MODELING AND COMPUTATIONS IN DYNAMICAL SYSTEMS

In commemoration of the 100th anniversary
of the birth of John von Neumann

WORLD SCIENTIFIC SERIES ON NONLINEAR SCIENCE

Editor: Leon O. Chua
University of California, Berkeley

WORLD SCIENTIFIC SERIES ON
NONLINEAR SCIENCE

Series Editor: Leon O. Chua

Series B   Vol. 13

MODELING AND COMPUTATIONS IN DYNAMICAL SYSTEMS

In commemoration of the 100th anniversary of the birth of John von Neumann

edited by

Eusebius J. Doedel
Concordia University, Canada

Gábor Domokos
Budapest University of Technology and Economics, Hungary

Ioannis G. Kevrekidis
Princeton University, USA

World Scientific

NEW JERSEY · LONDON · SINGAPORE · BEIJING · SHANGHAI · HONG KONG · TAIPEI · CHENNAI

Published by

World Scientific Publishing Co. Pte. Ltd.
5 Toh Tuck Link, Singapore 596224
USA office: 27 Warren Street, Suite 401-402, Hackensack, NJ 07601
UK office: 57 Shelton Street, Covent Garden, London WC2H 9HE

British Library Cataloguing-in-Publication Data
A catalogue record for this book is available from the British Library.

Cover Illustration: The image is an artistic rendering by Greg Jones (University of Bristol) of the Lorenz manifold as computed by the five different methods; see the chapter "A Survey of Methods for Computing (Un)Stable Manifolds of Vector Fields", by B. Krauskopf, H. M. Osinga, E. J. Doedel, M. E. Henderson, J. Guckenheimer, A. Vladimirsky, M. Dellnitz and O. Junge.

MODELING AND COMPUTATIONS IN DYNAMICAL SYSTEMS

ISBN   981-256-596-5

Typeset by Stallion Press
E-mail: enquiries@stallionpress.com

Printed by FuIsland Offset Printing (S) Pte Ltd, Singapore

CONTENTS

EDITORIAL

The papers in this issue are based on lectures presented at the October 2003 Budapest workshop on *Modeling and Computations in Dynamical Systems*, and complemented by selected additional contributions. The workshop, organized by G. Domokos, was held in commemoration of the 100th anniversary of the date of birth of John von Neumann, and made possible by generous support from The Thomas Cholnoky Foundation. Von Neumann made fundamental contributions to Computing, and he had a keen interest in Dynamical Systems, specifically, Hydrodynamic Turbulence. It was especially appropriate therefore, to dedicate the workshop (and this special issue) to the memory of von Neumann, one of the greatest and most influential mathematicians of the 20th century. While the topic of the Budapest workshop was rather well-defined, concentrating on modeling and computations in dynamical systems, the gathering attracted a diverse group of prominent researchers, theoreticians as well as computational scientists, with fields of expertise ranging from numerical techniques, including large scale computing, to fundamental aspects of dynamical systems. The papers in this special issue reflect these diverse interests, and, in fact, the wide-ranging nature of the field of Dynamical Systems. Applications of the work reported in this special issue include geometric integration, neural networks, linear programming, dynamical astronomy, chemical reaction models, and structural and fluid mechanics.

Eusebius Doedel,
Concordia University, Montreal, Canada

Gabor Domokos,
Budapest University of Technology and Economics, Hungary

Ioannis Kevrekidis,
Princeton University, USA

TRANSPORT IN DYNAMICAL ASTRONOMY AND MULTIBODY PROBLEMS

MICHAEL DELLNITZ*, OLIVER JUNGE*, WANG SANG KOON[†],
FRANCOIS LEKIEN[‡], MARTIN W. LO[§], JERROLD E. MARSDEN[†],
KATHRIN PADBERG*, ROBERT PREIS*, SHANE D. ROSS[†],
and BIANCA THIERE*

*Faculty of Computer Science, Electrical Engineering and Mathematics,
University of Paderborn, D-33095 Paderborn, Germany

[†]Control and Dynamical Systems, MC 107-81,
California Institute of Technology, Pasadena, CA 91125, USA

[‡]Department of Mechanical and Aerospace Engineering,
Princeton University Engineering Quad, Olden Street,
Princeton, NJ 08544-5263, USA

[§]Navigation and Mission Design, Jet Propulsion Laboratory,
California Institute of Technology, M/S 301-140L,
4800 Oak Grove Drive, Pasadena, CA 91109, USA

Received April 28, 2004; Revised July 5, 2004

We combine the techniques of almost invariant sets (using tree structured box elimination and graph partitioning algorithms) with invariant manifold and lobe dynamics techniques. The result is a new computational technique for computing key dynamical features, including almost invariant sets, resonance regions as well as transport rates and bottlenecks between regions in dynamical systems. This methodology can be applied to a variety of multibody problems, including those in molecular modeling, chemical reaction rates and dynamical astronomy. In this paper we focus on problems in dynamical astronomy to illustrate the power of the combination of these different numerical tools and their applicability. In particular, we compute transport rates between two resonance regions for the three-body system consisting of the Sun, Jupiter and a third body (such as an asteroid). These resonance regions are appropriate for certain comets and asteroids.

Keywords: Three-body problem; transport rates; dynamical systems; almost invariant sets; graph partitioning; set-oriented methods; invariant manifolds; lobe dynamics.

Contents

1. Introduction

The mathematical description of transport phenomena applies to a wide range of physical systems across many scales [Meiss, 1992; Wiggins, 1992; Rom-Kedar, 1999]. The recent and surprisingly effective application of methods combining dynamical systems ideas with those from chemistry to the transport of Mars impact ejecta underlines this point [Jaffé *et al.*, 2002]. In this paper, we develop computational methods to study transport based on the relationship between statistics and geometry in a nonlinear dynamical system with mixed regular and chaotic motion. Our focus is on the transport of material throughout the solar system. However, these methods are *fundamental and broad-based*; they may be applied to diverse areas of study, including fluid mixing [Rom-Kedar *et al.*, 1990; Malhotra & Wiggins, 1998; Poje & Haller, 1999; Coulliette & Wiggins, 2001; Lekien *et al.*, 2003], *N*-body problems in physical chemistry [Jaffé *et al.*, 2000; Lekien & Marsden, 2004] as well as other problems in dynamical astronomy. For example, the recent discovery of several binary pairs in the asteroid and Kuiper belts has stimulated interest in computing the formation and dissociation rates of such binary pairs (see, e.g. [Goldreich *et al.*, 2002; Scheeres, 2002; Scheeres *et al.*, 2002; Veillet *et al.*, 2002]).

Dynamical processes in the solar system

Our understanding of the solar system has changed dramatically in the past several decades with the realization that the orbits of the planets and some minor bodies are chaotic. In the case of planets, this chaos is of a sufficiently weak nature that their motion appears quite regular on relatively short time scales [Laskar, 1989]. In contrast, small bodies such as asteroids, comets, and Kuiper-belt objects can exhibit strongly chaotic motion through their interactions with the planets and the Sun, exhibiting Lyapunov times of only a few decades [Torbett & Smoluchowski, 1990; Tancredi, 1995].

The ability to predict the behavior of populations of these small but numerous objects is essential for understanding key transport phenomena in dynamical astronomy, such as the evolution of short period comets [Torbett & Smoluchowski, 1990], scattered Kuiper-belt objects [Malhotra *et al.*, 2000], and the intermediaries between these two populations [Tiscareno & Malhotra, 2003]. Furthermore, an understanding of how small bodies behave in *n*-body fields will aid in the gravitationally assisted transport of spacecraft using very little fuel [Koon *et al.*, 2000, 2001a, 2002; Gómez *et al.*, 2001; Dellnitz *et al.*, 2001a; Ross *et al.*, 2003; Yamato & Spencer, 2003]. This understanding also contributes to other fields such as astrobiology, for example, where comet

impact rates are key for determining the delivery of water to the Earth [Morbidelli *et al.*, 2000] and ejecta exchange rates are important for investigating the transportation of microbes between Mars and Earth [Gladman *et al.*, 1996; Mileikowsky *et al.*, 2000].

The recent discovery of several extrasolar planetary systems has stimulated interest in the morphological and dynamical features that may be present in *generic* planetary systems [Konacki *et al.*, 2003]. Some quantities of interest are the following: likely distributions of objects in the presence of dynamical sculpting due to planets and moons (e.g. generic circumstellar belts and circumsolar rings); rates of small body collision with a planet; and rates of capture and escape from one orbital resonance with a planet to another.

Short period comets

In order to develop a theory of chaotic transport that is computationally tractable, we will consider a physically relevant example from dynamical astronomy: the motion of (short period) comets in the gravitational field of the Sun and Jupiter. Our model, the planar circular restricted three-body problem (PCR3BP), will be described in a later section.

The role of the planar circular restricted three-body problem

The PCR3BP has long been considered an appropriate "baseline" model for providing a reasonable explanation for much of the dynamical behavior found in the large scale numerical experiments of solar system dynamics [Levison & Duncan, 1993; Malhotra *et al.*, 2000]. Malhotra's work [1996] provides a good recent example. Motivated by numerical studies of the stability of low-eccentricity and low-inclination orbits of small bodies in the trans-Neptunian Kuiper belt, Malhotra [1996] used the PCR3BP to describe the basic phase space structure in the neighborhood of Neptune's exterior mean motion resonances. The advantage of this simple model is that it allows the direct visualization, in two-dimensional surfaces-of-section, of a global mixed phase space structure of stable and chaotic zones. Much can be learned about populations of minor bodies from a semi-analytical study of the PCR3BP, i.e. careful numerics guided by dynamical systems theory.

1.1. *Need for modification of current transport calculations*

Several subjects make use of dynamical transport calculations. We indicate some of the reasons one would like to improve current techniques.

1.1.1. *Chemistry*

The transport of ensembles of points in phase space has been important for the theoretical determination of chemical reaction rates. One method, *transition state theory* (TST), has been a ubiquitous workhorse in the computational chemistry literature [Uzer *et al.*, 2002]. It is based on the identification of a transition state (TS) between large *realms* of phase space which correspond to either "reactants" or "products." If one assumes the phase space in each realm is structureless [Marston & De Leon, 1989], then the chemical reaction rate for the reaction under study can be estimated from the flux through the TS. However, rates given by TST can be off from the true rate by orders of magnitude [De Leon, 1992]. Modifications of transition state theory are necessary to calculate statistical quantities of interest [Hammes-Schiffer & Tully, 1995; Hammes-Schiffer, 2002; Agarwal *et al.*, 2002].

1.1.2. *Dynamical astronomy*

In principle, the computation of rates of mass transport can be accomplished by numerical simulations in which the orbits of vast numbers of test particles are propagated in time including as many gravitational interactions as desirable. Many investigators have used this approach successfully (cf. [Levison & Duncan, 1993]). However, such calculations are computationally demanding and it may be difficult to extract information from them about key dynamical mechanisms since the outcomes may depend sensitively on the initial conditions used for the simulation or may even be misleading. To obtain general features of planetary system evolution and morphology, which is a major goal of dynamical astronomy, other approaches may be necessary.

1.2. *Current methods for the study of transport in the PCR3BP*

Many of the important transport questions involve motion between different regions of the phase space. There have been a variety of approaches to deal with this question from various points of view. We recall some of them in this subsection.

1.2.1. *Analytical methods: single resonance theory and resonance overlap criterion*

One approach is to develop simple analytical models which provide answers to basic phase space transport questions. Much progress has been made in this area, but most of the work has focused on the study of the local dynamics around a single resonance, using a one-degree-of-freedom pendulum-like Hamiltonian with slowly varying parameters. Transport questions regarding capture into, and passage through resonance, have been addressed this way [Henrard, 1982; Neishtadt, 1996; Neishtadt *et al.*, 1997].

An important result regarding the interaction between resonances was obtained by Wisdom [1980], where the method of Chirikov [1979] was applied to the PCR3BP to determine a resonance overlap criterion for the onset of chaotic behavior for small mass parameter (ϵ). These analytical methods are still used today (see [Murray & Holman, 2001] and references therein).

1.2.2. *Toward a global picture of the phase space*

In [Koon *et al.*, 2000], dynamical systems techniques were applied to the problem of heteroclinic connections and interior–exterior transitions in the PCR3BP, laying the foundation for *tube dynamics*. In the point of view developed in [Koon *et al.*, 2000], the invariant manifold structures associated to L_1 and L_2, the (Conley-McGehee) *phase space tubes* [Conley, 1968; McGehee, 1969] play a key role. These tubes provide fundamental tools that can aid in understanding transport throughout the phase space, e.g. transport between the inside and outside of a planet's orbit, as seen in the comet P/Oterma [Carusi *et al.*, 1985], and chaotic trajectories leading to planetary impact, as in comet D/Shoemaker–Levy 9 [Benner & McKinnon, 1995].

The main new technical result in Koon, Lo, Marsden, and Ross [2000] is the numerical demonstration of the existence of a heteroclinic connection between pairs of periodic orbits, one around the libration point L_1 and the other around L_2, with the two periodic orbits having the same energy. This result is applied to the interior–exterior transition problem, providing insight into the "resonance hopping" of some short period comets (cf. [Tancredi *et al.*, 1990; Valsecchi, 1992; Belbruno & Marsden, 1997; Koon *et al.*, 2001]. Furthermore, an explicit numerical construction of interesting orbits with prescribed itineraries is developed, based on ideas from a proof of global motion in the PCR3BP.

For particles in the PCR3BP with energy slightly greater than that of L_2, the *interior, exterior* and *planetary realms* are connected by bottlenecks about L_1 and L_2 (see Fig. 1(c) in the next section). Particles can pass between realms only through these bottlenecks by being inside phase space tubes, regions bounded by pieces of the stable and unstable invariant manifolds of periodic orbits around L_1 and L_2. We can determine the flux between realms by monitoring the flux through these tubes.

1.2.3. *Mars escape rates*

Building on the ideas described in the preceding paragraph, the rate of escape of particles temporarily captured by Mars was computed in [Jaffé, et al., 2002; Ross, 2003]. The paper uses a statistical assumption that is common in transition state theory in chemistry, and which is appropriate for this problem. Theory and direct Monte Carlo simulations are shown to agree to within 1%, which showed the promise of a dynamical systems approach for the computation of interesting transport rates in dynamical astronomy.

The work of Rom-Kedar and Wiggins [1990], contains an investigation of the transport in the two-dimensional phase space of C^r diffeomorphisms ($r \geq 1$) of two-manifolds between regions of the phase space bounded by pieces of the stable and unstable manifolds of hyperbolic points. The transport mechanism is associated with the dynamics of homoclinic and heteroclinic tangles, and the study of this dynamics leads to a general formulation of the transport rates in terms of distributions of small phase space regions called "lobes". By following the evolution of these lobes, *lobe dynamics* supplies a method for theoretically computing short and long term transport rates. However, computational issues have limited its applications [Rom-Kedar & Wiggins, 1990, 1991; Meiss, 1992]. Important contributions to this effort were made by Lichtenberg and Lieberman [1983]; MacKay *et al.* [1984, 1987]; Meiss [1992]; Meiss and Ott [1986].

The manifolds computed in such problems are typically complicated because of the nature of homoclinic and heteroclinic tangles. Furthermore, the length of these complicated curves grows quickly with the size of the time window of interest. The number of points needed to describe long segments of manifolds can be prohibitively large if naive computational methods are used. One also needs to take into account the fine structure of the lobes and manifolds, and in particular, the effect of re-entrainment of the lobes, i.e. the implications of the lobes leaving and re-entering the specified regions on the transport rate. We show later on that this effect is in fact, important in the three-body problem and cannot be ignored.

Recent efforts made to incorporate lobe dynamics into geophysical, fluid and chemical transport calculations have brought new techniques to compute invariant manifolds (see [Coulliette & Wiggins, 2001; Lekien & Marsden, 2004; Lekien & Coulliette, 2004; Lekien *et al.*, 2003]). Using these techniques, one is able to compute very long segments of stable and unstable manifolds with high accuracy by *conditioning the manifolds adaptively*, for instance, by inserting more points along the manifold where the curvature is high (see [Hobson, 1993; Lekien, 2003]). As a result, the length and shape of the manifold is not an obstacle anymore and many more iterates of lobes than hitherto possible can be generated accurately. Using this approach, one keeps track of all the points throughout the computation, with the drawback that the resulting algorithms often require a great deal of memory. A related set of studies [You *et al.*, 1991; Kostelich *et al.*, 1996] describes a method for restricting the invariant manifold computation to specific regions of interest, thereby using significantly less memory, while rigorously guaranteeing that the computed manifold lies no further than a specified tolerance from the "true" manifold.

1.3. *Set oriented approach to transport*

In contrast to the geometric approach to the analysis of transport phenomena as described in the preceding paragraphs, the *set oriented approach* focuses on a global description of the dynamics on a coarse level. To this end, one considers a *transfer operator* associated to the underlying map. Roughly speaking, this operator describes how some initial distribution evolves under the dynamics. Via a partition of some interesting invariant part in phase space this operator can be discretized, yielding a stochastic matrix or, equivalently, a directed weighted graph, which may be viewed as a coarse-grain model of the global dynamics.

Transport rates between subsets of phase space can easily be computed using this matrix of *transition probabilities*. When these subsets are given as unions of partition elements, the computed rates are exact. However, in general the accuracy of the computed quantities is determined by the size of the partition elements.

In addition to computing transport rates, it is also possible to obtain insight about what "important" or interesting regions are in phase space. The idea is that the transfer operator encodes a macroscopic description of the dynamics. One way to reveal this information is to consider the corresponding graph, to which standard algorithms from graph theory can directly be applied for a further analysis. For example, we use algorithms for graph partitioning (see e.g. software-libraries such as CHACO [Hendrickson & Leland, 1995], JOSTLE [Walshaw, 2000], METIS [Karypsis & Kumar, 1999], SCOTCH [Pellegrini, 1996] or PARTY [Monien *et al.*, 2000] to find regions that are determined by (i) a high transport rate within the region and (ii) a small transport rate to other regions. In terms of dynamical systems, these sets are referred to as *almost invariant sets* [Dellnitz & Junge, 1999]. In particular, we use the PARTY library with extensions, which are explicitly developed for the analysis of almost invariant sets in dynamical systems [Dellnitz & Preis, 2003]. A key observation of this paper is that regions that we compute by this approach are actually those bounded by certain invariant manifolds.

1.4. *What is achieved in this paper*

The main results of this paper are

- Further development of the basic theory and application of computational techniques for transport. In particular, a comparison as well as a synthesis of tools from lobe dynamics and set oriented methods is presented. Error estimates are provided, which show, in particular, the convergence of the set-oriented methods.
- In regimes where the comparison makes sense, it is shown that the agreement is very good on a sample problem. Based on the initial information provided by the combination of the two methods, the set oriented methods are able to

carry out many more iterates than heretofore possible.

- As a concrete nontrivial example illustrating the methods, the transport rate from an interesting resonant region R_1 to a surrounding region R_2 in the Sun–Jupiter system, exterior to the orbit of Jupiter and at a particular energy value, are computed. It is computed that the probability (in the sense of the fractional area) that a transition from R_1 to R_2 occurs is about 28% in a period of about 1817 Earth years.
- The methods of this paper lay the foundation for many other computations of astrodynamical interest. In particular, in [Dellnitz *et al.*, 2004] we study the transport rate of asteroids from the Hilda region to a region defined by crossers of Mars' orbit as well as a remarkable relation between almost invariant sets associated with the Sun–Jupiter three-body system and the orbits of all the planets interior to Jupiter.

2. Description of the PCR3BP Global Dynamics

2.1. *Problem description*

The PCR3BP is a particular case of the general gravitational problem of three masses m_1, m_2, m_3 defined by the following restrictions: (a) the motion of all three bodies takes place in a common plane; (b) the masses m_1 and m_2 move on circular orbits about their common center of mass; and (c) the third body, m_3, has zero mass; therefore, it does not influence the motion of m_1 and m_2. In the context of this paper, m_1 represents the Sun and m_2 represents a planet, and we are concerned with the motion of the third body, the test particle m_3. The system is made nondimensional by the following choice of units: the unit of mass is taken to be $m_1 + m_2$; the unit of length is chosen to be a_P the constant separation between m_1 and m_2 (i.e. the mean separation of the Sun and planet); the unit of time is chosen such that the orbital period of m_1 and m_2 about their center of mass is 2π. Then the universal constant of gravitation, $G = 1$, and the masses of the Sun and planet are $1 - \epsilon$ and ϵ, where $\epsilon = m_2/(m_1 + m_2)$.

2.2. *Equations of motion*

Choosing a rotating coordinate system so that the origin is at the center of mass, the Sun and planet are on the x-axis at the points $(-\epsilon, 0)$ and $(1-\epsilon, 0)$ respectively. Let (x, y) be the position of the particle in the plane, then the equations of motion for the particle in this rotating frame are:

$$\ddot{x} - 2\dot{y} = -\overline{U}_x \quad \ddot{y} + 2\dot{x} = -\overline{U}_y, \tag{1}$$

where

$$\overline{U} = -\frac{x^2 + y^2}{2} - \frac{1-\epsilon}{r_S} - \frac{\epsilon}{r_P} - \frac{\epsilon(1-\epsilon)}{2}.$$

Here, the subscripts of $\overline{U}$ denote partial differentiation in the respective variable, and r_S, r_P are the distances from the particle to the Sun and planet, respectively. See [Szebehely, 1967] for more details on the derivation of this equation and [Koon et al., 2004] for its derivation using Lagrangian mechanics.

2.3. *Energy manifolds*

Equations (1) are autonomous and are in Euler–Lagrange form (and thus, using the Legendre transformation, can be put into Hamiltonian form as well). They have an energy integral

$$E = \frac{1}{2}(\dot{x}^2 + \dot{y}^2) + \overline{U}(x, y), \tag{2}$$

which is related to the Jacobi constant C by $C = -2E$. The motion of the test particle takes place on a three-dimensional energy manifold (defined by a particular value of E) embedded in the four-dimensional phase space, $(x, y, \dot{x}, \dot{y})$.

The value of the energy is an indicator of the type of global dynamics possible for a particle in the PCR3BP, which can be broken down into five cases (see Fig. 1). In case 1, shown in Fig. 1(a), the particle is trapped either exterior or interior to the planet's orbit, or around the planet itself (labeled the *exterior*, *interior*, and *planetary realms*, respectively). For energy values greater than that of L_2 (case 3), there is a bottleneck around L_1 and L_2, permitting particles to move between the three realms.

This paper considers case 1 to illustrate the techniques. It uses the Poincaré surface-of-section (s-o-s) defined by $y = 0, \dot{y} > 0$, and the coordinates $(x, \dot{x})$ on that section. The geometric interpretation is straightforward: we plot the x coordinate and velocity of the test particle at every conjunction with the planet. As a further restriction, we consider only the motion of test particles in the exterior realm (strictly speaking, with mean motion smaller than the planet's). For orbits exterior to the planet's, the s-o-s is crossed every time the test particle is aligned with the Sun and planet and is on the *opposite* side of the Sun from the planet, along the portion of the x-axis with $x < -1$, as shown in Fig. 2(a). Thus, the s-o-s becomes

$$y = 0, \quad \dot{y} > 0, \quad x < -1. \tag{3}$$

In the s-o-s so defined, *periodic* orbits of the test particle appear as a finite set of points. The successive crossings of the surface by a *quasiperiodic*

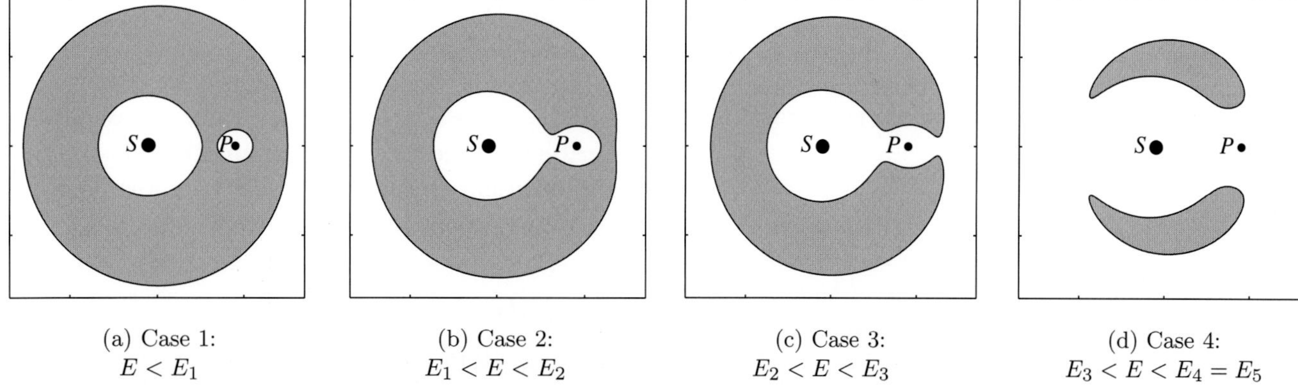

(a) Case 1: $E < E_1$	(b) Case 2: $E_1 < E < E_2$	(c) Case 3: $E_2 < E < E_3$	(d) Case 4: $E_3 < E < E_4 = E_5$

Fig. 1. **There are five cases of allowable motion.** The Sun and planet, denoted S and P, respectively, are fixed in this rotating frame. (a) In case 1, the particle is trapped either exterior or interior to the planet's orbit, or around the planet itself. It is energetically prohibited from crossing the *forbidden realm*, shown in gray. (b)–(d) As the energy E of the particle increases, the bottlenecks connecting the realms open. In case 5, not shown, the entire configuration space is energetically accessible.

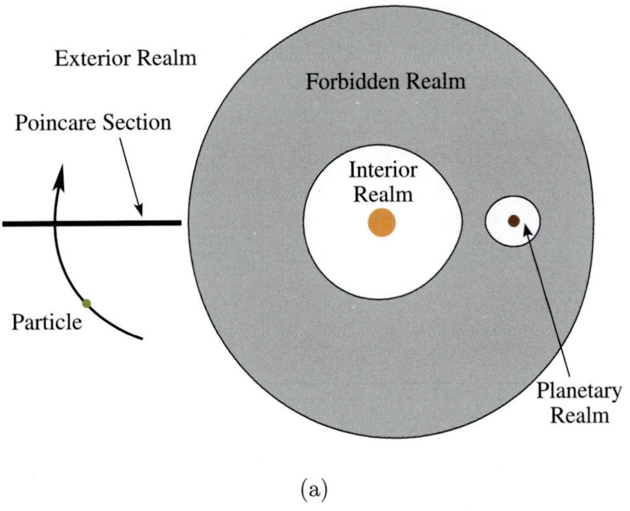

(a)

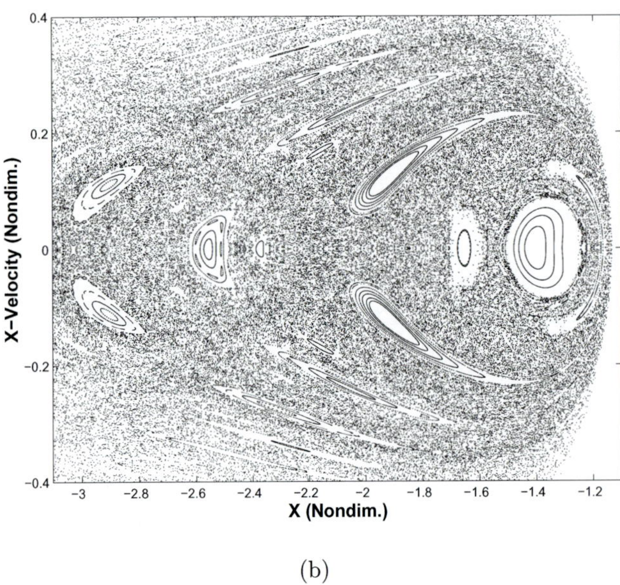

(b)

Fig. 2. **A Poincaré section of the flow in the restricted three-body problem.** (a) The location of the Poincaré surface-of-section (s-o-s) in this paper is shown in the configuration space for a case 1 energy, as in Fig. 2(a). (b) The mixed phase space structure of the PCR3BP is shown on this s-o-s. KAM tori and the chaotic sea are visible. Note that the Poincaré map of this s-o-s is area preserving.

orbit live on a set of closed smooth curves, such as the cross-section of a KAM torus. *Chaotic* orbits appear to approximately fill a two-dimensional area.

In general, by taking a grid of points on this s-o-s and integrating them forward for several iterates, one observes a mixed phase space structure of KAM tori embedded within a "chaotic sea", as in Fig. 2(b).

3. Computing Transport

As laid out in the previous section, our task is to compute the *transport* between *regions* in phase space. More precisely, we consider a volume- and orientation-preserving map $f : M \rightarrow M$ (e.g. the Poincaré map in the PCR3BP as described in the previous section) on some compact set $M \subset \mathbb{R}^d$ with volume-measure μ and ask for a suitable (i.e. depending on the application in mind) *partition* of M into compact *regions of interest* R_i, $i = 1, \ldots, N_R$, such that

$$M = \bigcup_{i=1}^{N_R} R_i \quad \text{and} \quad \mu(R_i \cap R_j) = 0 \quad \text{for } i \neq j. \quad (4)$$

Furthermore, we are interested in the following questions concerning the transport between the regions R_i (see [Wiggins, 1992]): "In order to keep track of the initial condition of a point as it moves throughout the regions we say that initially (i.e. at $t = 0$) region R_i is uniformly covered with species S_i. Thus, the species type of a point indicates the region in which it was located initially. Then we can generally state the transport problem as follows.

Describe the distribution of species $S_i, i = 1, \ldots, N_R$, throughout the regions R_j, $j = 1, \ldots, N_R$, for any time $t = n > 0$.

The quantity we want to compute is $T_{i,j}(n) \equiv$ *the total amount of species S_i contained in region R_j immediately after the nth iterate.*

The *flux* $\alpha_{i,j}(n)$ of species S_i into region R_j on the nth iterate is the change in the amount of species S_i in R_j on iteration n; namely, $\alpha_{i,j}(n) = T_{i,j}(n) - T_{i,j}(n-1)$. Since f is area-preserving, the flux is equal to the amount of species S_i entering region R_j at iteration n minus the amount of species S_i leaving R_j at iteration n.

Our goal is to determine $T_{i,j}(n), i, j = 1, \ldots, N_R$ for all n. Note, that $T_{i,i}(0) = \mu(R_i)$, and $T_{i,j}(0) = 0$ for $i \neq j$. In the following we briefly describe the theoretical background behind the two computational approaches to the transport problem that we are going to compare in Sec. 4.

3.1. *Lobe dynamics*

Following Rom-Kedar and Wiggins [1990], lobe dynamics theory states that the two-dimensional phase space M of the Poincaré map f can be divided as outlined above (see Eq. (4)), as illustrated in Fig. 3(a). A *region* is a connected subset of M with boundaries consisting of parts of the boundary of M

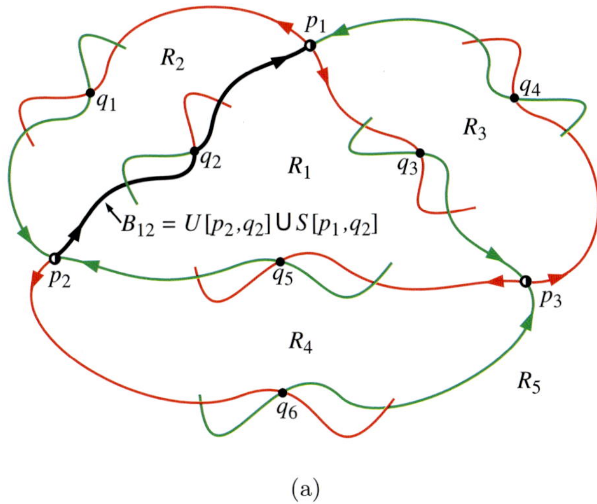

(a)

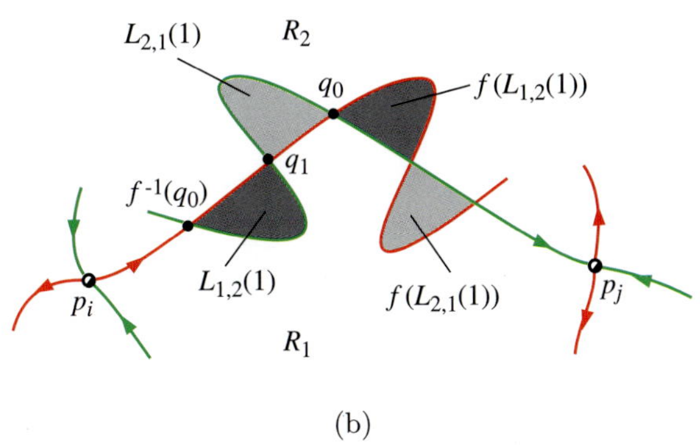

(b)

Fig. 3. **Transport between regions of the phase space**
M of a Poincaré map f. (a) The segment $S[p_1, q_2]$ of
the stable manifold $W^s(p_1)$ from p_1 to q_2 and the segment
$U[p_2, q_2]$ of the unstable manifold $W^u(p_2)$ from p_2 to q_2 intersect in the pip q_2. Therefore, the boundary B_{12} can be defined
as $B_{12} = U[p_2, q_2] \cup S[p_1, q_2]$. The region on one side of the
boundary may be labeled R_1 and the other side labeled R_2.
(b) q_1 is the only pip between the two pips q_0 and $f^{-1}(q_0)$
in $W^u(p_i) \cap W^s(p_j)$, thus $S[f^{-1}(q_0), q_0] \cup U[f^{-1}(q_0), q_0]$
forms the boundary of precisely two lobes; one in R_1, labeled
$L_{1,2}(1)$, and the other in R_2, labeled $L_{2,1}(1)$. Under one iteration of f, the only points that can move from R_1 into R_2
by crossing the boundary B are those in $L_{1,2}(1)$. Similarly,
under one iteration of f the only points that can move from
R_2 into R_1 by crossing B are those in $L_{2,1}(1)$.

3.1.1. *Boundaries, regions, pips, lobes,*
and turnstiles defined

To define a *boundary* between regions, one first
defines a *primary intersection point*, or *pip*. A point
q_k is called a pip if $S[p_i, q_k]$ intersects $U[p_j, q_k]$ only
at the point q_k, where $U[p_j, q_k]$ is a segment of the
unstable manifold $W^u(p_j)$ joining the unstable fixed
point p_j to q_k and similarly $S[p_i, q_k]$ is a segment
of the stable manifold $W^s(p_i)$ of the unstable fixed
point p_i joining p_i to q_k. The union of segments of
the unstable and stable manifolds naturally form
partial barriers, or *boundaries* $U[p_j, q_k] \cup S[p_i, q_k]$,
between *regions* of interest $R_i, i = 1, \ldots, N_R$, in
$M = \bigcup R_i$. In Fig. 3(a) several pips are shown
as well as the boundary B_{12}. Note that we could
have $p_i = p_j$, as will be the case studied in this
paper.

Consider Fig. 3(b). Let $q_0, q_1 \in W^u(p_i) \cap W^s$
(p_j) be two adjacent pips, i.e. there are no other pips
on $U[q_0, q_1]$ and $S[q_0, q_1]$, the segments of $W^u(p_i)$
and $W^s(p_j)$ connecting q_0 and q_1. We refer to the
region interior to $U[q_0, q_1] \cup S[q_0, q_1]$ as a *lobe*. Then
$S[f^{-1}(q_0), q_0] \cup U[f^{-1}(q_0), q_0]$ forms the boundary of precisely two lobes; one in R_1, defined by
$L_{1,2}(1) := \int(U[q_0, q_1] \cup S[q_0, q_1])$, where $\int$ denotes
the interior operation on sets, and the other in
R_2, $L_{2,1}(1) := \int \left(U[f^{-1}(q_0), q_1] \cup S[f^{-1}(q_0), q_1] \right)$.
Under one iteration of f, the only points that can
move from R_1 into R_2 by crossing B_{12} are those in
$L_{1,2}(1)$. Similarly, under one iteration of f the only
points that can move from R_2 into R_1 by crossing
B_{12} are those in $L_{2,1}(1)$. The two lobes $L_{1,2}(1)$ and
$L_{2,1}(1)$ are called a *turnstile*. It is important to note
that $f^{-n}(L_{1,2}(1)), n \geq 2$, need not be contained
entirely in R_1, i.e. the lobes can leave and re-enter
regions with strong implications for the dynamics.
As will be shown, the quantities of interest, $T_{i,j}(n)$,
can be expressed compactly in terms of intersection areas of images or preimages of turnstile
lobes.

3.1.2. *Multilobe, self-intersecting turnstiles*

Before we derive expressions for the $T_{i,j}(n)$, some
comments regarding technical points are in order
[Rom-Kedar & Wiggins, 1990]. In the previous
paragraph we assumed that there was only one pip
between q and $f^{-1}(q)$, but this is not the case for the
application to the PCR3BP in Sec. 4. Suppose that
there are k pips, $k \geq 1$, along $U[f^{-1}(q), q]$ besides q
and $f^{-1}(q)$. This gives rise to $k + 1$ lobes; m in R_2

(which may be at infinity) and/or segments of stable
and unstable manifolds of hyperbolic fixed points,
$p_i, i = 1, \ldots, N$. Moreover, the transport between
regions of phase space can be completely described
by the dynamical evolution of small regions of phase
space, "lobes" enclosed by segments of the stable
and unstable manifolds, as shown schematically in
Fig. 3(b), and defined below.

and $(k+1) - m$ in R_1. Suppose

$$L_0, L_1, \ldots, L_{k-m} \subset R_1,$$

$$L_{k-m+1}, L_{k-m+2}, \ldots, L_k \subset R_2.$$

Then we define

$$L_{1,2}(1) \equiv L_0 \cup L_1 \cup \cdots \cup L_{k-m},$$

$$L_{2,1}(1) \equiv L_{k-m+1} \cup L_{k-m+2} \cup \cdots \cup L_k,$$

and all the previous results hold.

Furthermore, we previously assumed that $L_{1,2}(1)$ and $L_{2,1}(1)$ lie entirely in R_1 and R_2, respectively. But $L_{1,2}(1)$ may intersect $L_{2,1}(1)$, as shown schematically in Fig. 4(a). We want $U[q, f^{-1}(q)]$ and $S[q, f^{-1}(q)]$ to intersect only in pips, so we must redefine our lobes, as shown in Fig. 4(b). Let

$$I = \text{int}\left(L_{1,2}(1) \cap L_{2,1}(1)\right).$$

The lobes defining the turnstile are redefined as

$$\begin{aligned}
\tilde{L}_{1,2}(1) &\equiv L_{1,2}(1) - I, \\
\tilde{L}_{2,1}(1) &\equiv L_{2,1}(1) - I,
\end{aligned} \tag{5}$$

and all our previous results hold. To the best of our knowledge, the PCR3BP is the first example of a physical system that has a multilobe turnstile, so the fact that it is a multilobe, self-intersecting turnstile is even more surprising. We believe this has a great effect on the dynamics.

3.1.3.   *Expressions for the transport of species*

In the application in the present paper, the phase space M is known to possess *resonance regions* whose boundaries have complicated lobe structures, which can lead to complicated transport properties (cf. [Meiss, 1992; Schroer & Ott, 1997; Koon *et al.*, 2000]. In this paper, we limit ourselves to the study of transport between just two regions. We suppose that our map f has a period-1 hyperbolic point p. We consider only one branch of the unstable manifold $W_+^u(p)$, and one branch of the stable manifold $W_+^s(p)$. We suppose that they intersect each other, as in Fig. 4, forming a boundary between two regions, R_1 and R_2. Using the lobe dynamics framework, the transport of species between the regions — $T_{i,j}(n), i, j = 1, 2$ — can be computed via the following formulas.

Let $L_{i,j}(m)$ denote the lobe that leaves R_i and enters R_j on the mth iterate, so that

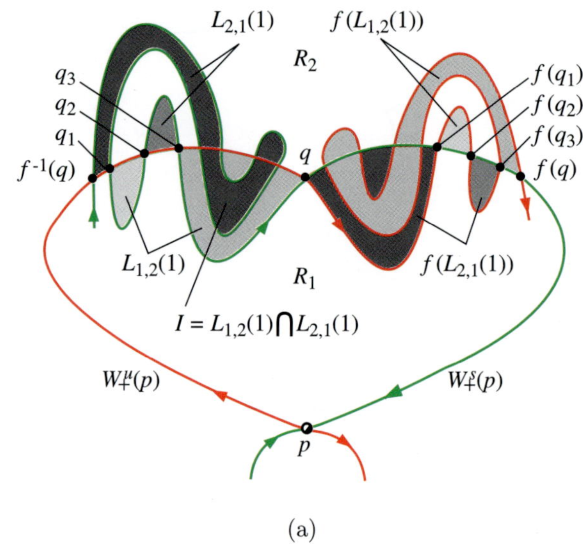

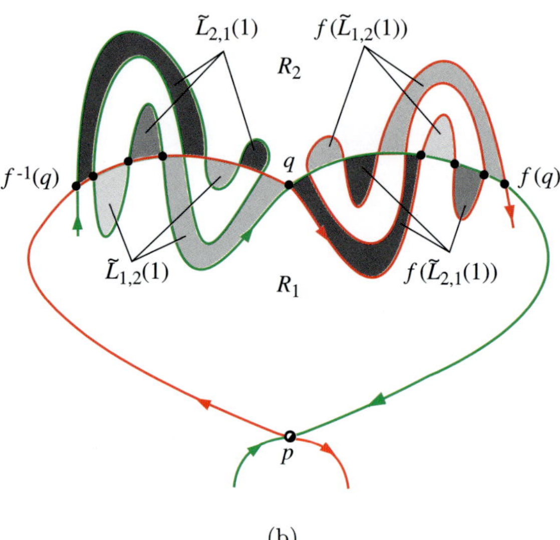

Fig. 4.   **A multilobe, self-intersecting turnstile.** The stable and unstable manifolds of the unstable fixed point p intersect in such a way that there are three pips between q and $f^{-1}(q)$, but our naively defined turnstile "lobes" have a nonempty intersection $I = \int \left(L_{1,2}(1) \cap L_{2,1}(1)\right) \neq \emptyset$. When we redefine the turnstile lobes such that $\tilde{L}_{1,2}(1) \equiv L_{1,2}(1) - I$ and $\tilde{L}_{2,1}(1) \equiv L_{2,1}(1) - I$, the result is a *multilobe, self-intersecting turnstile* consisting of a sequence of six regions; three defining $\tilde{L}_{1,2}(1)$ and three others defining $\tilde{L}_{2,1}(1)$.

$$f^{m-1}(L_{i,j}(m)) = L_{i,j}(1). \text{ Let } L_{i,j}^k(m) \equiv L_{i,j}(m) \cap R_k$$ denote the portion of lobe $L_{i,j}(m)$ that is in the region R_k. Then

$$T_{i,j}(n) - T_{i,j}(n-1)$$

$$= \sum_{k=1}^{2} \left[\mu\left(L_{k,j}^i(n)\right) - \mu\left(L_{j,k}^i(n)\right)\right] \tag{6}$$

where

$$\mu\left(L_{k,j}^i(n)\right) = \sum_{s=1}^{2}\sum_{m=0}^{n-1}\mu\left(L_{k,j}(1)\cap f^m(L_{i,s}(1))\right)$$

$$-\sum_{s=1}^{2}\sum_{m=1}^{n-1}\mu\left(L_{k,j}(1)\cap f^m(L_{s,i}(1))\right). \tag{7}$$

Thus, the dynamics associated with particles crossing B is reduced completely to a study of the dynamics of the turnstile lobes associated with B. The amount of computation necessary to obtain all the $T_{i,j}(n)$ can be reduced due to conservation of area and species, as well as symmetries of the map f (to be discussed in Sec. 4).

3.2. *Set oriented approach*

3.2.1. *The transfer operator*

Computing transport between regions in phase space is a question about the *global* dynamical behavior of the underlying dynamical system $f: M \to M$. One is interested in the evolution of sets or, more generally, of densities or measures on M instead of single trajectories. The evolution of e.g. a (signed) measure ν on M is compactly described in terms of the *transfer operator* (or *Perron–Frobenius operator*) associated with f, which is the linear operator $P : \mathcal{M} \to \mathcal{M}$,

$$(P\nu)(A) = \nu(f^{-1}(A)), \quad A \text{ measurable},$$

on the space $\mathcal{M}$ of signed measures on M.

To see how this operator relates to the transport quantities of interest, namely, the total amount $T_{i,j}(n)$ of species, consider the following observation.

Proposition 3.1. *Let $f: M \to M$ be an area preserving map, then*

$$T_{i,j}(n) = \mu(f^{-n}(R_j)\cap R_i)$$

(where, again, μ denotes the volume–measure on M).

Proof. By definition (see [Wiggins, 1992], p. 30 ff.), we have

$$T_{i,j}(n) = \mu\left(\bigcup_{k=1}^{N_R} f^n\left(L_{k,j}^i(n)\right)\right), \tag{8}$$

where $L_{k,j}^i(n)$ is the set of points that at time $t = n = 0$ is in R_i and is mapped from region R_k into region R_j on the nth iterate, i.e.

$$L_{k,j}^i(n) = f^{-n}(R_j)\cap f^{-(n-1)}(R_k)\cap R_i. \tag{9}$$

Combining (8) and (9) with the fact that f is a diffeomorphism yields

$$T_{i,j}(n) = \mu\left(\bigcup_{k=1}^{N_R} f^n(f^{-n}(R_j)\cap f^{-(n-1)}(R_k)\cap R_i)\right)$$

$$= \mu\left(\bigcup_{k=1}^{N_R} R_j \cap f(R_k)\cap f^n(R_i)\right)$$

$$= \mu\left(R_j \cap f^n(R_i)\cap \underbrace{\bigcup_{k=1}^{N_R} f(R_k)}_{=M}\right)$$

$$= \mu(f^{-n}(R_j)\cap R_i),$$

where the latter equality follows from the fact that f is area–preserving. $\blacksquare$

Since $\alpha_{i,j}(n) = T_{i,j}(n) - T_{i,j}(n-1)$, one obtains the formula

$$\alpha_{i,j}(n) = \mu(f^{-n}(R_j)\cap R_i) - \mu(f^{-(n-1)}(R_j)\cap R_i) \tag{10}$$

for the flux of species S_i into region R_j on the nth iterate.

The following consequence of Proposition 3.1 tells us how we can compute $\mu\left(f^{-n}(R_j)\cap R_i\right)$ using the transfer operator P (where, as usual, P^n refers to the n-fold application of P):

Corollary 3.2. *Let $\mu_i \in \mathcal{M}$ be the measure $\mu_i(A) = \mu(A\cap R_i) = \int_A \chi_{R_i}\,d\mu$, where χ_{R_i} denotes the indicator function on the region R_i. Then*

$$T_{i,j}(n) = (P^n\mu_i)(R_j). \tag{11}$$

Evidently, since we are interested in actually computing the quantities of interest for the PCR3BP, we need to explicitly deal with the transfer operator. Since an analytical expression for it will only be derivable for none but the most simple systems, we need to derive a finite-dimensional approximation to it. For more details on the following description see [Dellnitz *et al.*, 1997; Dellnitz & Junge, 1999; Dellnitz *et al.*, 2001b; Dellnitz & Junge, 2002].

3.2.2. *Discretization of the transfer operator*

Consider a covering of the phase space M by a finite collection $\mathcal{B} = \{B_1, \ldots, B_b\}$ of compact sets, i.e. a partition

$$M = \bigcup_{i=1}^{b} B_i \quad \text{and} \quad \mu(B_i \cap B_j) = 0 \quad \text{for } i \neq j.$$

In practice such a partition can be efficiently computed using a hierarchical multilevel approach as described in [Dellnitz & Hohmann, 1997].

As a finite dimensional space $\mathcal{M}_\mathcal{B}$ of measures on M we consider the space of absolutely continuous measures with density $h \in \Delta_\mathcal{B} := \text{span}\{\chi_B : B \in \mathcal{B}\}$, i.e. one which is piecewise constant on the elements of the partition $\mathcal{B}$. Let $Q_\mathcal{B} : L^1 \to \Delta_\mathcal{B}$ be the projection

$$Q_\mathcal{B} h = \sum_{B \in \mathcal{B}} \frac{1}{\mu(B)} \int_B h \, d\mu \, \chi_B,$$

then for every set A that is the union of partition elements we have

$$\int_A Q_\mathcal{B} h \, d\mu = \int_A h \, d\mu. \tag{12}$$

We define the discretized transfer operator $P_\mathcal{B} : \Delta_\mathcal{B} \to \Delta_\mathcal{B}$ as

$$P_\mathcal{B} = Q_\mathcal{B} P.$$

With respect to the basis $(\chi_B)_{B \in \mathcal{B}}$ it is represented by the matrix

$$P_\mathcal{B} = (p_{ij}), \quad \text{where } p_{ij} = \frac{\mu\left(f^{-1}(B_i) \cap B_j\right)}{\mu(B_j)},$$

$$1 \leq i, \, j \leq b. \tag{13}$$

For the computation of $\mu(f^{-1}(B_i) \cap B_j)$, that is, the measure of the subset of B_j that is mapped into B_i, one can use a Monte Carlo approach as described in [Hunt, 1993]:

$$\mu\left(f^{-1}(B_i) \cap B_j\right) \approx \frac{1}{K} \sum_{k=1}^{K} \chi_{B_i}\left(f(x_k)\right),$$

where the x_k's are selected at random in B_j from a uniform distribution. Evaluation of $\chi_{B_i}(f(x_k))$ only

means that we have to check whether or not the point $f(x_k)$ is contained in B_i. There are efficient ways to perform this check based on a hierarchical construction and storage of the collection $\mathcal{B}$ (see [Dellnitz & Hohmann, 1997; Dellnitz *et al.*, 1997]).

3.2.3. *Approximation of transport rates*

Note that we can write

$$T_{i,j}(n) = \int_{R_j} P^n \chi_{R_i} \, d\mu$$

For some (measurable) set A let

$$\underline{A} = \bigcup_{B \in \mathcal{B} : B \subset A} B \quad \text{and} \quad \overline{A} = \bigcup_{B \in \mathcal{B} : B \cap A \neq \emptyset} B.$$

Since P is positive, it follows that for two given regions R_i and R_j, $P^n(\chi_{R_i} - \chi_{\underline{R}_i}) \geq 0$, i.e. $P^n \chi_{R_i} \geq P^n \chi_{\underline{R}_i}$ and thus

$$\int_{\underline{R}_j} P^n \chi_{\underline{R}_i} \, d\mu \leq \int_{R_j} P^n \chi_{R_i} \, d\mu,$$

similarly, we can bound the term $\int_{R_j} P^n \chi_{R_i} \, d\mu$ from above and thus get the following estimate.

Proposition 3.3

$$\int_{\underline{R}_j} P^n \chi_{\underline{R}_i} \, d\mu \leq T_{i,j}(n) \leq \int_{\overline{R}_j} P^n \chi_{\overline{R}_i} \, d\mu. \tag{14}$$

The next step is to replace P^n by $P_\mathcal{B}^n$, since this is the operator we have at hand for computing. The error in making such a replacement is given by the estimate in the following Lemma.

Lemma 3.4. *Let* $R, S \subset M$ *and*

$$S_0 = S, \quad S_{k+1} = f^{-1}(\overline{S}_k), \quad k = 0, 1, 2, \ldots.$$

Then for $n = 1, 2, \ldots$

$$\left| \int_S P^n \chi_R \, d\mu - \int_S P_\mathcal{B}^n \chi_R \, d\mu \right|$$

$$\leq 2 \sum_{k=0}^{n-1} (n-k) \mu(R \cap \overline{S}_k \backslash S_k).$$

Proof. We proceed by induction on n. For $n = 1$ we use (12) and the fact that $\|I - Q_\mathcal{B}\| \leq 2$ and

$\|P\| = 1$ to obtain

$$\left| \int_S P\chi_R \, d\mu - \int_S P_{\mathcal{B}}\chi_R \, d\mu \right|$$

$$\leq \left| \int_{\overline{S}} (P - P_{\mathcal{B}})\chi_R \, d\mu - \int_{\overline{S}\backslash S} (P - P_{\mathcal{B}})\chi_R \, d\mu \right|$$

$$\leq \left| \int_{\overline{S}} (I - Q_{\mathcal{B}})P\chi_R \, d\mu \right|$$

$$+ \left| \int_{\overline{S}\backslash S} (I - Q_{\mathcal{B}})P\chi_R \, d\mu \right|$$

$$\leq 0 + 2\mu(R \cap \overline{S}\backslash S) = 2\mu(R \cap \overline{S}_0\backslash S_0).$$

Now note that since $\|I - Q_{\mathcal{B}}\| \leq 2$, $\|Q_{\mathcal{B}}\| = 1$ and $\|P\| = 1$,

$$\|P^n - (Q_{\mathcal{B}}P)^n\|$$

$$\leq \|P^n - Q_{\mathcal{B}}P^n\| + \|Q_{\mathcal{B}}P^n - (Q_{\mathcal{B}}P)^n\|$$

$$\leq 2 + \|Q_{\mathcal{B}}\|\|P\|\|P^{n-1} - (Q_{\mathcal{B}}P)^{n-1}\|$$

$$\leq 2n,$$

by induction. For $n > 1$ we get

$$\left| \int_S P^n\chi_R \, d\mu - \int_S P_{\mathcal{B}}^n\chi_R \, d\mu \right|$$

$$= \left| \int_{\overline{S}} (P^n - P_{\mathcal{B}}^n)\chi_R \, d\mu - \int_{\overline{S}\backslash S} (P^n - P_{\mathcal{B}}^n)\chi_R \, d\mu \right|$$

$$\leq \left| \int_{\overline{S}} (P^n - P_{\mathcal{B}}^n)\chi_R \, d\mu \right| + 2n\mu(R \cap \overline{S}\backslash S)$$

and, using (12) and the definition of P, the first term on the right-hand side can be estimated as

$$\int_{\overline{S}} (P^n - P_{\mathcal{B}}^n)\chi_R \, d\mu$$

$$= \int_{\overline{S}} (P^n - Q_{\mathcal{B}}P^n + Q_{\mathcal{B}}P^n - (Q_{\mathcal{B}}P)^n)\chi_R \, d\mu$$

$$= \int_{\overline{S}} (I - Q_{\mathcal{B}})P^n\chi_R \, d\mu$$

$$+ \int_{\overline{S}} (Q_{\mathcal{B}}P)(P^{n-1} - (Q_{\mathcal{B}}P)^{n-1})\chi_R \, d\mu$$

$$= \int_{\overline{S}} P(P^{n-1} - (Q_{\mathcal{B}}P)^{n-1})\chi_R \, d\mu,$$

$$= \int_{f^{-1}(\overline{S})} (P^{n-1} - (Q_{\mathcal{B}}P)^{n-1})\chi_R \, d\mu.$$

Thus, by induction, we obtain the claim. $\blacksquare$

Using Proposition 3.3 and Lemma 3.4 we obtain the following estimate on the error between the true transport rate $T_{i,j}(n)$ and its approximation. To abbreviate the notation, let $\underline{e}_i, \overline{e}_i, \underline{u}_i$ and $\overline{u}_i \in \mathbb{R}^b$ be defined by

$$(\underline{e}_i)_k = \begin{cases} 1, & \text{if } B_k \subset R_i, \\ 0, & \text{else} \end{cases},$$

$$(\overline{e}_i)_k = \begin{cases} 1, & \text{if } B_k \cap R_i \neq \emptyset, \\ 0, & \text{else} \end{cases}$$

and

$$(\underline{u}_i)_k = \begin{cases} \mu(B_k), & \text{if } B_k \subset R_i, \\ 0, & \text{else}, \end{cases},$$

$$(\overline{u}_i)_k = \begin{cases} \mu(B_k), & \text{if } B_k \cap R_i \neq \emptyset, \\ 0, & \text{else}, \end{cases},$$

where $k = 1, \ldots, b$.

Lemma 3.5.   *Let $R_i, R_j \subset M$ and*

$$R_0^j = R_j, \quad R_{k+1}^j = f^{-1}\left(\overline{R}_k^j\right), \quad k = 0, 1, 2, \ldots,$$

then for $n = 1, 2, \ldots$

$$\left| T_{i,j}(n) - \underline{e}_j^T P_{\mathcal{B}}^n \underline{u}_i \right|$$

$$\leq \underline{e}_j^T P_{\mathcal{B}}^n (\overline{u}_i - \underline{u}_i) + (\overline{e}_j - \underline{e}_j)^T P_{\mathcal{B}}^n \overline{u}_i$$

$$+ 2\sum_{k=0}^{n-1} (n - k)\mu\left(R_i \cap \overline{R}_k^j \backslash R_k^j \right).$$

Proof.

$$\left| T_{i,j}(n) - \underline{e}_j^T P_{\mathcal{B}}^n \underline{u}_i \right|$$

$$= \left| \int_{R_j} P^n\chi_{R_i} \, d\mu - \int_{\underline{R}_j} P_{\mathcal{B}}^n\chi_{\underline{R}_i} \, d\mu \right|$$

$$\leq \left| \int_{R_j} P^n\chi_{R_i} \, d\mu - \int_{R_j} P_{\mathcal{B}}^n\chi_{R_i} \, d\mu \right|$$

$$+ \left| \int_{R_j} P_{\mathcal{B}}^n\chi_{R_i} \, d\mu - \int_{\underline{R}_j} P_{\mathcal{B}}^n\chi_{\underline{R}_i} \, d\mu \right|$$

16   M. Dellnitz et al.

A bound on the first term on the right-hand side is given by Lemma 3.4. For the second term, we use the observation that led to Proposition 3.3 and get

$$\left| \int_{R_j} P_{\mathcal{B}}^n \chi_{R_i} d\mu - \int_{\underline{R}_j} P_{\mathcal{B}}^n \chi_{\underline{R}_i} d\mu \right|$$

$$\leq \int_{\overline{R}_j} P_{\mathcal{B}}^n \chi_{\overline{R}_i} d\mu - \int_{\underline{R}_j} P_{\mathcal{B}}^n \chi_{\underline{R}_i} d\mu$$

$$= \overline{e}_j^T P_{\mathcal{B}}^n \overline{u}_i - \underline{e}_j^T P_{\mathcal{B}}^n \underline{u}_i$$

$$= \underline{e}_j^T P_{\mathcal{B}}^n (\overline{u}_i - \underline{u}_i) + (\overline{e}_j - \underline{e}_j)^T P_{\mathcal{B}}^n \overline{u}_i,$$

which proves the claim. ■

This estimate gives a bound on the error between the true transport rate $T_{i,j}(n)$ and the one computed via the transition matrix, $\underline{e}_j^T P_{\mathcal{B}}^n \underline{u}_i$, in terms of those elements of the fine partition $\mathcal{B}$ that either intersect the boundary of R_i and are mapped into R_j or that intersect R_i at all and are mapped "onto" the boundary of R_j. In Fig. 5 we illustrate this idea by sketching two box-transitions that contribute to the error. So an obvious consequence of Lemma 3.5 is that in order to ensure a certain degree of accuracy of the transport rates for large n, these particular boxes need to be refined. Using (12) it also follows that for $n = 1$ the estimate in Proposition 3.3 holds for the discretized transfer operator, too, i.e.

$$\underline{e}_j^T P_{\mathcal{B}} \underline{u}_i \leq T_{i,j}(1) \leq \overline{e}_j^T P_{\mathcal{B}} \overline{u}_i. \tag{15}$$

Moreover, if $\underline{R}_i = \overline{R}_i$ for all sets R_i under consideration (i.e. the sets R_i are box collections) then

$$\underline{e}_j^T P_{\mathcal{B}}^n \underline{u}_i = \overline{e}_j^T P_{\mathcal{B}}^n \overline{u}_i$$

for all $n \in \mathbb{N}$. Notably the estimate in Lemma 3.5 reduces to

$$|T_{i,j}(n) - \underline{e}_j^T P_{\mathcal{B}}^n \underline{u}_i|$$

$$\leq 2 \sum_{k=0}^{n-1} (n-k)\mu\left(R_i \cap \overline{R}_k^j \backslash \underline{R}_k^j \right), \tag{16}$$

and for the special case of $n = 1$ we even get the exact transport rate:

$$\underline{e}_j^T P_{\mathcal{B}} \underline{u}_i = T_{i,j}(1) = \overline{e}_j^T P_{\mathcal{B}} \overline{u}_i. \tag{17}$$

Note in particular that the numerical effort to compute the approximate transport rate

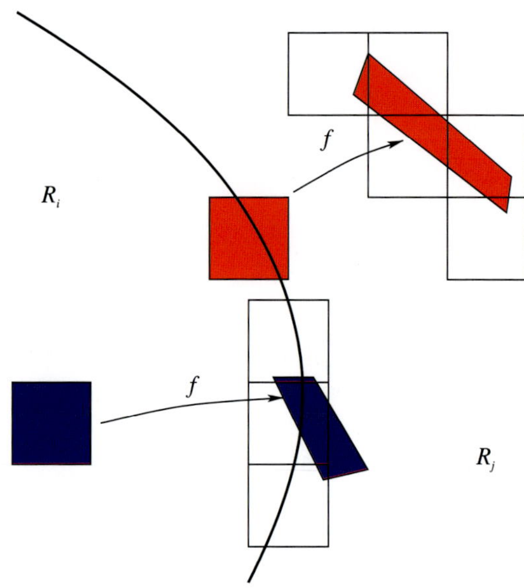

Fig. 5.   Two box transitions that contribute to the error between the computed and the actual value of the transport rate $T_{i,j}(1)$ from region R_i into region R_j after one iterate.

$\underline{e}_j^T P_{\mathcal{B}}^n \underline{u}_i$ essentially consists in n matrix-vector-multiplications — where the matrix $P_{\mathcal{B}}$ is sparse.

3.2.4.   Convergence

Lemma 3.5 yields the following convergence statement for the approximate transport rate $\underline{e}_j^T P_{\mathcal{B}}^n \underline{u}_i$ as the partition $\mathcal{B}$ is refined. Let $(\mathcal{B}_\ell)_\ell$ be a sequence of partitions such that

$$\max_{B \in \mathcal{B}_\ell} \operatorname{diam}(B) \to 0 \quad \text{as } \ell \to \infty. \tag{18}$$

Corollary 3.6. *If the regions $R_i, i = 1, \ldots, N_R$, are chosen such that for all i*

$$\mu\left(\bigcup_{\substack{B \in \mathcal{B}_\ell \\ B \cap \partial R_i \neq \emptyset}} B \right) \to 0 \quad \text{as } \ell \to \infty, \tag{19}$$

then for all n (fixed) and all i, j,

$$\underline{e}_j^T P_{\mathcal{B}_\ell}^n \underline{u}_i \to T_{i,j}(n) \tag{20}$$

as $\ell \to \infty$.

Clearly, under the assumption (18), the condition (20) will be satisfied if the boundaries of the regions R_i are piecewise smooth — as in our case, where the boundaries of the regions are composed of pieces of invariant manifolds.

3.2.5. *Almost invariant decompositions*

So far, we have discussed how to compute transport between two given regions. In the remainder of this section we will turn to the question of how to actually *find* regions of interest. In this context, a region will be of interest if it is almost invariant in the sense that typical points are mapped into the region itself with high probability. The problem of decomposing M into almost invariant sets can be formulated in graph theoretic notation and then solved by applying graph partitioning methods.

The transition probability for two measurable sets R_i and R_j is defined as

$$\rho(R_i, R_j) = \frac{\mu(f^{-1}(R_i) \cap R_j)}{\mu(R_j)}, \quad \mu(R_j) \neq 0. \quad (21)$$

If we consider the case $R_i = R_j = R$, then this transition probability measures which fraction (measured with respect to μ) of the points in R stays within R after one iteration of f. For an invariant set $R = f(R)$ with positive μ-measure, this ratio will be 1. We therefore define the *invariance ratio* of R as

$$\rho(R) = \rho(R, R). \quad (22)$$

For a given map $f : M \to M$ one can decompose its maximal invariant set into invariant parts, as e.g. chain recurrent sets and connecting orbits between them. For details on these concepts see e.g. [Easton, 1998]. But one may go one step further and ask for macroscopic dynamical structures within the chain recurrent sets themselves. One possible decomposition is given by an *almost invariant decomposition* of M (where for simplicity we assume M to be chain recurrent from now on) as defined in [Froyland & Dellnitz, 2003]: We ask for a measurable partition $\mathcal{R} = \{R_1, \ldots, R_{N_R}\}$ of M into N_R sets (with N_R fixed) with positive measure (i.e. $\mu(R_k) > 0$), such that the quantity

$$\rho(\mathcal{R}) = \frac{1}{N_R} \sum_{k=1}^{N_R} \rho(R_k) \quad (23)$$

is maximized over all such partitions.

Evidently the infinite dimensional optimization problem (23) needs to be discretized so it may be treated numerically. To this end we again restrict ourselves to sets within $\mathcal{C}_\mathcal{B}$, i.e. to sets that are unions of elements of the partition $\mathcal{B}$. Therefore, our goal is to look for partitions

$\tilde{\mathcal{R}} = \{\tilde{R}_1, \ldots, \tilde{R}_{N_R}\}, \tilde{R}_k \in \mathcal{C}_\mathcal{B}, \mu(\tilde{R}_k) > 0$, such that

$$\rho(\tilde{\mathcal{R}}) = \frac{1}{N_R} \sum_{k=1}^{N_R} \rho(\tilde{R}_k) = \frac{1}{N_R} \sum_{k=1}^{N_R} \rho(\tilde{R}_k, \tilde{R}_k) \quad (24)$$

is maximized over these special partitions.

3.2.6. *Graph formulation*

Consider the *transition matrix* $P_\mathcal{B}$ from (13). A transition matrix $P = (P_{ij})$ is called *reversible*, if for all i, j we have $p_j P_{ij} = p_i P_{ji}$, where p is the stationary distribution of P, i.e. $Pp = p$. The matrix $P_\mathcal{B}$ is not necessarily reversible. However, the matrix $Q_\mathcal{B}$ defined by

$$Q_\mathcal{B} = \frac{1}{2} \left(P_\mathcal{B} + D P_\mathcal{B}^T D^{-1} \right),$$

where $D = \text{diag}(\mu)$ denotes the diagonal matrix with the entries of μ on the diagonal and which has matrix entries

$$q_{ij} = \frac{\mu(B_j)p_{ij} + \mu(B_i)p_{ji}}{2\mu(B_j)}$$

is reversible. Let $\tilde{\mathcal{R}} = \{\tilde{R}_1, \ldots, \tilde{R}_{N_R}\}, \tilde{R}_k \in \mathcal{C}_\mathcal{B},$ $\mu(\tilde{R}_k) > 0$, be a partition of M into N_R sets. The function (24) to be optimized can be written as

$$\rho(\tilde{\mathcal{R}}) = \frac{1}{N_R} \sum_{k=1}^{N_R} \frac{\sum_{B_i, B_j \subset \tilde{R}_k} p_{ij} \cdot \mu(B_j)}{\sum_{B_j \subset \tilde{R}_k} \mu(B_j)}$$

$$= \frac{1}{N_R} \sum_{k=1}^{N_R} \frac{\sum_{B_i, B_j \subset \tilde{R}_k} q_{ij} \cdot \mu(B_j)}{\sum_{B_j \subset \tilde{R}_k} \mu(B_j)} \quad (25)$$

because of $\mu(B_j)p_{ij} + \mu(B_i)p_{ji} = 2\mu(B_j)q_{ij}$.

This optimization problem can be translated into the question of finding an optimal cut in a graph. Let $G = (V, E)$ be a graph with vertex set $V = \mathcal{B}$ and directed edge set

$$E = E(\mathcal{B}) = \{(B_1, B_2) \in \mathcal{B} \times \mathcal{B} \mid f(B_1) \cap B_2 \neq \emptyset\}.$$

The vertex weight function $vw \colon V \to \mathbb{R}$ with $vw(B_i) = \mu(B_i)$ assigns a weight to the vertices and the edge weight function $ew \colon E \to \mathbb{R}$ with $ew((B_i, B_j)) = \mu(B_i)p_{ji}$ assigns a weight to the edges. Furthermore, let

$$\overline{E} = \overline{E}(\mathcal{B})$$
$$= \{\{B_1, B_2\} \subset \mathcal{B} \mid$$
$$(f(B_1) \cap B_2) \cup (f(B_2) \cap B_1) \neq \emptyset\}.$$

This defines an undirected graph $\overline{G} = (V, \overline{E})$ with a weight function $\overline{ew}\colon \overline{E} \to \mathbb{R}$ with $\overline{ew}(\{B_i, B_j\}) = 2\mu(B_i)q_{ji} = 2\mu(B_j)q_{ij} = \mu(B_j)p_{ij} + \mu(B_i)p_{ji}$ on the edges. The difference between the graphs G and $\overline{G}$ is that in $\overline{G}$ the edge weight between two vertices is the sum of the edge weights of the two directed edges between the same vertices in G. Thus, the total edge weights of both graphs are identical.

The partition $\tilde{\mathcal{R}}$ corresponds to the partition of V into $\mathcal{V} = \{V_1, \ldots, V_{N_R}\}$ with $V_i = \{B_i; B_i \subset \tilde{R}_i\}$. For a set $W \subset V$ we denote

$$C_{\mathrm{int}}(W) = \frac{\sum_{(v,w) \in E; v,w \in W} ew(\{v,w\})}{\sum_{v \in W} vw(v)}$$

$$= \frac{\sum_{\{v,w\} \in \overline{E}; v,w \in W} \overline{ew}(\{v,w\})}{\sum_{v \in W} vw(v)},$$

called the *internal cost of W*. Note that the internal cost is independent from the choice between the directed graph G or the undirected graph $\overline{G}$. Thus, we are allowed to operate on undirected graphs, as we shall do in the following.

For a partition $\mathcal{V} = \{V_1, \ldots, V_{N_R}\}$ we denote

$$C_{\mathrm{int}}(\mathcal{V}) = \frac{1}{N_R} \sum_{i=1}^{N_R} C_{\mathrm{int}}(V_i) \qquad (26)$$

called the *internal cost of $\mathcal{V}$*. It is an easy task to check that $\rho(\tilde{\mathcal{R}}) = C_{\mathrm{int}}(\mathcal{V})$. Thus, the optimization of our cost function (24) is identical to the optimization of the internal costs of the partition $\mathcal{V}$ (26) written in graph notation and we have established the graph partitioning problem

$$C_{\mathrm{int}}(\mathcal{V}) \to \max. \qquad (27)$$

3.2.7.   *Heuristics and tools for the graph partitioning problem*

The optimization problem (27) is known to be NP-complete (even for constant weights, see [Garey & Johnson, 1979]), i.e. an efficient algorithm for solving this problem is not known. Efficient graph partitioning heuristics have been developed for a number of different applications. There are several software libraries, each of which provides a range of different methods. Examples are CHACO [Hendrickson & Leland, 1995], JOSTLE [Walshaw, 2000], METIS [Karypis & Kumar, 1999], SCOTCH [Pellegrini, 1996] or PARTY [Monien *et al.*, 2000]. These libraries are designed to create solutions to the balanced partitioning problem in which all parts

are restricted to have an equal (or almost equal) volume of the underlying measure. Therefore, we will use parts of the library PARTY and combine them with some new code which is specially designed to address our cost function (27).

PARTY, like other graph partitioning tools, follows the Multilevel Paradigm which has been proven to be a very powerful approach to efficient graph-partitioning. See e.g. [Gupta, 1997; Hendrickson & Leland, 1995; Karypis & Kumar, 1999; Monien *et al.*, 2000; Ponnusamy *et al.*, 1994; Preis, 2000] for a deeper discussion. The efficiency of this paradigm is dominated by two parts: graph coarsening and local improvement. The graph is coarsened down in several levels until a graph with a sufficiently small number of vertices is constructed. A single coarsening step between two levels can be performed by the use of graph matching (independent sets of vertex pairs).

Different methods for calculating the matching will result in different solutions of the partitioning problem. To achieve a selection of different results we will consider heuristics with the following graph matching algorithms:

1. Heavy Edge Matching (HEM): It is a simple, fast and widely used matching strategy in which the weight of the edges are considered.
2. Greedy Matching (GRM): The solution is within a factor of two from the optimal matching, but it requires the sorting of the edges in a preprocessing step.
3. Locally Heaviest Matching (LHM): It is a short algorithm which also guarantees a factor of at most two, but it runs in linear time [Preis, 1999].
4. Path Growing Matching (PGM): It has the same theoretical runtime and approximation quality as LHM but it follows a different strategy [Drake & Hogardy, 2002].

All these matching algorithms are implemented in PARTY and a discussion about their use in the graph partitioning context can be found in [Monien *et al.*, 2000; Preis, 2000].

The coarsening process is stopped when the number of vertices is equal to the desired number of parts N_R. Thus, each vertex of the coarse graph is one part of the partition. However, it is also possible to stop the coarsening process as soon as the number of vertices is sufficiently small. Then, any standard graph partitioning method can be used to calculate a partition of the coarse graph.

Finally, the partition of the smallest graph is projected back level-by-level to the initial graph and the partition is locally refined on each level. Standard methods for local improvement are Kernighan/Lin [Kernighan & Lin, 1970] type of algorithms with improvement ideas from Fiduccia/Mattheyses [Fiduccia & Mettheyses, 1982]. The algorithm moves single vertices between the parts to improve the cost function. The choice of the vertices to be moved depends on the cost function to be considered. Therefore, the Kernighan/Lin implementation has been modified in PARTY such that it optimizes the cost-function C_{int}.

The software environment GADS (Graph Algorithms for Dynamical Systems) has been established, which consists of a collection of graph algorithms which are useful for the analyses of dynamical systems. See [Dellnitz & Preis, 2003; Padberg *et al.*, 2004]. It has an interface to the graph partitioning library PARTY [Monien *et al.*, 2000] and is designed to work with the tool GAIO (Global Analysis of Invariant Objects, cf. [Dellnitz *et al.*, 2001b].

4. Example: The Sun–Jupiter–Asteroid System

We will compare and combine both methods from Sec. 3 within the example of the PCR3BP with the Sun and Jupiter as the main bodies, using $\epsilon = 9.5368 \times 10^{-4}$. We consider the motion of a particle (asteroid) that has an energy $E = -1.525$ (that is, the Jacobi constant is $C = 3.05$), case 1, as depicted in Fig. 1(a). We will study transport in the exterior realm, using the Poincaré section, $f: M \to M$ where $M \subset \mathbb{R}^2$, defined in Eq. (3), which is shown in Fig. 2(b).

4.1. *Lobe dynamics*

The only requirement to use lobe dynamics is being able to generate stable and unstable manifolds of the hyperbolic structures in phase space for the time window of interest. This has been done for many years using a simple principle. A small seed set near the hyperbolic point (positioned along the unstable eigenspace) will deform in time and stretch along the unstable invariant manifold. The same procedure performed backwards in time will render the stable manifold, but we can save computational effort by using symmetries of the map f.

4.1.1. *Symmetries of the Poincaré map f*

Using the following symmetry of the equations of motion (1),

$$y \mapsto -y, \quad t \mapsto -t, \quad \text{for all } x, \dot{y}$$

and therefore $\dot{x} \mapsto -\dot{x}$, the Poincaré map f on the surface of Sec. 3 has the corresponding symmetry

$$\text{sym}: M \times \mathbb{Z} \to M \times \mathbb{Z}, \quad (x, \dot{x}) \mapsto (x, -\dot{x}), \\ n \mapsto -n. \tag{28}$$

This symmetry implies that the Poincaré map f is symmetric with respect to reflection about the x-axis and time reversal. Note that this notion of symmetry with time reversal is very useful since it relates to stable and unstable manifolds.

4.1.2. *Finding a fixed point p of f*

Due to the symmetry (28), we may expect to find a fixed point for the Poincaré-map f along the x-axis. Using differential correction, we numerically find an unstable fixed point at $p = (\overline{x}, \dot{\overline{x}}) = (-2.029579567343744, 0)$, shown in Fig. 6(a).

4.1.3. *Finding the stable and unstable manifolds of p under f*

Denote the four branches of the stable and unstable manifolds of p by $W_+^u(p), W_-^u(p), W_+^s(p)$, and $W_-^s(p)$. We will consider only the "+" branches. Using the symmetry reduces the calculations by a factor of two, i.e. $W_+^s(p) = \text{sym}\left(W_+^u(p)\right)$. The local approximation to $W_+^u(p)$ can be obtained as given in [Parker & Chua, 1989]. The basic idea is to linearize the equations of motion about the periodic orbit in the energy surface and then use the monodromy matrix provided by Floquet theory to generate a linear approximation of $W_+^u(p)$. The linear approximation, in the form of a state vector, is numerically integrated in the nonlinear equations of motion to produce the approximation of $W_+^u(p)$.

In practice, we take a finite segment along this linear approximation described by an ordered array of points (the "seed"). Using a standard numerical integration scheme (in this case, RK78), we numerically integrate the equations of motion (1) to obtain the Poincaré map f. Under iterates of f, each point approaches the manifold at an exponential rate, reducing the positioning error. However, the seed also stretches in the direction of the manifold and

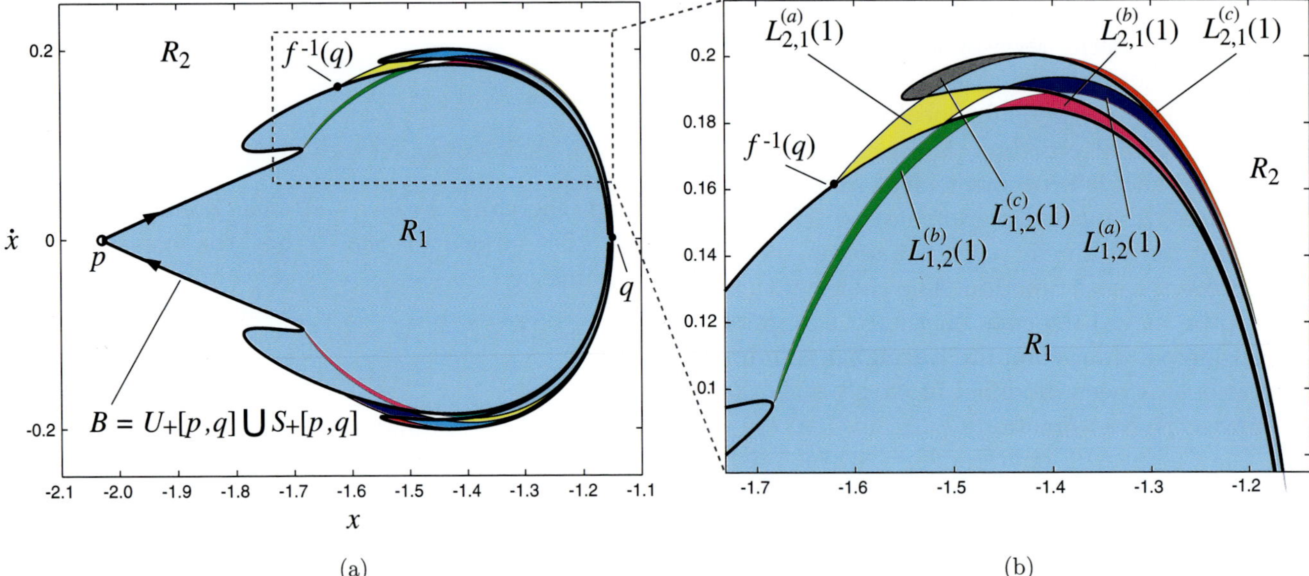

Fig. 6.   **Transport using lobe dynamics** for the same Poincaré surface of section shown in Fig. 2(b). (a) The boundary B between two regions is shown as the thick black line, formed by pieces of one branch of the stable and unstable manifolds of the unstable fixed point p. We can call the region inside of the boundary R_1 (in cyan) and the outside R_2 (in white). The pips q and $f^{-1}(q)$ are shown as black dots along the boundary and the turnstile lobes that will determine the transport between R_1 and R_2 are shown as colored regions. In (b), we see more details of the turnstile lobes. This is a case of a multilobe, self-intersecting turnstile discussed in Sec. 3.1. A schematic of this situation is shown in Fig. 4. In this case we define the turnstile lobes to be $L_{1,2}(1) = L_{1,2}^{(a)}(1) \cup L_{1,2}^{(b)}(1) \cup L_{1,2}^{(c)}(1)$ and $L_{2,1}(1) = L_{2,1}^{(a)}(1) \cup L_{2,1}^{(b)}(1) \cup L_{2,1}^{(c)}(1)$.

the distance between each point increases exponentially. Since the manifold experiences rapid stretching as it grows in length, it is necessary to check the distance between adjacent points and insert new points if necessary to insure that sufficient spatial resolution is maintained [Lekien, 2003]. The software package MANGEN is used to implement the adaptive conditioning of the mesh of points approximating the manifold [Lekien & Coulliette, 2004; Lekien, 2003]. More points are added where curvature or stretching is high.

4.1.4.   *Defining the regions and finding the relevant lobes*

The symmetry (28) is useful for defining the regions and lobes. The first intersection of $W_+^u(p)$ with the axis of symmetry is the natural choice for the pip q defining the boundary, shown in Fig. 6(a). We define R_1 (in cyan) to be the region bounded by $B = U_+[p,q] \cup S_+[p,q]$, where $U_+[p,q]$ and $S_+[p,q]$ are segments of $W_+^u(p)$ and $W_+^s(p)$, respectively, between p and q. We define R_2 (in white) to be the complement of R_1.

MANGEN can then be used to compute the turnstile lobes $L_{1,2}(1) \cup L_{2,1}(1)$. The turnstile lobes are shown as colored regions in the upper half plane of

Fig. 6(a). The first iterate of the turnstile lobes is shown in the lower half plane of Fig. 6(a) in corresponding colors. In the enlarged view, Fig. 6(b), the turnstile lobes are shown in greater detail. This is a case of a multilobe, self-intersecting turnstile, discussed in Sec. 3.1.

The area of the turnstile lobes, i.e. the flux of phase space across the boundary B (and the transport of species across B for *just the first iteration* of the map f), is summarized in Table 1.

4.1.5.   *Higher iterates of the map*

To compute all the transport quantities $T_{1,1}(n)$, $T_{1,2}(n)$, $T_{2,1}(n)$, and $T_{2,2}(n)$, it is only

Table 1.   **Flux of phase space across the boundary** in terms of canonical area per iterate. Note, $\mu(L_{1,2}(1))$ is the sum $\mu(L_{1,2}^{(a)}(1)) + \mu(L_{1,2}^{(b)}(1)) + \mu(L_{1,2}^{(c)}(1))$. This is the flux in *both directions*, i.e. $\mu(L_{1,2}(1)) = \mu(L_{2,1}(1))$, since the map f is area-preserving on M.

$\mu(L_{1,2}^{(a)}(1))$	$\mu(L_{1,2}^{(b)}(1))$	$\mu(L_{1,2}^{(c)}(1))$	$\mu(L_{1,2}(1))$
0.000956	0.000870	0.000399	**0.002225**

necessary to compute *one* of them. We compute $T_{1,2}(n)$. By area preservation of the map f, we have

$$T_{1,1}(n) = \mu(R_1) - T_{1,2}(n),$$

$$T_{2,1}(n) = T_{1,2}(n),$$

$$T_{2,2}(n) = \mu(R_2) - T_{1,2}(n).$$

The values for $T_{1,2}(n)$ up to $n = 5$ are given in Table 5. We cannot compute beyond $n = 5$ due to computer memory limitations of storing the windy boundaries of the lobes.

4.1.6. *Re-entrainment of the lobes*

We now illustrate the effect of re-entrainment of the lobes, i.e. lobes leaving and re-entering the specified regions. This geometric effect is believed to have important consequences for the behavior of $T_{i,j}(n)$ as n increases [Wiggins, 1992].

Consider Fig. 7. In this figure we show preimages and images of only the lobe labeled $L_{2,1}^{(b)}(1)$ in Fig. 6(b). Four preimages and five images of this lobe are shown in Fig. 7(a). By definition, we must have $f(L_{2,1}^{(b)}(1)) \subset R_1$, but the other images, i.e. $f^k(L_{2,1}^{(b)}(1))$ for $k > 1$, need not be contained entirely in R_1. In the specific geometry shown here, $f^k(L_{2,1}^{(b)}(1)) \cap R_2 \neq \emptyset$ for $k > k_f$, where $k_f = 3$. The boxed region in Fig. 7(a) is shown in more detail in Fig. 7(b). The area of the lobe which lies in R_1 or R_2 is shown in Fig. 7(c). We conclude that some particles in $L_{2,1}^{(b)}(1)$ which begins in R_2 will enter R_1 only to return to R_2 after just three iterates in R_1.

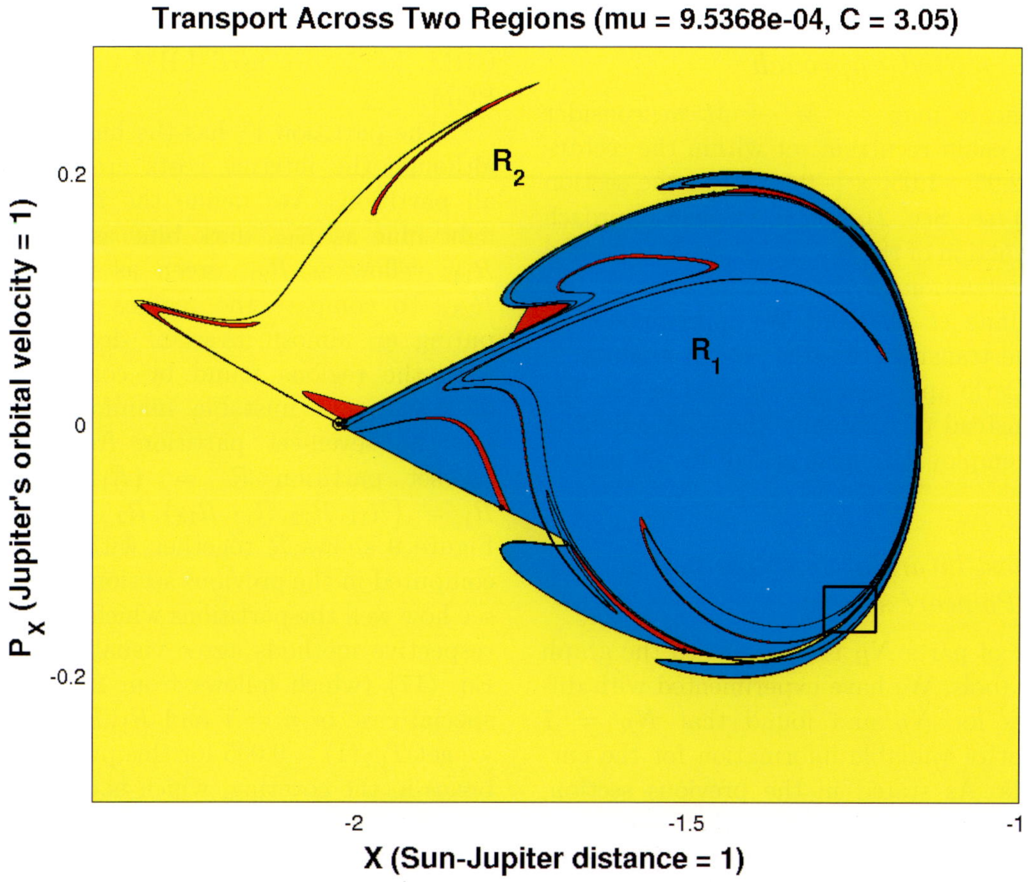

(a)

Fig. 7. **Re-entrainment of the lobes.** We show preimages and images of only the lobe labeled $L_{2,1}^{(b)}(1)$ in Fig. 6(b). (a) Four preimages and five images of this lobe are shown. Notice that the images are not contained entirely in R_1, i.e. $f^k(L_{2,1}^{(b)}(1)) \cap R_2 \neq \emptyset$ for $k > k_f$, where $k_f = 3$. (b) The boxed region in (a) is shown in more detail. (c) The area of the lobe which lies in R_1 or R_2 is shown.

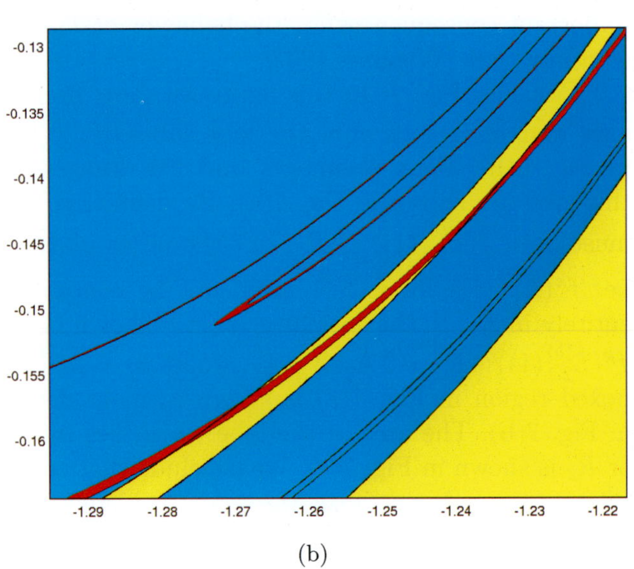

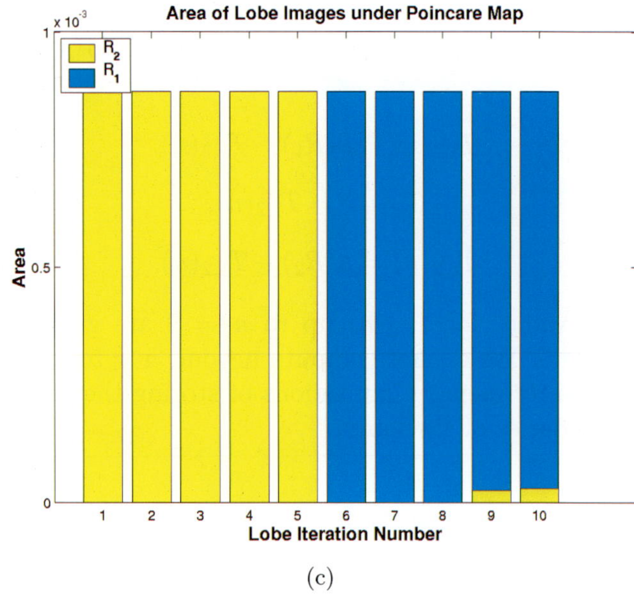

(b)   (c)

Fig. 7.   (*Continued*)

4.2. *Set oriented approach*

For the Poincaré map $f : M \to M$ we consider M to be the chain recurrent set within the rectangle $X = [-2.95, -1.05] \times [-0.5, 0.5]$ in the section $y = 0$, $\dot{y} > 0$ (see Sec. 2). For an efficient approach to the construction of the box coverings $\mathcal{B}$ as needed for the discretization of the transfer operator we refer to [Dellnitz *et al.*, 2000]. We approximate the entries of the transition matrix (13) in analogy to the Monte–Carlo approach as described in Sec. 3.2. Only here instead of randomly choosing points in each box we employ a uniform grid of 16×16 points.

4.2.1. *Almost invariant decomposition of the Poincaré section*

The number of parts N_R is an input to the graph partitioning tools. We have experimented with different values for N_R and found that $N_R = 7$ exhibits a lot of valuable information for the current example. As stated in the previous section, we would like to find a partition that maximizes our internal cost. Since the problem of computing an optimal solution is NP-complete, we apply some heuristics as described in Sec. 3.2. Figure 8 shows four different decompositions of M into seven almost invariant sets, obtained by different parameters for the coarsening step. For the partition $\mathcal{V}_1$ we used the matching strategy HEM, for $\mathcal{V}_2$ we used

GRM, for $\mathcal{V}_3$ we used LHM and for $\mathcal{V}_4$ we used PGM.

The partition $\mathcal{V}_2$ has the highest internal cost, although the internal costs are almost equal for all partitions. We define the red region as R_{11}, light blue as R_{12}, dark blue as R_{13}, magenta as R_{14}, yellow as R_{21}, green as R_{22} and white as R_{23}. To compare the regions obtained by computing an almost invariant decomposition of M with the regions found by considering branches of stable and unstable manifolds we agglomerate the seven-set partition from above into a two-set partition $\mathcal{R} = \{R_1, R_2\}$ by defining $R_1 = \{R_{11}, R_{12}, R_{13}, R_{14}\}, R_2 = \{R_{21}, R_{22}, R_{23}\}$. Figure 9 shows $\mathcal{R}$ together with the boundary as computed in the previous section. It is intriguing to see how well the partitions which were found by the respective methods agree visually. However, using Eq. (17) (which follows from Lemma 3.5 for the special case of $n = 1$ and R_1, R_2 box collections) we get $T_{1,2}(1) \approx 0.005$ for this particular size of the boxes in the covering, which is considerably larger than the value of 0.0022 as computed in the previous section for the corresponding partition given by the invariant manifolds.

4.2.2. *Transport for a two-set partition*

In this section we are going to compare the value for the quantity $T_{i,j}(1)$ as computed using lobe

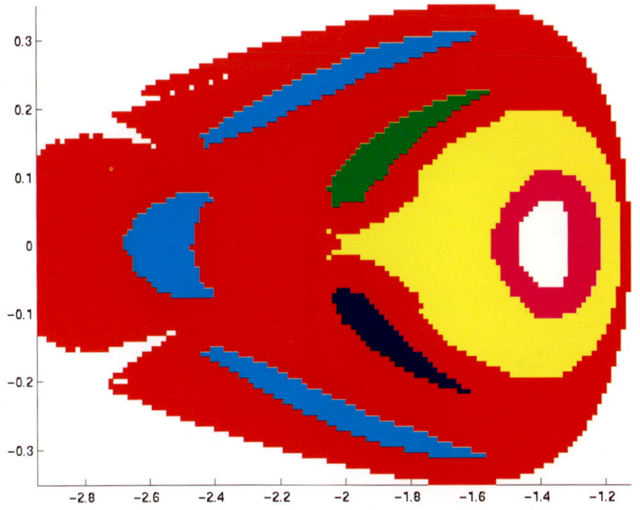

(a) The partition $\mathcal{V}_1$ obtained with the HEM coarsening strategy has an internal cost of 0.9453.

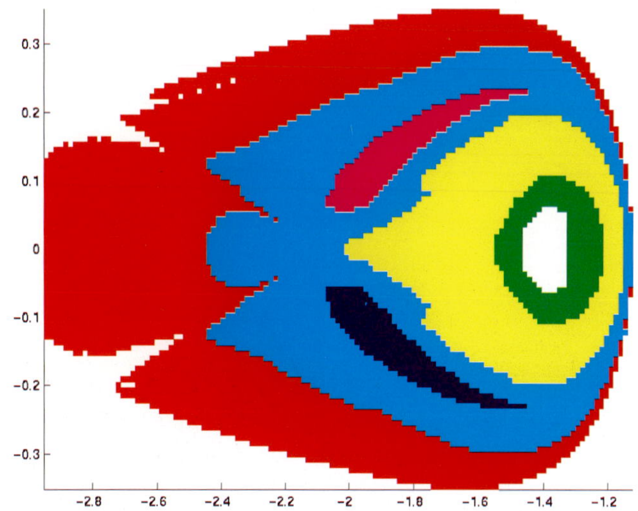

(b) The partition $\mathcal{V}_2$ obtained with the GRE coarsening strategy has an internal cost of 0.9493.

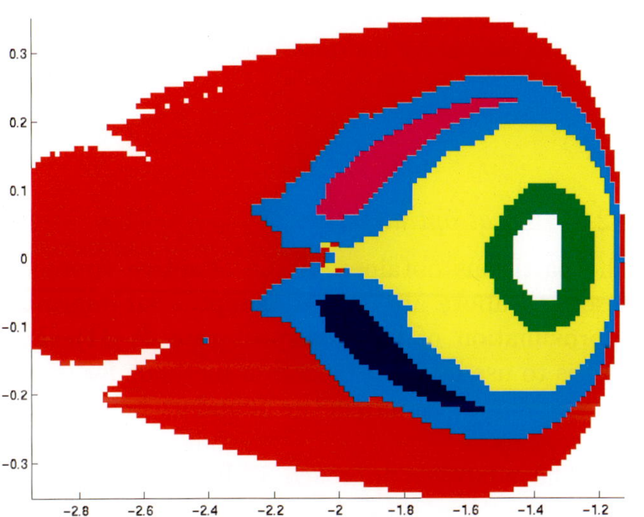

(c) The partition $\mathcal{V}_3$ obtained with the LHM coarsening strategy has an internal cost of 0.9472.

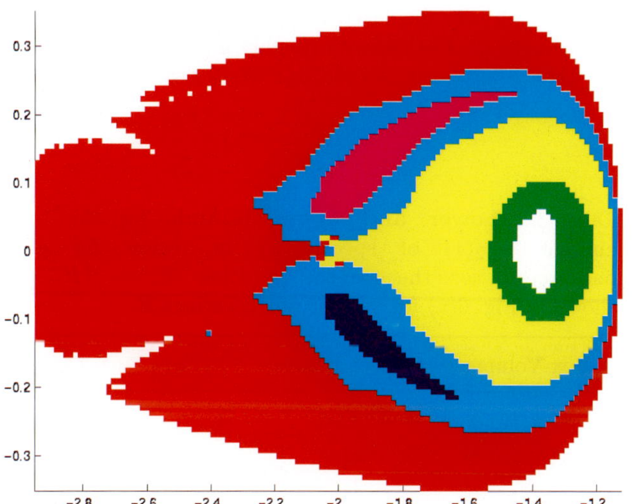

(d) The partition $\mathcal{V}_4$ obtained with the PGM coarsening strategy has an internal cost of 0.9458.

Fig. 8. *Almost invariant decomposition* of the chain recurrent set M into seven sets, indicated by different colors. We used different partitioning strategies to obtain the subfigures above.

dynamics with the one resulting from an application of the set oriented approach. For both computations we consider the two-set partition $\mathcal{R} = \{R_1, R_2\}$ defined by the two segments of stable and unstable manifolds of a certain fixed point as computed in Sec. 4.1, see Fig. 6(a).

The third and fourth columns of Table 2 show the values $\underline{e}_2^T P_\mathcal{B} \underline{u}_1$ and $\overline{e}_2^T P_\mathcal{B} \overline{u}_1$ for different partitions $\mathcal{B}$ of equally sized boxes. As suggested in Eq. (15) these values indeed sandwich the

value of 0.002225 for $T_{1,2}(1)$ as computed using lobe dynamics. However, the bounds are not very tight and seem to converge rather slowly towards the true value.

On the other hand, the error in computing $T_{i,j}(1)$ has to be related to a small subset of boxes of $\mathcal{B}$ only, see Lemma 3.5 and comments thereafter. It is therefore natural to consider an *adaptive* approach to the construction of the partitions $\mathcal{B}$ in the sense that one only refines boxes that

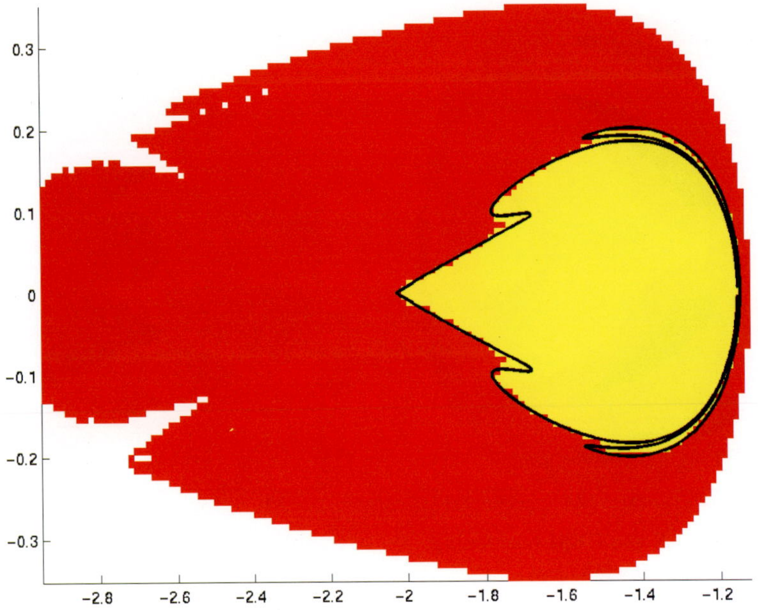

Fig. 9.   *Almost invariant decomposition into two sets.* We are interested in the transport between the two regions R_1 and R_2 which we already displayed in Fig. 7(a). The red and yellow areas are an almost invariant decomposition into two sets. The border between the two sets roughly matches the boundary formed by the branches of the stable and unstable manifolds of the fixed point $(-2.029579567343744, 0)$ drawn as a line.

Table 2. **Lower and upper bounds** for the total amount $T_{1,2}(1)$ of species S_1 in region R_2 after one iterate for the two-set partition $\mathcal{R} = \{R_1, R_2\}$ shown in Fig. 6(a) for various box coverings $\mathcal{B}$.

Box Volume	No. of Boxes	$\underline{e}_2^T P_\mathcal{B} \underline{u}_1$	$\overline{e}_2^T P_\mathcal{B} \overline{u}_1$
4.6387×10^{-4}	2238	0	0.067417
2.3193×10^{-4}	4436	0	0.058418
1.1597×10^{-4}	8673	0	0.041038
5.7983×10^{-5}	17216	0.000034	0.034708
2.8992×10^{-5}	32789	0.000258	0.022962

contribute to the error. To be able to identify these, we rely on the results from the previous section on the lobe dynamics approach, where we computed the pieces of the invariant manifolds bounding the two regions to high accuracy. Figure 10 shows the result of an implementation of this approach, the third and fourth columns of Table 3 show the corresponding lower and upper bounds. Note that for a comparable number of boxes in the partitions $\mathcal{B}$ these bounds are much tighter than those computed using equally sized boxes.

4.2.3.   *Local optimization*

The partitions obtained in the adaptive approach described above are used to compute an improved approximation of the transport rate $T_{1,2}(1)$. The idea is to use local optimization methods for graph partitioning to smoothen the boundary between the two regions.

The adaptive approach is based on a partition into three sets A_1, A_2 and A_b with $A_1 \cup A_2 \cup A_b = R_1 \cup R_2$ corresponding to an internal set $A_1 \subset R_1$, an external set $A_2 \subset R_2$ and a boundary set A_b, which is a box covering of the boundary between R_1 and R_2 provided by the results from the lobe dynamics approach. To get an approximation of the transport rates $T_{1,2}(1)$, we artificially construct a two-partition of the underlying graph corresponding to the sets A_1 and $A_2 \cup A_b$. This partition is then locally optimized (by maximizing the the internal costs (27) by the methods described in Sec. 3.2. In this way, one obtains approximations of the sets R_1 and R_2, which are given as box collections, so that for this particular setting we can again compute the transport rate $T_{1,2}(1)$ using Eq. (17). The fifth column of Table 3 presents the corresponding results.

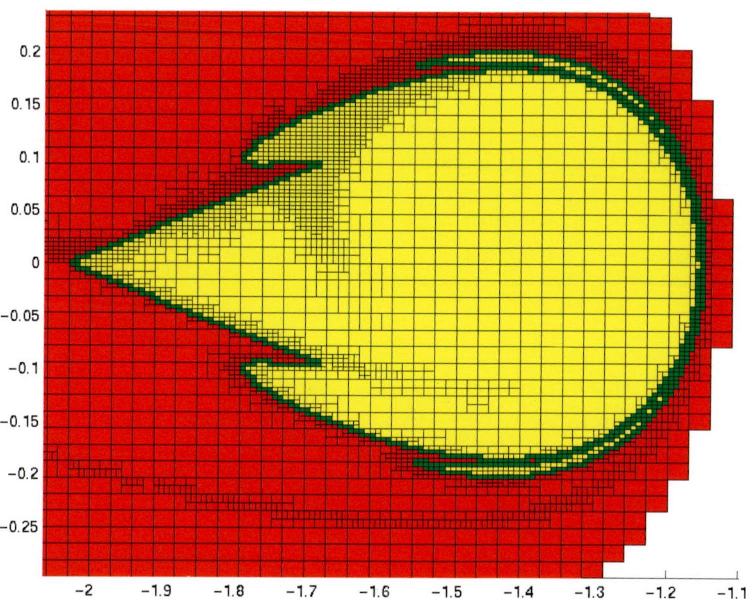

Fig. 10. Adaptive covering. Dynamical systems techniques have been used to identify locations in which box refinements are needed, i.e. where the lobes are located. This speeds up the computation considerably.

Table 3. Lower and upper bounds and the optimized value of the total amount of species S_1 contained in region R_2 after one iteration $T_{1,2}(1)$. The third and fourth columns present lower and upper bounds for the total amount $T_{1,2}(1)$ of species S_1 in region R_2 after one iterate for the two-set partition $\mathcal{R} = \{R_1, R_2\}$ shown in Fig. 6(a) for various adaptively refined box partitions $\mathcal{B}$. The fifth column lists the approximate value for $T_{1,2}(1)$, obtained by additionally locally optimizing the partition of $\mathcal{B}$ into two sets.

Box Volume (min)	No. of Boxes	$\underline{e}_2^T P_\mathcal{B} \underline{u}_1$	$e_2^T P_\mathcal{B} \overline{u}_1$	Optimized Value
4.6387×10^{-4}	2238	0	0.067417	0.008605
1.1597×10^{-4}	3269	0	0.041038	0.005166
2.8992×10^{-5}	5455	0.000258	0.022962	0.003497
7.2479×10^{-6}	10422	0.000790	0.012654	0.002622
1.8110×10^{-6}	21655	0.001362	0.007508	0.002324
4.5290×10^{-7}	45946	0.001722	0.004887	0.002314

4.2.4. *Extrapolation*

The results in Table 3 suggest that one should try to derive even better bounds by extrapolating the computed values. By Lemma 3.5 and comments thereafter, the error between $T_{i,j}(1)$ and its approximation $\underline{e}_j^T P_\mathcal{B} \underline{u}_i$ can be bounded in terms of the Lebesgue measure of a certain set of boxes that either intersect the boundary of region R_i or are mapped onto the boundary of region R_j. Roughly speaking this means that whenever those boxes are refined by bisection with respect to both coordinate directions, the Lebesgue measure of the set of boxes that contribute to the error will shrink by a factor of $1/2$. In view of this, we make the following Ansatz for an asymptotic expansion of the computed values in terms of the Lebesgue measure $\mu(\mathcal{B}) = \min_{B \in \mathcal{B}} \mu(B)$ of the relevant boxes:

$$\underline{e}_j^T P_\mathcal{B} \underline{u}_i \approx T_{i,j}(1) + C\sqrt{\mu(\mathcal{B})}, \qquad (29)$$

for some constant $C > 0$; similarly for $\overline{e}_j^T P_\mathcal{B} \overline{u}_i$ and the value as computed after locally optimizing the partition. Table 4 shows the results of extrapolating

Table 4. Extrapolation of the results in Table 3. Using the asymptotic expansion (29), the extrapolation is based on a linear interpolation of the values of two subsequent rows of Table 3.

$\mu(\mathcal{B})$	$\underline{e}_2^T P_{\mathcal{B}}\underline{u}_1$	$\overline{e}_2^T P_{\mathcal{B}}\overline{u}_1$	Linear Extrapolation
1.811×10^{-6}	0.0019337	0.0023648	0.002026
4.529×10^{-7}	0.0020821	0.0022651	0.002304

Table 5. Comparison of the two approaches for higher iterates. Approximate values for the amount $T_{1,2}(n)$ of species S_1 in region R_2 after n iterates.

n	$T_{1,2}(n)$ (lobe dynamics)	$\underline{e}_2^T P_{\mathcal{B}}^n \underline{u}_1$ (set oriented)
1	0.002230	0.002314
2	0.004461	0.004449
3	0.006692	0.006533
4	0.008898	0.008568
5	0.01110	0.01056
6	—	0.01250
7	—	0.01438
8	—	0.01623
9	—	0.01803
10	—	0.01978

the values in Table 3 based on the expansion (29). For the extrapolation we have been linearly interpolating the values of two subsequent rows of Table 3, respectively.

4.2.5. *Higher iterates of the map*

For the two-set partition $\mathcal{R} = \{R_1, R_2\}$ as employed in the previous section, Table 5 lists the approximate total amount $T_{i,j}(n)$ of species S_1 in region R_2 after n time steps. The values in the second column are based on the nth iterates of lobe volumes, whereas the values in the third column have been computed as $T_{i,j}(n) \approx \underline{e}_2^T P_{\mathcal{B}}^n \underline{u}_1$ using the approximations of R_1 and R_2 obtained by

the locally optimized partition on the finest box level in Table 3. Although the two methods do not use exactly the same two-set partition they agree to within 5% over their common domain. Table 5 and Fig. 11 show that using the set oriented approach one can efficiently approximate the quantities $T_{i,j}(n)$ for quite large n — every new iterate requires a single matrix-vector product (where the

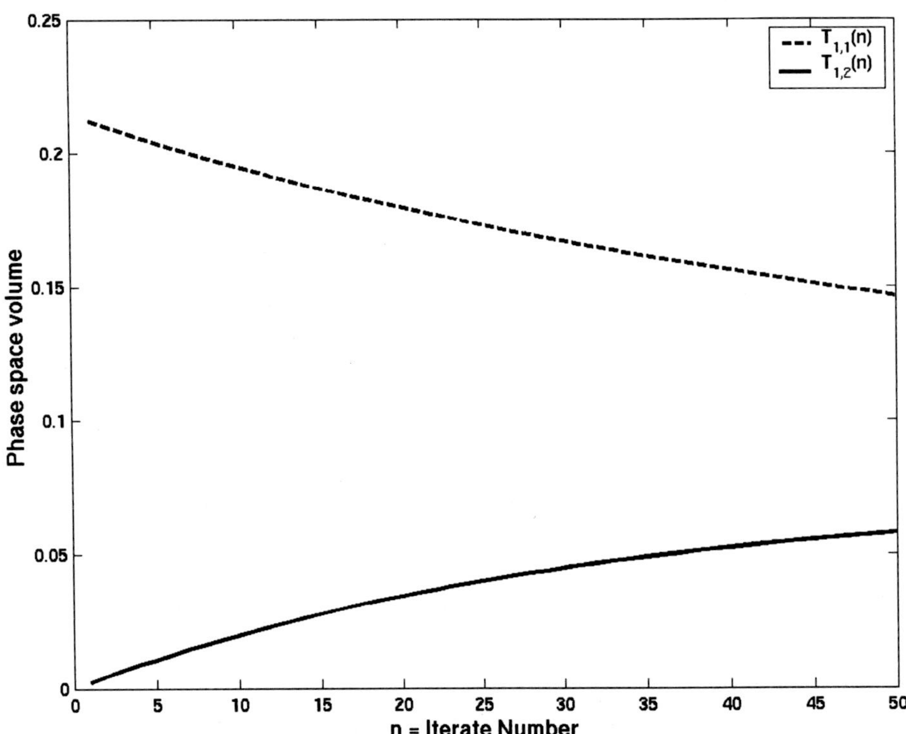

Fig. 11. Higher iterates using the set oriented method. Approximate values for $T_{1,2}(n)$ and $T_{1,1}(n)$ up to $n = 50$ iterates using the set oriented approach.

matrix is sparse) and a scalar product to be computed. Note however that since we are working with a covering consisting of boxes, typically there will be boxes that map outside the covering and thus the resulting transition matrix is not exactly stochastic. This will ultimately lead to $\underline{e}_2^T P_{\mathcal{B}}^n \underline{u}_1$ dropping to 0 with increasing n.

4.2.6. *Return times of the Poincaré map*

In terms of the time scale of the underlying differential equation, a species from $R = R_1 \cup R_2$ needs 13.02 to 36.34 years to return to R. In Fig. 11 the approximate values for $T_{1,2}(n)$ for 50 iterates are shown. Accordingly, the probability of the transition of a species from R_1 to R_2 is about 28% after 1817 years.

5. Conclusions and Future Directions

5.1. *Good agreement between approaches*

We have shown how invariant manifold techniques and the set oriented approach can work together in an important two degree of freedom example problem, reduced to a two-dimensional Poincaré map. For example, graph partitioning gives a coarse-grain global picture of the important regions and indicates where key unstable periodic points reside. The one-dimensional stable and unstable manifolds of those periodic points can then be computed and the lobe areas determined to yield highly accurate transport rates.

As one computes its extent of stable and unstable manifold curves from an initial seed, computer memory restrictions and the rapid stretching of the manifolds limits the length of the manifold which can be computed. This translates to a maximum iterate, $n_{\max}^{\text{lobe}}$, up to which transport can be accurately computed using the invariant manifold/lobe dynamics method.

Based on a coarse model of the underlying system in form of a finite-state Markov chain, the set oriented approach can compute transport quantities at higher iterates. Using a boundary between regions obtained from the invariant manifold method, one can implement adaptive refinement strategies for the underlying partition of phase space, which improves efficiency of the method. The good agreement between the set oriented approach with adaptive refinement and the lobe dynamics method over their common domain (up to $n_{\max}^{\text{lobe}}$)

leads us to believe that also for larger iterates $n > n_{\max}^{\text{lobe}}$ the computed transport rates are quite reliable. However, without proper modification the method is not yet suitable for very large iterates ($n > 100$). On the other hand, the method gives reliable transport rates for thousands of Earth years and with cautious extrapolation, one can conclude that the method is indeed of astrodynamical interest.

5.2. *Extension to higher dimensions and time dependent systems*

Some work has been done on transport in higher dimensions, for example, four-dimensional symplectic maps [Lekien & Marsden, 2004; Gillilan & Ezra, 1991]. In future work, we intend to use box methods and graph algorithms in conjunction with ideas from invariant manifold theory for studying phase space transport in higher dimensions.

Related to this, one can also consider an extension of lobe dynamics to the four-dimensional case [Lekien & Marsden, 2004; Lekien, 2003]. The four-dimensional phase space M of a volume- and orientation-preserving Poincaré map $f : M \to M$ can be divided into disjoint regions of interest, $R_i, i = 1, \ldots, N_R$, where the boundaries between regions are pieces of three-dimensional stable and unstable manifolds of two-dimensional normally hyperbolic invariant manifolds (NHIMs), $p_i, i = 1, \ldots, N_p$. Moreover, transport between regions of phase space can be completely described by the dynamical evolution of the higher dimensional turnstile lobes, volumes of the phase space enclosed by segments of the stable and unstable manifolds.

One way to approach this problem is to simply use the box subdivision and graph partitioning algorithms to partition M into its important regions and then compute the transport between them. However, this may not be computationally tractable. The complexity of box subdivision methods are proportional to the dimension of the object of interest, not the dimension of the embedding space. Thus, box subdivision methods could be used to (i) obtain the two-dimensional NHIMs, and then (ii) their stable and unstable three-dimensional manifolds which bound regions of M.

Another challenging but potentially very fruitful application is to put box subdivision methods and almost invariant sets into the time dependent context, such as occurs in, for instance, ocean dynamics [Lekien *et al.*, 2003]. For such systems, the idea of "fixed points" and "invariant manifolds" is problematic and one replaces them with notions such as those of Hallet [2002] involving *Lagrangian coherent structures*. Preliminary computations suggest that set oriented methods may be able to reveal such objects with similar properties.

5.3. *Merging techniques into a single software package*

The merging of statistical and geometric approaches yields a powerful tool. This could be reflected in the merging of the two software packages used in the current study, GAIO and MANGEN. We envision a software package for transport calculations using the box formulation along with adaptive strategies to reduce the computational effort based on highest transport and/or curvature of a low codimensional object.

Furthermore, we can make use of variational integration (VI) techniques, which are known to perform well when computing long time dynamics and chaotic invariant sets for mechanical systems, with and without dissipation. See, for instance [Kane *et al.*, 2000; Rowley & Marsden, 2002; Marsden West, 2001]. This also includes asynchronous VI techniques [Lew *et al.*, 2003] which are appropriate for taking different time steps in different *spatial regions* and yet maintaining all the conservation properties of variational integrators.

5.4. *Miscellany*

Other topics that warrant further investigation are the addition of dissipation and forcing to the problem, including the effect of other bodies, the Poynting Robertson drag and the Yarkovski effect. It would also be of interest to know how the choice of the coordinate system affects the results; specifically, lobe boundaries may be easier to handle in other coordinates, such as Delaunay (action-angle canonical) coordinates for the PCR3BP; in fact, they will not appear as convoluted in Delaunay coordinates.

5.5. *Progress towards the grand challenges in computational science*

In this paper, we seek to lay a foundation for significant progress toward some of the grand challenges in computational science, including computational astrodynamics, protein folding, and detailed predictive ocean dynamics. Our long term vision is to make a link between (i) the statistical methods which have been used to probe the dynamics in high dimensional systems and (ii) the geometric methods which provide detailed insight into the dynamics of low dimensional systems. This is a gap that we believe the work here begins to bridge.

We are ultimately interested in investigating whether the techniques described here will work for models of direct, practical interest. Thus we first work on simple models with an eye towards building more complex models using the results of simple models as building blocks.

Acknowledgments

This research was partly supported by the DAAD, DFG Priority Program 1095, NSF-ITR grant ACI-0204932, a Max Planck Research Award and the California Institute of Technology President's Fund. This work was carried out in part at the Jet Propulsion Laboratory and California Institute of Technology under a contract with National Aeronautics and Space Administration.

References

Agarwal, P. K., Billeter, S. R., Rajagopalan, P. T., Benkovic, S. J. & Hammes-Schiffer, S. [2002] "Network of coupled promoting motions in enzyme catalysis," *Proc. Nat. Acad. Sci. USA* **99**, 2794–2799.

Belbruno, E. & Marsden, B. [1997] "Resonance hopping in comets," *Astron. J.* **113**, 1433–1444.

Benner, L. A. M. & McKinnon, W. B. [1995] "On the orbital evolution and origin of comet Shoemaker–Levy 9," *Icarus* **118**, 155–168.

Carusi, A., Kresák, L., Pozzi, E. & Valsecchi, G. B. [1985] *Long Term Evolution of Short Period Comets* (Adam Hilger, Bristol, UK).

Chirikov, B. V. [1979] "A universal instability of many-dimensional oscillator systems," *Phys. Rep.* **52**, 263–379.

Conley, C. [1968] "Low energy transit orbits in the restricted three-body problem," *SIAM J. Appl. Math.* **16**, 732–746.

Coulliette, C. & Wiggins, S. [2001] "Intergyre transport in a wind-driven, quasigeostrophic double gyre: An application of lobe dynamics," *Nonlin. Process. Geophys.* **8**, 69–94.

De Leon, N., Mehta, M. A. & Topper, R. Q. [1991a] "Cylindrical manifolds in phase space as mediators of chemical reaction dynamics and kinetics. I. Theory," *J. Chem. Phys.* **94**, 8310–8328.

De Leon, N., Mehta, M. A. & Topper, R. Q. [1991b] "Cylindrical manifolds in phase space as mediators of chemical reaction dynamics and kinetics. II. Numerical considerations and applications to models with two degrees of freedom," *J. Chem. Phys.* **94**, 8329–8341.

De Leon, N. [1992] "Cylindrical manifolds and reactive island kinetic theory in the time domain," *J. Chem. Phys.* **96**, 285–297.

Dellnitz, M. & Hohmann, A. [1997] "A subdivision algorithm for the computation of unstable manifolds and global attractors," *Numer. Math.* **75**, 293–317.

Dellnitz, M., Hohmann, A., Junge, O. & Rumpf, M. [1997] "Exploring invariant sets and invariant measures," *Chaos* **7**, p. 221.

Dellnitz, M. & Junge, O. [1999] "On the approximation of complicated dynamical behavior," *SIAM J. Numer. Anal.* **36**, 491–515.

Dellnitz, M., Junge, O., Rumpf, M. & Strzodka, R. [2000] "The computation of an unstable invariant set inside a cylinder containing a knotted flow," in *Proc. Equadiff 99*, eds. Fiedler, B., Gröger, K. & Sprekels, J. (World Scientific, Singapore), pp. 1053–1059.

Dellnitz, M., Junge, O., Lo, M. & Thiere, B. [2001a] "On the detection of energetically efficient trajectories for spacecraft," *AAS/AIAA Astrodynamics Specialist Conf.*, Quebec City, Paper AAS 01-326.

Dellnitz, M., Froyland, G. & Junge, O. [2001b] "The algorithms behind GAIO — Set oriented numerical methods for dynamical systems," in *Ergodic Theory, Analysis, and Efficient Simulation of Dynamical Systems*, ed. Fiedler, B. (Springer), pp. 145–174.

Dellnitz, M. & Junge, O. [2002] "Set oriented numerical methods for dynamical systems," in *Handbook of Dynamical Systems II: Towards Applications*, eds. Fiedler, B., Iooss, G. & Kopell, N. (World Scientific, Singapore), pp. 221–264.

Dellnitz, M., Junge, O. Lo, M. W. Marsden, J. E., Padberg, K., Preis, R., Ross S. D. & Thiere B. [2004] "Transport of Mars-crossers from the quasi–Milda region," submitted for publication.

Dellnitz, M. & Preis, R. [2003] "Congestion and almost invariant sets in dynamical systems," in *Proc. SNSC'01*, ed. Winkler, F., LNCS, Vol. 2630 (Springer), pp. 183–209.

Drake, D. E. & Hougardy, S. [2002] "A simple approximation algorithm for the weighted matching problem," *Inform. Process. Lett.* **85**, 211–213.

Easton, R. W. [1998] *Geometric Methods for Discrete Dynamical Systems* (Oxford University Press, NY).

Froyland, G. & Dellnitz, M. [2003] "Detecting and locating near-optimal almost-invariant sets and cycles," *SIAM J. Sci. Comput.* **24**, 1839–1863.

Fiduccia, C. M. & Mattheyses, R. M. [1982] "A linear-time heuristic for improving network partitions," *Proc. IEEE Design Automation Conf.*, pp. 175–181.

Gillilan, R. E. & Ezra, G. S. [1991] "Transport and turnstiles in multidimensional Hamiltonian mappings for unimolecular fragmentation: Application to van del Waals predissociation," *J. Chem. Phys.* **94**, 2648–2668.

Gladman, B. J., Burns, J. A., Duncan, M., Lee, P. & Levison, H. F. [1996] "The exchange of impact ejecta between terrestrial planets," *Science* **271**, 1387–1392.

Goldreich, P., Lithwick, Y. & Sari, R. [2002] "Formation of Kuiper-belt binaries by dynamical friction and three-body encounters," *Nature* **240**, 643–646.

Gómez, G., Koon, W. S. Lo, M. W., Marsden, J. E., Masdemont, J. & Ross, S. D. [2001] "Invariant manifolds, the spatial three-body problem and space mission design," *Adv. Astronaut. Sci.* **109**, 3–22.

Gupta, A. [1997] "Fast and effective algorithms for graph partitioning and sparse matrix reordering," *IBM J. Res. Dev.* **41**, 171–183.

Haller, G. [2002] "Lagrangian coherent structures from approximate velocity data," *Phys. Fluids* **14**, 1851–1861.

Hammes-Schiffer, S. & Tully, J. C. [1995] "Nonadiabatic transition state theory and multiple potential energy surface molecular dynamics of infrequent events," *J. Chem. Phys.* **103**, 8528–8537.

Hammes-Schiffer, S. [2002] "Comparison of hydride, hydrogen atom, and proton-coupled electron transfer reactions," *Chem. Phys. Chem.* **3**, 33–42.

Henrard, J. [1982] "Capture into resonance: an extension of the use of adiabatic invariants," *Celest. Mech.* **27**, 3–22.

Hendrickson, B. & Leland, R. [1995] "A multilevel algorithm for partitioning graphs," *Proc. Supercomputing '95*, ACM.

Hobson, D. [1993] "An efficient method for computing invariant manifolds of planar maps," *J. Comput. Phys.* **104**, 14–22.

Hunt, F. Y. [1993] "A Monte Carlo approach to the approximation of invariant measures," National Insitute of Standards and Technology, *NISTIR* **4980**.

Jaffé, C., Farrelly, D. & Uzer, T. [2000] "Transition state theory without time-reversal symmetry: chaotic ionization of the hydrogen atom," *Phys. Rev. Lett.* **84**, 610–613.

Jaffé, C., Ross, S. D., Lo, M. W., Marsden, J. E., Farrelly, D. & Uzer, T. [2002] "Statistical theory of asteroid escape rates," *Phys. Rev. Lett.* **89**, 011101.

Johnson, M. R. & Johnson, D. S. [1979] *Computers and Intractability — A Guide to the Theory of NP-Completeness* (W.H. Freeman and Co).

Kane, C., Marsden, J. E., Ortiz, M. & West, M. [2000] "Variational integrators and the newmark algorithm for conservative and dissipative mechanical systems," *Int. J. Num. Math. Eng.* **49**, 1295–1325.

Karypis, G. & Kumar, V. [1999] "A fast and high quality multilevel scheme for partitioning irregular graphs," *SIAM J. Sci. Comput.* **20**, 359–392.

Kernighan, B. W. & Lin, S. [1970] "An effective heuristic procedure for partitioning graphs," *The Bell Syst. Tech. J.*, 291–307.

Konacki, M., Torres, G., Jha, S. & Sasselov, D. D. [2003] "An extrasolar planet that transits the disk of its parent star," *Nature* **421**, 507–509.

Koon, W. S., Lo, M. W., Marsden, J. E. & Ross, S. D. [2000] "Heteroclinic connections between periodic orbits and resonance transitions in celestial mechanics," *Chaos* **10**, 427–469.

Koon, W. S., Lo, M. W., Marsden, J. E. & Ross, S. D. [2001] "Resonance and capture of Jupiter comets," *Celest. Mech. Dyn. Astron.* **81**, 27–38.

Koon, W. S., Lo, M. W., Marsden, J. E. & Ross, S. D. [2001a] "Low energy transfer to the Moon," *Celest. Mech. Dyn. Astron.* **81**, 63–73.

Koon, W. S., Lo, M. W., Marsden, J. E. & Ross, S. D. [2002] "Constructing a low energy transfer between Jovian moons," *Contemp. Math.* **292**, 129–145.

Koon, W. S., Marsden, J. E., Ross, S., Lo, M. & Scheeres, D. J. [2004] "Geometric mechanics and the dynamics of asteroid pairs," *Ann. NY Acad. Sci.* **1017**, 11–38.

Kostelich, E. J., Yorke, J. A. & You, Z. [1996] "Plotting stable manifolds: Error estimates and noninvertible maps," *Physica* **D93**, 210–222.

Laskar, J. [1989] "A numerical experiment on the chaotic behaviour of the solar system," *Nature* **338**, 237–238.

Lekien, F. [2003] "Time-dependent dynamical systems and geophysical flows," Ph.D. thesis, California Institute of Technology.

Lekien, F., Coulliette, C. & Marsden, J. E. [2003] "Lagrangian structures in very high frequency radar data and optimal pollution timing," *7th Experimental Chaos Conf.* (AIP), pp. 162–168.

Lekien, F. & Coulliette, C. [2004] "MANGEN: Computation of hyperbolic trajectories, invariant manifolds and lobes of dynamical systems defined as 2D+1 data sets," in preparation.

Lekien, F. & Marsden, J. E. [2004] "Separatrices in high-dimensional phase spaces: Application to Van Der Waals dissociation," in preparation.

Levison, H. F. & Duncan, M. J. [1993] "The gravitational sculpting of the Kuiper belt," *Astrophys. J.* **406**, L35–L38.

Lew, A., Marsden, J. E., Ortiz, M. & West, M. [2003] "Asynchronous variational integrators," *Arch. Rat. Mech. An.* **167**, 85–146.

Lew, A., Marsden, J. E., Ortiz, M. & West, M. [2004] "Variational time integration for mechanical systems," *Int. J. Num. Meth. Engin.* **60**, 153–212.

Lichtenberg, A. J. & Lieberman, M. A. [1983] *Regular and Stochastic Motion* (Springer-Verlag, NY).

MacKay, R. S., Meiss, J. D. & Percival, I. C. [1984] "Transport in Hamiltonian systems," *Physica* **D13**, 55–81.

MacKay, R. S., Meiss, J. D. & Percival, I. C. [1987], "Resonances in area-preserving maps," *Physica* **D27**, 1–20.

Malhotra, N. & Wiggins, S. [1998] "Geometric structures, lobe dynamics, and Lagrangian transport in flows with aperiodic time dependence, with applications to Rossby wave flow," *J. Nonlin. Sci.* **8**, 401–456.

Malhotra, R. [1996] "The phase space structure near Neptune resonances in the Kuiper belt," *Astron. J.* **111**, 504–516.

Malhotra, R., Duncan, M. & Levison, H. [2000] "Dynamics of the Kuiper belt," in *Protostars and Planets IV*, eds. Mannings, V., Boss, A. P. & Russell S. S. (Univ. of Arizona Press, Tucson), pp. 1231–1254.

Marsden, J. E. & West, M. [2001] "Discrete mechanics and variational integrators," *Acta Numer.* **10**, 357–514.

Marston, C. C. & De Leon, N. [1989] "Reactive islands as essential mediators of unimolecular conformational isomerization: A dynamical study of 3-phospholene," *J. Chem. Phys.* **91**, 3392–3404.

McGehee, R. [1969] "Some homoclinic orbits for the restricted three body problem," Ph.D. thesis, University of Wisconsin, Madison, Wisconsin.

Meiss, J. D. & Ott, E. [1986] "Markov tree model of transport in area-preserving maps," *Physica* **D20**, 387–402.

Meiss, J. D. [1992] "Symplectic maps, variational principles, and transport," *Rev. Mod. Phys.* **64**, 795–848.

Mileikowsky, C., Cucinotta, F. A., Wilson, J. W., Gladman, B., Horneck, G., Lindegren, L., Melosh, J., Rickman, H., Valtonen M. & Zheng, J. Q. [2000] "Natural transfer of viable microbes in space — 1. From Mars to Earth and Earth to Mars," *Icarus* **145**, 391–427.

Monien, B., Preis, R. & Diekmann, R. [2000] "Quality matching and local improvement for multilevel graph-partitioning," *Parall. Comput.* **26**, 1609–1634.

Morbidelli, A., Chambers, J., Lunine, J. I., Petit, J. M., Robert, F., Valsecchi, G. B. & Cyr, K. E. [2000] "Source regions and timescales for the delivery of water to the Earth," *Meteor. Planet. Sci.* **35**, 1309–1320.

Murray, N. & Holman, M. [2001] "The role of chaotic resonances in the solar system," *Nature* **410**, 773–779.

Neishtadt, A. [1996] "Scattering by resonances," *Celest. Mech. Dyn. Astr.* **65**, 1–20.

Neishtadt, A. I., Sidorenko, V. V. & Treschev, D. V. [1997] "Stable periodic motions in the problem on passage through a separatrix," *Chaos* **7**, 2–11.

Ozorio de Almeida, A. M., De Leon, N., Mehta, M. A. & Marston, C. C. [1990] "Geometry and dynamics of stable and unstable cylinders in Hamiltonian systems," *Physica* **D46**, 265–285.

Padberg, K., Preis, R. & Dellnitz, M. [2004] "Integrating multilevel partitioning with hierarchical set-oriented methods for the analysis of dynamical systems," Technical Report, University of Paderborn.

Parker, T. S. & Chua, L. O. [1989] *Practical Numerical Algorithms for Chaotic Systems* (Springer-Verlag, NY).

Pellegrini, F. [1996] "SCOTCH 3.1 user's guide," Technical Report 1137-96, LaBRI, University of Bordeaux.

Perry, A. D. & Wiggins, S. [1994] "KAM tori are very sticky: Rigorous lower bounds on the time to move away from an invariant Lagrangian torus with linear flow," *Physica* **D71**, 102–121.

Poje, A. C., & Haller, G. [1999] "Geometry of cross-stream mixing in a double-gyre ocean model," *Phys. Oceanogr.* **29**, 1649–1665.

Ponnusamy, R., Mansour, N., Choudhary, A. & Fox, G. C. [1994] "Graph contraction for mapping data on parallel computers: A quality-cost tradeoff," *Sci. Program.* **3**, 73–82.

Preis, R. [1999] "Linear time 1/2-approximation algorithm for maximum weighted matching in general graphs," *Symp. Theoretical Aspects in Computer Science (STACS)*, pp. 259–269.

Preis, R. [2000] "Analyses and design of efficient graph partitioning methods," Dissertation. Heinz Nixdorf Institut Verlagsschriftenreihe, Universität Paderborn.

Rom-Kedar, V., Leonard, A. & Wiggins, S. [1990] "An analytical study of transport, mixing and chaos in an unsteady vortical flow," *J. Fluid Mech.* **214**, 347–394.

Rom-Kedar, V. & Wiggins, S. [1990] "Transport in two-dimensional maps," *Arch. Rat. Mech. Anal.* **109**, 239–298.

Rom-Kedar, V. & Wiggins, S. [1991] "Transport in two-dimensional maps: Concepts, examples, and a comparison of the theory of Rom-Kedar and Wiggins with the Markov model of MacKay, Meiss, Ott, and Percival," *Physica* **D51**, 248–266.

Rom-Kedar, V. [1999] "Transport in a class of *n*-d.o.f. systems," in *Hamiltonian Systems with Three or More Degrees of Freedom (S'Agaró, 1995)*, *NATO Adv. Sci. Inst. Ser. C Math. Phys. Sci.*, Vol. 533 (Kluwer Acad. Publ., Dordrecht), pp. 538–543.

Ross, S. D. [2003] "Statistical theory of interior-exterior transition and collision probabilities for minor bodies in the solar system," *Proc. Int. Conf. Libration Point Orbits and Applications*, Parador d'Aiguablava, Spain, June 10–14, 2002, eds. Gomez, G., Lo, M. W. & Masdemont, J. J. (World Scientific, Singapore), pp. 637–652.

Ross, S. D., Koon, W. S., Lo, M. W. & Marsden, J. E. [2003] "Design of a multi-moon orbiter,"

13th AAS/AIAA Space Flight Mechanics Meeting, Ponce, Puerto Rico, Paper AAS 03-143.

Rowley, C. W. & Marsden, J. E. [2002] "Variational integrators for point vortices," *Proc. CDC* **40**, 1521–1527.

Scheeres, D. J. [2002] "Stability of binary asteroids," *Icarus* **159**, 271–283.

Scheeres, D. J., Durda, D. D. & Geissler, P. E. [2002] "The fate of asteroid ejecta," in *Asteroids III*, eds. Bottk, W. M. *et al.*, University of Arizona, Tuscon, pp. 527–544.

Schroer, C. G. & Ott, E. [1997] "Targeting in Hamiltonian systems that have mixed regular/chaotic phase spaces," *Chaos* **7**, 512–519.

Szebehely, V. [1967] *Theory of Orbits* (Academic Press, NY–London).

Tancredi, G., Lindgren, M. & Rickman, H. [1990] "Temporary satellite capture and orbital evolution of comet P/Helin-Roman-Crockett," *Astron. Astrophys.* **239**, 375–380.

Tancredi, G. [1995] "The dynamical memory of Jupiter family comets," *Astron. Astrophys.* **299**, 288–292.

Tiscareno, M. & Malhotra, R. [2003] "The dynamics of known Centaurs," *Astron. J.* **126**, 3122–3131.

Torbett, M. V. & Smoluchowski, R. [1990] "Chaotic motion in a primordial comet disk beyond Neptune and comet influx to the Solar System," *Nature* **345**, 49–51.

Uzer, T., Jaffé, C., Palacián, J., Yanguas, P. & Wiggins, S. [2002] "The geometry of reaction dynamics," *Nonlinearity* **15**, 957–992.

Valsecchi, G. B. [1992] "Close encounters, planetary masses, and the evolution of cometary orbits," in *Periodic Comets*, eds. Fernández, J. A. & Rickman, H. Univ. de la Republica, Montevideo, Uruguay, pp. 143–157.

Veillet, C., Parker, J. W., Griffin, I., Marsden, B., Doressoundiram, A., Buie, M., Tholen, D. J., Connelley, M. & Holman, M. J. [2002] "The binary kuiper-belt object 1998 ww31," *Nature* **416**, 711–713.

Walshaw, C. [2000] "The Jostle user manual: Version 2.2," University of Greenwich.

Wiggins, S. [1992] *Chaotic transport in Dynamical Systems*, Interdisciplinary Appl. Math., Vol. 2 (Springer, Berlin–Heidelberg–NY).

Wisdom, J. [1980] "The resonance overlap criterion and the onset of stochastic behavior in the restricted three-body," *Astron. J.* **85**, 1122–1133.

Yamato, H. & Spencer, D. B. [2003] "Numerical investigation of perturbation effects on orbital classifications in the restricted three-body problem, *13th AAS/AIAA Space Flight Mechanics Meeting*, Ponce, Puerto Rico, Paper AAS 03-235.

You, Z., Kostelich, E. J. & Yorke, J. A. [1991] "Calculating stable and unstable manifolds," *Int. J. Bifurcation and Chaos* **1**, 605–623.

A BRIEF SURVEY ON THE NUMERICAL DYNAMICS FOR FUNCTIONAL DIFFERENTIAL EQUATIONS

BARNABAS M. GARAY

Department of Mathematics,
Budapest University of Technology and Economics,
H-1521 Budapest, Hungary

Received May 3, 2004; Revised June 16, 2004

GYULA FARKAS (1972–2002) IN MEMORIAM

This is a survey on discretizing delay equations from a geometric-qualitative view-point. Concepts like compact attractors, hyperbolic periodic orbits, the saddle structure around hyperbolic equilibria, center-unstable manifolds of equilibria, inertial manifolds, structural stability, and Kamke monotonicity are considered. Error estimates for smooth and nonsmooth initial data in various C^j topologies are provided. The emphasis is put on Runge–Kutta methods with natural interpolants. The paper ends with a collection of the related results on retarded functional differential equations with bounded delay.

Keywords: Delay equations; Runge–Kutta discretizations; invariant manifolds.

1. Introduction

The first monograph on numerical methods for delay differential equations was published by Bellen and Zennaro [2003]. Together with their own, they present relevant results of Baker, Brunner, Enright, Guglielmi, Hayashi, Hairer, in't Hout, Iserles, Jackiewicz, Koto, Maset, Tavernini, Torelli, Vermiglio and many other researchers. The book is almost 400 pages long and the bibliography contains 288 items.

The principle of organizing the material in [Bellen & Zennaro, 2003] is that — mutatis mutandis — all discretization methods used for ordinary differential equations can be applied for delay and neutral equations as well. Results for ordinary differential equations are followed by those on delay and neutral equations. Both similarities and differences (compared to the case of ordinary differential equations) are analyzed in details. In line with the mainstream tradition of presenting numerical analysis for ordinary diferential equations

[Butcher, 1987; Hairer *et al.*, 1993], the emphasis is put on convergence and stability properties of Runge–Kutta methods. The technicalities especially those related to the choice of the stepsize sequence depend on the type of the delay crucially. Delay equations of the form $\dot{x}(t) = f(t, x(t), x(t - \tau))$ and neutral equations of the form $\dot{x}(t) = f(t, x(t), x(t - \tau), \dot{x}(t - \tau))(t \geq t_0, t_0 \in \mathbf{R})$ are considered. With increasing complexity, the delay can be constant ($\tau = \tau_0 > 0$), bounded and time dependent ($\tau = \tau(t) \in [0, \tau_0]$) bounded and space dependent ($\tau = \tau(x(t)) \in [0, \tau_0]$), and proportional ($\tau = qt$ with some $q \in (0, 1)$ and $t_0 \geq 0$) — initial data are functions defined on the interval $[t_0 - \tau_0, t_0]$ and $[(1 - q)t_0, t_0]$, respectively. Multiple and distributed delays are discussed incidentally.

The monograph [Bellen & Zennaro, 2003] has grown out of *traditional numerical analysis*. Of course, the authors are well aware of the fact that the phase space of a delay equation is a function space. The approximating solution is computed first at the mesh points and then, via interpolation, on the intervals between mesh points. However, little attention is paid to the question if geometric-qualitative aspects of the solution dynamics are preserved under discretization. The main object of investigation is the relation between individual solution trajectories and their numerical approximation in $\mathbf{R}^n$. In other words, the much younger *tradition of numerical dynamics*, i.e. of handling numerical methods from the view-point of dynamical system theory plays a rather limited role in [Bellen & Zennaro, 2003].

Thus it is not without any reason to write a brief survey on delay equations placed within the general framework of numerical dynamics. When doing this, we reconsider some central topics discussed in [Stuart & Humphries, 1996], the first comprehensive presentation of numerical dynamics for ordinary differential equations:

- discretization methods as approximating semi-dynamical systems,
- compact attractors,
- hyperbolic periodic orbits,
- stable and unstable manifolds of hyperbolic equilibria.

Throughout this paper, we consider only autonomous equations with bounded delay and focus our attention to Runge–Kutta methods with polynomial interpolation. Aspects of

- inertial manifolds,
- structural stability,
- Kamke monotonicity

are also discussed. We refer frequently to papers of our late colleague *Gyula Farkas* who died in a car accident on February 27, 2002 — he was to receive his PhD Diploma at the end of the same week.

The development of numerical dynamics started for ordinary and parabolic partial differential equations simultaneously. Retarded equations followed with some delay and were influenced by the corresponding results on parabolic equations as well as by the general theory of semiflows in Banach spaces. An analysis of the related/underlying work on parabolic equations and general semiflows is beyond the scope of this paper. The most important contributions are cited in the literature we refer to. The monograph [Bellen & Zennaro, 2003] surveys connections to the numerics of Volterra integral equations.

2. Discretization as a Family of Approximating Discrete-Time Semidynamical Systems

In this section we collect some basic definitions and results on discretizing functional differential equations. This requires reformulation and restating within the framework of abstract dynamical systems theory. This level of abstractness — which is not needed for describing results on approximating individual solutions in [Bellen & Zennaro, 2003] — is absolutely essential when treating qualitative-geometric phenomena.

2.1. *Runge–Kutta discretization for delay equations*

For simplicity, take $t_0 = 0$, $\tau_0 = 1$, and consider first the initial value problem

$$\begin{cases} \dot{x}(t) = f(x(t), x(t-1)) & \text{for } t \geq 0 \\ x(t) = \eta(t) & \text{for } t \in [-1, 0] \end{cases} \tag{1}$$

where $f : \mathbf{R}^n \times \mathbf{R}^n \to \mathbf{R}^n$ is a bounded C^p function with bounded derivatives and $\eta \in C([-1,0], \mathbf{R}^n)$, the Banach space of continuous $\mathbf{R}^n$-valued functions on the interval $[0,1]$. The maximum norm on $C = C([-1,0], \mathbf{R}^n)$ is denoted by $\|\cdot\|$. The Euclidean norm on $\mathbf{R}^n$ is denoted by $|\cdot|$. The smoothness and boundedness assumptions on f imply that the initial value problem (1) has a unique solution $x = x_{\text{exact}}(\cdot, \eta) : [-1, \infty) \to \mathbf{R}^n$. Moreover, formula $(\Phi(t,\eta))(s) = x(t+s, \eta)$, $s \in [-1, 0]$ defines a semidynamical system $\Phi : \mathbf{R}^+ \times C \to C$. With respect to the second variable, Φ is of class C^p, $p = 1, 2, \ldots$. Note that $x_{\text{exact}}(\cdot, \eta)$ is differentiable at $t_0 = 0$ if and only if the left-hand side derivative of the initial function η exists at $t_0 = 0$ and $\dot{\eta}(0) = f(\eta(0), \eta(-1))$. It follows immediately that Φ is not differentiable with respect to its first variable. Since (1) defines a nonautonomous ordinary differential equation of the form $\dot{x} = f(x, \eta(t-1))$ on the interval $[0,1]$, $x_{\text{exact}}(\cdot, \eta)|_{[0,1]} = \Phi(1, \eta)$ is of class C^1 and, by a simple induction argument, $x_{\text{exact}}(\cdot, \eta)|_{[j, j+1]} = \Phi(j+1, \eta)$ is of class C^{j+1}, $j = 0, 1, \ldots, p$. Moreover, $x_{\text{exact}}(\cdot, \eta)|_{[j, \infty)}$ is of class C^{j+1}, $j = 0, 1, \ldots, p$ — this is the well-known *smoothing property* of the solution semidynamical system.

Fix $h_0 \in (0,1]$ and let $h \in (0,h_0]$. The *stepsize-h explicit Euler discretization operator with piecewise linear interpolant* is defined as $\varphi_{E,PLI} : (0,h_0] \times C \to C$, $(h,\eta) \to \varphi_{E,PLI}(h,\eta)$,

$$(\varphi_{E,PLI}(h,\eta))(s) = \begin{cases} \eta(h+s) & \text{if } s \in [-1,-h] \\ -\dfrac{s}{h}\eta(0) + \left(1 + \dfrac{s}{h}\right) X_E(h,\eta) & \text{if } s \in [-h,0] \end{cases} \tag{2}$$

where $X_E(h,\eta) = \eta(0) + hf(\eta(0),\eta(-1))$. The right-hand side of formula (2) defines the stepsize-h implicit Euler discretization operator $\varphi_{I,PLI}$ with piecewise linear interpolant if $X_E(h,\eta)$ is replaced by $X_I(h,\eta)$, the unique solution of equation $X = \eta(0) + hf(X,\eta(h-1))$, for $h \in (0,h_1]$, $h_1 > 0$ is sufficiently small.

Similarly, every Runge–Kutta method M (known from the numerics of ordinary differential equations [Butcher, 1987; Hairer *et al.*, 1993]) can be applied for Eq. (1). The *stepsize-h Runge–Kutta discretization operator with piecewise linear interpolant* is defined as $\varphi_{M,PLI} : (0,h_1^*] \times C \to C$, $(h,\eta) \to \varphi_{M,PLI}(h,\eta)$,

$$(\varphi_{M,PLI}(h,\eta))(s) = \begin{cases} \eta(h+s) & \text{if } s \in [-1,-h] \\ -\dfrac{s}{h}\eta(0) + \left(1 + \dfrac{s}{h}\right) X_M(h,\eta) & \text{if } s \in [-h,0] \end{cases} \tag{3}$$

where

$$X_M(h,\eta) = \eta(0) + h\sum_{i=1}^{\nu} b_i\, f(X^i, \eta(c_ih - 1)) \tag{4}$$

with

$$X^i = \eta(0) + h\sum_{j=1}^{\nu} a_{ij}\, f(X^j, \eta(c_jh - 1)),$$
$$i = 1,2,\ldots,\nu. \tag{5}$$

Here the positive integer ν and the real constants $\{a_{ij}\}_{i,j=1}^{\nu}$, $\{b_i\}_{i=1}^{\nu}$ and $\{c_i\}_{i=1}^{\nu}$ are the parameters of the Runge–Kutta method M. We leave them unspecified but assume that $c_i \in [0,1]$ for $i = 1,2,\ldots,\nu$. Note that for h sufficiently small, say $h \le h_1^* (\le h_0)$, the right-hand side of (5) defines a contraction operator on $\mathbf{R}^n \times \mathbf{R}^n \times \cdots \times \mathbf{R}^n$ (ν times).

The *stepsize-h Runge–Kutta discretization operator with a standard interpolant* is defined as $\varphi_{M,NFI} : (0,h_1^*] \times C \to C$, $(h,\eta) \to \varphi_{M,NFI}(h,\eta)$,

$$(\varphi_{M,NFI}(h,\eta))(s) = \begin{cases} \eta(h+s) & \text{if } s \in [-1,-h] \\ \eta(0) + h\sum_{i=1}^{\nu} \beta_i\left(-\dfrac{s}{h}\right) f(X^i, \eta(c_ih - 1)) & \text{if } s \in [-h,0] \end{cases} \tag{6}$$

where $\{X^i\}_{i=1}^{\nu}$ is determined by (5) and the polynomials $\beta_i : [0,1] \to \mathbf{R}$ satisfy $\beta_i(0) = b_i$, $\beta_i(1) = 0$, and $\beta_i(1 - c_j) = a_{ij}$, $i,j = 1,2,\ldots,\nu$. Note that the collection of the requirements on $\{\beta_i\}_{i=1}^{\nu}$ is equivalent to the collection of the properties

$$(\varphi_{M,NFI}(h,\eta))(-h) = \eta(0),$$

$$(\varphi_{M,NFI}(h,\eta))(0) = X_M(h,\eta) \quad \text{and}$$

$$(\varphi_{M,NFI}(h,\eta))(-h + c_ih) = X^i, \quad i = 1,2,\ldots,\nu.$$

The first two letters in *NFI* refer to the terminus technicus "natural" and "of the first class",

respectively. Throughout this paper, we assume that our Runge–Kutta method when applied for ordinary differential equations (like $\dot{x} = f(x, \eta(t-1))$ on the interval $[0,1]$) is of order p.

In most practical implementations, the approximating solution $x_{\text{approx}}(\cdot,\eta) : [-1,\infty) \to \mathbf{R}^n$ is computed only at the mesh points $\{k/N\}_{k\ge0}$. When keeping track on internal stage values, one arrives at the finite sequence of points $\{x_{\text{approx}}((k + c_i)/N,\eta)\}_{k\ge0;i=1,\ldots,\nu} \subset \mathbf{R}^n$. In particular, $x_{\text{approx}}(1/N,\eta) = X_M(h,\eta)$ and $x_{\text{approx}}(c_i/N,\eta) = X^i$ from (5)–(6), $i = 1,2,\ldots,\nu$.

It is immediate that $\varphi_{M,NFI} : (0, h_1^*] \times C \to C$ is continuous. With respect to the second variable, $\varphi_{M,NFI}$ is of class C^p. For $h \in (0, h_1^*]$ fixed, the iterates $\{\varphi_{M,NFI}^k(h, \eta)\}_{k=0}^{\infty}$ define a discrete-time semidynamical system on C. Our uniformity assumptions on f imply that, for any time $T \geq 1$ and for any ball $B \subset C$, the set

$$\{\varphi_{M,NFI}^k(h, \eta) \in C \,|\, 1 \leq kh \leq T, \, k \in \mathbf{N},$$
$$h \in (0, h_1^*], \, \eta \in B\}$$

consists of uniformly bounded and uniformly Lipschitz continuous functions. Note that the set $\Phi([1, T], B)$ consists of uniformly bounded and uniformly Lipschitz continuous functions, too — this is the well-known *compactifying property* of the solution semidynamical system.

From the view-point of a qualitative theory of discretizations, it is natural to define stepsize-h discretization operators as above i.e. as self-maps of the infinite-dimensional function space C. However, this is not quite satisfactory for practical purposes. In practice the initial function $\eta \in C$ is not always explicitly given but only its values on a uniform mesh are known. This leads to a parallel, more practical framework of establishing an abstract theory for discretizations.

Fix a positive integer N. By letting $\Pi_{1/N}(\eta)$ to be the piecewise linear continuous function with vertices $\{-1 + j/N, \eta(-1 + j/N)\}_{j=0}^{N}$, a linear projection $\Pi_{1/N} : C \to C$ is defined. The range of $\Pi_{1/N}$ is denoted by $C_{1/N} \subset C$. Obviously, $C_{1/N}$ can be identified with $\mathbf{R}^{n(N+1)}$ via the linear isomorphism $\eta \to \{\eta(-1 + j/N)\}_{j=0}^{N}$. Thus the stepsize-$1/N$ explicit Euler discretization method when applied to the delay equation (1) can be understood as a mapping $\varphi(1/N, \cdot) = \varphi_{P,E,PLI}(1/N, \cdot) : C_{1/N} \to C_{1/N}$ defined by

$$(\varphi(1/N, \cdot)) \left(\left\{ \eta \left(\frac{j-N}{N} \right) \right\}_{j=0}^{N} \right)$$

$$= \left(\eta \left(\frac{1-N}{N} \right), \eta \left(\frac{2-N}{N} \right), \ldots, \eta(0), X_E(\eta) \right). \tag{7}$$

Here the lower index $\mathcal{P}$ is an abbreviation for "practical". Note that the sequence of iterates $\{\varphi_{E,PLI}^k(1/N, \eta)\}_{k=0}^{\infty} \subset C$ depends solely on $(f$ and) $\{\eta(-1 + j/N)\}_{j=0}^{N}$, the restriction of η to the finite collection of the mesh points $\{-1 + j/N\}_{j=0}^{N}$

in $[-1, 0]$. By definition,

$$\varphi_{E,PLI}^k(1/N, \eta) = \varphi_{P,E,PLI}^k(1/N, \eta)$$
$$\text{whenever} \quad k \geq N \quad \text{and} \quad \eta \in C.$$

A similar construction is possible for general Runge–Kutta methods and leads to the definition of *stepsize-1/N practical Runge–Kutta discretization operators $\varphi_{P,M,PLI}$ with piecewise linear interpolant.*

2.2. *Error estimates for smooth initial functions*

For ordinary differential equations on a finite time interval $[0, T]$, it is well-known that stepsize-h Runge–Kutta approximating/discretized solutions converge to the exact solution as $h \to 0$. The order of a Runge–Kutta method M refers to the order of this convergence process

(a) on the set of the mesh points $\{kh\}_{k \geq 0}$ in $[0, T]$
(b) on the set of the stage points $\{kh\}_{k \geq 0} \cup \{(k + c_j)h\}_{k \geq 0; \, j=1, \ldots, \nu}$ in $[0, T]$ and, in case method M is combined with an interpolation operator $INTRP$,
(c) on the entire interval $[0, T]$.

The corresponding orders are called the classical (or nodal), the stage and the uniform order, respectively. A separate order can be defined for the interpolation operator $INTRP$ as well.

The monograph [Bellen & Zennaro, 2003] discusses all the order concepts above in the context of delay equations thoroughly. For delay equations of the form (1), their main result goes back to [Bellen, 1984] and can be restated as follows.

Lemma 1. *Let $\varphi_{M,NFI}$ be a Runge–Kutta discretization operator with standard interpolant. Given any finite interval $[0, T]$ and any C^p initial function η, there exists a positive constant K (depending only on f, M, T, as well as on $\|\eta'\|, \ldots, \|\eta^{(p)}\|$) such that*

$$|(\Phi(k/N, \eta))(0) - (\varphi_{M,NFI}^k(1/N, \eta))(0)| \leq K \cdot h^p \tag{8}$$

whenever $0 \neq N, k \in \mathbf{N}$, $k/N \leq T$ and $1/N \leq h_1^*$. Under some additional assumptions on the *interpolant NFI, also inequalities*

$$\left\| \frac{d^j}{ds^j}\, \Phi(k/N, \eta) - \frac{d^j}{ds^j}\, \varphi^k_{M,NFI}(1/N, \eta) \right\|$$

$$\leq K_j \cdot h^{q+1-j}, \quad j = 0, 1, \ldots, q \qquad (9)$$

hold true. Here $q \leq p - 1$ stays for the order of the interpolant NFI and constant K_j depends also on the underlying interpolation operator, $j = 0, 1, \ldots, q$. Derivatives at the mesh points in (9) are meant in the left/right sense.

Suppose that $\eta^{(p)}$ does not exist at some $s = -1 + \Delta$ with $h_0 < \Delta < 1$ but η is C^p on the interval $[-1, -1 + \Delta]$. Then the local approximation error satisfies inequality

$$|x_{\text{exact}}(h, \eta) - X_M(h, \eta)| \leq \text{const}(f, M, \eta) \cdot h^{p+1}$$

$$\text{for} \quad h \in (0, \Delta] \qquad (10)$$

where $\text{const}(f, M, \eta)$ depends only on f, M, as well as on the bounds for $|\eta'|, \ldots, |\eta^{(p)}|$. The restriction "for $h \in (0, \Delta]$" in (10) (which means that the inequality in (10) is not necessarily satisfied on the interval $(0, h_0]$) has important consequences to mesh point selection. In order to have local $\mathcal{O}(h^{p+1})$ error estimates, it implies that Δ has to be chosen for a mesh point. In view of the smoothing property of Φ, a similar argument shows that the mesh (still in order to have local $\mathcal{O}(h^{p+1})$ error estimates) should contain the points $0, 1, \ldots, p$ and also the points $1 + \Delta, 2 + \Delta$, etc. In general this leads to nonuniform mesh with a variable stepsize sequence $(h_1, h_2, \ldots)$, $0 < h_m \leq h_0$, $m = 1, 2, \ldots$. The discretization operator $\varphi_{M,NFI}$ gives rise to one with variable stepsize sequence by defining $\varphi_{M,NFI}(0, \eta) = \eta$ and then, inductively

$$\varphi_{M,NFI}(h_m, \ldots, h_1; \eta)$$

$$= \varphi_{M,NFI}(h_m, \varphi_{M,NFI}(h_{m-1}, \ldots, h_1; \eta))$$

$$\text{for} \quad m = 1, 2, \ldots.$$

Chapters 4 and 6 of [Bellen & Zennaro, 2003] contain several generalizations of what we called Lemma 1 above even for equations with state-dependent delay as well as for certain types of neutral equations where (still in order to have local $\mathcal{O}(h^{p+1})$ error estimates) stepsize selection is subject to various constraints. Also these results can be restated within the framework of a nonautonomous dynamical system theory. The

"additional assumptions on the interpolant *NFI*" from Lemma 1 (which go back to [Zennaro, 1986] and constitute one of the mostly involved part of [Bellen & Zennaro, 2003]) are discussed in Sec. 5.2.2. It is a challenging task to find such an interpolation operator that preserves the order of convergence in Lemma 1 and makes $x_{\text{approx}}(\cdot, \eta)$ to be of class C^j on the interval $[j, \infty)$, $j = 0, 1, \ldots, p$. The monograph [Bellen & Zennaro, 2003] refers to several results into this direction but none of them seems to ensure the same smoothness improvement for $x_{\text{approx}}(\cdot, \eta)$ along the intervals $\{[j, j+1]\}_{j=0}^{p}$ shared by $x_{\text{exact}}(\cdot, \eta)$.

There are various pro and contra arguments for variable stepsize sequences. A major pro argument has already been discussed. Though conflicting with higher order local approximation error estimates, further pro arguments are those behind adaptive error control in [Bellen & Zennaro, 2003, Chapter 7]. The major contra argument for variable stepsize sequences is the obvious pro argument for the uniform mesh $\{k/N\}_{k \geq 0}$ we outlined in the two last paragraphs of Sec. 2.1.

However, despite all efforts of putting stepsizes selection and the error control mechanism on a firm mathematical basis, heuristical aspects can hardly be avoided. This is particularly exemplified by considering a delay equation of the form

$$\dot{x}(t) = f(x(t), x(t - \rho), x(t - 1))$$

$$\text{where} \quad h_0 < \rho < 1.$$

Our first candidate is the uniform mesh $M_U = \{k/N\}_{k \geq 0}$. In order to go on with the explicit Euler method at a mesh point $k_0/N \leq T$, two earlier values of the approximate solution x_{approx} (at $k_0/N - \rho$ and $k_0/N - 1$ given or computed previously) are needed. It follows that the values of η at each point of the set

$$H_T = \{k/N - \ell\rho \in [-\rho, 0] \,|\, k \in \mathbf{Z},\, \ell \in \mathbf{N},\, k/N \leq T\}$$

are also needed. This is a pladoyee for interpolating and working within the $\varphi_{P,E,PLI}$ framework but, especially on moderate time intervals $[0, T]$, also the choice of $M_U + H_T$ (the algebraic sum of the two discrete sets M_U and H_T) as for a new, nonuniform mesh $M_{NU} = M_U + H_T$ is reasonable.

The subsection concludes with an example showing that local error estimates between exact and approximate solutions cannot be uniform in C. Nevertheless, it indicates that, on certain natural

38   B. M. Garay

subsets of C, uniform error estimates can be expected.

Example 1. If function f does not depend on the first n coordinates, then (1) simplifies to

$$\begin{cases} \dot{x}(t) = f(x(t-1)) & \text{for } t \geq 0 \\ x(t) = \eta(t) & \text{for } t \in [-1,0]. \end{cases} \quad (11)$$

Suppose that $f(0) = 0$ and $\eta(-jh) = 0$ whenever $h = 1/N$ and $j = 0,1,\ldots,N$. Then $(\Phi(1,\eta))(s) = \int_{-1}^{s} f(\eta(u))\,du$ but $(\varphi_{E,PLI}^{N}(1/N,\eta))(s) = 0$ for each $s \in [-1,0]$. In particular, $\|\Phi(1,\eta) - \varphi_{E,PLI}^{N}(1/N, \eta)\|$ can be arbitrarily large. Note that in our case,

$$\|\Phi(1,\eta) - \varphi_{E,PLI}^{N}(1/N,\eta)\|$$
$$\leq \int_{-1}^{0} |f(\eta(u))|du \leq \mathcal{L} \int_{-1}^{0} |\eta(u)|du$$

where $\mathcal{L}$ stays for the Lipschitz constant of f. Note that $\|\Phi(1,\eta) - \varphi_{E,PLI}^{N}(1/N,\eta)\| \to 0$ as $N \to \infty$ for each $\eta \in C$ because on $[0,1]$ ((1) is equivalent to a nonautonomous ordinary differential equation and thus) (11) simplifies to the integration problem $x(t) - \eta(0) = \int_0^t f(\eta(u-1))du$).

2.3. *Error estimates for nonsmooth initial functions*

If the initial function $\eta \in C$ is Lipschitz with constant $\text{Lip}(\eta) \leq L$, then the local approximation error satisfies inequality

$$|x_{\text{exact}}(h,\eta) - X_M(h,\eta)| \leq \text{const}(f,M,L) \cdot h^2$$
$$\text{for} \quad h \in (0,h_1^*]. \quad (12)$$

This is a consequence of (10) when applied to a sequence of C^1 Lipschitz functions $\{\eta_k\}_{k=1}^{\infty} \subset C$ with the properties that $\text{Lip}(\eta_k) \to \text{Lip}(\eta)$ and $\|\eta_k - \eta\| \to 0$ as $k \to \infty$.

Lemma 2. *Let $\varphi_{M,NFI}$ be a Runge–Kutta discretization operator with standard interpolant. Given any finite time interval $[0,T]$ and a finite collection of Lipschitz initial functions $\eta, \xi_1, \ldots, \xi_j$ with constants $\text{Lip}(\eta), \text{Lip}(\xi_1), \ldots, \text{Lip}(\xi_j) \leq L$ and $\|\xi_1\|, \ldots, \|\xi_j\| \leq 1$, the derivatives of the approximation error (as a j-linear operator between $C \times C \times \cdots \times C$ (j times) and C) satisfy the $j = 0,1,\ldots,p-1$*

chain of inequalities

$$\left\| \left[\frac{d^j}{d\eta^j} \, \Phi(k/N,\eta) \right](\xi_1,\ldots,\xi_j) \right.$$
$$\left. - \left[\frac{d^j}{d\eta^j} \, \varphi_{M,NFI}^{k}(1/N,\eta) \right](\xi_1,\ldots,\xi_j) \right\| \leq \kappa_j/N$$
$$(13)$$

whenever $0 \neq N, k \in \mathbf{N}$, $k/N \leq T$ and $1/N \leq h_1^*$. The positive constant κ_j, $j = 0,1,\ldots,p-1$ depends only on f, M, T, L, and on the underlying interpolation operator.

Proof. Case $j = 0$ is a direct consequence of inequality (12) via the standard Gronwall argument [Hairer *et al.*, 1993].

In order to prove case $j = 1$, we pass to a somewhat higher level of abstractness. Still with the initial value problem (1) in mind, we use the standard notation from the theory of retarded functional differential equations [Hale, 1977] and write $\dot{x}(t) = g(x_t)$, $x_0 = \eta$ instead. We consider also the initial value problem $\dot{y}(t) = [g'(x_t)]y_t$, $y_0 = \xi$ for the first variational equation. Define

$$G\begin{pmatrix} x_t \\ y_t \end{pmatrix} = \begin{pmatrix} g(x_t) \\ [g'(x_t)]y_t \end{pmatrix} \quad \text{and}$$

$$\Theta\left(h, \begin{pmatrix} \eta \\ \xi \end{pmatrix} \right) = \begin{pmatrix} \Phi(t,\eta) \\ \left[\dfrac{d}{d\eta} \Phi(t,\eta) \right] \xi \end{pmatrix}$$

where $t \geq 0$ and $\Phi : \mathbf{R}^+ \times C \to C$ denotes the solution semidynamical system for equation $\dot{x}(t) = g(x_t)$. It is clear that $\Theta : \mathbf{R}^+ \times (C \times C) \to C \times C$ is the solution semidynamical system for the retarded functional differential equation

$$\begin{pmatrix} \dot{x}(t) \\ \dot{y}(t) \end{pmatrix} = G\begin{pmatrix} x_t \\ y_t \end{pmatrix}, \quad t \geq 0. \quad (14)$$

Similarly, with $\varphi : (0,h_1^*] \times C \to C$ denoting an "approximation operator" for Φ, define

$$\psi\left(h, \begin{pmatrix} \eta \\ \xi \end{pmatrix} \right) = \begin{pmatrix} \varphi(h,x) \\ \left[\dfrac{d}{d\eta}\varphi(h,\eta) \right] \xi \end{pmatrix} \quad \text{whenever}$$

$$h \in (0,h_1^*] \quad \text{and} \quad \begin{pmatrix} \eta \\ \xi \end{pmatrix} \in C \times C.$$

Suppose that $\varphi = \varphi_{M,NFI,g}$ comes from a Runge–Kutta method with interpolant. Then the very same Runge–Kutta method applies to Eq. (14) and gives rise to operator $\varphi_{M,NFI,G}$. Analyzing (5)–(6), it is not hard to show that $\psi = \varphi_{M,NFI,G}$. We arrived at the conclusion that case $j = 1$ of (13) follows from inequality (12) — when applied to Eq. (14) instead of $\dot{x}(t) = g(x_t)$ — via the standard Gronwall argument. (Unfortunately, the uniformity assumptions we imposed on g remain no longer valid for G. The second coordinate of G is unbounded when $\|y_t\| \to \infty$. However, if our interest is reduced to a bounded subset of $C \times C$, then no difficulties arise.)

The remaining cases $j = 2, \ldots, p - 1$ follow by induction. ∎

Since $\Phi(0, \eta) = id_C \eta = \eta$ for each $\eta \in C$ and $(d^j/d\eta^j)\, \Phi(\cdot, \eta)$ is differentiable, we obtain immediately from (13) — or, by a direct analysis of the definition — that

$$\left\| \left[\frac{d^j}{d\eta^j}\, \varphi_{M,NFI}(1/N, \eta) \right](\xi_1, \ldots, \xi_j) \right.$$

$$\left. - \left[\frac{d^j}{d\eta^j}\, id_C \right](\xi_1, \ldots, \xi_j) \right\| \leq \tilde{\kappa}_j/N \qquad (15)$$

whenever $1/N \leq h_1^*$ and the initial functions $\eta, \xi_1, \ldots, \xi_j$ are Lipschitz with constants $\mathrm{Lip}(\eta)$, $\mathrm{Lip}(\xi_1), \ldots, \mathrm{Lip}(\xi_j) \leq L$ and $\|\xi_1\|, \ldots, \|\xi_j\| \leq 1$, $j = 0, 1, \ldots, p - 1$. Of course the positive constant $\tilde{\kappa}_j$ depends only on f, M, L, and on the underlying interpolation operator, $j = 0, 1, \ldots, p - 1$.

Recall that Lipschitz functions in C are of bounded variation and that the total variation $\mathrm{Tot}\, v(\eta)$ is not greater than $\mathrm{Lip}(\eta)$.

Hence our next result is an improvement over Lemma 2 for the explicit/implicit Euler method with piecewise linear interpolant.

Lemma 3. *Consider only the special cases $\varphi_{M,NFI} = \varphi_{E,PLI}$ or $\varphi_{I,PLI}$. Given any finite time interval $[0, T]$ and a finite collection of initial functions $\eta, \xi_1, \ldots, \xi_j$ of bounded variation with $\mathrm{Tot}\, v(\eta),\ \mathrm{Tot}\, v(\xi_1), \ldots, \mathrm{Tot}\, v(\xi_j) \leq L$, the $j = 0, 1, \ldots, p - 1$ chain of inequalities (13) still holds true.*

Proof. Applying the standard Gronwall argument [Hairer *et al.*, 1993] we have already referred to, we point out first that

$$|(\Phi(1, \eta))(0) - (\varphi_{E,PLI}^N(1/N, \eta))(0)|$$
$$\leq 2^{-1} e^{\mathcal{L}_1}(B + 4n\mathcal{L}_2 \cdot \mathrm{Tot}\, v(\eta))/N \qquad (16)$$

where B be is an upper bound for $|f|$ on $\mathbf{R}^n \times \mathbf{R}^n$ and $\mathcal{L}_i$ stays for the Lipschitz constant of f with respect to the ith variable, $i = 1, 2$. On the time interval $[0, 1]$, (1) is equivalent to the initial value problem $\dot{x} = f(x, q(t))$, $x(0) = x_0 = \eta(0)$ where $q(t) = \eta(t - 1)$. Let $\Psi(\cdot; t_*, x_*) : [t_*, 1] \to \mathbf{R}^n$ denote the right-hand side solution to the nonautonomous ordinary differential equation $\dot{x} = f(x, q(t))$ with initial data $(t_*, x_*) \in [0, 1] \times \mathbf{R}^n$. For brevity, we write $h = 1/N$,

$$t_k = kh, \quad x_k = (\varphi_{E,PLI}^N(1/N, \eta))(t_k - 1)$$
$$\text{for} \quad k = 0, 1, \ldots, N,$$

and $a = \mathcal{L}_1 B h^2/2$, $b = \mathcal{L}_2$, $d = e^{\mathcal{L}_1 h}$. Finally, for $k = 0, 1, \ldots, N - 1$, define

$$E_k = |\Psi(t_k; 0, x_0) - x_k| \quad \text{and}$$

$$c_k = \int_{t_k}^{t_{k+1}} |q(u) - q(t_k)|\, du$$

and observe that the right-hand side of inequality (16) is equal to $E_N = |\Psi(t_N; 0, x_0) - x_N|$.

We claim that $E_{k+1} \leq E_k d + a + b c_k$ for each $k = 0, 1, \ldots, N - 1$. In fact, we have for $k = 0, 1, \ldots, N - 1$ by the triangle inequality that

$$E_{k+1} \leq |\Psi(t_{k+1}; t_k, \Psi(t_k; 0, x_0)) - \Psi(t_{k+1}; t_k, x_k)|$$
$$+ |\Psi(t_{k+1}; t_k, x_k) - (x_k + h f(x_k, q(t_k)))|.$$

The first term can be estimated by using Gronwall lemma. In fact, for each $t \in [t_k, t_{k+1}]$, we have that

$$|\Psi(t; t_k, \Psi(t_k; 0, x_0)) - \Psi(t; t_k, x_k)|$$

$$= \left| \Psi(t_k; 0, x_0) \right.$$

$$+ \int_{t_k}^{t} f(\Psi(u; t_k, \Psi(t_k; 0, x_0)), q(u))\, du$$

$$\left. - \left(x_k + \int_{t_k}^{t} f(\Psi(u; t_k, x_k), q(u))\, du \right) \right|$$

$$\leq |\Psi(t_k; 0, x_0) - x_k|$$

$$+ \mathcal{L}_1 \int_{t_k}^{t} |\Psi(u; t_k, \Psi(t_k; 0, x_0)) - \Psi(u; t_k, x_k)|\, du$$

and, a fortiori, the first term is not greater than $e^{\mathcal{L}_1 h} E_k$. On the other hand, the second term is bounded by

$$\left| x_k + \int_{t_k}^{t} f(\Psi(u; t_k, x_k), q(u)) du \right.$$
$$\left. - \left(x_k + \int_{t_k}^{t_{k+1}} f(x_k, q(t_k)) du \right) \right|$$
$$\leq \mathcal{L}_1 \int_{t_k}^{t} |\Psi(u; t_k, x_k) - x_k| du$$
$$+ \mathcal{L}_2 \int_{t_k}^{t_{k+1}} |q(u) - q(t_k)| du$$

and the claim follows from observing that $x_k = \Psi(t_k; t_k, x_k)$ and $\Psi(\cdot; t_k, x_k)$ is Lipschitz with constant $\leq B$.

Starting from $E_0 = 0$, we conclude easily by induction that

$$E_N \leq a \frac{d^N - 1}{d - 1} + b \sum_{k=0}^{N-1} c_k d^{N-k-1}$$

$$\leq 2^{-1} e^{\mathcal{L}_1} \left(Bh + \mathcal{L}_2 \sum_{k=0}^{N-1} c_k \right).$$

It remains to prove that $\sum_{k=0}^{N-1} c_k \leq 2nh \cdot \operatorname{Tot} v(\eta)$. In fact, by the Jordan decomposition theorem, every coordinate function of q can be represented as $q_i = v_i - w_i$ where v_i and w_i are monotone increasing continuous real functions on $[0, 1]$ with the property that

$$\operatorname{Tot} v(v_i), \ \operatorname{Tot} v(w_i) \leq \operatorname{Tot} v(q_i) \leq \operatorname{Tot} v(q)$$
$$i = 1, \ldots, n.$$

It follows immediately that

$$\sum_{k=0}^{N-1} c_k \leq \sum_{k=0}^{N-1} \sum_{i=1}^{n} \int_{t_k}^{t_{k+1}} (|v_i(u) - v_i(t_k)|$$
$$+ |w_i(u) - w_i(t_k)|) du$$
$$\leq \sum_{i=1}^{n} \sum_{k=0}^{N-1} \int_{t_k}^{t_{k+1}} (|v_i(t_{k+1}) - v_i(t_k)|$$
$$+ |w_i(t_{k+1}) - w_i(t_k)|) du$$
$$\leq \sum_{i=1}^{n} h \cdot (\operatorname{Tot} v(v_i) + \operatorname{Tot} v(w_i))$$

$$\leq 2nh \cdot \operatorname{Tot} v(q)$$
$$= 2nh \operatorname{Tot} v(\eta).$$

(As a direct consequence of the uniform continuity of η, note that $\sum c_k \to 0$ as $N \to \infty$.)

By a repeated use of the standard Gronwall argument (when combined with piecewise linear interpolation), case $j = 0$ of the $\varphi_{M,NFI} = \varphi_{E,PLI}$, $L = \operatorname{Tot} v(\eta)$ version of inequality (13) follows with $\kappa_0 = \operatorname{const} \cdot e^{\mathcal{L}_1 T}(1 + L)$ easily.

The $\varphi_{M,NFI} = \varphi_{I,PLI}$ case can be reduced to the $\varphi_{M,NFI} = \varphi_{E,PLI}$ case already proven. The crucial point is to show that

$$|(\varphi_{I,PLI}^{N}(1/N, \eta))(0) - (\varphi_{E,PLI}^{N}(1/N, \eta))(0)|$$
$$\leq \operatorname{const} \cdot (1 + \operatorname{Tot} v(\eta))/N.$$

Without referring to the Jordan decomposition theorem any more, this follows via a simplified version of the recursion we used in deriving (16) above. (Having only Theorem 4.B in mind, we did not check if (16) holds true for a general discretization operator $\varphi_{M,NFI}$.)

The proof of the remaining cases $j = 1, 2, \ldots, p - 1$ is the same as in the proof of Lemma 2. ∎

Both for ordinary and delay differential equations, one-step and multistep methods, several versions of inequalities (8), (9), (13) and (15) are known from the literature. A weaker version of Lemma 3 has been stated in [Garay & Lóczi, 2004]. Inequality (16) is new.

The definition of an abstract discretization operator for Eq. (1) as well for the more abstract equation $\dot{x}(t) = g(x_t)$ we refer to in Sec. 4 below are based on case $j = 0, 1$ of inequalities (13) and (15).

In contrast to inequalities (8) and (9) which concern an individual exact and an individual approximating trajectory, the $j \geq 1$ cases of inequalities (13) and (15) relate to a collection of exact and approximating trajectories. Qualitative theory cannot live without differentiating with respect to initial data. This is why C^1 inequalities like case $j = 1$ of (13) (estimating the difference between exact and approximating solutions in C^1 topologies on the phase space) play a fundamental role in almost all papers on numerical dynamics. The typical result is that, for stepsizes sufficiently small, hyperbolic orbit configurations are preserved by discretization.

For ordinary differential equations, an abstract definition for discretization operators is based on C^j

properties. In one of the earliest papers on the qualitative theory of discretizations, [Beyn & Lorenz, 1987] suggest the following definition. Consider an ordinary differential equation $\dot{x} = f(x)$ where $f : \Omega \to \mathbf{R}^n$ is a C^{p+r+1} function. Let S be a compact subset of Ω. A mapping $\varphi : (0, h_0] \times S \to \mathbf{R}^n$ is an *abstract discretization operator of order p* if

(i) φ admits a C^{p+r+1} extension to an open neighborhood of $[0, h_0] \times S$ in $\mathbf{R} \times \Omega$

(ii) $|\Phi(h, x) - \varphi(h, x)| \leq K h^{p+1}$ whenever $(h, x) \in (0, h_0] \times S$

(iii) φ is locally determined by f. In other words, there exists a continuous function $\Delta : [0, h_0] \to \mathbf{R}^+$ with the properties that $\Delta(0) = 0$ and, for all $(h, x) \in (0, h_0] \times S$, $\varphi(h, x)$ is determined by the restriction of f to the set $\{z \in \mathbf{R}^n \mid |z - x| \leq \Delta(h)\}$.

As a simple consequence of assumptions (i)–(ii), $\varphi(h, \cdot)$ is a C^{p+r+1} diffeomorphism of S onto $\varphi(h, S)$ for h sufficiently small, and — on condition that $\Phi((k-1)h, x) \in S$ and $\varphi^{k-1}(h, x) \in S$ —

$$\left| \frac{d^j}{dx^j} \Phi(kh, x) - \frac{d^j}{dx^j} \varphi^k(h, x) \right|$$
$$\leq \kappa_j(T) \cdot h^{\min\{p, p+r-j\}}, \quad j = 0, 1, \ldots, p+r \tag{17}$$

whenever $k \in \mathbf{N}$, $h \in (0, h_0]$, $kh \leq T$ and $x \in S$. Clearly Runge–Kutta methods are subject to assumptions (i)–(iii). Moreover, for Runge–Kutta methods, inequality (17) is satisfied if f is chosen from the less smoother class of C^{p+r} functions. Mutatis mutandis, assumptions (i)–(iii) make sense if f is defined on a compact smooth manifold $\mathcal{M}$. This leads to the definition of abstract discretization operators on $\mathcal{M}$. For stepsize h small enough, $\varphi(h, \cdot)$ is a C^{p+r+1} self-diffeomorphism of $\mathcal{M}$. Also the C^j inequality (17) remains valid in the manifold setting. For details, see [Li, 1997; Garay, 2001].

Thus $\{\varphi(h, \cdot)\}_{h \in (0, h_0]}$ is a one-parameter family of diffeomorphisms approximating the one-parameter family of time-h diffeomorphism $\{\Phi(h, \cdot)\}_{h \in (0, h_0]}$ of the continuous-time solution dynamical system $\Phi : \mathbf{R} \times \mathbf{R}^n \to \mathbf{R}^n$ (or locally, $\Phi : \mathbf{R} \times \Omega \hookrightarrow \mathbf{R}^n$; or $\Phi : \mathbf{R} \times \mathcal{M} \to \mathcal{M}^n$). Consequently, for $h \in (0, h_0]$ fixed, discretization theory is part of perturbation theory for discrete-time dynamical systems. However, with $h \to 0$, both $\varphi(h, \cdot)$ and $\Phi(h, \cdot)$ approach the identity, an operator which behaves badly in perturbation theory: The behavior of h as of a small parameter is not entirely regular. We conclude that *the proof of a qualitative result in discretization theory requires a thorough reconsideration of the proof of the underlying abstract perturbation result in discrete dynamics with stepsize h as an additional small parameter, and the derivation of the accompanying error estimates.*

3. Qualitative Numerics for Delay Equations

What we described in the last paragraph for ordinary differential equations remains valid for delay equations, too. However, one is confronted with two major difficulties. These are the lack of uniform local error estimates and the lack of backward solvability. Fortunately, for L large enough say $L \geq L_*$, the closed set

$$C_{\mathrm{Lip}(L)} = \{\eta \in C | \eta \text{ is Lipschitz with}$$
$$\text{constant } \mathrm{Lip}(\eta) \leq L\}$$

is positively invariant with respect to the exact as well as to the discretized dynamics. By (13) and (15), C^j estimates on $C_{\mathrm{Lip}(L)}$ are uniform. Badly enough, $C_{\mathrm{Lip}(L_*)}$ is nowhere dense in C. What really helps is the smoothing/compactifying property of the exact and the discretized dynamics. As for the asymptotic theory, it implies that dynamical systems in finite and semidynamical systems in infinite dimension can be treated in a parallel way [Hale, 1988]. Distinguished subsets of the phases space like unstable manifolds of hyperbolic equilibria or of hyperbolic periodic orbits, inertial manifolds, and compact attractors consist of full trajectories i.e. trajectories defined on the entire real line $\mathbf{R}$. In particular, compact attractors and certain kinds of invariant manifolds of delay equations belong to $C_{\mathrm{Lip}(L_*)}$. This is why, in a final analysis, their qualitative discretization properties are (almost) the same as of their counterparts in ordinary differential equations.

From now on, let $p \geq 2$ and assume that all the regularity conditions we imposed on (1) in Sec. 2.1 are satisfied.

3.1. *The simplest hyperbolic orbit configurations*

The three major objects of the phase space investigated in [Stuart & Humphries, 1996] on numerical ordinary differential equations are

- compact attractors (i.e. asymptotically stable compact invariant sets)
- hyperbolic periodic orbits
- hyperbolic equilibria, together with their stable and unstable manifolds

In a well-defined technical sense, compact attractors, the saddle structure about hyperbolic equilibria, and periodic orbits are only slightly perturbed under discretization. As for compact attractors, the hyperbolic structure preserved is the transversal intersection structure between trajectories near the attractor and the level surfaces of suitable Liapunov functions. (The dynamics within the attractor itself is not assumed to be hyperbolic and can be changed dramatically under discretizaton.) The presentation in [Stuart & Humphries, 1996] is based on the original papers [Kloeden & Lorenz, 1986; Beyn, 1987a, 1987b]. No doubt these three papers belong to those few marking the birth of numerical dynamics as an independent field of research in the late eighties.

In what follows we present the corresponding results for delay equations.

Theorem 1. *Consider the delay equation* $\dot{x}(t) = f(x(t), x(t-1))$ *and let* $\varphi_{M,NFI} : (0, h_1^*] \times C \to C$ *be a Runge–Kutta discretization operator with standard interpolant.*

- A.) *[Kloeden & Schropp, 2004]: Let* $\emptyset \neq \mathcal{A}$ *be a compact attractor for the continuous-time solution semidynamical system* Φ. *Then, for stepsize-$1/N$ sufficiently small, the discrete-time semidynamical system* $\varphi_{M,NFI}(1/N, \cdot)$ *has a nonempty compact attractor* $\mathcal{A}_{1/N}$ *and the limiting process* $\mathcal{A}_{1/N} \to \mathcal{A}$ *as* $N \to \infty$ *(both in* $C_{\mathrm{Lip}(L_*)}$ *with nice Liapunov estimates and consequently, by using the general attraction results in Chapter 2 of [Hale, 1988], also in C) is upper semicontinuous.*
- B.1.) *[In't Hout & Lubich, 1998]: Let* Γ *be an exponentially stable periodic orbit for the continuous-time solution semidynamical system* Φ. *Then, for stepsize-$1/N$ sufficiently small, the discrete-time semidynamical system* $\varphi_{M,NFI}(1/N, \cdot)$ *has an exponentially stable invariant curve* $\Gamma_{1/N}$ *(both in* $\mathcal{N} \cap C_{\mathrm{Lip}(L_*)}$ *and in* $\mathcal{N}$ *where* $\mathcal{N}$ *is a suitable neighborhood of* Γ *in C but in the second case constant* κ *in the estimate*

$d(\varphi_{M,NFI}^k(1/N, \eta), \Gamma_{1/N}) \leq \kappa \cdot \mu^{k/N}$ *depends on* η; $\mu < 1$ *is fixed) such that* $d_{\mathrm{Hausdorff}}(\Gamma, \Gamma_{1/N}) \leq$ const$/N^p$.

- B.2.) *[Farkas, 2003]: Let* Γ *be a hyperbolic periodic orbit for the continuous-time solution semidynamical system* Φ *and assume that the period of* Γ *is at least two (i.e. two times the delay). Then, for stepsize-h sufficiently small (and not only for* $h = 1/N$ *with N large), the discrete-time semidynamical system* $\varphi_{M,NFI}(h, \cdot)$ *has a hyperbolic invariant curve* Γ_h *such that in normal coordinates around* Γ, *both* $|\Gamma_h|$ *and* $\mathrm{Lip}(\Gamma_h)$ *are of order h.*

In line with (17) and the general estimates for discretized normally hyperbolic compact invariant manifolds of ordinary differential equations [Garay, 2001], it seems plausible in Parts (B1) and (B2) that $\Gamma_{1/N} = \mathcal{F}_{1/N}(\Gamma)$ where $\mathcal{F}_{1/N}$ is a C^{p+r} embedding of Γ into $\mathbf{R}^n$ and the norm distance in $C^j(\Gamma, \mathbf{R}^n)$ between $\mathcal{F}_{1/N}$ and the inclusion of Γ in $\mathbf{R}^n$ is of order $1/N^{\min\{p, p+r-j\}}$, $j = 0, 1, \ldots, p + r$.

The unpublished PhD dissertation *Gyula Farkas: On Numerical Dynamics of Functional Differential Equations*, Budapest University of Technology, 2002, contains a lower semicontinuity result for discretized compact attractors of delay equations, the analogue of the one in [Stuart & Humphries, 1996] from the theory of discretized ordinary differential equations. As for upper semicontinuity, Farkas refers to [Gedeon & Hines, 1999] on upper semicontinuity of Morse sets under explicit ODE–Euler discretization of a one-dimensional delay equation (which results in a cyclic feedback system of ordinary differential equations). Though conceptually much easier, we note that Theorem 1.A is not a consequence of the results in [Gedeon & Hines, 1999].

For the rest of this subsection, we assume that $f(0,0) = 0$ or, equivalently, that $\eta_0 = 0 \in C$ is an equilibrium for Φ. The next result starts with a center-unstable versus strongly-stable $C = CU \times SS$ product decomposition of the phase space (invariant with respect to the linear semidynamical system generated by the solutions of the linearized equation $\dot{y}(t) = f_x'(0,0)y(t) + f_y'(0,0)y(t-1)$ and determined by its characteristic equation). Thus the equilibrium $0 \in C$ is not necessarily hyperbolic. As a consequence of basic spectral decomposition theory, note that the linear subspace CU is of finite dimension.

Theorem 2 [Farkas, 2002a]. *Consider the delay equation* $\dot{x}(t) = f(x(t), x(t-1))$ *again and let* $\varphi_{\mathcal{P},E,PLI}(1/N, \cdot) : C_{1/N} \to C_{1/N}$ *be the stepsize-*$1/N$ *practical explicit Euler discretization operator with piecewise linear interpolant. Assume that the equilibrium point* $0 \in C$ *has a center-unstable manifold of the form* $\mathrm{Graph}(G)$, *where* $G : CU \to SS$ *is a* C^2 *function. Then for* N *large enough (and under very mild additional technical conditions)* $C_{1/N}$ *admits a center-unstable versus strongly-stable* $C_{1/N} = CU_{1/N} \times SS_{1/N}$ *product decomposition with the properties as follows. Operator* $\varphi_{\mathcal{P},M,PLI}(1/N, \cdot)$ *has an invariant manifold of the form* $\mathrm{Graph}(G_{1/N})$, *where* $G_{1/N} : CU_{1/N} \to SS_{1/N}$ *is a* C^2 *function. In addition, there exists a linear isomorphism* $P_{1/N} : CU \to CU_{1/N}$ *such that, for* $j = 0$ *and* $j = 1$,

$$\left\| \frac{d^j}{d\eta^j} \Pi_{1/N} G - \frac{d^j}{d\eta^j} G_{1/N} P_{1/N} \right\| \to 0 \quad as \ N \to \infty. \tag{18}$$

Together with $\mathrm{Graph}(G)$, *also* $\mathrm{Graph}(G_{1/N})$ *is exponentially attractive, with asymptotic phase depending continuously on the stepsize as* $N \to \infty$.

Theorem 2 in [Farkas, 2002a] is accompanied by C^2 existence and C^1 approximation results for exact and discretized stable manifolds that correspond to $\overline{S}$ in the center-unstable versus stable product structure $CU \times \overline{S}$ where $\overline{S}$ is the finite-dimensional invariant subspace determined by a bounded set of the roots of the characteristic equation lying to the left of those belonging to CU. In the case of hyperbolic equilibria, also a numerical Grobman-Hartman lemma for partial linearizations [Farkas, 2001b] as well as C^1 shadowing results [Farkas, 2002a] are given.

As a preparation for the next subsection, we recall the simplest ordinary differential equation result on numerical structural stability [Garay, 1996]. The numerical saddle structure results in [Beyn, 1987a] can be interpreted as follows: Given a hyperbolic equilibrium $x_0 \in \mathbf{R}^n$ of an ordinary differential equation, there exist a neighborhood $\mathcal{U}$ of x_0 in $\mathbf{R}^n$, a constant $\kappa > 0$ and, for each $h \in (0, h_1^*]$, there exists a homeomorphism $\mathcal{H}_h$ of $\mathcal{U}$ into $\mathbf{R}^n$ with the properties that

$$\mathcal{H}_h(x_0) = x_0 \quad and \quad |\mathcal{H}_h(x) - x| \leq \kappa h^p$$
$$\text{for each } x \in \mathcal{U} \tag{19}$$

and, last but not least,

$$\mathcal{H}_h(\Phi(h, x)) = \varphi(h, \mathcal{H}_h(x))$$
$$whenever \quad x \in \mathcal{U}, \ \Phi(h, x) \in \mathcal{U}. \tag{20}$$

In other words, in the vicinity of hyperbolic equilibria, the exact and the discretized dynamics are conjugate and discretization is nothing else but an almost-identical coordinate transformation. With $\mathcal{H}_h$ being a C^{p+r} diffeomorphism, note that (19) and (20) can be proved in the vicinity of nonequilibria, too [Garay & Simon, 2001].

3.2. *Inertial manifolds and structural stability*

Throughout this subsection, we restrict ourselves to a certain type of delay equations with small delay. The smallness of the delay seems to be necessary for the C^2 smoothness of the inertial manifold. (The existence of C^1 inertial manifolds can be proved with moderate delay. However, if the delay is not small, then the gap condition (which is the basis for proving higher order smoothness) is violated and, for the time being, there is no way out of this difficulty. For details and references, see [Robinson, 1999; Farkas, 2002b, 2002c; Chicone, 2003]. Here we restrict ourselves to reminding the reader that inertial manifolds are global center-unstable invariant manifolds.) On the other hand, the C^2 smallness of the inertial manifold is necessary to apply [Li, 1997] on numerical structural stability for ordinary differential equations in proving Part B of the Theorem below. All numerical structural stability results we are aware of require at least C^2 smoothness assumptions.

Recall the definition of the practical stepsize-$1/N$ explicit Euler discretization operator $\varphi_{\mathcal{P},E,PLI}(1/N, \cdot) : C_{1/N} \to C_{1/N}$ from the last paragraph of Sec. 2.1. If the delay is $\varepsilon > 0$, then the phase space is $C^\varepsilon = C([-\varepsilon, 0], \mathbf{R}^n)$. A trivial modification of formula (7) gives rise to the definition of the stepsize-ε/N practical explicit Euler discretization operator $\varphi_{\mathcal{P},E,PLI}^\varepsilon(1/N, \cdot) : C_{1/N}^\varepsilon \to C_{1/N}^\varepsilon$.

Theorem 3. *Consider the delay equation* $\dot{x}(t) = Ax(t) + a(x(t)) + b(x(t - \varepsilon))$ *where* A *is an* $n \times n$ *real matrix,* $a, b : \mathbf{R}^n \to \mathbf{R}^n$ *are bounded* C^2 *functions with bounded derivatives, and* ε *is a positive parameter.*

- A. [Farkas, 2002c]: *Then there exists an* $\varepsilon_0 > 0$ *with the properties as follows. For every* $\varepsilon \in$

$(0, \varepsilon_0]$, *the delay equation has an invariant manifold of the form* $\mathrm{Graph}(J^\varepsilon)$ *where* $J^\varepsilon : CU^\varepsilon \to SS^\varepsilon$ *is of class* C^2, *the linear subspace* CU^ε *is finite-dimensional, and the* $C^\varepsilon = CU^\varepsilon \times SS^\varepsilon$ *product decomposition is given by* $CU^\varepsilon = \pi^\varepsilon(C^\varepsilon)$, $SS^\varepsilon = (id|_{C^\varepsilon} - \pi^\varepsilon)(C^\varepsilon)$ *with* $\pi^\varepsilon : C^\varepsilon \to C^\varepsilon$, $(\pi^\varepsilon(\eta))(s) = e^{As}\eta(0)$, $s \in [-\varepsilon, 0]$. *In addition, for* N *large enough (and under very mild additional technical conditions),* $C^\varepsilon_{1/N}$ *admits a* $C^\varepsilon_{1/N} = CU^\varepsilon_{1/N} \times SS^\varepsilon_{1/N}$ *product decomposition with the properties as follows. Operator* $\varphi^\varepsilon_{P,E,PLI}(1/N, \cdot)$ *has an invariant manifold of the form* $\mathrm{Graph}(J^\varepsilon_{1/N})$, *where* $J^\varepsilon_{1/N} : CU^\varepsilon_{1/N} \to SS^\varepsilon_{1/N}$ *is a* C^2 *function. In addition, there exists a linear isomorphism* $P^\varepsilon_{1/N} : CU^\varepsilon \to CU^\varepsilon_{1/N}$ *such that for* $j = 0$ *and* $j = 1$,

$$\left\| \frac{d^j}{d\eta^j} \, \Pi^\varepsilon_{1/N} J^\varepsilon - \frac{d^j}{d\eta^j} \, J^\varepsilon_{1/N} P^\varepsilon_{1/N} \right\| \to 0 \quad \text{as } N \to \infty.$$

$$(21)$$

Together with $\mathrm{Graph}(J^\varepsilon)$, *also* $\mathrm{Graph}(J^\varepsilon_{1/N})$ *is exponentially attractive, with asymptotic phase depending continuously on the stepsize as* $N \to \infty$.

- *B. [Farkas, 2002c]:* (CONTINUATION.) *Assume, in addition, that the solution flow* $\Psi : \mathbf{R} \times \mathbf{R}^n \to \mathbf{R}^n$ *of the limiting ordinary differential equation* $\dot{x} = Ax + a(x) + b(x)$ *is structurally stable and that the point at the* $\{\infty\}$ *of* $\mathbf{R}^n$ *is repulsive. Then, for* N *large enough, there*

exists a homeomorphism $\mathcal{H}^\varepsilon_{1/N}$ *of* $\mathbf{R}^n$ *onto* Graph $(J^\varepsilon_{1/N})$ *and a continuous time-reparametrization mapping* $r^\varepsilon_{1/N} : \mathbf{R}^n \to \mathbf{R}^+$ *such that*

$$\mathcal{H}^\varepsilon_{1/N}(\Psi(r^\varepsilon_{1/N}(x), x)) = \varphi^\varepsilon_{P,E,PLI}(1/N, H^\varepsilon_{1/N}(x))$$

$$\text{whenever} \quad x \in \mathbf{R}^n. \qquad (22)$$

If Ψ *is Morse–Smale and gradient-like, then* $r^\varepsilon_{1/N}(x) = 1/N$ *for each* $x \in \mathbf{R}^n$.

The reader is asked to make a comparison between (18) and (21) as well as between (20) and (22).

The proof of Theorem 3 requires a very careful handling of standard inertial manifold techniques like spectral decomposition, manipulations with cut-off functions on finite-dimensional subspaces, fixed-point equations in weighted sequences of Banach spaces, fiber contraction theorem, etc. extended for discretizations. As for numerical structural stability, it is just an application of the fundamental theorem on numerical structural stability in [Li, 1997], derived as by-product of the Moser–Robbin–Robinson approach to Smale's structural stability theorem.

3.3. *Kamke monotonicity*

Assume that, for some constant $\gamma \geq 0$, condition

$$\begin{cases} (f_i)'_{x_j}(x,y) \geq \gamma & \text{if } (x,y) \in \mathbf{R}^n \times \mathbf{R}^n, \ i,j = 1,2,\ldots,n \text{ and } i \neq j, \\ (f_i)'_{y_j}(x,y) \geq \gamma & \text{if } (x,y) \in \mathbf{R}^n \times \mathbf{R}^n, \ i,j = 1,2,\ldots,n \end{cases} \qquad (23)$$

Here of course f_i stands for the ith coordinate function of f, further $x = (x_1, x_2, \ldots, x_n)$ and $y = (y_1, y_2, \ldots, y_n)$ denote the first n and the last n coordinate variables of f_i, $i = 1, 2, \ldots, n$, respectively. By letting $x \leq \tilde{x}$ for $x, \tilde{x} \in \mathbf{R}^n$ if and only if $x_i \leq \tilde{x}_i$ for each $i = 1, 2, \ldots, n$, a closed partial order on $\mathbf{R}^n$ is defined. The closed partial order $\leq$ on $\mathbf{R}^n$ generates a closed partial order $\preceq$ on C. In particular, $\eta \preceq \tilde{\eta}$ holds if and only if $\eta(s) \leq \tilde{\eta}(s)$ for each $s \in [-1, 0]$. As an easy consequence of assumption (20) the semi-dynamical system Φ is Kamke monotone [Smith, 1995]. In other words, inequality $\Phi(t, \eta) \preceq \Phi(t, \tilde{\eta})$ holds true whenever $t \geq 0$ and $\eta, \tilde{\eta} \in C$ with $\eta \preceq \tilde{\eta}$.

Theorem 4. *Consider the initial value problem (1) under condition (23). Then*

- *A. [Garay & Lóczi, 2004]: Let* $\gamma > 0$. *Given any Runge–Kutta method* M *satisfying* $b_i \geq 0$ *for* $i = 1, \ldots, \nu$, *the discretization operator* $\varphi_{M,PLI}$ *(with piecewise linear interpolation) is monotone in the sense that, for sufficiently small stepsize-h and for any initial functions with* $\eta \preceq \tilde{\eta}$, *also the order relation* $\varphi_{M,PLI}(h, \eta) \preceq \varphi_{M,PLI}(h, \tilde{\eta})$ *holds true.*
- *B. [Garay & Lóczi, 2004]: Let* $\gamma > 0$. *Suppose that* $f_i(x, y) \geq 0$ *and* $(f_i)'_{x_i}(x, y) \geq 0$ *for*

$i = 1, 2, \ldots, n$. *Then, for sufficiently small stepsize-h and for any nondecreasing initial function η, we have that $\varphi_{E,PLI}(h, \eta) \preceq \Phi(h, \eta) \preceq \varphi_{I,PLI}(h, \eta)$.*

- *C. [Kloeden & Schropp, 2004]: Let $\gamma = 0$. Suppose we are given a Runge–Kutta method M with the properties that $a_{ij} > 0$ for $i, j = 1, \ldots, \nu$ and $b_i > 0$ for $i = 1, \ldots, \nu$. Then the discretization operator $\varphi_{M,PLI}$ is monotone. Moreover, the positivity assumption on the Runge–Kutta matrix $A = \{a_{ij}\}_{i,j=1}^{\nu}$ can be weakened to the nonnegativity assumption on the matrix function $\tau \to (I + \tau A)^{-1} A$, required on some nondegenerate τ-interval $[0, \tau_0]$.*

An iterative combination of Parts A and B formulates and generalizes the well-known observation that, given a one-dimensional ordinary differential equation with all solutions convex, then every solution curve is above the broken line determined by the explicit, and under the broken line determined by the implicit Euler method. Part C is entirely of different character. In a strong resemblance to results on contractivity in numerical ordinary differential equations [Hairer *et al.*, 1993], it provides a sufficient condition for a Runge–Kutta method to preserve monotonicity of the solution dynamics under discretization. In the light of the elegant counterexamples in [Kloeden & Schropp, 2003], this sufficient condition is almost necessary.

Even in the numerical contexts of differential equation theory, the word "monotonicity" can be used in a number of various ways. Monotonicity of iterative methods for delay equations has already been investigated in [Erbe & Liu, 1991].

4. Remarks on Functional Differential Equations

The previous considerations suggest that all the Theorems above are valid for retarded functional differential equations of the form $\dot{x}(t) = g(x_t)$ where $x_t(s) = x(t + s)$ for $s \in [0, 1]$, and $g : C \to \mathbf{R}^n$ is of class $C^p, p \geq 2$. Gyula Farkas has always formulated his results in this more general framework. However, occasionally, he carried out the proofs only for the special case $g(x_t) = f(x(t), x(t - 1))$ and indicated the technical modifications needed for a general g. He considered assumptions (2)–(4) in [Farkas, 2003] and assumptions (i)–(vi) of his Lemma 8 in [Farkas, 2002c] as a general definition of discretization operators for equations of the form $\dot{x}(t) = g(x_t)$ and $\dot{x}(t) = Lx_t + g(x_t)$, respectively. Here $L : C \to \mathbf{R}^n$ is a bounded linear operator (which, by a theorem of Riesz, can be represented as a Stieltjes integral $L\eta = \int_{-1}^{0} \eta(s)d\vartheta(s)$).

The general definitions above have little relevance to practical purposes. For example, in all numerical implementations we are aware of, $L\eta$ is replaced by a finite sum like $\sum_{i=1}^{N} N^{-1} \left(\vartheta((1 - j)/N) - \vartheta(-j/N)\right) \eta(-j/N)$. Several references on the numerics of retarded functional differential equations — chosen in the spirit of the Bellen–Zennaro monograph — are contained in [Maset, 2003].

The numerics of equations with infinite delay — partly because of the depth of the underlying functional analysis — is more complicated. There are only sporadic results into this direction [Liu, 1997]. We cite also the papers [Koto, 1999; Insperger & Stépán, 2002] representing those devoted to some qualitative aspects of numerical bifurcation and numerical stability of retarded/delay equations. We are not aware of any papers on the numerics of delay equations with computer-assisted proofs.

All in all, we conclude by emphasizing that the large gap that characterized the relation of abstract dynamical systems theory and the numerical practice of solving differential equations until the early nineties of the last century, has been considerably filled in the last ten years. Having read papers like [Shub, 1986] or [Matijasevich, 1985] on numerical methods, it is clear to us that many of the most distinguished mathematicians have (i) hoped for (ii) guessed (iii) foreseen (iv) worked for this development. Among them John von-Neumann is pioneer number one.

Acknowledgments

Parts of this paper were written during a stay of the author at the University of Padova. Hospitality of the Department of Mathematics is gratefully acknowledged. The author is indebted to Wolf-Jürgen Beyn, Giovanni Colombo, Peter Kloeden, and Johannes Schropp for valuable discussions during the preparation of the paper. The paper is supported by the Hungarian National Science Foundation OTKA No. T037491.

References

Bellen, A. [1984] "One-step collocation for delay differential equations," *J. Comput. Appl. Math.* **10**, 275–283.

Bellen, A. & Zennaro, M. [2003] *Numerical Methods for Delay Differential Equations* (Oxford University Press, Oxford).

Beyn, W. J. [1987a] "On the numerical approximation of phase portraits near stationary points," *SIAM J. Numer. Anal.* **24**, 1095–1113.

Beyn, W. J. [1987b] "On invariant closed curves of one-step methods," *Numer. Math.* **51**, 103–122.

Beyn, W. J. & Lorenz, J. [1987] "Center manifolds of dynamical systems under discretization," *Num. Funct. Anal. Optimiz.* **9**, 318–414.

Butcher, J. F. [1987] *The Numerical Analysis of Ordinary Differential Equations* (Wiley, London).

Chicone, C. [2003] "Inertial and slow manifolds for delay equations with small delays," *J. Diff. Eq.* **190**, 364–406.

Erbe, L. & Liu, X. [1991] "Monotone iterative methods for differential systems with finite delay," *Appl. Math. Comput.* **43**, 43–64.

Farkas, G. [2001a] "Unstable manifolds for RFDEs under discretization: The Euler method," *Comput. Math. Appl.* **42**, 1069–1081.

Farkas, G. [2001b] "A Grobman–Hartman result for retarded functional differential equations with an application to the numerics of hyperbolic equilibria," *Z. Angew. Math. Phys.* **52**, 421–432.

Farkas, G. [2002a] "A numerical C^1-shadowing result for retarded functional differential equations," *J. Comput. Appl. Math.* **45**, 269–289.

Farkas, G. [2002b] "Nonexistence of uniform exponential dichotomies for delay equations," *J. Diff. Eq.* **182**, 266–268.

Farkas, G. [2002c] "Small delay inertial manifolds under numerics: A numerical structural stability result," *J. Dyn. Diff. Eq.* **14**, 549–588.

Farkas, G. [2003] "Discretizing hyperbolic periodic orbits of delay differential equations," *Z. Angew. Math. Mech.* **83**, 38–49.

Garay, B. M. [1996] "On structural stability of ordinary differential equations with respect to discretization methods," *Numer. Math.* **72**, 449–479.

Garay, B. M. [2001] "Estimates in discretizing normally hyperbolic compact invariant manifolds of ordinary differential equations," *Comput. Math. Appl.* **42**, 1103–1122.

Garay, B. M. & Simon, P. L. [2001] "Numerical flow-box theorems under structural assumptions," *IMA J. Numer. Anal.* **21**, 733–749.

Garay, B. M. & Lóczi, L. [2004] "Monotone delay equations and Runge–Kutta discretizations," *Funct. Diff. Eq.* **11**, 59–67.

Gedeon, T. & Hines, G. [1999] "Upper semicontinuity of Morse sets of a discretization of a delay-differential equation," *J. Diff. Eq.* **151**, 36–78.

Hairer, E., Norsett, S. P. & Wanner, G. [1993] *Solving Ordinary Differential Equations I. Nonstiff Problems* (Springer, Berlin).

Hale, J. K. [1977] *Theory of Functional Differential Equations* (Springer, Berlin).

Hale, J. K. [1988] *Asymptotic Behaviour of Dissipative Systems* (AMS, Providence).

Insperger, T. & Stépán, G. [2002] "Stability chart for the delayed Mathieu equation," *Roy. Soc. London Proc. Ser. A. Math. Phys. Eng. Sci.* **458**, 1989–1998.

In't Hout, K. & Lubich, Ch. [1998] "Periodic orbits of delay differential equations under discretization," *BIT* **38**, 72–91.

Kloeden, P. E. & Lorenz, J. [1986] "Stable attracting sets in dynamical systems and their one-step discretization," *SIAM J. Num. Anal.* **23**, 986–995.

Kloeden, P. E. & Schropp, J. [2003] "Runge–Kutta methods for monotone differential and delay equations," *BIT* **43**, 571–586.

Kloeden, P. E. & Schropp, J. [2004] "Stable attracting sets in delay differential equations and in their Runge–Kutta discretization," submitted.

Koto, T. [1999] "Neumark–Sacker bifurcation in the Euler method for a delay differential equations," *BIT* **39**, 110–115.

Li, M. C. [1997] "Structural stability of flows under numerics," *J. Diff. Eq.* **141**, 1–12.

Liu, Y. [1997] "On the θ-method for delay equations with infinite time lag," *J. Comput. Appl. Math.* **71**, 177–190.

Maset, S. [2003] "Numerical solution of retarded functional differential equations as abstract Cauchy problems," *J. Comput. Appl. Math.* **16**, 259–282.

Matijasevich, Yu. V. [1985] "*A posteriori* interval analysis," *EUROCAL*, Vol. 2., ed. Caviness, B. F. (Springer, Berlin), pp. 328–334.

Robinson, J. C. [1999] "Inertial manifolds with and without delay," *Discr. Cont. Dyn. Syst.* **5**, 813–824.

Shub, M. [1986] "Some remarks on dynamical systems and numerical analysis," *Dynamical Systems and Partial Differential Equations*, eds. Lara–Carrero, L. & Lewowicz, J. (Univ. Simon Bolivar, Caracas), pp. 69–91.

Smith, H. L. [1995] *Monotone Dynamical Systems* (AMS, Providence).

Stuart, A. M. & Humphries, A. R. [1996] *Dynamical Systems and Numerical Analysis* (Cambridge University Press, Cambridge).

Zennaro, M. [1986] "Natural continuous extensions of Runge–Kutta methods," *Math. Comput.* **46**, 119–133.

BIFURCATIONS AND CONTINUOUS TRANSITIONS OF ATTRACTORS IN AUTONOMOUS AND NONAUTONOMOUS SYSTEMS

P. E. KLOEDEN and S. SIEGMUND

Fachbereich Mathematik, Johann Wolfgang Goethe Universität,
D-60054 Frankfurt am Main, Germany

Received February 16, 2004; Revised June 8, 2004

Nonautonomous bifurcation theory studies the change of attractors of nonautonomous systems which are introduced here with the process formalism as well as the skew product formalism.

We present a total stability theorem ensuring the existence of nearby attractors of perturbed systems. They depend continuously on a parameter if and only if the attraction is uniform w.r.t. parameter, i.e. the attractors are equiattracting.

We apply these principles to explicit systems to clarify the meaning of continuous and abrupt transitions of attractors in contrast to bifurcations, i.e. splitting of minimal invariant subsets into others within the attractor. Several examples are treated, including a nonautonomous pitchfork bifurcation.

Keywords: Total stability; attractor transition; attractor bifurcation; subcritical bifurcation; supercritical bifurcation; nonautonomous pitchfork bifurcation; nonautonomous dynamical system; process; skew product flow.

1. Introduction

We have several aims in writing this article, which is really more an essay than a research, survey or tutorial paper, although it contains elements of all three. Two main aims of particular long term interest are to understand what is meant by

1. a bifurcation or transition of a nontrivial attractor set in an autonomous system (e.g. such as a Lorenz attractor in its chaotic regime or a Chua attractor),
2. a bifurcation in a nonautonomous system.

As we shall see, both are closely related through the skew product flow representation of nonautonomous dynamical systems.

Our discussion is by no means complete. Nevertheless, we hope that our comments will provide the reader with some insight into the issues that are involved and will stimulate further investigations.

Underlying our considerations are two general principles. The first is the concept of total stability: if a system has a uniformly asymptotically stable compact set, such as a global attractor, then so do all nearby systems. In the autonomous case, the perturbed systems then have attractors, which converge in general upper semicontinuously to that of the original system. In the nonautonomous case, the perturbed systems also have nearby compact absorbing or attracting sets, but the existence of attractors is complicated by the ambiguity of just how an attractor should actually be defined in nonautonomous systems — we will introduce the reader to several possibilities below.

The idea of total stability underlies a rarely mentioned fact in autonomous bifurcation theory. Although subcritical and supercritical bifurcations are commonly encountered in such systems, it is not always easy to determine which of them actually occurs. However, if — at the bifurcation point — an equilibrium point remains asymptotically stable for the nonlinear system when it loses stability in the linearized system, then the bifurcation is supercritical (e.g. see [Arrowsmith & Place, 1990, Theorem 4.2.1] in connection with Hopf bifurcations). In this case the global (or maximal if only a local) attractor of the nonlinear system in fact depends continuously on the bifurcation parameter. We will see below that this also holds for a subcritical bifurcation at the point of loss of linear stability, but not where the nonlocal bifurcating equilibrium points first arise. This is a consequence of a second general principle: as recently shown in [Li & Kloeden, 2004a] (see also [Li & Kloeden, 2004b; Wang *et al.*, 2004]), attractors depend continuously on a parameter if and only if they are equiattracting, i.e. uniformly attracting with respect to the parameter.

The term "bifurcation" usually refers to situations when a system linearized about a minimal invariant set such as an equilibrium point, a periodic solution or an almost periodic solution loses asymptotic stability and several new invariant sets come into existence. Can one really talk about the bifurcation of a general global attractor? Firstly, it is not clear about which solutions within a global attractor one should linearize and, secondly, a global attractor is both unique and connected, if it exists, so cannot split into new disjoint invariant sets. Obviously, one needs to think in terms of changes of the dynamics within the global attractor rather than of the attractor itself. We will use the term *transition* when discussing changes to the attractors as system parameters vary and reserve the term "bifurcation" for the usual situations mentioned above, i.e. the splitting of minimal invariant subsets into others within the attractors. In particular, we will refer to a *continuous transition* when the global (or maximal if only local) attractors depend continuously on the parameters and to an *abrupt* or *discontinuous transition* when the attractors depend only upper semicontinuously (and not continuously) on the parameter. We will see in many examples that although transitions, in general, need only be abrupt, they are in fact typically continuous for most parameter values.

Next we describe the structure of this article, followed by some notation at the end of this section.

Section 2 deals with autonomous systems and some of their bifurcations. It contains scalar examples of a supercritical and a subcritical bifurcation and an explanation why e.g. the transcritical

bifurcation does not fit into our total stability scenario. We also give examples of supercritical, subcritical and saddle-node bifurcations for triangular autonomous systems.

In Sec. 3 we introduce nonautonomous systems from two different points of views using the process formalism and the skew product formalism. Our examples are a nonautonomous version of the autonomous example for the supercritical pitchfork bifurcation and triangular autonomous systems, now interpreted differently.

Section 4 contains our theorem on total stability of nonautonomous systems under a uniform parametric dependence condition. We apply it in Sec. 5 to derive conditions for continuous bifurcations, e.g. supercritical bifurcations.

Section 6 concludes the discussion with some remarks and open questions. The proof of the total stability theorem is contained in an Appendix.

The Hausdorff semi-metric $H_X^*(A, B)$ of nonempty compact subsets A and B of a metric space (X, d) is defined as

$$H_X^*(A, B) := \max_{a \in A} \operatorname{dist}(a, B),$$

$$\text{where } \operatorname{dist}(a, B) := \min_{b \in B} d(a, b),$$

and $H_X(A, B) = \max\left\{ H_X^*(A, B), H_X^*(B, A) \right\}$ is a metric, called the Hausdorff metric, on the space of nonempty compact subsets of (X, d).

Remark 1. To simplify the exposition we will always assume that we have global bounds and constants (e.g. in Lipschitz conditions and approximation estimates). Our theorems and proofs are in fact valid for locally defined bounds and constants, but require technical modifications (e.g. see [Kloeden & Lorenz, 1986]), which, we feel, distract from our emphasis here on the dynamical behavior.

2. Autonomous Systems

We begin with several examples of well-known bifurcations in scalar autonomous systems and in two-dimensional triangular autonomous systems, which provide simple but useful insights into the topics we wish to discuss.

2.1. *Scalar autonomous systems*

2.1.1. *Supercritical bifurcation*

The autonomous differential equation

$$\frac{dx}{dt} = \nu x - bx^3 \tag{1}$$

with $b > 0$ has a global attractor $A_\nu = \{0\}$ for $\nu < 0$. For $\nu = 0$, the equilibrium point 0 loses asymptotic stability for the linearized equation (it remains stable there), but the set $A_0 = \{0\}$ is still a global attractor for the nonlinear system. To see this we use the Lyapunov function $V(x) = x^2$ to obtain

$$\frac{d}{dt} V(x_0(t)) = -2bx_0(t)^4 = -2bV(x_0(t))^2$$

and hence

$$V(x_0(t)) = \frac{V(x_0(0))}{1 + 2bV(x_0(0))t} \to 0 \quad \text{as } t \to \infty,$$

where $x_0(t)$ is any solution of the differential equation (1) for $\nu = 0$. (This Lyapunov function can also be used for $\nu < 0$.)

For $\nu > 0$, there are three equilibrium points 0 and $\pm\sqrt{\nu/b_0}$, and the global attractor is now

$$A_\nu = \left[-\sqrt{\nu/b_0}, \ \sqrt{\nu/b_0} \right].$$

Here $H_{\mathbb{R}}(A_\nu, A_0) = H_{\mathbb{R}}^*(A_\nu, \{0\}) = \sqrt{\nu/b_0} \to 0$ as $\nu \to 0$, i.e. the set-valued mapping $\nu \mapsto A_\nu$ is continuous (in the Hausdorff metric) at the bifurcation point $\nu = 0$.

This classical example of a supercritical pitchfork bifurcation in an autonomous system is our first example of what we have called above a "continuous transition" of the global attractor.

2.1.2. *Subcritical bifurcation*

We now consider a subcritical bifurcation arising in the scalar autonomous differential equation

$$\frac{dx}{dt} = -x \left(x^4 - 2x^2 + 1 - \nu \right), \tag{2}$$

for which there are three parameter regimes for equilibrium solutions $\overline{x}_\nu$:

(i) $\overline{x}_\nu = 0$ for $\nu < 0$
(ii) $\overline{x}_\nu = 0, \pm\sqrt{1 + \sqrt{\nu}}, \pm\sqrt{1 - \sqrt{\nu}}$ for $0 \leq \nu < 1$, and
(iii) $\overline{x}_\nu = 0, \pm\sqrt{1 + \sqrt{\nu}}$ for $\nu \geq 1$.

The zero solution here loses linear stability at $\nu = 1$ in a subcritical bifurcation to the nonlocal solutions $\pm\sqrt{1 + \sqrt{\nu}}$. Note, however, that these equilibria as well as $\pm\sqrt{1 - \sqrt{\nu}}$ first appear at $\nu = 0$. The equilibria $\pm\sqrt{1 + \sqrt{\nu}}$ are asymptotically stable for $\nu > 0$, whereas the equilibria $\pm\sqrt{1 - \sqrt{\nu}}$ are unstable in their existence interval $0 \leq \nu < 1$.

The global attractors here are $A_\nu = \{0\}$ for $\nu < 0$ and

$$A_\nu = \left[-\sqrt{1 + \sqrt{\nu}}, \ \sqrt{1 + \sqrt{\nu}} \right]$$

for $\nu \geq 0$. In particular, the set-valued mapping $\nu \mapsto A_\nu$ is not continuous at $\nu = 0$ (being only

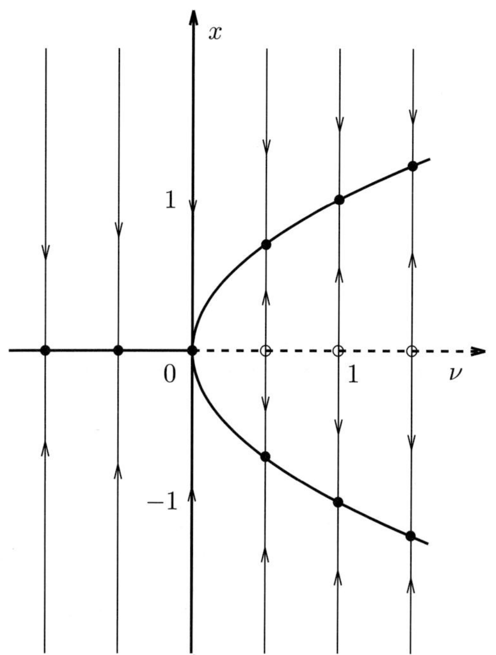

Fig. 1.   Supercritical pitchfork bifurcation.

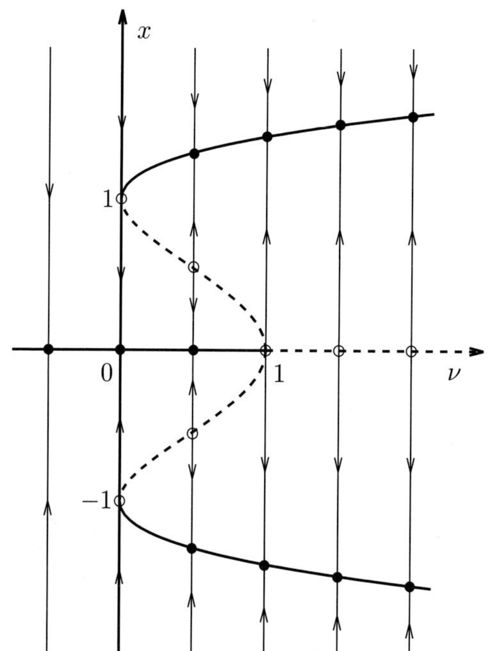

Fig. 2.   Subcritical bifurcation.

upper semicontinuous there), but is continuous at $\nu = 1$. The attractor thus undergoes a discontinuous transition at $\nu = 0$ and a continuous transition at $\nu = 1$ (and, in fact, at any $\nu \neq 0$). An inspection of the direction fields shows that the attractors are clearly not equiattracting for parameter values in a neighborhood of $\nu = 0$, since the attraction is not uniform there.

2.1.3.   *Some nonapplicable situations*

We mention for completeness that a transcritical bifurcation does not fit into our total stability scenario since there is no nonlinear attractor at the bifurcation point, e.g. the autonomous equation

$$\frac{dx}{dt} = \nu x - x^2$$

has (nonglobal) local attractors $A_\nu = \{0\}$ for $\nu < 0$ and $A_\nu = \{\nu\}$ for $\nu > 0$, but $A_0 = \{0\}$ is not attracting from any neighborhood, being attracting on one side and repelling on the other.

A similar situation occurs in a small neighborhood of the nonlocal subcritically bifurcating equilibria ± 1 in system (2) at $\nu = 0$, these equilibria also being attracting on one side and repelling on the other. Unlike in the transcritical bifurcation above, the equilibria here exist only on one side of the critical parameter value $\nu = 0$.

2.2.   *Triangular autonomous systems*

We consider some autonomous differential equations in $\mathbb{R}^2$ with the triangular form

$$\frac{dx}{dt} = f(x,p), \quad \frac{dp}{dt} = g(p), \quad (x,p) \in \mathbb{R}^2.$$

Such triangular systems are examples of skew product flows with the uncoupled component for p being considered as "driving" the coupled or "driven" system for x.

In the following three examples we consider bifurcations of the triangular system due to bifurcations in either the "driving" equation (for p) or the "driven" equation (for x), the first two involving a supercritical and subcritical bifurcation, respectively, of the driving system and the third a saddle-node bifurcation in the driven system. The systems are all Morse–Smale systems with global attractors consisting of a finite number of equilibria and their heteroclinic trajectories. Since the systems are

two-dimensional their global attractors can be easily determined by elementary algebra and direction field arguments.

2.2.1.   *Supercritical bifurcation*

The driving system of the triangular system

$$\frac{dx}{dt} = -x + p, \quad \frac{dp}{dt} = \nu p - p^3, \qquad (x,p) \in \mathbb{R}^2 \quad (3)$$

is, in fact, the scalar autonomous differential equation (1) with $b = 1$, thus with equilibria $\overline{p}_\nu = 0$ for $\nu \leq 0$ and $\overline{p}_\nu = 0, \pm\sqrt{\nu}$ for $\nu > 0$. The equilibria $(\overline{x}_\nu, \overline{p}_\nu)$ of the triangular system (3) satisfy $\overline{x}_\nu = \overline{p}_\nu$, where the $\overline{p}_\nu$ are equilibria of the driving system. Thus there are two cases

(i) $(\overline{x}_\nu, \overline{p}_\nu) = (0,0)$ for $\nu \leq 0$, and
(ii) $(\overline{x}_\nu, \overline{p}_\nu) = (0,0), (\pm\sqrt{\nu}, \pm\sqrt{\nu})$ for $\nu > 0$.

The global attractor of the coupled system is thus $A_\nu = \{(0,0)\}$ for $\nu \leq 0$ and

$$A_\nu = \{(0,0), (\pm\sqrt{\nu}, \pm\sqrt{\nu})\}$$
$$\cup \ \{\text{heteroclinic trajectories}\}$$

for $\nu > 0$.

A heteroclinic trajectory lies below the $x = p$ line in the (p, x) plane if $p' = \nu p - p^3$ is positive there, and above if it is negative, see Fig. 3.

The set-valued mapping $\nu \to A_\nu$ is thus continuous for all ν, including the supercritical bifurcation point $\nu = 0$ of the driving system, where the single equilibria $(0,0)$ in A_0 (actually, $A_0 = \{(0,0)\}$ here) undergoes a supercritical bifurcation to yield the new equilibria $(\pm\sqrt{\nu}, \pm\sqrt{\nu})$ in A_ν for $\nu > 0$.

2.2.2.   *Subcritical bifurcation*

The situation is analogous for the triangular system

$$\frac{dx}{dt} = -x + p, \quad \frac{dp}{dt} = -p\left(p^4 - 2p^2 + 1 - \nu\right),$$
$$(x,p) \in \mathbb{R}^2, \qquad (4)$$

where the driving system is now the subcritically bifurcating scalar differential equation (2). The triangular system (4) has equilibria $(\overline{x}_\nu, \overline{p}_\nu)$ with $\overline{x}_\nu = \overline{p}_\nu$, where the $\overline{p}_\nu$ are equilibria of the driving

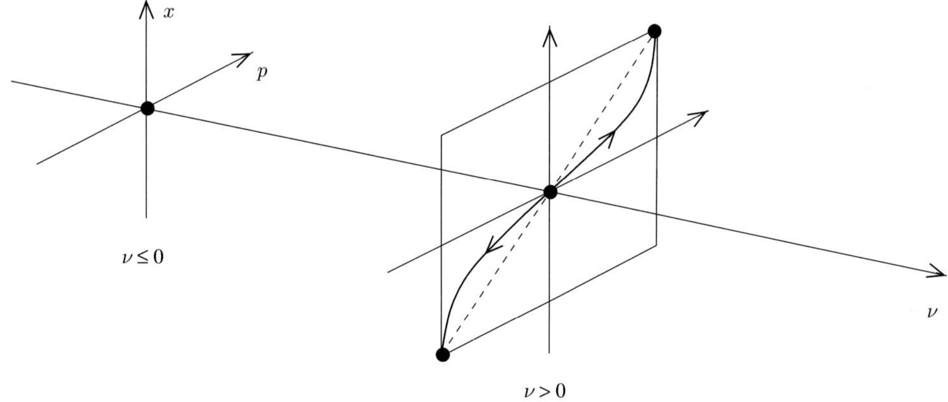

Fig. 3.   The attractor for $\nu \leq 0$ and $\nu > 0$: Supercritical case.

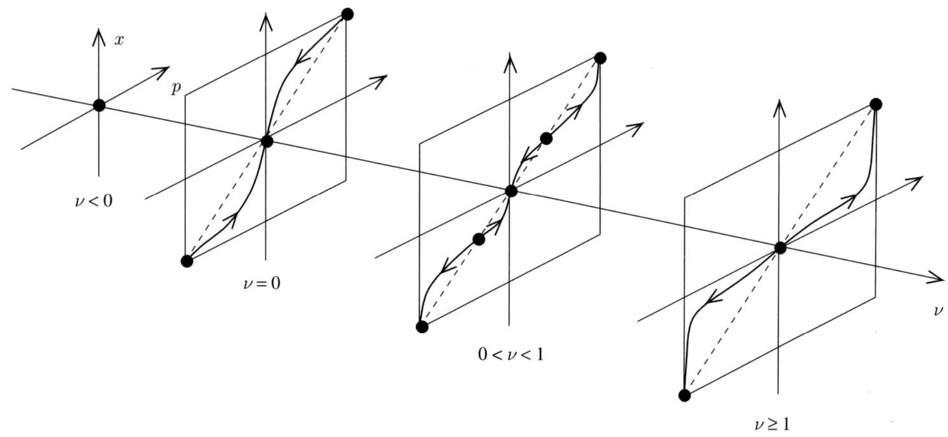

Fig. 4.   The attractor for different values of ν: Subcritical case.

system, i.e. with the three cases

(i) $\overline{p}_\nu = 0$ for $\nu < 0$,

(ii) $\overline{p}_\nu = 0, \pm\sqrt{1 - \sqrt{\nu}}, \pm\sqrt{1 + \sqrt{\nu}}$ for $0 \leq \nu < 1$, and

(iii) $\overline{p}_\nu = 0, \pm\sqrt{1 + \sqrt{\nu}}$ for $\nu \geq 1$.

The global attractor of the coupled system is thus $A_\nu = \{(0,0)\}$ for $\nu < 0$ with

$$A_\nu = \left\{(0,0), \left(\pm\sqrt{1 \pm \sqrt{\nu}}, \pm\sqrt{1 \pm \sqrt{\nu}}\right)\right\}$$

$$\cup \{\text{heteroclinic trajectories}\}$$

for $0 \leq \nu < 1$, and

$$A_\nu = \left\{(0,0), \left(\pm\sqrt{1 + \sqrt{\nu}}, \pm\sqrt{1 + \sqrt{\nu}}\right)\right\}$$

$$\cup \{\text{heteroclinic trajectories}\}$$

for $\nu \geq 1$. A heteroclinic trajectory lies below the $x = p$ line in the (p, x) plane if $p' = -p(p^4 -$

$2p^2 + 1 - \nu)$ is positive there, and above if it is negative, see Fig. 4. The transition in A_ν is thus continuous for all $\nu \neq 0$ and discontinuous only for $\nu = 0$.

2.2.3. *Saddle-node bifurcation*

The autonomous differential equation in $\mathbb{R}^2$,

$$\frac{dx}{dt} = x\left(1 - x^2 - y^2\right) + y\left(1 + \nu + x\right),$$

$$\frac{dy}{dt} = -x\left(1 + \nu + x\right) + y\left(1 - x^2 - y^2\right),$$

is obviously not of triangular form. However, if we change from cartesian to polar coordinates, then we obtain an equivalent system in triangular form,

$$\frac{d\theta}{dt} = -(1 + \nu + r\cos\theta), \quad \frac{dr}{dt} = r - r^3,$$

$$(\theta, r) \in [0, 2\pi] \times \mathbb{R}^+, \tag{5}$$

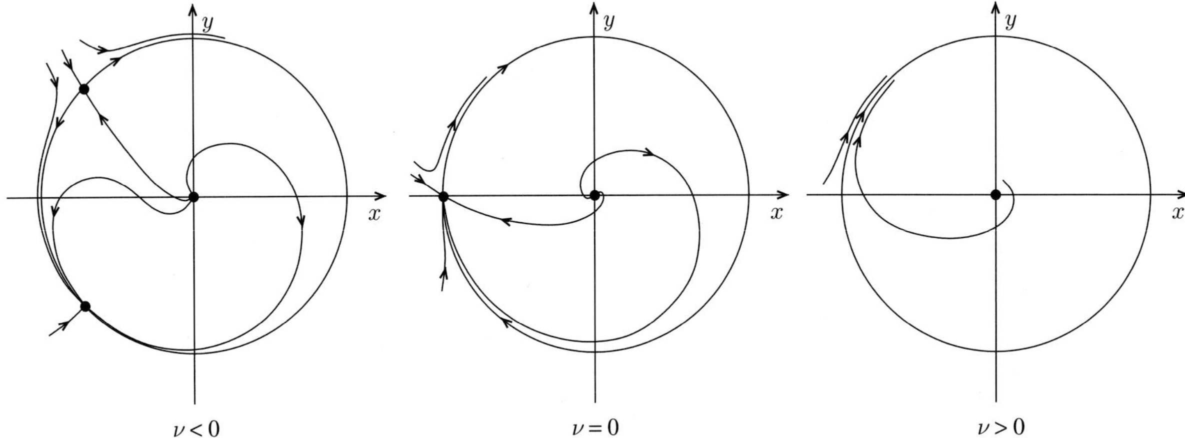

Fig. 5.   Bifurcation diagram, cartesian coordinates.

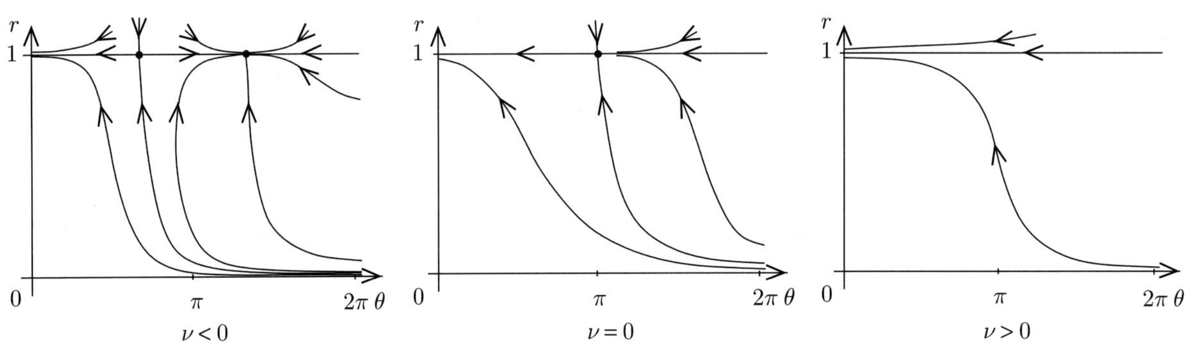

Fig. 6.   Bifurcation diagram, polar coordinates.

for which the dynamical behavior is more transparent [Arrowsmith & Place, 1990; Glendinning, 1994; Reitmann, 1996].

The global attractor A_ν here is the unit disk $\{(x, y) \in \mathbb{R}^2 \;:\; r^2 = x^2 + y^2 \leq 1\}$ for all ν, i.e. it does not change at all as ν changes (so it obviously depends continuously on ν). However, the dynamics within and near A_ν do change significantly as ν passes through zero. The origin is an equilibrium point and the unit circle is invariant for all values of ν. For negative values of ν, there are two equilibrium points on the unit circle, one a saddle point and the other a stable node, which coalesce at $\nu = 0$ to form a single saddle-node. For positive ν this saddle node disappears. This is easily visualized in Cartesian coordinates (see Fig. 5), but it is interesting to also use polar coordinates (see Fig. 6).

This example also differs from the previous two on triangular systems in that the bifurcation occurs in the coupled "state" space equation (θ here) rather than in the "driving" system equation (r here).

3.   Nonautonomous Systems

We consider two abstract formalisms of nonautonomous dynamical systems through simple examples: the process formalism, which at first sight seems more natural, and the skew product formalism in which a state space system is driven by an inputed autonomous system. Various definitions of nonautonomous attractors will then be introduced and illustrated with examples.

3.1.   *Process formalism of a nonautonomous system*

Solution mappings are one of the main motivations for the process definition [Dafermos, 1971] (see also

[Hale, 1988]) of an abstract nonautonomous dynamical system on a state space X. A *process* is a continuous mapping $(t, t_0, x_0) \mapsto x(t, t_0, x_0) \in X$ for $t \geq t_0$, $t_0 \in \mathbb{R}$ and $x_0 \in X$, with the initial value and evolution properties

(i) $x(t_0, t_0, x_0) = x_0$ for all $t_0 \in \mathbb{R}$ and $x_0 \in X$,
(ii) $x(t_2, t_0, x_0) = x(t_2, t_1, x(t_1, t_0, x_0))$ for all $t_0 \leq t_1 \leq t_2$ in $\mathbb{R}$ and $x_0 \in X$.

It is often also called a *two-parameter semigroup* on X in contrast with the one-parameter semigroup of an autonomous dynamical system.

A nonautonomous attractor A now consists of a family of nonempty compact sets $\{A(t), t \in \mathbb{R}\}$ which is invariant in the sense that

$$x(t, t_0, A(t_0)) = A(t) \quad \text{for all } t \geq t_0, t_0 \in \mathbb{R}$$

(from which it follows that the set-valued mapping $t \mapsto A(t)$ is continuous in $t \in \mathbb{R}$). There are now two ways to define attraction of A, which are equivalent in the autonomous case. The first, and perhaps more obvious, corresponds to Lyapunov asymptotic stability, i.e. with *forward attraction* in the sense of

$$\lim_{t \to \infty} H_{\mathbb{R}}^{*}(x(t, t_0, x_0), A(t)) = 0. \tag{6}$$

Note that the target set $A(t)$ is changing in time.

The other attraction, called *pullback attraction* involves a fixed target set with progressively earlier starting time, i.e.

$$\lim_{t_0 \to -\infty} H_{\mathbb{R}}^{*}(x(t, t_0, x_0), A(t)) = 0.$$

See Fig. 7 for the case where the family $\{A(t), t \in \mathbb{R}\}$ corresponds to a single trajectory $\overline{\varphi}(t)$, i.e. $A(t) = \{\overline{\varphi}(t)\}$ for all $t \in \mathbb{R}$. For an extensive, but elementary introduction to forward and pullback attractors see [Caraballo *et al.*, 2003; Grüne & Kloeden, 2001].

We now consider a nonautonomous analogue of the pitchfork bifurcation of the differential equation (1), namely

$$\frac{dx}{dt} = \nu x - b(t)x^3, \tag{7}$$

with continuous $b : \mathbb{R} \to \mathbb{R}$, $b(t) \in [b_0, b_1]$ for all $t \in \mathbb{R}$ where $0 < b_0 < b_1 < \infty$. See [Kloeden, 2004] for the bifurcatory analysis of a multidimensional version of this equation.

Assuming $b(t)$ is nonconstant, the nonautonomous differential equation has only one equilibrium point 0 and this exists for all values of ν. Supposing that $\nu \leq 0$, we see for the Lyapunov function $V(x) = x^2$ that

$$\frac{d}{dt}V(x_\nu(t)) = 2\nu x_\nu(t)^2 - 2b(t)x_\nu(t)^4$$

$$\leq 2\nu V(x_\nu(t)) - 2b_0 V(x_\nu(t))^2$$

$$\leq -2b_0 V(x_\nu(t))^2,$$

so

$$V(x_\nu(t)) \leq \frac{V(x_\nu(t_0))}{1 + 2b_0\, V(x_\nu(t_0))(t - t_0)} \to 0$$

$$\text{as } t \to \infty,$$

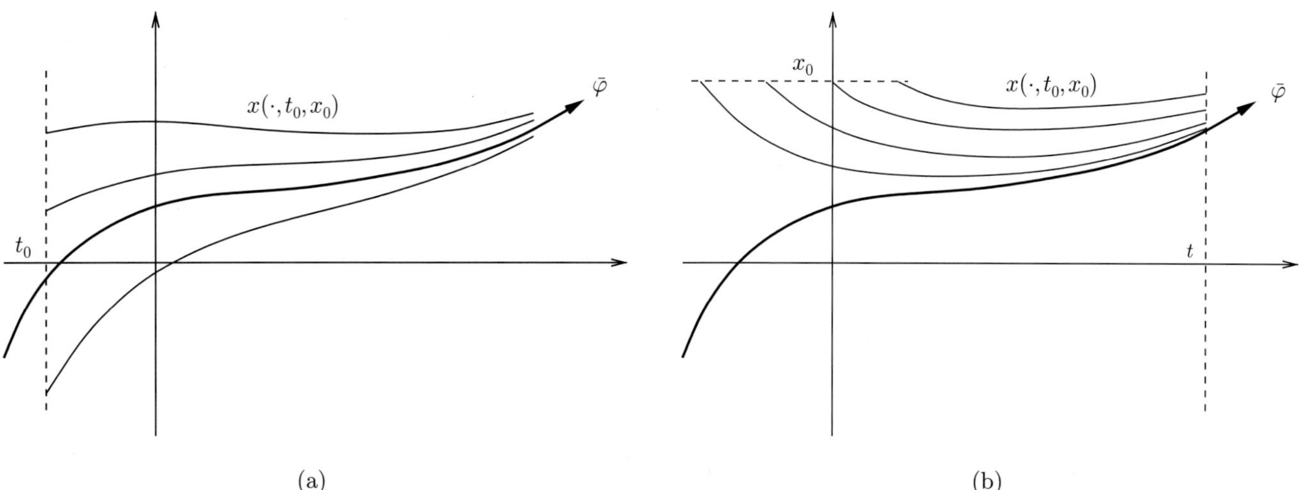

(a)   (b)

Fig. 7. (a) Forward and (b) pullback attraction.

where $x_\nu(t)$ is any solution of the differential equation (7) for $\nu \leq 0$. Moreover, the equilibrium point 0 loses stability in the linearized equation at $\nu = 0$ and is unstable for $\nu > 0$.

Unlike the autonomous case above, no new equilibrium points come into existence when $\nu > 0$. Instead we will show that there is a family of time-dependent sets

$$A_\nu(t) = [-\overline{\varphi}_\nu(t), \overline{\varphi}_\nu(t)], \quad t \in \mathbb{R}, \tag{8}$$

which are uniformly Lyapunov asymptotically stable (see Eq. (12) below). Here $\pm\overline{\varphi}_\nu$ are solutions of the nonautonomous differential equation (7) given by

$$\overline{\varphi}_\nu(t) = \frac{1}{\sqrt{2 \displaystyle\int_{-\infty}^{t} b(s) e^{-2\nu(t-s)} \, ds}} \tag{9}$$

and satisfies $\overline{\varphi}_\nu(t) \in \left[\sqrt{\nu/b_1}, \sqrt{\nu/b_0}\right]$ for all $t \in \mathbb{R}$, which means, in particular that

$$H_\mathbb{R}(A_\nu(t), A_0) = H_\mathbb{R}(A_\nu(t), \{0\})$$
$$\leq \sqrt{\nu/b_0} \to 0 \quad \text{as } \nu \to 0,$$

for each $t \in \mathbb{R}$, see Fig. 8. (Note that $\overline{\varphi}_\nu$ is almost periodic when the coefficient b is almost periodic [Fink, 1974; Sell, 1971].)

Thus we have another example of a continuous transition of the attractor, this time nonautonomous.

The proofs of the above assertions are instructive. We first note that Eq. (7) is a Bernoulli equation, so can be converted into a linear differential equation

$$\frac{dv}{dt} + 2\nu v = 2b(t)$$

with the substitution $v = x^{-2}$ (recall that $x = 0$ is an equilibrium of the Bernoulli equation (7)), which we integrate to obtain

$$\frac{1}{x_\nu(t, t_0, x_0)^2} = \frac{1}{x_0^2} e^{-2\nu(t-t_0)}$$
$$+ 2 \int_{t_0}^{t} b(s) e^{-2\nu(t-s)} \, ds. \tag{10}$$

We hold t and x_0 fixed in (10) and take the *pullback* limit as $t_0 \to -\infty$, to obtain the pullback limit solution

$$\overline{\varphi}_\nu(t) = \lim_{t_0 \to -\infty} x_\nu(t, t_0, x_0)$$

given by (9). This is itself a solution of the differential equation (7) and thus satisfies (10) with $x_0 = \overline{\varphi}_\nu(t_0)$, specifically

$$\frac{1}{\overline{\varphi}_\nu(t)^2} = \frac{1}{\overline{\varphi}_\nu(t_0)^2} e^{-2\nu(t-t_0)}$$
$$+ 2 \int_{t_0}^{t} b(s) e^{-2\nu(t-s)} \, ds. \tag{11}$$

To show that $\overline{\varphi}_\nu(t)$ is asymptotically stable for all $x_0 > 0$ (cf. [Langa *et al.*, 2002; Langa *et al.*, 2004;

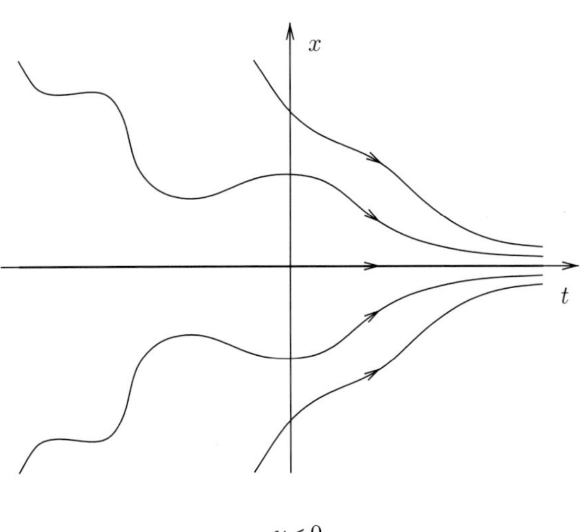

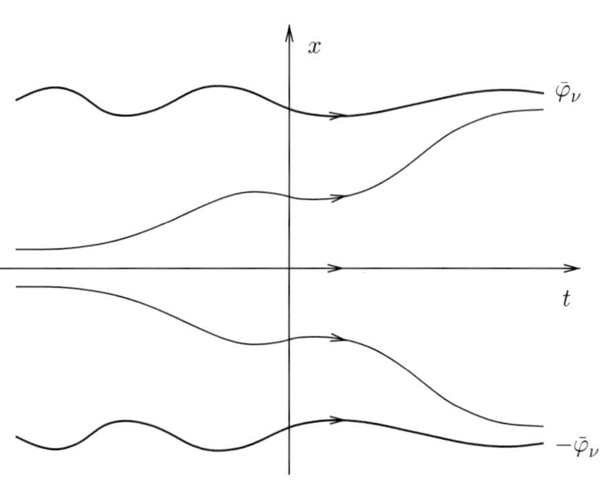

Fig. 8.   Nonautonomous pitchfork bifurcation.

Langa & Suárez, 2002]), we subtract (10) from (11) to obtain

$$\frac{1}{\overline{\varphi}_\nu(t)^2} - \frac{1}{x_\nu(t,t_0,x_0)^2}$$

$$= \left(\frac{1}{\overline{\varphi}_\nu(t_0)^2} - \frac{1}{x_0^2}\right) e^{-2\nu(t-t_0)} \to 0 \quad \text{as } t \to \infty,$$

from which the result follows. It is in fact uniformly asymptotically stable because using the fact that $\overline{\varphi}_\nu(t) \in \left[\sqrt{\nu/b_1}, \sqrt{\nu/b_0}\right]$, we can find for every $\varepsilon > 0$ and $x_0 > 0$ a $T(x_0, \varepsilon) \geq 0$ such that

$$\left(\frac{1}{\overline{\varphi}_\nu(t_0)^2} - \frac{1}{x_0^2}\right) e^{-2\nu(t-t_0)}$$

$$\leq \left(\frac{b_1}{\nu} - \frac{1}{x_0^2}\right) e^{-2\nu(t-t_0)} < \varepsilon \qquad (12)$$

for $t \geq T(x_0, \varepsilon) + t_0$. The corresponding result holds for the solution $-\overline{\varphi}_\nu(t)$ for all $x_0 < 0$.

We note here that the nonzero solutions $x_\nu(t, t_0, x_0)$ of the differential equation (7) converge to $\mathrm{sgn}(x_0)\overline{\varphi}_\nu(t)$ in both the usual forward sense (i.e. $t \to \infty$ with t_0 fixed) as well as in the pullback sense (i.e. $t_0 \to -\infty$ with t fixed). In general, forward and pullback convergence are independent concepts and neither one implies the other [Cheban *et al.*, 2002; Wang *et al.*, 2004], although in autonomous systems or in uniform systems, as in this example, they are equivalent. Pullback convergence is useful in constructing limiting objects, since the limit is fixed and not changing in time as in forward convergence.

3.2. *Triangular systems as nonautonomous systems*

Here we will reconsider our examples in Sec. 2.2 of bifurcations in autonomous differential equations of the triangular form

$$\frac{dx}{dt} = f(x,p), \quad \frac{dp}{dt} = g(p), \qquad (x,p) \in \mathbb{R}^2, \quad (13)$$

which we now interpret as nonautonomous differential equations

$$\frac{dx}{dt} = \tilde{f}_{p_0}(x,t) := f(x,p(t,p_0)), \qquad x \in \mathbb{R}^1, \quad (14)$$

where the solution $p(t,p_0)$ with initial value $p(0,p_0) = p_0$ of the second uncoupled component of the triangular system (13) is considered as an external driving force. In general, $p(t,p_0)$ need not be the solution of an autonomous differential equation, but just a function or a trajectory of an autonomous dynamical system, examples of which will be given in Sec. 3.3 and later.

We will reinterpret the bifurcations in these examples as nonautonomous bifurcations, using the pullback convergence method to construct explicitly the heteroclinic trajectories inside the previously determined global attractors.

In particular, we consider the supercritical bifurcation in the system

$$\frac{dx}{dt} = -x + p, \quad \frac{dp}{dt} = \nu p - p^3, \qquad (x,p) \in \mathbb{R}^2. \quad (15)$$

As seen earlier, the global attractor of the coupled system (15) is $A_\nu = \{(0,0)\}$ for $\nu \leq 0$ and

$$A_\nu = \{(0,0), (\pm\sqrt{\nu}, \pm\sqrt{\nu})\}$$
$$\cup \{\text{heteroclinic trajectories}\}$$

for $\nu > 0$.

We know that the global attractor of the uncoupled equation for p when $\nu > 0$ is $P_\nu = [-\sqrt{\nu}, \sqrt{\nu}]$ and, moreover, that the solution $p_\nu(t,p_0)$ exists for all $t \in \mathbb{R}$ when $p_0 \in P_\nu$, since P_ν is a compact invariant set for the p-dynamics. For such a solution, the nonautonomous differential equation (14) for the first component,

$$\frac{dx}{dt} = -x + p_\nu(t,p_0),$$

with initial value $x(t_0) = x_0$ has the solution

$$x_\nu(t,t_0,x_0,p_0) = x_0 e^{-(t-t_0)}$$
$$+ e^{-t}\int_{t_0}^t e^s p_\nu(s,p_0)\, ds. \quad (16)$$

Holding t fixed and taking the limit as $t_0 \to -\infty$, we obtain the pullback limit

$$\lim_{t_0 \to -\infty} x_\nu(t,t_0,x_0,p_0) = \overline{\varphi}_\nu(t,p_0)$$

$$:= e^{-t}\int_{-\infty}^t e^s p_\nu(s,p_0)\, ds. \quad (17)$$

Obviously, we have $-\sqrt{\nu} \leq \overline{x}_\nu(t,p_0) \leq \sqrt{\nu}$ for all $t \in \mathbb{R}$ with, in particular, $\overline{\varphi}_\nu(t,\overline{p}_\nu) \equiv \overline{p}_\nu$, when $p_0 = \overline{p}_\nu$ is one of the equilibria $0, \pm\sqrt{\nu}$ of the p-equation. In fact, $(\overline{\varphi}_\nu(t,p_0), p_\nu(t,p_0)) \in A_\nu$ for all $t \in \mathbb{R}$ and the heteroclinic trajectories in the global attractor A_ν are the curves

$$p_0 \mapsto \overline{\Phi}_\nu(p_0), \quad p_0 \in P_\nu = [-\sqrt{\nu}, \sqrt{\nu}],$$

where $\overline{\Phi}_\nu(p_0) := \overline{\varphi}_\nu(0, p_0)$. In this case we can solve the Bernoulli equation for p explicitly (cf. (10)) to obtain

$$\overline{\Phi}_\nu(p_0) = \int_{-\infty}^0 \frac{e^{(\nu+1)s}\, ds}{\sqrt{e^{2\nu s} + \left(\dfrac{\nu}{p_0^2} - 1\right)}}, \quad p_0^2 \le \nu.$$

The analysis is similar in the other two examples, so we will not present the details (which are somewhat more complicated) here.

3.3. *Skew product flow formalism of nonautonomous systems*

Following [Kloeden *et al.*, 1999] (see also [Berger & Siegmund, 2003; Cheban *et al.*, 2002; Grüne & Kloeden, 2001; Kloeden & Kozyakin, 2001; Kloeden & Stonier, 1998; Langa *et al.*, 2002; Li & Kloeden, 2004b; Wang *et al.*, 2004; Wiggins, 2003]) we define a *nonautonomous dynamical system* (θ, φ), abbreviated NDS, in terms of a cocycle mapping φ on a state space X (a metric space) which is driven by an autonomous dynamical system θ acting on a base or parameter space P (also a metric space). Specifically, $\theta = \{\theta_t : t \in \mathbb{R}\}$ is an *autonomous dynamical system* on P, i.e. a group of homeomorphisms under composition on P with the properties that

1. $\theta_0(p) = p$ for all $p \in P$;
2. $\theta_{s+t} = \theta_s(\theta_t(p))$ for all $s,\, t \in \mathbb{R}$;
3. the mapping $(t, p) \mapsto \theta_t(p)$ is continuous,

and the *cocycle* mapping $\varphi : \mathbb{R}^+ \times P \times X \to X$ satisfies

1. $\varphi(0, p, x) = x$ for all $(p, x) \in P \times X$;
2. $\varphi(s + t, p, x) = \varphi(s, \theta_t(p), \varphi(t, p, x))$ for all $s,\, t,\, \in \mathbb{R}^+$, $(p, x) \in P \times X$;
3. the mapping $(t, p, x) \mapsto \varphi(t, p, x)$ is continuous.

Then the mapping $\pi : \mathbb{R}^+ \times Y \to Y$ defined by

$$\pi(t, (p, x)) := (\theta_t(p), \varphi(t, p, x))$$

forms an autonomous semidynamical system on $Y = P \times X$ over $\mathbb{R}^+$, which is called the *skew product flow* associated with the nonautonomous dynamical system (θ, φ).

Triangular autonomous systems (13) are special cases of skew product flows. In general, the driving system is not generated by an autonomous differential equation, so the autonomous global attractor of the skew product flow may not always be physically meaningful or, at least, not as physically meaningful as the dynamics in the state space variable x. For example, consider the nonautonomous Bernoulli differential equation

$$\frac{dx}{dt} = \nu x - b(t) x^3, \tag{18}$$

with $b : \mathbb{R} \to \mathbb{R}$ is an almost periodic function [Fink, 1974; Sell, 1971], which, in general, will not be the solution of an autonomous differential equation. However, we can use a construction of Bebutov [1940] to formulate changes in b as an autonomous dynamical system, namely for the shift operators $\theta_t b(\cdot) := b(t + \cdot)$ for all $t \in \mathbb{R}$ which determine an autonomous dynamical system on the space

$$P := \mathrm{cl}\,\{b(t + \cdot) : t \in \mathbb{R}\},$$

where the closure is with respect to the norm $\|f\|_\infty := \sup_{t \in \mathbb{R}} |f(t)|$. Since b is almost periodic, $P \subset C(\mathbb{R}, \mathbb{R})$ is a compact metric space with the metric corresponding to this norm (however, if b is only bounded we need the weak* topology for P to be a compact metric space). The x variable may be a physical quantity such as a chemical concentration or population density, but it is not clear how we should interpret a point (p, x) in the global attractor $A_\nu \subset P \times \mathbb{R}$ of the corresponding skew product flow.[1]

We note that the quasi-periodically forced linear systems of [Grebogi *et al.*, 1984] (see also [Glendinning, 2004]) with their intriguing nonchaotic strange attractors can also be formulated as nonautonomous dynamical systems in this way.

The global attractor $\mathfrak{A}$ (if it exists) of a skew product flow $\pi = (\theta, \varphi)$ has the form [Cheban *et al.*, 2002]

$$\mathfrak{A} = \bigcup_{p \in P^*} (\{p\} \times A_p),$$

where P^* is the global attractor of the driving system θ, thus a nonempty compact invariant

---

[1]A similar situation occurs for *random* dynamical systems [Arnold, 1998; Ashwin & Ochs, 2003], as when the function b here is a stochastic process. Then we write $b(t, \omega) = \overline{b}(\theta_t(\omega))$ where $\theta_t : \Omega \to \Omega$ is a metric dynamical system, i.e. with $\omega \to \theta_t(\omega)$ measurable rather than continuous. Here Ω is the sample space in an appropriately chosen probability space $(\Omega, \mathcal{F}, \mathbb{P})$. The physical interpretation of a point (ω, x) is also not obvious here.

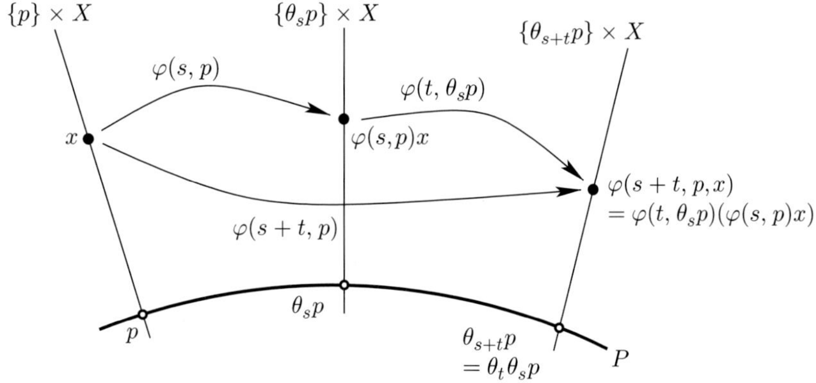

Fig. 9.   The cocycle property.

subset of P, and A_p are nonempty compact subsets of X for each $p \in P^*$, which are φ-invariant in the sense that

$$\varphi(t, p, A_p) = A_{\theta_t(p)} \quad \text{for all } t \geq 0, \; p \in P^* \quad (19)$$

and *pullback attracting* in the sense that

$$\lim_{t \to \infty} H_X^*(\varphi(t, \theta_{-t}(p), B), A_p) = 0 \quad (20)$$

for every nonempty bounded subset B of X and every $p \in P^*$. However, the A_p sets here need not be *forward attracting*, i.e. in the sense that

$$\lim_{t \to \infty} H_X^*(\varphi(t, p, B), A_{\theta_t(p)}) = 0 \quad (21)$$

for every nonempty bounded subset B of X and every $p \in P^*$.

This suggests several possible definitions of a nonautonomous attractor of an NDS (θ, φ) when one wishes to focus attention on the state space X and the dynamics therein. A family $\hat{A} = \{A_p : p \in P\}$ of nonempty compact subsets of X is said to be a *pullback attractor* of an NDS (θ, φ) if it is φ-invariant and pullback attracting as in (19) and (20) on P. (Usually one restricts attention to $P = P^*$, which we will do henceforth.) Such a family $\hat{A}$ is said to be a *forward attractor* of the NDS (θ, φ) if it is φ-invariant and forward attracting as in (19) and (21) on P.

The existence of a pullback attractor follows from that of an absorbing set (or family of absorbing sets) and a compactness property of the cocycle mapping [Cheban *et al.*, 2002; Grüne & Kloeden, 2001; Wiggins, 2003]. Obviously, any uniform pullback attractor is also a uniform forward attractor, and vice versa, where uniformity is with respect to $p \in P$, but in general neither of pullback and forward convergence implies the other.

The relationship between pullback and forward attractors and the subset $\mathcal{A}$ of $P \times X$ defined by $\bigcup_{p \in P}(\{p\} \times A_p)$ with component sets from such pullback or forward attractors and a possible global attractor of the associated autonomous skew product flow π is discussed in [Cheban *et al.*, 2002; Wang *et al.*, 2004].

As a simple example, we note that the family of singleton sets $A_p = \{\overline{\Phi}_\nu(p)\}$ for $p \in P_\nu = [-\sqrt{\nu}, \sqrt{\nu}]$ where $\overline{\Phi}_\nu(p)$ in the previous subsection is a pullback attractor on $X = \mathbb{R}$ for $p \in P_\nu$. In this case, the family is also a forward attractor and the subset $\bigcup_{p \in P_\nu}(\{p\} \times A_p)$ of $P_\nu \times \mathbb{R}$ is the global attractor for the associated skew product flow, i.e. the triangular system (15).

The above skew product formalism is particularly advantageous when the base space P of the driving system is compact, as for example, in almost periodically forced differential equations. However, P need not be compact. In fact, with $P = \mathbb{R}$ and $\theta_t(t_0) := t + t_0$ for all t and $t_0 \in \mathbb{R}$, the above formalism reduces to the process formalism of a nonautonomous dynamical system with $x(t + t_0, t_0, x_0) := \varphi(t, t_0, x_0)$ for all $t \geq 0$, $t_0 \in \mathbb{R}$ and $x_0 \in X$. We have already seen counterparts of above definitions of pullback or forward attractors for such processes in Sec. 3.1 above. We observe that an attractor does not exist for the skew product flow in this context.

4.   A Total Stability Theorem

For the total stability theorem and its application in the next section we consider parametrized nonautonomous differential equations

$$\frac{dx}{dt} = f_\nu(t, x), \quad \nu \in [-\nu^*, \nu^*], \quad (22)$$

where the vector fields $f_\nu : \mathbb{R} \times \mathbb{R}^d \to \mathbb{R}^d$ satisfy the following *standard assumptions*: The functions f_ν are continuous in $(t, x) \in \mathbb{R} \times \mathbb{R}^d$, globally Lipschitz in $x \in \mathbb{R}^d$ uniformly in $t \in \mathbb{R}$ with Lipschitz constant L_ν, and satisfy the *parametric dependence condition*[2]:

There exists $\omega(\nu)$ for $\nu \in [-\nu^*, \nu^*]$ with $\omega(\nu) \to 0$ as $\nu \to 0$ such that

$$\sup_{t \in \mathbb{R}, x \in \mathbb{R}^d} |f_\nu(t, x) - f_0(t, x)| \le \omega(\nu) \qquad (23)$$

for $\nu \in [-\nu^*, \nu^*]$.

We will denote the solution of (22) with initial value $x(t_0) = x_0$ by $x_\nu(t, t_0, x_0)$. We will also assume that for $\nu = 0$ the system (22) has a uniformly asymptotically stable compact connected set A_0 (which may be an attractor, but need not be). Our next theorem on total stability is a consequence of the *uniform* asymptotic stability of A_0, the uniformity being essential here.

Theorem 1. *Consider Eq. (22) satisfying the standard assumptions and suppose that there is a nonempty compact set A_0 which is uniformly asymptotically stable (possibly only locally) for (22) with $\nu = 0$.*

Then there is a $\nu^{**} \in (0, \nu^*]$ such that for each ν with $0 < |\nu| \le \nu^{**}$ there exists a family $\mathcal{A}_\nu = \{A_\nu(t_0), t_0 \in \mathbb{R}\}$ of nonempty compact connected subsets of $\mathbb{R}^d$, which are invariant with respect to (22), i.e.

$$x_\nu(t, t_0, A_\nu(t_0)) = A_\nu(t), \quad t \ge t_0, \qquad (24)$$

and converge upper semicontinuously to A_0 uniformly in $t \in \mathbb{R}$, i.e.

$$\sup_{t \in \mathbb{R}} H^*_{\mathbb{R}^d}(A_\nu(t), A_0) \to 0 \quad as \; \nu \to 0. \qquad (25)$$

The family $\mathcal{A}_\nu$ is a *pullback attractor* for the nonautonomous process or two-parameter semigroup $\{x_\nu(t, t_0, \cdot)\}$ defined by the solutions of the differential equation (22). Note that the set-valued mapping $t \mapsto A_\nu(t)$ is continuous due to the invariance property (24) and the continuity of the process.

The proof of Theorem 1 is given in the Appendix. It is based on the following Lyapunov inequality (see Lemma 3 in the Appendix)

$$V(t, x_\nu(t, t_0, x_0)) \le e^{-(t-t_0)} V(t_0, x_0) + KL\omega(\nu),$$

$$0 \le t - t_0 \le 1, \qquad (26)$$

for all $t_0 \in \mathbb{R}$ and $x_0 \in \mathbb{R}^d$, where V is a Lyapunov function with Lipschitz constant L (and K is another positive constant), which characterizes the uniform asymptotical stability of the set A_0 for the differential equation (22) with $\nu = 0$. The existence of V is ensured by Theorem 2 in the Appendix. We use the Lyapunov inequality (26) to establish the existence of a family of nonempty compact subsets $\{\Lambda^*_\nu(t_0), t_0 \in \mathbb{R}\}$ of $\mathbb{R}^d$, provided ν is sufficiently small, which is positively invariant, i.e.

$$x_\nu(t, t_0, \Lambda^*_\nu(t_0)) \subseteq \Lambda^*_\nu(t), \quad \forall \, t \ge t_0,$$

and absorbing uniformly in $t_0 \in \mathbb{R}$, i.e. for every compact subset D of $\mathbb{R}$ there exists $T_{D,\nu} \ge 0$ such that

$$x_\nu(t, t_0, D) \subseteq \Lambda^*_\nu(t), \qquad t \ge t_0 + T_{D,\nu}, \quad \forall \, t_0 \in \mathbb{R}.$$

The component sets of the pullback attractor are then determined by

$$A_\nu(t_0) = \bigcap_{\tau \ge 0} x_\nu(t_0, t_0 - \tau, \Lambda^*_\nu(t_0 - \tau))$$

for each $t_0 \in \mathbb{R}$. These are *pullback* attracting in the sense that

$$\mathrm{dist}\,(x_\nu(t_0, t_0 - \tau, x_0), A_\nu(t_0)) \to 0 \quad as \; \tau \to \infty,$$

$$\forall \, t_0 \in \mathbb{R}, \; x_0 \in \mathbb{R}^d, \qquad (27)$$

but they need not be *forward* attracting in the sense that

$$\mathrm{dist}\,(x_\nu(t, t_0, x_0), A_\nu(t)) \to 0 \quad as \; t \to \infty,$$

$$\forall \, t_0 \in \mathbb{R}, \; x_0 \in \mathbb{R}^d. \qquad (28)$$

Note that the pullback attractor $\mathcal{A}_\nu$ consists of a single set A_ν if the differential equation (22) for this value of ν is autonomous. In this case $x_\nu(t, t_0, x_0) \equiv x_\nu(t - t_0, 0, x_0)$, so pullback and forward convergences are equivalent. For a detailed discussion of the relationship between pullback and forward convergences see [Cheban *et al.*, 2002; Wang *et al.*, 2004].

The upper semicontinuous convergence (25) follows from the fact that $A_\nu(t_0) \subset \Lambda^*_\nu(t_0)$ and the

---

[2]Unlike the uniformity in x here (see Remark 1), the uniformity in $t \in \mathbb{R}$ is a strong restriction but is essential for total stability. However it does hold for almost periodic functions.

construction of the $\Lambda_\nu^*(t_0)$ sets, which leads to

$$H_{\mathbb{R}^d}^*(A_\nu(t_0), A_0) \leq H_{\mathbb{R}^d}^*(\Lambda_\nu^*(t_0), A_0) \to 0 \text{ as } \nu \to 0$$

for all $t_0 \in \mathbb{R}$. In general, it cannot be strengthened to continuous convergence (i.e. with $H_{\mathbb{R}^d}^*$ replaced by the Hausdorff metric $H_{\mathbb{R}^d}$). A simple counterexample is given by the example of a subcritical bifurcation in the autonomous differential equation (2) at $\nu = 0$ (i.e. where the nonlocal equilibria first arise). As mentioned above, equiattraction ensures continuous convergence of attractors in the autonomous case. An analogous result also holds in the nonautonomous case [Li & Kloeden, 2004b].

Under suitable assumptions (see [Wang *et al.*, 2004]) it can be shown that the set-valued mapping $t \mapsto A_\nu(t)$ is periodic, respectively almost periodic, when the functions $f_\nu(t, x)$ are so. In particular, this holds for the Bernoulli equation (7) when the function b is periodic, respectively almost periodic, for which $A_\nu(t)$ are given by (8).

Remark 2. It is possible to generalize the above theorem to assume that the system has a uniformly Lyapunov asymptotically stable (in the forward sense) family of nonempty compact invariant sets $\{A(t) : t \in \mathbb{R}\}$ instead of the attracting set A_0 (see [Yoshizawa, 1966]) or to assume that the right-hand side of the differential equation (22) is of the form $f(x, p)$ with a uniform pullback attractor $\{A_p : p \in P\}$ with P compact [Kloeden & Kozyakin, 2001].

Remark 3. The following example due to Li Desheng (private communication) shows that the perturbed attracting set need not be globally attracting even when the unperturbed system is globally attracting.

The autonomous scalar ordinary differential equation $x'(t) = -xe^{-x^2/2} + \nu$ has a global attractor $A_0 = \{0\}$ when $\nu = 0$. For $0 < \nu^2 < e^{-1}$, there are two equilibria with the one closer to zero being locally asymptotically stable and thus forming

a singleton set local attractor. The other equilibrium is unstable and coalesces with the first as ν^2 approaches e^{-1} from below, then both equilibria disappear for $\nu^2 > e^{-1}$. This example also shows that the parameter interval in which the perturbed attractor exists can be very small, e.g. when we apply the total stability theorem to the above differential equation with a ν value such that $e^{-\frac{1}{2}} - \nu > 0$ is very small.

5. Applications of the Total Stability Theorem

As a first comment, if to emphasize the obvious, we mention that uniform asymptotic stability is concerned solely with what happens *outside* of the set A_0 and says nothing about what may happen inside of A_0. Indeed the Lyapunov function vanishes on A_0. Nevertheless the internal dynamics of A_0 may have a significant effect as the above example of a saddle-node bifurcation shows. The importance of the total stability theorem is that it ensures the existence of nearby attracting objects in perturbed systems. The theorem of [Li & Kloeden, 2004a] (see [Li & Kloeden, 2004b] for nonautonomous dynamical systems) then says that the convergence of these perturbed attractors is in fact continuous rather than just upper semicontinuous when the attractors are equiattracting.

For a supercritical bifurcation one needs to assume more about the attractor A_0 of the reference system and the perturbed systems: for example, that the differential equations (22) with f_ν satisfying in addition to the standard assumptions the following properties

1. $f_\nu(t, 0) = 0$ for all $t \in \mathbb{R}$ and all $\nu \in [-\nu^*, \nu^*]$;
2. the equilibrium solution $x_\nu(t) \equiv 0$ of (22) is stable for $\nu < 0$ and unstable for $\nu \geq 0$ for the linearized system.
3. the equilibrium solution $x_0(t) \equiv 0$ of (22) with $\nu = 0$ is uniformly asymptotically stable.

Thus Theorem 1 is applicable and we get $A_0 = \{0\}$ with $0 \in A_\nu \neq \{0\}$ for $\nu > 0$, so

$$H_{\mathbb{R}^d}(A_\nu(t_0), \{0\}) \equiv H_{\mathbb{R}^d}^*(A_\nu(t_0), \{0\}) \to 0$$

$$\text{as } \nu \to 0$$

for all $t_0 \in \mathbb{R}$. In this case we have a continuous bifurcation at $\nu = 0$. The supercritical bifurcations of the autonomous differential equation (1) and the

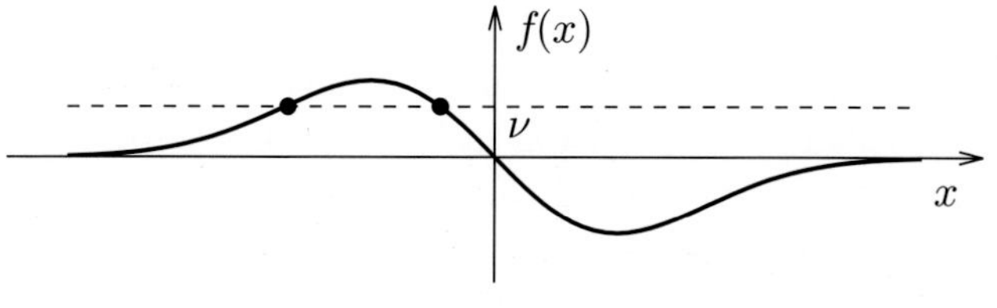

Fig. 10. Graph of $f(x) = -xe^{-x^2/2}$.

nonautonomous differential equation (7) are examples of this result. We note that a similar analysis is possible if the equilibrium solution 0 here is replaced by a periodic or almost periodic solution.

Two features are required here for a supercritical bifurcation:

(i) the bifurcating solution is uniformly asymptotically stable for the nonlinear system at the parameter value where it loses linear stability, and

(ii) the bifurcating solution (or at least a continuation of it) exists and is unstable after the bifurcation point.

The first of these ensures the existence of an attractor after the bifurcation point and the second that these attractors contain something more than just the continuation of the bifurcating solution and thus the occurence of not just a bifurcation, but a supercritical bifurcation.

Usually the supercriticality of a bifurcation is determined by the sign of a coefficient in an expansion of the new solution in what is essentially a normal form expression [Glendinning, 1994]. As can be seen from the examples in [Glendinning, 1994; Marsden & McCracken, 1976], such expressions are difficult enough to determine for specific examples of codimension-one bifurcations and thus cannot be expected to be any easier for higher codimensions. The nonlinear asymptotic stability of the bifurcating solution at the point of loss of linear stability provides an alternative test for supercriticality in these cases.

As a final point, we note that we are not restricted here to applying the total stability theorem to a global or maximal attractor, but could equally well apply it *locally* to the bifurcation of an equilibrium solution, say, inside such an attractor. This would verify the continued existence of the solution or a continuation of it of some form after a bifurcation and the existence of other nearby minimal solutions if the continued solution is unstable after the bifurcation point.

6. Concluding Remarks and Questions

We have restricted our attention here to well-known basic bifurcations, partly because these already illustrate our ideas clearly in the autonomous case and partly because investigations of bifurcations in the general nonautonomous case have not progressed much beyond these elementary

bifurcations. More complicated types of bifurcations, e.g. homoclinic bifurcations, certainly might occur inside autonomous attractors and their nonautonomous counterparts. On the other hand, it is known that a bifurcation at infinity may completely destroy a pullback attractor [Kloeden & Kozyakin, 2001]. Less dramatically, as we have seen in the autonomous differential equation $x'(t) = -xe^{-x^2/2} + \nu$ above (see Remark 3), the bifurcation at infinity for $\nu = 0$ destroys the global attractivity of the perturbed attractor but not the attractor itself.

In all of our examples the bifurcation parameter appears in an autonomous linear part of the differential equation. Johnson *et al.* [2002] considered the analog of Hopf bifurcations for equations where the bifurcation parameter has a time-dependent coefficient, as e.g. in the Duffing–van der Pol equation with almost periodic coefficient $b(t)$

$$\frac{d}{dt}\begin{pmatrix} x \\ y \end{pmatrix} = \begin{pmatrix} 0 & 1 \\ -\alpha + \sigma b(t) & \beta \end{pmatrix}\begin{pmatrix} x \\ y \end{pmatrix}$$

$$- \begin{pmatrix} 0 \\ x^2 y + x^3 \end{pmatrix}. \tag{29}$$

The bifurcation then appears to occur in two stages, perhaps because the corresponding Sacker–Sell spectrum consists typically of intervals rather than single points, with the bifurcation being only complete when the whole spectral interval has crossed over into the positive part of the real line (see [Siegmund, 2002a] for spectral theory and [Siegmund, 2001, 2002b] for a nonautonomous normal form of (29)). Should one consider such bifurcations as the authentic nonautonomous bifurcations and the ones that we have focussed on in this article only as special cases? In any case, the above discussion on total stability and supercriticality remains valid here too.

Our autonomous triangular systems were all Morse–Smale systems. They provided us with examples of nonautonomous systems with particularly robust attractors for the case that the driving systems were generated by differential equations. However, we also considered examples of nonautonomous systems for which the driving system was not generated by a differential equation, but by, say, the shift operator on the hull of an almost periodic function. What then is the counterpart of a Morse–Smale system in such situations?

In our examples, a discontinuous transition of an attractor was always associated with a

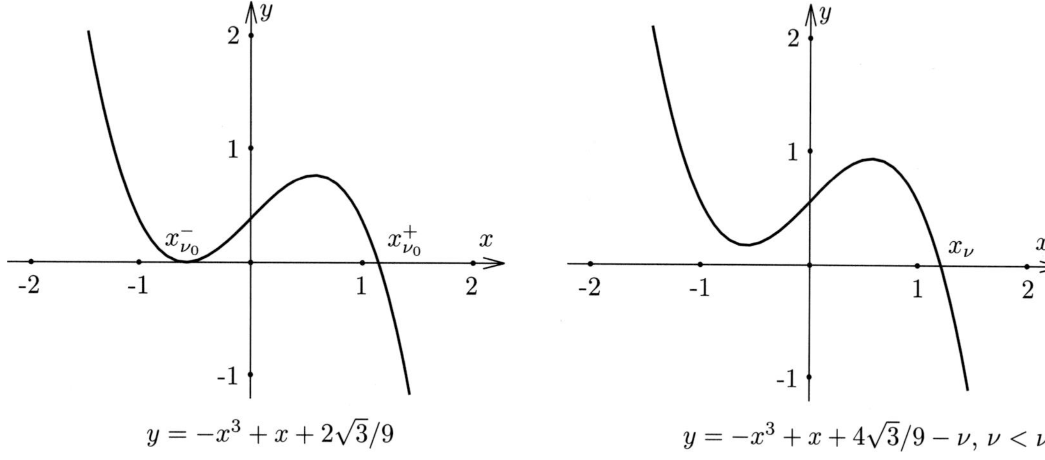

Fig. 11.   Two equilibria at ν_0 and one equilibrium for $\nu < \nu_0$.

subcritical bifurcation within the attractor. Is this the only cause of a discontinuous transition? Do such discontinuous transitions only occur at isolated parameter values? In fact, a subcritical bifurcation is not necessary but rather the sudden vanishing or rising of nonlocal equilibria (or other minimal invariant sets), i.e. which exist only on one side of a critical parameter value. This can be seen in Example 3.2 in [Li & Kloeden, 2004a] for the scalar equation

$$\dot{x} = -x^3 + x + 4\sqrt{3}/9 - \nu$$

with $\nu \in [0, \nu_0]$ where $\nu := 2\sqrt{3}/9$. For $\nu = \nu_0$ the equation has two distinct equilibria $x_{\nu_0}^-$ and $x_{\nu_0}^+$ and for $\nu < \nu_0$ one equilibrium x_ν (see Fig. 11).

Note that the equilibrium near 1 is asymptotically stable for all ν. In particular, there is no subcritical bifurcation.

References

Arnold, L. [1998] *Random Dynamical Systems* (Springer–Verlag, Heidelberg).

Arrowsmith, D. K. & Place, C. M. [1990] *An Introduction to Dynamical Systems* (Cambridge University Press, Cambridge).

Ashwin, P. & Ochs, G. [2003] "Convergence to local random attractors," *Dyn. Syst.* **18**, 139–158.

Bebutov, M. V. [1940] "Sur les systèmes dynamiques dans l'espace des fonctions continues," *Doklady Akad Nauk SSSR* **27**, 904–906.

Berger, A. & Siegmund, S. [2003] "On the gap between random dynamical systems and continuous skew products," *J. Dyn. Diff. Eqs.* **15**, 237–279.

Caraballo, T., Kloeden, P. E. & Langa, J. [2003] "Atractores globales para sistemas diferenciales no autónomos," *Cubo Matemática Educacional* **5**, 305–329.

Cheban, D., Kloeden, P. E. & Schmalfuß, B. [2002] "The relationship between pullback, forward and global attractors of nonautonomous dynamical systems," *Nonlin. Dyn. Syst. Th.* **2**, 9–28.

Dafermos, C. M. [1971] "An invariance principle for compact processes," *J. Diff. Eqs.* **9**, 239–252.

Fink, A. M. [1974] *Almost Periodic Differential Equations*, Springer Lecture Notes in Mathematics, Vol. 377 (Springer-Verlag, Heidelberg).

Glendinning, P. [1994] *Stability, Instability and Chaos* (Cambridge University Press, Cambridge).

Glendinning, P. [2004] "The non-smooth pitchfork bifurcation," *Discr. Contin. Dyn. Syst. Ser.* **B4**, 457–464.

Grebogi, C., Ott, E., Pelikan, S. & Yorke, J. A. [1984] "Strange attractors that are not chaotic," *Physica* **D13**, 261–268.

Grüne, L. & Kloeden, P. E. [2001] "Discretization, inflation and perturbation of attractors," in *Ergodic Theory: Analysis and Efficient Simulation of Dynamical Systems*, ed. Fiedler, B. (Springer-Verlag), pp. 399–416.

Hale, J. [1988] *Asymptotic Behavior of Dissipative Dynamical Systems* (Amer. Math. Soc., Providence).

Johnson, R. A., Kloeden, P. E. & Pavani, R. [2002] "Two–step transition in nonautonomous bifurcations: An explanation," *Stoch. Dyn.* **2**, 67–92.

Kloeden, P. E. & Lorenz, J. [1986] "Stable attracting sets in dynamical systems and in their one-step discretizations," *SIAM J. Numer. Anal.* **23**, 986–995.

Kloeden, P. E. & Stonier, D. J. [1998] "Cocycle attractors in nonautonomously perturbed differential equations," *Dyn. Contin. Discr. Impuls. Syst.* **4**, 211–226.

Kloeden, P. E., Keller, H. & Schmalfuß, B. [1999] "Towards a theory of random numerical dynamics," in *Stochastic Dynamics*, eds. Crauel, H. & Gundlach, V. M. (Springer-Verlag, Heidelberg), pp. 259–282.

Kloeden, P. E. & Kozyakin, V. S. [2000] "The inflation of attractors and discretization: The autonomous case," *Nonlin. Anal. TMA* **40**, 333–343.

Kloeden, P. E. & Kozyakin, V. S. [2001] "The perturbation of attractors of skew-product flows with a shadowing driving system," *Discr. Contin. Dyn. Syst.* **7**, 883–893.

Kloeden, P. E. [2004] "Pitchfork and transcritical bifurcations in systems with homogeneous nonlinearities and an almost periodic time coefficient," *Commun. Pure Appl. Anal.* **3**, 161–173.

Kloeden, P. E. & Kozyakin, V. S. [2004] "Uniform nonautonomous attractors under discretization," *Discr. Contin. Dyn. Syst.* **10**, 423–433.

Koksch, N. & Siegmund, S. [2002] "Pullback attracting inertial manifolds for nonautonomous dynamical systems," *J. Dyn. Diff. Eqs.* **14**, 889–941.

Krasnosel'skii, M. A., Burd, V. Sh. & Kolesov, Yu. S. [1973] *Nonlinear Almost Periodic Solutions* (John Wiley & Sons, NY).

Langa, J. A., Robinson, J. C. & Suárez, A. [2002] "Stability, instability and bifurcation phenomena in non-autonomous differential equations," *Nonlinearity* **15**, 887–903.

Langa, J. A. & Suárez, A. [2002] "Bifurcation phenomena for a non autonomous logistic equation," *Electron. J. Diff. Eqs.* **72**, 1–20.

Langa, J. A., Robinson, J. C. & Suárez, A. [2004] "Bifurcations in non-autonomous scalar equations," submitted.

Li Desheng & Kloeden, P. E. [2004a] "Equi-attraction and the continuous dependence of attractors on parameters," *Glasgow Math. J.* **46**, 131–141.

Li Desheng & Kloeden, P. E. [2004b] "Equi-attraction and the continuous dependence of pullback attractors on parameters," *Stoch. Dyn.* **4**, 373–384.

Marsden, J. & McCracken, M. [1976] *The Hopf Bifurcation and its Applications* (Springer-Verlag, NY).

Reitmann, V. [1996] *Reguläre und Chaotische Dynamik* (B.G. Teubner, Stuttgart).

Sell, G. R. [1971] *Lectures on Topological Dynamics and Differential Equations* (Van Nostrand–Reinbold, London).

Siegmund, S. [2001] "Normal form of Duffing–van der Pol oscillator under nonautonomous parametric perturbations," *Discr. Contin. Dyn. Syst.*, 357–361, Kennesaw conference issue available from http://AIMSciences.org/

Siegmund, S. [2002a] "Dichotomy spectrum for nonautonomous differential equations," *J. Dyn. Diff. Eqs.* **14**, 243–258.

Siegmund, S. [2002b] "Normal forms for nonautonomous differential equations," *J. Diff. Eqs.* **178**, 541–573.

Wang Yejuan, Li Desheng & Kloeden, P. E. [2004] "Uniform attractors of almost periodic nonautonomous dynamical systems," *Nonlin. Anal. TMA* **59**, 35–53.

Wiggins, S. [2003] *Introduction to Applied Nonlinear Dynamical Systems and Chaos*, 2nd edition. (Springer-Verlag, Heidelberg).

Yoshizawa, T. [1966] *Stability Theory by Lyapunov's Second Method* (Mathematical Society of Japan, Tokyo).

A.   Appendix: Proof of Theorem 1

The following theorem, based on Theorem 22.5 in [Yoshizawa, 1966], provides the existence of a Lyapunov function which characterizes the uniform asymptotical stability of a globally uniformly asymptotically stable compact set A_0 of a nonautonomous differential equation

$$\frac{dx}{dt} = f(t, x). \qquad (A.1)$$

Theorem 2. *Suppose that $f : \mathbb{R} \times \mathbb{R}^d \to \mathbb{R}^d$ in $(A.1)$ is continuous in (t, x) and globally Lipschitz in $x \in \mathbb{R}^d$ uniformly in $t \in \mathbb{R}$ and suppose that $(A.1)$ has a globally uniformly asymptotically stable compact set A_0.*

Then there exists a Lyapunov function $V : \mathbb{R} \times \mathbb{R}^d \to [0, \infty)$ for which:

1. *V is globally Lipschitz in $x \in \mathbb{R}^d$ uniformly in $t \in \mathbb{R}$, i.e. there exists a constant $L > 0$ such that*

$$|V(t, x) - V(t, y)| \leq L\,|x - y|$$
$$\textit{for all} \quad x, y \in \mathbb{R}^d, \quad t \in \mathbb{R}; \qquad (A.2)$$

2. *there exist continuous strictly increasing functions $\alpha, \beta : \mathbb{R}^d \mapsto [0, \infty)$ with $\alpha(0) = \beta(0) = 0$ and $0 < \alpha(r) < \beta(r)$ for all $r > 0$ such that*

$$\alpha(\mathrm{dist}(x, A_0)) \leq V(t, x) \leq \beta(\mathrm{dist}(x, A_0))$$
$$\textit{for all} \quad x \in \mathbb{R}^d;$$

3. *V decreases exponentially fast along trajectories of $(A.1)$ uniformly in $t_0 \in \mathbb{R}$, i.e. we have for $t_0 \in \mathbb{R}, x_0 \in \mathbb{R}^d$*

$$V(t, x(t, t_0, x_0)) \leq e^{-(t-t_0)}\, V(t_0, x_0)$$
$$\textit{for all} \quad t \geq t_0. \qquad (A.3)$$

In fact, if differential equation (A.1) has a uniformly asymptotically stable equilibrium solution and the function f in (A.1) is almost periodic uniformly in $x \in B[0; R]$ for each $R > 0$ (respectively, periodic or autonomous), then, from Theorem 19.8 of [Yoshizawa, 1966], the Lyapunov function V can be chosen to be almost periodic in t (respectively, periodic or autonomous).

The following Lyapunov inequality, which is similar to inequalities in [Kloeden & Kozyakin, 2000; Kloeden & Lorenz, 1986], will be one of the key tools in the proof of Theorem 1.

Lemma 3. *Under the assumptions of Theorem 1 there is a Lyapunov function V with*

$$V(t, x_\nu(t, t_0, x_0)) \leq e^{-(t-t_0)} V(t_0, x_0) + KL\omega(\nu),$$
$$0 \leq t - t_0 \leq 1, \tag{A.4}$$

for the solution $x_\nu(t, t_0, x_0)$ with initial value $x_\nu(t_0, t_0, x_0) = x_0$ of the differential equation (22) with parameter $\nu \neq 0$ for any $t_0 \in \mathbb{R}$ and $x_0 \in \mathbb{R}^d$.

Proof. Apply Theorem 2 to (22) for $\nu \neq 0$. Using the Lipschitz property (A.2) and the exponential decay inequality (A.2) of the Lyapunov function V, we obtain

$$V(t, x_\nu(t, t_0, x_0))$$
$$\leq V(t, x_0(t, t_0, x_0))$$
$$\quad + |V(t, x_\nu(t, t_0, x_0)) - V(t, x_0(t, t_0, x_0))|$$
$$\leq e^{-(t-t_0)} V(t_0, x_0)$$
$$\quad + L \, |x_\nu(t, t_0, x_0) - x_0(t, t_0, x_0)|.$$

To estimate $|x_\nu(t, t_0, x_0) - x_0(t, t_0, x_0)| =: |x_\nu(t) - x_0(t)|$ for $0 \leq t - t_0 \leq 1$, we use the integral equation representation of the differential equation (22), the Lipschitz constant L_0 of (22) for $\nu = 0$ and the parametric dependence condition (23) to get

$$|x_\nu(t) - x_0(t)|$$
$$= \left| x_0 + \int_{t_0}^{t} f_\nu(s, x_\nu(s)) \, ds - x_0 \right.$$
$$\quad \left. - \int_{t_0}^{t} f_0(s, x_0(s)) \, ds \right|$$
$$\leq \int_{t_0}^{t} |f_\nu(s, x_\nu(s)) - f_0(s, x_0(s))| \, ds$$
$$\leq \int_{t_0}^{t} |f_\nu(s, x_\nu(s)) - f_0(s, x_\nu(s))| \, ds$$
$$\quad + \int_{t_0}^{t} |f_0(s, x_\nu(s)) - f_0(s, x_0(s))| \, ds$$
$$\leq \int_{t_0}^{t} \omega(\nu) \, ds + L_0 \int_{t_0}^{t} |x_\nu(s) - x_0(s)| \, ds.$$

Thus we have

$$|x_\nu(t) - x_0(t)| \leq \omega(\nu)(t - t_0)$$
$$+ L_0 \int_{t_0}^{t} |x_\nu(s) - x_0(s)| \, ds$$

for $t - t_0 \geq 0$. Hence by the Gronwall inequality we obtain

$$|x_\nu(t) - x_0(t)| \leq e^{L_0(t-t_0)}\omega(\nu)(t - t_0)$$

for any $t - t_0 \geq 0$. If we restrict to $0 \leq t - t_0 \leq 1$, then we have

$$|x_\nu(t) - x_0(t)| \leq e^{L_0}\omega(\nu) =: K\omega(\nu),$$

proving the lemma. ∎

A.1. *Existence of a positively invariant family of absorbing sets for the perturbed dynamics*

Since $\omega(\nu) \to 0$ as $\nu \to 0$ (see (23)), we can choose $\nu^{**} \in (0, \nu^*]$ such that for each ν with $0 < |\nu| \leq \nu^{**}$ we have $\omega(\nu) < ((e-1)/Le)^2$. Then $\Delta_\nu := \ln\left(1/[1 - L\sqrt{\omega(\nu)}]\right) < 1$ and

$$1 - e^{-\Delta_\nu} = L\sqrt{\omega(\nu)} \quad \text{and}$$
$$\frac{1}{2}\left(1 + e^{-\Delta_\nu}\right) \leq e^{-\frac{1}{4}\Delta_\nu} \tag{A.5}$$

(the reason for the last inequality will become apparent in the proofs of Lemmas 5 and 6, cf. Lemma 3.4 of [Kloeden & Lorenz, 1986]), and

$$\eta(\nu) := 2K\sqrt{\omega(\nu)} \to 0+ \quad \text{and} \quad \Delta_\nu \to 0+$$
$$\text{as } \nu \to 0.$$

Then define

$$\Lambda_\nu(t_0) := \{x \in \mathbb{R}^d \ : \ V(t_0, x) \leq \eta(\nu)\}$$

for each $t_0 \in \mathbb{R}$.

Lemma 4. *$\Lambda_\nu(t_0)$ is a nonempty compact subset of $\mathbb{R}^d$ for each $t_0 \in \mathbb{R}$ with*

$$H^*_{\mathbb{R}^d}(\Lambda_\nu(t_0), A_0) \leq \alpha^{-1}(\eta(\nu)) \to 0 \quad \text{as } \nu \to 0.$$
$$\tag{A.6}$$

Proof. Since $V(t_0, x) = 0$ for $x \in A_0$, so $A_0 \subset \Lambda_\nu(t_0)$, hence $\Lambda_\nu(t_0)$ is nonempty. It is compact by the continuity of $x \mapsto V(t_0, x)$ and the fact that $\Lambda_\nu(t_0) = V(t_0, \cdot)^{-1}([0, \eta(\nu)])$. The inequality (A.6) follows from the inequalities

$$\alpha(\mathrm{dist}(x, A_0)) \leq V(t_0, x) \leq \eta(\nu) \quad \text{for all } x \in \Lambda_\nu(t_0).$$

∎

The family of sets $\{\Lambda_\nu(t_0), t_0 \in \mathbb{R}\}$ is positively invariant with respect to the discrete time process $x_\nu(t_0 + n\Delta_\nu, t_0, x_0)$, in the sense that

Lemma 5. $x_\nu(t_0 + n\Delta_\nu, t_0, \Lambda_\nu(t_0)) \subseteq \Lambda_\nu(t_0 + n\Delta_\nu)$ *for all* $n \geq 0$, $t_0 \in \mathbb{R}$.

Proof. It suffices to consider the case $n = 1$. Take any $x_0 \in \Lambda_\nu(t_0)$. Then $V(t_0, x_0) \leq \eta(\nu)$. By the key Lyapunov inequality (A.4), the definition of $\eta(\nu)$ and (A.5) we have

$$V(t_0 + \Delta_\nu, x_\nu(t_0 + \Delta_\nu, t_0, x_0))$$
$$\leq e^{-\Delta_\nu} V(t_0, x_0) + KL\omega(\nu)$$
$$\leq e^{-\Delta_\nu} \eta(\nu) + \frac{1}{2}\left(1 - e^{-\Delta_\nu}\right)\eta(\nu)$$
$$= \frac{1}{2}\left(1 + e^{-\Delta_\nu}\right)\eta(\nu)$$
$$\leq \eta(\nu),$$

so $x_\nu(t_0 + \Delta_\nu, t_0, x_0) \in \Lambda_\nu(t_0 + \Delta_\nu)$. ∎

The family of sets $\{\Lambda_\nu(t_0), t_0 \in \mathbb{R}\}$ is in fact absorbing for the discrete time process $x_\nu(t_0 + n\Delta_\nu, t_0, x_0)$ uniformly in $t_0 \in \mathbb{R}$, provided ν (and hence Δ_ν) is sufficiently small.

Lemma 6. *For each ν such that $0 < |\nu| \leq \nu^{**}$ and each compact subset D of $\mathbb{R}^d$ there exists an integer $N_{D,\nu} \geq 0$, for which*

$$x_\nu(t_0 + n\Delta_\nu, t_0, x_0) \in \Lambda_\nu(t_0 + n\Delta_\nu)$$

for all $n \geq N_{D,\nu}$, $x_0 \in D$ and $t_0 \in \mathbb{R}$.

Proof. Choose x_0 in a compact subset D of $\mathbb{R}^d$, let us write $t_n = t_0 + n\Delta_\nu$, $x_n = x_\nu(t_0 + n\Delta_\nu, t_0, x_0)$. If $x_0 \in \Lambda_\nu(t_0)$ we have nothing to prove. Now assume that $x_0 \notin \Lambda_\nu(t_0)$. Then, by the Lyapunov inequality

(A.4) and the definition of $\eta(\nu)$ we have

$$V(t_1, x_1) \leq e^{-\Delta_\nu} V(t_0, x_0) + KL\omega(\nu)$$
$$= e^{-\Delta_\nu} V(t_0, x_0) + \frac{1}{2}\left(1 - e^{-\Delta_\nu}\right)\eta(\nu)$$
$$< \frac{1}{2}\left(1 + e^{-\Delta_\nu}\right) V(t_0, x_0)$$
$$\leq e^{-\frac{1}{4}\Delta_\nu} V(t_0, x_0)$$

since $V(t_0, x_0) > \eta(\nu)$. Repeating this argument, we have

$$V(t_n, x_n) < e^{-\frac{n}{4}\Delta_\nu} V(t_0, x_0)$$

as long as $x_j \notin \Lambda_\nu(t_0 + j\Delta_\nu)$ for $j = 0, \ldots, n - 1$. Now

$$V(t_0, x_0) \leq \beta(\mathrm{dist}(x_0, A_0)) \leq \beta(H^*_{\mathbb{R}^d}(D, A_0)) < \infty$$

for all $x_0 \in D$, so

$$V(t_n, x_n) < e^{-\frac{n}{4}\Delta_\nu} \beta(H^*_{\mathbb{R}^d}(D, A_0))$$

as long as $x_j \notin \Lambda_\nu(t_0 + j\Delta_\nu)$ for $j = 0, \ldots, n - 1$. Define $N_{D,\nu}$ to be the smallest integer n for which

$$e^{-\frac{n+1}{4}\Delta_\nu} \beta(H^*_{\mathbb{R}^d}(D, A_0))$$
$$\leq \eta(\nu) < e^{-\frac{n}{4}\Delta_\nu} \beta(H^*_{\mathbb{R}^d}(D, A_0)).$$

Thus for each $x_0 \in D$ there exists an integer $n_0 \leq N_{D,\nu}$, possibly 0, such that $x_{n_0} = x_\nu(t_0 + n_0\Delta_\nu, t_0, x_0) \in \Lambda_\nu(t_0 + n_0\Delta_\nu)$. By the positive invariance of the family of sets $\{\Lambda_\nu(t_0), t_0 \in \mathbb{R}\}$ proved in Lemma 5 all successive values x_n remain in $\Lambda_\nu(t_0 + n\Delta_\nu)$, so the proof of Lemma 6 is complete. ∎

However, we need a family of nonempty compact subsets of $\mathbb{R}^d$ which is positively invariant and uniformly absorbing for the continuous time process $x_\nu(t, t_0, x_0)$. For this we define

$$\Lambda^*_\nu(t_0) = \bigcup_{t_0 - \Delta_\nu \leq \tau \leq t_0} x_\nu(t_0, \tau, \Lambda_\nu(\tau))$$

for each $t_0 \in \mathbb{R}$ (see Fig. 12). These sets are obviously nonempty and compact. Note that

$$x^* \in \Lambda^*_\nu(t_0) \Rightarrow x^* = x_\nu(t_0, \tau^*, z^*) \quad \text{with}$$
$$\tau^* \in [t_0 - \Delta_\nu, t_0], z^* \in \Lambda_\nu(\tau^*). \qquad \text{(A.7)}$$

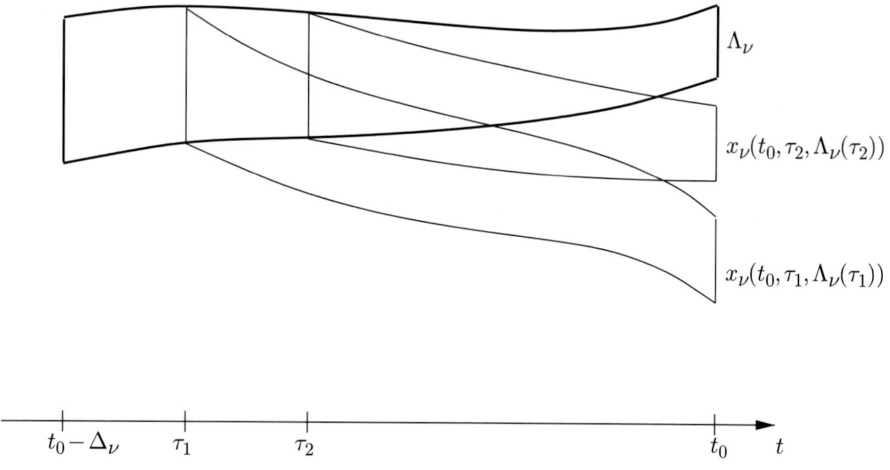

Fig. 12.   The definition of $\Lambda_\nu^*(t_0)$.

We will show that the family of sets $\{\Lambda_\nu^*(t_0),\ t_0 \in \mathbb{R}\}$ is positively invariant and absorbing for the continuous time process $x_\nu(t, t_0, x_0)$ uniformly in $t_0 \in \mathbb{R}$, provided ν (and hence Δ_ν) is sufficiently small.

Lemma 7. $x_\nu(t, t_0, \Lambda_\nu^*(t_0)) \subseteq \Lambda_\nu^*(t)$ *for all* $t \geq t_0$.

Proof.   Consider an arbitrary point $x^* \in \Lambda_\nu^*(t_0)$, using (A.7) we get $x^* = x_\nu(t_0, \tau^*, z^*)$. We prove that $x_\nu(t, t_0, x^*) \in \Lambda_\nu^*(t)$ for $t \in [t_0, t_0 + \Delta_\nu]$ in two steps by showing it for $t \in [t_0, \tau^* + \Delta_\nu]$ and then for $t \in [\tau^* + \Delta_\nu, t_0 + \Delta_\nu]$.

Step 1.   Using the fact that

$$x_\nu(t, t_0, x^*) = x_\nu(t, t_0, x_\nu(t_0, \tau^*, z^*))$$

$$= x_\nu(t, \tau^*, z^*) \in x_\nu(t, \tau^*, \Lambda_\nu(\tau^*))$$

for $t \geq t_0$, we get $x_\nu(t, t_0, x^*) \in \Lambda_\nu^*(t)$ for $t \in [t_0, \tau^* + \Delta_\nu]$.

Step 2.   From Lemma 5, we have $x_\nu(\tau^* + \Delta_\nu, \tau^*, \Lambda_\nu(\tau^*)) \subseteq \Lambda_\nu(\tau^* + \Delta_\nu)$, so

$$x_\nu(\tau^* + \Delta_\nu, t_0, x^*)$$

$$= x_\nu(\tau^* + \Delta_\nu, \tau^*, z^*) \in \Lambda_\nu(\tau^* + \Delta_\nu).$$

Hence $x_\nu(t, t_0, x^*) \in x_\nu(t, \tau^* + \Delta_\nu, \Lambda_\nu(\tau^* + \Delta_\nu))$ for all $t \geq \tau^* + \Delta_\nu$, from which it follows that $x_\nu(t, t_0, x^*) \in \Lambda_\nu^*(t)$ for at least $t \in [\tau^* + \Delta_\nu, t_0 + \Delta_\nu]$. Combining Steps 1 and 2, we have

$$x_\nu(t, t_0, \Lambda_\nu^*(t_0)) \subseteq \Lambda_\nu^*(t)$$

for at least $t \in [t_0, t_0 + \Delta_\nu]$.

We can repeat the above argument on the intervals $[t_0 + n\Delta_\nu, t_0 + (n+1)\Delta_\nu]$ for $n = 1, 2, \ldots$ to obtain the inclusion for all $t \geq t_0$.   ∎

The proof of the absorbing property is easier.

Lemma 8.   *For each compact subset* D *of* $\mathbb{R}^d$ *there exists a time* $T_{D,\nu} \geq 0$ *such that*

$$x_\nu(t, t_0, D) \subseteq \Lambda_\nu^*(t)$$

for all $t \geq t_0 + T_{D,\nu}$ *and each* $t_0 \in \mathbb{R}$.

Proof.   We note from Lemma 6 that

$$x_\nu(t_0 + n\Delta_\nu, t_0, D) \subseteq \Lambda_\nu(t_0 + n\Delta_\nu) \subseteq \Lambda_\nu^*(t_0 + n\Delta_\nu)$$

for $n = N_{D,\nu} \geq 0$, so by the positive invariance property we then obtain

$$x_\nu(t, t_0, D) \subseteq \Lambda_\nu^*(t)$$

for $t \geq t_0 + T_{D,\nu}$, where $T_{D,\nu} := N_{D,\nu}\Delta_\nu$.   ∎

We notice that the time elapsed until being absorbed $T_{D,\nu}$ does not depend on t_0. From this, we conclude that the family $\{\Lambda_\nu^*(t_0), t_0 \in \mathbb{R}\}$ is also absorbing in the pullback sense of the following Lemma.

Lemma 9.   *For each compact subset* D *of* $\mathbb{R}^d$ *there exists a time* $T_{D,\nu} \geq 0$ *such that*

$$x_\nu(t_0, t_0 - \tau, D) \subseteq \Lambda_\nu^*(t_0)$$

for all $\tau \geq T_{D,\nu}$ *and each* $t_0 \in \mathbb{R}$.

Finally the pullback attracting component sets converge upper semicontinuously to A_0 uniformly in $t_0 \in \mathbb{R}$.

Lemma 10. $H^*_{\mathbb{R}^d}(\Lambda^*_\nu(t_0), A_0) \leq \alpha^{-1}(\eta(\nu) + KL\omega(\nu)) \to 0$ *as* $\nu \to 0$.

Proof. We apply the Lyapunov inequality (A.4) to $x^* \in \Lambda^*_\nu(t_0)$ given by $x^* = x_\nu(t_0, \tau^*, z^*)$ for $\tau^* \in [t_0 - \Delta_\nu, t_0]$ and $z^* \in \Lambda_\nu(\tau^*)$, see (A.7), to obtain

$$V(t_0, x^*)$$
$$= V(t_0, x_\nu(t_0, \tau^*, z^*))$$
$$\leq e^{-(t_0 - \tau^*)} V(\tau^*, z^*) + KL\omega(\nu)$$
$$\leq e^{-(t_0 - \tau^*)} \eta(\nu) + KL\omega(\nu) \leq \eta(\nu) + KL\omega(\nu).$$

The result then follows from the fact that

$$\alpha(\mathrm{dist}(x^*, A_0)) \leq V(t_0, x^*). \qquad \blacksquare$$

A.2. *Existence and convergence of pullback attractors*

We apply standard theoretic methods for nonautonomous dynamical systems to the continuous time process $x_\nu(t, t_0, x_0)$ and the family $\Lambda^*_\nu = \{\Lambda^*_\nu(t_0), t_0 \in \mathbb{R}\}$ of pullback absorbing sets defined in the previous subsection to obtain the existence of a pullback attractor $\mathcal{A}_\nu = \{A_\nu(t_0), t_0 \in \mathbb{R}\}$ defined through

$$A_\nu(t_0) = \bigcap_{\tau \geq 0} x_\nu(t_0, t_0 - \tau, \Lambda^*_\nu(t_0 - \tau))$$

which is a nonempty and compact set, since the intersecting sets are nonempty compact and

nested, i.e.

$$x_\nu(t_0, t_0 - \tau_2, \Lambda^*_\nu(t_0 - \tau_2))$$
$$\subseteq x_\nu(t_0, t_0 - \tau_1, \Lambda^*_\nu(t_0 - \tau_1))$$

for $\tau_1 \leq \tau_2$. This follows from the two-parameter evolution property and the fact that the family of absorbing sets is positively invariant, thus

$$x_\nu(t_0, t_0 - \tau_2, \Lambda^*_\nu(t_0 - \tau_2))$$
$$= x_\nu(t_0, t_0 - \tau_1, x_\nu(t_0 - \tau_1, t_0 - \tau_2, \Lambda^*_\nu(t_0 - \tau_2)))$$
$$\subseteq x_\nu(t_0, t_0 - \tau_1, \Lambda^*_\nu(t_0 - \tau_1)).$$

The invariance

$$x_\nu(t, t_0, A_\nu(t_0)) = A_\nu(t)$$

follows from the above construction and the continuity $t \mapsto A_\nu(t)$ from the invariance and continuity of the process, since

$$H_{\mathbb{R}^d}(A_\nu(t), A_\nu(t_0))$$
$$= H_{\mathbb{R}^d}(x_\nu(t, t_0, A_\nu(t_0)), A_\nu(t_0)) \to 0 \quad \text{as } t \to t_0.$$

In addition, the uniform upper semicontinuous convergence follows from the fact that $A_\nu(t_0) \subset \Lambda^*_\nu(t_0)$, so

$$H^*_{\mathbb{R}^d}(A_\nu(t_0), A_0) \leq H^*_{\mathbb{R}^d}(\Lambda^*_\nu(t_0), A_0)$$

for all $t_0 \in \mathbb{R}$ and the result follows from Lemma 10.

We also note that the $A_\nu(t)$ are connected sets, since the $\Lambda^*_\nu(t_0)$ are connected and the $A_\nu(t)$ are an intersecting family of nested connected sets.

A SURVEY OF METHODS FOR COMPUTING (UN)STABLE MANIFOLDS OF VECTOR FIELDS

B. KRAUSKOPF and H. M. OSINGA
Department of Engineering Mathematics, University of Bristol,
Queen's Building, Bristol BS8 1TR, UK

E. J. DOEDEL
Department of Computer Science, Concordia University,
1455 Boulevard de Maisonneuve O., Montréal Québec, H3G 1M8 Canada

M. E. HENDERSON
IBM Research, PO Box 218, Yorktown Heights, NY 10598, USA

J. GUCKENHEIMER and A. VLADIMIRSKY
Department of Mathematics, Cornell University,
Malott Hall, Ithaca, NY 14853–4201, USA

M. DELLNITZ and O. JUNGE
Institute for Mathematics, University of Paderborn,
D-33095 Paderborn, Germany

Received May 14, 2004; Revised June 16, 2004

The computation of global invariant manifolds has seen renewed interest in recent years. We survey different approaches for computing a global stable or unstable manifold of a vector field, where we concentrate on the case of a two-dimensional manifold. All methods are illustrated with the same example — the two-dimensional stable manifold of the origin in the Lorenz system.

Keywords: Stable and unstable manifolds; numerical methods; Lorenz equations.

1. Introduction

Many applications give rise to mathematical models in the form of a system of ordinary differential equations. Well-known examples are periodically forced oscillators and the Lorenz system (introduced in Sec. 1.1); see, for example [Guckenheimer & Holmes, 1986; Kuznetsov, 1998; Strogatz, 1994] for further references. Such a dynamical system can be written in the general form

$$\frac{d\mathbf{x}}{dt} = f(\mathbf{x}), \tag{1}$$

where $\mathbf{x} \in \mathbb{R}^n$ and the map $f : \mathbb{R}^n \mapsto \mathbb{R}^n$ is sufficiently smooth. We remark that, in general, the function f will depend on parameters. However, we assume that all parameters are fixed and use (1) as the appropriate setting for the discussion of global manifolds.

The goal is to understand the overall dynamics of system (1). To this end, one needs to find special invariant sets, namely the equilibria, periodic orbits and possibly invariant tori. Furthermore, if these invariant sets are of saddle type then they come with global stable and unstable manifolds. For example, the stable and unstable manifolds $W^s(\mathbf{x}_0)$ and $W^u(\mathbf{x}_0)$ of a saddle equilibrium $\mathbf{x}_0$ are defined as

$$W^s(\mathbf{x}_0) := \left\{ \mathbf{x} \in \mathbb{R}^n | \lim_{t \to \infty} \phi^t(\mathbf{x}) = \mathbf{x}_0 \right\}$$

$$W^u(\mathbf{x}_0) := \left\{ \mathbf{x} \in \mathbb{R}^n | \lim_{t \to \infty} \phi^{-t}(\mathbf{x}) = \mathbf{x}_0 \right\},$$

68   *B. Krauskopf et al.*

respectively, where ϕ^t is the flow of (1). Hence, trajectories on the stable (unstable) manifold converge to $\mathbf{x}_0$ in forward (backward) time. Knowing these manifolds is crucial as they organize the dynamics on a global scale. For example, stable manifolds may form boundaries of basins of attraction, and it is well known that intersections of stable and unstable manifolds lead to complicated dynamics and chaos.

Generally, global stable and unstable manifolds cannot be found analytically. Furthermore, they are not implicitly defined, meaning that it is not possible to find them as the zero-set of some function of the phase space variables. Hence, points on global invariant manifolds cannot be found "locally". Instead, these manifolds must be "grown" from local knowledge, for example from linear information, near a fixed point $\mathbf{x}_0$.

It is the purpose of this paper to review different numerical techniques that have recently become available to compute these global objects. We review five algorithms in detail and characterize their properties using a common test-case example, namely, the *Lorenz manifold* which is introduced now.

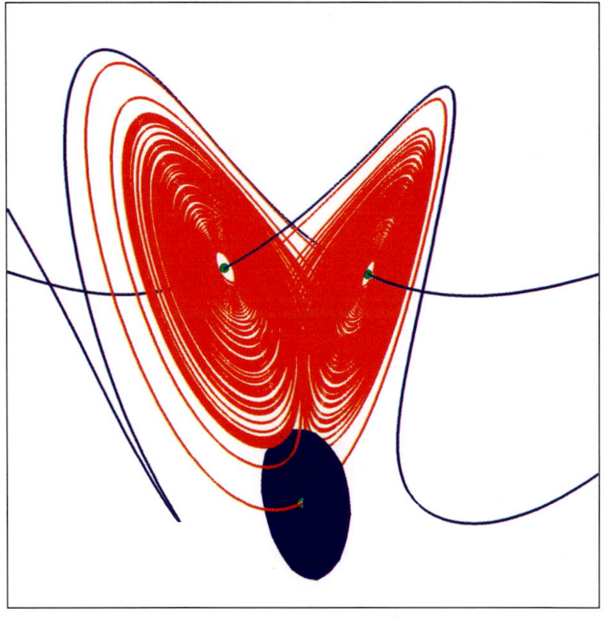

Fig. 1.   The unstable manifold $W^u(\mathbf{0})$ (red curve) accumulates on the butterfly-shaped Lorenz attractor. The blue disk is the linear approximation $E^s(\mathbf{0})$ of the Lorenz manifold $W^s(\mathbf{0})$. Also shown are the two equilibria at the centers of the "wings" of the butterfly and their one-dimensional stable manifolds (blue curves).

1.1. *The Lorenz manifold*

The Lorenz system [Lorenz, 1963] is a classic example of a vector field with a chaotic attractor. It is given as

$$\begin{cases} \dot{x} = \sigma(y - x), \\ \dot{y} = \varrho x - y - xz, \\ \dot{z} = xy - \beta z, \end{cases} \qquad (2)$$

where we fix the parameters at the standard choice $\sigma = 10$, $\varrho = 28$ and $\beta = 8/3$, for which one finds the famous butterfly-shaped Lorenz attractor. Note that the Lorenz system (2) has the symmetry $(x, y, z) \mapsto (-x, -y, z)$ of rotation by π about the z-axis. In particular, the z-axis is invariant under the flow.

The origin is a saddle point of (2) with real eigenvalues $-\beta$ and $-(\sigma + 1)/2 \pm (1/2)\sqrt{(\sigma + 1)^2 + 4\sigma(\rho - 1)}$, that is, approximately -22.828, -2.667 and 11.828. The origin is contained in the Lorenz attractor, so that its one-dimensional unstable manifold $W^u(\mathbf{0})$ can be used to approximate the Lorenz attractor; this is illustrated in Fig. 1 where $W^u(\mathbf{0})$ is shown in red. At the centers of the "wings" of the butterfly are two more equilibria of (2), approximately at $(\pm 8.485, \pm 8.485, 27)$, which are each other's image under the symmetry

of (2). Each of these equilibria has one negative real eigenvalue, giving rise to a one-dimensional stable manifold, and an unstable pair of complex conjugate eigenvalues with positive real part. Figure 1 shows all equilibria of (2) in green, together with their one-dimensional global manifolds. As mentioned, the red curve is the unstable manifold $W^u(\mathbf{0})$ of the origin, whose closure is the Lorenz attractor. The blue curves are the stable manifolds of the two other equilibria. The blue disk lies in the linear eigenspace $E^s(\mathbf{0})$ of the origin.

The Lorenz attractor, that is, the red curve in Fig. 1 conveys the chaotic nature of the system, but does not give any information on the overall organization of the phase space of (2). This role is played by the two-dimensional stable manifold $W^s(\mathbf{0})$ of the origin — which we refer to as the *Lorenz manifold* from now on. The Lorenz manifold $W^s(\mathbf{0})$ is tangent at 0 to the eigenspace $E^s(\mathbf{0})$ spanned by the eigenvectors associated with the eigenvalues -22.828 and -2.667. This is a generic property of stable and unstable manifolds; see Sec. 1.2. Note the large difference in magnitude between the two stable eigenvalues, leading to a dominance of the strong stable manifold, which is tangent to the eigenspace of the eigenvalue -22.828.

The Lorenz manifold has a number of astonishing properties. Imagine that the little blue disk in Fig. 1 "grows" to become the Lorenz manifold $W^s(\mathbf{0})$, but without ever intersecting the red unstable manifold $W^u(\mathbf{0})$. In other words, the Lorenz manifold stays "in between" trajectories on the Lorenz attractor, but "spirals" simultaneously into both wings of the butterfly. Now imagine how trajectories on this manifold must be able to pass from one wing to the other. Any finitely grown part of $W^s(\mathbf{0})$ is topologically still a two-dimensional disk, but one with a particularily intriguing embedding into $\mathbb{R}^3$. The geometry of $W^s(\mathbf{0})$ can only truly be appreciated if one can draw an image of it.

Some early work on the geometry of the Lorenz manifold can be found in [Perelló, 1979]. Using "a desktop computer with a plotter" Perello studied the embedding of the stable manifold of the origin as a function of the parameter ρ and, in particular, provides a sketch for ρ close to 24.74. Pioneering efforts to visualize the Lorenz system are due to Stewart. Trajectories that illustrate the (local) stable manifold can be found in [Thompson & Stewart, 1986, Fig. 11.6], while [Stewart, 1986] is an extended abstract of a movie that visualizes the dynamics and global bifurcations (as a function of R) of the Lorenz system with computer graphics in the three-dimensional phase space. The first, hand-drawn image of (the structure of) the Lorenz manifold (that is, for the standard parameter values also used here) appeared in the book [Abraham & Shaw, 1985]. The first published computer-generated image of the Lorenz manifold is that in [Guckenheimer & Worfolk, 1993].

Not in the least due to its intriguing nature, the Lorenz manifold has become a much-used test-case example for evaluating algorithms that compute two-dimensional (un)stable manifolds of vector fields. For each of the methods discussed in this paper we present an image of the computed Lorenz manifold that is always taken from a viewpoint along the line spanned by the vector $(\sqrt{3}, 1, 0)$ in the (x, y)-plane.

1.2. *Stable and unstable manifolds*

In order to explain the different methods for computing two-dimensional (un)stable manifolds, we need to introduce some notation. To keep the exposition simple, we consider here the case of a global (un)stable manifold of a hyperbolic saddle point $\mathbf{x}_0 \in \mathbb{R}^n$ of (1). Furthermore, we present all theory and the different methods for the case of an unstable manifold. This is not a restriction, because a stable manifold can be computed as an unstable manifold when time is reversed in system (1).

Suppose now that $f(\mathbf{x}_0) = \mathbf{0}$ and for some $1 < k < n$ the Jacobian $Df(\mathbf{x}_0)$ of f at $\mathbf{x}_0$ has k eigenvalues with positive real parts and $(n - k)$ eigenvalues with negative real parts (counted with multiplicity). The Stable and Unstable Manifold Theorem (see, e.g. [Guckenheimer & Holmes, 1986; Kuznetsov, 1998]) states that a local unstable manifold $W^u_{\mathrm{loc}}(\mathbf{x}_0)$ exists in a neighborhood of $\mathbf{x}_0$. Furthermore, $W^u_{\mathrm{loc}}(\mathbf{x}_0)$ is as smooth as f and tangent to the unstable (generalized) eigenspace $E^u(\mathbf{x}_0)$ of $Df(\mathbf{x}_0)$ at $\mathbf{x}_0$. This means that we may define the global unstable manifold $W^u(\mathbf{x}_0)$ as

$$W^u(\mathbf{x}_0) = \left\{ \mathbf{x} \in \mathbb{R}^n \mid \lim_{t \to -\infty} \phi^t(\mathbf{x}) = \mathbf{x}_0 \right\}$$
$$= \bigcup_{t>0} \phi^t(W^u_{\mathrm{loc}}(\mathbf{x}_0)). \tag{3}$$

Hence, $W^u(\mathbf{x}_0)$ is a k-dimensional (immersed) manifold, defined as the globalization of $W^u_{\mathrm{loc}}(\mathbf{x}_0)$ under the flow ϕ^t. Note that the local stable manifold $W^s_{\mathrm{loc}}(\mathbf{x}_0)$ and the stable manifold $W^s(\mathbf{x}_0)$ are similarly related with respect to the reversed direction of time, namely

$$W^s(\mathbf{x}_0) = \left\{ \mathbf{x} \in \mathbb{R}^n \mid \lim_{t \to \infty} \phi^t(\mathbf{x}) = \mathbf{x}_0 \right\}$$
$$= \bigcup_{t<0} \phi^t(W^s_{\mathrm{loc}}(\mathbf{x}_0)). \tag{4}$$

This indeed shows that it is sufficient to consider only the case of an unstable manifold, possibly after reversing time.

Definition (3) already suggests a method for computing $W^u(\mathbf{x}_0)$: take a small $(k - 1)$-sphere (or other "outflow boundary" such as an ellipsoid) $S_\delta \subset W^u_{\mathrm{loc}}(\mathbf{x}_0)$ with radius δ around $\mathbf{x}_0$ and grow the manifold $W^u(\mathbf{x}_0)$ by evolving S_δ under the flow ϕ^t. As starting data, one can take $S_\delta \subset E^u(\mathbf{x}_0)$ or a higher-order approximation of $W^u_{\mathrm{loc}}(\mathbf{x}_0)$.

In the special case $k = 1$ of computing a one-dimensional manifold, this method works well, because it boils down to evolving two points at distance δ from $\mathbf{x}_0$ under the flow. This can be done reliably by numerical integration of (1), so that computing one-dimensional unstable manifolds is straightforward. The one-dimensional manifolds in Fig. 1 were computed in this way.

However, the above method of evolving a $(k-1)$-sphere S_δ with $k \geq 2$ under the flow ϕ^t generally gives very poor results. This is so because S_δ will typically deform very rapidly under ϕ^t. In particular, it will stretch out along the strong unstable directions (if present). Furthermore, S_δ is a continuous object that will have to be discretized by some mesh. Any mesh on S_δ will deteriorate rapidly under the flow ϕ^t, so that it will not be a good representation of $W^u(\mathbf{x}_0)$ as a k-dimensional manifold.

1.3. *Different approaches to computing $W^u(\mathbf{x}_0)$*

It is quite a challenge to compute a global unstable manifold $W^u(\mathbf{x}_0)$ of dimension at least two. Indeed simple numerical integration of the flow is not sufficient (except in very special cases) — dedicated algorithms are needed for this task. Before we describe some recent methods in more detail, we first explain the underlying approaches in general terms. It is useful to consider for this purpose different parametrizations of $W^u(\mathbf{x}_0)$.

We concentrate in this survey on the first nontrivial case $k = 2$ of a two-dimensional unstable manifold. While all methods could be used in principle to compute higher-dimensional manifolds, almost all implementations are for $k = 2$. Furthermore, visualizing higher-dimensional manifolds remains a serious challenge. The different methods use the idea of growing $W^u(\mathbf{x}_0)$ from a local neighborhood of $\mathbf{x}_0$. They differ in how they ensure that a good mesh representing $W^u(\mathbf{x}_0)$ is computed during this growth process.

Consider as starting data a small smooth closed curve $S_\delta \subset W^u_{\text{loc}}(\mathbf{x}_0)$, also referred to as a (topological) circle in what follows, of points that all lie within a distance δ from $\mathbf{x}_0$. (As was mentioned, one can take $S_\delta \subset E^u(\mathbf{x}_0)$ if δ is small enough.) The goal is to find a "nice" parametrization of $W^u(\mathbf{x}_0)$ in terms of the starting data S_δ.

As we have seen above, the parametrization

$$W^u(\mathbf{x}_0) = \{\phi^t(S_\delta)\}_{t \in \mathbb{R}} \tag{5}$$

is not practical. While the $\phi^t(S_\delta)$ are smooth closed curves for all t, they are typically not "nice" and "round". Indeed the curvature along these curves typically varies dramatically, and they soon tend to look like very elongated ellipses.

In order to define the parametrization of $W^u(\mathbf{x}_0)$ as a family of the nicest possible topological circles, recall that the geodesic distance $d_g(\mathbf{x}, \boldsymbol{y})$ is defined as the arclength of the shortest path in $W^u(\mathbf{x}_0)$ connecting $\mathbf{x}$ and $\boldsymbol{y}$, called a *geodesic*. Consider now the geodesic parametrization of $W^u(\mathbf{x}_0)$ given by

$$W^u(\mathbf{x}_0) = \{S_\eta\}_{\eta > 0}$$

$$\text{where } S_\eta := \{\mathbf{x} \in W^u(\mathbf{x}_0) | d_g(\mathbf{x}, \mathbf{x}_0) = \eta\}. \tag{6}$$

The geodesic parametrization (6) is entirely in terms of the geometry of $W^u(\mathbf{x}_0)$, and not in terms of the dynamics on the manifold. Since $W^u(\mathbf{x}_0)$ is a smooth manifold tangent to $E^u(\mathbf{x}_0)$ at $\mathbf{x}_0$, there must be some $\eta_{\text{max}} > 0$ so that the geodesic level sets S_η for $0 < \eta \leq \eta_{\text{max}}$ are all smooth closed curves without self-intersection, that is, topological circles; see, for example, [Spivak, 1979]. We also refer to geodesic level sets for $\eta \leq \eta_{\text{max}}$ as *geodesic circles*. Up until η_{max}, the geodesic parametrization (6) is geometrically the nicest parametrization, because its elements, the geodesic circles, are the nicest possible topological circles on $W^u(\mathbf{x}_0)$. (This means here that the metric is exactly the identity.) For the Lorenz manifold, apparently $\eta_{\text{max}} = \infty$. However, the case of a finite η_{max} is possible and it typically involves a non-smooth geodesic circle; see [Krauskopf & Osinga, 2003] for details.

The idea of computing $W^u(\mathbf{x}_0)$ as a sequence of geodesic circles goes back to [Guckenheimer & Worfolk, 1993]. Starting with a small geodesic circle (or ellipse) S_δ around $\mathbf{x}_0$, they modify the vector field so that the component tangential to the last computed geodesic level set is practically zero, retaining only the radial part. Then the flow of the rescaled radial vector field is used to evolve (a sufficient number of points on) this geodesic circle by integration over a suitably small and fixed integration time (now corresponding to geodesic distance up to a rescaling of the radial part of the vector field). Figure 2 shows 36 approximate geodesic circles of the Lorenz manifold computed with this method up to geodesic distance 180. The output was produced in the DsTool software environment [Back *et al.*, 1992], the manifold could be rendered as a two-dimensional surface by post-processing the data. When the vector field f is largely tangential to the geodesic circles, the computation of that vector field's radial component becomes unstable unless the integration time τ is sufficiently small (see the ripples on the last few geodesic circles near the helix at the

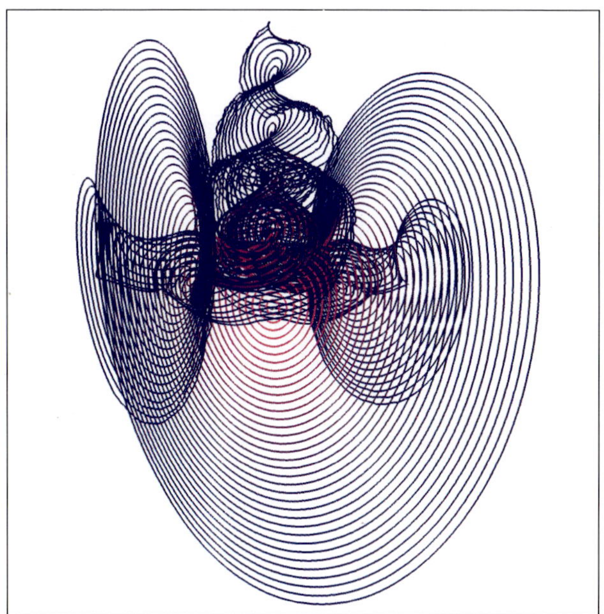

Fig. 2. The Lorenz manifold computed with the method of [Guckenheimer & Worfolk, 1993] up to geodesic distance 180; the computed approximate geodesic level sets are at increasing radial distances from the origin with steps of 5.0 in between, which are indicated by a color change from magenta (small) to blue (large).

center of Fig. 2). This CFL-type stability condition becomes increasingly restrictive as the angle between the trajectories and geodesic circles decreases. More generally, the method from [Guckenheimer & Worfolk, 1993] can approximate stably only a part of the manifold, on which the vector field remains transverse to each geodesic circle.

The method by [Krauskopf & Osinga, 1999, 2003], discussed in detail in Sec. 2, also computes $W^u(\mathbf{x}_0)$ as a sequence of geodesic circles, but does not rescale the vector field. Instead, the idea is to find the next geodesic circle in a local (and changing) coordinate system given by hyperplanes perpendicular to the present geodesic circle. Determined by certain accuracy parameters, a suitable number of mesh points on the next geodesic circle is computed by solving appropriate boundary value problems. During the computation the interpolation error stays bounded, so that the overall quality of the mesh is guaranteed.

A different approach is to reparametrize time so that the flow with respect to the new time progresses with the same speed along all trajectories through S_δ, meaning that the same arclength is covered per unit time along all trajectories. One also speaks of arclength integration. We then have the

new parametrization of $W^u(\mathbf{x}_0)$ given by

$$W^u(\mathbf{x}_0) = \{A_\eta\}_{\eta>0}$$

$$\text{where } A_\eta := \{\mathbf{x} \in W^u(\mathbf{x}_0) | d_a(\mathbf{x}, \mathbf{x}_0) = \eta\}, \qquad (7)$$

where $d_a(\mathbf{x}, \mathbf{y})$ denotes the arclength distance between two points $\mathbf{x}$ and $\mathbf{y}$ on the same trajectory; we set $d_a(\mathbf{x}, \mathbf{y}) = \infty$ if $\mathbf{x}$ and $\mathbf{y}$ are not on the same trajectory. This parametrization can be considered as the best in terms of dynamically defined topological circles on $W^u(\mathbf{x}_0)$.

Johnson *et al.* [1997] used essentially this parametrization by trajectory arclength, but considered integration in the product of time and phase space. They started with a uniform mesh on a first small circle $A_\delta \in E^u(\mathbf{x}_0)$ and then integrated at each step the present mesh points up to a specified arclength. This leads to a new circle, on which a uniform mesh is then constructed by interpolation between the integration points. Figure 3 shows the Lorenz manifold computed with this method up to an approximate arclength distance of 200. The method is quite fast since it involves only direct integration and redistribution of points by interpolation. On the other hand, it is difficult to control the interpolation error, which is determined by the (unknown) dynamics on $W^u(\mathbf{x}_0)$.

An altogether different parametrization of $W^u(\mathbf{x}_0)$ is the dual parametrization to (5) and (7) that consists of the individual trajectories through a fixed $S_\delta \subset E^u(\mathbf{x}_0)$. It is formally given as

$$W^u(\mathbf{x}_0) = \{B_p\}_{p\in S_\delta}$$

$$\text{where } B_p := \{\phi^t(p) | t \in \mathbb{R}\}. \qquad (8)$$

Notice that, in the case of a two-dimensional manifold $W^u(\mathbf{x}_0)$ considered here, parametrization (8) is a one-parameter family of trajectories, while (5) and (7) are one-parameter families of closed curves.

The method by Doedel, discussed in detail in Sec. 3, computes two-dimensional (un)stable manifolds by following trajectories B_p as a boundary value problem where the initial condition $p \in S_\delta$ is parametrized with one of the free continuation parameters. This method is very accurate and flexible by allowing for different boundary conditions at the other end point of the trajectory B_p, which includes specifying a fixed arclength L of the trajectory. During a computation, mesh points are distributed along the trajectories to maintain the accuracy of the computation.

Fig. 3.   The Lorenz manifold computed with the method of [Johnson *et al.*, 1997] up to a total trajectory arclength of about 200.

The method of [Guckenheimer & Vladimirsky, 2004], discussed in detail in Sec. 5, locally models $W^u(\mathbf{x}_0)$ as the graph of a function g that satisfies a quasilinear partial differential equation (PDE) expressing the tangency of the vector field f to the graph of g. The PDE is discretized in an Eulerian framework and the manifold is approximated by a triangulated mesh. At each step one new point is added to the mesh, leading to a new simplex whose other vertices are previously known mesh points. An Ordered Upwind Method determines where the next point/simplex is added and the ordering of new simplices is based on the arclength of the trajectories.

The method of Dellnitz and Hohmann [1996, 1997], discussed in detail in Sec. 6, is complementary to the previous methods in that it computes an outer approximation of the manifold by boxes of the same dimension n as the phase space of (1). This method uses the time-τ map of the flow ϕ^t for some fixed τ. A subdivision algorithm first finds a covering of $W^u_{\mathrm{loc}}(\mathbf{x}_0)$ with n-dimensional boxes of suitably small diameter. This local box covering is then globalized in steps by adding new boxes (of the same small size) that are "hit" under the time-τ map by the present collection of boxes. The practical problem is to detect reliably when the image of one box intersects another box (for example, by using test points). If *a priori* bounds on the local growth rate of the vector field are known then it is possible to compute a rigorous box covering of $W^u(\mathbf{x}_0)$; see [Junge, 2000a].

In the following sections we present the different algorithms in more detail, again illustrated with the computation of the Lorenz manifold $W^s(\mathbf{0})$.

The method of [Henderson, 2003], discussed in detail in Sec. 4, also considers parametrization (8) of $W^u(\mathbf{x}_0)$ by orbits. However, the manifold is constructed directly as a two-dimensional object by computing fat trajectories. A fat trajectory is a string of polyhedral patches along a trajectory, where the size of each patch is given by local curvature information. When a fat trajectory reaches the prescribed total arclength L, the boundary of the computed part of the manifold is determined. Then a suitable starting point for the next fat trajectory is found and the computation continues. When no more possible starting points exist, the computation stops.

2. Approximation by Geodesic Level Sets

The method of Krauskopf and Osinga [1999, 2003] approximates a global (un)stable manifold as a sequence of geodesic circles of the parametrization (6). Only the case of a two-dimensional unstable manifold of a saddle point in a three-dimensional space is presented here. However, the method can be formulated in terms of computing a k-dimensional manifold of a vector field in $\mathbb{R}^n$, and has been implemented to compute two-dimensional (un)stable manifolds of saddle points and saddle periodic orbits in a phase space of any dimension;

see the examples in [Krauskopf & Osinga, 1999, 2003] and also in [Osinga, 2000, 2003]. Variants of this method exist to compute global manifolds of maps; see [Krauskopf & Osinga, 1998a, 1998b].

The method completely steps away from evolving an existing mesh. Instead, new mesh points are computed by means of solving appropriate boundary value problems; see Sec. 2.1. The boundary conditions predetermine where the new mesh points need to be added in order to achieve a prescribed mesh quality. This method is as independent of the dynamics as possible and it grows the manifold as a sequence of discretized geodesic circles until $\eta_{\max}$ is reached where the geodesic level sets are no longer smooth circles; see Sec. 1.3.

To be more specific, let M_i denote a circular list of mesh points from which a continuous topological circle C_i is formed by connecting neighboring points of M_i by line segments. The mesh points in M_i are computed to ensure that C_i is a good approximation (according to prespecified accuracy parameters) of an appropriate geodesic circle S_{η_i}. The manifold $W^u(\mathbf{x}_0)$ is then approximated up to a prescribed geodesic distance L by the triangulation formed by the total mesh $\mathcal{M} = \bigcup_{0 \leq i \leq l} M_i$, where $l \in \mathbb{N}$ depends on L and the accuracy parameters.

The start data is a uniform mesh M_0 on an initial small geodesic circle $S_{\eta_0} = S_\delta \subset E^u(\mathbf{x}_0)$ at some prescribed distance δ from $\mathbf{x}_0$. The method then computes at each step i a new circular list M_{i+1} that approximates the next level set $S_{\eta_{i+1}}$. In other words, at every step a new band is added to $W^u(\mathbf{x}_0)$; the width of this band is determined by the curvature of geodesics. The method stops when the prespecified fixed geodesic distance L from $\mathbf{x}_0$ is reached.

2.1. *Finding a new point in M_{i+1}*

Let us consider the task of finding M_{i+1} at some prescribed increment Δ_i from a known circular list M_i representing S_{η_i}. The circular list M_{i+1} is constructed pointwise. Let $r \in M_i$ and consider the (half)plane $\mathcal{F}_r$ through r that is (approximately) perpendicular to C_i at r. (In the implementation the normal to $\mathcal{F}_r$ is defined as the average of the two unit vectors through r and its immediate left and right neighbors.) Then $W^u(\mathbf{x}_0) \cap \mathcal{F}_r$ is a well-defined one-dimensional curve locally near r, which is parametrized by the time it takes to reach

$W^u(\mathbf{x}_0) \cap \mathcal{F}_r$ by integration from C_i. Points in $W^u(\mathbf{x}_0) \cap \mathcal{F}_r$ can be found by solving the two-point boundary value problem

$$q_r(t) \in C_i, \tag{9}$$

$$b_r(t) := \phi^t(q_r(t)) \in \mathcal{F}_r, \tag{10}$$

where the integration time t is a free parameter. The situation is shown in Fig. 4 with actual data for the Lorenz manifold $W^s(\mathbf{0})$ presented in Sec. 2.3. Note that for an unstable manifold $t \geq 0$ and for a stable manifold $t \leq 0$.

The point $b_r(t_r) \in \mathcal{F}_r$ is uniquely defined by the property that t_r is the smallest integration time (in absolute value) for which $\|b_r(t_r) - r\| = \Delta_i$. If Δ_i is small enough then $b_r(t_r)$ exists and can be found by continuation of the trivial solution $b_r(0) = q_r(0) = r$ for $t = 0$ while checking for the first zero of the test function

$$\Delta_i - \|b_r(t) - r\|. \tag{11}$$

When the first zero is found then $b_r(t_r) = b_r(t)$ is the candidate for a point in M_{i+1}; see Fig. 4.

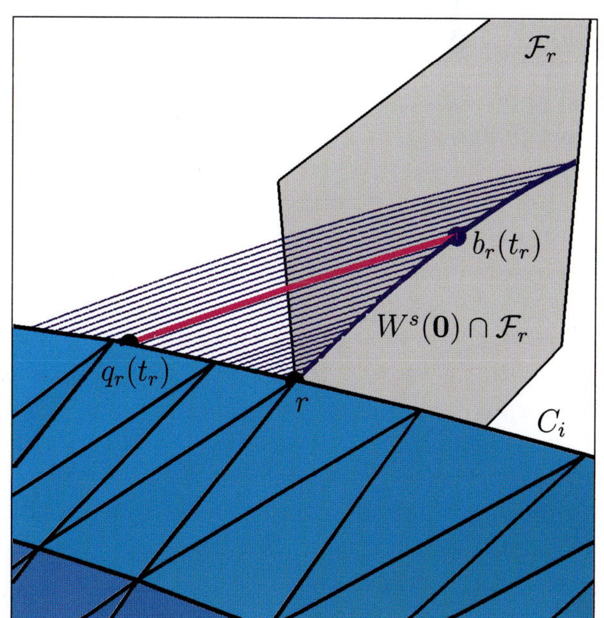

Fig. 4. The boundary value problem formulated for a mesh point r on the geodesic level set C_i is solved by a family of trajectories, starting at $q_r(t)$ on C_i and ending at $b_r(t)$ in $\mathcal{F}_r$, that is parametrized by integration time t. There is a unique first orbit such that $\|b_r(t_r) - r\| = \Delta_i$. The image shows actual data for the Lorenz manifold $W^s(\mathbf{0})$ of Fig. 5 where $C_i \approx S_{\eta_i}$ with $\eta_i = 32.75$ and $\Delta_i = 4.0$.

2.2. *Mesh adaptation*

Once all candidate points in M_{i+1} have been found, all for the same Δ_i, then it is decided whether the step size Δ_i was appropriate. To this end, it is checked that the curvature of (approximate) geodesics through all points $r \in M_i$ was not too large. This is done with a criterion that was originally introduced for one-dimensional global manifolds of maps [Hobson, 1993]. Let α_r denote the angle between the line through r and $b_r(t_r)$ and the line through p_r and r, where $p_r \in M_{i-1}$ is the associated point of M_{i-1} on the approximate geodesic. The step of geodesic distance Δ_i was acceptable if both

$$\alpha_r < \alpha_{\max}, \quad \text{and} \tag{12}$$

$$\Delta_i \cdot \alpha_r < (\Delta\alpha)_{\max} \tag{13}$$

hold for all $r \in M_i$. In this case M_{i+1} is accepted and step i is complete. If there is some $r \in M_i$ that fails either (12) or (13) then Δ_i is halved and step i is repeated with this smaller Δ_i. Similarly, Δ_i may be doubled if for every $r \in M_i$ both α_r and $\Delta_i \cdot \alpha_r$ are well below the respective upper bounds in (12) or (13), say, below $\alpha_{\min}$ and $(\Delta\alpha)_{\min}$ respectively. The parameters $\alpha_{\min}$, $\alpha_{\max}$, $(\Delta\alpha)_{\min}$, and $(\Delta\alpha)_{\max}$ implicitly determine the mesh adaptation along geodesics and are fixed by the user before a computation.

It is important to ensure that C_{i+1} is also a good approximation of $S_{\eta_{i+1}}$. In other words, neighboring points of M_{i+1} may not be too close or too far from each other. When two neighboring points of M_i lead to two neighboring points of M_{i+1} at more than the prespecified distance $\Delta_{\mathcal{F}}$ from each other, then a new point is added in between. This is not done by interpolating between points of M_{i+1} but by applying step i of Sec. 2.2 for finding a new point in M_{i+1} to the middle point on C_i. In other words, no interpolation is ever performed between points that are more than $\Delta_{\mathcal{F}}$ distance apart. In order to ensure proper order relations between directly neighboring points of M_{i+1} a point is removed if two neighboring points in M_{i+1} lie closer together than a prespecified distance $\delta_{\mathcal{F}}$.

The mesh adaptation as decribed ensures that the overall error of a computation up to a prescribed geodesic distance L is bounded. This means that the computed piece of the manifold lies in an ε-neighborhood of $W^u(\mathbf{x}_0)$, provided the accuray parameters are chosen small enough; see [Krauskopf & Osinga, 2003] for the proof.

2.3. *The Lorenz manifold approximated by geodesic circles*

Figure 5 shows the Lorenz manifold $W^s(\mathbf{0})$ represented by a total of 75 bands and with total geodesic distance 154.75. The manifold was computed starting with a mesh M_0 of 20 points on $S_\delta \subset E^s(\mathbf{0})$ with $\delta = 1.0$. The computation was initiated with $\Delta_1 = 0.25$ and the mesh was generated using the accuracy parameters $\alpha_{\min} = 0.3$, $\alpha_{\max} = 0.4$, $(\Delta\alpha)_{\min} = 0.1$, $(\Delta\alpha)_{\max} = 1.0$, $\Delta_{\mathcal{F}} = 2.0$, and $\delta_{\mathcal{F}} = 0.67$. The coloring illustrates the geodesic distance from the origin, where blue is small, green is intermediate and red is large. The manifold was rendered as a two-dimensional surface with the visualization package Geomview [Phillips *et al.*, 1993]; other illustrations of the Lorenz manifold can be found in [Krauskopf & Osinga, 2003, 2004; Osinga & Krauskopf, 2002] and animations with [Krauskopf & Osinga, 2003, 2004].

Figure 5(a) shows the entire computed part of the Lorenz manifold from the common viewpoint; notice the similarity with the geodesic level sets in Fig. 2. Figure 5(b) shows an enlargement of the Lorenz manifold where the manifold is now transparent. This brings out the detail of the manifold, in particular, the development of a pair of extra helices that follow the main helix along the z-axis. Notice that points of the same color are on the same geodesic circle, which shows that points on $W^s(\mathbf{0})$ that are close to the origin in Euclidean distance need not be close to the origin in geodesic distance. Figure 5(c) shows a further enlargement near the Lorenz attractor, which is illustrated in magenta by plotting the unstable manifold $W^u(\mathbf{0})$. In this image only every second band is shown to obtain a see-through effect, showing clearly how the Lorenz manifold "rolls" into the Lorenz attractor.

Figure 5(d) gives an impression of the computed mesh with an enlargement looking into one of the outer scrolls. Geodesic circles can be seen as spiraling curves (between bands of the same color). The approximate geodesics are the curves that point approximately radially out in the image. They are perpendicular to the geodesic circles, and locations where points were added can be identified as starting points of new approximate geodesics. Notice that the last six bands are closer together. The image illustrates how the distance between geodesic circles is determined by the curvature along geodesics, while the mesh distribution on the

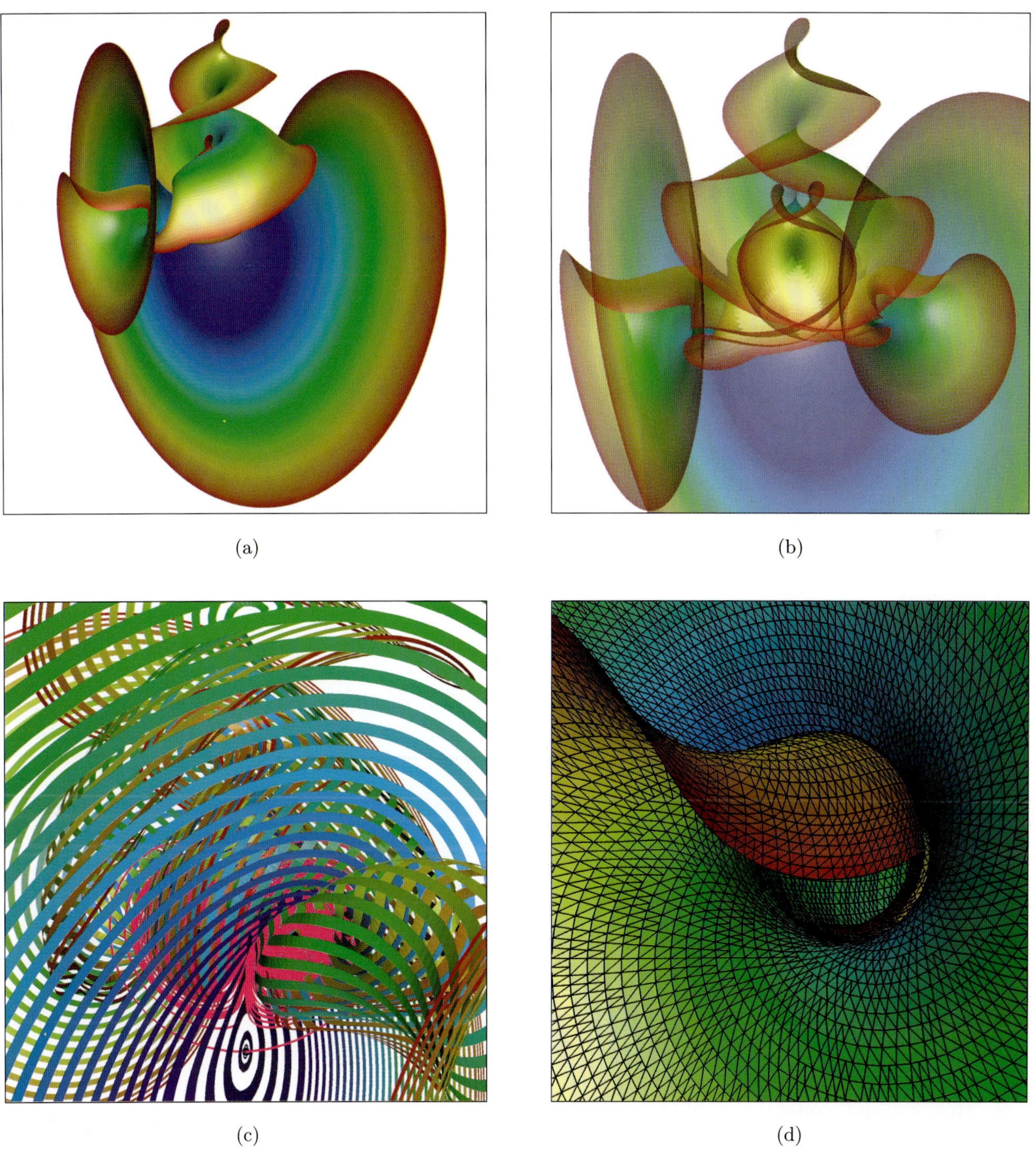

(a)　　　　　　　　　　　　　(b)

(c)　　　　　　　　　　　　　(d)

Fig. 5.   The Lorenz manifold computed with the method of Krauskopf and Osinga up to geodesic distance 154.75. Panel (a) shows the entire manifold, panel (b) an enlargement where the manifold is transparent, panel (c) a further enlargement near the Lorenz attractor (in magenta) where only every second band is shown, and panel (d) the computed mesh when looking into the outer scroll.

geodesic circles is allowed to vary between $\delta_{\mathcal{F}} = 0.67$ and $\Delta_{\mathcal{F}} = 2.0$.

3. BVP Continuation of Trajectories

It seems very natural to use parametrization (8) for defining a one-parameter family that describes the unstable manifold $W^u(\mathbf{x}_0)$ of a saddle equilibrium $\mathbf{x}_0$ of (1). An approximation to $W^u(\mathbf{x}_0)$ could then be attempted by simple integration of Eq. (1) for a sufficient number of initial conditions that lie on the circle (or ellipse) S_δ of small radius δ in the stable eigenspace $E^u(\mathbf{x}_0)$ centered at $\mathbf{x}_0$. However, as was already explained in Sec. 1.3, this procedure does not generally produce $W^u(\mathbf{x}_0)$ as a surface. The main task is to properly space the initial conditions around the circle, so that the result gives a reasonable distribution of the computed trajectories along the stable manifold. This is a major problem because the entire calculated trajectory (e.g. of a fixed finite length) depends very sensitively on the initial condition.

The method of Doedel uses numerical continuation to solve this problem. The basic idea of continuation is to follow a (one-dimensional) branch of solutions that exists according to the Implicit Function Theorem around a regular root of a system of m equations with $m+1$ unknowns. The step size in the continuation procedure (see Sec. 3.1 for details) measures the change of the *entire computed trajectory* (and various parameters), and not just the change in the initial condition. It is this key property of continuation that generally results in a reasonable distribution of trajectories along the stable manifold.

In this section we only consider the computation of one-parameter families of trajectories, which together describe a two-dimensional (un)stable manifold of a fixed point. Most existing continuation algorithms can handle the computation of such one-dimensional families (also called *solution branches*); see, for example [Beyn *et al.*, 2002; Doedel *et al.*, 1991a; Doedel *et al.*, 1991b; Keller, 1977; Rheinboldt, 1986; Seydel, 1995], and [Kuznetsov, 1998, Chapter 10]. The continuation method described here was implemented in the continuation package AUTO [Doedel, 1981; Doedel *et al.*, 1997; Doedel *et al.*, 2000] by specifying the respective driver files.

Continuation algorithms have also been developed for the higher-dimensional case; see, for example [Allgower & Georg, 1996; Henderson, 2002]. Hence, this method could be applied, in principle, equally well to compute manifolds of dimension larger than two.

3.1. *Pseudo-arclength continuation*

Let us begin with a discussion of some basic notions of continuation. Consider the finite-dimensional equation

$$F(X) = 0, \quad F : \mathbb{R}^{m+1} \to \mathbb{R}^m, \qquad (14)$$

where F is assumed to be sufficiently smooth. This equation has one more variable than it has equations. Given a solution X_0, one has, generically, a locally unique solution branch that passes through X_0. To compute a next point, say, X_1, on this branch, one can use Newton's method to solve the extended system

$$F(X_1) = 0, \qquad (15)$$

$$(X_1 - X_0)^* \dot{X}_0 = \Delta s. \qquad (16)$$

Here $\dot{X}_0$ is the unit tangent to the path of solutions at X_0, the symbol $*$ denotes transpose, and Δs is a step size in the continuation procedure. The vector $\dot{X}_0$ is a null vector of the $m \times (m+1)$-dimensional Jacobian matrix $F_X(X_0)$, and it can be computed at little cost [Doedel *et al.*, 1991a]. This continuation method is known as Keller's *pseudo-arclength method* [Keller, 1977]. The size of the pseudo-arclength step Δs is normally adapted along the branch, depending, for example, on the convergence history of Newton's method. It is very important to note that the stepsize is measured with respect to all components of the solution, and not just one.

The continuation procedure is well posed near a *regular solution* X_0, that is, if the null space of $F_X(X_0)$ is one-dimensional. Namely, in this case the Jacobian of the entire system (15)–(16) at X_0, that is, the $(m+1) \times (m+1)$ matrix

$$\begin{pmatrix} F_X(X_0) \\ \dot{X}_0^* \end{pmatrix} \qquad (17)$$

is nonsingular. The Implicit Function Theorem then guarantees that a locally unique solution branch passes through X_0. This branch can be parametrized locally by Δs. Moreover, for Δs sufficiently small, and for sufficiently accurate initial approximation (for example, when taking $X_1^{(0)} = X_0 + \Delta s \dot{X}_0$), Newton's method for solving Eqs. (15)–(16) converges.

3.2. *Boundary value problem formulation*

When computing a branch of solutions to an ODE of the form (1), parametrized by initial conditions and the integration time T, one must keep in mind that (1) has infinitely many solutions and boundary or integral constraints must be imposed. Furthermore, the pseudo-arclength constraint (16) is then typically given in functional form; more details can be found in [Doedel *et al.*, 1991b]. This means that the possibly unknown total integration time T is embedded in the equations. To this end, the vector field (1) is rescaled so that integration always takes place over the interval $[0, 1]$, and the actual integration time T appears as a parameter. Hence, in this context, Eqs. (15)–(16) take the form

$$\mathbf{x}_1'(t) = \hat{f}(\mathbf{x}_1(t), \lambda_1), \tag{18}$$

$$b(\mathbf{x}_1(0), \mathbf{x}_1(1), \lambda_1) = 0, \tag{19}$$

$$\int_0^1 q(\mathbf{x}_1(s), \lambda_1)\, ds = 0, \tag{20}$$

$$\int_0^1 (\mathbf{x}_1(\tau) - \mathbf{x}_p(\tau))^* \dot{\mathbf{x}}_p(\tau)\, d\tau$$
$$+ (\lambda_1 - \lambda_0)^* \dot{\lambda}_0 = \Delta s, \tag{21}$$

where the dimension of λ_1 must be chosen consistently with the dimensions of the boundary conditions (19) and the integral constraints (20) in order to ensure a one-dimensional family of solutions. Again, we stress that the continuation stepsize is for the entire solution X, and not just for the parameter vector λ_1. Equations (15)–(16) must be solved for $X_1 = (\mathbf{x}_1(\cdot), \lambda_1)$, given a previous solution $X_0 = (\mathbf{x}_p(\cdot), \lambda_0)$ of the ODE and the path tangent $\dot{X}_0 = (\dot{\mathbf{x}}_p(\cdot), \dot{\lambda}_0)$. That is, in a function space setting, Eqs. (18)–(20) correspond to the equation $F(X) = 0$, as in Eq. (14). Note that the dimension $(m + 1)$ of $X = (x(\cdot), \lambda)$ may be much larger than the dimension n of the phase space of (1). In particular, λ always contains the parameter T, which may or may not vary during the continuation; see Sec. 3.3 for specific examples. If $\lambda = T$ then $\hat{f}(\mathbf{x}_1(t), \lambda_1) = T f(\mathbf{x}_1(t))$.

In each continuation step, Eqs. (18)–(21) are solved by a numerical boundary value algorithm. Here, the package AUTO [Doedel, 1981; Doedel *et al.*, 1997; Doedel *et al.*, 2000] is used, which uses piecewise polynomial collocation with Gauss–Legendre collocation points (also called *orthogonal collocation*), similar to COLSYS with adaptive mesh selection [Ascher *et al.*, 1995; De Boor & Swartz, 1973; Russell & Christiansen, 1978]. In combination with continuation, this allows the numerical solution of "difficult" orbits. Moreover, for the case of periodic solutions, AUTO determines the characteristic multipliers (or Floquet multipliers) that determine asymptotic stability and bifurcation properties, as a by-product of the decomposition of the Jacobian of the boundary value collocation system [Doedel *et al.*, 1991b; Fairgrieve & Jepson, 1991]; see also [Lust, 2001].

3.3. *BVP continuation of the (un)stable manifold of an equilibrium*

Consider now the situation that (1) has a saddle equilibrium $\mathbf{x}_0$ with a two-dimensional unstable manifold, meaning that the Jacobian $Df(\mathbf{x}_0)$ has exactly two eigenvalues μ_1 and μ_2 with positive real part. Suppose further that $\mathbf{v}_1$ and $\mathbf{v}_2$ are the associated (generalized) eigenvectors. We are looking for solutions of the system

$$\mathbf{x}'(t) = T f(\mathbf{x}(t)), \tag{22}$$

$$\mathbf{x}(0) = \mathbf{x}_0 + \delta(\cos(\theta)\mathbf{v}_1 + \sin(\theta)\mathbf{v}_2), \tag{23}$$

which is a combination of Eqs. (18) and (19) with $\lambda = (\theta, T)$. Note that in Eqs. (22)–(23) the continuation equation corresponding to Eq. (21) (or Eq. (16)) has been omitted, even though it is an essential part of the continuation procedure. The continuation equation will also not be written explicitly in subsequent continuation systems.

If the eigenvalues μ_1 and μ_2 are real, then it is advantageous to choose the initial condition on the ellipse that is given by the ratio of the eigenvalues as

$$\mathbf{x}(0) = \mathbf{x}_0 + \delta \left(\cos(\theta)\frac{\mathbf{v}_1}{|\mu_1|} + \sin(\theta)\frac{\mathbf{v}_2}{|\mu_2|} \right). \tag{24}$$

In other words, in the continuation Eq. (23) is replaced by Eq. (24).

Obvious starting data for the system (22)–(23) consist of a value of θ $(0 \leq \theta < 2\pi)$, $T = 0$, and $\mathbf{x}(t) = \mathbf{x}_0 + \delta(\cos(\theta)\mathbf{v}_1 + \sin(\theta)\mathbf{v}_2)$, that is, $\mathbf{x}(t)$ is constant. An actual trajectory for a specific value of θ can now be obtained using continuation as well. While this may seem superfluous, it has the added benefit that the output files of this first step in AUTO are then compatible with subsequent continuation steps. In this continuation step, system (22)–(23) is solved for $X = (\mathbf{x}(\cdot), T)$, keeping the angle θ fixed. Here, $T > 0$ for an unstable manifold and $T < 0$ for a stable manifold since then integration is backward (or negative) in time.

Once a single orbit is obtained up to a desired length, defined by a suitable end-point condition, then this orbit is continued numerically as a boundary value problem where the initial condition on the small circle (or ellipse) is now a component of the continuation variable. In this way, the family (8) of such orbits on (part of) the stable manifold $W^u(\mathbf{x}_0)$ is approximated. The simplest way to do this is to fix T in the continuation system (22)–(23) after the first step and allow θ, the angle of the starting point on S_δ to vary freely. It is important to note that θ is not used as the sole continuation parameter. Instead each continuation step is taken in the full continuation variable $X = (\mathbf{x}(\cdot), \theta)$, so that the continuation stepsize includes variations along the entire orbit. Also, θ is one of the variables solved for in each continuation step and it is not fixed *a priori*.

Instead of keeping T fixed, there are other ways to perform the continuation. For example, one can constrain the end point $\mathbf{x}(1)$ as one wishes. This is done by adding to system (22)–(23) the equation

$$g(\mathbf{x}(1), \theta, T) - \alpha = 0. \tag{25}$$

Here g is an appropriate functional, chosen to control the end point in a desirable manner, for example, by requiring one coordinate to have a particular fixed value. The continuation variable can now be taken as $X = (\mathbf{x}(\cdot), \theta, T)$, while α is kept fixed.

Another possibility is to impose an integral constraint along the orbit, namely adding to (22)–(23) the equation

$$\int_0^1 h(\mathbf{x}(s), \theta, T)\, ds - L = 0. \tag{26}$$

Now h is an appropriate functional, chosen to control the orbit in a desirable manner. The continuation variable can again be taken as $X = (\mathbf{x}(\cdot), \theta, T)$, but now keeping L fixed. A particularly useful choice is $h(\mathbf{x}, \lambda, T) = T\|f(\mathbf{x}, \lambda)\|$, which results in the total arclength of the orbit being kept fixed during the continuation. Finally, it is entirely possible to use a combination of end-point conditions and integral constraints, but this will not be used here.

3.4. *The Lorenz manifold as a family of trajectories*

Figure 6 shows an enlargement near the origin of the orbits that were continued on the Lorenz manifold $W^s(\mathbf{0})$ (for negative T). The angle θ is allowed to vary from 0 to 2π, so that the initial condition varies along the ellipse in the middle of the image, which is defined by (24) with $\delta = 5.0$, $\mu_1 = -22.828$ and $\mu_1 = -2.667$. All orbits have the same arclength and the coloring is in terms of the total integration time T along each trajectory. In other words, the coloring gives an indication of the speed of the flow along trajectories, where red is fast and green is slower. The flow is fastest along the strong stable

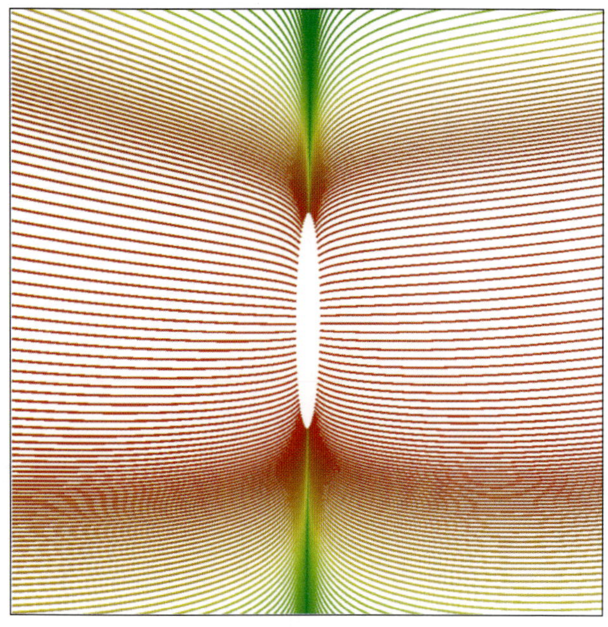

Fig. 6. Continued trajectories on $W^s(\mathbf{0})$ near the origin starting from the ellipse (24) with $\delta = 5.0$, $\mu_1 = -22.828$ and $\mu_2 = -2.667$; the coloring is according to integration time T, where red indicates faster and green slower flow.

manifold, which is located in the middle of the red region. Note that the distribution of points is much denser near the top and bottom of the ellipse, that is, near the invariant z-axis, which ensures a good distribution of orbits over the Lorenz manifold $W^s(\mathbf{0})$.

Figures 7(a)–7(c) show the Lorenz manifold $W^s(\mathbf{0})$ covered by 2284 trajectories of arclength 250, where the ellipse of initial conditions is as in Fig. 6. The number of mesh points along each trajectory was NTST = 75, with NCOL = 4 collocation points in each mesh interval. Figure 7(a) shows the entire computed part of the Lorenz manifold from the common viewpoint. The coloring changes from blue to red according to the mesh point number along a trajectory, which gives an impression of the arclength of trajectories. Figures 7(b) and 7(c) show enlargements where the coloring shows the total integration time T along trajectories. As in Fig. 6, this indicates the speed of the flow; the strong stable manifold is located in the red region of fast flow. In Fig. 7(b) every fourth trajectory is rendered as a thin tube. This results in a better sense of depth so that an impression is given of how trajectories lie in phase space to form $W^s(\mathbf{0})$. Figure 7(c) is an enlargement of Fig. 7(a) (though with a different color scheme) showing how the manifold forms a scroll.

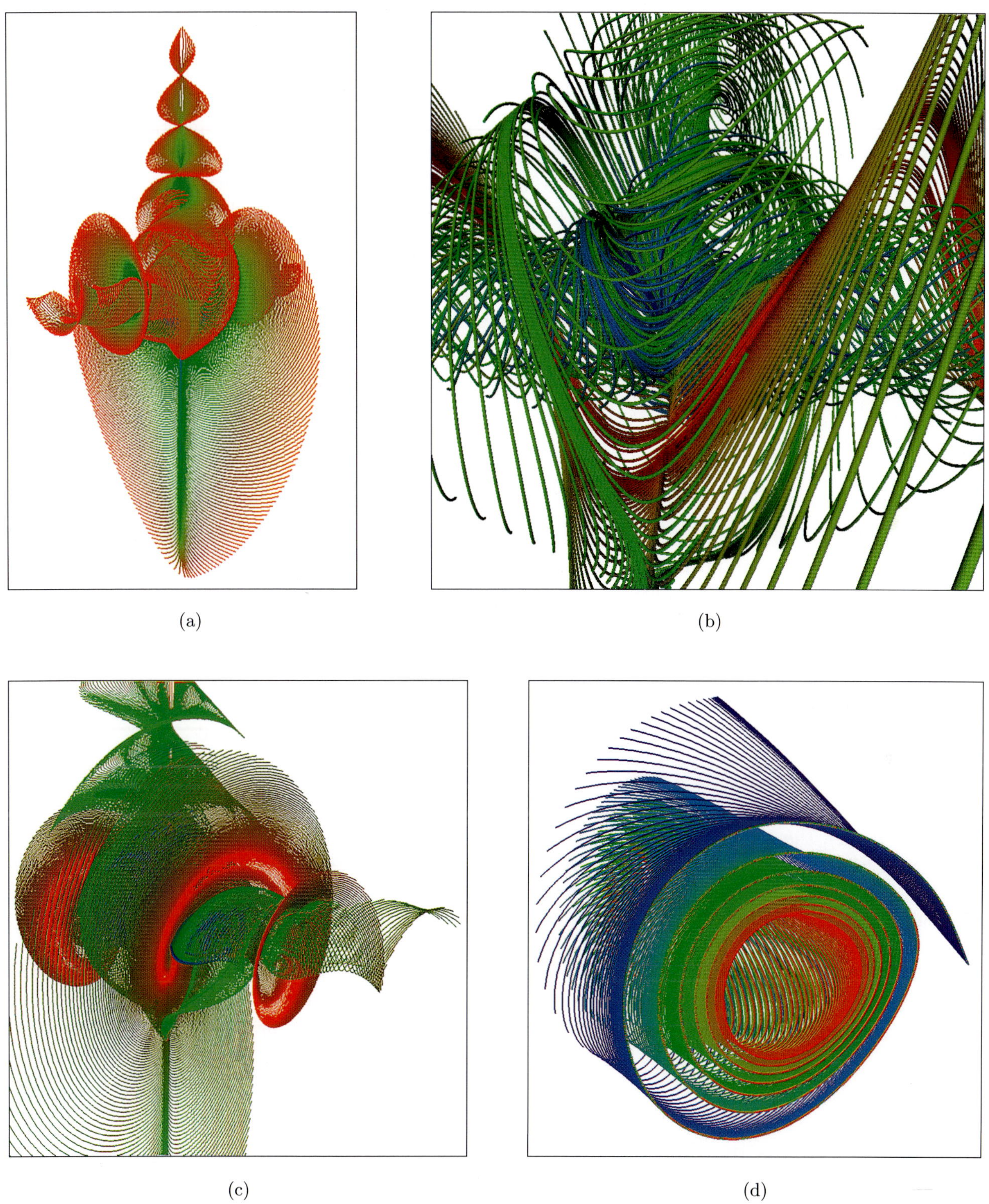

Fig. 7. The Lorenz manifold computed with the continuation method of Doedel. Panels (a)–(c) show the manifold where the arclength of the trajectories is fixed at $L = 250$. In panel (a) the coloring indicates the arclength along trajectories and in panels (b) and (c) the coloring is according to the total integration time T of trajectories; the strong stable manifold lies inside the red region. Panels (a) and (c) show all trajectories, while panel (b) shows only every fourth trajectory as a tube. Panel (d) demonstrates that only a part of interest of the stable manifold may be computed, such as a part of the main scroll; this was done by fixing $x = -25$ at the end point of trajectories.

Figure 7(d) illustrates the flexibility of the method by showing part of the Lorenz manifold computed by numerical continuation of solutions to the boundary value problem (22)–(23) and (25) for the choice $g(\mathbf{x}, \lambda, T) = x$. This results in the x-coordinate of the end point $\mathbf{x}(1)$ being kept fixed during the continuation, and it was set to $x = -25$ in the computation. For an appropriate choice of α, for which some trajectories intersect this plane several times, the continuation procedure then naturally leads to nonmonotonous variation of θ, thereby allowing the computation of a scroll-like structure on the stable manifold. In Fig. 7(d) the origin is the point on the right from which all trajectories emerge.

4. Computation of Fat Trajectories

The method of Henderson [2003] computes a compact piece of a k-dimensional invariant manifold by covering it with k-dimensional spherical balls in the tangent space, centered at a set of well-distributed points. This set is found by computing so-called fat trajectories, which are trajectories augmented with tangent and curvature information at each point. The centers of the balls are points on the fat trajectory, and the radius is determined by the curvature.

For the implemented case of computing a two-dimensional unstable manifold $W^u(\mathbf{x}_0)$ of a saddle point of (1), the method starts with a small circle $S_\delta \subset E^u(\mathbf{x}_0)$ and at every step circular disks are added along a fat trajectory with a fixed total arclength (from $\mathbf{x}_0$) of L. Initially all fat trajectories start on S_δ, but at later stages fat trajectories begin at points interpolated where two fat trajectories move too far from each other. The method stops when $W^u(\mathbf{x}_0)$ has been covered up to the prescribed arclength L.

4.1. *Fat trajectories on the global stable manifold*

The method requires a basis for the tangent space and the curvatures in that basis to construct the disks. As was mentioned in the introduction, invariant manifolds are not defined locally, so that there is no local way of determining the tangent space or curvature for a given point on the invariant manifold. This information is known at points on the initial curve $S_\delta \subset E^u(\mathbf{x}_0)$, for example, the tangent

to S_δ is known, and if the flow is transverse to the initial curve S_δ, f can be used as the second tangent. The circle S_δ (or possibly an ellipse) may be chosen to be transverse to the flow for sufficiently small δ. The curvature information can be obtained using the second derivative tensor.

The tangent and curvature can be "transported" over $W^u(\mathbf{x}_0)$ by deriving and solving evolution equations along a trajectory. To this end, one writes the parametrization (5) in the form

$$x(t, \sigma) = c(\sigma) + \int_0^t f(\mathbf{x}(s, \sigma))ds, \qquad (27)$$

where $c(\sigma)$ parametrizes S_δ with the one-dimensional parameter σ. (An example of such a parametrization is (24).) Then the tangent space at $\mathbf{x}(t, \sigma)$ is spanned by $\mathbf{x}_\sigma$ and $\mathbf{x}_t = f$, and the corresponding curvatures are given by the second derivatives $\mathbf{x}_{\sigma\sigma}$, $\mathbf{x}_{t\sigma} = f_\mathbf{x}\mathbf{x}_\sigma$ and $\mathbf{x}_{tt} = f_\mathbf{x}f$. Evolution equations for the unknown quantities can be found by differentiating (27)

$$\frac{\mathrm{d}}{\mathrm{d}t}\mathbf{x} = f, \qquad (28)$$

$$\frac{\mathrm{d}}{\mathrm{d}t}\mathbf{x}_\sigma = f_\mathbf{x}\mathbf{x}_\sigma, \qquad (29)$$

$$\frac{\mathrm{d}}{\mathrm{d}t}\mathbf{x}_{\sigma\sigma} = f_\mathbf{x}\mathbf{x}_{\sigma\sigma} + f_{\mathbf{x}\mathbf{x}}\mathbf{x}_\sigma\mathbf{x}_\sigma. \qquad (30)$$

Note that, even if $\mathbf{x}_\sigma$ is orthogonal to f at the initial point, there is no reason to expect the basis to remain orthogonal. In [Henderson, 2003], equations are derived for the evolution of a local parametrization which does remain orthonormal and has minimal change in the basis along the trajectory. (This is analogous to finding Riemannian normal coordinates in gravitation, where trajectories play the role of geodesics [Misner *et al.*, 1970].) If the tangents in the local parametrization are $\mathbf{u}_0$ and $\mathbf{u}_1$, they evolve according to

$$\frac{\mathrm{d}}{\mathrm{d}t}\mathbf{u}_0 = f_\mathbf{x}\mathbf{u}_0 - \mathbf{u}_0^T f_\mathbf{x}\mathbf{u}_0\,\mathbf{u}_0 - \mathbf{u}_1^T f_\mathbf{x}\mathbf{u}_0\,\mathbf{u}_1, \qquad (31)$$

$$\frac{\mathrm{d}}{\mathrm{d}t}\mathbf{u}_1 = f_\mathbf{x}\mathbf{u}_1 - \mathbf{u}_0^T f_\mathbf{x}\mathbf{u}_1\,\mathbf{u}_0 - \mathbf{u}_1^T f_\mathbf{x}\mathbf{u}_1\,\mathbf{u}_1. \qquad (32)$$

4.2. *Interpolation points on the invariant manifold*

The method starts with a set of well-distributed points on the initial curve S_δ, which can be found using the algorithm described in [Henderson, 2002]. At each such point on S_δ an orthonormal basis for the invariant manifold and second derivatives of the manifold in that basis are computed, and used as initial conditions for finding a set of disks along a fat trajectory. Because trajectories may move apart from each other, these disks will generally not cover $W^u(\mathbf{x}_0)$; see Fig. 8. This means that additional fat trajectories must be started at suitable points until $W^u(\mathbf{x}_0)$ is covered. In order to generate a well-spaced set of points on $W^u(\mathbf{x}_0)$, one chooses a starting point from the boundary of the computed part of the manifold.

The method in [Henderson, 2002] represents the boundary of the union of disks $\{D_i\}$ using polygons related to the Voronoi regions of the centers of the disks. A disk D_i consists of a center $\mathbf{x}(t_i, \sigma_i)$ (a point on a fat trajectory), the orthonormal basis for the tangent space of the manifold $\mathbf{u}_0(t_i, \sigma_i)$ and $\mathbf{u}_1(t_i, \sigma_i)$, a radius R_i, and polygon P_i. The polygon P_i is represented by a list of vertices in the tangent space and edges joining them (which actually works in arbitrary dimensions). The polygons are constructed in such a way that each edge of P_i that crosses the boundary of D_i corresponds to a neighboring disk D_j. The situation is sketched in Fig. 8.

Suppose that part of $W^u(\mathbf{x}_0)$ is represented in this way, and a new disk D_i is to be added. P_i is initially a square centered at the origin with sides $2R_i$, and for each disk D_j that intersects the new disk D_i complementary half spaces are subtracted from P_i and P_j. The projection of D_j into the tangent space at $\mathbf{x}_i$ is approximated by a disk of radius R_j centered at the projection of $\mathbf{x}_j$. If R_i and R_j are small enough so that the distance between the tangent space and the manifold is small (this depends on the curvature of $W^u(\mathbf{x}_0)$), then this is a good approximation. This pair of disks in the tangent space at $\mathbf{x}_i$ defines a line containing the intersection of the circles bounding the disks, and one subtracts from P_i a half space bounded by this line. The same approach is used to update P_j by projecting $\mathbf{x}_i$ into the tangent space at $\mathbf{x}_j$.

With these polygons a point on the boundary of the union can easily be found. Any point on $\delta D_i \cap P_i$ is near the boundary of the union (the distance to the boundary is controlled by the distance between the tangent space and the manifold at the radius). Points on the boundary where two disks meet correspond to points where an edge of P_i crosses δD_i (the point obtained is in the tangent space of the manifold and must be projected onto the manifold).

If one considers the part of the invariant manifold that is not yet covered (that is, the exterior of the union of neighborhoods, $t < T$), one can define something resembling a constrained minimization problem (it lacks a global objective function), which looks for a point in this region that lies furthest back in time under the flow. With a mild assumption about the shape of the region (it must be a topological ball), such a minimal point must exist. It must lie on the boundary of the region at the intersection of two disks. This point is a "minimum" if the flow vector extended backwards intersects the interior of the edge joining the centers of the intersecting disks. (This is, in fact, Guckenheimer and Vladimirksy's upwinding criterion; see Sec. 5.) One can easily find candidate points on the boundary from the edges of the polygons, and checking the upwinding criterion is a matter of computing a projection. One can then either interpolate tangents and curvatures from the disks' centers (the method used in the computations shown in Fig. 9) or use a homotopy (as Doedel uses in AUTO [Doedel, 1981; Doedel *et al.*, 1997; Doedel *et al.*, 2000]) to move from the fat trajectory from S_δ through the center of one of the disks to the fat trajectory which

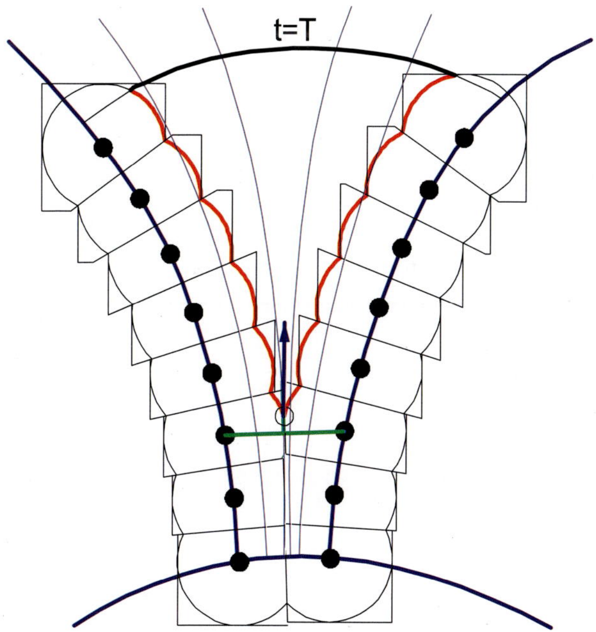

Fig. 8.   Two adjacent fat trajectories starting from S_δ. A new fat trajectory starts from the point where the two fat trajectories separate. This point can be found by interpolation between two suitable mesh points, indicated by the green lines.

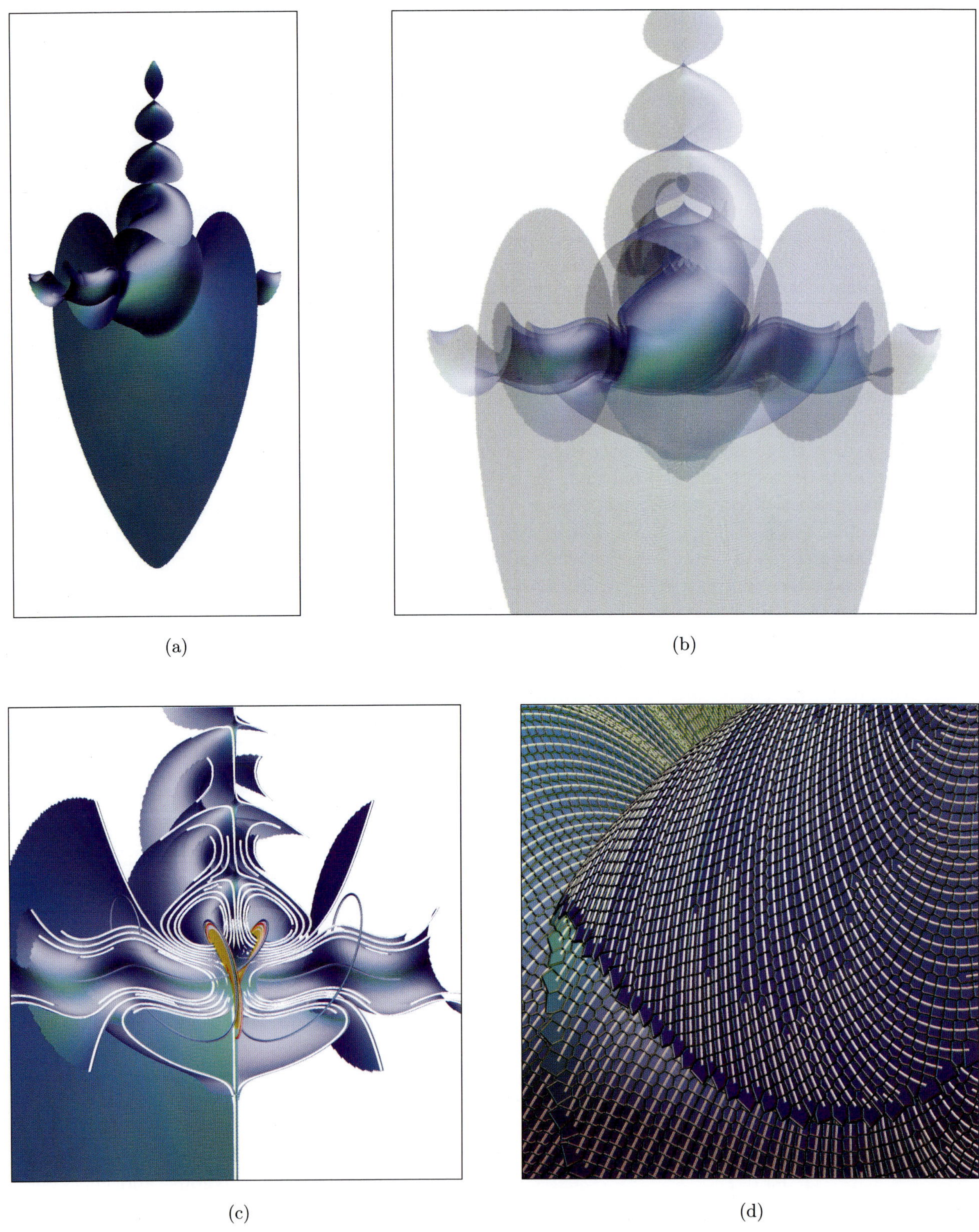

Fig. 9.   The Lorenz manifold computed with the method of Henderson up to a total trajectory arclength of 250. Panel (a) shows a view of the entire manifold, panel (b) a transparent enlargement near the main scroll, panel (c) shows the part of the manifold for $x < 0$ together with the Lorenz attractor and the one-dimensional stable manifolds of the two other equilibria, and panel (d) gives an impression of the computed mesh.

starts on S_δ and passes underneath the interpolation point.

This interpolation to find new starting points for fat trajectories completes the algorithm. It computes a covering of the manifold $W^u(\mathbf{x}_0)$ with disks centered at well-spaced points. Provided the disks are sufficiently small compared to the curvature, the algorithm is guaranteed to terminate, and all points lie on trajectories that originate on the initial curve S_δ or at points interpolated between nearby trajectories.

The fat trajectory, with its string of disks and polygons, is integrated until a prespecified total arclength L is reached. This is repeated for all the points on the initial curve. (The total integration time T of fat trajectories varies with the initial condition.)

4.3. *The Lorenz manifold covered by fat trajectories*

Figure 9 shows the Lorenz manifold $W^s(\mathbf{0})$ computed (using integration backward in time) up to a total trajectory arclength of 250. The step was controlled so that the distance between the tangent space and $W^u(\mathbf{x}_0)$ over each disk was less than 0.5. The scaled time step along trajectories was 0.01 (many more than one time step is taken between successive points on a fat trajectory), and no radius is greater than 2.0. The result was a total of 221,210 disks. Figure 9(a) shows the entire computed part of the Lorenz manifold from the common viewpoint. Figure 9(b) shows an enlargement of the Lorenz manifold near the central region where the manifold is now transparent. Notice the different "sheets" of manifold in the scroll and the extra helices forming around the z-axis. This complicated structure of the Lorenz manifold is further illustrated in Fig. 9(c) where only the half of $W^s(\mathbf{0})$ with negative x-coordinate is shown. The intersection curves of the manifold with the plane $\{x = 0\}$ are shown in white. Also shown is the one-dimensional unstable manifold $W^u(\mathbf{0})$ (red curve) accumulating on the Lorenz attractor (yellow) and the stable manifolds (blue curves) of the other two equilibria.

Figure 9(d) gives an impression of the computed mesh. The fat trajectories are the white curves and they are surrounded by the polygons that make up the Lorenz manifold. Clearly visible are points where new fat trajectories are started from

interpolated data. The boundary of the manifold at termination simply consists of the disks that are of distance L from $\mathbf{x}_0$ (measured along trajectories).

5. PDE Formulation

Another method for approximating invariant manifolds of hyperbolic equilibria was introduced by Guckenheimer and Vladimirsky [2004]. Their approach locally models a codimension-one invariant manifold as the graph of a function g satisfying a quasi-linear PDE that expresses the tangency of the vector field f of (1) to the graph of g. The PDE is then discretized in an Eulerian framework and the manifold is approximated by a triangulated mesh. We denote by $\mathcal{M}$ the triangulated approximation of the "known" part of the manifold. It can be extended by adding simplices at the current polygonal boundary $\partial\mathcal{M}$ in a locally-outward direction in the tangent plane. The discretized version of the PDE is then solved to obtain the correct slope for the newly added simplices. To avoid solving the discretized equations simultaneously, an Ordered Upwind Method (OUM) is used to decouple the system: the causality is ensured by ordering the addition/recomputation of new simplices based on the lengths Λ of the vector field's trajectories.

Two key ideas provide for the method's efficiency:

1. The use of Eulerian discretization ensures that geometric stiffness, a high nonuniformity of separation rates for nearby trajectories on different parts of the manifold, does not affect the quality of the produced approximation: new simplices constructed at the current boundary $\partial\mathcal{M}$ are as regular as is compatible with the previously constructed mesh.
2. Since OUM is noniterative, the PDE-solving step of the method is quite fast.

5.1. *Tangency condition*

The method is explained here for a two-dimensional manifold $W^u(\mathbf{x}_0)$ of a saddle point $\mathbf{x}_0$ in $\mathbb{R}^3$; see [Guckenheimer & Vladimirsky, 2004] for more details. Let $(\boldsymbol{u}, g(\boldsymbol{u})) = (u_1, u_2, g(u_1, u_2))$ be a local parametrization of the manifold of (1). Then the vector field f should be tangential to the graph of

$g(u_1, u_2)$, that is, the dot product

$$\left[\frac{\partial}{\partial u_1} g(u_1, u_2),\ \frac{\partial}{\partial u_2} g(u_1, u_2),\ -1\right]$$

$$\cdot f(u_1, u_2, g(u_1, u_2)) = 0. \tag{33}$$

The above first-order quasi-linear PDE can be solved to grow the manifold in steps, because the Dirichlet boundary condition is specified on the boundary $\partial \mathcal{M}$ of the piece of the manifold computed in previous steps. The initial boundary is chosen by discretizing a small circle or ellipse $S_\delta \subset E^u(\mathbf{x}_0)$ that is transverse to f, so that the vector field is outward-pointing everywhere.

Unlike a general quasi-linear PDE, Eq. (33) always has a smooth solution as long as the chosen parametrization remains valid. Thus, switching to local coordinates when solving the PDE avoids checking the continued validity of the parametrization.

In [Guckenheimer & Vladimirsky, 2004] the PDE formulation (33) is extended to approximate two-dimensional manifolds in $\mathbb{R}^n$. A similar characterization can be used for general k-dimensional invariant manifolds in $\mathbb{R}^n$, but the current numerical implementation relies on $k = 2$.

The PDE approach for characterizing invariant surfaces goes back to at least the 1960s. The existence and smoothness of solutions for equations equivalent to (33) were the subjects of Sacker's analytical perturbation theory [Sacker, 1965] and later served as a basis for several numerical methods, for example, those in [Dieci & Lorenz, 1995; Dieci *et al.*, 1991; Edoh *et al.*, 1995]. However, all this work was done for the computation of invariant tori. There are two very important distinctions between the PDE methods for tori and the method presented in this section:

1. These prior methods assume the existence of a coordinate system in which the invariant torus is indeed *globally* a graph of a function $g : \mathbb{T}^k \mapsto \mathbb{R}^{n-k}$. This implies the availability of a global mesh, on which the PDE can be solved. For invariant manifolds of hyperbolic equilibria such a mesh is not available *a priori* and has to be constructed in the process of growing the approximation $\mathcal{M}$.
2. For the invariant tori computations, the solution function g has periodic boundary conditions; hence, the discretized equations are inherently coupled and have to be solved simultaneously.

For the approximation of $W^u(\mathbf{x}_0)$ all characteristics of the PDE start at the initial boundary (chosen in $E^u(\mathbf{x}_0)$) and run "outward". Knowledge of the direction of information flow can be used to decouple the discretized system, resulting in a much faster computational method.

5.2. *Eulerian discretization*

To enable decoupling of the discretized system, our discretization of Eq. (33) at a "new" mesh point $\boldsymbol{y}$ has to be "upwinding", i.e. it should use only previously-computed mesh points straddling $\boldsymbol{y}$'s approximate trajectory. For a two-dimensional invariant manifold in $\mathbb{R}^3$, let $G(u_1, u_2)$ be a piecewise-linear numerical approximation of the local parameterization $g(u_1, u_2)$. Consider a simplex $\boldsymbol{y}\boldsymbol{y}^1\boldsymbol{y}^2$, where $\boldsymbol{y}^i = (u_1^i, u_2^i, G(u_1^i, u_2^i)) = (\boldsymbol{u}^i, G(\boldsymbol{u}^i))$ and $\boldsymbol{y} = (u_1, u_2, G(u_1, u_2)) = (\boldsymbol{u}, G(\boldsymbol{u}))$. Suppose that the vertices $\boldsymbol{y}^1$ and $\boldsymbol{y}^2$ are two adjacent mesh points on the discretization of the current manifold boundary, called *AcceptedFront* (thus, $G(\boldsymbol{u}^1)$ and $G(\boldsymbol{u}^2)$ are known and can be used in computing $G(\boldsymbol{u})$). If $\boldsymbol{u}$ is chosen so that the simplex $\boldsymbol{u}\boldsymbol{u}^1\boldsymbol{u}^2$ is well-conditioned, then $\boldsymbol{y} = (\boldsymbol{u}, G(\boldsymbol{u}))$ can be determined from the PDE. Define the unit vectors $\boldsymbol{P_i} = (\boldsymbol{u} - \boldsymbol{u}^i/\|\boldsymbol{u} - \boldsymbol{u}^i\|)$ and let P be the square invertible matrix with the $\boldsymbol{P_i}$'s as its rows. The directional derivative of G in the direction $\boldsymbol{P_i}$ can be computed as $v_i(\boldsymbol{u}) = (G(\boldsymbol{u}) - G(\boldsymbol{u}^i))/\|\boldsymbol{u} - \boldsymbol{u}^i\|$, for $i = 1, 2$. Therefore, $\nabla g(\boldsymbol{u}) \approx \nabla G(\boldsymbol{u}) = P^{-1}\boldsymbol{v}$, where $\boldsymbol{v} = \begin{bmatrix} v_1 \\ v_2 \end{bmatrix}$. This yields the discretized version of Eq. (33) as

$$\left[P^{-1}\boldsymbol{v}(\boldsymbol{u})\right]_1 f_1(\boldsymbol{u}, G(\boldsymbol{u}))$$
$$+ \left[P^{-1}\boldsymbol{v}(\boldsymbol{u})\right]_2 f_2(\boldsymbol{u}, G(\boldsymbol{u})) = f_3(\boldsymbol{u}, G(\boldsymbol{u})). \tag{34}$$

This nonlinear equation can be solved for $G(\boldsymbol{u})$ by the Newton–Raphson method or any other robust zero-solver. In addition, it has an especially simple geometric interpretation if the local coordinates are chosen so that $G(\boldsymbol{u}^1) = G(\boldsymbol{u}^2) = 0$. Namely, we reduce the problem to finding the correct "tilt" of the simplex $\boldsymbol{y}\boldsymbol{y}^1\boldsymbol{y}^2$ with respect to the simplex $\hat{\boldsymbol{y}}\boldsymbol{y}^1\boldsymbol{y}^2$ where $\hat{\boldsymbol{y}} = (\boldsymbol{u}, 0)$ can be interpreted as a preliminary position (predictor) of $\boldsymbol{y}$. (As discussed in Sec. 5.3 below, when $\hat{\boldsymbol{y}}$ is first added to the mesh, $\boldsymbol{u}$ is chosen so that $\hat{\boldsymbol{y}}\boldsymbol{y}^1\boldsymbol{y}^2$ is a well-conditioned simplex in a tangent plane.) Hence,

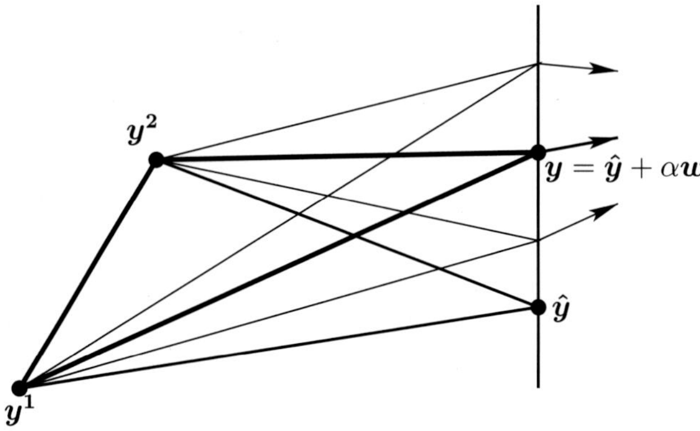

Fig. 10. Geometric interpretation of Eq. (34). The search space for $\boldsymbol{y}$ is the normal subspace, here corresponding to the line spanned by $\boldsymbol{w}$. The segment $\boldsymbol{y}^1\boldsymbol{y}^2$ is a part of the *AcceptedFront*, and $\hat{\boldsymbol{y}}$ is a *Considered* point.

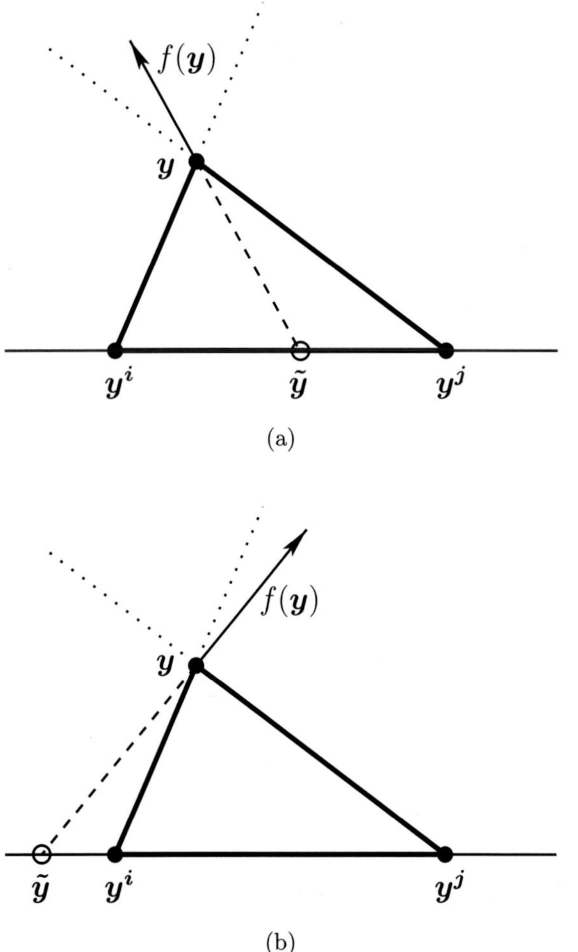

(a)

(b)

Fig. 11. (a) An acceptable and (b) an unacceptable approximation of $f(\boldsymbol{y})$; the range of upwinding directions is shown by dotted lines; the local linear approximation to the trajectory is shown by a dashed line; $\tilde{\boldsymbol{y}}$ is its intersection with the line $\boldsymbol{y}^i\boldsymbol{y}^j$. In the second case the upwinding criterion is not satisfied and the update for $\boldsymbol{y}$ should be computed using another segment of *AcceptedFront*.

solving Eq. (34) is equivalent to finding $\alpha \in \mathbb{R}$ such that $f(\hat{\boldsymbol{y}} + \alpha\boldsymbol{w})$ lies in the plane defined by $\boldsymbol{y}^1$, $\boldsymbol{y}^2$, and $\boldsymbol{y} = \hat{\boldsymbol{y}} + \alpha\boldsymbol{w}$, where $\boldsymbol{w}$ is the unit vector normal to $\hat{\boldsymbol{y}}\boldsymbol{y}^1\boldsymbol{y}^2$; see Fig. 10. A similar discretization and geometric interpretation can be derived for the general case of $k \geq 2$ and $n \geq 3$ [Guckenheimer & Vladimirsky, 2004].

The described discretization procedure is similar in spirit to an *implicit Euler's method* for solving initial value problems since $\boldsymbol{y}^1$ and $\boldsymbol{y}^2$ are assumed to be known and the vector field is computed at the to-be-determined point $\boldsymbol{y}$. In solving first-order PDEs, a fundamental condition for the numerical stability requires that the mathematical domain of dependence should be included in the numerical domain of dependence. Since the characteristics of PDE (33) coincide with the trajectories of the vector field, $G(\boldsymbol{u})$ should be computed using the triangle through which the corresponding (approximate) trajectory runs. Thus, having computed $\boldsymbol{y} = (\boldsymbol{u}, G(\boldsymbol{u}))$ by (34) using two adjacent mesh points $\boldsymbol{y}^i$ and $\boldsymbol{y}^j$, we need to verify an additional *upwinding condition*: the linear approximation to the trajectory of $\boldsymbol{y}$ should intersect the line $\boldsymbol{y}^i\boldsymbol{y}^j$ at a point $\tilde{\boldsymbol{y}} = (\tilde{\boldsymbol{u}}, G(\tilde{\boldsymbol{u}}))$ that lies between $\boldsymbol{y}^i$ and $\boldsymbol{y}^j$; see Fig. 11. An equivalent formulation is that $f(\boldsymbol{y})$ should point *from* the newly computed simplex $\boldsymbol{y}\boldsymbol{y}^i\boldsymbol{y}^j$.

Algebraically, if $\boldsymbol{y}$ solves (34), then $f(\boldsymbol{y}) = \beta_1(\boldsymbol{y} - \boldsymbol{y}^i) + \beta_2(\boldsymbol{y} - \boldsymbol{y}^j)$; thus, the upwinding criterion above simply requires $\beta_1, \beta_2 \geq 0$. In this case the discretization is locally second-order accurate and the arclength $\Lambda(\boldsymbol{y})$ of the trajectory up to the point $\boldsymbol{y}$ can be approximated as

$$\Lambda(\boldsymbol{y}) \approx \|\boldsymbol{y} - \tilde{\boldsymbol{y}}\| + \Lambda(\tilde{\boldsymbol{y}})$$

$$\approx \frac{\|f(\boldsymbol{y})\|}{\beta_1 + \beta_2} + \beta_1\Lambda(\boldsymbol{y}^i) + \beta_2\Lambda(\boldsymbol{y}^j)$$

$$\approx d_a(\boldsymbol{0}, \boldsymbol{y}). \tag{35}$$

Numerical evidence indicates that the resulting method is globally first-order accurate [Guckenheimer & Vladimirsky, 2004].

5.3. *Ordered Upwind Method*

Ordered Upwind Methods (OUMs) were originally introduced for static Hamilton–Jacobi–Bellman PDEs [Sethian & Vladimirsky, 2003]. In [Guckenheimer & Vladimirsky, 2004] the same idea of *space-marching* for boundary value problems is used to solve Eq. (33). All mesh points are divided into those that are *Accepted*, that is, already fixed as belonging to the approximation $\mathcal{M}$, and those *Considered*, which are in a tentative position adjacent to the current polygonal manifold boundary $\partial\mathcal{M}$, called the *AcceptedFront*. A tentative position can be computed for each *Considered* mesh point $\boldsymbol{y}$ under the assumption that its trajectory intersects $\partial\mathcal{M}$ in some neighborhood $\mathcal{N}(\boldsymbol{y})$ of that point. In other words, $\boldsymbol{y}$ is updated by solving Eq. (34) for a "virtual simplex" $\boldsymbol{y}\boldsymbol{y^i}\boldsymbol{y^j}$ such that $\boldsymbol{y^i}\boldsymbol{y^j} \in \partial\mathcal{M} \cap \mathcal{N}(\boldsymbol{y})$ and the upwinding criterion is satisfied. All *Considered* points are sorted based on the approximate trajectory arclengths $\Lambda(\boldsymbol{y})$ defined by (35). The method starts with $\partial\mathcal{M}$ discretizing a small ellipse in $E^u(\boldsymbol{x_0})$. That initial boundary is surrounded by a single "layer" of *Considered* mesh points (also in $E^u(\boldsymbol{x_0})$).

A typical step of the algorithm consists of picking the *Considered* point $\overline{\boldsymbol{y}}$ with the smallest Λ and making it *Accepted*. This operation modifies $\partial\mathcal{M}$ ($\overline{\boldsymbol{y}}$ is included, and the mesh points that are no longer on the boundary are removed) and causes a possible recomputation of all the not-yet-*Accepted* mesh points near $\overline{\boldsymbol{y}}$. If $\boldsymbol{y^i}$ is adjacent to $\overline{\boldsymbol{y}}$ and $\boldsymbol{y^i}\overline{\boldsymbol{y}}$ is on the boundary, then the mesh is locally extended by adding a new *Considered* mesh point $\boldsymbol{y}$ connected to $\boldsymbol{y^i}\overline{\boldsymbol{y}}$ in a tangent plane. To maintain good aspect ratios of newly-created simplices, the current implementation relies on an "advancing front mesh generation" method similar to [Peraire *et al.*, 1999]. Other local mesh-extension strategies can be implemented similarly to methods in [Rebay, 1993] or [Henderson, 2002].

The vector field near $\partial\mathcal{M}$ determines the order in which the correct "tilts" for tentative simplex-patches are computed and the *Considered* mesh points are *Accepted*. This ordering has the effect of reducing the approximation error (since a mesh point $\boldsymbol{y}$ first computed from a relatively far part of $\mathcal{N}(\boldsymbol{y})$ is likely to be recomputed before it gets *Accepted*). The default stopping criterion is to enforce $\Lambda(\overline{\boldsymbol{y}}) \leq L$, so that the algorithm terminates when the maximal approximate arclength L is reached. Other stopping criteria (for example, based on Euclidean or geodesic distance or the maximum number of simplices) can be used as well. Current algorithmic parameters include L, the radius R_N of the neighborhood $\mathcal{N}(\boldsymbol{y})$, and the desired simplex size Δ. (The simplex size is fixed in the present implementation; it could be adapted according to curvature information.) As in the original OUMs, the computational complexity of the algorithm is $O(M \log M)$, where $M = O(L^2/\Delta^2)$ is the total number of mesh points and the $(\log M)$ factor results from the necessity to maintain a sorted list of *Considered* mesh points. A detailed discussion of the algorithmic issues can be found in [Guckenheimer & Vladimirsky, 2004].

5.4.   *The Lorenz manifold computed with the PDE formulation*

Figure 12 shows the Lorenz manifold $W^s(\boldsymbol{0})$ computed up to an approximate total arclength of $L = 174$. The computation started from $S_\delta \subset E^s(\boldsymbol{0})$ with $\delta = 2.0$, $\Delta = 0.6$ and $R_N = 4\Delta$, which resulted in the total of $271\,469$ mesh points. The coloring shows arclength along trajectories where blue is small and red is large. The manifold was rendered as a two-dimensional surface with MATLAB; other illustrations and associated animations can be found in [Guckenheimer & Vladimirsky, 2004].

Figure 12(a) shows the entire computed part of the Lorenz manifold from the common viewpoint. Figure 12(b) is an enlargement near the central scrolls where the manifold is now shown transparent. Clearly visible are two secondary spirals forming near the positive z-axis. The coloring is such that points of the same color are equally far away from the origin in arclength along trajectories. Figure 12(c) is a further enlargement near the unstable manifold $W^u(\boldsymbol{0})$ accumulating on the Lorenz attractor. This clearly shows how the Lorenz manifold "rolls" into both wings of the Lorenz attractor, creating different sheets that do not actually intersect the shown trajectories representing the unstable manifold $W^u(\boldsymbol{0})$.

Figure 12(d) gives an enlarged impression of the computed mesh looking into one of the outer scrolls. The simplices of the mesh are sufficiently uniform in spite of the complicated geometry of the manifold they represent. The red boundary of the computed manifold is not a smooth curve, because it is formed simply by the last simplices that were added locally.

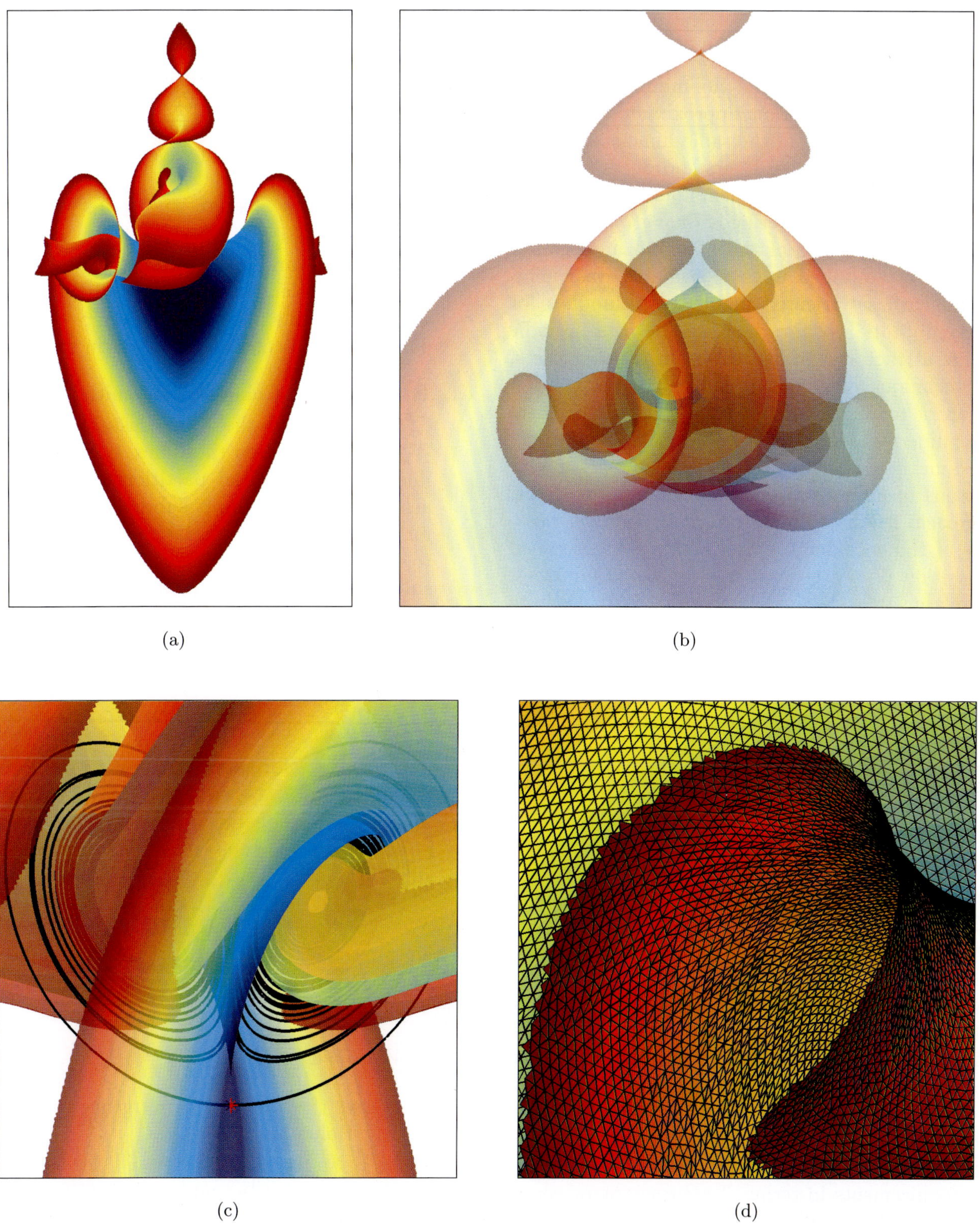

Fig. 12. The Lorenz manifold computed with the method of Guckenheimer and Vladimirsky up to a total trajectory arclength of about 174. Panel (a) shows a view of the entire manifold, panel (b) an enlargement near the main scroll where the manifold is shown transparent, panel (c) shows how the manifold interacts with the Lorenz attractor, and panel (d) gives an impression of the computed mesh.

6. Box Covering

In contrast to the techniques described so far, the method of Dellnitz and Hohmann [1996, 1997] presented in this section approximates invariant manifolds by objects of the same dimension as the underlying phase space. It first produces an *outer covering* of a local unstable manifold by a finite collection of sets. This covering is then grown in order to cover larger parts of the manifold analogously to what is described in Secs. 2 and 5. In combination with set-oriented multilevel techniques for the computation of invariant sets, such as periodic orbits, attractors and general chain recurrent sets, the technique allows, in principle, for the computation of manifolds of arbitrary dimension, where the numerical effort is essentially determined by the dimension of the manifold. In combination with rigorous techniques for the implementation of this approach, it is possible to compute *rigorous coverings* of the considered object. For a more detailed exposition of the general method see [Dellnitz & Hohmann, 1996, 1997; Dellnitz *et al.*, 2001; Dellnitz & Junge, 2002]. The algorithm is implemented in the software package GAIO [Dellnitz *et al.*, 2001].

6.1.   *The box covering algorithm*

The box covering algorithm applies to a discrete-time dynamical system, that is, to a diffeomorphism D. In the context of approximating global manifolds, it can compute the unstable manifold of an (unstable) invariant set of D in a compact region of interest Q. In this section, we explain how this method can be used for the computation of a two-dimensional (un)stable manifold of a saddle $\mathbf{x}_0$ in $\mathbb{R}^3$.

Here, the diffeomorphism $D : \mathbb{R}^3 \to \mathbb{R}^3$ is given by the time-τ map of the vector field (1). For an unstable manifold $\tau > 0$, while for a stable manifold $\tau < 0$ to account for reversing time. Numerically, the map D may be realized by classical one-step integration schemes. Since the algorithm involves integration over short time intervals only, typically the requirements in terms of accuracy or preservation of structures of the underlying vector field f are rather mild. The diffeomorphism D then has a hyperbolic saddle fixed point $\bar{\mathbf{x}} = \mathbf{x}_0$ and, in the case $\tau < 0$, $\bar{\mathbf{x}}$ has a two-dimensional unstable manifold $W^u(\bar{\mathbf{x}})$, which is identical to the stable manifold of $\mathbf{x}_0$.

The idea of the algorithm is as follows. Imagine a finite partition $\mathcal{P}$ of Q. The method first finds a (small) collection $\mathcal{C}_0 \subset \mathcal{P}$ that covers the local unstable manifold $W^u_{\mathrm{loc}}(\bar{\mathbf{x}})$. This local covering of $W^u(\bar{\mathbf{x}})$ is extended in steps, where in each step the sets in the current collection $\mathcal{C}_k$ are mapped forward under D. All sets in $\mathcal{P}$ that have an intersection with the images of $\mathcal{C}_k$ are added to the current collection of sets, yielding $\mathcal{C}_{k+1}$.

More formally, let $\mathcal{P}_0, \mathcal{P}_1, \ldots$ be a nested sequence of successively finer partitions of Q: We take $\mathcal{P}_0 = \{Q\}$ and each element $P \in \mathcal{P}_{\ell+1}$ is contained in an element $P' \in \mathcal{P}_\ell$ and $\mathrm{diam}(P) \le \gamma\,\mathrm{diam}(P')$ for some fixed number $0 < \gamma < 1$.

The algorithm consists of two main steps:

1. *Initialization*: Compute an initial covering $\mathcal{C}_0^{(k)} \subset \mathcal{P}_{\ell+k}$ of the local unstable manifold $W^u_{\mathrm{loc}}(\bar{x})$ of $\bar{x}$. (Here the index k indicates the fineness of the initial partition.) This can be achieved by applying a subdivision algorithm for the computation of relative global attractors to the element $P \in \mathcal{P}_\ell$ containing $\bar{\mathbf{x}}$ for some suitable ℓ; see [Dellnitz & Hohmann, 1996].

2. *Growth*: From the collection $\mathcal{C}_j^{(k)}$ the next collection $\mathcal{C}_{j+1}^{(k)}$ is obtained by setting

$$\mathcal{C}_{j+1}^{(k)} = \Big\{ P \in \mathcal{P}_{\ell+k} : D(\tilde{P}) \cap P \ne \emptyset$$
$$\text{for some set } \tilde{P} \in \mathcal{C}_j^{(k)} \Big\}.$$

This step is repeated until no more sets are added to the current collection, that is, until $\mathcal{C}_j^{(k)} = \mathcal{C}_{j+1}^{(k)}$.

We can show that this method converges to a certain subset of $W^u(\bar{\mathbf{x}})$ in Q. Namely, let $W_0 = W^u_{\mathrm{loc}}(\bar{\mathbf{x}}) \cap P$, where P is the element in $\mathcal{P}_\ell$ containing $\bar{\mathbf{x}}$ and define

$$W_{j+1} = D(W_j) \cap Q, \quad j = 0, 1, 2, \ldots.$$

Then we have the following convergence result (see [Dellnitz & Hohmann, 1996]):

1. the sets $C_j^{(k)} = \bigcup_{P \in \mathcal{C}_j^{(k)}} P$ are coverings of W_j for all $j, k = 0, 1, \ldots$;

2. for fixed j and $k \to \infty$, the covering $C_j^{(k)}$ converges to W_j in Hausdorff distance.

In general, one cannot guarantee that the algorithm leads to an approximation of the entire set $W^u(\bar{\mathbf{x}}) \cap Q$. This is due to the fact that parts of $W^u(\bar{\mathbf{x}})$ that do not lie in Q may map into Q.

In this case, the method will indeed not cover all of $W^u(\overline{\mathbf{x}}) \cap Q$.

Under certain hyperbolicity assumptions on $W^u(\overline{\mathbf{x}})$ it is possible to obtain statements about the speed of convergence in terms of how the Hausdorff distance between the covering and the approximated subset of $W^u(\overline{\mathbf{x}})$ depends on the diameter of the sets in the covering collection; see [Dellnitz & Junge, 2002] for details.

6.2. *Realization of the method*

The efficiency of the growth part of the algorithm significantly depends on the realization of the collections $\mathcal{P}_\ell$. In the implementation the $\mathcal{P}_\ell$ are partitions of Q into *boxes*

$$B(c,r) = \{y \in \mathbb{R}^n : |y_i - c_i| \le r_i \text{ for } i = 1, \ldots, n\},$$

where $c, r \in \mathbb{R}^n$, $r_i > 0$, are the center and the sizes of the box $B(c,r)$, respectively. Moreover, only partitions are used that result from bisecting the initial box Q repeatedly, where in this process of bisecting the relevant coordinate direction is changed systematically (typically, the bisected coordinate direction is varied cyclically).

Starting with $\mathcal{P}_0 = \{Q\}$, this process yields a sequence $\mathcal{P}_\ell$ of partitions of Q, that can efficiently be stored in a binary tree. Note that it is easy to store arbitrary subsets of the full partition $\mathcal{P}_\ell$ just by storing the corresponding part of the tree. In fact, in the initialization of the algorithm one starts with a single box on a given level ℓ, so that the stored tree consists of a single leaf. Whenever sets are added to the current collection, the corresponding paths are added to the tree. Figure 13 illustrates the first three growth steps for the computation of

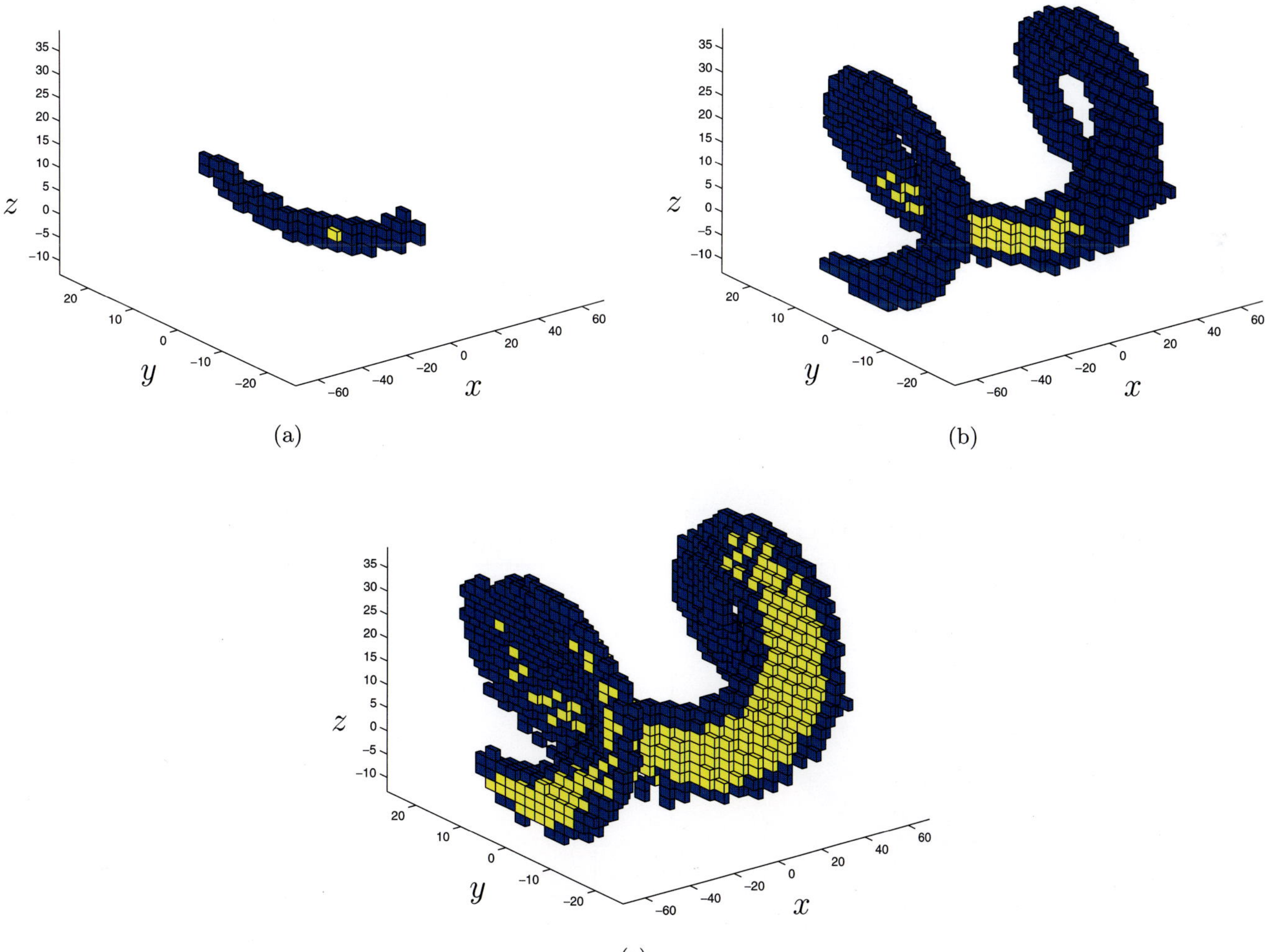

(a)

(b)

(c)

Fig. 13. Coverings of the Lorenz manifold during the first three growth steps are shown in panels (a)–(c), where the covering of the previous step (the initialization box in the case of (a)) is shown in yellow.

a covering of the Lorenz manifold on level 18 of the tree (all other parameters are as described in Sec. 6.3 below). The yellow box in Fig. 13(a) was created in the initialization step and then grown in one step to obtain the blue boxes. Panels (b) and (c) show two further growth steps, where the covering of the previous step is again shown in yellow.

The hierarchical storage scheme has another crucial computational advantage in that it is easier to decide which boxes are "hit" by mapping the boxes that were added in the previous step of the continuation algorithm. Namely, for each of these boxes $B \in \mathcal{P}_{\ell+k}$ one needs to compute the set $\mathcal{F}(B) = \{B' \in \mathcal{P}_{\ell+k} | D(B) \cap B' \neq \emptyset\}$. Since B contains an uncountable number of points, this problem must be discretized. The obvious approach is to choose a finite set T of *test points* in B and to approximate $\mathcal{F}(B)$ by $\tilde{\mathcal{F}}(B) = \{B' \in \mathcal{P}_{\ell+k} | D(\mathsf{T}) \cap B' \neq \emptyset\}$. Using the tree structure, the determination of the box that contains the image of a test point can be accomplished with a complexity that only depends logarithmically on the number of boxes in $\mathcal{P}_\ell$ [Dellnitz & Hohmann, 1997].

6.3. *Box covering of the Lorenz manifold*

Figure 14 shows a box covering of the Lorenz manifold $W^s(\mathbf{0})$. For the computation the time-τ map of the Lorenz system (2) was considered with $\tau = -0.1$. This map is realized by the classical Runge–Kutta scheme of fourth order with a fixed step size of -0.01. The region of interest Q is a box with radius $(70, 70, 70)$ and center $(10^{-1}, 10^{-1}, 10^{-1})$; this offset centering is for a practical reason: it avoids having the origin on the edge of a box. Level $\ell = 27$ of the tree was used and 16 growth steps were performed, starting from a single box containing the origin (i.e. $k = 0$). In each growth step, an equidistant grid of 125 test points in each box was mapped forward. The resulting object contains more than 4 million boxes.

Figure 14(a) shows the entire computed part of the Lorenz manifold from the common viewpoint. The same view is shown in Fig. 14(b) but now the manifold is transparent. Figure 14(c) shows an enlargement of the transparent rendering near the central region. Because the method is using the time-τ map of the Lorenz system (2), the Lorenz

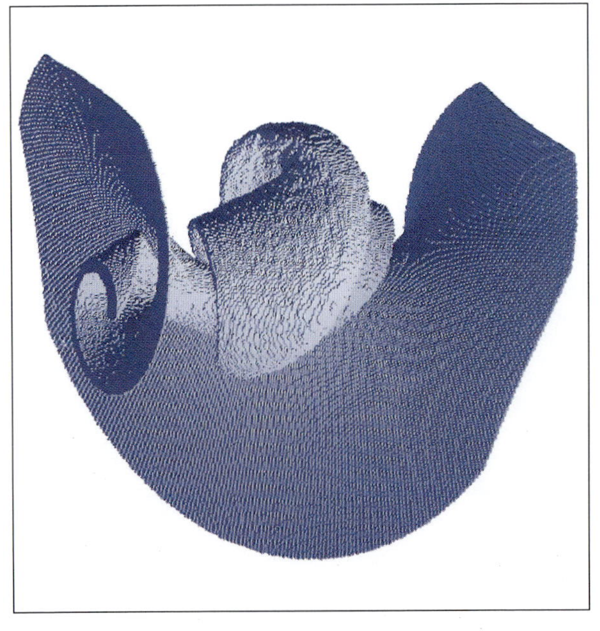

(a)

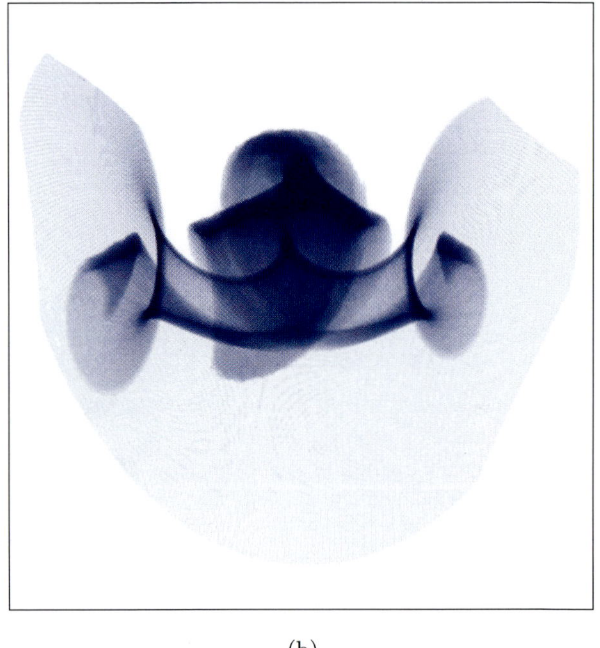

(b)

Fig. 14.   The Lorenz manifold computed with the box covering method of Dellnitz and Hohmann seen from the common viewpoint (a). In panels (b) and (c) the manifold is rendered transparently. Panel (c) shows an enlargement near the z-axis, and panel (d) gives a closer look at the computed boxes.

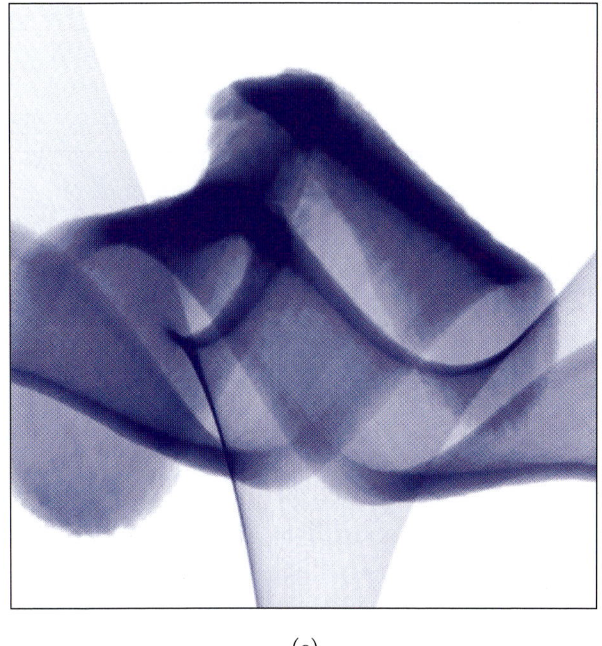

(c)

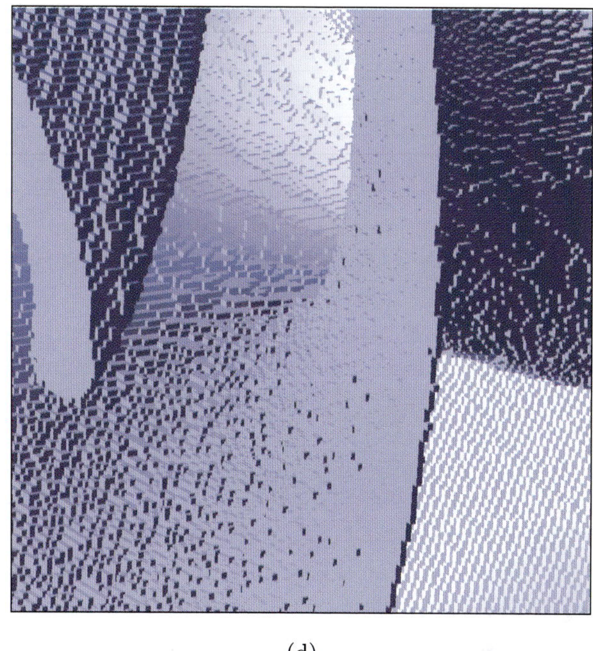

(d)

Fig. 14. (*Continued*)

manifold first grows initially mainly in the direction of the strong unstable direction until the boundary of the box of interest is reached. This can be seen nicely in Fig. 13. Later steps of the growth process then start to build up the other part of the manifold, resulting in the images in Figs. 14(a) and 14(b). The further enlargement near the scroll of the manifold in Fig. 14(d) gives a local impression of the box covering. Notice that the covering of the manifold has a thickness of several box diameters at the end of the scroll.

7. Discussion

After a recent flurry of research activity, several complementary methods are available today to compute global (un)stable manifolds in applications. While these methods are still somewhat under development and testing, we hope that this survey will encourage the reader to consider computing such global objects in systems arising in applications.

Each of the methods presented in the previous sections is based on a particular point of view of characterizing a global (un)stable manifold. Common to all is the idea that the manifold must be grown from local information near the saddle point, and the difference is in how this is done. The choice

of method will generally depend on the application one has in mind and on the particular questions one wants to answer. This discussion is intended to give an indication of the specific properties of the different approaches.

7.1. *Approximation by geodesic level sets*

The method of Krauskopf and Osinga [1999, 2003] is presently implemented for two-dimensional manifolds of saddle points and saddle periodic orbits in a phase space of arbitrary dimension; see also [Osinga, 2000, 2003] This implementation approximates the manifold linearly between mesh points, while the boundary value problems (9)–(10) are solved by single shooting. It would be possible to use higher order interpolation between mesh points and collocation for solving the boundary value problems. The method produces a very regular mesh that consists of (approximate) geodesic circles and approximate geodesics. This means that the manifold is rendered as a geometric object, independently of the dynamics on it. The mesh is, in fact, constructed so regularly that it can be interpreted as a crochet pattern. This allows one to produce a real-life model of the Lorenz manifold; see [Osinga & Krauskopf, 2004] for details. During a computation, the interpolation error is controlled by prescribed mesh quality parameters, so that the correctness of the method can be proved; see [Krauskopf & Osinga, 2003] for details.

The price one has to pay for obtaining a guaranteed "geometric mesh" is that one needs to set up and continue a boundary value problem for each new mesh point. This makes the method more expensive compared to other methods. With the nonoptimized present implementation and the accuracy parameters as in Sec. 2.3, computing the Lorenz manifold up to geodesic distance 140 takes about 10 minutes, while the larger image in Fig. 5 with 69900 mesh points took 40 minutes and 47 seconds on an 800 MHz Pentium III machine.

Because it is based on the geodesic parametrization (6), the method works as long as the geodesic level sets of this parametrization remain smooth circles. While this is not an obstruction for computing the Lorenz manifold, there are examples where the computation stops when a geodesic circle ceases to be smooth; see [Krauskopf & Osinga, 2003]. Furthermore, the method stops when it encounters an equilibrium or a periodic orbit on (the closure of) the (un)stable manifold.

An implementation for global (un)stable manifolds of dimension three would already be quite challenging. First of all, geodesic level sets are spheres in this case, on which one needs to compute a regular mesh. Secondly, the method would require multiparameter continuation to continue the boundary value problems (9)–(10).

7.2. *BVP continuation of trajectories*

The method by Doedel is arguably the most straightforward one. The continuation calculations can be carried out using the standard boundary value continuation capabilities of AUTO. This means that all that is required are rather standard AUTO equations and parameter files. The orbits that make up the manifold are computed very accurately, due to the high accuracy of the orthogonal collocation method, which is superconvergent for the solution at the mesh points and for scalar variables. Furthermore, the boundary value continuation algorithms in AUTO, written in the f77 or C programming language, are rather efficient, so that the calculations can generally be done in relatively little computer time. For example, computing the Lorenz manifold up to a trajectory arclength of 250 with a high resolution of NTST = 75, as in

Figs. 7(a)–7(c), takes 30 seconds on a 1.6 MHz Pentium M laptop; for NTST = 25, which still gives good resulution, the computation time (including writing the output) drops to just over 10 seconds.

The method is very flexible in that it allows for different boundary conditions at the endpoint of a trajectory. This means that one can compute only a part of interest of the manifold, as was illustrated in Fig. 7(d). However, the manifold cannot be grown, so that the continuation must be repeated if a larger part of the manifold is desired.

While visualizing or even animating the computed trajectories gives much insight into the geometry of the manifold, it would require substantial post-processing to produce a nice mesh representation of the manifold as a two-dimensional object. In particular, the density of the orbits may be high in areas where the further evolution of the trajectories depends sensitively on the current state. For example, in Figs. 7(a) and 7(c) the density of the orbits is high along a curve in the direction of the z-axis, that is, the direction of the weakly stable eigenvector.

7.3. *Computation of fat trajectories*

While also essentially computing trajectories, the method of [Henderson, 2003] does produce a nice mesh representation by "fattening" the trajectories with a string of polygonal patches. The method tends to minimize the need for interpolation. When interpolation is needed there is a guarantee that appropriate points exist, and at those points information is available which allows higher order interpolation or the generation of an interpolating trajectory. The algorithms for computing fat trajectories, for finding a third-order approximation to the manifold, and for finding interpolation points are implemented for any dimension k of the manifold. The interpolation itself is presently limited to $k = 2$. The code used to compute the Lorenz manifold is available as OpenSource; see [Henderson, 2003].

The computation of a fat trajectory is more expensive than straighforward integration, because it adds equations for the evolution of the tangent space and curvatures. However, the implementation of updating the computed boundary is quite efficient; see also [Henderson, 2002]. The overall algorithm is relatively fast. For example, the Lorenz manifold in Fig. 9 was computed on a 375 MHz Power3 processor in about 7.3 hours.

Finally, the algorithm may encounter a geometric problem. It must be able to distinguish between mesh points on different sheets of the invariant manifold, for example, where a trajectory returns close to itself. This can be done by checking the values of t and σ at the centers of the disks, but it demands sufficiently small disks so that those quantities vary only a little across each disk. This requirement may result in many more mesh points being computed than is necessary to obtain a geometrically smooth manifold. This geometric problem occurs when trajectories spiral tightly, as is the case, for example, on the unstable manifolds of the two equilibria on the wings of the Lorenz attractor.

7.4. *PDE formulation*

The PDE approach by [Guckenheimer & Vladimirsky, 2004] leads to a very efficient numerical method for computing a mesh representation of a global (un)stable manifold. The computational cost of this method is largely independent of the geometric stiffness present in the system. For example, the Lorenz manifold in Fig. 12 was computed in under 90 seconds on a Pentium III 850 MHz processor.

The constructed approximation $\mathcal{M}$ is "causal", that is, it contains approximate trajectories for all the mesh points on $\partial\mathcal{M}$. The method is not restricted to manifolds where the level curves of the geodesic distance remain smooth. In particular, the method can be used for approximating manifolds containing homoclinic and heteroclinic trajectories; see [Guckenheimer & Vladimirsky, 2004] for examples.

The computational cost of adding each mesh point is proportional to the codimension $(n - k)$ of the manifold. When approximating manifolds of high codimension, this is clearly a disadvantage compared to other methods for which this cost is proportional to the dimension k of the manifold. A second limitation of the method is that the constructed approximation is globally only first-order accurate, in contrast with, for example, the second-order accuracy of computing fat trajectories.

A variant of the code exists that uses a global coordinate system defined by a triangulated mesh. This means that the PDE method could be used in a continuation framework, where an approximation of the manifold for one parameter value is used to build a global parametrization for nearby parameter values. This would reduce the cost of locally extending the mesh near $\partial\mathcal{M}$ at every step of the continuation.

The current implementation of the PDE approach works for two-dimensional manifolds in a phase space of arbitrary dimension. An adaptive implementation for $k \geq 3$ will have to employ a robust algorithm for a higher-dimensional local mesh extension, which remains a challenge.

7.5. *Box covering*

The box covering algorithm of [Dellnitz & Hohmann, 1996, 1997; Dellnitz *et al.*, 2001; Dellnitz & Junge, 2002] constructs a covering of (part of) the global invariant manifold. This covering consists of a collection of small boxes. The method is formulated for discrete-time systems, and differential equations can be handled by considering a corresponding time-τ map. It allows for the computation of (un)stable manifolds of arbitrary invariant sets. It is possible (and implemented in GAIO) to compute manifolds of arbitrary dimension. The "thickness" of the covering depends on the contraction rate transverse to the manifold. The stronger the contraction, the fewer "box-layers" along the manifold will be produced. In particular, the algorithm needs to be modified in order to apply it to Hamiltonian systems [Junge, 2000b].

The key implementational issue, namely how to compute the image of a given box, is typically discretized by mapping a (finite) set of test points in each box. Evidently, depending on the properties of the underlying map, the choice of these points determines the quality of the resulting covering. Using too few points may lead to missing boxes, while using too many slows down the computation. There exist strategies for a near-optimal choice of these points. In the case that Lipschitz estimates of the dynamical system are available, one may compute rigorous coverings. In this case, it can be ensured that the manifold is contained inside the union of the sets in the constructed covering [Dellnitz *et al.*, 2001; Junge, 2000a].

The overall computational cost is quite high when good resolution, that is, many boxes are required. For example, the Lorenz manifold in Fig. 14 of more that 4 million boxes took about 120 minutes on a 1.25 GHz G4 processor. Since the numerical cost depends on the dimension of the manifold, for manifolds of dimension larger than two it may only be feasible to compute rather coarse approximations.

Acknowledgment

The authors thank Mike Jolly for providing the image in Fig. 3 of the Lorenz manifold computed with the method in [Johnson *et al.*, 1997], and seen from the common viewpoint.

References

Abraham, R. H. & Shaw, C. D. [1985] *Dynamics — The Geometry of Behavior, Part Three: Global Behavior* (Aerial Press, Santa Cruz).

Allgower, E. L. & Georg, K. [1996] "Numerical path following," *Handbook of Numerical Analysis*, Vol. 5, eds. Ciarlet, P. G. & Lions, J. L. (North Holland Publishing), pp. 3–207.

Ascher, U. M., Mattheij, R. M. M. & Russell, R. D. [1995] *Numerical Solution of Boundary Value Problems for Ordinary Differential Equations* (SIAM).

Back, A., Guckenheimer, J., Myers, M. R., Wicklin, F. J. & Worfolk, P. A. [1992] "DsTool: Computer assisted exploration of dynamical systems," *Notices Amer. Math. Soc.* **39**, p. 303.

Beyn, W.-J., Champneys, A., Doedel, E. J., Govaerts, W., Sandstede, B. & Kuznetov, Yu. A. [2002] "Numerical continuation and computation of normal forms," *Handbook of Dynamical Systems*, Vol. 2, ed. Fiedler, B. (Elsevier Science), pp. 149–219.

De Boor, C. & Swartz, B. [1973] "Collocation at Gaussian points," *SIAM J. Numer. Anal.* **10**, 582–606.

Dellnitz, M. & Hohmann, A. [1996] "The computation of unstable manifolds using subdivision and continuation," *Nonlinear Dynamical Systems and Chaos PNLDE* 19, eds. Broer, H. W., Van Gils, S. A., Hoveijn, I. & Takens, F. (Birkhäuser, Basel), pp. 449–459.

Dellnitz, M. & Hohmann, A. [1997] "A subdivision algorithm for the computation of unstable manifolds and global attractors," *Numer. Math.* **75**, 293–317.

Dellnitz, M., Hohmann, A., Junge, O. & Rumpf, M. [1997] "Exploring invariant sets and invariant measures," *Chaos* **7**, 221–228.

Dellnitz, M., Froyland, G. & Junge, O. [2001] "The algorithms behind GAIO — Set oriented numerical methods for dynamical systems," *Ergodic Theory, Analysis, and Efficient Simulation of Dynamical Systems*, ed. Fiedler, B. (Springer-Verlag, Berlin), pp. 145–174; software available at http://www.dynamicalsystems.org/sw/sw/detail?item=30.

Dellnitz, M. & Junge, O. [2002] "Set oriented numerical methods for dynamical systems," *Handbook of Dynamical Systems II: Towards Applications*, eds. Fiedler, B., Iooss, G. & Kopell, N. (World Scientific, Singapore), pp. 221–264.

Dieci, L. & Lorenz, J. [1995] "Computation of invariant tori by the method of characteristics," *SIAM J. Num. Anal.* **32**, 1436–1474.

Dieci, L., Lorenz, J. & Russell, R. D. [1991] "Numerical calculation of invariant tori," *SIAM J. Sci. Stat. Comput.* **12**, 607–647.

Doedel, E. J. [1981] "AUTO, a program for the automatic bifurcation analysis of autonomous systems," *Congr. Numer.* **30**, 265–384.

Doedel, E. J., Keller, H. B. & Kernévez, J. P. [1991a] "Numerical analysis and control of bifurcation problems: I," *Int. J. Bifurcation and Chaos* **1**, 493–520.

Doedel, E. J., Keller, H. B. & Kernévez, J. P. [1991b] "Numerical analysis and control of bifurcation problems: II," *Int. J. Bifurcation and Chaos* **1**, 745–772.

Doedel, E. J., Champneys, A. R., Fairgrieve, T. F., Kuznetsov, Yu. A., Sandstede, B. & Wang, X. J. [1997] "AUTO97: Continuation and bifurcation software for ordinary differential equations," available via http://cmvl.cs.concordia.ca/.

Doedel, E. J., Paffenroth, R. C., Champneys, A. R., Fairgrieve, T. F., Kuznetsov, Yu. A., Oldeman, B. E., Sandstede, B. & Wang, X. J. [2000] "AUTO2000: Continuation and bifurcation software for ordinary differential equations," available via http://cmvl.cs.concordia.ca/.

Edoh, K. D., Russell, R. D. & Sun, W. [1995] "Orthogonal collocation for hyperbolic PDEs & computation of invariant tori," Australian National Univ., Mathematics Research Report No. MRR 060-95.

Fairgrieve, T. F. & Jepson, A. D. [1991] "O. K. Floquet multipliers," *SIAM J. Numer. Anal.* **28**, 1446–1462.

Guckenheimer, J. & Holmes, P. [1986] *Nonlinear Oscillations, Dynamical Systems and Bifurcations of Vector Fields*, 2nd edition (Springer-Verlag, NY).

Guckenheimer, J. & Worfolk, P. [1993] "Dynamical systems: Some computational problems," *Bifurcations and Periodic Orbits of Vector Fields*, ed. Schlomiuk, D. (Kluwer Academic Publishers), pp. 241–277.

Guckenheimer, J. & Vladimirsky, A. [2004] "A fast method for approximating invariant manifolds," *SIAM J. Appl. Dyn. Syst.* **3**, 232–260; animations available at http://epubs.siam.org/sambin/dbq/article/60017.

Henderson, M. E. [2002] "Multiple parameter continuation: Computing implicitly defined k-manifolds," *Int. J. Bifurcation and Chaos* **12**, 451–476.

Henderson, M. E. [2003] "Computing invariant manifolds by integrating fat trajectories," *SIAM J. Appl. Dyn. Syst.*, in press.

Hobson, D. [1993] "An efficient method for computing invariant manifolds of planar maps," *J. Comput. Phys.* **104**, 14–22.

Johnson, M. E., Jolly, M. S. & Kevrekidis, I. G. [1997] "Two-dimensional invariant manifolds and global bifurcations: Some approximation and visualization studies," *Numer. Alg.* **14**, 125–140.

Johnson, M. E., Jolly, M. S. & Kevrekidis, I. G. [2001] "The Oseberg transition: Visualization of global bifurcations for the Kuramoto–Sivashinsky equation," *Int. J. Bifurcation and Chaos* **11**, 1–18.

Junge, O. [2000a] "Rigorous discretization of subdivision techniques," in *Proc. Int. Conf. Diff. Eqs.* Vol. 2, eds. Fiedler, B., Gröger, K. & Sprekels, J. (World Scientific, Singapore), pp. 916–918.

Junge, O. [2000b] *Mengenorientierte Methoden zur Numerischen Analyse Dynamischer Systeme* (Shaker, Aachen).

Keller, H. B. [1977] "Numerical solution of bifurcation and nonlinear eigenvalue problems," *Applications of Bifurcation Theory*, ed. Rabinowitz, P. H. (Academic Press), pp. 359–384.

Krauskopf, B. & Osinga, H. M. [1998a] "Globalizing two-dimensional unstable manifolds of maps," *Int. J. Bifurcation and Chaos* **8**, 483–503.

Krauskopf, B. & Osinga, H. M. [1998b] "Growing 1D and quasi 2D unstable manifolds of maps," *J. Comp. Phys.* **146**, 404–419.

Krauskopf, B. & Osinga, H. M. [1999] "Two-dimensional global manifolds of vector fields," *Chaos* **9**, 768–774.

Krauskopf, B. & Osinga, H. M. [2003] "Computing geodesic level sets on global (un)stable manifolds of vector fields," *SIAM J. Appl. Dyn. Syst.* **4**, 546–569.

Krauskopf, B. & Osinga, H. M. [2004] "The Lorenz manifold as a collection of geodesic level sets," *Nonlinearity* **17**, C1–C6.

Kuznetsov, Yu. A. [1998] *Elements of Applied Bifurcation Theory*, 2nd edition (Springer Verlag, NY).

Lorenz, E. N. [1963] "Deterministic nonperiodic flow," *J. Atmosph. Sci.* **20**, 130–141.

Lust, K. [2001] "Improved numerical Floquet multipliers," *Int. J. Bifurcation and Chaos* **11**, 2389–2410.

Misner, C. W., Thorne, K. S. & Wheeler, J. A. [1970] *Gravitation* (W. H. Freeman and Company, San Francisco).

Osinga, H. M. [2000] "Non-orientable manifolds of periodic orbits," in *Proc. Int. Conf. Differential Eqations, Equadiff 99 (Berlin)* Vol. 2, eds. Fiedler, B., Gröger, K. & Sprekels, J. (World Scientific, Singapore), pp. 922–924.

Osinga, H. M. & Krauskopf, B. [2002] "Visualizing the structure of chaos in the Lorenz system," *Comput. Graph.* **26**, 815–823.

Osinga, H. M. [2003] "Non-orientable manifolds in three-dimensional vector fields," *Int. J. Bifurcation and Chaos* **13**, 553–570.

Osinga, H. M. & Krauskopf, B. [2004] "Crocheting the Lorenz manifold," *The Math. Intell.* **26**, 25–37.

Peraire, J., Peiro, J. & Morgan, K. [1999] "Advancing front grid generation," *Handbook of Grid Generation*, eds. Thompson, J. F., Soni, B. K. & Weatherill, N. P. (CRC Press), Chap. 17.

Perelló, C. [1979] "Intertwining invariant manifolds and Lorenz attractor," in *Global Theory of Dynamical Systems (Proc. Internat. Conf., Northwestern Univ., Evanston, Ill., 1979)*, *Lecture Notes in Mathematics*, Vol. 819 (Springer-Verlag, Berlin), pp. 375–378.

Phillips, M., Levy, S. & Munzner, T. [1993] "Geomview: An interactive geometry viewer," *Not. Amer. Math. Soc.* **40**, 985–988.

Rebay, S. [1993] "Efficient unstructured mesh generation by means of Delaunay triangulation and Bowyer–Watson algorithm," *J. Comp. Phys.* **106**, 125–138.

Rheinboldt, W. C. [1986] *Numerical Analysis of Parametrized Nonlinear Equations*, University of Arkansas Lecture Notes in the Mathematical Sciences (Wiley-Interscience).

Russell, R. D. & Christiansen, J. [1978] "Adaptive mesh selection strategies for solving boundary value problems," *SIAM J. Numer. Anal.* **15**, 59–80.

Sacker, R. J. [1965] "A new approach to the perturbation theory of invariant surfaces," *Comm. Pure Appl. Math.* **18**, 717–732.

Sethian, J. A. & Vladimirsky, A. [2003] "Ordered upwind methods for static Hamilton-Jacobi equations: Theory & applications," *SIAM J. Numer. Anal.* **41**, 325–363.

Seydel, R. [1995] *From Equilibrium to Chaos. Practical Bifurcation and Stability Analysis*, 2nd edition (Springer-Verlag, NY).

Spivak, M. [1979] *Differential Geometry*, 2nd edition (Publish or Perish, Houston, Texas).

Stewart, H. B. [1986] "Visualization of the Lorenz system," *Physica* **D18**, 479–480.

Strogatz, S. H. [1994] *Nonlinear Dynamics and Chaos* (Addison-Wesley, Reading, MA).

Thompson, J. M. T. & Stewart, H. B. [1986] *Nonlinear Dynamics and Chaos* (John Wiley, Chichester/NY).

COMMUTATORS OF SKEW-SYMMETRIC MATRICES

ANTHONY M. BLOCH

Department of Mathematics, University of Michigan,
Ann Arbor, MI 48109, USA

ARIEH ISERLES

Department of Applied Mathematics and Theoretical Physics,
Centre for Mathematical Sciences, University of Cambridge,
Wilberforce Road, Cambridge CB3 0WA, England

Received March 26, 2004; Revised June 8, 2004

In this paper we develop a theory for analysing the "radius" of the Lie algebra of a matrix Lie group, which is a measure of the size of its commutators. Complete details are given for the Lie algebra $\mathfrak{so}(n)$ of skew symmetric matrices where we prove $\|[X, Y]\| \le \sqrt{2}\|X\| \cdot \|Y\|$, $X, Y \in \mathfrak{so}(n)$, for the Frobenius norm. We indicate how these ideas might be extended to other matrix Lie algebras. We discuss why these ideas are of interest in applications such as geometric integration and optimal control.

Keywords: Lie algebras; symmetric gauges; commutator matrices.

1. Norms and Commutators in $\mathbf{M}_n[\mathbb{R}]$ and $\mathfrak{so}(n)$

This paper is concerned with the following question. Let $\mathfrak{g}$ be a matrix *Lie algebra*. (We refer the reader to [Carter *et al.*, 1995; Humphreys, 1978; Varadarajan, 1984] for elements of Lie-algebraic theory of relevance to this paper.) Given $X, Y \in \mathfrak{g}$ and a norm $\|\cdot\| : \mathfrak{g} \to \mathbb{R}_+$, what is the size of $\|[X, Y]\|$ in comparison with $\|X\| \cdot \|Y\|$? We assume that the norm satisfies the Banach inequality $\|XY\| \le \|X\| \cdot \|Y\|$.

On the face of this, the question has little merit since the elementary inequality

$$\|[X, Y]\| \le 2\|X\| \cdot \|Y\| \tag{1}$$

always holds for $X, Y \in \mathrm{M}_n[\mathbb{R}]$, the set of all $n \times n$ real matrices and an arbitrary matrix norm $\|\cdot\|$. (This follows purely from the additive and multiplicative properties of norms, writing $[X, Y] = XY - YX$.) Moreover, it is easy to prove that the bound (1) can be attained for most norms of practical interest. In particular, this is the case

for two types of norms closely associated with a remarkable paper of von Neumann [1937].

We recall that a *symmetric gauge* is a vector norm $|\cdot|$ which is both symmetric and positive. In other words, for every $\mathbf{x} \in \mathbb{R}^n$ it is true that $|\mathbf{x}_\pi| = |\mathbf{x}|$ and $\||\mathbf{x}|\| = |\mathbf{x}|$, where π is a permutation of $\{1, 2, \ldots, n\}$, $\mathbf{x}_\pi^\top = [x_{\pi_1}, x_{\pi_2}, \ldots, x_{\pi_n}]$ and $|\mathbf{x}|^\top = [|x_1|, |x_2|, \ldots, |x_n|]$. We consider two norms, firstly the *operator norm*

$$|X| = \max_{\mathbf{v} \neq \mathbf{0}} \frac{|X\mathbf{v}|}{|\mathbf{v}|}$$

and secondly the norm

$$\|X\| = |\boldsymbol{\sigma}(X)|, \tag{2}$$

where $\boldsymbol{\sigma}(X)$ are the *singular values* of X, arbitrarily ordered. While it is easy to see that (2) is a unitary norm (i.e. invariant under multiplication by a unitary matrix), von Neumann proved that *all* unitary norms are of this form. We remark that the standard $\ell_p[\mathbb{R}^n]$ vector norm, $1 \le p \le \infty$, is a symmetric gauge. Therefore it gives rise to a unitarily-invariant norm, the *Schatten p-norm* $\|\cdot\|_p = |\boldsymbol{\sigma}(\cdot)|_p$ [Horn & Johnson, 1994].

We consider just the case $n = 2$, since it can be embedded in $M_n[\mathbb{R}]$ for any $n \geq 2$ and this is sufficient for analysing the upper bound for general $n \geq 2$. Let

$$X = \begin{bmatrix} 1 & 0 \\ 0 & -1 \end{bmatrix}, \quad Y = \begin{bmatrix} 0 & 1 \\ -1 & 0 \end{bmatrix},$$

$$Z = [X, Y] = \begin{bmatrix} 0 & 2 \\ 2 & 0 \end{bmatrix}.$$

It is easy to verify that $|X| = |Y| = 1$ and $|Z| = 2$. Moreover, since $\boldsymbol{\sigma}(X) = \boldsymbol{\sigma}(Y) = [1, 1]$ and $\boldsymbol{\sigma}(Z) = [2, 2]$, it is also true that $\|X\|, \|Y\| = |\mathbf{1}|$ and $\|Z\| = 2|\mathbf{1}|$, where $\mathbf{1}^\top = [1, 1]$. In both cases the upper bound in (1) is attained.

Yet, there is a basic difference between $M_n[\mathbb{R}]$ and a Lie algebra $\mathfrak{g} \subset M_n[\mathbb{R}]$: while $\dim M_n[\mathbb{R}] = n^2$, the Lie algebra typically has a lower dimension: for example, $\dim \mathfrak{so}(n) = (1/2)(n-1)n$. Thus, it makes sense to pose the question whether, *once X and Y are restricted to $\mathfrak{g}$*, the inequality (1) might still be obeyed as an equality or whether 2 might be replaced by a smaller constant for all $X, Y \in \mathfrak{g}$. Thus, given a norm $\|\cdot\|$, we say that the *radius* of a Lie algebra $\mathfrak{g}$ is the least number $\omega(\mathfrak{g}) \in [0, 2]$ such that

$$\|[X, Y]\| \leq \omega(X)\|X\| \cdot \|Y\|, \quad X, Y \in \mathfrak{g}.$$

In other words,

$$\omega(\mathfrak{g}) = \max \left\{ \frac{\|[X, Y]\|}{\|X\| \cdot \|Y\|} : \right.$$
$$\left. X, Y \in \mathfrak{g}, \ X, Y \neq O \right\}, \tag{3}$$

the operator norm of the commutator.

It is important when defining ω to keep in mind which underlying norm we are using. In the following we shall denote by $\|v\|_p$, $v \in \mathbb{R}^n$, the vector p-norm and by

$$\|X\|_p = \max \left\{ \frac{\|Xv\|_p}{\|v\|_p} : v \neq 0 \right\}$$

the corresponding operator norm as above. In the case $p = 2$ we shall call this the Euclidean norm.

We denote by

$$\|X\|_\mathsf{F} = \left(\sum_{k,l=1}^{n} X_{kl}^2 \right)^{\frac{1}{2}}$$

the *Frobenius* norm.

We recall also the important facts to be used below, namely that $\|X\|_2$ is equal to the magnitude of the largest singular value of X while $\|X\|_\mathsf{F} = \|\sigma(X)\|_2$.

When the context is not clear we will label ω by a subscript denoting which norm is being used.

Trivially, the Lie algebra $\mathfrak{g}$ is commutative if and only if $\omega(\mathfrak{g}) = 0$, but this observation is devoid of any insight. More interestingly, consider $\mathfrak{so}(3)$ and the Euclidean norm. Letting

$$X = \begin{bmatrix} 0 & x_1 & x_2 \\ -x_1 & 0 & x_3 \\ -x_2 & -x_3 & 0 \end{bmatrix}, \quad Y = \begin{bmatrix} 0 & y_1 & y_2 \\ -y_1 & 0 & y_3 \\ -y_2 & -y_3 & 0 \end{bmatrix}$$

and observing that in $\mathfrak{so}(n)$ the Euclidean norm coincides with the spectral radius, we commence by noting that

$$\|X\| = \|\mathbf{x}\|, \quad \|Y\| = \|\mathbf{y}\|.$$

Moreover, if $Z = [X, Y]$ then, by an easy direct calculation,

$$\|\mathbf{z}\|^2 = \|\mathbf{x}\| \cdot \|\mathbf{y}\| - (\mathbf{x}^\top \mathbf{y})^2. \tag{4}$$

Therefore

$$\|[X, Y]\| = \left[\|X\|^2 \|Y\|^2 - (\mathbf{x}^\top \mathbf{y})^2 \right]^{\frac{1}{2}} \leq \|X\| \cdot \|Y\|,$$

with the upper bound holding as an equality when $\mathbf{x}$ is orthogonal to $\mathbf{y}$. We thus deduce that $\omega(\mathfrak{so}(3)) = 1$.

Remark 1. There is a natural Lie algebra homomorphism between $\mathfrak{so}(3)$ and $\mathbb{R}^3$ endowed with the cross product. The above computation may be repeated with this in mind and (4) is a standard vector identity. One could of course use the "hat" notation (see e.g. [Marsden & Ratiu, 1999]) for this homomorphism but we prefer our notation here because we require below a more general relationship between vectors and matrices.

Remark 2. It is also of interest to repeat the above computation for the Frobenius norm. One determines immediately that $\omega_\mathsf{F}(\mathfrak{so}(3)) = 1/\sqrt{2}$. However for Lie algebraic reasons that will become apparent below it is more natural to scale the Frobenius norm by a factor of $\sqrt{2}$. With this scaling we also have $\omega_\mathsf{F}(\mathfrak{so}(3)) = 1$. Strikingly this result does not hold for n larger than 3.

The Main Result. In this paper we determine $\omega(\mathfrak{so}(n))$ for all $n \geq 3$ ($\mathfrak{so}(2)$ is a commutative algebra, hence $\omega(\mathfrak{so}(2)) = 0$) with respect to the (scaled) Frobenius norm. Specifically, we prove that $\omega_{\mathsf{F}}(\mathfrak{so}(n)) = \sqrt{2}$ for $n \geq 4$ (with the above-mentioned scaling). Note that $\|X\|_{\mathsf{F}} = -\langle X, X\rangle$, where $\langle \cdot, \cdot \rangle$ is a multiple of the *Killing form* in $\mathfrak{so}(n)$, hence it has deeper Lie-algebraic significance. The Killing form evaluated on a pair of $n \times n$ skew-symmetric matrices A, B is actually $(n-2)$ trace AB (see [Kobayashi & Nomizu, 1969]). (Of course for a noncompact Lie algebra the Killing form does not provide a norm since it is not definite.) Another reason why the Frobenius norm is of interest is that the radius of $\mathfrak{so}(n)$, $n \geq 4$, is just equal to 2 for most other norms of interest. Consider for example the following analysis.

Thus, again, let $|\cdot|$ be a symmetric gauge and

$$X = \begin{bmatrix} 0 & 1 & 0 & 0 \\ -1 & 0 & 0 & 0 \\ 0 & 0 & 0 & 1 \\ 0 & 0 & -1 & 0 \end{bmatrix}, \quad Y = \begin{bmatrix} 0 & 0 & 1 & 0 \\ 0 & 0 & 0 & -1 \\ -1 & 0 & 0 & 0 \\ 0 & 1 & 0 & 0 \end{bmatrix},$$

therefore

$$[X, Y] = \begin{bmatrix} 0 & 0 & 0 & -2 \\ 0 & 0 & -2 & 0 \\ 0 & 2 & 0 & 0 \\ 2 & 0 & 0 & 0 \end{bmatrix}.$$

Note that for every $\mathbf{v} \in \mathbb{R}^4$ positivity and symmetry of the symmetric gauge imply that

$$|X\mathbf{v}| = \left| \begin{bmatrix} v_2 \\ -v_1 \\ v_4 \\ -v_3 \end{bmatrix} \right| = |\mathbf{v}|$$

and, similarly, $|Y\mathbf{v}| = |\mathbf{v}|$ and $|[X,Y]\mathbf{v}| = 2|\mathbf{v}|$. Therefore, in the underlying operator norm $|X|$, $|Y| = 1$ and $|[X, Y]| = 2$. Consequently $\omega(\mathfrak{so}(4)) = 2$ and this can be extended to all Lie algebras $\mathfrak{so}(n)$, $n \geq 4$, since they form a flag. This example cannot, however, be extended to unitary norms (2), unless $\|1\| = 1$. Note that the latter condition holds when $|\cdot| = |\cdot|_\infty$ (the ∞-Schatten norm [Horn & Johnson, 1994], which is equivalent to the operator Euclidean norm), but in that instance $\|\cdot\| = \|\cdot\|_2$ and we are back to the area covered earlier in this paragraph. On the other hand, $|\cdot| = |\cdot|_2$, whence $\|1\| = 2$, results in $\|\cdot\| = \|\cdot\|_{\mathsf{F}}$. This, of course, does not necessarily mean that $\omega(\mathfrak{so}(4)) < 2$ in the Frobenius norm.

In Sec. 2 we discuss the structure of the commutator operator, considered as a linear transformation from $\mathfrak{so}(n)$ to itself. We prove that, subject to an appropriate representation of $\mathfrak{so}(n)$, the commutator matrix in the $(1/2)(n-1)n$-dimensional linear space can be read explicitly from a certain directed graph and investigate its eigenstructure. Section 3 is devoted to the proof of the main result of this paper, namely that, once we use the (scaled) Frobenius norm, $\omega(\mathfrak{so}(n)) = \sqrt{2}$ for all $n \geq 4$.

The subject matter of this paper is motivated by a raft of issues arising from *geometric numerical integration*. The simplest such problem is the convergence of the sum

$$f(t; X, Y) = \sum_{m=0}^{\infty} a_m t^m \mathrm{ad}_X^m Y, \quad X, Y \in \mathfrak{g},$$

where $\{a_m\}_{m \in \mathbb{Z}_+}$ is a given sequence and ad_X is the *adjoint operator* of the Lie algebra $\mathfrak{g}$,

$$\mathrm{ad}_X^0 Y = Y, \quad \mathrm{ad}_X^m Y = [X, \mathrm{ad}_X^{m-1} Y], \quad m \in \mathbb{N}.$$

It is trivial to deduce from the triangle inequality that

$$\|f(t; X, Y)\| \leq \|Y\| \sum_{m=0}^{\infty} a_m [t\omega(\mathfrak{g})\|X\|]^m,$$

thereby relating the convergence of F to the domain of analyticity of the generating function $\sum_{m=0}^{\infty} a_m z^m$. The benefit of smaller $\omega(\mathfrak{g})$ in the convergence of such a function is clear. Similar and more complicated problems abound in the analysis of Lie-group methods [Iserles *et al.*, 2000].

The norm of a bracket is also important in determining the maximum allowable step size in certain minimization problems on adjoint orbits, see the work of [Brockett, 1993].

A related problem of interest in analysing certain systems of differential equations is that of finding a bound on the norm of the bracket $[X, N]$ where N is fixed and X varies over the adjoint orbit of a group. This problem is discussed in [Brockett, 1994]. In that setting for $\mathfrak{so}(n)$ one has to solve the problem of maximizing $\|[X, N]\|$ over all $X = \theta^T \Lambda \theta$ for N, Λ fixed in $\mathfrak{so}(n)$ and θ in the group SO(n).

2. The Reduced Commutator Matrix in $\mathfrak{so}(n)$

2.1. *The reduced commutator matrix*

Let $\mathfrak{g} \subseteq \mathrm{M}_n[\mathbb{R}]$ be an m-dimensional matrix Lie algebra, $1 \leq m \leq n^2$. An obvious means to explore the norm of the commutator in $\mathfrak{g}$ is by means of the *natural embedding* $\boldsymbol{\theta} : \mathfrak{g} \to \mathbb{R}^{n^2}$ that "stretches" a matrix X into a vector, e.g. by letting $\boldsymbol{\theta}_{(l-1)n+k}(X) = X_{k,l}$, $k, l = 1, 2, \ldots, n$ (columnwise ordering). Since commutation is a linear transformation, it follows that for every $X \in \mathfrak{g}$ there exists a matrix $\tilde{C}_X \in \mathrm{M}_{n^2}[\mathbb{R}]$ such that

$$\boldsymbol{\theta}([X,Y]) = \tilde{C}_X \boldsymbol{\theta}(Y), \quad Y \in \mathfrak{g}.$$

It is known that

$$\sigma(\tilde{C}_X) = \{\lambda_k - \lambda_l \ : \ \lambda_k, \lambda_l \in \sigma(X), \ k, l = 1, 2, \ldots, n\}$$

[Hille, 1969], and this provides a useful tool to explore commutators in a classical setting. Yet, this line of reasoning disregards the fact that $\mathfrak{g}$ is a Lie algebra, typically of much smaller dimension than that of $\mathrm{M}_n[\mathbb{R}]$. Thus, in place of $\boldsymbol{\theta}$, we propose a *restricted embedding* $\boldsymbol{\nu} : \mathfrak{g} \to \mathbb{R}^m$. Let $\mathcal{Q} = \{Q_1, Q_2, \ldots, Q_m\}$ be a basis of $\mathfrak{g}$. We define an isomorphism $\boldsymbol{\nu}$ from $\mathfrak{g}$ on $\mathbb{R}^m$ through

$$\mathfrak{g} \ni X = \sum_{k=1}^{m} x_k Q_k \ \Leftrightarrow \ \boldsymbol{\nu}(X) = \mathbf{x} = \begin{bmatrix} x_1 \\ x_2 \\ \vdots \\ x_m \end{bmatrix}.$$

Remark. We note that this is just a vector space isomorphism and not in general a Lie algebra homomorphism and there is in general no natural cross-product operation in $\mathbb{R}^m$. Thus one cannot use the earlier argument for $\mathfrak{so}(3)$.

The *restricted commutator matrix* $C_X \in \mathrm{M}_m[\mathbb{R}]$ is then defined by the identity

$$\boldsymbol{\nu}([X,Y]) = C_X \boldsymbol{\nu}(Y), \quad Y \in \mathfrak{g}.$$

Spectral information on C_X is no longer readily and explicitly available, yet the procedure has the great virtue of reducing the dimension and allowing for a more natural incorporation of Lie-algebraic information. Specifically, let $\{c_{k,l}^j\}_{k,l,j=1,2,\ldots,m}$ be the *structure constants* of $\mathfrak{g}$ with respect to $\mathcal{Q}$,

$$[Q_k, Q_l] = \sum_{j=1}^{m} c_{k,l}^j Q_j.$$

It is an elementary exercise that

$$(C_X)_{j,l} = \sum_{k=1}^{m} x_k c_{k,l}^j, \quad j, l = 1, 2, \ldots, m. \qquad (5)$$

Recalling that the definition of the *radius* $\omega(\mathfrak{g})$ of the Lie algebra is

$$\max \left\{ \frac{\|[X,Y]\|}{\|X\| \cdot \|Y\|} \ : \ X, Y \in \mathfrak{g}, \ X, Y \neq O \right\},$$

where $\|\cdot\|$ is a given norm *induced by the vector norm* on $\mathbb{R}^m$, i.e. $\|X\| = \|\boldsymbol{\nu}(X)\|$, we observe that

Proposition 1.

$$\omega(\mathfrak{g}) = \max \left\{ \frac{\|C_X\|}{\|X\|} \ : \ X \in \mathfrak{g}, \ X \neq O \right\}, \qquad (6)$$

where the Lie algebra norm is that induced by the map $\boldsymbol{\nu}$.

Proof. We have

$$\|[X,Y]\| = \|\boldsymbol{\nu}([X,Y])\| = \|C_X \boldsymbol{\nu}(Y)\|.$$

Hence

$$\omega(\mathfrak{g}) = \max_{X \in \mathfrak{g}\backslash\{O\}} \max_{Y \in \mathfrak{g}\backslash\{O\}} \frac{\|C_X \boldsymbol{\nu}(Y)\|}{\|\boldsymbol{\nu}(X)\| \cdot \|\boldsymbol{\nu}(Y)\|}$$

$$= \max_{X \in \mathfrak{g}\backslash\{O\}} \frac{\|C_X\|}{\|\boldsymbol{\nu}(X)\|}. \qquad \blacksquare$$

We conclude this section by addressing the question of multiple representations. Suppose thus that we have two bases of $\mathfrak{g}$, $\mathcal{Q} = \{Q_1, Q_2, \ldots, Q_m\}$ and $\mathcal{P} = \{P_1, P_2, \ldots, P_m\}$, say. Set

$$P = [\mathbf{p}_1 \ \ \mathbf{p}_2 \ \ \cdots \ \ \mathbf{p}_m], \quad Q = [\mathbf{q}_1 \ \ \mathbf{q}_2 \ \ \cdots \ \ \mathbf{q}_m],$$

where $\mathbf{p}_k = \boldsymbol{\nu}(P_k)$, $\mathbf{q}_k = \boldsymbol{\nu}(Q_k)$, $k = 1, 2, \ldots, m$. Then

$$\mathfrak{g} \ni X = \sum_{k=1}^{m} x_k P_k = \sum_{k=1}^{m} \tilde{x}_k Q_k$$

implies at once that $\mathbf{x} = Q\tilde{\mathbf{x}}$. Therefore $C_X^Q = Q^{-1} C_X^P Q$, where C_X^P and C_X^Q are reduced commutators with respect to the two bases. In particular, if Q is an orthogonal matrix and the bases are orthogonally similar then the radius of $\mathfrak{g}$ does not depend on the choice of the basis.

Example. We calculate $\omega(\mathfrak{so}(3))$ using this formalism.
    If

$$X = \begin{bmatrix} 0 & a & b \\ -a & 0 & c \\ -b & -c & 0 \end{bmatrix}$$

then we easily compute

$$C_X = \begin{bmatrix} 0 & c & -b \\ -c & 0 & a \\ b & -a & 0 \end{bmatrix}.$$

Now, $\boldsymbol{\nu}(X) = [x, y, z]^T$, and hence $\|\boldsymbol{\nu}(X)\|_2 = (a^2 + b^2 + c^2)^{1/2}$. On the other hand, $\sigma(C_X) = \{0, \pm(a^2 + b^2 + c^2)^{1/2}\}$, hence $\|C_X\|_2 = (a^2 + b^2 + c^2)^{1/2}$ and we obtain $\omega(\mathfrak{so}(3)) = 1$.

2.2. *The reduced commutator matrix in $\mathfrak{so}(n)$ and directed graphs*

We denote by $E_{k,l} \in \mathrm{M}_n[\mathbb{R}]$ the matrix whose (k,l)th component is $+1$ and otherwise is zero, $k, l = 1, 2, \ldots, n$, and choose the basis

$$\mathcal{Q} = \{Q_{k,l} = E_{k,l} - E_{l,k} : 1 \leq k < l \leq n\}$$

of $\mathfrak{so}(n)$. The restricted embedding $\boldsymbol{\nu}$ takes each $Q_{k,l}$ to $\mathbf{e}_{\mu(k,l)} \in \mathbb{R}^m$, where $m = (1/2)(n-1)n$ and $\boldsymbol{\mu}$ is an arbitrary isomorphism mapping pairs $\mathcal{I} = \{(k, l) : 1 \leq k < l \leq n\}$, into $\{1, 2, \ldots, m\}$. However, it is more convenient to discuss restricted commutator matrices in the formalism of the $Q_{k,l}$, bypassing $\boldsymbol{\nu}$ altogether. Thus, we index the structure constants and the entries of the restricted commutator matrix by pairs $(i, j) \in \mathcal{I}$.

For ease of notation we let $Q_{k,l} = -Q_{l,k}$ for $k > l$ and $Q_{k,k} = O$. Since

$$[Q_{k,l}, Q_{r,s}] = \delta_{l,r}Q_{k,s} - \delta_{k,r}Q_{l,s} - \delta_{l,s}Q_{k,r} + \delta_{k,s}Q_{l,r},$$

the structure constants are

$$c^{(i,j)}_{(k,l),(r,s)}$$
$$= \begin{cases} -1, & k = r, \ l \neq s, \ i = l, \ j = s, \\ +1, & k \neq r, \ l = r, \ i = k, \ j = s, \\ -1, & k \neq r, \ l = s, \ i = k, \ j = r, \quad (i,j) \in \mathcal{I}. \\ +1, & k = s, \ l \neq r, \ i = l, \ j = r, \\ 0, & \text{otherwise.} \end{cases}$$

$$(7)$$

In other words, most structure constants vanish: not surprising, given that our basis is consistent with the root space decomposition of $\mathfrak{so}(n)$, hence

likely to lend itself to a sparse set of structure constants. Specifically, the nonzero structure constants are precisely

$$c^{(l,s)}_{(k,l),(k,s)} = -1, \ l < s, \qquad c^{(l,s)}_{(k,l),(k,s)} = +1, \ l > s,$$

$$c^{(k,s)}_{(k,l),(l,s)} = +1, \ k < s, \qquad c^{(k,s)}_{(k,l),(l,s)} = -1, \ k > s,$$

$$c^{(k,r)}_{(k,l),(r,l)} = -1, \ k < r, \qquad c^{(k,r)}_{(k,l),(r,l)} = +1, \ k > r,$$

$$c^{(l,r)}_{(k,l),(r,k)} = +1, \ l < r, \qquad c^{(l,r)}_{(k,l),(r,k)} = -1, \ l > r.$$

Given

$$\mathfrak{so}(n) \ni X = \sum_{k=1}^{n-1} \sum_{l=k+1}^{n} x_{k,l} Q_{k,l},$$

(5) implies that

$$(C_X)_{(k,l),(i,j)} = \sum_{(r,s) \in \mathcal{I}} x_{r,s} c^{(i,j)}_{(r,s),(k,l)}$$

$$= -\sum_{(r,s) \in \mathcal{I}} c^{(r,s)}_{(k,l),(i,j)}, \quad (k,l), (i,j) \in \mathcal{I}.$$

We observe that $C_X \in \mathfrak{so}(m)$ and that it is a very sparse matrix. Specifically, for any $(k, l) \in \mathcal{I}$ the only nonzero entries are

$$(C_X)_{(k,l),(l,j)} = x_{k,j}, \quad j = l+1, l+2, \ldots, n,$$

$$(C_X)_{(k,l),(k,j)} = x_{j,l}, \quad j = k+1, k+2, \ldots, n,$$
$$j \neq l,$$

$$(C_X)_{(k,l),(i,l)} = x_{i,k}, \quad i = 1, 2, \ldots, l-1, \ i \neq k,$$

$$(C_X)_{(k,l),(i,k)} = x_{l,i}, \quad i = 1, 2, \ldots, k-1.$$

$$(8)$$

Altogether, just $(n-2)(n-1)$, out of $(1/2)(n-1)^2 n^2$, entries of C_X are nonzero.

The elements of C_X lend themselves to a very convenient representation in terms of *labeled digraphs*. Any matrix $A \in \mathrm{M}_m[\mathbb{R}]$ can be represented by a digraph with m vertices, adopting the convention that, once $A_{k,l} \neq 0$, then there is a directed edge from vertex k to vertex l with the label $A_{k,l}$. As an example, let us examine the digraph corresponding to C_X for $n = 4$ (hence $m = 6$):

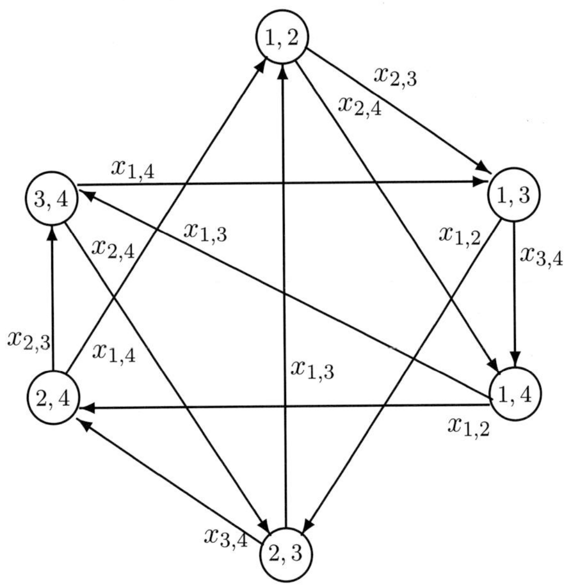

We commence by noting that the graph is 4-regular [Chartrand & Lesniak, 1986]: each vertex is of degree 4. Moreover, two of the edges at each vertex commence and two terminate there.

A generalization for all $n \geq 3$ is clear from (8). For every $1 \leq k < l < j \leq n$ we have

$$(C_X)_{(k,l),(l,j)} = x_{k,j},$$

$$(C_X)_{(l,j),(k,j)} = x_{k,l},$$

$$(C_X)_{(k,j),(k,l)} = x_{l,j}.$$

In the notation of labeled digraphs this corresponds to the 3-cycle

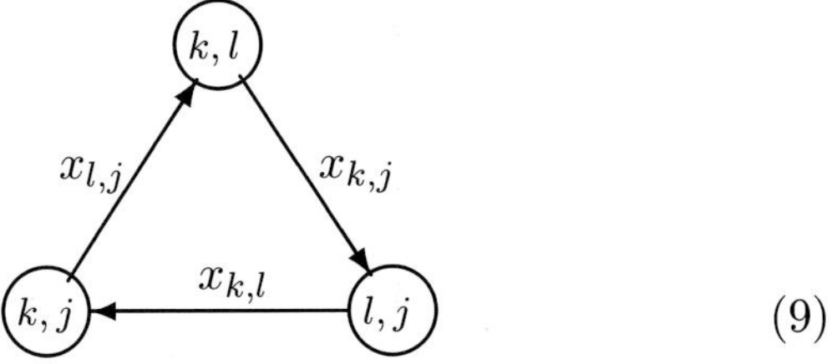

$$(9)$$

Needless to say, (9) can be read "backwards": an arrow from (l,j) to (k,j) with a label $x_{k,l}$ is the same as an arrow from (k,j) to (l,j) with the label $-x_{k,l}$.

Lemma 2. *For all $n \geq 3$ the directed graph of C_X is the sum of all $\binom{n}{3}$ 3-cycles (9) for all $1 \leq k < l < j \leq n$. It is r-regular, where $r = 2(n-2)$.*

Proof. The first statement of the lemma follows at once from our analysis. Because of symmetry, clearly the graph must be r-regular for some $r \geq 1$. Therefore, the sum of all the degrees of all the vertices is mr. Since each 3-cycle (9) accounts for exactly six degrees and $m = (1/2)(n-1)n$, we have

$$r = \frac{6\binom{n}{3}}{\frac{1}{2}(n-1)n} = 2(n-2). \quad \blacksquare$$

Given $X, Y \in \mathfrak{so}(n)$, $n \geq 3$, we can reconstruct the representation of $[X, Y]$ in the basis $\mathcal{Q}$ directly from the digraph in Lemma 2. Since

$$[X,Y] = \sum_{(k,l)\in\mathcal{I}} \sum_{(r,s)\in\mathcal{I}} x_{k,l} y_{r,s} (\delta_{l,r} Q_{k,s} - \delta_{k,r} Q_{l,s}$$

$$- \delta_{l,s} Q_{k,r} + \delta_{k,s} Q_{l,r}),$$

we have the following contributions to the $(i,j) \in \mathcal{I}$ component of the commutator,

$$
\begin{aligned}
(k,s) = (i,j), \quad & r = l \;\Rightarrow\; +x_{i,l} y_{l,j}, \quad & i+1 \leq l \leq j-1, \\
(l,s) = (i,j), \quad & r = k \;\Rightarrow\; -x_{k,i} y_{k,j}, \quad & 1 \leq k \leq \min\{i,j\}-1, \\
(k,r) = (i,j), \quad & s = l \;\Rightarrow\; -x_{i,l} y_{j,l}, \quad & \max\{i,j\}+1 \leq l \leq n, \\
(l,r) = (i,j), \quad & s = k \;\Rightarrow\; +x_{k,i} y_{j,k}, \quad & j+1 \leq k \leq i-1, \\
(k,s) = (j,i), \quad & r = l \;\Rightarrow\; -x_{j,l} y_{l,i}, \quad & j+1 \leq l \leq i-1, \\
(l,s) = (j,i), \quad & r = k \;\Rightarrow\; +x_{k,j} y_{k,i}, \quad & 1 \leq k \leq \min\{i,j\}-1, \\
(k,r) = (j,i), \quad & s = l \;\Rightarrow\; +x_{j,l} y_{i,l}, \quad & \max\{i,j\}+1 \leq l \leq n, \\
(l,r) = (j,i), \quad & s = k \;\Rightarrow\; -x_{k,j} y_{i,k}, \quad & i+1 \leq k \leq n.
\end{aligned}
$$

For $n = 4$ and $(i,j) = (1,3)$ just four terms survive from the above list,

$$[X,Y]_{1,3} = x_{1,2}y_{2,3} - x_{1,4}y_{3,4} + x_{3,4}y_{1,4} - x_{2,3}y_{1,2}.$$

Examine now the digraph for $n = 4$: $(1,3)$ is connected to $(2,3)$ with label $x_{1,2}$ and outgoing arrow, to $(3,4)$ with label $x_{1,4}$ and incoming arrow, etc. In general, it is easy to confirm the following general rule for the reconstruction of the commutator in our basis.

Lemma 3. *Let $n \geq 3$. Then, for every $(i,j) \in \mathcal{I}$ the element $[X,Y]_{i,j}$ is the sum of terms of the form $\pm x_{k,l}y_{r,s}$ over all the $2(n-2)$ edges adjoining the vertices (i,j) and (r,s) with the weight $x_{k,l}$, and with the sign being $+1$ if the arrow is outgoing from (i,j), -1 otherwise.*

Lemma 3 becomes very useful when the matrix Y is sparse, since the algorithm therein lends itself handily to the exploitation of structure and sparsity.

3. The Radius of $\mathfrak{so}(n)$ for $n \geq 4$

3.1. *The eigenstructure of $\mathfrak{so}(n)$ in $\mathbb{R}^m$*

The evaluation of the Frobenius norm of a commutator comes as something of an anticlimax, since the spectrum of the *restricted* commutator operator can be evaluated with relative ease. We have already noted that the eigenvalues of the *full* commutator operator, acting in $\mathbb{R}^{n^2}$, are $\{i(\lambda_k - \lambda_l) \, : \, k,l = 1,2,\ldots,n\}$, where $\sigma(X) = \{i\lambda_1, i\lambda_2, \ldots, i\lambda_n\}$. Our contention is that $m = (1/2)(n-1)n$ of these eigenvalues survive intact once we consider the restricted commutator.

To this end, we commence by revisiting the classical analysis of the eigenstructure of the full commutator. Thus, suppose that $X \in \mathrm{M}_n[\mathbb{R}]$ has a full set of eigenvectors, therefore $X = VDV^{-1}$, where $D = \mathrm{diag}\,\boldsymbol{\lambda}$. For every $k,l = 1,2,\ldots,n$, $k \neq l$, we set $E_{k,l} \in \mathrm{M}_n[\mathbb{R}]$ as a zero matrix, except for a unit element at the (k,l) entry. Therefore

$$V^{-1}[X, E_{k,l}]V = DE_{k,l} - E_{k,l}D = (\lambda_k - \lambda_l)E_{k,l}$$

and $[X, W_{k,l}] = (\lambda_k - \lambda_l)W_{k,l}$, where $W_{k,l} = VE_{k,l}V^{-1}$, $k,l = 1,2,\ldots,n$. However, if X resides in a Lie algebra $\mathfrak{g}$, we cannot expect $W_{k,l}$ to belong to $\mathfrak{g}$: If $\mathfrak{g} = \mathfrak{so}(n)$ then this is in general false.

Suppose that $X \in \mathfrak{so}(n)$ and assume that $n = 2N$ — the case of an odd n will be addressed

briefly in the sequel. We set $J = \left[\begin{smallmatrix} 0 & 1 \\ -1 & 0 \end{smallmatrix}\right]$. Then there exists a matrix $Q \in \mathrm{SO}(n)$ such that

$$QXQ^\top = A = \begin{bmatrix} A_1 & O & \cdots & O \\ O & A_2 & \ddots & \vdots \\ \vdots & \ddots & \ddots & O \\ O & \cdots & O & A_N \end{bmatrix},$$

where $A_k = \alpha_k J$, $k = 1,2,\ldots,N$. Note that the eigenvalues of X are $\pm i\alpha_k$, $k = 1,2,\ldots,N$.

Choose $1 \leq k < l \leq N$ and let $V \in \mathfrak{so}(n,\mathbb{C})$ be a zero matrix, except that

$$\begin{bmatrix} V_{2k-1,2l-1} & V_{2k-1,2l} \\ V_{2k,2l-1} & V_{2k,2l} \end{bmatrix} = U,$$

$$\begin{bmatrix} V_{2l-1,2k-1} & V_{2l-1,2k} \\ V_{2l,2k-1} & V_{2l,2k} \end{bmatrix} = -U,$$

where $U = \left[\begin{smallmatrix} u_1 & u_2 \\ u_3 & u_4 \end{smallmatrix}\right]$. Letting $Z = [A, V]$, we observe that all the entries of Z vanish, except for

$$\begin{bmatrix} Z_{2k-1,2l-1} & Z_{2k-1,2l} \\ Z_{2k,2l-1} & Z_{2k,2l} \end{bmatrix} = A_k U - UA_l,$$

$$\begin{bmatrix} Z_{2l-1,2k-1} & Z_{2l-1,2k} \\ Z_{2l,2k-1} & Z_{2l,2k} \end{bmatrix} = UA_k - A_lU.$$

Assume that $\gamma \in \mathbb{C}$ and $\mathbf{u} \neq \mathbf{0}$ are an eigenvalue and an eigenvector, respectively, of the matrix

$$\begin{bmatrix} 0 & \alpha_l & \alpha_k & 0 \\ -\alpha_l & 0 & 0 & \alpha_k \\ -\alpha_k & 0 & 0 & \alpha_l \\ 0 & -\alpha_k & -\alpha_l & 0 \end{bmatrix}.$$

Then $A_kU - UA_l = \gamma U$ and it follows that $\boldsymbol{\nu}(V)$ is an eigenvector of C_A, corresponding to the eigenvalue γ. This results for each $k < l$ in *four* eigenvalue/eigenvector pairs,

$$\gamma = i(\alpha_k + \alpha_l), \quad U = \begin{bmatrix} 1 & i \\ i & 1 \end{bmatrix};$$

$$\gamma = i(-\alpha_k + \alpha_l), \quad U = \begin{bmatrix} 1 & i \\ -i & 1 \end{bmatrix};$$

$$\gamma = i(\alpha_k - \alpha_l), \quad U = \begin{bmatrix} 1 & -i \\ i & 1 \end{bmatrix};$$

$$\gamma = i(-\alpha_k - \alpha_l), \quad U = \begin{bmatrix} 1 & -i \\ -i & 1 \end{bmatrix}.$$

Altogether, this results in $(1/2)(N-1)N = (1/2)(n-1)n - (1/2)n$ eigenvalues of C_A.

The remaining $N = (1/2)n$ eigenvalues of C_A are zero. This is easy to verify by letting, for any $k = 1, 2, \ldots, N$, $V \in \mathfrak{so}(n, \mathbb{C})$ be zero, except that

$$\begin{bmatrix} V_{2k-1,2k-1} & V_{2k-1,2k} \\ V_{2k,2k-1} & V_{2k,2k} \end{bmatrix} = J,$$

whence $[A, V] = O$.

Once we have determined $\sigma(C_A)$, we note that $\sigma(C_X) = \sigma(C_A)$ whenever X and A are similar, since $X = QAQ^{-1}$ means that

$$C_X \boldsymbol{\nu}(Y) = \boldsymbol{\nu}(Z) \Leftrightarrow C_A \boldsymbol{\nu}(Q^{-1}YQ) = \boldsymbol{\nu}(Q^{-1}ZQ).$$

Lemma 4. *Suppose that $n = 2N$ and that the eigenvalues of $X \in \mathfrak{so}(n)$ are $\pm i\alpha_k$, $k = 1, 2, \ldots, N$. Then the eigenvalues of the restricted commutator C_X are*

$$i(\pm\alpha_k \pm \alpha_l), \quad 1 \leq k < l \leq N, \tag{10}$$

as well as a zero eigenvalue of multiplicity N.

We note as an aside that we have just determined that the *centralizer* of $X \in \mathfrak{so}(n)$ is $(n/2)$-dimensional, as well as presenting its basis.

Lemma 5. *Suppose that $n = 2N + 1$ and that the eigenvalues of $X \in \mathfrak{so}(n)$ are $\pm i\alpha_k$, $k = 1, 2, \ldots, N$ and zero. Then the eigenvalues of the restricted commutator C_X are*

$$i(\pm\alpha_k \pm \alpha_l), \quad 1 \leq k < l \leq N, \tag{11}$$
$$\pm i\alpha_k, \quad 1 \leq k \leq N, \tag{12}$$

as well as a zero eigenvalue of multiplicity N.

Proof. Since $X \in \mathfrak{so}(n)$ is necessarily singular, we need to add to A a bottom row and rightmost column of zeros: We denote the new, $(2N+1)\times(2N+1)$ matrix by $\tilde{A}$. All the eigenvectors of C_A, suitably padded by zeros, can be extended to $C_{\tilde{A}}$. Moreover, let $\mathbf{v} \in \mathbb{C}^{2N}$ be a nonzero eigenvector of $\tilde{A}$ with an eigenvalue $i\gamma$ and set

$$V = \begin{bmatrix} O & \mathbf{v} \\ -\mathbf{v}^T & 0 \end{bmatrix}.$$

Then we can easily verify that $[\tilde{A}, V] = i\gamma V$. Hence we recover the eigenvalues (12). Altogether we have $N(2N+1)$ eigenvalues, hence the full spectrum of $C_{\tilde{A}}$. Since the spectrum of the restricted commutator is invariant under similarity transformation, the proof is complete. ∎

3.2. *The radius of $\mathfrak{so}(n)$*

Up to $\sqrt{2}$, measuring $\mathfrak{so}(n)$ in the Frobenius norm is the same as using the Euclidean norm in $\mathbb{R}^{\frac{1}{2}(n-1)n}$, $\|X\|_{\mathsf{F}} = \sqrt{2}\|\boldsymbol{\nu}(X)\|_2$. Moreover, C_X is skew symmetric, therefore normal, and its Euclidean norm coincides with its spectral radius.

Theorem 6. *For every $n \geq 4$ it is true that*

$$\omega(\mathfrak{so}(n)) = \sqrt{2}. \tag{13}$$

Proof. We commence with even $n = 2N$ and assume that, without loss of generality,

$$|\alpha_1| \geq |\alpha_2| \geq \cdots \geq |\alpha_N|.$$

Therefore, according to (10),

$$\|C_X\|_2 = \rho(C_X) = |\alpha_1| + |\alpha_2|.$$

Since $\|X\|_{\mathsf{F}}^2 = \sum_{\lambda \in \sigma(X)} |\lambda|^2$, we deduce that

$$\frac{\|C_X\|_2}{\|\boldsymbol{\nu}(X)\|_2} = \frac{|\alpha_1| + |\alpha_2|}{\frac{1}{\sqrt{2}}\|X\|_{\mathsf{F}}} = \frac{|\alpha_1| + |\alpha_2|}{\sqrt{\sum_{k=1}^{N} |\alpha_k|^2}}$$

$$\leq \frac{|\alpha_1| + |\alpha_2|}{\sqrt{|\alpha_1|^2 + |\alpha_2|^2}} \leq \sqrt{2},$$

with the upper bound attainable when $\alpha_1 = \alpha_2 > 0$, $\alpha_k = 0$, $k \geq 3$, which corresponds to an embedding of $\mathfrak{so}(4)$ in the algebra. Note that the inequality above holds by Young's inequality for $p = 2$, i.e. we have $2|\alpha_1||\alpha_2| \leq |\alpha_1|^2 + |\alpha_2|^2$.

Therefore $\omega(\mathfrak{so}(2N)) = \sqrt{2}$.

The proof for $n = 2N + 1$ is virtually identical, since

$$\rho(C_X) = \max\left\{ \max_{1 \leq k < l \leq N} |\alpha_k| + |\alpha_l|, \max_{k=1,2,\ldots,N} |\alpha_k| \right\}$$

$$= |\alpha_1| + |\alpha_2|,$$

and we again obtain the radius (13). ∎

Example. It is instructive to analyse the special case $\mathfrak{so}(4)$. Using the structure constants one can compute that for

$$X = \begin{bmatrix} 0 & x_1 & x_2 & x_3 \\ -x_1 & 0 & x_4 & x_5 \\ -x_2 & -x_4 & 0 & x_6 \\ -x_3 & -x_5 & -x_6 & 0 \end{bmatrix},$$

we have

$$C_X = \begin{bmatrix} 0 & x_4 & x_5 & -x_2 & -x_3 & 0 \\ -x_4 & 0 & x_6 & x_1 & 0 & -x_3 \\ -x_5 & -x_6 & 0 & 0 & x_1 & x_2 \\ x_2 & -x_1 & 0 & 0 & x_6 & -x_5 \\ x_3 & 0 & -x_1 & -x_6 & 0 & x_4 \\ 0 & x_3 & -x_2 & x_5 & -x_4 & 0 \end{bmatrix}.$$

The 2-norm of C_X may then be computed to be $\|x\|^2 + 2|x_1 x_6 - x_2 x_5 + x_3 x_4|$. Using Lagrange multipliers to maximize this subject to $\|x\| = 1$ yields indeed that $\omega(\mathfrak{so}(4)) \le \sqrt{2}$.

It is of interest in fact to characterize all $\mathfrak{so}(4)$ matrices whose restricted commutator has the norm $\sqrt{2}$. These take either the form

$$X = \begin{bmatrix} 0 & a & b & c \\ -a & 0 & c & -b \\ -b & -c & 0 & a \\ -c & b & -a & 0 \end{bmatrix},$$

or

$$X = \begin{bmatrix} 0 & a & b & c \\ -a & 0 & -c & b \\ -b & c & 0 & -a \\ -c & -b & a & 0 \end{bmatrix}$$

for arbitrary $a, b, c \in \mathbb{R}$ which are not all zero.

In each case the spectrum of C_X consists of four zero eigenvalues and $\pm i 2(a^2 + b^2 + c^2)^{1/2}$. Hence $\|X\|_F = 2(a^2 + b^2 + c^2)^{1/2}$ and thus $\|\boldsymbol{\nu}(X)\|_2 = \sqrt{2}(a^2 + b^2 + c^2)^{1/2}$ and $\|C_X\|_2 = 2(a^2 + b^2 + c^2)^{1/2}$ and it follows that $\|C_X\|_2 / \|\boldsymbol{\nu}(X)\|_2 = \sqrt{2}$.

4. Conclusion

We have defined the radius of a Lie algebra and computed its value for $\mathfrak{so}(n)$ and the Frobenius norm. It is of interest to compute the radius for other Lie algebras. We intend to do this in a future publication. In generalizing the work here one needs to distinguish between compact and noncompact Lie algebras (where the Killing form is definite and indefinite, respectively) and between real and complex algebras. The compact real form of a complex Lie algebra is natural to look at — for example $\mathfrak{su}(n)$, the compact real form of $\mathfrak{sl}(n, \mathbb{C})$. In the case of $\mathfrak{su}(2)$ one has of course a Lie algebra isomorphism

between $\mathfrak{su}(2)$ and $\mathfrak{so}(3)$ and $\mathbb{R}^3$ endowed with the cross product. The map in this case is given by (see e.g. [Marsden & Ratiu, 1999])

$$\begin{bmatrix} x_1 \\ x_2 \\ x_3 \end{bmatrix} \to X = \frac{1}{2} \begin{bmatrix} -\mathrm{i} x_3 & -\mathrm{i} x_1 - x_2 \\ -\mathrm{i} x_1 + x_2 & \mathrm{i} x_3 \end{bmatrix}.$$

Thus our earlier argument for $\mathfrak{so}(3)$ shows that $\omega(\mathfrak{su}(2)) = 1$ with respect to the norm induced by the vector norm on $\mathbb{R}^3$.

Acknowledgments

We would like to thank Brad Baxter, Reng-Cang Li, Elizabeth Mansfield, Alexei Shadrin and Mike Shub for useful comments. We would like also to thank the referee whose comments greatly improved the exposition. The research of AMB was supported in part by the National Science Foundation.

References

Brockett, R. [1993] "Differential geometry and the design of gradient algorithms," *Proc. Symp. Pure Math.* **54**, 69–92.

Brockett, R. [1994] "Differential equations and matrix inequalities on isospectral families," *Lin. Alg. Appl.* **203/204**, 189–207.

Carter, R., Segal, G. & Macdonald, I. [1995] *Lectures on Lie Groups and Lie Algebras* (Cambridge University Press, Cambridge).

Chartrand, G. & Lesniak, L. [1986] *Graphs and Digraphs* (Wadsworth & Brooks/Cole).

Hille, E. [1969] *Lectures on Ordinary Differential Equations* (Addison-Wesley, Reading, MA).

Horn, R. A. & Johnson, C. R. [1994] *Topics in Matrix Analysis* (Cambridge University Press, Cambridge).

Humphreys, J. E. [1978] *Introduction to Lie Algebras and Representation Theory* (Springer-Verlag, Berlin).

Iserles, A., Munthe-Kaas, H. Z., Nørsett, S. P. & Zanna, A. [2000] "Lie-group methods," *Acta Numerica* **9**, 215–365.

Kobayashi, S. & Nomizu, K. [1969] *Foundations of Differential Geometry* (John Wiley, NY).

Marsden, J. E. & Ratiu, T. [1984] *Introduction to Mechanics and Symmetry* (Springer-Verlag, NY).

Varadarajan, V. S. [1984] *Lie Groups, Lie Algebras, and Their Representations* (Springer-Verlag, NY).

von Neumann, J. [1937] "Some matrix inequalities and metrization of matrix space," *Tomsk Univ. Rev.* **1**, 286–300; Reprinted [1962] in *Collected Works* (Pergamon, Oxford), Vol. IV, pp. 205–218.

SIMPLE NEURAL NETWORKS THAT OPTIMIZE DECISIONS

ERIC BROWN*, JUAN GAO[†], PHILIP HOLMES*[,†], RAFAL BOGACZ*[,‡],

MARK GILZENRAT[‡], and JONATHAN D. COHEN[‡]

*Program in Applied and Computational Mathematics,
[†]Department of Mechanical and Aerospace Engineering,
[‡]Department of Psychology,
Princeton University, Princeton, NJ 08544, USA

Received April 2, 2004; Revised July 7, 2004

We review simple connectionist and firing rate models for mutually inhibiting pools of neurons that discriminate between pairs of stimuli. Both are two-dimensional nonlinear stochastic ordinary differential equations, and although they differ in how inputs and stimuli enter, we show that they are equivalent under state variable and parameter coordinate changes. A key parameter is gain: the maximum slope of the sigmoidal activation function. We develop piecewise-linear and purely linear models, and one-dimensional reductions to Ornstein–Uhlenbeck processes that can be viewed as linear filters, and show that reaction time and error rate statistics are well approximated by these simpler models. We then pose and solve the optimal gain problem for the Ornstein–Uhlenbeck processes, finding explicit gain schedules that minimize error rates for time-varying stimuli. We relate these to time courses of norepinephrine release in cortical areas, and argue that transient firing rate changes in the brainstem nucleus locus coeruleus may be responsible for approximate gain optimization.

Keywords: Gain; neural network model; decision task; stochastic differential equation; reaction time; optimal speed and accuracy; matched filter; locus coeruleus.

1. Introduction

The psychological and neural bases of decision making are active areas of inquiry in cognitive science [Schall, 2001; Gold & Shadlen, 2001; Schall et al., 2002; Gold & Shadlen, 2002; Shadlen & Newsome, 2001; Platt & Glimcher, 1999; Stone, 1960; Laming, 1968; Ratcliff, 1978; Ratcliff et al., 1999; Usher & McClelland, 2001; Roitman & Shadlen, 2002; Wang, 2002]. There is a wealth of data on simple decision tasks which require discrimination among alternative stimuli as quickly and accurately as possible. Typically, this discriminatory process has been modeled as a competition among different neural populations, each representing alternate interpretations of the current stimulus [Cohen et al., 1990; Usher & McClelland, 2001]. Recent direct recordings in visual and motor areas of monkeys performing sensory discrimination tasks support this interpretation by revealing that, following training, certain "decision" neurons become selective for different stimulus alternatives, and upon presentation of the relevant stimulus their firing rates gradually increase accordingly; when these rates cross thresholds, the corresponding behavioral response is initiated (e.g. [Schall, 2001; Gold & Shadlen, 2001; Schall et al., 2002; Roitman & Shadlen, 2002; Gold & Shadlen, 2002]). This neural evidence adds to behavioral evidence noted below, suggesting that decisions are made by comparing integrated "weights of evidence", encoded by the firing rates of neural groups. Here, we explore the computational mechanisms required to optimize such a process.

The stimuli relevant to making a decision are often not static: their saliences may change over time. In the simplest case, a change occurs only at the moment when the stimulus itself appears. This is typically modeled in simulations of decision tasks (e.g. in [Cohen & Huston, 1994; Brown & Holmes, 2001; Cho *et al.*, 2002], cf. [Laming, 1968]) by dividing the task into two distinct periods: a preparatory period, in which no stimulus is present, and a trial period, in which a stimulus of constant discriminability is presented. Alternatively, stimulus discriminability may change in a stepwise manner or vary continuously.

The following specific example motivates our analysis of two specific cases in Sec. 2.5. In the "moving dots" paradigm of the two alternative forced choice task [Britten *et al.*, 1993; Shadlen & Newsome, 2001; Gold & Shadlen, 2002] a display of moving dots is presented, and the subject must indicate whether a majority of dots is moving to the right or the left. In the simplest case, the subject focuses on a neutral fixation point during the preparatory period, after which the dots appear, with a certain "coherent" fraction moving either left or right, and the rest moving randomly. A variant is obtained by showing a zero coherence display of dots during the preparatory period, and suddenly increasing coherence to a fixed value.

Even if external stimuli have constant strengths, their representations in neural populations that decide between alternative hypotheses may *gradually* rise, due to accumulating activity in input layers, fluctuations in attention, or both [Mozer, 1988; Cohen *et al.*, 1992; Usher *et al.*, 1999; Gilzenrat *et al.*, 2002]. Another possible source of time varying salience is the increasing noise levels that may accompany higher firing rates. A richer situation, in which the stimulus salience increases and decreases over time, is explored in [Huk *et al.*, 2002]. A focus of the present paper is how stimuli with time-dependent salience can be *optimally* processed in simple neurally-based models of decision networks. We study the reduction of such networks to linearized, one-dimensional approximations (cf. [Usher & McClelland, 2001; Brown & Holmes, 2001; Bogacz *et al.*, 2004]) for which optimality conditions can be fully characterized, and identify two distinct mechanisms, one involving intrinsic properties of decision networks and the other involving external modulation, that

can implement optimal processing of time-varying stimuli.

Optimality principles have found wide application in psychology and neuroscience (e.g. [Bialek *et al.*, 1991; Anderson, 1990; Fairhall *et al.*, 2001]). In particular, Stone [1960] applied the optimal Sequential Probability Ratio Test (SPRT) to model behavioral data in a two-alternative forced choice task. This was followed by the extensive work of Laming [1968]. The SPRT computes time-dependent likelihood ratios between the probabilities of two competing hypotheses, a procedure equivalent to the signal processing strategy that maximizes signal-to-noise ratio in the difference between two incoming stimuli. For stimuli with constant signal-to-noise ratios, the SPRT is equivalent, in an appropriate continuum limit, to the constant-drift diffusion model, which has been shown by Ratcliff and others to fit a wide variety of behavioral data (see [Ratcliff, 1978; Ratcliff *et al.*, 1999] and references therein) and also to describe the dynamics of neural firing rates in sensori-motor brain areas [Schall *et al.*, 2002; Gold & Shadlen, 2002], cf. [Smith & Ratcliff, 2004]. Specifically, in [Gold & Shadlen, 2002], the notion of reward rate is introduced for the constant-drift diffusion model, and [Bogacz *et al.*, 2004] shows that higher performing subjects do optimize this quantity in a specific behavioral task. However, although [Laming, 1968] does allow for accumulation of noise to have occurred before stimulus presentation (see Laming's Appendix A7), in all these studies the decision process is modeled only *after* presentation of a stimulus having constant signal-to-noise ratio; furthermore, the parameters describing processing of incoming information are not explicitly allowed to vary in time.

In this paper we show how models of mutually inhibiting neural populations can make nearly optimal decisions about the identity of *time-varying* stimuli. This is accomplished via dynamical adjustments in an *effective* gain parameter for the linearized population dynamics. The gain determines the sensitivity of (equilibrium) population firing rates to changes in averaged input currents to the population, and the word "effective" is used here because these changes can result either from transient variations in the gain parameter describing this sensitivity or directly from the nonlinearities of neural input–output functions. There is much current research into neural mechanisms for the modulation of gain in neural populations, identifying such factors as levels of norepinephrine [Usher *et al.*, 1999] and the strength of fluctuations in individual neurons comprising the population (e.g. [Chance *et al.*, 2002; Amit & Tsodyks, 1991; Brunel *et al.*, 2001]). In particular, Shin *et al.* [1999] proposes a mechanism in which frequency-current curves of individual neurons adapt to match operating ranges to neural inputs, via intracellular calcium signals. This may be viewed as a biophysical implementation of the earlier "automatic gain control" (see Eq. (9) of [Grossberg, 1988] and references therein), which is implemented via multiplicative "shunting" terms in neural network models and also keeps neural units in the sensitive regimes of their input–output functions. Gain plays a different role in the present paper: we identify, for three different models, the distinct time-dependent (effective) gain schedules which implement optimal processing strategies for time-dependent signals. These provide predictions for gain manipulations that diverse neural mechanisms may implement to improve task performance.

The balance of the paper proceeds as follows. In Sec. 2 we introduce the forced and free response decision tasks, and three types of stochastic differential equation (SDE) models for these tasks. We show that two of these are related via a coordinate transformation, and discuss linearized and one-dimensional reductions of them, exploring the accuracy of these reductions in two rather general cases. In the following Sec. 3, we compute time-dependent values of gain that optimize signal processing in the one-dimensional models. This involves calculating gain functions that enable them to implement the classical signal processing notion of matched filters. Section 4 interprets these results in terms of cortical norepinephrine (NE) release mediated by the brainstem nucleus locus coeruleus (LC), showing that LC and NE dynamics indeed appear to approximate optimal time courses. Section 5 concludes the paper with a brief discussion.

Although we only consider simple models of a prototypical cognitive task, we believe that this paper is appropriate for a volume celebrating the centenary of John von Neumann's birth. Early in 1956 von Neumann was working on a manuscript in preparation for the Silliman memorial lectures at Yale, which he had been invited to deliver that Spring. Unfortunately, his final illness intervened and he entered the Walter Reed Hospital in April, where he remained until his death in February 1957. The lectures were never given, but his remarkable book, *The Computer and the Brain* [von Neumann, 1958], remains among his final work. In it, he makes elegant and simple estimates of human neural computational capacity based on notions drawn from the theory of analog and digital automata (which he had largely developed), and from information theory. Although neuronal spikes appear as 1's (and their absence as 0's), he argues that neural computation is necessarily inaccurate and noisy, and hence must be "statistical" rather than "digital." He points out that firing rates in sensory neurons tend to be monotone functions of stimulus strength and, as an early proponent of rate coding, he can be seen as pioneering the class of firing rate models treated here.

2. Models of Decision Tasks

2.1. *Decision tasks: The forced and free response protocols*

We consider two distinct tasks, both widely used in cognitive neuroscience, in each of which a decision maker must discriminate between two alternatives, henceforth denoted "1" and "2". The sensory information itself, as well as its neural representation, is assumed to be noisy, so that discrimination errors occur. The first task is the *forced-response* paradigm, in which subjects must respond at a fixed time T following stimulus onset with their best estimate of which alternative (1 or 2) was presented. Performance on this task is measured by the error rate, or one minus the fraction of correct responses. We will also refer to this as the *interrogation protocol*, noting that it is distinct from deadlining (not considered further here), in which subjects are apprised in advance of a fixed, maximal time *before* which all responses must be made.

In the second, *free-response* paradigm, decisions are not demanded at a preset time, but are given when the subject feels that sufficient evidence in favor of one alternative has accumulated. Since the sensory evidence is noisy, response times vary from trial to trial and performance under the free-response condition is characterized by both reaction times and error rates. Here, optimality requires an appropriate balance of speed and accuracy [Wickelgren, 1977; Gold & Shadlen, 2002; Bogacz *et al.*, 2004].

Following [Usher & McClelland, 2001] and others, we shall model both these tasks by a pair of competing (mutually inhibitory) neural populations, each of which is selectively responsive to sensory input corresponding to one of the two alternatives. In the forced-response protocol, the neural population with the highest firing rate at time T determines the decision. For free responses, the first of the two populations to cross a firing rate threshold establishes the choice. We do not address the (interesting) question of how thresholds are set or threshold crossings are detected.

2.2. *Two-dimensional nonlinear models and the neural gain parameter*

In this section we consider the dynamics of two mutually inhibiting neural populations, each of which receives noisy sensory input from components of the stimulus representing one of the alternatives. We describe two models for such populations, both in wide use, and both in the form of systems of stochastic ordinary differential equations (SDEs) [Arnold, 1974].

The first of these, the leaky integrator *connectionist* model [McClelland, 1979; Usher & McClelland, 2001], is:

$$\tau_c \frac{dx_1}{dt} = -x_1 - \beta f_{g(t)}(x_2) + a_1(t) + \frac{c(t)}{\sqrt{2}} \eta_t^1, \quad (1)$$

$$\tau_c \frac{dx_2}{dt} = -x_2 - \beta f_{g(t)}(x_1) + a_2(t) + \frac{c(t)}{\sqrt{2}} \eta_t^2, \quad (2)$$

where the state variables $x_j(t)$ denote the mean input currents to cell bodies of the jth neural population, the integration implicit in the differential equations modeling temporal summation of dendritic synaptic inputs ([Grossberg, 1988] and references therein). Additionally, the parameter β sets the strength of mutual inhibition via population firing rates $f_{g(t)}(x_j(t))$, where $f_{g(t)}(\cdot)$ is the sigmoidal "activation" (or "frequency-current" or neural "input–output") function to be described shortly. The stimulus signal received by each population is $a_j(t)$, and the noise terms polluting this signal are $c(t)\eta_t^j$, where $c(t)$ sets r.m.s. noise strength and the η_t^j are (independent) white noise processes with variance $E(\eta_t^j - \eta_{t'}^j)^2 = \delta(t-t')$. The time constant τ_c reflects the rate at which neural activities decay in the absence of inputs and respond to

input changes. Under the free-response paradigm a decision is made and the response initiated when the firing rate $f_{g(t)}(x_j)$ of either population first exceeds a preset threshold θ_j; it is normally assumed that $\theta_1 = \theta_2 = \theta$. For the interrogation protocol, the population with greatest activity (and also firing rate) at time T determines the decision. We also assume that activities decay to zero after response and prior to the next trial, so that the initial conditions for (1)–(2) are $x_j(0) = 0$.

The subscript in $f_{g(t)}(\cdot)$ indicates dependence on the time-varying gain, or sensitivity, $g(t)$ of the neural populations: gain sets the slope of the activation function. For example, the logistic function

$$f_{g(t)}(x) = \frac{1}{1 + \exp(-4g(t)(x - b))}$$

$$= \frac{1}{2}[1 + \tanh(2g(t)(x - b))] \quad (3)$$

has maximal slope $g(t)$ (see Fig. 1, left). While this specific form is not required for the results derived below, we do assume that f_g takes its time-dependent maximal slope $g(t)$ at some time-independent point, as for (3).

As already mentioned, the connectionist model describes the time evolution of current inputs. A second model is derived in [Wilson & Cowan, 1972], cf. [Hopfield, 1984; Abbott, 1991; Gerstner & Kistler, 2002], in which the firing rates of neural populations are themselves integrated over time. First we give the linearized version of this *firing rate* model:

$$\tau_c \frac{dy_1}{dt} = -y_1 + f_{g(t)}^l \left(-\beta y_2 + a_1(t) + \frac{c(t)}{\sqrt{2}} \eta_t^1 \right), \quad (4)$$

$$\tau_c \frac{dy_2}{dt} = -y_2 + f_{g(t)}^l \left(-\beta y_1 + a_2(t) + \frac{c(t)}{\sqrt{2}} \eta_t^2 \right). \quad (5)$$

Here, the y_j are the firing rates of population j and other terms are as above. The linear function

$$f_{g(t)}^l(x) = \frac{1}{2} + g(t)(x - b), \quad (6)$$

derives from replacing the logistic (or any similar monotonic) function by the linear approximation $f_{g(t)}^l(\cdot)$ around its point of maximal slope. Note that the firing rate y_j of the jth population approaches an equilibrium set by the input currents to this population, passed through the (linearized) frequency-current function. This model must be reformulated to allow for nonlinear functions $f_{g(t)}$, because white noise does not make sense as an argument in such a function, cf. [Gardiner, 1985]. In particular, we assume that, as in (4)–(5), the strength of firing rate fluctuations in response to noise in inputs scales with $g(t)$ (i.e. with the maximal sensitivity of firing rates to the deterministic component of the input). This yields

$$\tau_c \frac{dy_1}{dt} = -y_1 + f_{g(t)}(-\beta y_2 + a_1(t)) + g(t)\frac{c(t)}{\sqrt{2}}\eta_t^1,$$
$$(7)$$

$$\tau_c \frac{dy_2}{dt} = -y_2 + f_{g(t)}(-\beta y_1 + a_2(t)) + g(t)\frac{c(t)}{\sqrt{2}}\eta_t^2,$$
$$(8)$$

which is valid for all $f(\cdot)$ and reduces to the form (4)–(5) for linear $f(\cdot)$. Note that the firing rate model (7)–(8) is a standard two-unit recurrent neural network with additive noise [Hertz *et al.*, 1991]. As above, we take initial conditions $y_j(0) = 0$, and note that threshold-crossing in the free-response case is detected directly via $y_j = \theta_j$.

For the questions of optimal stimulus processing addressed here, the most important distinction between the connectionist (1)–(2) and firing rate (4–5)–(7–8) models is whether the inputs $a_j(t) + c(t)/\sqrt{2}\eta_t^j$ enter as separate additive terms, as in the former, or as arguments to the activation function $f_g(t)$, as in the latter. As explained at the end of Sec. 3, this determines whether changes in gain directly adjust the sensitivity of neural units to all inputs or just to feedback from the competing unit, and it results in qualitatively different predictions for optimal gain schedules in the two models. While we expect that future work on low-dimensional descriptions of the population dynamics of spiking neurons (extending, e.g. [Brunel *et al.*, 2001; Wang, 2002; Omurtag *et al.*, 2000; Shelley & McLaughlin, 2002; Ermentrout, 1994] to include neurotransmitter effects) will result in more refined models, here we study the "simple" connectionist and firing rate descriptions. Throughout, we use variables x_j in referring to the former and y_j to the latter.

2.3. *Equivalence of the firing rate and connectionist models*

We now show that the firing rate and connectionist models are equivalent under a (generally time-dependent) coordinate change and corresponding adjustment of parameters, initial conditions, and thresholds. Specifically, for any activation function that is odd around some input value, such as (3), (7)–(8) can be written in the form (1)–(2). Hence, for every parameterization of the firing rate model, there is a connectionist model that produces identical trajectories as well as error rate and reaction time statistics, and vice-versa. This shows that the two models are effectively equivalent, up to parameterization. However, in Sec. 3 below we demonstrate that, because of the different ways that gain $g(t)$ enters them, their optimal gain trajectories differ significantly.

Starting with Eqs. (7)–(8), we extend the S–Σ exchange transformation of Grossberg [1988] to define the new coordinates

$$\tilde{y}_1 = 2b + \beta y_1 - a_2, \quad \tilde{y}_2 = 2b + \beta y_2 - a_1, \quad (9)$$

so that $-\beta y_1 + a_2 = -\tilde{y}_1 + 2b$ and $-\beta y_2 + a_1 = -\tilde{y}_2 + 2b$. In terms of these (7)–(8) become

$$\tau_c \frac{d\tilde{y}_1}{dt} = \beta\left[-\frac{1}{\beta}(a_2 + \tilde{y}_1 - 2b) + f_{g(t)}(-\tilde{y}_2 + 2b)\right.$$
$$\left. + g(t)\frac{c(t)}{\sqrt{2}}\eta_t^1\right] - \frac{da_2}{dt}, \quad (10)$$

$$\tau_c \frac{d\tilde{y}_2}{dt} = \beta\left[-\frac{1}{\beta}(a_1 + \tilde{y}_2 - 2b) + f_{g(t)}(-\tilde{y}_1 + 2b)\right.$$
$$\left. + g(t)\frac{c(t)}{\sqrt{2}}\eta_t^2\right] - \frac{da_1}{dt}, \quad (11)$$

and using the following property of the logistic activation function (3):

$$f_{g(t)}(-\xi + 2b) = \frac{1}{2}[1 + \tanh(2g(t)[-\xi + 2b - b])]$$

$$= \frac{1}{2}[1 + \tanh(-2g(t)[\xi - b])]$$

$$= \frac{1}{2}[1 - \tanh(2g(t)[\xi - b])]$$

$$= 1 - \frac{1}{2}[1 + \tanh(2g(t)[\xi - b])]$$

$$= 1 - f_{g(t)}(\xi), \quad (12)$$

(10)–(11) become

$$\tau_c \frac{d\tilde{y}_1}{dt} = -\tilde{y}_1 - \beta f_{g(t)}(\tilde{y}_2) - a_2 - \dot{a}_2$$

$$+ 2b + \beta + \beta g(t)\frac{c(t)}{\sqrt{2}}\eta_t^1,$$

$$\tau_c \frac{d\tilde{y}_2}{dt} = -\tilde{y}_2 - \beta f_{g(t)}(\tilde{y}_1) - a_1 - \dot{a}_1$$

$$+ 2b + \beta + \beta g(t)\frac{c(t)}{\sqrt{2}}\eta_t^2.$$

This SDE has the same form as (1)–(2) with parameters mapped as follows:

$$a_1 \mapsto 2b + \beta - a_2 - \dot{a}_2,$$
$$a_2 \mapsto 2b + \beta - a_1 - \dot{a}_1. \tag{13}$$

The firing rate model (7)–(8) therefore produces identical statistics to the connectionist model (1)–(2) with appropriately remapped parameters and state variables. Note that thresholds and initial conditions for the firing rate variables y_1, y_2 must be transformed under (9) to apply to the equivalent connectionist model, that a_1 and a_2 are interchanged in the inputs, and that the noise terms are multiplied by gain $g(t)$.

2.4.   *Piecewise-linear approximations*

As in [Usher & McClelland, 2001; Brown & Holmes, 2001], Eq. (3) may be approximated by a piecewise-linear function:

$$f_{g(t)}(\xi)$$
$$\approx f_{g(t)}^{pw}(\xi)$$
$$= \begin{cases} 0 & \text{for } \xi \in \left(-\infty, b - \dfrac{1}{2g}\right] \\[2ex] \dfrac{1}{2} + g(t)(\xi - b) & \text{for } \xi \in \left[b - \dfrac{1}{2g}, b + \dfrac{1}{2g}\right], \\[2ex] 1 & \text{for } \xi \in \left[b + \dfrac{1}{2g}, \infty\right) \end{cases}$$
$$\tag{14}$$

as illustrated in Fig. 1. Note that our choice to set the slope of $f_{g(t)}^{pw}$ in its central domain equal to the maximal slope $g(t)$ of the nonlinear function $f_{g(t)}$ does not minimize the distance between the two functions in the L^∞ or L^2 norms. The best L^∞ match is obtained by setting the maximal slope of $f_{g(t)}^{pw}$ equal to $0.71g(t)$, and in L^2 by a $g(t)$-dependent value ranging between $0.72g(t)$ and $0.76g(t)$ (for $g(t)$ between 0.25 and 3). However, all

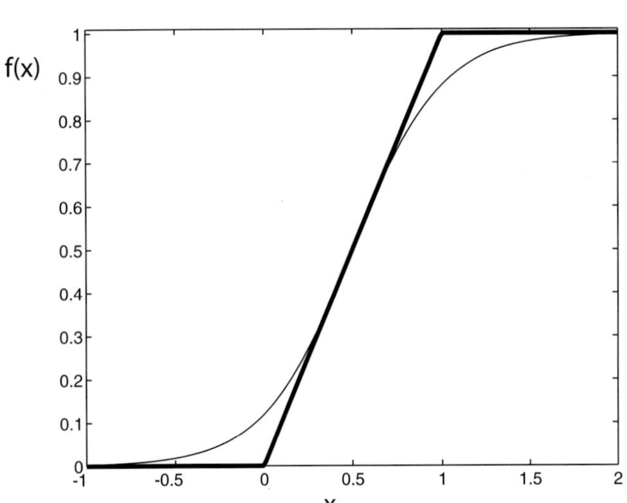
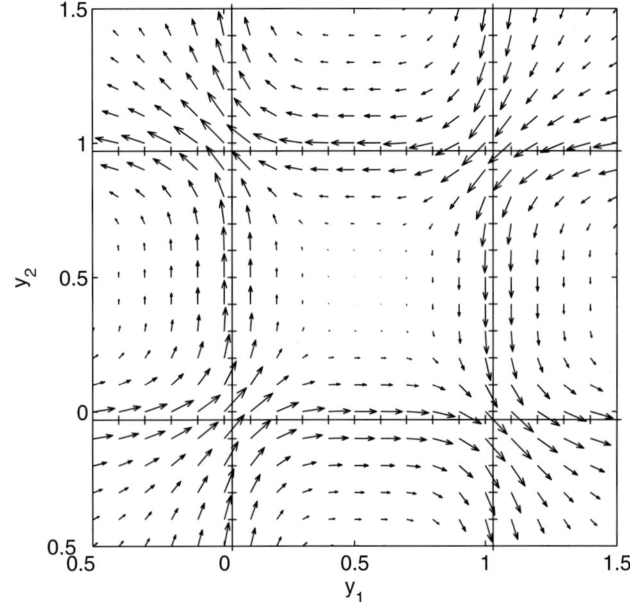

Fig. 1.   (Left) Comparison of logistic and piecewise-linear activation functions; $g = 1, b = 0.5$. (Right) Comparison of logistic and piecewise-linear vectorfields $F(y_1, y_2)$ and $F^{pw}(y_1, y_2)$ for the piecewise-linear firing rate model (15)–(16): the difference $F(y_1, y_2) - F_{pw}(y_1, y_2)$ is plotted. Also shown for reference are the nine phase space tiles described in Fig. 2. Here additionally $\tau_c = 1, \beta = 1, a_1 = 1.03, a_2 = 0.97$.

these choices result in similar error rate and reaction time statistics, and we use (14) in what follows.

For ease of reference, we rewrite Eqs. (7)–(8) following piecewise linearization:

$$\tau_c \frac{dy_1}{dt} = -y_1 + f_{g(t)}^{pw}(-\beta y_2 + a_1(t)) + g(t)\frac{c(t)}{\sqrt{2}}\eta_t^1, \tag{15}$$

$$\tau_c \frac{dy_2}{dt} = -y_2 + f_{g(t)}^{pw}(-\beta y_1 + a_2(t)) + g(t)\frac{c(t)}{\sqrt{2}}\eta_t^2. \tag{16}$$

The difference between the vectorfield of the fully nonlinear model (7)–(8) and that of (15)–(16) is illustrated in Fig. 1 (right) for a specific parameter choice. In Sec. 2.6 below, we shall explicitly compare reaction times and error rates predicted by these two models.

The (y_1, y_2) phase space of the piecewise-linear firing rate model (and of the analogous connectionist model) is tiled by nine regions divided by pairs of horizontal and vertical lines at the break points of f_g^{pw}, each having a distinct linear vectorfield: see Fig. 2. In the following section, we will describe two cases in which this tiled structure can be used to reduce Eqs. (7)–(8) to a one-dimensional system.

2.5. *Representing decision dynamics in one dimension*

As discussed above and in [Usher & McClelland, 2001], in the forced response protocol, the choice $j = 1$ or 2 is made according to which of the two neural populations has the greatest activity or firing rate at interrogation time T. Therefore, knowledge of the difference

$$y(T) \stackrel{\triangle}{=} y_1(T) - y_2(T) \quad \text{or} \tag{17}$$
$$x(T) \stackrel{\triangle}{=} x_1(T) - x_2(T)$$

determines the outcome and reduction of the original two-dimensional problem to a single variable does not *inherently* imply any loss in accuracy. For example, if the difference in firing rates is described by a time-dependent probability density $p(y, t)$ (whose distribution represents variability across behavioral trials), then the error rate at

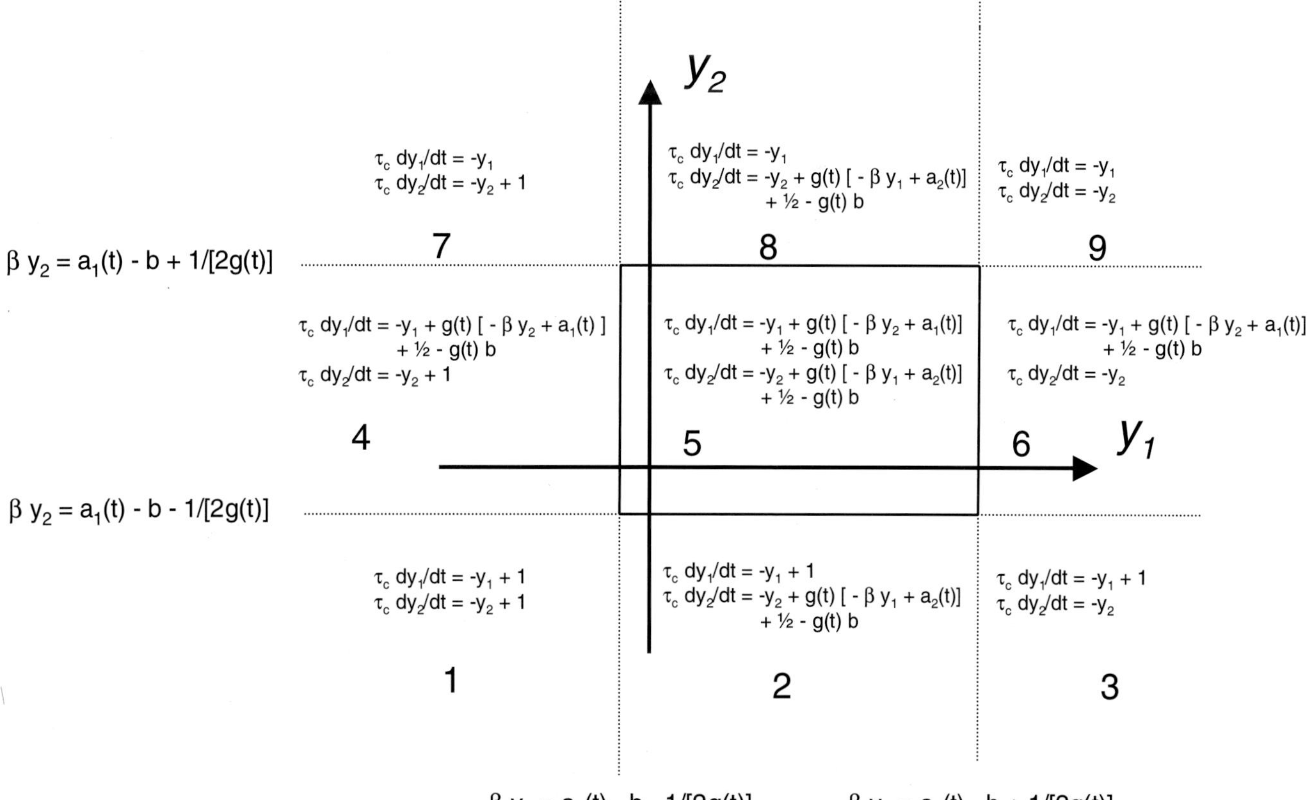

Fig. 2.   The piecewise-linear vectorfield of the firing rate model (15)–(16). The central tile is surrounded by a solid box.

interrogation time T is

$$ER = \int_0^\infty p(y, T)dy \tag{18}$$

if alternative 2 was presented (i.e. if $a_2 > a_1$ for $t > t_s$), and

$$ER = \int_{-\infty}^0 p(y, T)dy \tag{19}$$

if alternative 1 was presented. Similar conclusions hold for the connectionist model.

For the free choice protocol the situation is more subtle. The single variable x or y is sufficient to characterize the decision only if the probability density of solutions to (7)–(8) or (1)–(2) has approximately collapsed along a one-dimensional "decision manifold" $\mathcal{M}$ by the time the threshold is crossed; see Fig. 3. In this sketch, the decision manifold, parameterized by y, is the unstable, center or weak stable manifold [Guckenheimer & Holmes, 1983] of the indicated fixed point, which, for the linearized system, coincides with its eigenspace.

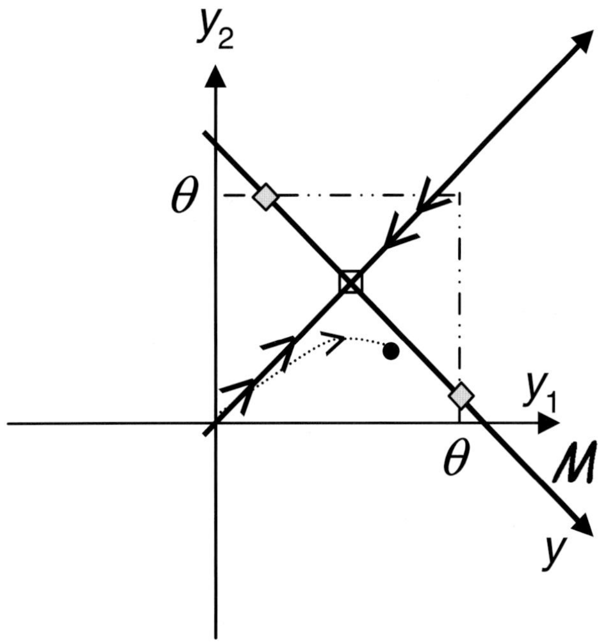

Fig. 3. Reduction to one dimension. The coordinate y (or x) of Eq. (17) parameterizes the decision manifold $\mathcal{M}$ (see text): the invariant manifold containing the fixed point indicated by the square. In the free response protocol, collapse of noisy solutions along $\mathcal{M}$ is required for accurate description in one dimension (cf. Figs. 4 and 5 (right)) so that sample paths (dotted line and point) cross thresholds arbitrarily close to the intersections of $\mathcal{M}$ with the thresholds $y_j = 0$. This is not required for the forced response/interrogation protocol, in which the probability density $p(y, t)$ is simply cut along $y_1 = y_2$ at $t = T$.

The existence of center manifolds $\mathcal{M}$ for SDEs with additive noise, such as those considered here, has been proven rather generally: see [Boxler, 1991] and [Arnold, 1998, Chap. 7]; also [Knobloch & Wiesenfeld, 1983] for an early analysis and explicit examples. However, here we consider only the fully linear and piecewise linear systems, for which the "diagonal" coordinates $y = y_1 - y_2$, $\tilde{y} = y_1 + y_2$ and assumption of independent white noise processes decouple the components of (7)–(8) (and analogously of (1)–(2)) [Bogacz et al., 2004], and so we do not need the full power of these results.

For collapse to $\mathcal{M}$ to occur, the eigenvalue characterizing dynamics normal to the manifold must be sufficiently negative compared with the other eigenvalue and the noise strength c, so that the joint probability density $p(y_1, y_2, t)$ rapidly concentrates near $\mathcal{M}$ and a substantial majority of sample paths crosses the thresholds $x_j = \theta$ (or $y_j = \theta$) near their intersections with $\mathcal{M}$ [Usher & McClelland, 2001; Brown & Holmes, 2001; Bogacz et al., 2004]. These requirements are met by two distinct parameter sets to be introduced below, and in Sec. 2.6 we compare the resulting reaction times and error rates determined from one-dimensional reductions with those of the original two-dimensional models.

2.5.1. *Dimension reduction and transient gain in two simple cases*

In two cases, a simple equation for the evolution of $x(t)$ or $y(t)$ may be derived. These cases are characterized by a dominant proportion of solutions to (15)–(16) (i.e. for "most" realizations of the noise processes $\eta_j(t)$) (i) being confined to a single tile for the duration of the decision process or (ii) "jumping" together between tiles. The first of these situations occurs for Case 1 parameter sets, in which, for example, the onset of salience (i.e. $a_1 \neq a_2$) in input currents is accompanied by large transients in the magnitude of these inputs. The second Case 2 occurs for stimuli in which salience appears without such transients in magnitude. We now consider these cases in detail for the firing rate model.

Case 1. **Trajectories confined to the central tile, gain parameter directly modulated**

The central tile of the firing rate phase plane, where both functions $f_{g(t)}^{pw}(\cdot)$ appearing in Eqs. (15)–(16) are linearly increasing, is defined by $\beta y_1 \in [a_2(t)$

$-b - (1/2g(t)), a_2(t) - b + (1/2g(t))]$ and $\beta y_2 \in$ $[a_1(t) - b - (1/2g(t)), a_1(t) - b + (1/2g(t))]$. If

$$|a_1(t) - b| < \frac{1}{2g(t)}, \quad |a_2(t) - b| < \frac{1}{2g(t)}, \quad (20)$$

then the central tile always contains the origin and some part of the first quadrant (note that this quadrant is invariant under the deterministic part of Eqs. (7)–(8) if f is non-negative) so that decision dynamics starting at the origin may (for suitable choices of other parameters) take place entirely within the central tile. For example, if $b = 0.5$ and $0 < g(t) \leq 1$, then $a_1(t)$, $a_2(t)$ may take values between 0 and 1 while still satisfying (20).

Figure 4 shows a sample of solutions of the piecewise-linearized firing rate model for the piecewise constant parameters $g(t) = \{0.3, \ t < t_s; 1, \ t \geq t_s\}$, $a_1(t) = \{1, t \leq t_s; 1.03, t > t_s\}$, $a_2(t) = \{1, t \leq t_s; 0.97, t > t_s\}$, $c(t) \equiv 0.09\sqrt{2}$, $b = 0.5$, $\tau_c = 1$, $\theta = 0.725$, $t_s = 10$ and $\beta = 1$. Note that stimuli

$a_j(t) \neq 0$ are present throughout, but that coherence $(a_1(t) \neq a_2(t))$ appears in the inputs a_j only at $t = t_s$, so that times $t < t_s$ make up the preparatory phase mentioned in the introduction and the situation corresponds to the introduction of coherence into an entirely random pattern. Assuming that decision thresholds are set within the boundaries of the central tile or that the interrogation time T is sufficiently small so that only a negligible proportion of solutions have left this tile, solutions are effectively confined to the central tile for all times of interest. This behavior characterizes Case 1 parameter sets, for which subtraction of Eqs. (15)–(16) yields the one-dimensional SDE

$$\tau_c \frac{dy}{dt} = -y + g(t)(\beta y + a(t)) + g(t)c(t)\eta_t$$

$$\text{(firing rate model)}, \quad (21)$$

where we define the net rate of incoming evidence as

$$a(t) = a_1(t) - a_2(t). \quad (22)$$

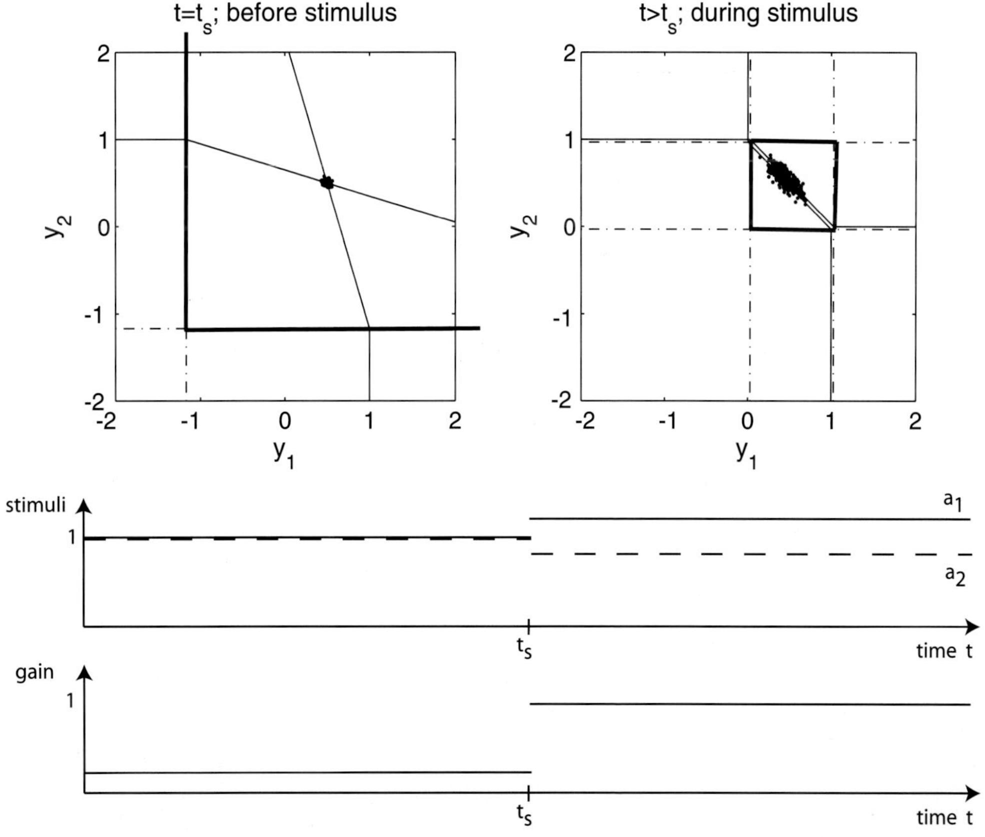

Fig. 4. Case 1: solutions confined to central tile. Scatter plot of trajectories both at the end of the preparatory period and hence at the moment of stimulus onset t_s (left) and during the stimulus ($t = t_s + 2$, right). The tiling of the plane is shown with dot-dashed lines; cf. Fig. 2; the central tile is outlined in solid and extends outside the plotted domain in the left panel. Parameter values are given in text. Also shown are nullclines for Eqs. (15)–(16) as thin solid lines. The lower panels show stimuli $a_j(t)$ and gain $g(t)$ as functions of time.

We note that transient gain values in this case result from modifications to the firing rate function itself, as solutions explore only the central region of this function in which it is practically linear. This is the "external" mechanism of dynamic gain change discussed in the Introduction.

For future reference, we also note that an analytical expression for the density of reaction times may be derived if the parameters in (21) are constant (i.e. $a(t) \equiv a$, $c(t) \equiv c$) and the gain "balances" the decay: e.g. $g(t) \equiv g = 1$ in (21) (see, e.g. [Ratcliff *et al.*, 1999]). In this case, (21) simplifies to a constant drift diffusion process and the probability that a trajectory first escapes the interval $[-\bar{\theta}, \bar{\theta}]$ at a time $RT = \inf\{t : |y(t)| > \bar{\theta}\}$ from initial condition $y(0) = 0$ has density

$$p(RT) = \frac{\pi c^2}{\bar{\theta}^2} e^{-\frac{a^2\,RT}{2c^2}} \left(e^{-\frac{\bar{\theta}a}{c^2}} + e^{\frac{\bar{\theta}a}{c^2}} \right)$$
$$\times \sum_{k=1}^{\infty} k \sin\left(\frac{k\pi}{2}\right) \exp\left(\frac{-k^2\pi^2 c^2\,RT}{8\bar{\theta}^2}\right). \tag{23}$$

Here $\pm\bar{\theta}$ correspond to the intersections of the decision manifold $\mathcal{M}$ with the thresholds $y_j = \theta$ of the two-dimensional process (Fig. 3). Equation (23) may be extended to account for distributed initial conditions $y(0) \neq 0$ and other generalizations [Ratcliff *et al.*, 1999], but we do not use such extensions here.

Similar considerations yield the reduction of the connectionist model restricted to its respective central tile:

$$\tau_c \frac{dx}{dt} = -x + \beta g(t)x + a(t) + c(t)\eta_t$$
$$\text{(connectionist model).} \tag{24}$$

Note that gain multiplies the last three terms in (21), but only the second in (24).

Case 2. Trajectories switch tiles, changing effective gain

We now consider the case of stimuli $a_j(t)$ that "suddenly" turns on from zero at time t_s while the gain parameter $g(t) \equiv g$ remains constant, and show how stimulus onset itself can give rise to a time-dependent one-dimensional reduction that resembles the reduction to (21) obtained above. This corresponds to appearance of a partially coherent stimulus replacing a fixation spot. Since $a_1(t) = a_2(t) = 0$ for $t \leq t_s$, in this period there is a stable

fixed point at $(0,0)$ if $b \geq 1/2g$. If $b = 1/2g$, the situation simplifies: while $t \leq t_s$, $(0,0)$ lies exactly at the corner of tile 9 (see Fig. 2), to which tile solutions are confined (modulo noise effects). At stimulus onset t_s, tile boundaries shift, so that, for appropriate choices of $a_1(t), a_2(t) > 1/2g(t) - b$ for $t > t_s$, the origin and the cluster of solutions in its neighborhood at time $t = t_s^+$, suddenly finds itself in the central tile 5. For concreteness, we fix parameters meeting the requirements $b = 1/2g$ and $a_1(t) = a_2(t) = 0$ for $t \leq t_s$ as follows: $a_1(t) = \{0, t \leq t_s; 1.03, t > t_s\}$, $a_2(t) = \{0, t \leq t_s; 0.97, t > t_s\}$, $g = 1$ and all other parameters as for the example in Case 1. See Fig. 5.

To determine the appropriate linear (two- and one-dimensional) reductions for these parameters, we use Eqs. (15)–(16) restricted to tile 9 for the preparatory phase $t \leq t_s$, and restricted to tile 5 for times $t > t_s$ during stimulus presentation (we make the same assumptions about the interrogation time or thresholds as for Case 1, so that solutions remain in the central tile 5 for all times $t > t_s$ of relevance to the decision). This yields the one-dimensional equation

$$\tau_c \frac{dy}{dt} = -y + \begin{cases} gc(t)\eta_t & \text{for } t \leq t_s \\ g[\beta y + a(t)] + gc(t)\eta_t & \text{for } t > t_s \end{cases}, \tag{25}$$

(and an analogous reduction to a linear two-dimensional model).

Equation (25) is similar to the reduction (21), if the stimulus and gain functions in the latter are piecewise constant, as for the example parameters of Case 1. The major difference is that the noise coefficient remains constant for (25). As we see in the next section, the statistics produced by the one-dimensional models (21) and (25) can nevertheless agree rather well. Thus, transient gain strategies to be derived for the more general (21) in Sec. 3 can be approximately implemented for stimuli undergoing large steps, with no changes in the gain of the activation functions *per se*.

Similar considerations hold for Cases 1 and 2 reductions of the connectionist model, but we do not pursue this here.

2.6. *Accuracy of the reduced models*

Figure 6 demonstrates that our simplifications of the nonlinear firing rate model (7)–(8) accurately capture reaction time statistics for Case 1

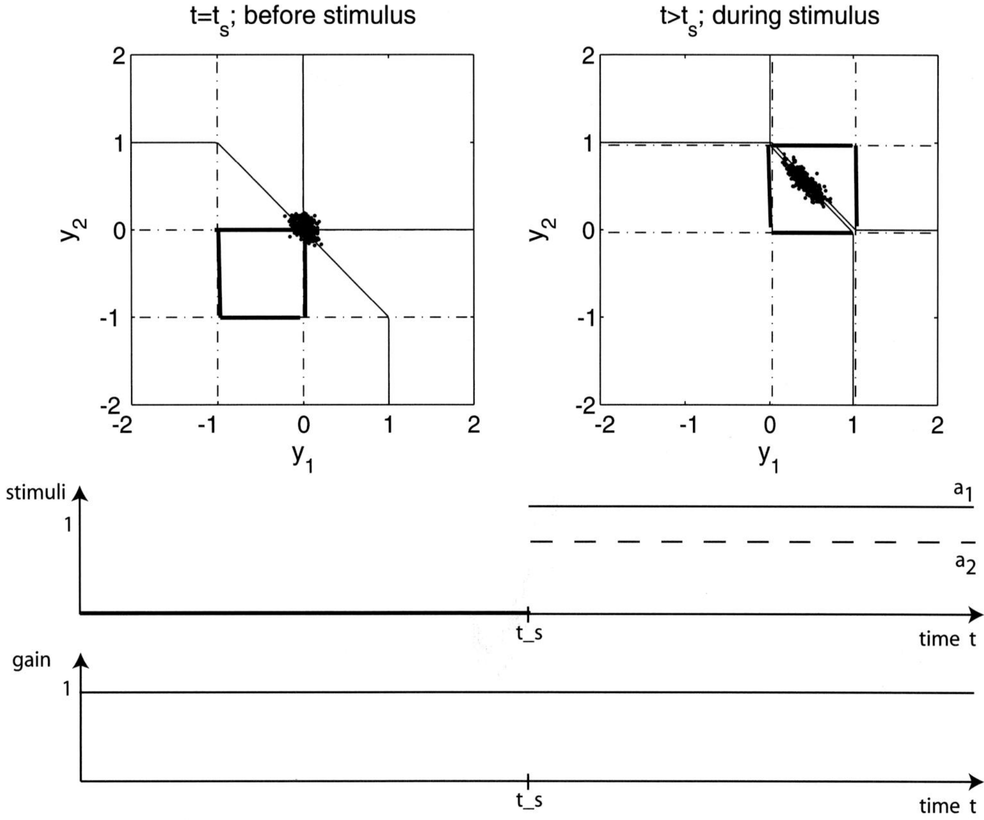

Fig. 5. Case 2: trajectories switch tiles. Scatter plot of trajectories both at the end of the preparatory period and hence at the moment of stimulus onset t_s (left) and during the stimulus ($t = t_s + 2$, right). The tiling of the plane is shown with dot-dashed lines; cf. Fig. 2; the central tile is outlined in solid. Parameter values are given in text. Also shown are nullclines for Eqs. (15)–(16) as thin solid lines. The lower panels show stimuli $a_j(t)$ and gain $g(t)$ as functions of time.

parameters. For the one- and two-dimensional linear reductions, linearized activation functions take piecewise constant (in time) values appropriate to the tiles containing the dominant proportion of solution trajectories during the preparatory and trial periods, exactly as in (21). That is: for Case 1, $f^l_{g(t)}(x) = 1/2 + (x - b)$ for all t, as solutions remain in the central tile 5. For Case 2, $f^l_{g(t)}(x) = 0$ for all $t < t_s$ (when solutions are in tile 9) and $f^l_{g(t)}(x) = (1/2) + (x - b)$ for $t > t_s$, when solutions are in tile 5.

For Case 1, the error rates corresponding to the reaction time distributions of Fig. 6 are 0.050, 0.051, 0.051, 0.035, and 0.034 respectively for the two-dimensional firing rate model with logistic activation functions $f_{g(t)}$, the two-dimensional model with piecewise-linear activation functions $f^{pw}_{g(t)}$ (15)–(16), the two-dimensional model with linear activation functions, the one-dimensional reduction (21), and the expression (23), which describes the one-dimensional reduction with initial condition

$y(t_s) = 0$ at the time of stimulus presentation (keeping the first 10 terms of sum). For Case 2, these error rates are 0.060, 0.065, 0.059, 0.042, 0.034. Thus, in both cases while the different two-dimensional models are in close agreement, the one-dimensional reductions produce significantly lower error rates. Figures 4 and 5 show why: the distribution of solutions is not entirely collapsed along the attracting decision manifold, and the spatially extended "incorrect" thresholds of the two-dimensional models require smaller (and hence more probable) excursions to cross. Closer agreement between one- and two-dimensional models can be achieved with, for example, higher values of β or lower values of noise strength c: see [Bogacz *et al.*, 2004].

As an additional comparison among the various models, we separately computed error rates for interrogation at a time $T = t_s + 1$ (see [Usher & McClelland, 2001] for an earlier, related comparison between the nonlinear two-dimensional and linear one-dimensional models). For Case 1,

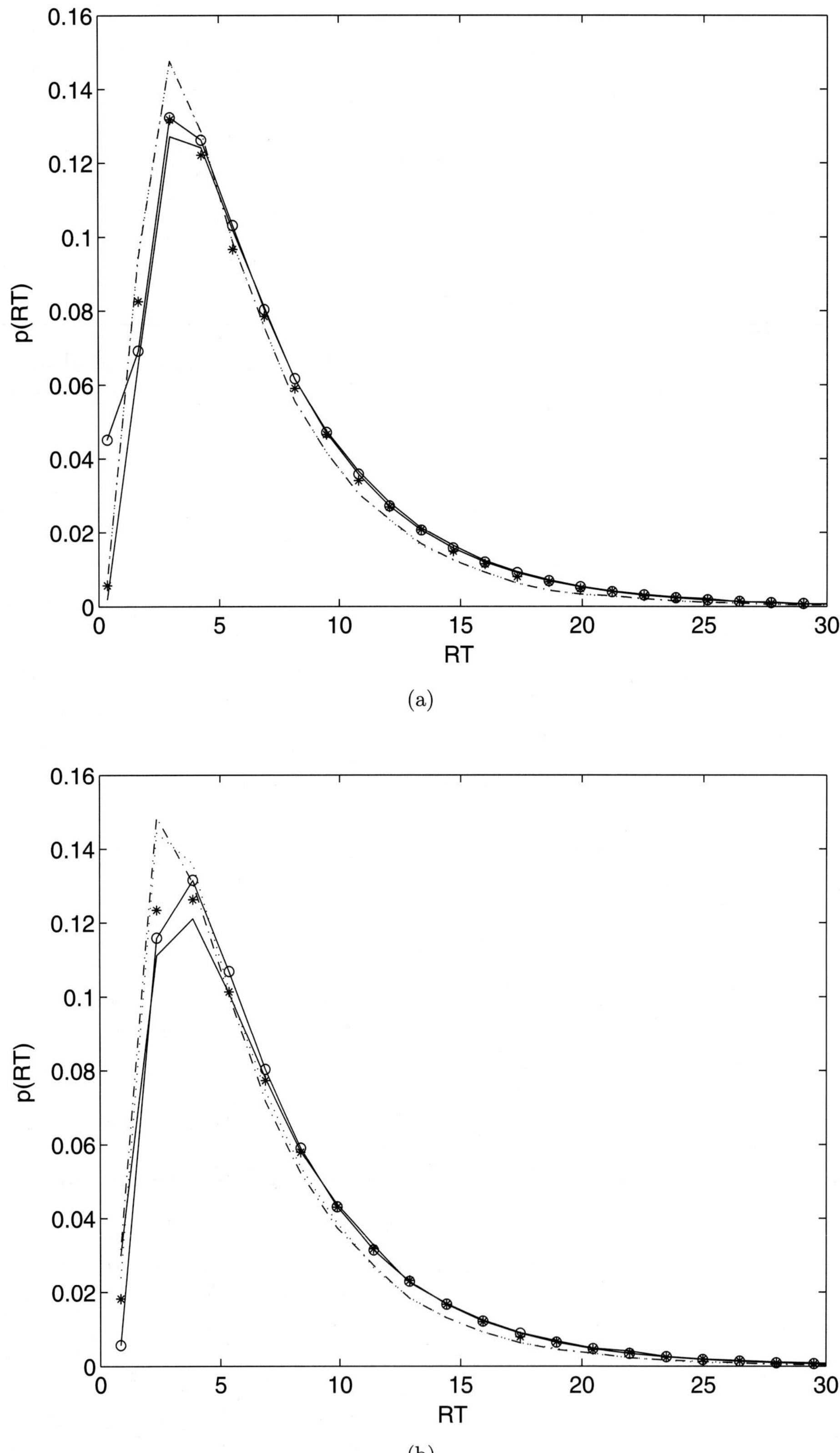

Fig. 6. Reaction time densities for the nonlinear firing rate model of Eqs. (7)–(8) (stars) and its various reductions, with thresholds $\theta = 0.725$: dot-dashed line, two-dimensional model with piecewise-linear activation functions $f^{pw}_{g(t)}$; dotted line, two-dimensional model with linear activation functions $f^{l}_{g(t)}$; solid line, linear one-dimensional reduction, solid line with circles, analytic 1-D expression with zero variance at trial onset (see text). (a) Case 1: solutions confined to central tile. (b) Case 2: trajectories switch tiles. Parameter values are given in main text.

the interrogation error rates are (in the same order as above) 0.323, 0.321, 0.321, 0.324, and 0.319. For Case 2, these error rates are 0.374, 0.363, 0.354, 0.350, and 0.319. For both cases interrogation error rates are more similar for the various model reductions than the free response error rates reported in the previous paragraph. This is expected from the discussion in Sec. 2.5, since accurate description of the interrogation protocol by a one-dimensional model does not require that solutions are confined near the decision manifold.

2.7. Drift-diffusion and the one-dimensional models as linear filters

We introduce a third one-dimensional SDE, an extension of the drift-diffusion model of [Laming, 1968; Ratcliff, 1978] in which both drift and diffusion terms are multiplied by a common gain factor $g(t)$:

$$\tau_c \frac{dz}{dt} = g(t)[a(t) + c(t)\eta_t] \quad \text{((pure) drift-diffusion}$$
$$\text{model).} \quad (26)$$

Equation (26) and the one-dimensional reductions of the firing rate and connectionist equations (21) and (24) are Ornstein–Uhlenbeck processes, (affine-) linear in the activities x, y and z and in the input

$$\underbrace{I(t)}_{\text{input}} = \underbrace{a(t)}_{\text{signal}} + \underbrace{c(t)\eta_t}_{\text{noise}} . \quad (27)$$

We may explicitly solve all these SDEs, for a given realization of the white noise process η_s, $s \in [0,t]$, to obtain respectively

$$z(t) = \int_0^t \frac{g(s)a(s)}{\tau_c} ds + \int_0^t \frac{g(s)c(s)}{\tau_c} dW_s \quad (28)$$

for the drift diffusion model,

$$x(t) = \int_0^t \frac{a(s)}{\tau_c} \exp\left(\frac{1}{\tau_c}\int_s^t [\beta g(s') - 1]\, ds'\right) ds$$
$$+ \int_0^t \frac{c(s)}{\tau_c} \exp\left(\frac{1}{\tau_c}\int_s^t [\beta g(s') - 1]\, ds'\right) dW_s \quad (29)$$

for the connectionist model, and

$$y(t) = \int_0^t \frac{a(s)g(s)}{\tau_c} \exp\left(\frac{1}{\tau_c}\int_s^t [\beta g(s') - 1]\, ds'\right) ds$$
$$+ \int_0^t \frac{c(s)g(s)}{\tau_c} \exp\left(\frac{1}{\tau_c}\int_s^t [\beta g(s') - 1]\, ds'\right) dW_s \quad (30)$$

for the firing rate model. Here, dW_s is an increment of a Wiener process, of which the white noise process η_s is the formal time derivative, and we have assumed unbiased initial data $x(0) = y(0) = z(0) = 0$. These expressions all take the form

$$w(t) = \int_0^t K(t,s)a(s)ds + \int_0^t K(t,s)c(s)dW_s , \quad (31)$$

and so we conclude that (28)–(30) all compute linear filters of their inputs.

At any fixed time t, $w(t)$ is a Gaussian-distributed random variable with mean $\int_0^t K(t,s)a(s)$ and variance $\int_0^t K^2(t,s)c^2(s)ds$. Using this fact, after a change of variables the error rate expression (19) becomes

$$\text{ER} = \frac{1}{2}\left[1 - \text{erf}\left(\frac{\left|\int_0^t K(t,s)a(s)\right|}{\sqrt{\int_0^t K^2(t,s)c^2(s)ds}}\right)\right] . \quad (32)$$

3. Optimal Signal Discrimination in the One-Dimensional Models

We now ask what functional form of $g(t)$ optimizes performance for Eqs. (28)–(30), thereby computing optimal gain trajectories for the (reduced) drift-diffusion, connectionist, and firing rate models.

3.1. Optimal statistical tests

Given only the noisy input function (27), consider the task of deciding whether $I(t)$ was generated by time-dependent signals $a_0(t)$ or $a_1(t)$: hypotheses 0 and 1, resp. This can be accomplished in two distinct ways, mirroring the interrogation and free response protocols of Sec. 2. In the first, the decision is made at a fixed time T; in the second, it is made when some preset level of confidence is reached. Optimal performance in the first version of the task implies that as few errors as possible are made; in the second, it implies that the decision must be made as quickly as possible for a fixed error tolerance, timed from stimulus onset at time $t = 0$. The best strategy in the first version is the (continuum limit of the) Neyman–Pearson test; in the second version it is the sequential probability ratio test (SPRT) [Wald, 1947; Lehmann, 1959]. Both tests compute an evolving estimate of the log likelihood ratio:

$$l(t) = \log \left[\frac{p(\{I(s)|a_0(s), s \in [0,t]\})}{p(\{I(s)|a_1(s), s \in [0,t]\})} \right]$$

$$\triangleq \log \left[\frac{p_0(\{I(s), s \in [0,t]\})}{p_1(\{I(s), s \in [0,t]\})} \right]. \tag{33}$$

(the base of the logarithm is arbitrary). In the Neyman–Pearson test, hypothesis 0 is chosen if $l(T) > 0$ and hypothesis 1 if $l(T) < 0$; in the SPRT, hypothesis 0 (resp. 1) is chosen when $l(t)$ first crosses threshold θ (resp. $-\theta$), θ being determined by the error tolerance.

Writing the input $I(t)$ (27) as a sum of its increments for an appropriate discretization of time $\{t^j\}$:

$$I(t) = \sum_j dI^j = \sum_j a(t^j)dt + c(t^j)\,dW_t^j, \tag{34}$$

we obtain

$$l(t) = \sum_j \log \left[\frac{p_0(dI^j)}{p_1(dI^j)} \right]. \tag{35}$$

Now restrict to the special case in which $a_0(t) = -a_1(t) = a(t)$ and consider the likelihood distributions (now themselves time-dependent) that correspond to an increment $dI(t) = a(t)dt + c(t)dW_t$. Since the dW_t are normally distributed with mean 0 and variance dt, we have

$$p_0(t)(dI(t)) = \frac{1}{\sqrt{2\pi c^2(t)dt}}\, e^{-(dI(t)+a(t)dt))^2 / (2c^2(t)dt)}, \tag{36}$$

$$p_1(t)(dI(t)) = \frac{1}{\sqrt{2\pi c^2(t)dt}}\, e^{-(dI(t)-a(t)dt))^2 / (2c^2(t)dt)}. \tag{37}$$

The corresponding increment of likelihood evidence to (33) is

$$dl_t = \log \left(\frac{p_1(dI_t)}{p_0(dI_t)} \right) = k\frac{a(t)}{c^2(t)}dI_t, \tag{38}$$

where $k = 2\log(e)$ depends on the base of the logarithm. Substituting for dI_t, we obtain a differential equation for the total evidence l_t accumulated at

time t,

$$dl_t = k\left[\frac{a^2(t)}{c^2(t)}\,dt + \frac{a(t)}{c(t)}\,dW_t \right], \tag{39}$$

which may be integrated to yield:

$$l(t) = \int_0^t k\frac{a^2(s)}{c^2(s)}ds + \int_0^t k\frac{a(s)}{c(s)}dW_s. \tag{40}$$

Comparing with Eq. (31) shows that the optimal filter is

$$K(t,s) = k\frac{a(s)}{c^2(s)} : \tag{41}$$

this is the matched filter for white noise which is fundamental in signal processing [Papoulis, 1977]. Note that, in (39)–(40) only the signal-to-noise ratio (a/c) appears.

3.2. *A direct proof that the kernel $K(t,s) = k(a(s)/c^2(s))$ is optimal in the interrogation paradigm*

As follows from its matched filter property, the linear filter $K(t,s) = k(a(s)/c^2(s))$ which computes log likelihood $l(t)$ for inputs with white noise also produces, for all times t, a filtered (and Gaussian) version $w(t)$ of the input [Eq. (31)] with a maximal integrated signal-to-noise ratio

$$F[K; a, c](t) = \frac{\left| \int_0^t K(t,s)a(s)ds \right|}{\sqrt{\mathbb{E}\left(\int_0^t K(t,s)c(s)dW_s \right)^2}}$$

$$= \frac{\left| \int_0^t K(t,s)a(s)ds \right|}{\sqrt{\int_0^t K^2(t,s)c^2(s)ds}}. \tag{42}$$

For completeness, we now demonstrate this directly.

Minimization of the error rate (18) or (19) for (fixed) interrogation at time $t = T$ is achieved by maximizing F over all possible kernels $K(s)$. This problem in the calculus of variations is solved by computing the first and second variations, with respect to K, of the functional F, setting the first to zero to determine a candidate $\overline{K}$ for the optimal K, and evaluating the second at $\overline{K}$ to check that $D_K^2 F$ is negative (semi-) definite. Henceforth we drop explicit reference to the (fixed, arbitrary) interrogation time $t = T$ in the function K and write $K(T, s) = K(s)$. We compute:

$$\frac{\delta F}{\delta K} = \lim_{\epsilon \to 0} \frac{d}{d\epsilon} F[K + \epsilon\gamma; a, c](T) = \lim_{\epsilon \to 0} \frac{d}{d\epsilon} \left\{ \frac{\int_0^T a(s)[K(s) + \epsilon\gamma(s)]\, ds}{\left[2 \int_0^T c^2(s)[K^2(s) + 2\epsilon g(s)\gamma(s) + \epsilon^2\gamma^2(s)]\, ds\right]^{\frac{1}{2}}} \right\}$$

$$= \lim_{\epsilon \to 0} \frac{1}{\sqrt{2}} \left\{ \frac{\int_0^T a(s)\gamma(s)\, ds}{[H(T,\epsilon)]^{\frac{1}{2}}} - \frac{\int_0^T a(s)[K(s) + \epsilon\gamma(s)]\, ds \int_0^T c^2(s)[K(s)\gamma(s) + \epsilon\gamma^2(s)]\, ds}{[H(T,\epsilon)]^{\frac{3}{2}}} \right\}$$

$$= \frac{\int_0^T a(s)\gamma(s)\, ds \int_0^T c^2(s)K^2(s)\, ds - \int_0^T a(s)K(s)\, ds \int_0^T c^2(s)K(s)\gamma(s)\, ds}{\sqrt{2} \left[\int_0^T c^2(s)K^2(s)\, ds\right]^{\frac{3}{2}}}, \tag{43}$$

where $H(T, \epsilon) = \int_0^T c^2(s)[K^2(s) + 2\epsilon K(s)\gamma(s) + \epsilon^2\gamma^2(s)]\, ds$. Setting (43) equal to zero and using the fact that the variation $\gamma(s)$ is arbitrary, we conclude that the critical point indeed occurs at $\overline{K}(s) = k(a(s)/c^2(s))$, as given by (41).

To compute the second derivative we differentiate the expression within braces in the penultimate step of (43) with respect to ϵ once more, set $\epsilon = 0$, and evaluate the resulting expression at the critical point (41), obtaining:

$$\left. \frac{\delta^2 F}{\delta K^2} \right|_{K=\overline{K}} = -\frac{\int_0^T c^2(s)\overline{K}^2(s)\, ds \int_0^T c^2(s)\gamma^2(s)\, ds - \left(\int_0^T c^2(s)\overline{K}(s)\gamma(s)\, ds\right)^2}{\sqrt{2} \left[\int_0^T c^2(s)\overline{K}^2(s)\, ds\right]^{\frac{3}{2}}} \leq 0. \tag{44}$$

In the last step we appeal to Schwarz's inequality. This proves that the second variation is negative semidefinite, and vanishes identically only for variations $\gamma(s) = \kappa\overline{K}(s)$ in the direction of $\overline{K}$ (as expected from (41), which contains the arbitrary "scaling" parameter k).

Substituting (41) into (42) we obtain

$$F[\overline{g}; a, c](T) = \sqrt{\frac{1}{2} \int_0^T \frac{a^2(s)}{c^2(s)}\, ds}, \tag{45}$$

and using (32), we obtain the minimum possible error rate for interrogation at time t:

$$\mathrm{ER} = \frac{1}{2} \left[1 - \mathrm{erf}\left(\sqrt{\frac{1}{2} \int_0^T \frac{a^2(s)}{c^2(s)}\, ds}\right)\right]. \tag{46}$$

Since the integrand $(a/c)^2$ is non-negative, the error rate continues to decrease or at worst remains constant as T increases.

3.3. Optimal gains for the three models

We may now extract explicit expressions for optimal gains by setting $K(s) = \overline{K}(s)$ in (31) and comparing the resulting integrands with those in the SDE solutions (28)–(30).

3.3.1. Pure drift-diffusion model

Comparing (31) with (28), we see that the optimal gain is simply $\overline{K}$:

$$\overline{g}_{dd}(s) = \tau_c \overline{K}(s) = \tau_c k \frac{a(s)}{c^2(s)}; \tag{47}$$

thus, there is a continuum of optimal schedules differing only by a multiplicative scale factor.

3.3.2. *Connectionist model*

Equations (31) and (29) give

$$\tau_c \overline{K}(s) = \tau_c k \frac{a(s)}{c^2(s)}$$

$$= \exp\left(\frac{1}{\tau_c} \int_s^T [\beta \overline{g}_c(s') - 1]\, ds'\right), \quad (48)$$

where $\overline{g}_c$ is the optimal gain for the connectionist model. Taking the log of this expression, differentiating with respect to s, and solving for $\overline{g}_c(s)$, we obtain:

$$\overline{g}_c(s) = \frac{1}{\beta}\left[1 - \tau_c \frac{d}{ds} \log\left(\frac{a(s)}{c^2(s)}\right)\right]. \quad (49)$$

Note that $\overline{g}_c$ is unique and in particular, independent of k and the interrogation time T. However, $\overline{g}_c$ is not required to be positive, so may not always be physically admissable. The form of $\overline{g}_c$ may be interpreted as follows. When $(a(s)/c^2(s))$ is decreasing, $\overline{g}_c(s) > 1/\beta$ and the O–U process (24) is unstable; hence solutions "run away," in the direction $x(s)$, emphasizing higher-fidelity information that was previously collected. When $(a(s)/c^2(s))$ is increasing, $\overline{g}_c(s) < 1/\beta$, the O–U process is stable, and the linear term in (24) is attractive, thereby discounting previously integrated information in favor of the higher-fidelity input currently arriving.

We note that, because the "output" neural activity is determined by a gain-dependent function of the dynamical variable x in the connectionist model (see text following Eqs. (1)–(2)), transient gain schedules also adjust the position of free-response thresholds with respect to x. We leave an exploration of this effect, which does not enter the interrogation protocol or affect the firing rate model, for future studies.

3.3.3. *Firing rate model*

Equations (31) and (30) give

$$\tau_c \overline{K}(s) = \tau_c k \frac{a(s)}{c^2(s)}$$

$$= \overline{g}_f(s) \exp\left(\frac{1}{\tau_c} \int_s^T [\beta \overline{g}_f(s') - 1]\, ds'\right). \quad (50)$$

Defining $f(s) = \tau_c k(a(s)/c^2(s))e^{(1/\tau_c)(T-s)}$, differentiating with respect to s, and restricting to positive functions $\overline{g}_f$, a and c^2 (which we justify below), (50) yields

$$f'(s) = \frac{d}{ds}\left[\overline{g}_f(s) \exp\left(\frac{1}{\tau_c} \int_s^T \beta \overline{g}_f(s')\, ds'\right)\right]$$

$$= \overline{g}_f'(s) \exp\left(\frac{1}{\tau_c} \int_s^T \beta \overline{g}_f(s')\, ds'\right)$$

$$\quad - \frac{\beta}{\tau_c}\overline{g}_f^2(s) \exp\left(\frac{1}{\tau_c} \int_s^T \beta \overline{g}_f(s')\, ds'\right)$$

$$= \overline{g}_f'(s)\frac{f(s)}{\overline{g}_f(s)} - \frac{\beta}{\tau_c}\overline{g}_f(s)f(s). \quad (51)$$

Rewriting (51), we obtain

$$\frac{d\overline{g}_f(s)}{ds} = \frac{\beta}{\tau_c}\overline{g}_f^2(s) + \overline{g}_f(s)\frac{f'(s)}{f(s)}$$

$$= \frac{\beta}{\tau_c}\overline{g}_f^2(s) + \overline{g}_f(s)\frac{d}{ds}\log(f(s))$$

$$= \frac{\beta}{\tau_c}\overline{g}_f^2(s) + \overline{g}_f(s)\left[\frac{d}{ds}\log\left(\frac{a(s)}{c^2(s)}\right) - \frac{1}{\tau_c}\right]. \quad (52)$$

Thus, the condition for optimal gain in the linearized firing rate model is a differential equation, unlike the algebraic relationships for the drift-diffusion and connectionist cases. Note that solutions to (52) initialized at positive values remain positive for all times, since the equation has an equilibrium at $\overline{g}_f = 0$, preventing passage through this point. This justifies our assumption of positive $\overline{g}_f$ above and ensures that the optimum gain is "physical" this sense. In fact, (52) may be solved explicitly using the integrating factor $I(s) = \exp\left(\int_0^s l(s')ds'\right)$, where $l(s') \triangleq (d/ds')\log(a(s')/c^2(s')) - 1/\tau_c$, yielding

$$\overline{g}_f(s) = \frac{\exp\left(\displaystyle\int_0^s l(s')ds'\right)}{\dfrac{\beta}{\tau_c}\displaystyle\int_0^s \left[\exp\left(\int_0^{s'} l(s'')ds''\right)\right] ds' + \dfrac{1}{g(0)}}. \quad (53)$$

The integral equation (50) specifies only an *arbitrary*, positive final condition $\overline{g}_f(T) = k(a(T)/c^2(T))$ for (52), since k is itself arbitrary. Any solution of (52) with positive initial condition (as long as it is defined) therefore delivers a member of the continuum of optimal gain functions for the linearized firing rate model. This is in striking contrast to the unique optimal gain (49) in the connectionist model, and, since the different $\bar{g}_f$ generally have different forms (see below), it also contrasts with the multiplicity of "scaled" optimal drift-diffusion gain functions (47). The optimality of $\bar{g}_f$ schedules with such different forms follows from the fact that gain multiplies the inputs to the firing rate model (21). For example, optimal gain schedules with $(\beta\bar{g}_f(s) - 1) < 0$ may implement the SPRT even when the signal-to-noise-ratio is constant (see Example 1 below), because discounting of previously integrated evidence is compensated for via weighting incoming evidence by a decreasing function $\bar{g}_f(s)$.

3.3.4. *Numerical examples*

Example 1. We first take constant signal $a(s) \equiv a = 0.06$ and constant noise strength $c(s) \equiv 0.09$

with $\tau_c = \beta = 1$. Then, Eq. (47) gives the family of optimal constant gain functions for the pure drift-diffusion model,

$$\bar{g}_{dd}(s) \equiv \tau_c k a, \tag{54}$$

and Eq. (49) gives the unique optimal gain for the connectionist model, again a constant:

$$\bar{g}_c(s) \equiv \frac{1}{\beta}. \tag{55}$$

For the same parameter values, the firing rate model gain ODE (52) becomes

$$\frac{d}{ds}\bar{g}_f(s) = \frac{\beta}{\tau_c}\bar{g}_f^2(s) - \frac{1}{\tau_c}\bar{g}_f(s). \tag{56}$$

Initial conditions $\bar{g}_f(0) \in [0, 1/\beta]$ decay to the fixed point at $\bar{g}_f = 0$, while for $\bar{g}_f(0) > 1/\beta$, gain functions increase to ∞ in finite time. The initial condition $\bar{g}_f(0) = 1/\beta$ yields the constant gain function $\bar{g}_f(s) \equiv 1/\beta$, for which the linearized firing rate model again becomes constant drift Brownian motion: see Fig. 7. As expected, all gain profiles

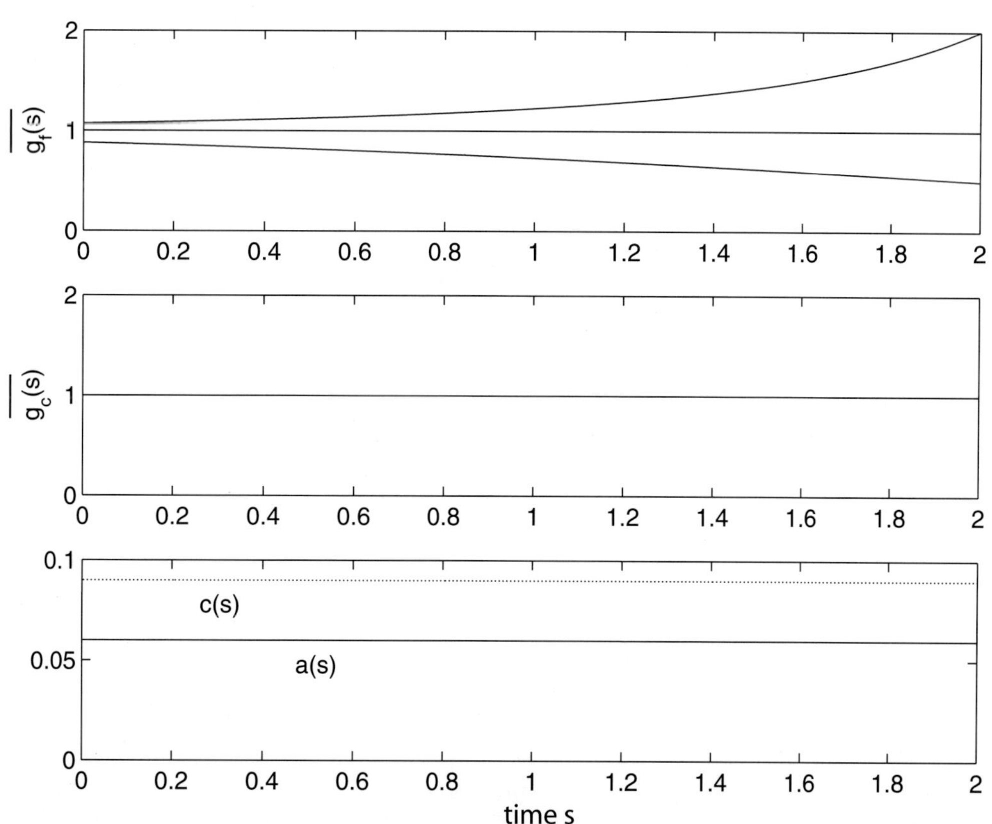

Fig. 7. Optimal gains for constant signal strength $a(s) \equiv 0.06$ (solid line in bottom panel) and constant noise amplitude $c(s) \equiv 0.09$ (dotted line). Top panel: three optimal gain schedules $\bar{g}_f$ solving (52); note that these include, but are not limited to, $\bar{g}_f(s) \equiv 1/\beta$ (here $\beta = 1$). Central panel: the unique optimal gain function $\bar{g}_c(s) \equiv 1/\beta$ for the connectionist model, given by Eq. (49).

produced optimal performance (with 82.7% correct responses returned at interrogation time $T = 2$).

Example 2. We now assume that signal amplitude is zero up to stimulus presentation at time t_s and rises exponentially toward $\bar{a}$ thereafter: $a(s) = \bar{a}[1 - e^{-r(s-t_s)}]$ for $s > t_s$. This form is motivated by the saturating dynamics of input layers which feed forward to decision units in simple connectionist models. We set $\bar{a} = 0.06$, $r = 10$, $t_s = 1$ and take constant noise strength $c(s) \equiv 0.09$ and $\tau_c = \beta = 1$ as previously: see Fig. 8 (bottom). As $r \to \infty$, $a(s)$ approaches the piecewise constant functions of Secs. 2.5.1–2.6, for which the one-dimensional reduction was shown to be an adequate model.

For the pure drift-diffusion model, Eq. (47) gives

$$\bar{g}_{dd}(s) = \tau_c k a(s), \tag{57}$$

so that, as above, optimal gain trajectories are scaled versions of the signal strength and, in

particular, $\bar{g}(s) = 0$ for $s \leq t_s$. For the connectionist and firing rate models, however, the formulae (49) and (52) are valid only while $a(s) > 0$, and additional reasoning is needed to determine optimal gain values in the pre-stimulus period $s < t_s$. For the connectionist model, the integral equation (48) is clearly satisfied for $a(s) = 0$ if $g_c(s) = -\infty$, so we set $\bar{g}_c(s) = -\infty$, $s \leq t_s$. Since for a "physical" neural network, activation functions $f_{g(t)}(\cdot)$ are nondecreasing, such negative gain values are not directly relevant to biological applications, but illustrate the demand that relative activation x be clamped at zero before the stimulus arrives. As before, we define $\bar{g}_c(s)$ via (49) for $s > t_s$. That is, for $t > t_s$,

$$\bar{g}_c(s) = \frac{1}{\beta}[1 - \tau_c l(s)], \tag{58}$$

where $l(s) = (d/ds)\log(a(s)/c^2(s)) = r/e^{r(s-t_s)} - 1$ decays from ∞ to 0 as time s increases.

For the firing rate model, we also appeal directly to the integral equation (50) to define $g_f(s)$ when $a(s) = 0$. Since (50) is satisfied by $\bar{g}_f(s) = 0$,

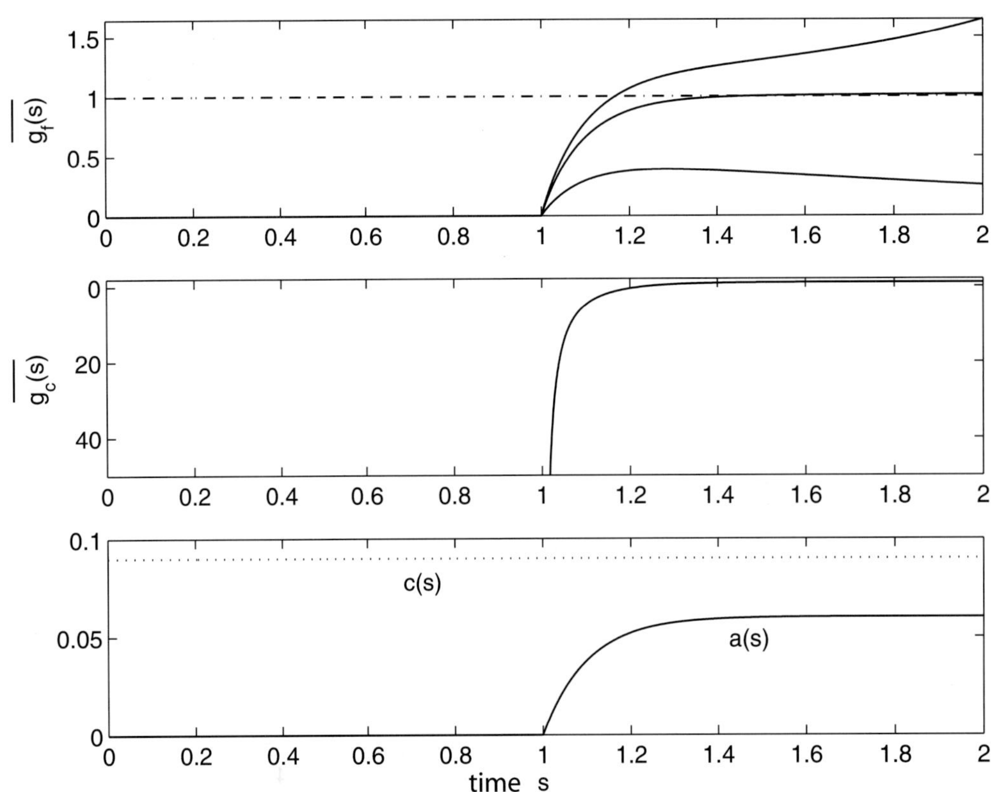

Fig. 8.   Optimal gains for exponentially asymptoting signal strength $a(s)$ (solid line in bottom panel) and constant noise amplitude $c(s) \equiv 0.09$ (dotted line). Top panel: three optimal gain schedules $\bar{g}_f$ for the firing rate model solving (52) (solid curves); the nonoptimal constant gain $g \equiv 1/\beta$ is shown as dot-dashed for reference. The lowest of the solid $\bar{g}_f$'s displays the rise-decay form discussed in the text. Central panel: the unique optimal gain function for the connectionist model, given by Eq. (49); $\bar{g}_c(s) = -\infty$ for $s \leq t_s$.

we assume this for $s \leq t_s$. We then determine $\overline{g}_f(s)$ for $s > t_s$ from (52), allowing a discontinuity at t_s and taking arbitrary "initial" conditions $\overline{g}_f(t_s)$. Figure 8 illustrates several optimal functions arising from different choices of $\overline{g}_f(t_s)$. The following fact is helpful in understanding positive solutions of (52): orbits lying below $(1/\beta)[1 - \tau_c \, l(s)]$ at any time s decrease toward 0; those above this value increase. Since $(1/\beta)[1 - \tau_c \, l(s)] \to 1/\beta$ as $s \to \infty$, $1/\beta$ asymptotically forms a separatrix between optimal gain trajectories that decay and those that diverge to ∞. Also, note that Case 2 parameters for the two-dimensional firing rate model of Sec. 2.5.1 implement a step in effective gain values up to $1/\beta = 1$, so that in this case nearly optimal signal processing occurs with no explicit adjustment of the gain parameter. The performance resulting from optimal gain trajectories in all models is 73.1% correct responses at interrogation at time $T = 2$; for comparison, the (*nonoptimal*) constant gain $\overline{g}_f(s) \equiv 1/\beta$ produces only 66.4% correct.

Gains must remain bounded for all time to be of practical interest. A family of optimal gain schedules of this form, determined by their (sufficiently small) initial conditions, will always exist for monotonically rising and bounded stimuli $a(s)$ such as that chosen here. As we elaborate in Sec. 4, their "rise-decay" pattern resembles the gain produced by dissipating pulses of the neuromodulator norepinephrine delivered to cortical decision areas via the locus coeruleus, hence providing a clue that this brainstem organ may be assisting near-optimal decision making.

Example 3. We finally assume that $a(s)$ smoothly increases from a low to a higher level and then returns to its original level, corresponding to a transient increase in stimulus salience. We model this as a difference of two sigmoids: $a(s) = a_0 + (\overline{a}/(1 + \exp(-4r(t_{s,1}-s)))) - (\overline{a}/(1 + \exp(4r(t_{s,2}-s))))$, with parameters $a_0 = -0.04$, $\overline{a} = 0.045$, $t_{s,1} = 0.75$, $t_{s,2} = 1.25$, and $r = 20$: see Fig. 9. Additionally, we take constant noise strength $c(s) \equiv 0.06$ and $\tau_c = \beta = 1$.

For the pure drift-diffusion model, Eq. (47) again gives $\overline{g}_{dd}(s) = \tau_c k a(s)$, and for the

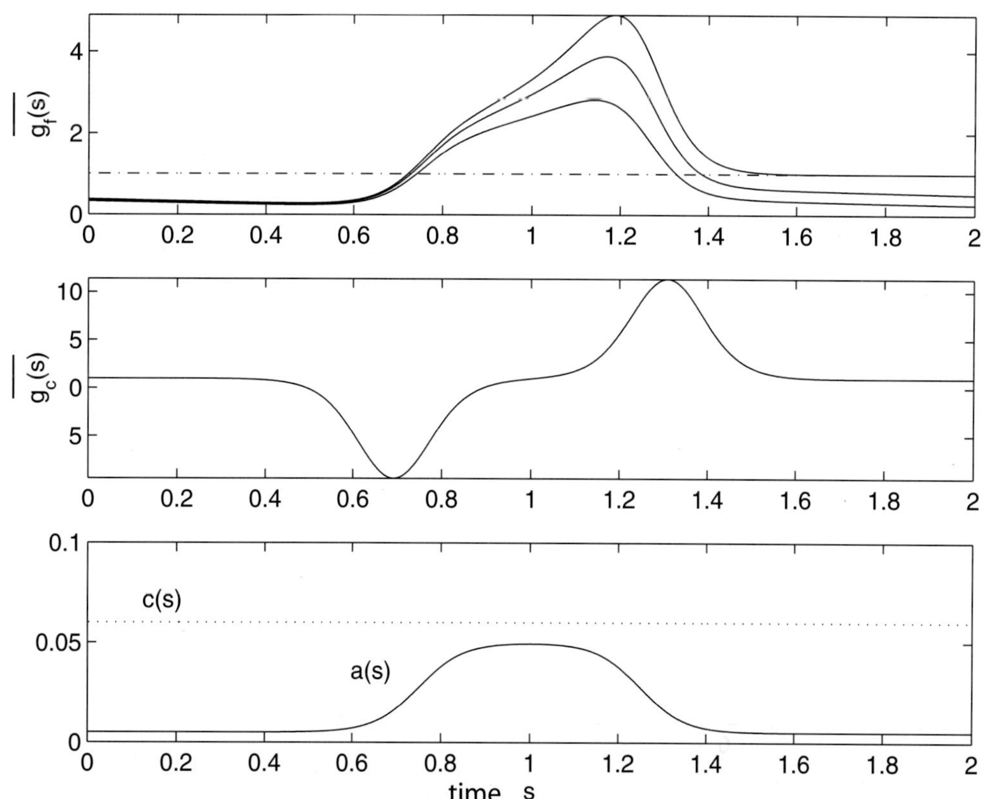

Fig. 9.   Optimal gains for pulsed signal strength $a(s)$ (solid line in bottom panel) and constant noise amplitude $c(s) \equiv 0.06$ (dotted line). Top panel: three optimal gain schedules $\overline{g}_f$ for the firing rate model solving (52) (solid curves); the nonoptimal constant gain function $g \equiv 1/\beta$ is shown dot-dashed for reference. Central panel: the unique optimal gain function for the connectionist model, given by Eq. (49).

connectionist and firing rate models, we may use (49) and (52) for the entire time interval of interest since $a(s)$ is strictly positive. The resulting optimal gain trajectories, shown in Fig. 9, yield 70.8% correct responses at interrogation time $T = 2$, compared with 64.9% correct obtained for constant gain $g_f(s) \equiv 1/\beta$ in the firing rate model. Note that the form of the optimal $\overline{g}_c(s)$ illustrates the intuitive explanation given in Sec. 3.3.2: when the signal-to-noise ratio increases, $\overline{g}_c(s)$ *decreases*, suppressing previously integrated information, and vice-versa.

In summary, we have shown in this section that gain schedules yielding optimal performance in (reduced) models of decision tasks depend strongly on the time course of task stimuli as well as the structure of the underlying model, although they all implement matched filters and maximize the signal-to-noise ratio in the difference between activities of neural populations representing competing alternatives. For systems well-described by connectionist models, neural mechanisms may be expected to depress the gain (i.e. strength of inhibitory feedback) below the "balanced" level of $1/\beta$ when stimulus salience is increasing, and enhance it above this level when salience is decreasing. However, for the firing rate model an optimal network can "choose" among a variety of gain schedules of qualitatively different forms. One neurobiological implication of this flexibility is explored in the following section.

4. The Locus Coeruleus Brainstem Area and Optimal Gain Trajectories

Neurons comprising the brainstem nucleus locus coeruleus (LC) emit the neurotransmitter norepinephrine (NE) to targets widely distributed throughout the brain, including cortical areas involved in decision tasks. While NE has disparate and complex effects on different brain regions, a dominant cortical role is believed to be modulation of neuronal gain at both the single cell and population levels [Usher *et al.*, 1999; Servan-Schreiber *et al.*, 1990]. Recordings of cortical neuron responses to stereotyped inputs at various latencies following activation of LC reveal these gain effects: responses to a fixed input are larger (in certain experimental ranges) following LC activation than in control recordings without LC, and this elevated sensitivity decays with a time constant $\tau_{NE} \approx 0.2$ sec [Waterhouse *et al.*, 1998].

Since the firing rate of LC neurons governs NE release rate, we propose the following simple model for cortical gain $g(t)$:

$$\tau_{NE}\,\dot{g}(t) = k_{LC}\,LC(t) - g(t)\,. \tag{59}$$

Here, $LC(t)$ denotes the time-dependent rate of LC firing and k_{LC} is a constant relating this rate to equilibrium values of cortical gain. This model's limitations in describing the underlying biology include the fact that $g(t)$ decays to zero in the absence of LC firing (this could be rectified by adding a constant "gain floor" g_{base}). Nevertheless, it allows us to make an interesting qualitative point in relating recent data on LC firing rates to optimal strategies for the processing of noisy sensory stimuli. Inverting (59) and inserting an optimal gain trajectory yields a prediction for the optimal time course of LC activity:

$$\overline{LC}(t) = \frac{1}{k_{LC}}\left(\tau_{NE}\,\dot{\overline{g}}(t) + \overline{g}(t)\right). \tag{60}$$

Figure 10(d) shows histograms of LC firing rates recorded from monkeys performing two different psychological tasks: target identification, in which a horizontal or vertical bar must be detected, and the Eriksen flanker task, in which a central cue must be identified while an array of distractors is ignored. Since the second task involves more complex stimulus processing, we assume as in [Brown *et al.*, 2004b] that the onset of stimulus representation in cortical decision areas is more gradual in this than in the target identification task. Specifically, for t greater than the time t_s of stimulus arrival we take $a(t) = \overline{a}(1 - e^{-r(t-t_s)})$ with $r = 50$ (time constant 0.02 sec) for target identification and $r = 10$ (time constant 0.1 sec) for the Eriksen task; also, we set $\overline{a} = 0.06$; and $\tau_c = 0.5$ sec: see Fig. 10(b). Additionally, we assume that t_s follows presentation of the sensory cue by a processing time lag of 0.1 sec (cf. [Aston-Jones *et al.*, 1994]). Optimal gain schedules $\overline{g}_f(t)$ for the firing rate model with these stimuli, computed as in the preceding section, are shown in Fig. 10(a). To produce panel (c), these gain functions were inserted into Eq. (60) to yield corresponding optimal LC firing rates, the discontinuity in $\overline{g}_f(t)$ at stimulus onset having negligible effect. (Also note that assuming a smoother profile for $a(t)$ would eliminate the jump in $LC(t)$.) The similarity between overall form and decay rates of optimal gain functions $\overline{LC}(t)$ and the empirical data of Fig. 10(d) supports the

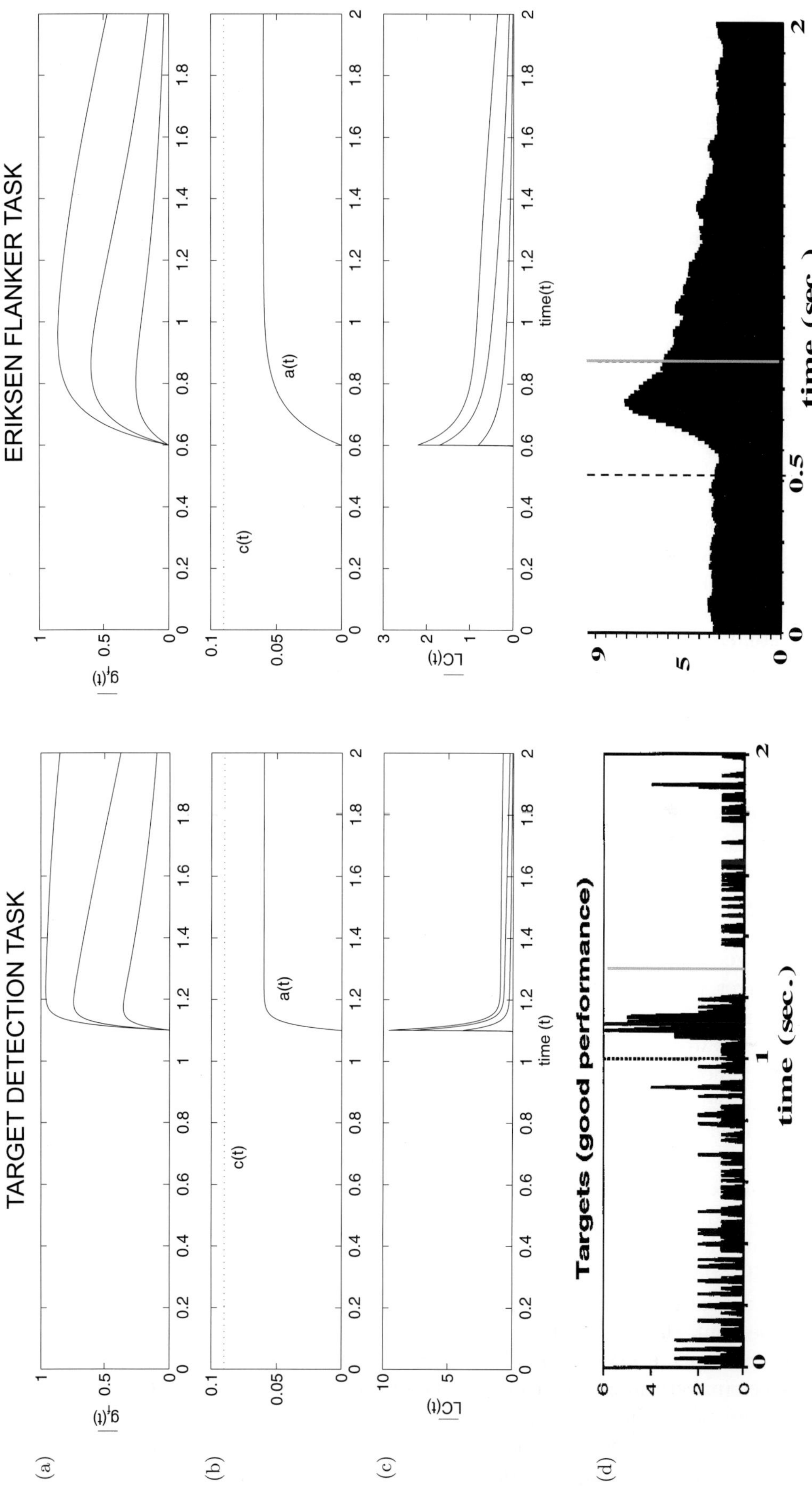

Fig. 10. Comparison of optimal gain theory with empirical data for two psychological tasks. (a) Optimal gain schedules for the firing rate model, for rapid (left) and gradual (right) onset of stimulus $a(t)$ to neural units (with a processing time lag of 0.1 sec following sensory cue), as shown in (b). (c) The corresponding optimal time courses of LC firing rate. (d) Histograms of LC firing rates recorded in the two tasks: (left) the target detection task [Usher *et al.*, 1999] (right) the Eriksen flanker task, with data kindly provided by the authors of [Clayton *et al.*, 2004]. Vertical dashed lines indicate onset of sensory stimuli, and vertical gray (solid) lines indicate mean behavioral reaction time (standard deviations are $\approx$ 34 and 114 msec for the target detection and Eriksen tasks, respectively).

hypothesis that the LC may affect near-optimal processing of sensory stimuli. This is true even though LC firing rates are not sustained at the initial high values that follow stimulus onset; in fact, both LC firing rate relaxation and NE time constants are compatible with optimal gain schedules.

We note that the optimal gains, and hence $LC(t)$ time courses, are computed assuming prior knowledge of the stimulus $a(t)$ and signal-to-noise ratio $a(t)/c(t)$. If this were the case, LC firing patterns should be well-correlated with stimulus onset. However, experimental data of [Clayton *et al.*, 2004], which involved variable stimulus onset times, indicates tighter correlations with behavioral responses. Here, the function $a(t)$ is perhaps better interpreted as input to motor neurons, the onsets of sensory stimuli having been detected earlier in decision layers. Thus, the most appropriate LC data for use in Fig. 10 would be aligned with transients in firing rates in intermediate processing layers; here we provide data aligned with sensory stimuli as the closest available surrogate. Explicit models of multi-layer decision/response dynamics with variable gain are studied in [Brown *et al.*, 2004a].

5. Discussion and Conclusions

In this paper we explicitly compute optimal gain trajectories for one-dimensional, linearized reductions of simplified models for competing neural groups involved in decisions between two alternatives. We also demonstrate via simulations that such reductions provide good approximations for the reaction time and error rate statistics of the nonlinear two-dimensional connectionist and firing rate models from which they were derived.

We first show that the nonlinear connectionist and firing rate models are equivalent, under suitable variable and parameter coordinate changes. We then develop a piecewise-linear approximation to the canonical sigmoidal activation or firing rate function. The resulting two-dimensional piecewise-linear SDEs (15)–(16) introduced in Sec. 2.4 form a midpoint in our simplification process. This system can be easily solved on each of nine "tiles" forming its phase plane, but solutions must be assembled by matching constants of integration. To illustrate this, we focus on two specific cases in Sec. 2.5.1, motivated by the moving dots' paradigm [Britten *et al.*, 1993; Shadlen & Newsome, 2001; Gold & Shadlen, 2002], that correspond to differing stimulus presentation conditions and rely on

different neural mechanisms to implement transient effective gain values.

In Case 1, the development of salience (i.e. $a_1 \neq a_2$), in sensory stimuli at time t_s is not accompanied by large changes in the stimulus magnitudes; in fact the summed magnitude is unchanged. This mild stimulus onset is insufficient to move solutions between tiles, so variations in gain must result from modulation of the gain of the neural activation function itself, presumably via influence of other brain areas such as the locus coeruleus. However, in Case 2, the appearance of salience is accompanied by large changes in stimulus magnitude, either due to properties of the stimulus itself or due to additive biases that shift the activation function to the left, as has been proposed in connectionist models that address the effects of attention [Mozer, 1988; Cohen *et al.*, 1992]. In this case, no external modulation of gain is required, since the decision dynamics themselves move the system between regions of the activation function where desired sensitivities (and hence gains) are achieved. The possibility that neural systems are tuned so that the presence of target stimuli causes solutions to move into sensitive regions of their activation functions has been previously suggested in behavioral neuroscience [Servan-Schreiber *et al.*, 1990]; here we reformulate this idea in terms of optimal signal processing.

We end by showing that the (nonunique) optimal gain schedules for the firing rate model include time courses that are consistent with release of norepinephrine due to transient increases in the activity of neurons in locus coeruleus.

The external modification of gain considered in Case 1 assumes prior knowledge of the time course of the absolute values of sensory inputs $a_j(t)$, the task of the decision maker being merely to identify their signs. In [Brown *et al.*, 2004a] the more general case in which this information is not available is treated, and strategies must additionally include a mechanism for detecting increases in signal-to-noise ratio of sensory inputs.

Acknowledgments

This work was partially supported by DoE grant DE-FG02-95ER25238 and PHS grants MH58480 and MH62196 (Cognitive and Neural Mechanisms of Conflict and Control, Silvio M. Conte Center). E. Brown was supported under a National Science Foundation Graduate Fellowship and a Burroughs–Wellcome Training Grant in Biological Dynamics.

The authors thank Josh Gold and Jaime Cisternas for useful contributions and discussions, as well as Ed Clayton and Gary Aston-Jones for providing the data of Fig. 10 and for their insights into the role of the LC in modulating decisions.

References

Abbott, L. [1991] "Firing-rate models for neural populations," in *Neural Networks: From Biology to High-Energy Physics*, eds. Benhar, O., Bosio, C., Del Giudice, P. & Tabat, E. (ETS Editrice, Pisa), pp. 179–196.

Amit, D. & Tsodyks, M. [1991] "Quantitative study of attractor neural network retrieving at low spike rates: I. Substrate–spikes, rates, and neuronal gain," *Network* **2**, 259–273.

Anderson, J. [1990] *The Adaptive Character of Thought* (Lawrence Erlbaum, Hillsdale, NJ).

Arnold, L. [1974] *Stochastic Differential Equations* (John Wiley, NY).

Arnold, L. [1998] *Random Dynamical Systems* (Springer, Heidelberg).

Aston-Jones, G., Rajkowski, J., Kubiak, P. & Alexinsky, T. [1994] "Locus coeruleus neurons in the monkey are selectively activated by attended stimuli in a vigilance task," *J. Neurosci.* **14**, 4467–4480.

Bialek, W., Rieke, F., de Reuter van Steveninck, R. & Warland, D. [1991] "Reading a neural code," *Science* **252**, 1854–1857.

Bogacz, R., Brown, E., Moehlis, J., Hu, P., Holmes, P. & Cohen, J. D. [2004] "The physics of optimal decision making: A formal analysis of models of performance in two alternative forced choice tasks," *Psych. Rev.*, in review.

Boxler, P. [1991] "How to construct stochastic center manifolds on the level of vector fields," in *Lyapunov Exponents*, eds. Arnold, L., Crauel, H. & Eckmann, J.-P., Lecture Notes in Mathematics, Vol. 1486 (Springer, Heidelberg), pp. 141–158.

Britten, K. H., Shadlen, M. N., Newsome, W. T. & Movshon, J. A. [1993] "Responses of neurons in macaque MT to stochastic motion signals," *Vis. Neurosci.* **10**, 1157–1169.

Brown, E. & Holmes, P. [2001] "Modeling a simple choice task: stochastic dynamics of mutually inhibitory neural groups," *Stochast. Dyn.* **1**, 159–191.

Brown, E., Gilzenrat, M. S. & Cohen, J. D. [2004a] "The locus coeruleus, adpative gain, and the optimization of simple decision tasks," Technical Report #04–02, Center for the Study of Mind, Brain, and Behavior, Princeton University.

Brown, E., Moehlis, J., Holmes, P., Clayton, E., Rajkowski, J. & Aston-Jones, G. [2004b] "The influence of spike rate and stimulus duration on noradrenergic neurons," *J. Comput. Neurosci* **17**, 5–21.

Brunel, N., Chance, F., Fourcaud, N. & Abbott, L. F. [2001] "Effects of synaptic noise and filtering on the frequency response of spiking neurons," *Phys. Rev. Lett.* **86**, 2186–2189.

Chance, F. S., Abbott, L. F. & Reyes, A. D. [2002] "Gain modulation from background synaptic input," *Neuron* **35**, 773–782.

Cho, R., Nystrom, L., Brown, E., Jones, A., Braver, T., Holmes, P. & Cohen, J. D. [2002] "Mechanisms underlying performance dependencies on stimulus history in a two-alternative forced choice task," *Cogn. Affect. Behav. Neurosci.* **2**, 283–299.

Clayton, E., Rajkowski, J., Cohen, J. D. & Aston-Jones, G. [2004] "Decision-related activation of monkey locus coeruleus neurons in a forced choice task," under preparation.

Cohen, J. D., Dunbar, K. & McClelland, J. L. [1990] "On the control of automatic processes: A parallel distributed processing model of the Stroop effect," *Psychol. Rev.* **97**, 332–361.

Cohen, J. D., Servan-Schreiber, D. & McClelland, J. L. [1992] "A parallel distributed processing approach to automaticity," *Amer. J. Psychol.* **105**, 239–269.

Cohen, J. D. & Huston, T. A. [1994] "Progress in the use of interactive models for understanding attention and performance," in *Attention and Performance XV*, eds. Umilta, C. & Moscovitch, M. (MIT Press, Cambridge), pp. 453–476.

Ermentrout, G. B. [1994] "Reduction of conductance-based models with slow synapses to neural nets," *Neur. Comput.* **6**, 679–695.

Fairhall, A., Lewen, G., Bialek, W. & de Ruyter van Steveninck, R. [2001] "Effciency and ambiguity in an adaptive neural code," *Nature* **412**, 787–792.

Gardiner, C. W. [1985] *Handbook of Stochastic Methods*, 2nd edition (Springer, NY).

Gerstner, W. & Kistler, W. [2002] *Spiking Neuron Models* (Cambridge University Press, Cambridge, UK).

Gilzenrat, M. S., Holmes, B. D., Rajkowski, J., Aston-Jones, G. & Cohen, J. D. [2002] "Simplified dynamics in a model of noradrenergic modulation of cognitive performance," *Neural Networks* **15**, 647–663.

Gold, J. I. & Shadlen, M. N. [2001] "Neural computations that underlie decisions about sensory stimuli," *Trends Cogn. Sci.* **5**, 10–16.

Gold, J. I. & Shadlen, M. N. [2002] "Banburismus and the brain: Decoding the relationship between sensory stimuli, decisions, and reward," *Neuron* **36**, 299–308.

Grossberg, S. [1988] "Nonlinear neural networks: Principles, mechanisms, and architectures," *Neural Networks* **1**, 17–61.

Guckenheimer, J. & Holmes, P. J. [1983] *Nonlinear Oscillations, Dynamical Systems and Bifurcations of Vector Fields* (Springer-Verlag, NY).

Hertz, J., Krough, A. & Palmer, R. [1991] *Introduction to the Theory of Neural Computation* (Perseus Book Group, NY).

Hopfield, J. J. [1984] "Neurons with graded response have collective computational properties like those of two-state neurons," *Proc. Natl. Acad. Sci. USA* **82**, 3088–3092.

Huk, A., Palmer, J. & Shadlen, M. [2002] "Temporal integration of motion energy underlies perceptual decisions and response times," *Annual Society for Neuroscience Meeting*, Orlando, FL, Nov 2–7, 2002, Abstract No. 353.5.

Knobloch, E. & Weisenfeld, K. A. [1983] "Bifurcations in fluctuating systems: The center manifold approach," *J. Stat. Phys.* **33**, 611–637.

Laming, D. R. J. [1968] *Information Theory of Choice-Reaction Times* (Academic Press, NY).

Lehmann, E. L. [1959] *Testing Statistical Hypotheses* (John Wiley, NY).

McClelland, J. L. [1979] "On the time relations of mental processes: An examination of systems of processes in cascade," *Psychol. Rev.* **86**, 287–330.

Mozer, M. [1998] "A connectionist model of selective attention in visual perception," in *Proc. Tenth Ann. Conf. Cognitive Science Society* (Erlbaum, Hillsdale, NJ), pp. 195–201.

Omurtag, A., Kaplan, E., Knight, B. W. & Sirovich, L. [2000] "A population approach to cortical dynamics with an application to orientation tuning," *Network* **11**, 247–260.

Papoulis, A. [1977] *Signal Analysis* (McGraw-Hill, NY).

Platt, M. L. & Glimcher, P. W. [2001] "Neural correlates of decision variable in parietal cortex," *Nature* **400**, 233–238.

Ratcliff, R. [1978] "A theory of memory retrieval," *Psych. Rev.* **85**, 59–108.

Ratcliff, R., Van Zandt, T. & McKoon, G. [1999] "Connectionist and diffusion models of reaction time," *Psych. Rev.* **106**, 261–300.

Roitman, J. & Shadlen, M. [2002] "Response of neurons in the lateral intraparietal area during a combined visual discrimination reaction time task," *J. Neurosci.* **22**, 9475–9489.

Schall, J. D. [2001] "Neural basis of deciding, choosing, and acting," *Nature Rev.: Neurosci.* **2**, 33–42.

Schall, J., Stuphorn, V. & Brown, J. [2002] "Monitoring and control of action by the frontal lobes," *Neuron* **36**, 309–322.

Servan-Schreiber, D., Printz, H. & Cohen, J. D. [1990] "A network model of catecholamine effects: Gain, signal-to-noise ratio, and behavior," *Science* **249**, 892–895.

Shadlen, M. N. & Newsome, W. T. [2001] "Neural basis of a perceptual decision in the parietal cortex (area LIP) of the rhesus monkey," *J. Neurophysiol.* **86**, 1916–1936.

Shelley, M. & McLaughlin, D. [2002] "Coarse-grained reduction and analysis of a network model of cortical response. I. drifting grating stimuli," *J. Comput. Neurosci.* **12**, 97–122.

Shin, J., Koch, C. & Douglas, R. [1999] "Adaptive neural coding dependent on the time varying statistics of the somatic input current," *Neural Comput.* **11**, 1083–1913.

Smith, P. L. & Ratcliff, R. [2004] "Psychology and neurobiology of simple decisions," *Trends in Neurosci.* **27**, 161–168.

Stone, M. [1960] "Models for choice-reaction time," *Psychometrika* **25**, 251–260.

Usher, M., Cohen, J. D., Servan-Schreiber, D., Rajkowsky, J. & Aston-Jones, G. [1999] "The role of locus coeruleus in the regulation of cognitive performance," *Science* **283**, 549–554.

Usher, M. & McClelland, J. L. [2001] "On the time course of perceptual choice: The leaky competing accumulator model," *Psych. Rev.* **108**, 550–592.

von Neumann, J. [1958] *The Computer and the Brain* (Yale University Press, New Haven, CT); 2nd edition [2000], with a foreword by Paul and Patricia Churchland.

Wald, A. [1947] *Sequential Analysis* (John Wiley, NY).

Wang, X.-J. [2002] "Probabilistic decision making by slow reverberation in cortical circuits," *Neuron* **36**, 955–968.

Waterhouse, B., Moises, H. & Woodward, D. [1998] "Phasic activation of the locus coeruleus enhances responses of primary sensory cortical neurons to peripheral receptive field stimulation," *Brain Res.* **790**, 33–44.

Wickelgren, W. A. [1977] "Speed-accuracy tradeoff and information processing dynamics," *Acta Psychol.* **41**, 67–85.

Wilson, H. & Cowan, J. [1972] "Excitatory and inhibitory interactions in localized populations of model neurons," *Biophys. J.* **12**, 1–24.

NEWTON FLOW AND INTERIOR POINT METHODS IN LINEAR PROGRAMMING

JEAN-PIERRE DEDIEU

MIP. Département de Mathématique, Université Paul Sabatier,
31062 Toulouse cedex 04, France

MIKE SHUB

Department of Mathematics, University of Toronto,
100 St. George Street, Toronto, Ontario M5S 3G3, Canada

Received January 12, 2004; Revised June 17, 2004

We study the geometry of the central paths of linear programming theory. These paths are the solution curves of the Newton vector field of the logarithmic barrier function. This vector field extends to the boundary of the polytope and we study the main properties of this extension: continuity, analyticity, singularities.

Keywords: Linear programming; interior point method; central path; Newton vector field; extension.

1. Introduction

In this paper we take up once again the subject of the geometry of the central paths of linear programming theory. We study the boundary behavior of these paths as in [Megiddo & Shub, 1989], but from a different perspective and with a different emphasis. Our main goal will be to give a global picture of the central paths even for degenerate problems as solution curves of the Newton vector field, $N(x)$, of the logarithmic barrier function which we describe below. See also [Bayer & Lagarias, 1989a, 1989b, 1991]. The Newton vector field extends to the boundary of the polytope. It has the properties that it is tangent to the boundary and restricted to any face of dimension i it has a unique source with unstable manifold dimension equal to i, the rest of the orbits tending to the boundary of the face. Every orbit tends either to a vertex or one of these sources in a face. See Corollary 4.1. This highly cellular structure of the flow lends itself to the conjecture that the total curvature of these central paths may be linearly bounded by the dimension n of the polytope. The orbits may be relatively straight, except for orbits which come close to an orbit in a face of dimension i which itself comes close to a singularity in a boundary face of dimension less than i. This orbit is then forced to turn almost parallel to the lower dimensional face so its tangent vector may be forced to turn as well. See the two figures at the end of this paper. As this process involves a reduction of the dimension of the face it can only happen for the dimension of the polytopetimes. So our optimistic conjecture is that the total curvature of a central path is $O(n)$. We have verified the conjecture in an average sense in [Dedieu *et al.*]. It is not difficult to give an example showing that $O(n)$ is the best possible for the worst case. Such an example is worked out in [Megiddo & Shub, 1989]. The average behavior may be however much better. Ultimately we hope that an understanding of the curvature of the central paths may contribute to the analysis of algorithms which use them. In [Vavasis & Ye, 1996] the authors explore similar structure to give an algorithm whose running time depends only on the polytope.

We prove in Corollary 4.1 that the extended vector field is Lipschitz on the closed polytope.

Under a genericity hypothesis we prove in Theorem 5.1 that it extends to be real analytic on a neighborhood of the polytope. Under the same genericity hypothesis we prove in Theorem 5.2 that the singularities are all hyperbolic. The eigenvalues of $-N(x)$ at the singularities are all $+1$ tangent to the face and -1 transversal to the face. In dynamical systems terminology the vector field is Morse–Smale. The vertices are the sinks. Finally, we mention that in order to prove that $N(x)$ always extends continuously to the boundary of the polytope we prove Lemma 4.2 which may be of independent interest about the continuity of the Moore–Penrose inverse of a family of linear maps of variable rank.

2. The Central Path is a Trajectory of the Newton Vector Field

Linear programming problems are frequently presented in different formats. We will work with one of them here which we find convenient. The polytopes defined in one format are usually affinely equivalent to the polytopes defined in another. So we begin with a discussion of Newton vector fields and how they transform under affine equivalence. This material is quite standard. An excellent source for this fact and linear programming in general is [Renegar, 2001].

Let $\mathbb{Q}$ be an affine subspace of $\mathbb{R}^n$ (or a Hilbert space if you prefer, in which case, assume $\mathbb{Q}$ is closed). Denote the tangent space of $\mathbb{Q}$ by $\mathbb{V}$. Suppose that U is an open subset of $\mathbb{Q}$. Let $f : U \to \mathbb{R}$ be twice continuously differentiable. The derivative $Df(x)$ belongs to $L(\mathbb{V}, \mathbb{R})$, the linear maps from $\mathbb{V}$ to $\mathbb{R}$. So $Df(x)$ defines a map from U to $L(\mathbb{V}, \mathbb{R})$. The second derivative $D^2f(x)$ is an element of $L(\mathbb{V}, L(\mathbb{V}, \mathbb{R}))$. Thus $D^2f(x)$ is a linear map from a vector space to another isomorphic space and $D^2f(x)$ may be invertible.

Definition 2.1. If f is as above and $D^2f(x)$ is invertible we define the Newton vector field, $N_f(x)$ by

$$N_f(x) = -(D^2f(x))^{-1}Df(x).$$

Note that if $\mathbb{V}$ has a nondegenerate inner product $\langle \, , \, \rangle$ then the gradient of f, grad $f(x) \in \mathbb{V}$, and Hessian, hess $f(x) \in L(\mathbb{V}, \mathbb{V})$, are defined by

$$Df(x)u = \langle u, \text{grad } f(x) \rangle$$

and

$$D^2f(x)(u, v) = \langle u, (\text{hess } f(x))v \rangle.$$

It follows then that $N_f(x) = -(\text{hess } f(x))^{-1}$ grad $f(x)$.

Now let A be an affine map from $\mathbb{P}$ to $\mathbb{Q}$ whose linear part L is an isomorphism. Suppose U_1 is open in $\mathbb{P}$ and $A(U_1) \subseteq U$. Let $g = f \circ A$.

Proposition 2.1. *A maps the solution curves of N_g to the solution curves of N_f.*

Proof. By the chain rule $Dg(y) = Df(A(y))L$ and

$$D^2g(y)(u, v) = D^2f(A(y))(Lu, Lv).$$

So $u = N_g(y)$ if and only if $D^2g(y)(u, v) = -Dg(y)(v)$ for all v if and only if $D^2f(A(y))(Lu, Lv) = -Df(A(y))Lv$ for all v, i.e. $N_f(A(y)) = L(u)$ or $LN_g(y) = N_f A(y)$. This last is the equation expressing that the vector field N_f is the push forward by the map of the vector field N_g and hence the solution curves of the N_g field are mapped by A to the solution curves of N_f. ∎

Now we make explicit the linear programming format used in this paper, define the central paths and relate them to the Newton vector field of the logarithmic barrier function.

Let $\mathcal{P}$ be a compact polytope in $\mathbb{R}^n$ defined by m affine inequalities

$$A_i x \geq b_i, \quad 1 \leq i \leq m.$$

Here $A_i x$ denotes the matrix product of the row vector $A_i = (a_{i1}, \ldots, a_{in})$ by the column vector $x = (x_1, \ldots, x_n)^T$, A is the $m \times n$ matrix with rows A_i and we assume rank $A = n$. Given $c \in \mathbb{R}^n$, we consider the linear programming problem

$$(LP) \quad \min_{\substack{A_i x \geq b_i \\ 1 \leq i \leq m}} \langle c, x \rangle.$$

Let us denote by

$$f(x) = \sum_{i=1}^{m} \ln(A_i x - b_i)$$

($\ln(s) = -\infty$ when $s \leq 0$) the logarithmic barrier function associated with the description $Ax \geq b$ of $\mathcal{P}$. The barrier technique considers the family of nonlinear convex optimization problems

$$(LP(t)) \quad \min_{x \in \mathbb{R}^n} t\langle c, x \rangle - f(x)$$

with $t > 0$. The objective function

$$f_t(x) = t \langle c, x \rangle - f(x)$$

is strictly convex, smooth, and satisfies

$$\lim_{\substack{x \to \partial \mathcal{P} \\ x \in \mathrm{Int}\, \mathcal{P}}} f_t(x) = \infty.$$

Thus, there exists a unique optimal solution $\gamma(t)$ to $(LP(t))$ for any $t > 0$. This curve is called the central path of our problem. Let us denote as D_x the $m \times m$ diagonal matrix $D_x = \mathrm{Diag}(A_i x - b_i)$. This matrix is nonsingular for any $x \in \mathrm{Int}\, \mathcal{P}$. We also let $e = (1, \ldots, 1)^T \in \mathbb{R}^m$,

$$g(x) = \mathrm{grad}\, f(x) = \sum_{i=1}^m \frac{A_i^T}{A_i x - b_i} = A^T D_x^{-1} e$$

and

$$h(x) = \mathrm{hess}\, f(x) = -A^T D_x^{-2} A.$$

Since f_t is smooth and strictly convex the central path is given by the equation $\mathrm{grad}\, f_t(\gamma(t)) = 0$ i.e.

$$g(\gamma(t)) = tc, \quad t > 0.$$

When $t \to 0$, the limit of $\gamma(t)$ is given by

$$-f(\gamma(0)) = \min_{x \in \mathbb{R}^n} -f(x).$$

It is called the analytic center of $\mathcal{P}$ and denoted by $c_{\mathcal{P}}$.

Lemma 2.1. *$g: \mathrm{Int}\, \mathcal{P} \to \mathbb{R}^n$ is real analytic and invertible. Its inverse is also real analytic.*

Proof. For any $c \in \mathbb{R}^n$ the optimization problem

$$\min_{x \in \mathbb{R}^n} \langle c, x \rangle - f(x)$$

has a unique solution in $\mathrm{Int}\, \mathcal{P}$ because the objective function is smooth, strictly convex and $\mathcal{P}$ is compact. Thus $g(x) = c$ has a unique solution that is g bijective. We also notice that, for any x, $Dg(x)$ is nonsingular. Thus g^{-1} is real analytic by the inverse function theorem. ∎

According to this lemma, the central path is the inverse image by g of the ray $c\mathbb{R}_+$. When c varies

in $\mathbb{R}^n$ we obtain a family of curves. Our aim in this paper is to investigate the structure of this family.

For a subspace $\mathcal{B} \subset \mathbb{R}^m$ we denote by $\Pi_{\mathcal{B}}$ the orthogonal projection $\mathbb{R}^m \to \mathcal{B}$. Let $b_1, \ldots, b_r$ be a basis of $\mathcal{B}$ and let us denote by B the $m \times r$ matrix with columns of the vectors b_i. Then $\Pi_{\mathcal{B}}$, also denoted Π_B, is given by $\Pi_B = B(B^T B)^{-1} B^T = BB^\dagger$ ($B^\dagger$ is the generalized inverse of B equal to $(B^T B)^{-1} B^T$ because B is injective).

Definition 2.2. The Newton vector field associated with g is

$$\begin{aligned} N(x) &= -Dg(x)^{-1} g(x) \\ &= (A^T D_x^{-2} A)^{-1} A^T D_x^{-1} e \\ &= A^\dagger D_x \Pi_{D_x^{-1} A} e. \end{aligned}$$

It is defined and analytic on $\mathrm{Int}\, \mathcal{P}$.

Note that the expression $A^\dagger D_x \Pi_{D_x^{-1} A} e$ is defined for all $x \in \mathbb{R}^n$ for which $A_i x - b_i$ is not equal to 0 for all i. Thus $N(x)$ is defined by the rational expression in Definition 2.2 for almost all $x \in \mathbb{R}^n$. Later we will prove that this rational expression has a continuous extension to all $\mathbb{R}^n$.

Lemma 2.2. *The central paths $\gamma(t)$, $c \in \mathbb{R}^n$, are the trajectories of the vector field $-N(x)$.*

Proof. A central path is given by

$$g(\gamma(t)) = tc, \quad t > 0,$$

for a given $c \in \mathbb{R}^n$. Let us change variable: $t = \exp s$ and $d(s) = \gamma(t)$ with $s \in \mathbb{R}$. Then

$$g(d(s)) = \exp(s)c, \quad s \in \mathbb{R},$$

so that

$$\frac{d}{ds} g(d(s)) = \exp(s)c = g(d(s)).$$

Let us denote $\dot{d}(s) = (d/ds)d(s)$. We have

$$\frac{d}{ds} g(d(s)) = Dg(d(s))\dot{d}(s)$$

thus

$$\dot{d}(s) = Dg(d(s))^{-1} g(d(s)) = -N(d(s))$$

and $d(s)$ is a trajectory of the Newton vector field. Conversely, if $\dot{d}(s) = -N(d(s)) = Dg(d(s))^{-1} g(d(s))$, $s \in \mathbb{R}$, then

$$\frac{d}{ds} g(d(s)) = Dg(d(s))\dot{d}(s) = g(d(s))$$

so that

$$g(d(s)) = \exp(s)g(d(0))$$

which is the central path related to $c = g(d(0))$. ∎

Remark 2.1. The trajectories of $N(x)$ and $-N(x)$ are the same with time reversed. As $t \to \infty$, $\gamma(t)$ tends to the optimal points of the linear programming problem. So we are interested in the positive time trajectories of $-N(x)$.

Lemma 2.3. *The analytic center $c_{\mathcal{P}}$ is the unique singular point of the Newton vector field $N(x)$, $x \in Int\ \mathcal{P}$.*

Proof. $N(x) = 0$ if and only if $g(x) = 0$, that is $x = c_{\mathcal{P}}$. ∎

3. An Analytic Expression for the Newton Vector Field

In this section we compute an analytic expression for $N(x)$ which will be useful later. For any subset $K_n \subset \{1, \ldots, m\}$, $K_n = \{k_1 < \cdots < k_n\}$, we denote by A_{K_n} the $n \times n$ submatrix of A with rows $A_{k_1}, \ldots, A_{k_n}$, by b_{K_n} the vector in $\mathbb{R}^n$ with coordinates $b_{k_1}, \ldots, b_{k_n}$, and by u_{K_n} the unique solution of the system $A_{K_n} u_{K_n} = b_{K_n}$ when the matrix A_{K_n} is nonsingular. With these notations we have:

Proposition 3.1. *For any $x \in Int\ \mathcal{P}$,*

$$N(x) = \frac{\displaystyle\sum_{\substack{K_n \subset \{1,\ldots,m\} \\ \det A_{K_n} \neq 0}} (x - u_{K_n})(\det A_{K_n})^2 \prod_{l \notin K_n} (A_l x - b_l)^2}{\displaystyle\sum_{\substack{K_n \subset \{1,\ldots,m\} \\ \det A_{K_n} \neq 0}} (\det A_{K_n})^2 \prod_{l \notin K_n} (A_l x - b_l)^2}$$

Proof. Let us denote $\Pi = \prod_{l=1}^{m} (A_l x - b_l)$ and $\Pi_k = \prod_{l \neq k} (A_l x - b_l)$. We already know (Definition 2.2) that $N(x) = (A^T D_x^{-2} A)^{-1} A^T D_x^{-1} e$ with

$$(A^T D_x^{-2} A)_{ij} = \sum_{k=1}^{m} \frac{a_{ki} a_{kj}}{(A_k x - b_k)^2}$$

$$= \frac{1}{\Pi^2} \sum_{k=1}^{m} a_{ki} a_{kj} \Pi_k^2$$

$$= \frac{1}{\Pi^2} X_{ij}$$

where X is the $n \times n$ matrix given by $X_{ij} = \sum_{k=1}^{m} a_{ki} a_{kj} \Pi_k^2$. Moreover

$$(A^T D_x^{-1} e)_i = \sum_{k=1}^{m} \frac{a_{ki}}{A_k x - b_k}$$

$$= \frac{1}{\Pi} \sum_{k=1}^{m} a_{ki} \Pi_k$$

$$= \frac{1}{\Pi} V_i$$

where V is the n vector given by $V_i = \sum_{k=1}^{m} a_{ki} \Pi_k$. This gives

$$N(x) = \Pi X^{-1} V.$$

To compute X^{-1} we use Cramer's formula: $X^{-1} = \operatorname{cof}(X)^T / \det(X)$ where $\operatorname{cof}(X)$ denotes the matrix of cofactors: $\operatorname{cof}(X)_{ij} = (-1)^{i+j} \det(X^{ij})$ with X^{ij} the $n-1 \times n-1$ matrix obtained by deleting in X the ith row and jth column. We first compute $\det X$. We have

$$\det X = \sum_{\sigma \in \mathbb{S}_n} \epsilon(\sigma) X_{1\sigma(1)} \cdots X_{n\sigma(n)}$$

where $\mathbb{S}_n$ is the group of permutations of $\{1, \ldots, n\}$ and $\epsilon(\sigma)$ the signature of σ. Thus

$$\det X = \sum_{\sigma \in \mathbb{S}_n} \epsilon(\sigma) \prod_{j=1}^{n} \sum_{k_j=1}^{m} a_{k_j j} a_{k_j \sigma(j)} \Pi_{k_j}^2$$

$$= \sum_{\substack{1 \leq k_j \leq m \\ 1 \leq j \leq n}} \Pi_{k_1}^2 \cdots \Pi_{k_n}^2 a_{k_1 1} \cdots a_{k_n n}$$

$$\times \sum_{\sigma \in \mathbb{S}_n} \epsilon(\sigma) a_{k_1 \sigma(1)} \cdots a_{k_n \sigma(n)}$$

$$= \sum_{\substack{1 \leq k_j \leq m \\ 1 \leq j \leq n}} \Pi_{k_1}^2 \cdots \Pi_{k_n}^2 a_{k_1 1} \cdots a_{k_n n} \det A_{k_1 \cdots k_n}$$

where $A_{k_1 \cdots k_n}$ is the matrix with rows $A_{k_1} \cdots A_{k_n}$. When two or more indices k_j are equal the corresponding coefficient $\det A_{k_1 \cdots k_n}$ is zero. For this reason, instead of this sum taken for n independent indices k_j we consider a set $K_n \subset \{1, \ldots, m\}$, $K_n = \{k_1 < \cdots < k_n\}$, and all the possible permutations $\sigma \in \mathbb{S}(K_n)$. We obtain

$$\det X = \sum_{\substack{K_n \subset \{1,\ldots,m\} \\ \sigma \in \mathbb{S}(K_n)}} \Pi^2_{\sigma(k_1)} \cdots \Pi^2_{\sigma(k_n)}$$

$$\times \, a_{\sigma(k_1)1} \cdots a_{\sigma(k_n)n} \det A_{\sigma(k_1) \cdots \sigma(k_n)}$$

$$= \sum_{K_n \subset \{1,\ldots,m\}} \Pi^2_{k_1} \cdots \Pi^2_{k_n}$$

$$\times \sum_{\sigma \in \mathbb{S}(K_n)} \epsilon(\sigma) a_{\sigma(k_1)1} \cdots a_{\sigma(k_n)n} \det A_{k_1 \cdots k_n}$$

$$= \sum_{K_n \subset \{1,\ldots,m\}} \Pi^2_{k_1} \cdots \Pi^2_{k_n} \left(\det A_{K_n} \right)^2.$$

Note that, for any $l = 1, \ldots, m$, the product $\Pi^2_{k_1} \cdots \Pi^2_{k_n}$ contains $(A_l x - b_l)^{2n}$ if $l \notin K_n$ and $(A_l x - b_l)^{2n-2}$ otherwise. For this reason

$$\det X = \Pi^{2n-2} \sum_{K_n \subset \{1,\ldots,m\}} (\det A_{K_n})^2$$

$$\times \prod_{l \notin K_n} (A_l x - b_l)^2.$$

Let us now compute $Y = \operatorname{cof}(X)^T V$. We have

$$Y_i = \sum_{j=1}^{n} (-1)^{i+j} \det(X^{ji}) V_j$$

$$= \sum_{j=1}^{n} (-1)^{i+j} \det(X^{ji}) \sum_{k=1}^{m} a_{kj} \Pi_k$$

$$= \sum_{k=1}^{m} \Pi_k \sum_{j=1}^{n} (-1)^{i+j} \det(X^{ij}) a_{kj}$$

because X is symmetric. This last sum is the determinant of the matrix with rows $X_1 \cdots X_{i-1} A_k X_{i+1} \cdots X_n$ so that

$$Y_i = \sum_{k=1}^{m} \Pi_k \sum_{\sigma \in \mathbb{S}_n} \epsilon(\sigma) X_{1\sigma(1)} \cdots X_{i-1\sigma(i-1)} a_{k\sigma(i)}$$

$$\times X_{i+1\sigma(i+1)} \cdots X_{n\sigma(n)}$$

$$= \sum_{k=1}^{m} \Pi_k \sum_{\sigma \in \mathbb{S}_n} \epsilon(\sigma) a_{k\sigma(i)} \prod_{\substack{j=1 \\ j \neq i}}^{n} \sum_{k_j=1}^{m} a_{k_j j} a_{k_j \sigma(j)} \Pi^2_{k_j}$$

$$= \sum_{k=1}^{m} \Pi_k \sum_{\sigma \in \mathbb{S}_n} \epsilon(\sigma) a_{k\sigma(i)}$$

$$\times \sum_{\substack{k_j=1 \\ 1 \leq j \leq n \\ j \neq i}}^{m} a_{k_1 1} a_{k_1 \sigma(1)} \Pi^2_{k_1} \cdots a_{k_n n} a_{k_n \sigma(n)} \Pi^2_{k_n}$$

$$= \sum_{k=1}^{m} \Pi_k \sum_{\substack{k_j=1 \\ 1 \leq j \leq n \\ j \neq i}}^{m} a_{k_1 1} \cdots a_{k_n n} \Pi^2_{k_1} \cdots \Pi^2_{k_n}$$

$$\times \sum_{\sigma \in \mathbb{S}_n} \epsilon(\sigma) a_{k_1 \sigma(1)} \cdots a_{k\sigma(i)} \cdots a_{k_n \sigma(n)}$$

which gives

$$Y_i = \sum_{k=1}^{m} \Pi_k \sum_{\substack{k_j=1 \\ 1 \leq j \leq n \\ j \neq i}}^{m} a_{k_1 1} \cdots a_{k_n n} \Pi^2_{k_1} \cdots \Pi^2_{k_n}$$

$$\times \det A_{k_1 \cdots k_{i-1} k k_{i+1} \cdots k_n}.$$

By a similar argument as before we sum up for any set with $n - 1$ elements $K_{n-1} \subset \{1, \ldots, m\}$, $K_{n-1} = \{k_1 < \cdots < k_{i-1} < k_{i+1} < \cdots < k_n\}$ and for any permutation $\sigma \in \mathbb{S}(K_{n-1})$. We obtain as previously

$$Y_i = \sum_{k=1}^{m} \Pi_k \sum_{K_{n-1} \subset \{1,\ldots,m\}} \Pi^2_{k_1} \cdots \Pi^2_{k_n}$$

$$\times \det A^i_{k_1 \cdots k_{i-1} k_{i+1} \cdots k_n} \det A_{k_1 \cdots k_{i-1} k k_{i+1} \cdots k_n}$$

with $A^i_{k_1 \cdots k_{i-1} k_{i+1} \cdots k_n}$ the matrix with rows A_{k_j}, $j \in K_{n-1}$ and the ith column removed. The quantity $A_l x - b_l$ appears in the product $\Pi_k \Pi^2_{k_1} \cdots \Pi^2_{k_n}$ with an exponent equal to

- $2n - 1$ when $l \neq k$ and $l \notin K_{n-1}$,
- $2n - 2$ when $l = k$ and $l \notin K_{n-1}$,
- $2n - 3$ when $l \neq k$ and $l \in K_{n-1}$,
- $2n - 4$ when $l = k$ and $l \in K_{n-1}$.

In this latter case, two rows of the matrix $A_{k_1 \cdots k_{i-1} k k_{i+1} \cdots k_n}$ are equal and its determinant is zero. Thus, each term $A_l x - b_l$ appears at least $2n-3$

times so that

$$Y_i = \Pi^{2n-3} \sum_{\substack{k=1 \\ K_{n-1}}}^{m} (A_k x - b_k) \prod_{\substack{l \neq k \\ l \notin K_{n-1}}} (A_l x - b_l)^2 \det A^i_{k_1 \cdots k_{i-1} k_{i+1} \cdots k_n} \det A_{k_1 \cdots k_{i-1} k k_{i+1} \cdots k_n}.$$

The ith component of the Newton vector field is equal to $N(x)_i = \Pi Y_i / \det X$ so that

$$N(x)_i = \frac{\displaystyle\sum_{\substack{k=1 \\ K_{n-1}}}^{m} (A_k x - b_k) \prod_{\substack{l \neq k \\ l \notin K_{n-1}}} (A_l x - b_l)^2 \det A^i_{k_1 \cdots k_{i-1} k_{i+1} \cdots k_n} \det A_{k_1 \cdots k_{i-1} k k_{i+1} \cdots k_n}}{\displaystyle\sum_{K_n} (\det A_{K_n})^2 \prod_{l \notin K_n} (A_l x - b_l)^2}.$$

Instead of a sum taken for k and K_{n-1} in the numerator we use a subset $K_n \subset \{1, \ldots, m\}$ equal to the union of k and K_{n-1}. Notice that $\det A_{k_1 \cdots k_{i-1} k k_{i+1} \cdots k_n} = 0$ when $k \in K_{n-1}$ so that this case is not considered. Conversely, for a given $K_n = \{k_1 \cdots k_n\}$, we can write it in n different ways as a union of $k = k_j$ and $K_{n-1} = K_n \backslash \{k_j\}$. For these reasons we get

$$N(x)_i = \frac{\displaystyle\sum_{K_n} \left(\sum_{j=1}^{n} (A_{k_j} x - b_{k_j}) \det A^{ji}_{K_n} \det A_{K_n, i, j} \right) \prod_{l \notin K_n} (A_l x - b_l)^2}{\displaystyle\sum_{K_n} (\det A_{K_n})^2 \prod_{l \notin K_n} (A_l x - b_l)^2}$$

with $A^{ji}_{K_n}$ the matrix obtained from A_{K_n} in deleting the jth row and ith column, and $A_{K_n, i, j}$ obtained from A_{K_n} in removing the line A_j, and in reinserting it as the ith line, the other lines remaining with the same ordering. Note that $\det A_{K_n, i, j} = (-1)^{i+j} \det A_{K_n}$ thus

$$N(x)_i = \frac{\displaystyle\sum_{K_n} \left(\sum_{j=1}^{n} (A_{k_j} x - b_{k_j})(-1)^{i+j} \det A^{ji}_{K_n} \right) \det A_{K_n} \prod_{l \notin K_n} (A_l x - b_l)^2}{\displaystyle\sum_{K_n} (\det A_{K_n})^2 \prod_{l \notin K_n} (A_l x - b_l)^2}.$$

In fact this sum is taken for the sets K_n such that A_{K_n} is nonsingular, otherwise, the coefficient $\det A_{K_n}$ vanishes and the corresponding term is zero.

According to Cramer's formulas, the expression $(-1)^{i+j} \det A^{ji}_{K_n} / \det A_{K_n}$ is equal to $\left(A^{-1}_{K_n} \right)_{ij}$. Thus

$$\sum_{j=1}^{n} (A_{k_j} x - b_{k_j})(-1)^{i+j} \det A^{ji}_{K_n}$$

$$= (A^{-1}_{K_n}(A_{K_n} x - b_{K_n}))_i$$

$$= x_i - \left(A^{-1}_{K_n} b_{K_n} \right)_i = x_i - u_{K_n, i}.$$

We get

$$N(x)_i = \frac{\displaystyle\sum_{K_n} (x_i - u_{K_n, i})(\det A_{K_n})^2 \prod_{l \notin K_n} (A_l x - b_l)^2}{\displaystyle\sum_{K_n} (\det A_{K_n})^2 \prod_{l \notin K_n} (A_l x - b_l)^2}$$

and we are done. ∎

4. Extension to the Faces of $\mathcal{P}$

Our aim is to extend the Newton vector field, defined in the interior of $\mathcal{P}$, to its different faces.

Let $\mathcal{P}_J$ be the face of $\mathcal{P}$ defined by

$$\mathcal{P}_J = \{x \in \mathbb{R}^n : A_i x = b_i \text{ for any } i \in I \\ \text{and } A_i x \geq b_i \text{ for any } i \in J\}.$$

Here I is a subset of $\{1, 2, \ldots, m\}$ containing m_I integers, $J = \{1, 2, \ldots, m\} \backslash I$ and $m_J = m - m_I$.

Definition 4.1. The face $\mathcal{P}_J$ is regularly described when the relative interior of the face is given by

$$ri - \mathcal{P}_J = \{x \in \mathbb{R}^n : A_i x = b_i \text{ for any } i \in I \\ \text{and } A_i x > b_i \text{ for any } i \in J\}.$$

The polytope is regularly described when all its faces have this property.

We assume here that $\mathcal{P}$ is regularly described. This definition avoids, for example, in the description of a $\mathcal{P}_J$ a hyperplane defined by two inequalities: $A_i x \geq b_i$ and $A_i x \leq b_i$ instead of $A_i x = b_i$. Note that every face of a regularly described $\mathcal{P}$ has a unique regular description, the set I consists of all indices i such that $A_i x = b_i$ on the face. The affine hull of $\mathcal{P}_J$ is denoted by

$$F_J = \{x = (x_1, \ldots, x_n)^T \in \mathbb{R}^n : \\ A_i x = b_i \text{ for any } i \in I\}$$

which is parallel to the vector subspace

$$G_J = \{x = (x_1, \ldots, x_n)^T \in \mathbb{R}^n : \\ A_i x = 0 \text{ for any } i \in I\}.$$

We also let

$$E_J = \{y = (y_1, \ldots, y_m)^T \in \mathbb{R}^m : \\ y_i = 0 \text{ for any } i \in I\}.$$

E_I is defined similarly.

Let us denote by A_J (resp. A_I) the $m_J \times n$ (resp. $m_I \times n$) matrix whose ith row is A_i, $i \in J$ (resp. $i \in I$). A_J defines a linear operator $A_J : \mathbb{R}^n \to \mathbb{R}^{m_J}$. We also let

$$b_J : G_J \to \mathbb{R}^{m_J}, \quad b_J = A_J|_{G_J}$$

so that

$$b_J^T : \mathbb{R}^{m_J} \to G_J, \quad b_J^T = \Pi_{G_J} A_J.$$

Here, for a vector subspace E, Π_E denotes the orthogonal projection onto E. Let $D_{x,J}$ (resp. $D_{x,I}$) be the diagonal matrix with diagonal entries

$A_i x - b_i$, $i \in J$ (resp. $i \in I$). It defines a linear operator $D_{x,J} : \mathbb{R}^{m_J} \to \mathbb{R}^{m_J}$.

Since the faces of the polytope are regularly described, for any $x \in ri - \mathcal{P}_J$, $D_{x,J}$ is nonsingular. $\mathcal{P}_J$ is associated with the linear program

$$(LP_J) \quad \min_{x \in \mathcal{P}_J} \langle c, x \rangle.$$

The barrier function

$$f_J(x) = \sum_{i \in J} \ln(A_i x - b_i)$$

is defined for any $x \in F_J$ and finite in $ri - \mathcal{P}_J$ the relative interior of $\mathcal{P}_J$. The barrier technique considers the family of nonlinear convex optimization problems $(LP_J(t))$

$$\min_{x \in F_J} t\langle c, x \rangle - f_J(x)$$

with $t > 0$. The objective function

$$f_{t,J}(x) = t\langle c, x \rangle - f_J(x)$$

is smooth, strictly convex and

$$\lim_{x \to \partial \mathcal{P}_J} f_{t,J}(x) = \infty,$$

thus $(LP_J(t))$ has a unique solution $\gamma_J(t) \in ri - \mathcal{P}_J$ given by

$$Df_{t,J}(\gamma_J(t)) = 0.$$

For any $x \in ri - \mathcal{P}_J$, the first derivative of f_J is given by

$$Df_J(x)u = \sum_{i \in J} \frac{A_i u}{A_i x - b_i} = \langle A_J^T D_{x,J}^{-1} e_J, u \rangle$$

with $u \in G_J$ and $e_J = (1, \ldots, 1)^T \in \mathbb{R}^{m_J}$. We have

$$g_J(x) = \text{grad } f_J(x) = \Pi_{G_J} A_J^T D_{x,J}^{-1} e_J \\ = b_J^T D_{x,J}^{-1} e_J.$$

The second derivative of f_J at $x \in ri - \mathcal{P}_J$ is given by

$$D^2 f_J(x)(u, v) = -\sum_{i \in J} \frac{(A_i u)(A_i v)}{(A_i x - b_i)^2}$$

$$= -\langle A_J^T D_{x,J}^{-2} A_J v, u \rangle$$

$$= -\langle b_J^T D_{x,J}^{-2} b_J v, u \rangle$$

for any u, $v \in G_J$ so that

$$Dg_J(x) = \text{hess } f_J(x) = -b_J^T D_{x,J}^{-2} b_J.$$

To $\mathcal{P}_J$ we associate the Newton vector field given by

$$N_J(x) = -Dg_J(x)^{-1} g_J(x), \quad x \in ri - \mathcal{P}_J.$$

We have:

Lemma 4.1. *For any $x \in ri - \mathcal{P}_J$ this vector field is defined and*

$$N_J(x) = (b_J^T D_{x,J}^{-2} b_J)^{-1} b_J^T D_{x,J}^{-1} e_J$$
$$= b_J^\dagger D_{x,J} \Pi_{\text{im}(D_{x,J}^{-1} b_J)} e_J \in G_J.$$

Proof. We first have to prove that $Dg_J(x)$ is nonsingular and that $N_J(x) \in G_J$. This second point is clear. For the first, we take $u \in G_J$ such that $Dg_J(x)u = 0$. This gives $A_J u = b_J u = 0$ which implies $Au = 0$ because $u \in G_J$ that is $A_I u = 0$. Since A is injective we get $u = 0$. By the same argument we see that b_J is injective so that $b_J^T b_J$ is nonsingular. The first expression for $N_J(x)$ comes from the description of g_J and Dg_J. We have

$$N_J(x) = (b_J^T D_{x,J}^{-2} b_J)^{-1} b_J^T D_{x,J}^{-1} e_J$$
$$= (b_J^T b_J)^{-1} b_J^T D_{x,J} D_{x,J}^{-1} b_J$$
$$\times (b_J^T D_{x,J}^{-2} b_J)^{-1} b_J^T D_{x,J}^{-1} e_J$$
$$= b_J^\dagger D_{x,J} \Pi_{\text{im}(D_{x,J}^{-1} b_J)} e_J \in G_J. \quad \blacksquare$$

The curve $\gamma_J(t)$, $0 < t < \infty$, is the central path of the face $\mathcal{P}_J$. It is given by

$$\gamma_J(t) \in F_J \quad \text{and} \quad Df_J(\gamma_J(t)) - tc = 0$$

that is

$$x \in F_J, \quad A_J^T D_{x,J}^{-1} e_J - tc \in G_J^\perp \quad \text{and} \quad \gamma_J(t) = x$$

or, projecting on G_J,

$$A_i x = b_i, \quad i \in I, \quad b_J^T D_{x,J}^{-1} e_J - t\Pi_{G_J} c = 0$$

$$\text{and} \quad \gamma_J(t) = x.$$

When $t \to 0$, $\gamma_J(t)$ tends to the analytic center $\gamma_J(0)$ of $\mathcal{P}_J$ defined as the unique solution of the convex program

$$-f_J(\gamma_J(0)) = \min_{x \in F_J} -f_J(x).$$

The analytic center is also given by

$$A_i x = b_i, \quad i \in I, \quad b_J^T D_{x,J}^{-1} e_J = 0 \quad \text{and} \quad \gamma_J(0) = x$$

so that $\gamma_J(0)$ is the unique singular point of N_J in the face $\mathcal{P}_J$.

We now investigate the properties of this extended vector field: continuity, derivability and so on. We shall investigate the following abstract problem: for any $y \in \mathbb{R}^m$ we consider the linear operator

$$\mathcal{D}_y : \mathbb{R}^m \to \mathbb{R}^m$$

given by the $m \times m$ diagonal matrix $\mathcal{D}_y = \text{Diag}(y_i)$. Let P be a vector subspace in $\mathbb{R}^m$. Then, for any $y \in \mathbb{R}^m$ with nonzero coordinates, the operator

$$\mathcal{D}_y \circ \Pi_{\mathcal{D}_y^{-1}(P)} : \mathbb{R}^m \to \mathbb{R}^m$$

is well defined. Can we extend its definition to any $y \in \mathbb{R}^m$? The answer is yes and proved in the following

Lemma 4.2. *Let $\bar{y} \in E_J$ be such that $\bar{y}_i \neq 0$ for any $i \in J$.*

Then $\mathcal{D}_{\bar{y}}|_{E_J} : E_J \to E_J$ is nonsingular and

$$\lim_{y \to \bar{y}} \mathcal{D}_y \circ \Pi_{\mathcal{D}_y^{-1}(P)} = \mathcal{D}_{\bar{y}}|_{E_J} \circ \Pi_{(\mathcal{D}_{\bar{y}}|_{E_J})^{-1}(P \cap E_J)}.$$

Proof. To prove this lemma we suppose that $I = \{1, 2, \ldots, m_1\}$ and $J = \{m_1 + 1, \ldots, m_1 + m_2 = m\}$.

Let us denote $p = \dim P$. P is identified to an $n \times p$ matrix with rank $P = p$. We also introduce the following matrices:

$$\mathcal{D}_y = \begin{pmatrix} D_{y,1} & 0 \\ 0 & D_{y,2} \end{pmatrix}, \quad P = \begin{pmatrix} U & 0 \\ V & W \end{pmatrix}.$$

The different blocks appearing in these two matrices have the following dimensions: $D_{y,1} : m_1 \times m_1$, $D_{y,2} : m_2 \times m_2$, $U : m_1 \times p_1$, $V : m_2 \times p_1$, $W : m_2 \times p_2$. We also suppose that the columns of $\begin{pmatrix} 0 \\ W \end{pmatrix}$ are a basis for $P \cap E_J$ and those of $\begin{pmatrix} U \\ V \end{pmatrix}$ a basis of the orthogonal complement of $P \cap E_J$ in P that is $(P \cap E_J)^\perp \cap P$. Let us notice that $p_2 \leq m_2$ and rank $W = p_2$ and also that $p_1 \leq m_1$ and rank $U = p_1$. Let us prove this last assertion. Let U_i, $1 \leq i \leq p_1$ be the columns of U. If $\alpha_1 U_1 + \cdots + \alpha_{p_1} U_{p_1} = 0$, we have

$$\alpha_1 \begin{pmatrix} U_1 \\ V_1 \end{pmatrix} + \cdots + \alpha_{p_1} \begin{pmatrix} U_{p_1} \\ V_{p_1} \end{pmatrix}$$

$$= \begin{pmatrix} 0 \\ \alpha_1 V_1 + \cdots + \alpha_{p_1} V_{p_1} \end{pmatrix}.$$

The left-hand side of this equation is in $(P \cap E_J)^\perp \cap P$ and the right-hand side in $P \cap E_J$. Thus this vector is equal to 0 and since rank $P = p$ we get $\alpha_1 = \cdots = \alpha_{p_1} = 0$.

For every subspace X in $\mathbb{R}^m$ with $\dim X = p$ identified with an $m \times p$ rank p matrix we have

$$\Pi_X = X(X^T X)^{-1} X^T.$$

This gives here

$$\Pi_{\mathcal{D}_y^{-1} P}$$

$$= \begin{pmatrix} D_{y,1}^{-1} U & 0 \\ D_{y,2}^{-1} V & D_{y,2}^{-1} W \end{pmatrix}$$

$$\times \begin{pmatrix} U^T D_{y,1}^{-2} U + V^T D_{y,2}^{-2} V & V^T D_{y,2}^{-2} W \\ W^T D_{y,2}^{-2} V & W^T D_{y,2}^{-2} W \end{pmatrix}^{-1}$$

$$\times \begin{pmatrix} U^T D_{y,1}^{-1} & V^T D_{y,2}^{-1} \\ 0 & W^T D_{y,2}^{-1} \end{pmatrix}$$

and

$$\Pi_{E_J} = \begin{pmatrix} 0 & 0 \\ 0 & I_{m_2} \end{pmatrix}.$$

We also notice that

$$\mathcal{D}_y \Pi_{\mathcal{D}_y^{-1} P} = \mathcal{D}_y \Pi_{E_I} \Pi_{\mathcal{D}_y^{-1} P} + \mathcal{D}_y \Pi_{E_J} \Pi_{\mathcal{D}_y^{-1} P}.$$

We have

$$\lim_{y \to \bar{y}} \mathcal{D}_y \Pi_{E_I} \Pi_{\mathcal{D}_y^{-1} P} = 0.$$

This is a consequence of the two following

$$\left\| \Pi_{E_I} \Pi_{\mathcal{D}_y^{-1} P} \right\| \le 1$$

because it is the product of two orthogonal projections and

$$\lim_{y \to \bar{y}} \mathcal{D}_y \Pi_{E_I} = \mathcal{D}_{\bar{y}} \Pi_{E_I} = 0.$$

We now have to study the limit

$$\lim_{y \to \bar{y}} \mathcal{D}_y \Pi_{E_J} \Pi_{\mathcal{D}_y^{-1} P}.$$

Let us denote $A = U^T D_{y,1}^{-2} U$. The following identities hold:

$$\mathcal{D}_y \Pi_{E_J} \Pi_{\mathcal{D}_y^{-1} P}$$

$$= \begin{pmatrix} D_{y,1} & 0 \\ 0 & D_{y,2} \end{pmatrix} \begin{pmatrix} 0 & 0 \\ 0 & I_{m_2} \end{pmatrix} \begin{pmatrix} D_{y,1}^{-1} U & 0 \\ D_{y,2}^{-1} V & D_{y,2}^{-1} W \end{pmatrix}$$

$$\times \begin{pmatrix} U^T D_{y,1}^{-2} U + V^T D_{y,2}^{-2} V & V^T D_{y,2}^{-2} W \\ W^T D_{y,2}^{-2} V & W^T D_{y,2}^{-2} W \end{pmatrix}^{-1} \begin{pmatrix} U^T D_{y,1}^{-1} & V^T D_{y,2}^{-1} \\ 0 & W^T D_{y,2}^{-1} \end{pmatrix}$$

$$= \begin{pmatrix} 0 & 0 \\ V & W \end{pmatrix} \left[\begin{pmatrix} A & 0 \\ 0 & I_{m_2} \end{pmatrix} \begin{pmatrix} I_{m_1} + A^{-1}(V^T D_{y,2}^{-2} V) & A^{-1}(V^T D_{y,2}^{-2} W) \\ W^T D_{y,2}^{-2} V & W^T D_{y,2}^{-2} W \end{pmatrix} \right]^{-1} \begin{pmatrix} U^T D_{y,1}^{-1} & V^T D_{y,2}^{-1} \\ 0 & W^T D_{y,2}^{-1} \end{pmatrix}$$

$$= \begin{pmatrix} 0 & 0 \\ V & W \end{pmatrix} \begin{pmatrix} I_{m_1} + A^{-1}(V^T D_{y,2}^{-2} V) & A^{-1}(V^T D_{y,2}^{-2} W) \\ W^T D_{y,2}^{-2} V & W^T D_{y,2}^{-2} W \end{pmatrix}^{-1} \begin{pmatrix} A^{-1}(U^T D_{y,1}^{-1}) & A^{-1}(V^T D_{y,2}^{-1}) \\ 0 & W^T D_{y,2}^{-1} \end{pmatrix}.$$

We will prove later that

$$\lim A^{-1} = \lim A^{-1} \left(U^T D_{y,1}^{-1} \right) = 0$$

when $y \to \bar{y}$. Since

$$\lim_{y \to \bar{y}} D_{y,2} = D_{\bar{y},2}$$

is a nonsingular matrix we get

$$\lim_{y \to \overline{y}} \mathcal{D}_y \Pi_{E_J} \Pi_{\mathcal{D}_y^{-1} P}$$

$$= \begin{pmatrix} 0 & 0 \\ V & W \end{pmatrix} \begin{pmatrix} I_{m_1} & 0 \\ W^T D_{\overline{y},2}^{-2} V & W^T D_{\overline{y},2}^{-2} W \end{pmatrix}^{-1}$$

$$\times \begin{pmatrix} 0 & 0 \\ 0 & W^T D_{\overline{y},2}^{-1} \end{pmatrix}$$

$$= \begin{pmatrix} 0 & 0 \\ 0 & W(W^T D_{\overline{y},2}^{-2} W)^{-1} W^T D_{\overline{y},2}^{-1} \end{pmatrix}$$

and this last matrix represents the operator

$$\mathcal{D}_{\overline{y}}|_{E_J} \circ \Pi_{(\mathcal{D}_{\overline{y}}|_{E_J})^{-1}(P \cap E_J)}$$

as announced in this lemma.

To achieve the proof of this lemma we have to show that

$$\lim A^{-1} = \lim A^{-1} \left(U^T D_{y,1}^{-1} \right) = 0$$

with $A = U^T D_{y,1}^{-2} U$. In fact it suffices to prove $\lim A^{-1} = 0$ because

$$\| A^{-1}(U^T D_{y,1}^{-1}) \|^2$$

$$= \| A^{-1}(U^T D_{y,1}^{-1})(A^{-1}(U^T D_{y,1}^{-1}))^T \|$$

$$= \| A^{-1} \|.$$

Since U is full rank, the matrix $U^T U$ is positive definite so that

$$\mu = \min \ \mathrm{Spec}(U^T U) > 0.$$

Let us denote $\succ$ the ordering on square matrices given by the cone of non-negative matrices. We have

$$D_{y,1}^{-2} \succ \frac{1}{\max y_i^2} I_{m_1} \succ \frac{1}{\|y\|^2} I_{m_1}$$

so that

$$U^T D_{y,1}^{-2} U \succ \frac{1}{\|y\|^2} U^T U \succ \frac{\mu}{\|y\|^2} I_{p_1}.$$

Taking the inverses changes this inequality in the following

$$0 \prec (U^T D_{y,1}^{-2} U)^{-1} \prec \frac{\|y\|^2}{\mu} I_{p_1} \to 0$$

when $y \to \overline{y}$ and we are done. ∎

Corollary 4.1. *The vector field $N(x)$ extends continuously to all of R^n. Moreover it is Lipschitz on compact sets. When all the faces of the polytope $\mathcal{P}$ are regularly described, the continuous extension of $N(x)$ to the face $\mathcal{P}_J$ of $\mathcal{P}$ equals $N_J(x)$. Consequently any orbit of $N(x)$ in the polytope $\mathcal{P}$ tends to one of the singularities of the extended vector field, i.e. either to a vertex or an analytic center of one of the faces.*

Proof. It is a consequence of Definition 2.2, Lemmas 4.1 and 4.2 and the equality $A_J^\dagger y = A^\dagger y$ for any $y \in E_J$ that $N(x)$ extends continuously to all of R^n and equals $N_J(x)$ on $\mathcal{P}_J$. Moreover a rational function which is continuous on R^n has bounded partial derivatives on compact sets and hence is Lipschitz. Now we use the characterization of the vectorfield restricted to the face to see that any orbit which is not the analytic center of a face tends to the boundary of the face and any orbit which enters a small enough neighborhood of a vertex tends to that vertex. ∎

Remark 4.1. We have shown that $N(x)$ is Lipschitz. We do not know an example where it is not analytic and wonder as to what its order of smoothness is, in general. In the next section we will show it is analytic generically.

5. Analyticity and Derivatives

In Sec. 3 we gave the following expression for the Newton vector field:

$$N(x)$$

$$= \frac{\displaystyle\sum_{\substack{K_n \subset \{1,\dots,m\} \\ \det K_n \neq 0}} (x - u_{K_n})(\det A_{K_n})^2 \prod_{l \notin K_n} (A_l x - b_l)^2}{\displaystyle\sum_{\substack{K_n \subset \{1,\dots,m\} \\ \det K_n \neq 0}} (\det A_{K_n})^2 \prod_{l \notin K_n} (A_l x - b_l)^2}$$

for any $x \in \mathrm{Int} \ \mathcal{P}$. Under a mild geometric assumption, the denominator of this fraction never vanishes so that $N(x)$ may be extended in a real analytic vector field.

Theorem 5.1. *Suppose that for any $x \in \partial\mathcal{P}$ contained in the relative interior of a codimension d face of $\mathcal{P}$, we have $A_{k_i} x = b_{k_i}$ for exactly d indices in $\{1, \dots, m\}$ and $A_l x > b_l$ for the other indices. In that case the line vectors $A_{k_i}, 1 \le i \le d$, are linearly* *independent. Moreover, for such an x*

$$\sum_{\substack{K_n \subset \{1,\dots,m\} \\ \det K_n \neq 0}} (\det A_{K_n})^2 \prod_{l \notin K_n} (A_l x - b_l)^2 \neq 0$$

so that $N(x)$ extends analytically to a neighborhood of $\mathcal{P}$.

Proof. Under this assumption, for any $x \in \mathcal{P}$, there exists a subset $K_n \subset \{1, \dots, m\}$ such that the submatrix A_{K_n} is nonsingular and $A_l x - b_l > 0$ for any $x \notin K_n$. $\blacksquare$

Our next objective is to describe the singular points of this extended vector field.

Theorem 5.2. *Under the previous geometric assumption, the singularities of the extended vector field are: the analytic center of the polytope and the analytic centers of the different faces of the polytope, including the vertices. Each of them is hyperbolic: if $x \in \partial \mathcal{P}$ is the analytic center of a codimension d face $\mathcal{F}$ of $\mathcal{P}$, then the derivative $DN(x)$ has $n - d$ eigenvalues equal to -1 with corresponding eigenvectors contained in the linear space $\mathcal{F}_0$ parallel to $\mathcal{F}$ and d eigenvalues equal to 1 with corresponding eigenvectors contained in a complement of $\mathcal{F}_0$.*

Proof. The first part of this theorem, about the -1 eigenvalues, is the consequence of two facts.

The first one is a well-known fact about the Newton operator: its derivative is equal to $-$id at a zero (if $N(x) = 0$, then $DN(x) = D(-Dg(x)^{-1}g(x)) = D(-Dg(x)^{-1})g(x) - Dg(x)^{-1}Dg(x) = -$id). The second fact is proved in Sec. 4: the restriction of $N(x)$ to a face is the Newton vector field associated with the restriction of $g(x)$ to this face.

We have now take care of the 1 eigenvalues. To simplify the notations we suppose that $A_i x = b_i$ for $1 \leq i \leq d$, $A_i x > b_i$ when $i + 1 \leq i \leq m$, and $N(x) = 0$. N is analytic and its derivative in the direction v is given by

$$DN(x)v = \frac{\text{Num}}{\displaystyle\sum_{\substack{K_n \subset \{1,\dots,m\} \\ \det K_n \neq 0}} (\det A_{K_n})^2 \prod_{l \notin K_n} (A_l x - b_l)^2}$$

with

$$\text{Num} = \sum_{\substack{K_n \subset \{1,\dots,m\} \\ \det K_n \neq 0}} v(\det A_{K_n})^2 \prod_{l \notin K_n} (A_l x - b_l)^2$$

$$+ \sum_{\substack{K_n \subset \{1,\dots,m\} \\ \det K_n \neq 0}} (x - u_{K_n})(\det A_{K_n})^2$$

$$\times \sum_{l_0 \notin K_n} 2 A_{l_0} v (A_{l_0} x - b_{l_0}) \prod_{\substack{l \notin K_n \\ l \neq l_0}} (A_l x - b_l)^2$$

which gives

$$DN(x)v = v + \frac{\displaystyle\sum_{\substack{K_n \\ \det K_n \neq 0}} (x - u_{K_n})(\det A_{K_n})^2 \sum_{l_0 \notin K_n} 2 A_{l_0} v (A_{l_0} x - b_{l_0}) \prod_{\substack{l \notin K_n \\ l \neq l_0}} (A_l x - b_l)^2}{\displaystyle\sum_{\substack{K_n \subset \{1,\dots,m\} \\ \det K_n \neq 0}} (\det A_{K_n})^2 \prod_{l \notin K_n} (A_l x - b_l)^2} = v + Mv$$

where M is, up to a constant factor (i.e. constant in v), the $n \times n$ matrix equal to

$$\sum_{\substack{K_n \\ \det K_n \neq 0 \\ l_0 \notin K_n}} (\det A_{K_n})^2 (A_{l_0} x - b_{l_0}) \left(\prod_{\substack{l \notin K_n \\ l \neq l_0}} (A_l x - b_l)^2 \right) (x - u_{K_n}) A_{l_0}$$

which is also equal to

$$\sum_{\substack{\{1,\dots,d\} \subset K_n \\ \det K_n \neq 0 \\ d+1 \leq l_0 \leq m \\ l_0 \notin K_n}} \frac{(\det A_{K_n})^2}{A_{l_0} x - b_{l_0}} \left(\prod_{l \notin K_n} (A_l x - b_l)^2 \right) (x - u_{K_n}) A_{l_0}$$

because $A_i x = b_i$ when $1 \leq i \leq d$ and $A_i x > b_i$ otherwise.

To prove our theorem we have to show that $\dim \ker M \geq d$. This gives at least d independent vectors v_i such that $M v_i = 0$, that is, $DN(x)v_i = v_i$; thus 1 is an eigenvalue of $DN(x)$ and its multiplicity is $\geq d$. In fact it is exactly d because we already have the eigenvalue -1 with multiplicity $n - d$. The inequality $\dim \ker M \geq d$ is given by $\operatorname{rank} M \leq n - d$. Why is it true? M is a linear combination of rank 1 matrices $(x - u_{K_n})A_{l_0}$ so that the rank of M is less than or equal to the dimension of the system of vectors $x - u_{K_n}$ with K_n as before. Since $\{1, \ldots, d\} \subset K_n$, $A_i x = b_i$ when $1 \leq i \leq d$, and $A u_{K_n} = b_{K_n}$ we have $A(x - u_{K_n}) = (0, \ldots, 0, y_{d+1}, \ldots, y_m)^T$. From the hypothesis, the line vectors $A_1, \ldots, A_d$ defining the face $\mathcal{F}$ are independent, thus the set of vectors $u \in \mathbb{R}^n$ such that the vector $Au \in \mathbb{R}^m$ begins by d zeros has dimension $n - d$ and we are done. $\blacksquare$

Remark 5.1. The last theorem implies that $N(x)$ is Morse–Smale in the terminology of dynamical systems. Recall also that we are really interested in the positive time trajectories of $-N(x)$. For $-N(x)$ the eigenvalues at the critical points are multiplied by -1 so in the faces the critical points of $-N(x)$ are sources and their stable manifolds are transverse to the faces.

6. Example

Let us consider the case of a triangle in the plane. Since the Newton vector field is affinely invariant (Proposition 2.1) we may only consider the triangle with vertices $(0,0)$, $(1,0)$ and $(0,1)$. A dual description is given by the three inequalities $x \geq 0$, $y \geq 0$, $-x - y \geq -1$ which correspond to the following data:

$$A = \begin{pmatrix} 1 & 0 \\ 0 & 1 \\ -1 & -1 \end{pmatrix}, \quad B = \begin{pmatrix} 0 \\ 0 \\ -1 \end{pmatrix},$$

$$D_{(x,y)} = \begin{pmatrix} x & 0 & 0 \\ 0 & y & 0 \\ 0 & 0 & 1 - x - y \end{pmatrix}.$$

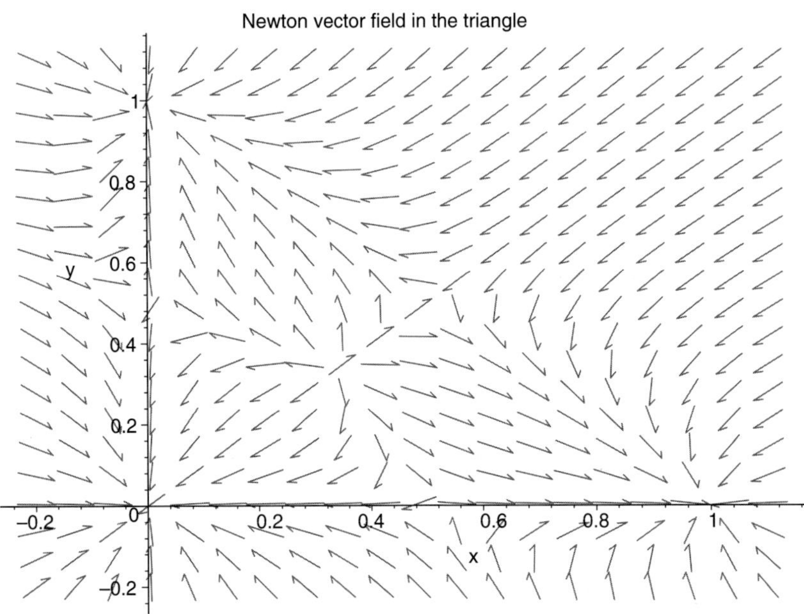

The corresponding Newton vector field is given by the rational expressions

$$N(x,y) = \begin{bmatrix} \dfrac{xz^2 - x^2 z + xy^2 - x^2 y}{z^2 + y^2 + x^2} \\[2ex] \dfrac{x^2 y - xy^2 + yz^2 - y^2 z}{z^2 + y^2 + x^2} \end{bmatrix}$$

with $z = 1 - x - y$. This vector field is analytic on the whole plane. The singular points are the three vertices, the midpoints of the three sides and the center of gravity. The arrows in the figure are for $-N(x)$ and the critical points are clearly sources in their faces.

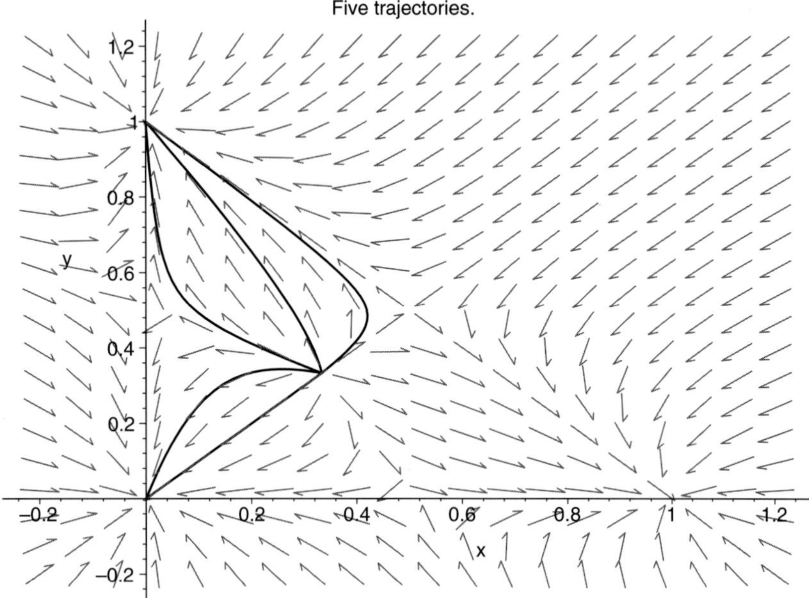

References

Bayer, D. & Lagarias, J. [1989a] "The non-linear geometry of linear programming I: Affine and projective scaling trajectories," *Trans. Amer. Math. Soc.* **314**, 499–526.

Bayer, D. & Lagarias, J. [1989b] "The non-linear geometry of linear programming II: Legendre transform coordinates and central trajectories," *Trans. Amer. Math. Soc.* **314**, 527–581.

Bayer, D. & Lagarias, J. [1991] "Karmarkar's linear programming algorithm and Newton's method," *Math. Progr.* **A50**, 291–330.

Dedieu, J.-P., Malajovich, G. & Shub, M. "On the curvature of the central path of linear programming theory," to appear.

Meggido, N. & Shub, M. [1989] "Boundary behaviour of interior point algorithms in linear programming," *Math. Oper. Res.* **14**, 97–146.

Renegar, J. [2001] *A Mathematical View of Interior-Point Methods in Convex Optimization* (SIAM, Philadelphia).

Vavasis, S. & Ye, Y. [1996] "A primal-dual accelerated interior point method whose running time depends only on *A*," *Math. Progr.* **A74**, 79–120.

NUMERICAL CONTINUATION OF BRANCH POINTS OF EQUILIBRIA AND PERIODIC ORBITS

E. J. DOEDEL

Department of Computer Science, Concordia University,
1455 Boulevard de Maisonneuve O., Montréal Québec, H3G 1M8, Canada

W. GOVAERTS

Department of Applied Mathematics and Computer Science,
Ghent University, Krijgslaan 281-S9, B-9000 Gent, Belgium

YU. A. KUZNETSOV

Mathematisch Instituut, Universiteit Utrecht,
Boedapestlaan 6, 3584 CD Utrecht, The Netherlands

A. DHOOGE

Department of Applied Mathematics and Computer Science,
Ghent University, Krijgslaan 281-S9, B-9000 Gent, Belgium

Received March 10, 2004; Revised June 14, 2004

We consider the three-parameter numerical continuation of branch points in dynamical systems, with emphasis on the continuation of branch points of periodic orbits. We consider both, the case of branch points along one-parameter families of limit cycles (typical in systems with symmetry), and the case of branch points along fold-curves of limit cycles (typical in generic systems). We discuss new algorithms based on bordered matrices for both detection and continuation. We apply the techniques to a model of a chemical reactor, a model of an electronic circuit, and a model from celestial mechanics. Our algorithms have been implemented in freely available software.

Keywords: Rank defect; test function; bifurcation.

1. Introduction

We deal with the numerical continuation of special solution families associated with a smooth dynamical system of the form

$$\frac{dx}{dt} = f(x, \alpha), \qquad (1)$$

with $x \in \mathbb{R}^n, f(x, \alpha) \in \mathbb{R}^n$, and α a vector of parameters.

Software packages such as AUTO [Doedel *et al.*, 2001], CONTENT [Kuznetsov & Levitin, 1997] and MATCONT [Dhooge *et al.*, 2003] can compute families ("branches") of equilibria and periodic orbits, as well as families of fold and Hopf bifurcation points, and fold-, flip- and torus bifurcations of periodic orbits, Bogdanov–Takens points, etc. The singular points can be classified in terms of the codimension of point type and the number of free parameters required for continuation, cf. [Beyn *et al.*, 2002; Govaerts, 2000; Kuznetsov, 1998].

Branch points disturb this nice picture. First, unlike other bifurcation points, a branch point is defined with respect to a particular parameter. If the equilibrium or periodic orbit of (1) is defined by

$$F(X, \alpha_0) = 0, \qquad (2)$$

where α_0 is a component of α, and X and $F(X, \alpha_0)$ are in compatible state spaces, then a branch point is characterized by the fact that $[F_X \; F_{\alpha_0}]$ is rank deficient. This is a codimension-2 phenomenon, and therefore not generic in one-parameter problems. In generic systems, branch points can be expected only in two-parameter problems, and their continuation requires three free parameters. Second, branch points depend more intimately on the parameters of the problem than, say, limit points. Unlike the previously mentioned bifurcation points, their numerical continuation uses second-order derivatives with respect to parameters.

There are reasons for not considering branch points at all in generic systems. However, problems arising in applications often have a special structure (e.g. equivariant, Hamiltonian, etc.). For this reason, standard software packages do provide the option to detect and accurately locate branch points, as well as branch switching, although they do not provide for their numerical continuation. One of the first standard codes that supported detection of equilibrium branch points and allowed for branch switching was STAFF [Borisyuk, 1981]. Similar facilities are provided by AUTO [Doedel *et al.*, 2001], CONTENT [Kuznetsov & Levitin, 1997], and several other bifurcation programs.

Generic software can often deal with structured problems if an artificial "unfolding" parameter is introduced to break the special structure, thereby embedding the problem in a generic class; see, for example [Doedel *et al.*, 2003] and [Muñoz-Almarez *et al.*, 2003]. Thus, three-parameter continuation of branch points for generic systems is also useful for structured problems.

In this paper we describe a mathematical framework for the three-parameter continuation of branch points of equilibria and branch points of periodic solutions, concentrating on the use of minimal extended systems. The adjective "minimal" reflects the fact that we append only two scalar equations to the system that defines the equilibria or the periodic solutions, much in the spirit of the computation of fold, flip and torus bifurcations of periodic orbits in [Doedel *et al.*, 2003].

We note that the computation of branches bifurcating at branch points of equilibria is studied intensively in many papers; we refer in particular to the Proceedings volumes [Küpper *et al.*, 1984] and [Mittelmann & Weber, 1980], as well as to the book [Allgower & Georg, 1990]. Fully extended systems for branch points of systems of equations (including boundary value problems) are discussed in [Moore, 1980] and [Mei, 1989, 2000]. These systems can be compared to our system (28) except that in (28) $[p_1^* \; p_2^* \; p_3]^*$ is a fixed vector, while the corresponding entities in [Moore, 1980] and [Mei, 1989] are unknowns of the problem. Therefore the systems that we obtain are essentially smaller. Minimal extended systems for branch points of equibria were proposed in [Griewank *et al.*, 1984] and [Allgower & Schwetlick, 1997] and ours are equivalent to these.

On the other hand, the fact that branch points generically appear in families of limit points of equilibria and periodic orbits seem to have received little attention in the numerical literature. Our example in Sec. 5 shows how it helps to understand how connections between various objects can switch if parameters change.

In Secs. 2 and 3 we discuss mathematical features of branch points of equilibria and periodic orbits, respectively. In Sec. 4 we deal with the numerical implementation, i.e. the detection, computation and continuation of branch points of periodic orbits. In Secs. 5–7 we give numerical examples, using an implementation of the algorithms in the MATLAB-based software MATCONT.

2. Equilibria and their Branch Points

An equilibrium is a constant solution of (1), i.e. a solution of

$$f(x, \alpha) = 0. \tag{3}$$

Let (x^0, α^0) be an equilibrium point, and assume that a component β of α is free. By the implicit function theorem there is a unique solution family of (3) passing through (x^0, β^0) in (x, β)-space, if the $(n, n+1)$-dimensional Jacobian matrix

$$[f_x(x^0, \alpha^0) \quad f_\beta(x^0, \alpha^0)], \tag{4}$$

has full rank n. For a generic $(n, n+1)$ matrix, corank 1 is a codimension-2 phenomenon, and corank 2 is a codimension-6 phenomenon; see, e.g. [Govaerts, 2000], Proposition 3.4.2. We restrict to the corank 1 case, and simply call it a *branch point*. In this case, the existence of a unique family of equilibria is not guaranteed. In fact, the behavior of the equilibrium solutions near (x^0, α^0) can be quite complicated; this is the subject of singularity theory

[Golubitsky & Schaeffer, 1985; Golubitsky *et al.*, 1988; Govaerts, 2000]. However, the most common situation is that of a transcritical or pitchfork bifurcation.

A local characterization of the manifold of $(n, n+1)$ — corank 1 matrices near (4) follows from [Govaerts, 2000, Propositions 3.4.1–2]. We recall the main facts, using "$*$" to denote transposed matrices.

Proposition 1. *Let* $\phi_{11}, \phi_{21} \in \mathbb{R}^n$, $\phi_{12}, \phi_{22} \in \mathbb{R}$, *be such that* $\phi_1 = (\phi_{11}^*, \phi_{12})^*$, $\phi_2 = (\phi_{21}^*, \phi_{22})^*$, *together with the rows of (4) span* $\mathbb{R}^{n+1}$. *Also, let* $\psi \in \mathbb{R}^n$ *be a vector that together with the columns of (4) spans* $\mathbb{R}^n$. *Then the bordered matrix*

$$B_J = \begin{bmatrix} f_x(x^0, \alpha^0) & f_\beta(x^0, \alpha^0) & \psi \\ \phi_{11}^* & \phi_{12} & 0 \\ \phi_{21}^* & \phi_{22} & 0 \end{bmatrix}, \qquad (5)$$

is nonsingular.

Proof. This follows from [Govaerts, 2000, Proposition 3.2.1]. ∎

Proposition 2. *Let* $\phi_{11}, \phi_{21} \in \mathbb{R}^n$, $\phi_{12}, \phi_{22} \in \mathbb{R}$ *and* $\psi \in \mathbb{R}^n$ *be as in Proposition 1. Let* B *be a matrix having the same structure as the matrix* B_J *in (5):*

$$B = \begin{bmatrix} B_1 & B_2 & \psi \\ \phi_{11}^* & \phi_{12} & 0 \\ \phi_{21}^* & \phi_{22} & 0 \end{bmatrix}, \qquad (6)$$

and define $v_{11}, v_{21} \in \mathbb{R}^n$, $v_{12}, v_{22}, g_1, g_2 \in \mathbb{R}$ *by requiring*

$$B \begin{bmatrix} v_{11} & v_{21} \\ v_{12} & v_{22} \\ g_1 & g_2 \end{bmatrix} = \begin{bmatrix} 0_n & 0_n \\ 1 & 0 \\ 0 & 1 \end{bmatrix}. \qquad (7)$$

If B_1, B_2 *are sufficiently close to* $f_x(x^0, \alpha^0)$, $f_\beta(x^0, \alpha^0)$, *respectively, then* B *is nonsingular. Furthermore,*

$$\begin{bmatrix} B_1 & B_2 \end{bmatrix}$$

has corank 1 if and only if

$$\begin{cases} g_1 = 0, \\ g_2 = 0. \end{cases} \qquad (8)$$

Moreover, (8) is locally a regular defining system for the manifold of corank 1 matrices.

Proof. This follows from [Govaerts, 2000, Proposition 3.4.2]. ∎

By the previous results, the branch points of the equilibria of (1) near (x^0, α^0) are defined by the system consisting of (3) and (8), where B_1, B_2 are replaced by $f_x(x, \alpha)$, $f_\beta(x, \alpha)$, respectively, so that g_1, g_2 are functions of x, α. If three components of α are freed, then, generically, a family of branch points can be computed. "Generically" means that the Jacobian matrix of (3), (8) with respect to the components of x and the free parameters has full rank; geometrically this can be related to a transversal intersection of manifolds.

We note that the definition of branch points depends on the choice of the component β of α; a solution of (3) may be a branch point with respect to one component of α but not with respect to another component. Also, the three free parameters generically needed to compute a family of branch points with respect to β, may or may not include β. In applications they usually do.

In the numerical continuation of a family of branch points it is highly desirable to have explicit expressions for the derivatives of g_1, g_2 with respect to the state variables and the free parameters. To this end we solve the adjoint system corresponding to (7)

$$B^* \begin{bmatrix} w \\ g_1 \\ g_2 \end{bmatrix} = \begin{bmatrix} 0_{n+1} \\ 1 \end{bmatrix}, \qquad (9)$$

where $w \in \mathbb{R}^n$. If z is one of the components of x, or a free parameter, then by taking derivatives of (7) and multiplying from the left with

$$\begin{bmatrix} w^* & g_1 & g_2 \end{bmatrix},$$

we obtain

$$g_{iz} = -w^* f_{xz} v_{i1} - w^* f_{\beta z} v_{i2}, \qquad (i = 1, 2). \qquad (10)$$

If z is a parameter component, then these expressions involve the second derivatives of f with respect to parameters.

3. Periodic Solutions and their Branch Points

3.1. *Periodic solutions*

A periodic solution is a nonconstant solution of (1) with finite period $T > 0$, i.e. $x(0) = x(T)$. Since T is not known in advance, we use an equivalent system defined on the fixed interval $[0, 1]$, by rescaling time. Then the system, with $\dot{x} = dx/dt$, reads

$$\begin{cases} \dot{x} - Tf(x, \alpha) = 0, \\ x(0) - x(1) = 0. \end{cases} \tag{11}$$

The phase shifted function $\phi(t) = x(t + s)$ is also a solution of (11), for any value of s. In order to have a unique solution, an extra constraint is needed. The following integral constraint is often used [Doedel et al., 2001; Kuznetsov & Levitin, 1997]:

$$\langle x, \dot{x}_{\text{old}} \rangle = 0, \tag{12}$$

where $\dot{x}_{\text{old}}$ is the time derivative of a previously calculated periodic solution, and therefore known. For given $x, y \in \mathcal{C}^0([0, 1], \mathbb{R}^n)$, we denote

$$\langle x, y \rangle = \text{Int}_y(x) = \int_0^1 x^*(t)y(t)\, dt.$$

The phase condition (12) selects the periodic solution x with the smallest phase difference compared to the previous solution x_{old}. The most appropriate choice for x_{old} is the preceding periodic solution computed in the continuation process.

The complete boundary value problem (BVP) defining a periodic solution now consists of (11) and (12).

Consider the variational equation

$$\dot{X} - Tf_x(x(t), \alpha)X = 0, \tag{13}$$

and the adjoint variational equation

$$\dot{X} + Tf_x^*(x(t), \alpha)X = 0. \tag{14}$$

Denote by $\Phi(t)$ the fundamental matrix solution of (13), for which $\Phi(0) = I$, where $I = I_n$ is the n-dimensional identity matrix. Then $\Phi(1)$ is the *monodromy matrix* of the periodic solution. The eigenvalues of $\Phi(1)$ are the *Floquet multipliers*. There is always at least one multiplier that is equal to 1, with corresponding eigenvector $\dot{x}(0)$, where $x(t)$ is a solution of (11). For a *regular periodic solution*, the multiplier 1 has geometric multiplicity 1. Similarly denote by $\Psi(t)$ the fundamental matrix solution of (14), for which $\Psi(0) = I$. One has $\Psi(t) = [(\Phi(t))^{-1}]^*$.

If $v(t)$ is a vector solution of (13), with initial values $v(0) = v_0$, and $w(t)$ is a vector solution to (14), with initial values $w(0) = w_0$, then the inner product satisfies $w^*(t)v(t) = w_0^*v_0$, i.e. it is independent of time t.

The left and right eigenvectors of the monodromy matrix $\Phi(1)$ for a geometrically simple eigenvalue 1 will be denoted p_0, q_0 respectively. It is easily seen that p_0 (respectively, q_0) is also the right (respectively, left) eigenvector of $\Psi(1)$ for the eigenvalue 1. Furthermore, q_0 is a scalar multiple of $\dot{x}(0)$.

3.2. *Branch points of periodic solutions*

If we select a component β of the parameter vector α, then the periodic solution equations (11), (12) admit a smooth solution family in $(x(t), T, \beta)$-space, passing through a given periodic solution, if the Jacobian operator

$$J = \begin{bmatrix} D - Tf_x(x(t), \alpha) & -f(x(t), \alpha) & -Tf_\beta(x(t), \alpha) \\ \delta_0 - \delta_1 & 0 & 0 \\ \text{Int}_{\dot{x}_{\text{old}}(t)} & 0 & 0 \end{bmatrix} \tag{15}$$

is onto and has a one-dimensional kernel. If this condition is violated, then the periodic solution is called a *branch point*, and, as in the case of equilibria, more information is needed to decide about the behavior of nearby periodic solutions.

We call J the *branching operator*. To study it in more detail we first recall some basic facts about the operator that is implicitly defined by (11), when linearized about a regular solution $(x(t), T, \alpha)$.

Proposition 3.   *If $(x(t), T, \alpha)$ is a regular solution of (11), (12), then the operator*

$$\begin{bmatrix} D - Tf_x(x(t), \alpha) \\ \delta_0 - \delta_1 \end{bmatrix} :$$

$$\mathcal{C}^1([0, 1], \mathbb{R}^n) \to \mathcal{C}^0([0, 1], \mathbb{R}^n) \times \mathbb{R}^n \tag{16}$$

has a one-dimensional kernel spanned by Φq_0. Its range has codimension 1; if $\zeta \in C^0([0,1], \mathbb{R}^n)$, $r \in \mathbb{R}^n$ then $(\zeta, r)^*$ is in the range if and only if $\langle \Psi p_0, \zeta \rangle = p_0^* r$. In particular, if $r = 0$ then $(\zeta, 0)^*$ is in the range if and only if $\langle \Psi p_0, \zeta \rangle = 0$.

Proof. See [Doedel *et al.*, 2003b, Proposition 1]. ∎

Proposition 4. *Let $(x(t), T, \alpha)$ be a regular solution of (11), (12), and assume that $x_{\text{old}}(t)$ is close enough to $x(t)$, so that $\langle \dot{x}, \dot{x}_{\text{old}} \rangle \neq 0$. Then the operator*

$$\begin{bmatrix} D - Tf_x(x(t), \alpha) \\ \delta_0 - \delta_1 \\ \text{Int}_{\dot{x}_{\text{old}}(t)} \end{bmatrix} : C^1([0,1], \mathbb{R}^n)$$

$$\to C^0([0,1], \mathbb{R}^n) \times \mathbb{R}^n \times \mathbb{R} \qquad (17)$$

is one-to-one. Its range has codimension 1; if $\zeta \in C^0([0,1], \mathbb{R}^n)$, $r \in \mathbb{R}^n$, $s \in \mathbb{R}$ then $(\zeta, r, s)^*$ is in the range if and only if $\langle \Psi p_0, \zeta \rangle = p_0^* r$. In particular, if $r = 0, s = 0$ then $(\zeta, 0, 0)^*$ is in the range if and only if $\langle \Psi p_0, \zeta \rangle = 0$.

Proof. First assume that $y(t)$ is in the kernel of (17). Then it is also in the kernel of (16), i.e. $y(t)$ is a multiple of $\dot{x}(t)$ by Proposition 3. By the assumption on the closeness of x_{old}, this implies that $y(t) \equiv 0$.

Next, consider any $\zeta \in C^0([0,1], \mathbb{R}^n)$, $r \in \mathbb{R}^n$, $s \in \mathbb{R}$. By Proposition 3 this implies $\langle \Psi p_0, \zeta \rangle = p_0^* r$. Conversely, if this condition holds then, again by Proposition 3, there exists an $s' \in \mathbb{R}$ such that $(\zeta, r, s')^*$ is in the range of (17). On the other hand, this range contains also a vector of the form $(0, 0, s - s')^*$ (the image of a multiple of $\dot{x}(t)$). ∎

Proposition 5. *Let $(x(t), T, \alpha)$ be a regular solution of (11), (12) and assume that $x_{\text{old}}(t)$ is close enough to $x(t)$, so that $\langle \dot{x}, \dot{x}_{\text{old}} \rangle \neq 0$. Let $k_1, k_2 \in C^0([0,1], \mathbb{R})$ and consider the operator*

$$J_1 = \begin{bmatrix} D - Tf_x(x(t), \alpha) & k_1(t) & k_2(t) \\ \delta_0 - \delta_1 & 0 & 0 \\ \text{Int}_{\dot{x}_{\text{old}}(t)} & 0 & 0 \end{bmatrix}, \quad (18)$$

where $J_1 : C^1([0,1], \mathbb{R}^n) \times \mathbb{R} \times \mathbb{R} \to C^0([0,1], \mathbb{R}^n) \times \mathbb{R}^n \times \mathbb{R}$. Then the following statements are equivalent:

(1) One of $\langle \Psi p_0, k_1 \rangle$, $\langle \Psi p_0, k_2 \rangle$ is nonzero.
(2) J_1 is onto and has a one-dimensional kernel.
(3) J_1 is onto.

Proof. To prove that (1) implies (2), we may assume that $\langle \Psi p_0, k_1 \rangle \neq 0$. By Proposition 4, the second column of J_1 is not in the range of (17), and so J_1 is an onto operator. Next, the first two block columns of J_1 span the range of J_1, so the third column is a linear combination of the first two block columns; this implies that the kernel of J_1 is at least one-dimensional. To prove that the kernel is one-dimensional, suppose that $(y_1(t), u_1, v_1)$, $(y_2(t), u_2, v_2)$ are both in the kernel of J_1. Then there exist $a, b \in \mathbb{R}$, not both equal to zero, such that $av_1 + bv_2 = 0$. Hence

$$\begin{bmatrix} (au_1 + bu_2)k_1(t) \\ 0 \\ 0 \end{bmatrix}$$

is in the range of (17). By Proposition 4 and the assumption on $k_1(t)$, this implies that $au_1 + bu_2 = 0$. Hence $ay_1(t) + by_2(t)$ is in the kernel of (17). However, this operator is one-to-one, so $ay_1(t) + by_2(t) = 0$. This proves that $(y_1(t), u_1, v_1)$ and $(y_2(t), u_2, v_2)$ are linearly dependent, so the kernel of J_1 is one-dimensional.

The implication (2) $\Rightarrow$ (3) is trivial. Now assume that J_1 is onto. If $\langle \Psi p_0, k_1 \rangle = \langle \Psi p_0, k_2 \rangle = 0$ then, by Proposition 4, the range of J_1 is the range of (17), so J_1 is not onto. ∎

Corollary 1. *Let $(x(t), T, \alpha)$ be a regular solution of (11), (12) and assume that $x_{\text{old}}(t)$ is close enough to $x(t)$, so that $\langle \dot{x}, \dot{x}_{\text{old}} \rangle \neq 0$. Let β be a component of α. Then $(x(t), T, \alpha)$ is a branch point of the periodic solutions in (x, β)-space if and only if $\langle \Psi p_0, f(x, \alpha) \rangle = \langle \Psi p_0, f_\beta(x, \alpha) \rangle = 0$.*

From [Doedel *et al.*, 2003b], Proposition 5 we recall that the condition $\langle \Psi p_0, f(x, \alpha) \rangle = 0$ generically characterizes fold bifurcations of limit cycles. This is in accordance with the fact that in generic systems BPC points appear as special points in families of LPC points.

4. Numerical Detection, Computation and Continuation of Branch Points of Periodic Solutions

4.1. *Time discretization*

We concentrate on the orthogonal collocation method [Ascher *et al.*, 1995] to discretize the periodic solutions, because of its good convergence properties [De Boor & Swartz, 1973], and its widespread use, e.g. in COLSYS [Ascher *et al.*, 1981], AUTO [Doedel *et al.*, 2001], CONTENT [Kuznetsov & Levitin, 1997] and MATCONT [Dhooge *et al.*, 2003]. We recall the basic features.

First the interval $[0, 1]$ is subdivided into N smaller intervals.

$$0 = \tau_0 < \tau_1 < \cdots < \tau_N = 1.$$

In each of these intervals the solution $x(\tau)$ is approximated by an order m vector valued polynomial $x^{(i)}(\tau)$. This is done by defining $m+1$ equidistant points on each interval:

$$\tau_{i,j} = \tau_i + \frac{j}{m}(\tau_{i+1} - \tau_i) \ (j = 0, 1, \ldots, m),$$

and defining the polynomials $x^{(i)}(\tau)$ as

$$x^{(i)}(\tau) = \sum_{j=0}^{m} x^{i,j} \ell_{i,j}(\tau).$$

Here $x^{i,j}$ is the discretization of $x(\tau)$ at $\tau = \tau_{i,j}$ (we note that $x^{i,m} = x^{i+1,0}$), and the $\ell_{i,j}(\tau)$'s are the Lagrange basis polynomials

$$\ell_{i,j}(\tau) = \prod_{k=0, k \neq j}^{m} \frac{\tau - \tau_{i,k}}{\tau_{i,j} - \tau_{i,k}}.$$

In each interval $[\tau_i, \tau_{i+1}]$ we require that the polynomials $x^{(i)}(\tau)$ satisfy the BVP exactly at m collocation points $\zeta_{i,j}$ $(j = 1, \ldots, m)$. It can be proved that the best choice for the collocation points are the Gauss points [De Boor & Swartz, 1973], i.e. the roots of the Legendre polynomial of degree m, relative to the interval $[\tau_i, \tau_{i+1}]$.

Now let $x(t)$ be a function defined in $[0, 1]$, and assume that we want to integrate it over $[0, 1]$. If, for example, $N = 3$ (mesh intervals), and $m = 2$ (collocation points), then the following data are associated with the discretized interval $[0, 1]$:

τ_0		τ_1		τ_2		τ_3
○	○	●	○	●	○	○
$\tau_{0,0}$	$\tau_{0,1}$	$\tau_{0,2}$		$\tau_{2,0}$	$\tau_{2,1}$	$\tau_{2,2}$
		$\tau_{1,0}$	$\tau_{1,1}$	$\tau_{1,2}$		$\tau_{3,0}$
$t_1 w_1$	$t_1 w_2$	$t_1 w_3 + t_2 w_1$	$t_2 w_2$	$t_2 w_3 + t_3 w_1$	$t_3 w_2$	$t_3 w_3$
$\sigma_{0,0}$	$\sigma_{0,1}$	$\sigma_{1,0}$	$\sigma_{1,1}$	$\sigma_{2,0}$	$\sigma_{2,1}$	$\sigma_{3,0}$

The total number of mesh points (tps) is $N \times m + 1$, the total number of points $(ncoords)$ is $tps \times n$. Each mesh point $\tau_{i,j}$ in a mesh interval $[\tau_i, \tau_{i+1}]$ has a particular weight w_{j+1}, the Gauss–Lagrange quadrature coefficient. Some mesh points (the black bullets) belong to two mesh intervals. We set $t_i = \tau_i - \tau_{i-1}$, $(i = 1, \ldots, N)$. The integration weight $\sigma_{i,j}$ of $\tau_{i,j}$ is given by $w_{j+1} t_{i+1}$, for $0 \leq i \leq N - 1$ and $0 < j < m$. For $i = 0, \ldots, N - 2$, the integration weight of $\tau_{i,m} = \tau_{i+1,0}$ is given by $\sigma_{i,m} = w_{m+1} t_{i+1} + w_1 t_{i+2}$, and the integration weights of τ_0 and τ_N are given by $w_1 t_1$ and $w_{m+1} t_N$, respectively. The integral $\int_0^1 x(t)\, dt$ is approximated by $\sum_{i=0}^{N-1} \sum_{j=0}^{m-1} x(\tau_{i,j}) \sigma_{i,j} + x(1) \sigma_{N,0}$.

4.2. *Discretization of the BVP*

Using the discretization described in Sec. 4.1 we obtain the discretized BVP

$$\begin{cases} \left(\sum_{j=0}^{m} x^{i,j} \dot{\ell}_{i,j}(\zeta_{i,k}) \right) - T f \left(\sum_{j=0}^{m} x^{i,j} \ell_{i,j}(\zeta_{i,k}), \alpha \right) = 0, \\[2mm] x^{0,0} - x^{N-1,m} = 0, \\[2mm] \sum_{i=0}^{N-1} \sum_{j=0}^{m-1} \sigma_{i,j} \left[x^{i,j} \right]^* \dot{x}_{\text{old}}^{i,j} + \sigma_{N,0} \left[x^{N,0} \right]^* \dot{x}_{\text{old}}^{N,0} = 0. \end{cases}$$

$$(19)$$

The first equation actually represents Nm equations, one for each combination of $i = 0, 1, 2, \ldots, N - 1$ and $k = 1, 2, \ldots, m$.

The Jacobian of the discretized system is sparse. During the continuation process, each Newton iteration requires the numerical solution of a linear system consisting of this Jacobian matrix, with a extra row that corresponds to the tangent vector to the solution branch. For example, if $N = 3$ (mesh intervals), $m = 2$ (collocation points), and $n = 2$, this matrix has the following sparsity structure [Doedel *et al.*, 1991]:

$$
\begin{pmatrix}
x^{0,0} & x^{0,1} & x^{1,0} & x^{1,1} & x^{2,0} & x^{2,1} & x^{3,0} & T & \alpha \\
\bullet & \bullet & \bullet & \bullet & \bullet & \bullet & & & \bullet & \bullet \\
\bullet & \bullet & \bullet & \bullet & \bullet & \bullet & & & \bullet & \bullet \\
\bullet & \bullet & \bullet & \bullet & \bullet & \bullet & & & \bullet & \bullet \\
\bullet & \bullet & \bullet & \bullet & \bullet & \bullet & & & \bullet & \bullet \\
& & \bullet & \bullet & \bullet & \bullet & \bullet & \bullet & \bullet & \bullet \\
& & \bullet & \bullet & \bullet & \bullet & \bullet & \bullet & \bullet & \bullet \\
& & \bullet & \bullet & \bullet & \bullet & \bullet & \bullet & \bullet & \bullet \\
& & \bullet & \bullet & \bullet & \bullet & \bullet & \bullet & \bullet & \bullet \\
& & & & \bullet & \bullet & \bullet & \bullet & \bullet & \bullet & \bullet & \bullet \\
& & & & \bullet & \bullet & \bullet & \bullet & \bullet & \bullet & \bullet & \bullet \\
& & & & \bullet & \bullet & \bullet & \bullet & \bullet & \bullet & \bullet & \bullet \\
& & & & \bullet & \bullet & \bullet & \bullet & \bullet & \bullet & \bullet & \bullet \\
\bullet & & & & & & & \bullet \\
& \bullet & & & & & & \bullet \\
\bullet & \bullet & \bullet & \bullet & \bullet & \bullet & \bullet & \bullet & \bullet & \bullet & \bullet & \bullet & \bullet & \bullet \\
\bullet & \bullet & \bullet & \bullet & \bullet & \bullet & \bullet & \bullet & \bullet & \bullet & \bullet & \bullet & \bullet & \bullet
\end{pmatrix}, \qquad (20)
$$

where the $\bullet$'s denote elements that are generally nonzero. The columns of (20) label the unknowns of the discretized problem. The first $n = 2$ rows correspond to the first collocation point, etc. In (11) and (12) there are three unknown quantities: the orbit x, the period T and a parameter α. The part of the Jacobian that corresponds to the first equation in (11), has the following form:

$$[D - T f_x(x, \alpha) \quad -f(x, \alpha) \quad -T f_\alpha(x, \alpha)].$$

In (20), $D - T f_x(x, \alpha)$ corresponds to $N = 3$ blocks, of dimension $nm \times n(m+1)$, i.e. 4×6. The part of (20) that defines the periodic boundary conditions has the form:

$$[I_n \quad 0_{n \times (Nm-1)n} \quad -I_n \quad 0_n].$$

In (20), these are $n = 2$ rows following the 4×6 blocks. These rows contain two nonzero parts, corresponding to $x^{0,0}$ and $x^{N,0}$ (i.e. $\pm I_2$). The next to last row in (20) is the derivative of the discretization of the phase condition (12). The last row, which basically corresponds to Keller's pseudo-arclength continuation equation [Keller, 1977], is automatically added in our implementation.

4.3. *Numerical detection of BPC cycles in generic systems*

For generic systems, Corollary 1 provides the key to detect and compute BPC cycles. First we consider the test functions

$$T_{LPC} = \langle \Psi p_0, f(x, \alpha) \rangle$$

$$= \int_0^1 [\Psi(t) p_0]^* f(x(t), \alpha) \, dt$$

and

$$T_{BPC} = \langle \Psi p_0, f_\beta(x, \alpha) \rangle$$

$$= \int_0^1 [\Psi(t)p_0]^* f_\beta(x(t), \alpha)\, dt$$

from Corollary 1. By Proposition 3

$$\begin{bmatrix} \Psi(t)p_0 \\ -p_0 \end{bmatrix}$$

is orthogonal to the range of

$$M = \begin{bmatrix} D - Tf_x(x(t), \alpha) \\ \delta_0 - \delta_1 \end{bmatrix}. \tag{21}$$

Thus the conditions $T_{LPC} = 0, T_{BPC} = 0$, are equivalent to the statements that

$$\begin{bmatrix} f(x(t), \alpha) \\ 0_n \end{bmatrix},$$

and

$$\begin{bmatrix} f_\beta(x(t), \alpha) \\ 0_n \end{bmatrix},$$

are in the range of (21). Now the discretized form of (21) is the square matrix M_D, obtained from (20) by removing the last two rows and columns. To be precise, if $h \in C^1([0, 1], \mathbb{R}^n)$, then

$$Mh = \begin{bmatrix} \dot{h} - Tf_x(x(t), \alpha)h \\ h(0) - h(1) \end{bmatrix},$$

and

$$M_D(h)_{dm} = \begin{bmatrix} (\dot{h} - Tf_x(x(t), \alpha)h)_{dc} \\ h(0) - h(1) \end{bmatrix},$$

where $()_{dm}$ and $()_{dc}$ denote discretization in mesh points and in collocation points, respectively.

M_D has rank defect one, and its right singular vector is $(\Phi(t)q_0)_{dm}$, or, equivalently, $(\dot{x}(t))_{dm}$.

For numerical purposes we therefore replace the conditions $T_{LPC} = 0$ and $T_{BPC} = 0$ by the requirements that

$$\begin{bmatrix} (f(x(t), \alpha))_{dc} \\ 0_n \end{bmatrix},$$

respectively

$$\begin{bmatrix} (f_\beta(x(t), \alpha))_{dc} \\ 0_n \end{bmatrix},$$

are in the range space of M_D, or, equivalently, that they are orthogonal to the left singular vector of M_D. To compute this vector, we expand M by adding a column w and a row v^* so that

$$M_{Db} = \begin{bmatrix} M_d & w \\ v^* & 0 \end{bmatrix}$$

is nonsingular, and we solve the system

$$M_{Db}^* \begin{bmatrix} \psi_1 \\ \psi_2 \\ \psi_3 \end{bmatrix} = \begin{bmatrix} 0_{(Nm+1)n} \\ 1 \end{bmatrix},$$

where ψ_1, ψ_2 have Nmn, n components, respectively, and ψ_3 is a scalar. We note that in exact arithmetic $\psi_3 = 0$. So the numerical test functions are

$$T_{LPCd} = \psi_1^*(f(x(t), \alpha))_{dc}$$

and

$$T_{BPCd} = \psi_1^*(f_\beta(x(t), \alpha))_{dc},$$

respectively.

4.4.   *Numerical detection of BPC cycles in families of limit cycles*

BPC cycles are not generic in families of limit cycles, but they are common in the case of symmetries, if the branching parameter is also the continuation parameter; examples are given in Secs. 6 and 7. The test function in AUTO and CONTENT is the determinant of a small matrix obtained from (20), by an elimination that preserves the rank of the matrix. MATCONT uses a strategy that requires only the solution of linear systems; it is based on the fact that in a symmetry-breaking BPC cycle M_D has rank defect two. Therefore we border M_D with two additional rows and columns to obtain

$$M_{Dbb} = \begin{bmatrix} M_D & w_1 & w_2 \\ v_1^* & 0 & 0 \\ v_2^* & 0 & 0 \end{bmatrix},$$

so that M_{Dbb} is nonsingular in the BPC cycle. Then we solve the systems

$$M_{Dbb} \begin{bmatrix} \psi_{11} & \psi_{12} \\ g_{BPC11} & g_{BPC12} \\ g_{BPC21} & g_{BPC22} \end{bmatrix} = \begin{bmatrix} 0_{(Nm+1)n} & 0_{(Nm+1)n} \\ 1 & 0 \\ 0 & 1 \end{bmatrix},$$

where ψ_{11}, ψ_{12} have $(Nm+1)n$ components, and g_{BPC11}, g_{BPC12}, g_{BPC21}, and g_{BPC22}, are scalar test functions for the BPC. In the BPC cycle they all vanish. In the examples in Secs. 6 and 7 we show that they indeed change sign, and that they can therefore detect the BPC cycles.

4.5. *Numerical computation of BPC cycles in families of LPC cycles*

In the generic case, T_{BPCd} can be used to locate a BPC cycle on a curve of LPC cycles exactly. This method is implemented in MATCONT. We note that the branching parameter may be different from the continuation parameters. An example is given in Sec. 5.

4.6. *Numerical continuation of BPC cycles*

The discretization of the branching operator is a matrix J_D, that we formally obtain by removing the last row of (20), and replacing α by a specific component. Thus J_D is an N_D by $N_D + 1$ matrix where $N_D = (Nm + 1) \times n + 1$. As in Sec. 2, we shall express that this matrix has rank defect 1. We formulate the essential results, omitting the proofs.

Proposition 6. *Let $\phi_1, \phi_2 \in \mathbb{R}^{N_D+1}$ be such that together with the rows of J_D they span $\mathbb{R}^{N_D+1}$. Also, let $\psi \in \mathbb{R}^{N_D}$ be a vector that together with the columns of J_D spans $\mathbb{R}^{N_D}$. Then the bordered matrix*

$$B_{JD} = \begin{bmatrix} J_D & \psi \\ \phi_1^* & 0 \\ \phi_2^* & 0 \end{bmatrix} \tag{22}$$

is nonsingular.

Proposition 7. *Let $\phi_1, \phi_2 \in \mathbb{R}^{N_D+1}$, and $\psi \in \mathbb{R}^{N_D}$ be as in Proposition 6. Let B_D be a matrix with the structure of B_{JD}*

$$B_D = \begin{bmatrix} B & \psi \\ \phi_1^* & 0 \\ \phi_2^* & 0 \end{bmatrix} \tag{23}$$

and define $v_{D1}, v_{D2} \in \mathbb{R}^{N_D+1}$, $g_{D1}, g_{D2} \in \mathbb{R}$ by requiring

$$B_D \begin{bmatrix} v_{D1} & v_{D2} \\ g_{D1} & g_{D2} \end{bmatrix} = \begin{bmatrix} 0_{N_D} & 0_{N_D} \\ 1 & 0 \\ 0 & 1 \end{bmatrix}. \tag{24}$$

If B is sufficiently close to J_D, then B is nonsingular. Furthermore, B has corank 1 if and only if

$$\begin{aligned} g_{D1} &= 0, \\ g_{D2} &= 0. \end{aligned} \tag{25}$$

The numerical equations for a branch point of a periodic solution of (1), near a given $(x^0(t), T^0, \alpha^0)$, are defined by the system consisting of (19) and (25), where B is replaced by J_D, so that g_{D1}, g_{D2} are functions of the discretized orbit $x(t)$, and of T and α. If three components of α are freed, then, generically, a family of branch points can be computed.

In the computations we also need the derivatives of g_{D1}, g_{D2}, with respect to the components of $x(t), T$ and α. This can be done as in Sec. 2. We solve the adjoint system corresponding to (24)

$$B_D^* \begin{bmatrix} w_D \\ g_{D1} \\ g_{D2} \end{bmatrix} = \begin{bmatrix} 0_{N_D+1} \\ 1 \end{bmatrix}, \tag{26}$$

where $w_D \in \mathbb{R}^{N_D}$. If z is one of the components of x, T, or a free parameter, then by taking derivatives of (24), and multiplying from the left with

$$\begin{bmatrix} w_D^* & g_{D1} & g_{D2} \end{bmatrix},$$

we obtain

$$g_{Diz} = -w_D^* B_{Dz} v_{Di}, \quad (i = 1, 2). \tag{27}$$

The continuation of BPC cycles is supported by MATCONT. We note that the second-order partial derivatives (the *Hessian*) of f with respect to x and α are required.

4.7. *Numerical computation of BPC cycles in families of limit cycles*

In the case of symmetries, where a BPC point occurs along a family of limit cycles, AUTO and CONTENT locate the BPC by a rootfinding procedure on the test function for detection. MATCONT, on the other hand, uses a specific locator algorithm that guarantees quadratic convergence. This locator has many features in common with the numerical continuation described in Sec. 4.6.

The idea is to set up a system based on (19), that contains an artificial scalar unknown β, and two additional equations:

$$
\begin{cases}
\left(\displaystyle\sum_{j=0}^{m} x^{i,j} \dot{\ell}_{i,j}(\zeta_{i,k}) \right) \\[2ex]
\quad - Tf\left(\displaystyle\sum_{j=0}^{m} x^{i,j} \ell_{i,j}(\zeta_{i,k}), \alpha \right) + \beta p_1 = 0, \\[2ex]
\qquad x^{0,0} - x^{N-1,m} + \beta p_2 = 0, \\[2ex]
\displaystyle\sum_{i=0}^{N-1}\sum_{j=0}^{m-1} \sigma_{i,j} \left[x^{i,j} \right]^* \dot{x}_{\text{old}}^{i,j} \\[2ex]
\qquad + \sigma_{N,0} \left[x^{N,0} \right]^* \dot{x}_{\text{old}}^{N,0} + \beta p_3 = 0, \\[2ex]
\qquad\qquad g_{D1}(x, T, \alpha) = 0, \\[1ex]
\qquad\qquad g_{D2}(x, T, \alpha) = 0,
\end{cases}
\tag{28}
$$

where g_{D1}, g_{D2} are defined as in (23), and $[p_1^* \, p_2^* \, p_3]^*$ is the bordering vector ψ that appears in (22).

We solve this system with respect to x, T, α, β by Newton's method with initial $\beta = 0$. A branch point (x, T, α) corresponds to a regular solution $(x, T, \alpha, 0)$ of system (28) (see [Beyn *et al.*, 2002, p. 165]). We note again that the second-order partial derivatives (Hessian) of f with respect to x and α are required.

5. Example 1: The $A \to B \to C$ Reaction

In this section we discuss a generic example, i.e. a model without symmetries. The continuation

of BPC points then involves three effective parameters.

The model is that of a continuous stirred tank reactor, with consecutive $A \to B \to C$ reactions, as studied by [Doedel & Heinemann, 1983]. It has three state variables, u_1, u_2, u_3, and five parameters, p_1, p_2, p_3, p_4, p_5:

$$
\begin{cases}
\dot{u}_1 = -u_1 + p_1(1 - u_1)e^{u_3}, \\[1ex]
\dot{u}_2 = -u_2 + p_1 e^{u_3}(1 - u_1 - p_5 u_2), \\[1ex]
\dot{u}_3 = -u_3 - p_3 u_3 + p_1 p_4 e^{u_3}(1 - u_1 + p_2 p_5 u_2).
\end{cases}
\tag{29}
$$

This model is used as a demo in the AUTO manual [Doedel *et al.*, 2001]. In the notation of [Doedel & Heinemann, 1983], we have $u_1 = y$, where $1 - y$ is the concentration of reactant A, $u_2 = z$, the concentration of reactant B, $u_3 = \theta$, the temperature, $p_1 = D$, the Damkohler number, $p_2 = \alpha$, the ratio of reaction heats, $p_3 = \beta$, the heat transfer coefficient, $p_4 = B$, the adiabatic temperature rise, and $p_5 = \sigma$, the selectivity ratio.

Figure 1(a) reproduces the equilibria found in [Doedel & Heinemann, 1983], recomputed with MATCONT. The parameter values are $p_2 = 1, p_3 = 1.5, p_4 = 8, p_5 = 0.04$, with free parameter p_1, starting from the equilibrium at $p_1 = 0.1$, for which $u_1 = 0.13304, u_2 = 0.13223, u_3 = 0.42833$. The curve of equilibria contains four Hopf points, denoted, from left to right, H_1, H_2, H_3, H_4, respectively.

Using MATCONT, we reproduce in Fig. 1(b) another equilibrium curve, for the same parameter values, except with $p_2 = 0.9$. This curve looks qualitatively similar to that in Fig. 1(a), and it also has four Hopf points. As shown in [Doedel & Heinemann, 1983], in the case $p_2 = 1$, the Hopf points H_1 and H_4 are connected by a family of periodic solutions, and H_2 and H_3 are similarly connected. Figure 2(a) shows the family of periodic solutions that connects H_1 to H_4.

Interestingly, the situation is different when $p_2 = 0.9$. In this case H_1 and H_2 are connected by a family of periodic solutions, and so are H_3 and H_4. Figure 2(b) shows the family of periodic solutions that connects H_1 to H_2.

As shown in Fig. 2(a) ($p_2 = 1$), the family of solutions that connects H_1 to H_4 contains three fold bifurcations of periodic solutions, as also observed in [Doedel & Heinemann, 1983]. In Fig. 5(a) we

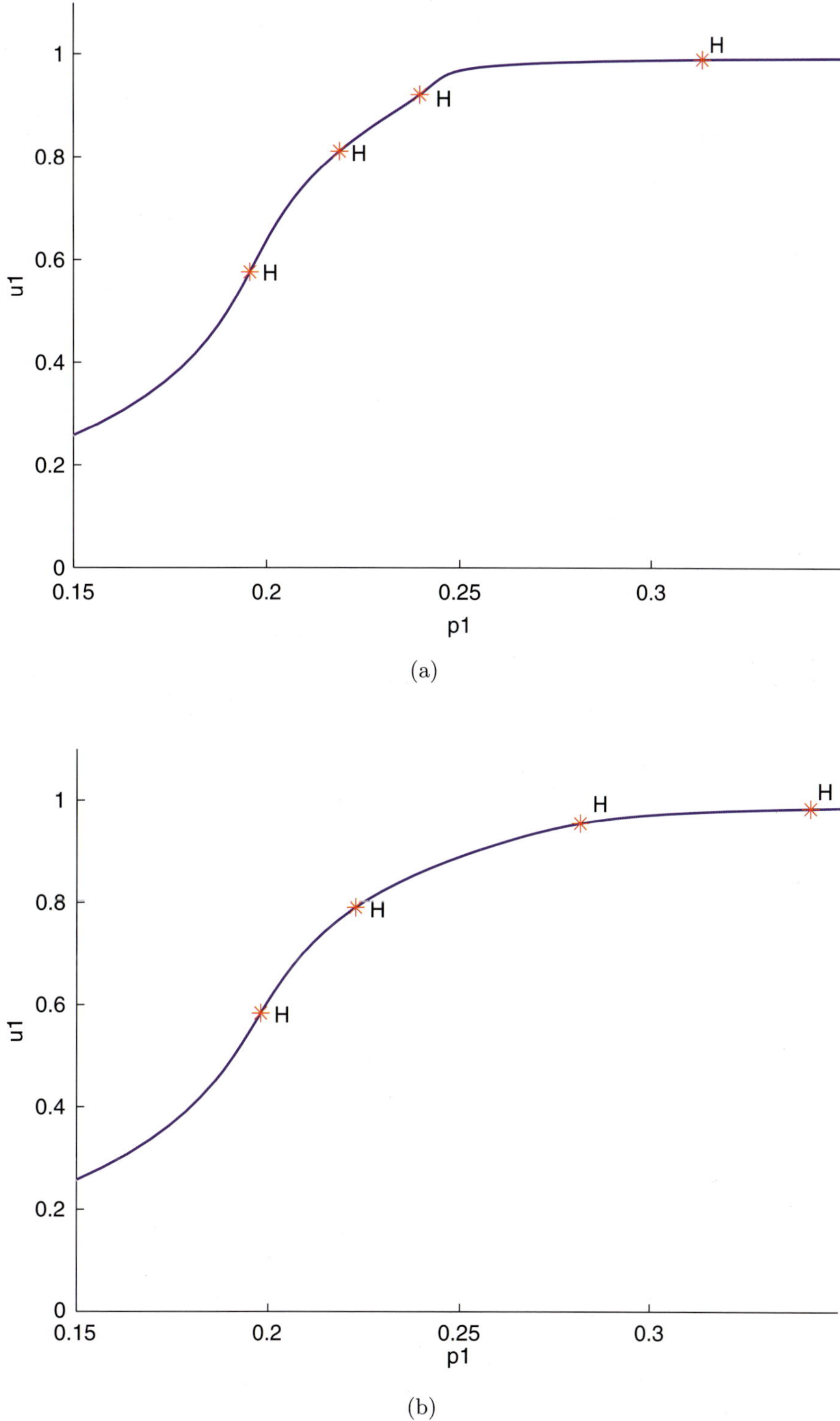

Fig. 1. (a) Equilibrium curve of the $A \to B \to C$ reaction, for $p_2 = 1$. (b) Equilibrium curve of the $A \to B \to C$ reaction for $p_2 = 0.9$.

plot T_{LPCd} versus p_1; T_{LPCd} clearly vanishes (and changes sign) at the fold bifurcations.

The observations above imply that, for certain nearby values of other parameters, we can expect an exchange of connections, i.e. a branch point of periodic orbits with respect to p_1. In order to locate it, we continue the first fold bifurcation of periodic solutions numerically; freeing both p_1 and p_2. This

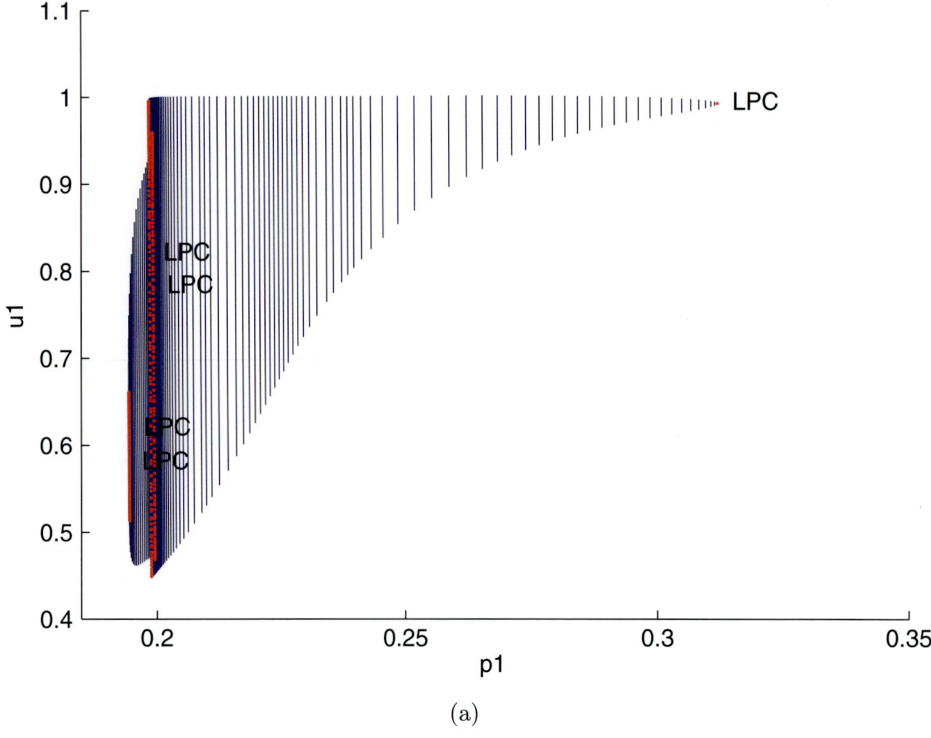

(a)

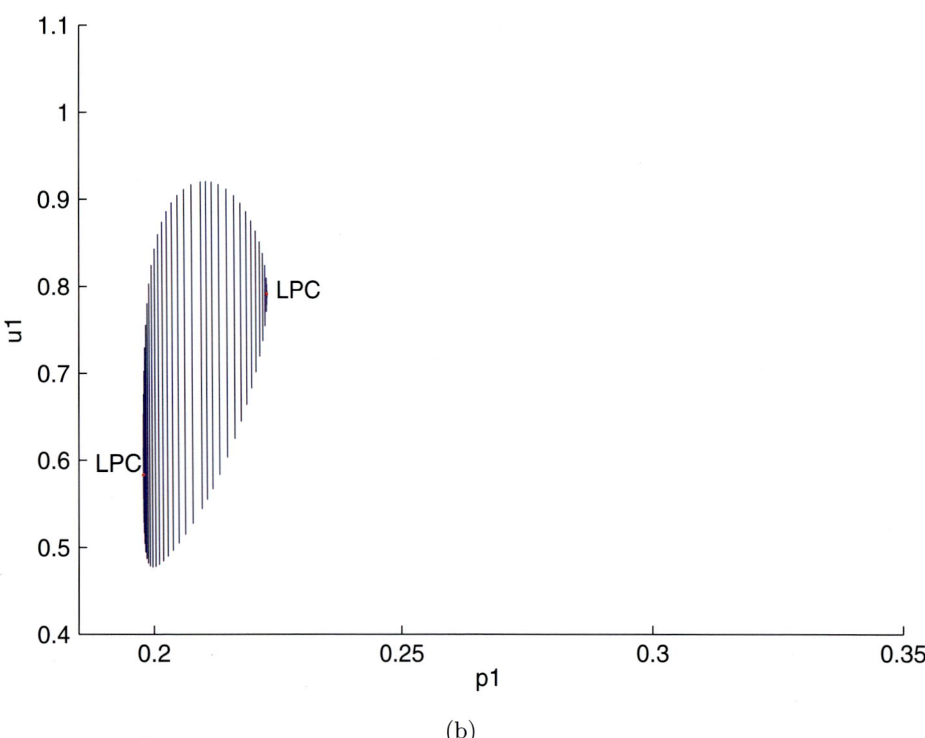

(b)

Fig. 2.   (a) Family of periodic orbits connecting the first and fourth Hopf points in Fig. 1(a). (b) Family of periodic orbits connecting the first and second Hopf points in Fig. 1(b).

family contains indeed a BPC point with respect to p_1; see Figs. 3 and 4. This BPC point was detected as a zero of T_{BPCd}; the symbol BPC1 in Figs. 3 and 4 reminds us that the branching is with respect to the first parameter of the system p_1. The critical parameter values are $p_1 = 0.211201156173, p_2 = 0.940211847478$. We note that the local extremum with respect to p_1 in Fig. 3 corresponds to a cusp in

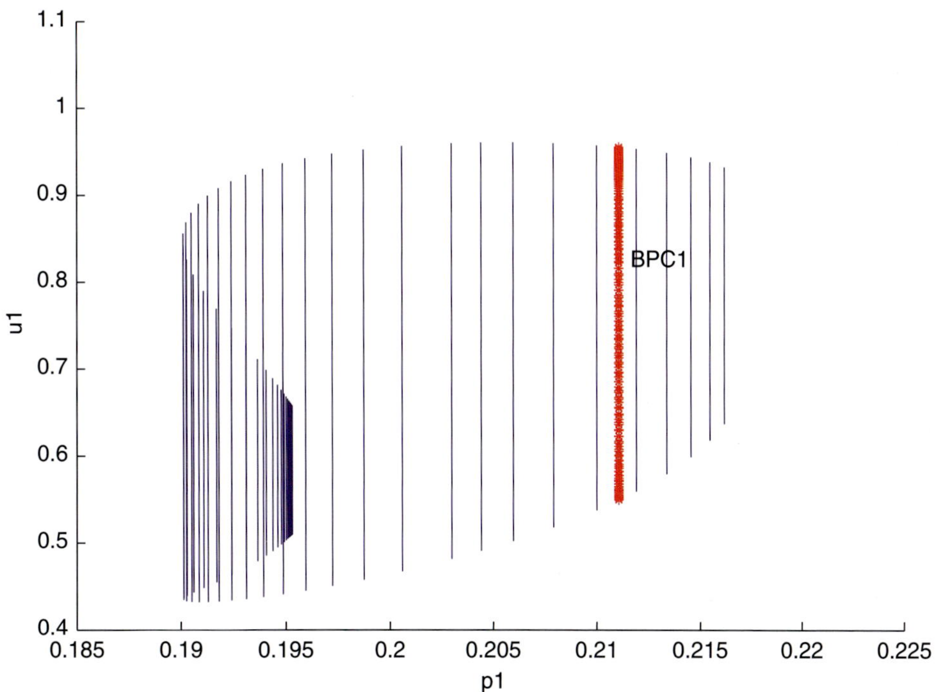

Fig. 3.   LPC curve, with a BPC point with respect to p_1, for the $A \to B \to C$ reaction.

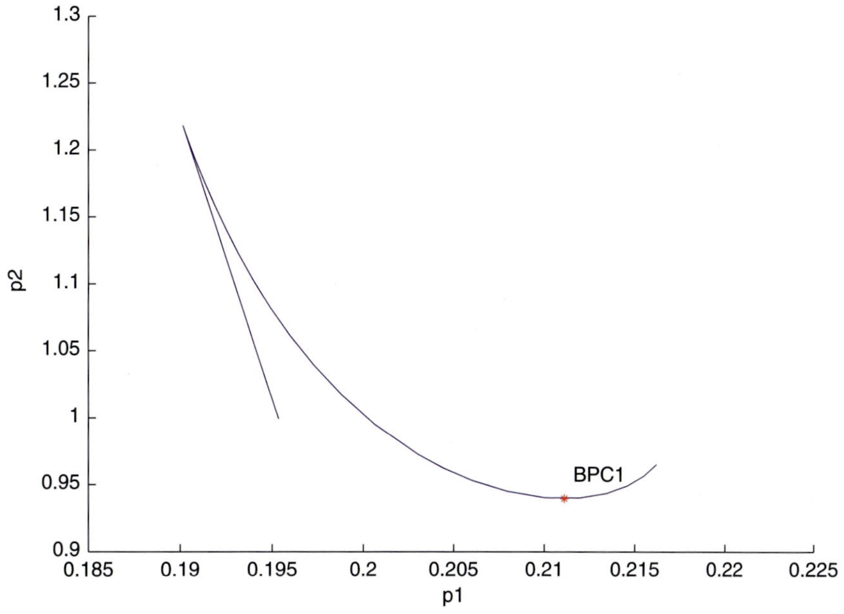

Fig. 4.   The family from Fig. 3 in (p_1, p_2)-space.

the parameter plane in Fig. 4. Also, in the parameter plane, the branch point with respect to p_1 corresponds to a local extremum with respect to p_2. In Fig. 5(b) we plot T_{LPCd} and T_{BPCd} versus p_1. As expected, T_{LPCd} vanishes at all points of the curve, while T_{BPCd} vanishes (and changes sign) at the BPC point only.

It is now possible to continue the BPC point by freeing a third parameter. Selecting another point on this family, and freezing again the third

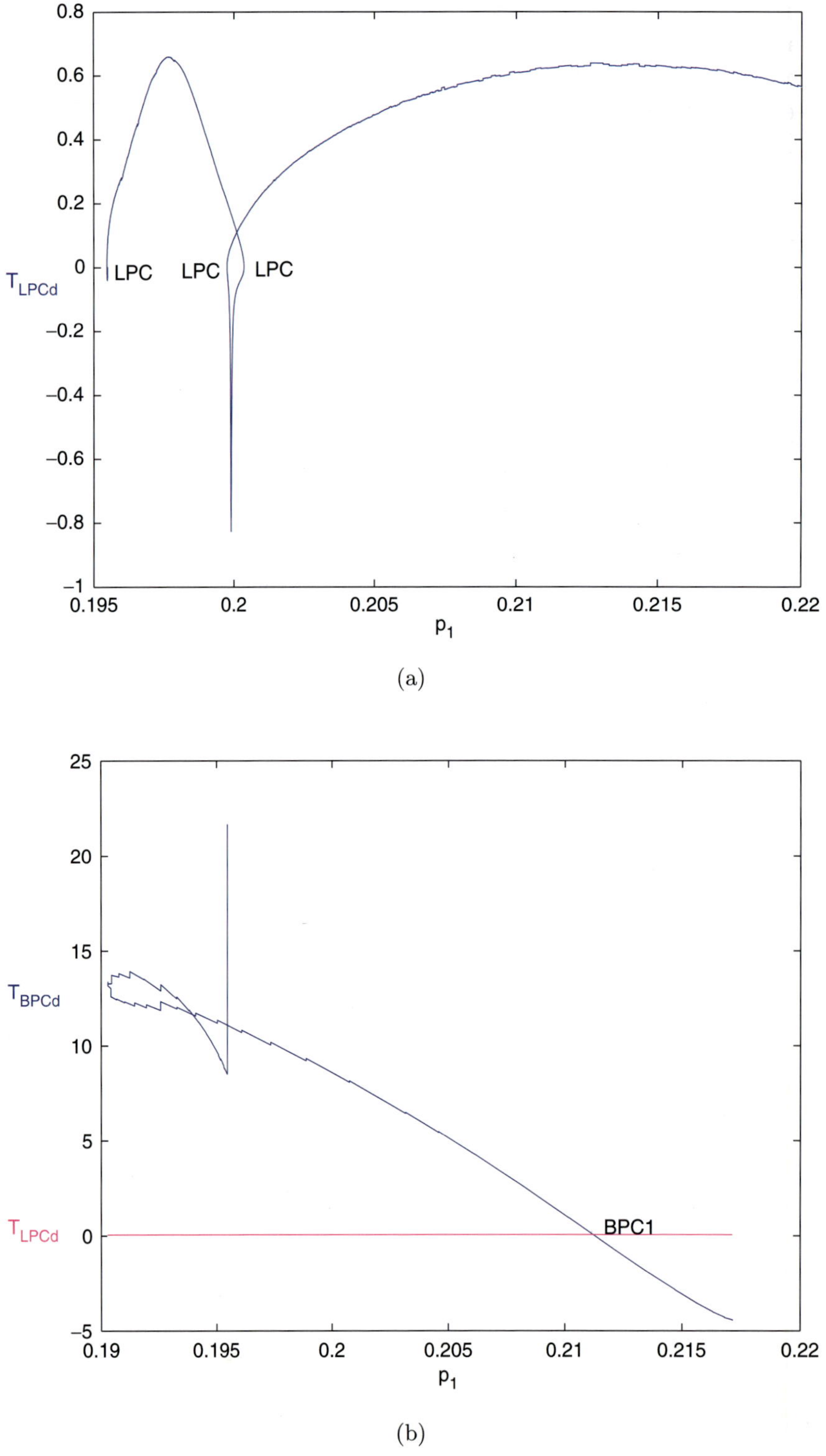

(a)

(b)

Fig. 5. (a) T_{LPCd} on a LC curve. (b) T_{BPCd} and T_{LPCd} on a LPC curve.

parameter we can produce pictures qualitatively similar to Fig. 4.

6. Example 2: An Electronic Circuit

In this section we discuss a nongeneric situation, i.e. a problem with a symmetry, where the continuation of BPC points includes two effective parameters, and one artificial parameter.

The model is that of an autonomous electronic circuit, studied in [Freire *et al.*, 1983]. It has three state variables x, y, z, and six parameters $\gamma, r, a_3, b_3, \nu, \beta$:

$$\begin{cases} \dot{x} = \Big(-(\beta + \nu)x + \beta y - a_3 x^3 + b_3(y - x)^3 \Big) r^{-1}, \\ \dot{y} = \beta x - (\beta + \gamma)y - z - b_3(y - x)^3, \\ \dot{z} = y. \end{cases} \tag{30}$$

This model is also used as a demo in AUTO2000 [Doedel *et al.*, 2001]. It has the trivial solution family, where $x = y = z = 0$, for all parameter values. Moreover, it has the Z_2-symmetry $x \mapsto -x, y \mapsto -y, z \mapsto -z$.

We start by computing the trivial family, with fixed parameters $\gamma = -0.6, r = 0.6, a_3 = 0.328578$, $b_3 = 0.933578, \beta = 0.5$, and with ν as free parameter, with initially $\nu = -0.9$. Along this family a Hopf point is detected at $\nu = -0.58933644$, and a branch point of equilibria at $\nu = -0.5$. From the Hopf point we start the computation of a family of periodic solutions, using 25 mesh intervals and four collocation points. This is a family of symmetric solutions of (30); we detect one LPC and two BPC, see Fig. 6.

In Fig. 7 we show how the first BPC point is detected by the simultaneous sign changes of the test functions g_{BPC11}, g_{BPC12} g_{BPC21}, g_{BPC22} discussed in Sec. 4.4.

To compute the family of BPC points with respect to ν, through the first BPC, with free parameters ν, β, we need to introduce an additional free parameter that breaks the symmetry. There are many choices for this; we choose to introduce a parameter ϵ, and extend the system (30) by simply adding a term $+\epsilon$ to the first right-hand-side. For $\epsilon = 0$ this reduces to (30), while for $\epsilon \neq 0$ the symmetry is broken. Using the algorithm for the continuation of generic BPC points, with three

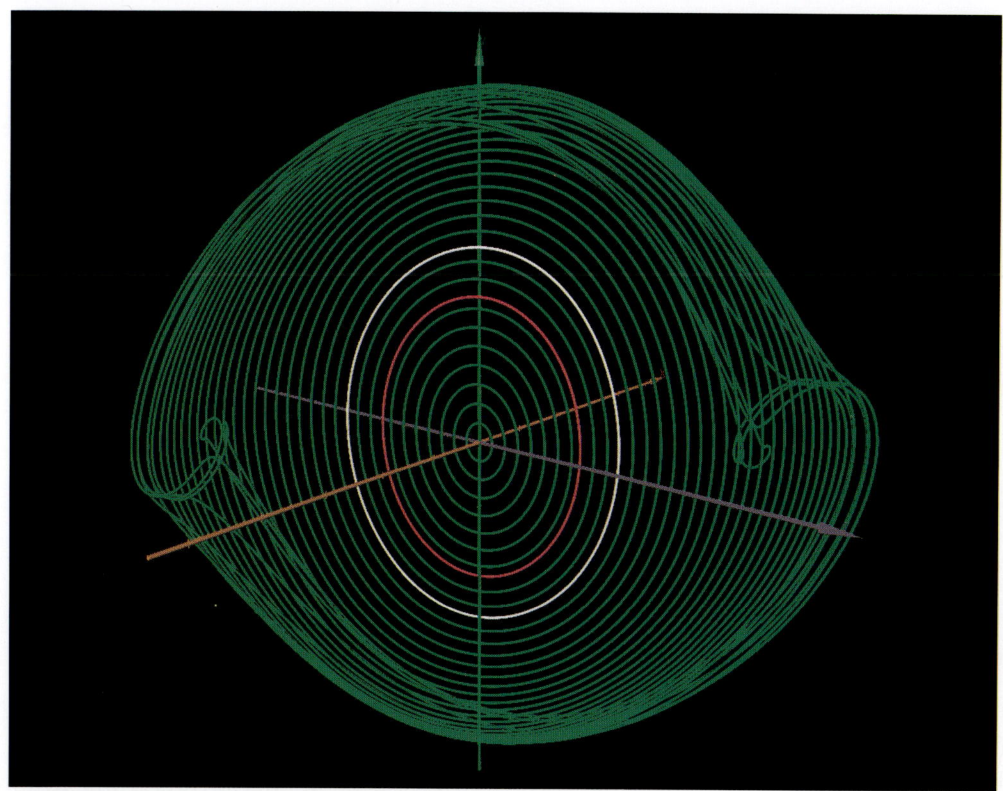

Fig. 6.   Family of periodic solutions, with LPC and branch points, in the circuit example.

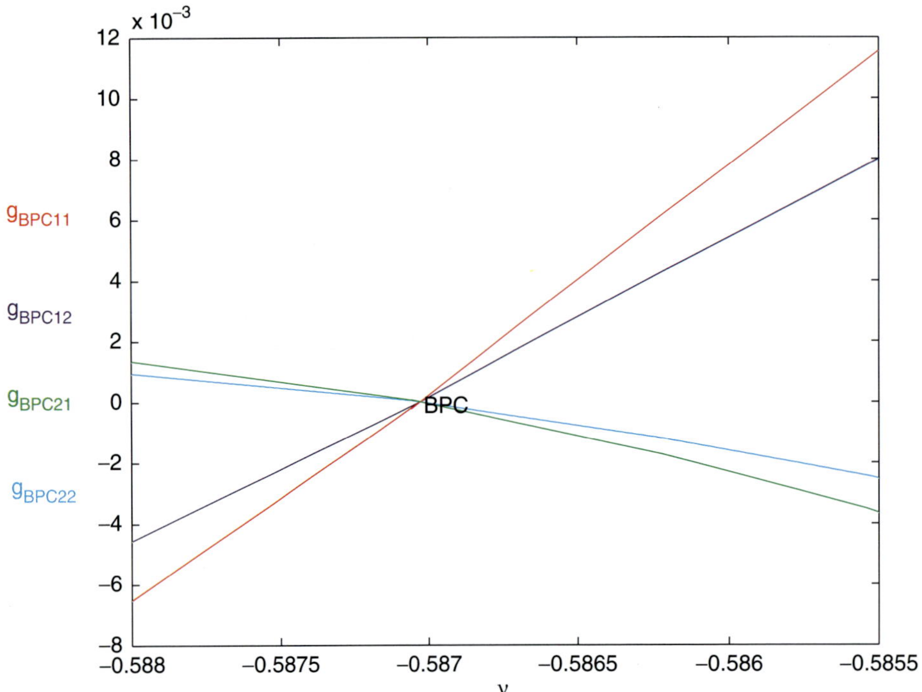

Fig. 7. Evolution of the test functions for branch points in a BPC in the circuit example.

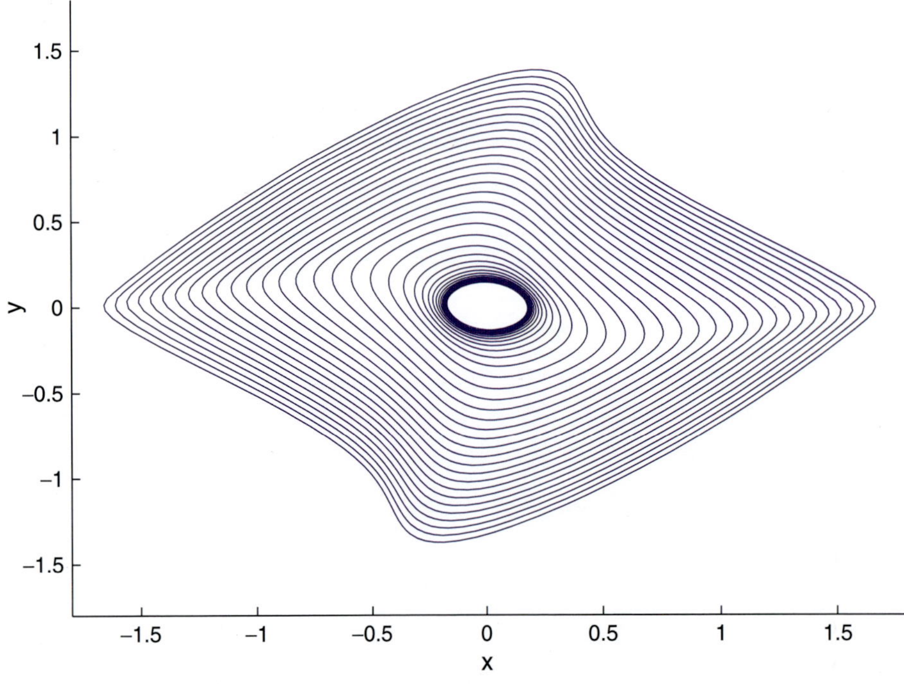

Fig. 8. Curve of BPC points in the circuit example.

free parameters ν, β, ϵ, we continue the curve of nongeneric BPC points, where ϵ remains zero, up to numerical precision. Figure 8 illustrates the fact that the symmetry is preserved.

7. Example 3

In this section we discuss an even less generic situation than in the preceding example, namely, an equation with several symmetries, where the continuation of BPC points involves one effective parameter, and two artificial parameters.

Our model is that of the circular restricted three-body problem (CR3BP), as studied, for example, in [Doedel *et al.*, 2003a; Doedel *et al.*, 2003c]:

$$\begin{cases} \ddot{x} = 2\dot{y} + x - (1-\mu)(x+\mu)r_1^{-3} \\ \quad - \mu(x-1+\mu)r_2^{-3}, \\ \ddot{y} = -2\dot{x} + y - (1-\mu)yr_1^{-3} - \mu yr_2^{-3}, \\ \ddot{z} = -(1-\mu)zr_1^{-3} - \mu zr_2^{-3}. \end{cases} \quad (31)$$

The larger and smaller primary bodies (say, the Earth and the Moon) are located at $(-\mu,0,0)$ and $(1-\mu,0,0)$ respectively, where μ is the mass-ratio parameter. In the Earth–Moon case, $\mu = 0.01215$. In (31) $r_1 = \sqrt{(x+\mu)^2 + y^2 + z^2}$ and $r_2 = \sqrt{(x-1+\mu)^2 + y^2 + z^2}$ are the distances of the negligible-mass satellite (x,y,z) to the two primary bodies, respectively.

We convert (31) to a first-order system in the usual way, by introducing the velocities v_x, v_y, v_z as additional variables; furthermore we introduce artificial parameters λ_1, λ_2, to obtain:

$$\begin{cases} \dot{x} = v_x + \lambda_1 \dfrac{\partial E}{\partial x}, \\[2mm] \dot{y} = v_y + \lambda_1 \dfrac{\partial E}{\partial y}, \\[2mm] \dot{z} = v_z + \lambda_1 \dfrac{\partial E}{\partial z}, \\[2mm] \dot{v}_x = 2v_y + x - (1-\mu)(x+\mu)r_1^{-3} - \mu(x-1+\mu)r_2^{-3} + \lambda_1 \dfrac{\partial E}{\partial v_x}, \\[2mm] \dot{v}_y = -2v_x + y - (1-\mu)yr_1^{-3} - \mu yr_2^{-3} + \lambda_1 \dfrac{\partial E}{\partial v_y}, \\[2mm] \dot{v}_z = -(1-\mu)zr_1^{-3} - \mu zr_2^{-3} + \lambda_1 \dfrac{\partial E}{\partial v_z} + \lambda_2, \end{cases} \quad (32)$$

where

$$E = \frac{1}{2}\left(v_x^2 + v_y^2 + v_z^2\right) \\ - \frac{1}{2}(x^2+y^2) - \frac{1-\mu}{r_1} - \frac{\mu}{r_2} - \frac{\mu(1-\mu)}{2}$$

is the *Jacobi constant* (the *energy*), which is preserved along orbits.

We are only interested in the case where $\lambda_1 = \lambda_2 = 0$. It is known that in this case, for each value of $\mu, (0 < \mu < 1)$, the (x,y) plane contains five equilibria, the so-called *libration points*. Three of them, L1, L2, L3, are collinear with the primary bodies, while both L4 and L5 form an equilateral triangle with the primaries. Here we compute the planar periodic orbits that arise from L1, the libration point between the two primary bodies. L1 is a solution of the equilibrium equations of (32), with $\lambda_1 = \lambda_2 = y = z = v_x = v_y = v_z = 0$. We detect it by a continuation of the equilibrium equations of (32), with μ as the free parameter. Since MATCONT

can locate zeroes of user functions, we detect L1 as a zero of the function $\mu - 0.01215$.

L1 is a (degenerate) Hopf bifurcation point of (32); we compute the family of planar periodic solutions (with $z = 0$) born there, using 40 mesh intervals and four collocation points, with λ_1 as the free parameter. Along this family λ_1 remains zero, up to numerical precision, and a BPC is detected, see Fig. 9.

In Fig. 10, we show how the BPC point is detected by the simultaneous sign changes of the test functions $g_{BPC11}, g_{BPC12}, g_{BPC21}, g_{BPC22}$ discussed in Sec. 4.4.

Selecting this BPC point as starting point, we compute the locus of branch points, using $\mu, \lambda_1, \lambda_2$ as the free parameters. The resulting curve is shown in Fig. 11. We note that the curve that represents the BPC point, shrinks to a single point, $x = -1, y = 0$, if μ tends to 1 (left side of Fig. 11), and also to a single point, $x = 1, y = 0$, if μ tends to 0 (right side of Fig. 11).

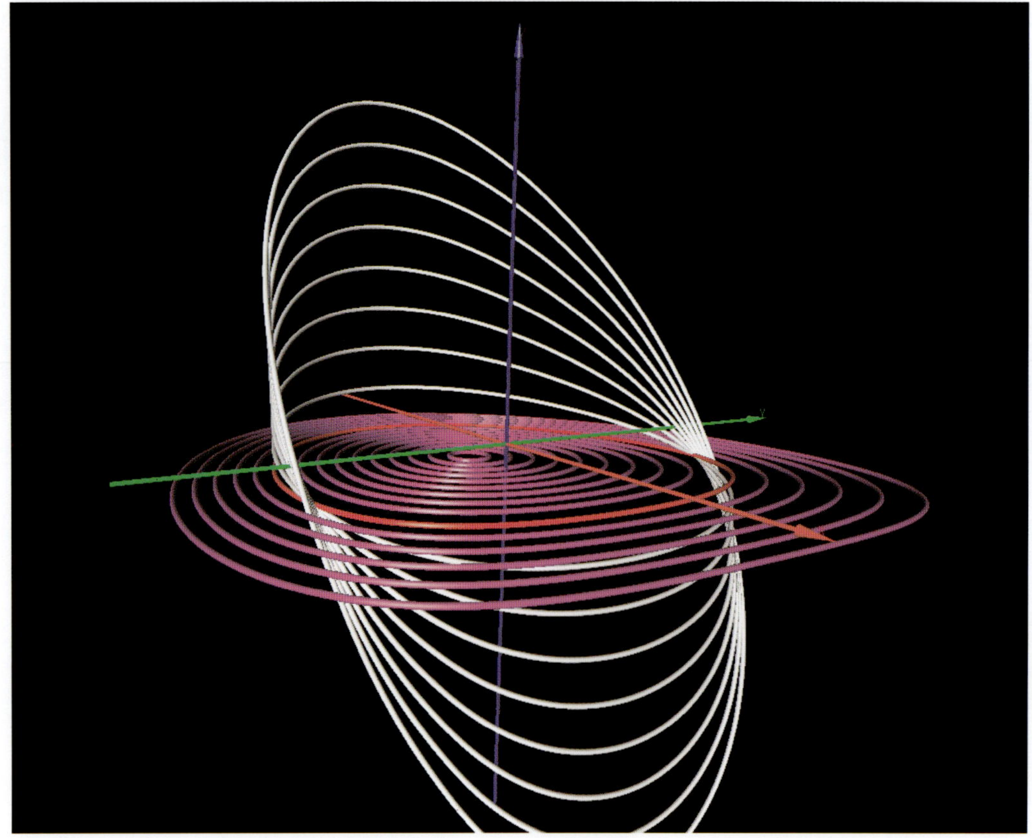

Fig. 9.   Some planar Lyapunov orbits, with BPC point, and some bifurcating orbits in the CR3BP.

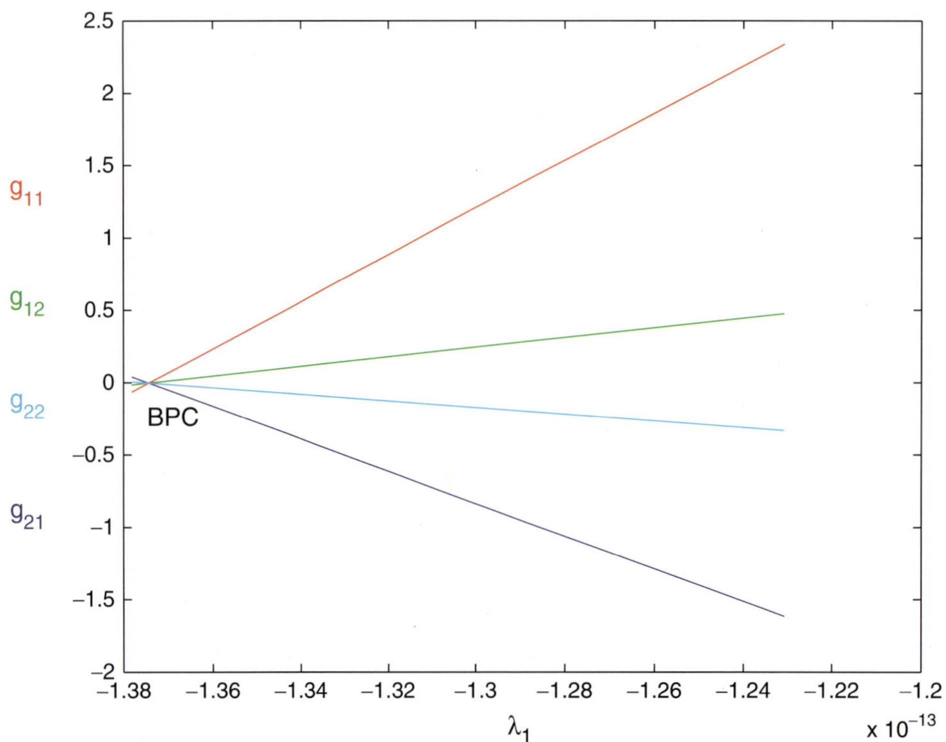

Fig. 10.   Evolution of the test functions for branch points in a BPC in the CR3BP example.

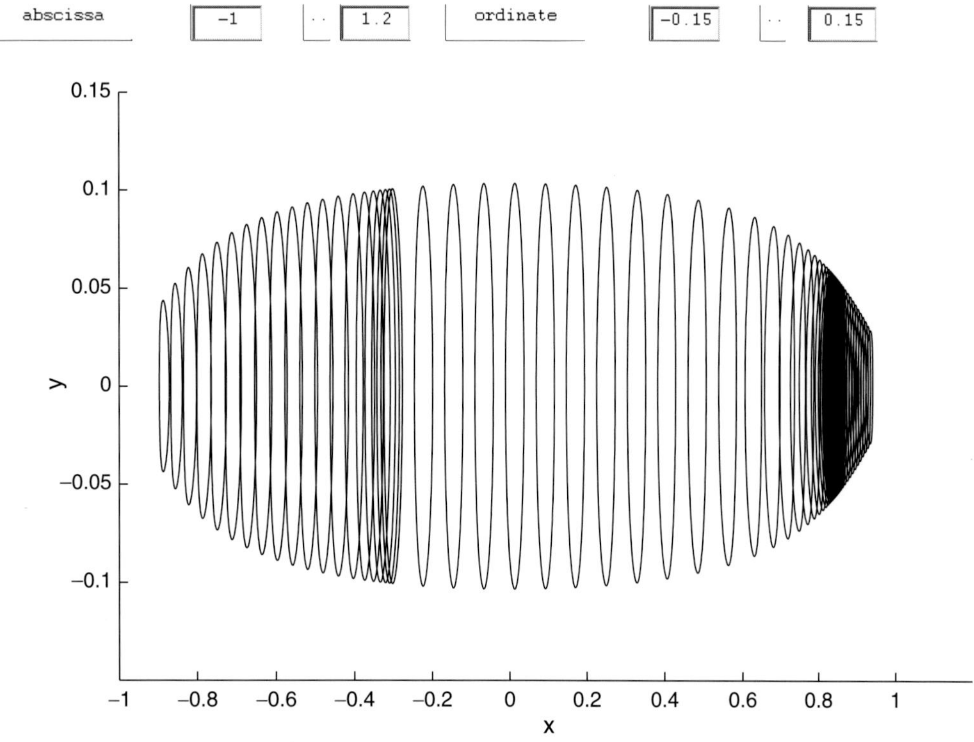

Fig. 11.   Family of BPC points in the CR3BP.

References

Allgower, E. L. & Georg, K. [1990] *Numerical Continuation Methods. An Introduction* (Springer-Verlag, NY).

Allgower, E. L. & Schwetlick, H. [1997] "A general view of minimally extended systems for simple bifurcation points," *Z. Angew. Math. Mech.* **77**, 83–97.

Ascher, U. M., Christiansen, J. & Russell, R. D. [1981] "Collocation software for boundary value ODEs," *ACM Trans. Math. Softw.* **7**, 209–222.

Ascher, U. M., Mattheij, R. M. M. & Russell, R. D. [1995] *Numerical Solution of Boundary Value Problems for Ordinary Differential Equations* (SIAM, Philadelphia).

Beyn, W.-J., Champneys, A., Doedel, E. J., Govaerts, W., Sandstede, B. & Kuznetov, Yu. A. [2002] "Numerical continuation and computation of normal forms," *Handbook of Dynamical Systems*, Vol. 2, ed. Fiedler, B. (Elsevier Science), pp. 149–219.

Borisyuk, R. M. [1981] *Stationary Solutions of a System of Ordinary Differential Equations Depending upon a Parameter*, FORTRAN Software Series, Vol. 6 (Research Computing Centre, USSR Academy of Sciences, Pushchino) [in Russian].

De Boor, C. & Swartz, B. [1973] "Collocation at Gaussian points," *SIAM J. Numer. Anal.* **10**, 582–606.

Dhooge, A., Govaerts, W., Kuznetsov, Yu. A., Mestrom, W. & Riet, A. M. [2003] "CL_MATCONT: A continuation toolbox in MATLAB," *Proc. 2003 ACM Symp. Applied Computing* (Melbourne, Florida, March 2003), pp. 161–166.

Dhooge, A., Govaerts, W. & Kuznetsov, Yu. A. [2003] "MATCONT: A MATLAB package for numerical bifurcation analysis of ODEs," *ACM Trans. Math. Softw.* **29**, 141–164.

Doedel, E. J. & Heinemann, R. F. [1983] "Numerical computation of periodic solution branches and oscillatory dynamics of the stirred tank reactor with A → B → C reactions," *Chem. Engin. Sci.* **38**, 1493–1499.

Doedel, E. J., Keller, H. B. & Kernévez, J. P. [1991] "Numerical analysis and control of bifurcation problems: (II)," *Int. J. Bifurcation and Chaos* **1**, 745–772.

Doedel, E. J., Champneys, A. R., Fairgrieve, T. F., Kuznetsov, Yu. A., Sandstede, B. & Wang, X. J. [2001] AUTO97-AUTO2000: "Continuation and Bifurcation Software for Ordinary Differential Equations (with HomCont)," *User's Guide*, Concordia University, Montreal, Canada, http://cmvl.cs.concordia.ca.

Doedel, E. J., Dichmann, D. J., Galán-Vioque, J., Keller, H. B., Paffenroth, R. C. & Vanderbauwhede, A. [2003a] "Elemental periodic orbits of the CR3BP: A brief selection of computational results," in *Proc. Equadiff Conf.*, Maastricht, to appear.

Doedel, E. J., Govaerts, W. & Kuznetsov, Yu. A. [2003b] "Computation of periodic solution bifurcations in ODEs using bordered systems," *SIAM J. Numer. Anal.* **41**, 401–435.

Doedel, E. J., Paffenroth, R. C., Keller, H. B., Dichmann, D. J., Galán, J. & Vanderbauwhede, A. [2003c] "Continuation of periodic solutions in conservative systems with application to the 3-Body problem," *Int. J. Bifurcation and Chaos* **13**, 1–29.

Freire, E., Rodríguez-Luis, A., Gamero, E. & Ponce, E. [1993] "A case study for homoclinic chaos in an autonomous electronic circuit: A trip from Takens–Bogdanov to Hopf–Shilnikov," *Physica* **D62**, 230–253.

Golubitsky, M. & Schaeffer, D. G. [1985] *Singularities and Groups in Bifurcation Theory*, Vol. I (Springer Verlag, NY).

Golubitsky, M., Stewart, I. & Schaeffer, D. G. [1988] *Singularities and Groups in Bifurcation Theory*, Vol. II (Springer Verlag, NY).

Govaerts, W. [2000] *Numerical Methods for Bifurcations of Dynamical Equilibria* (SIAM, Philadelphia).

Griewank, A. & Reddien, G. W. [1984] "Characterization and computation of generalized turning points," *SIAM J. Numer. Anal.* **21**, 176–185.

Keller, H. B. [1977] "Numerical solution of bifurcation and nonlinear eigenvalue problems," in *Applications of Bifurcation Theory*, ed. Rabinowitz, P. H. (Academic Press), pp. 359–384.

Küpper, T., Mittelmann, H. D. & Weber, H. (eds.) [1984] *Numerical Methods for Bifurcation Problems*, ISNM Vol. 70 (Birkhäuser Verlag, Boston).

Kuznetsov, Yu. A. & Levitin, V. V. [1997] "CONTENT: Integrated environment for analysis of dynamical systems," CWI, Amsterdam: ftp://ftp.cwi.nl/pub/CONTENT

Kuznetsov, Yu. A. [1998] *Elements of Applied Bifurcation Theory*, 2nd Ed. (Springer-Verlag, NY).

Mei, Z. [1989] "A numerical approximation for the simple bifurcation problems," *Numer. Funct. Anal. Optim.* **10**, 383–400.

Mei, Z. [2000] *Numerical Bifurcation Analysis for Reaction-Diffusion Equations* (Springer-Verlag, Berlin).

Mittelmann, H. D. & Weber, H. (eds.) [1980] *Bifurcation Problems and their Numerical Solutions* (Birkhäuser Verlag, Boston).

Moore, G. [1980] "The numerical treatment of nontrivial bifurcation points," *Numer. Funct. Anal. Optim.* **2**, 441–472.

Muñoz-Almaraz, J., Freire, E., Galán, J., Doedel, E. J. & Vanderbauwhede, A. [2003] "Continuation of periodic orbits in conservative and Hamiltonian systems," *Physica* **D181**, 1–38.

COARSE-GRAINED OBSERVATION OF DISCRETIZED MAPS

GÁBOR DOMOKOS

Department of Mechanics, Materials and Structures and
Center for Applied Mathematics and Computational Physics,
Budapest University of Technology and Economics, H-1521 Budapest, Hungary

Received April 30, 2004; Revised June 15, 2004

We investigate why discretized versions f_N of one-dimensional ergodic maps $f : I \to I$ behave in many ways similarly to their continuous counterparts. We propose to register observations of the $N \times N$ discretization f_N on a coarse $M \times M$ grid, with $N = cM$, c being an integer. We prove that rounding errors behave like uniformly distributed random variables, and by assuming their independence, the $M \times M$ incidence matrix A^M associated with the continuous map (indicating which of the M equal subintervals is mapped onto which) can be expected to be identical to the incidence matrix $B^{N,M}$ associated with the aforementioned coarse grid, if $c \geq \sqrt{\deg(f)N}$, where $\deg(f)$ denotes the degree of f. We show how coarse-grained registration can be used as a "digital" definition of an unstable orbit and how this can be applied in real computations. Combination of these results with ideas from the random map model suggests an intuitive explanation for the statistical similarity between f and f_N. Our approach is not a rigorous one, however, we hope that the results will be useful for the computational community and may facilitate a rigourous mathematical description.

Keywords: Coarse-grained model; discretization; chaos.

1. Introduction

Computer simulations of ergodic maps $f : I \to I$ necessarily distort the map due to roundoff errors. These simulations have many surprising qualities.

For physicists and engineers it may appear as self-evident that by applying sufficiently high arithmetic precision, the aforementioned distortion becomes negligible. This is, however, not always the case. Regard, for example, the "diadic" doubling map $f : x \to 2x \bmod 1$ which, when interpreted as an iteration, exhibits for any typical (irrational) x_0 random-like, ergodic behavior. Similar to many other mixing and expanding maps, the diadic map posseses a unique, absolutely continuous, invariant density function $\rho(x)$, and for this map ρ happens to be uniform, so the statistical distribution of the iterated points is uniform on I. By applying any

finite arithmetic which subdivides the unit interval into $N = 2^k$ subintervals, the iteration ends after a maximum of k steps at the fixed-point at $x = 0$. (Since computers do use discretizations of this type, the reader is encouraged to test this statement by writing a 3-line code and run it.) By changing N to $N = 2^k + 1$ we will see radically different dynamical behavior, with several, finite cycles. Each of these n cycles is associated with a discrete density $\rho_{N,i}, i = 1, 2, \ldots n$, so the discrete map typically possesses *several* densities as opposed to the single one associated with the continuous map. Depending on the choice of initial condition the statistics converges to one of the densities $\rho_{N,i}, i = 1, 2, \ldots n$. Although the diadic map is rather special, it is true for any ergodic map f that the exact dynamical behavior of the discretization f_N (i.e. the number and length of the cycles) depends sensitively on N. One central computational problem is how to restore on the basis of f_N the original density ρ. Apparently, the first person to think about this problem was Stanislaw Ulam, who proposed in his book [Ulam, 1960] an averaging scheme for the computation of invariant densities. Later on, several authors used the idea of *random perturbations* to restore the invariant density ρ from the discrete map f_N, fundamental convergence results for expanding maps are due to [Kifer, 1997; Liverani, 2001] and [Keller, 1982], see also [Benettin *et al.*, 1978; Blank, 1988; Gora & Boyarski, 1988]. In [Domokos & Szász, 2003] the authors slightly improve previous results to identify the *minimally necessary* random perturbation; also they show that the mentioned "minimally perturbed" scheme is equivalent to a special version of Ulam's original scheme. So, the "engineering puzzle" can be regarded as settled as far as the reconstruction of ρ is concerned, at least in case of expanding maps. However, discretized ergodic maps can surprise not only engineers or physicists, but mathematicians as well.

For the latter, the radical difference between a piecewise continuous map and a discrete one is almost evident. What may come as a surprise is that discrete maps can reproduce several relevant properties of their continuous counterparts in a relatively robust manner. For example, by applying double-precision arithmetic (i.e. $N \approx 2^{53}$) physicists are able to compute unstable cycles of the continuous map with reasonable precision (cf. [Grebogi & Yorke, 1988; Cvitanovic *et al.*, 1999]), and the invariant densities $\rho_{N,i}$ associated with the discrete cycles tend to approximate fairly accurately the unique invariant density ρ of the continuous map [Lanford, 1998]. These quantitative agreements may appear as rather baffling from the mathematical point of view and, as Lanford points out [Lanford, 1998], there is no simple explanation at hand. One possible approach to predict the qualitative behavior (i.e. the length and number of cycles) of discrete maps f_N for very large N is to investigate *random maps*, where in each iterative step a different map is chosen randomly from the set of all possible discrete maps. In several respects, this approach is equivalent to the drawing of numbered balls from an urn with replacement, until the same number appears the second time. The number of draws between the first and second appearances of this number can be regarded as an approximation of the cycle length. The number of draws before the first appearance may be regarded as the length of the *transient* leading into a cycle. According to [Lanford, 1998], this *random map* approach has been first proposed by D. Ruelle in order to explain some features of numerical experiments performed in [Levy, 1982]. This model has been developed further in [Grebogi & Yorke, 1988]. More recent results on the random map model are summarized in [Lanford, 1998] and [Kloeden *et al.*, 1996]. Independently, the same idea appears in [Domokos, 1990]. The random map model predicts cycles of length $\approx \sqrt{N}$ and transients of the same length [Domokos, 1990; Lanford, 1998]. The number of cycles is predicted to grow approximately as $\log N$.

Both previously described approaches introduce randomness into the discrete map (either by adding random perturbation or by regarding the whole system as random) in order to give meaningful predictions for large N. In this paper we take a different approach and regard directly the discrete map. Our goal is also to describe asymptotic behavior for large N, in particular, we would like to demonstrate that if we regard the discrete map on a coarse-grained level (i.e. on a $M \times M$ grid, with $M \ll N$), then the discrete map does approximate the continuous one. We will show that the *incidence matrices* (indicating which subinterval is mapped onto which one), associated with the coarse-grained system and the continuous map can be expected to be identical if $M < \sqrt{N}$. In Sec. 2 we give the basic definitions, Sec. 3 proves the main result. In Sec. 4 we outline applications and in Sec. 5 we summarize our results and survey other ways to investigate the discrete to continuous convergence for ergodic maps. Our approach is not a rigorous one, rather, we hope that as a plausible argument it will be useful for the computational community and it also facilitates a rigorous mathematical description.

2.   Definitions and Assumptions

2.1.   *Definitions and assumptions about the maps f and f_N*

We investigate maps of the unit interval $f : I \to I$ with independent variable $x \in I$. We will assume that $f(x)$ is piecewise continuous so its derivative $f'(x)$ exists for almost all values of x. The map f defines the iterated sequence

$$x_{i+1} = f(x_i), \quad x_i \in [0, 1). \tag{1}$$

The discretized map f_N will be interpreted on the $N \times N$ quadratic lattice, where N is the integer describing the arithmetic precision of the computer in the independent variable x (the smallest value of x different from zero is $1/N$). We define the discretized map as

$$f_N\left(\frac{i}{N}\right) = \frac{1}{N}\left[Nf\left(\frac{i}{N}\right)\right], \quad i = 0, 1, 2, \ldots, N, \tag{2}$$

where $[\ ]$ denotes the integer part of a real number. The discrete map f_N defines an iterated sequence X_i:

$$X_{i+1} = f_N(X_i),$$
$$X_i \in \{0, 1/N, 2/N, \ldots, (N-1)/N\}. \tag{3}$$

For convenience, the equivalent, "inflated" sequence NX_i of integers can be used as well.

Now we can proceed to define the incidence matrix A^N associated with the continuous map $f(x)$. We denote the jth subinterval $[(j-1)/N, j/N)$ by I_j^N, and using this notation the incidence matrix is defined as

$$A_{i,j}^N = \begin{cases} 0 & \text{if } f(I_i^N) \cap I_j^N = \emptyset \\ 1 & \text{otherwise.} \end{cases} \tag{4}$$

The incidence matrix $B^{N,N}$ associated with the discrete map $f_N(i/N)$ is easy to define as

$$B_{i,j}^{N,N} = \begin{cases} 0 & \text{if } f_N\left(\dfrac{i}{N}\right) \neq \dfrac{j}{N} \\ 1 & \text{if } f_N\left(\dfrac{i}{N}\right) = \dfrac{j}{N}. \end{cases} \tag{5}$$

2.2. Definitions for the coarse-grained map $f_{M,N}$

The basic idea behind the coarse-grained approach is that we deliberately ignore a certain amount of information contained in the discrete trajectories of f_N by *registering* the iterated values on a coarse $M \times M$ mesh with

$$N = cM, \quad c \in Z^+, \tag{6}$$

so that instead of the original iterated sequence X_i we regard the modified sequence

$$\overline{X}_i = [MX_i]/M,$$
$$\overline{X}_i \in \{0, 1/M, 2/M, \ldots, (M-1)/M\}, \tag{7}$$

where $[\ ]$ denotes the integer part of a real number. Here again, the equivalent, "inflated" sequence MX_i of integers may be more convenient to use.

In an analogous manner, one can also look for the coarse-grained $\overline{x}_i$ version of the x_i sequence (1) associated with the continuous map:

$$\overline{x}_i = [Mx_i]/M,$$
$$\overline{x}_i \in \{0, 1/M, 2/M, \ldots, (M-1)/M\}, \tag{8}$$

or at the inflated version $M\overline{x}_i$.

From now on the N-mesh will refer to the original, $N \times N$ mesh, the M-mesh to the coarse one. Although it would be misleading to speak about a coarse-grained *map*, we can readily define a coarse-grained incidence matrix $B^{M,N}$ by using the subintervals $I_j^M = [(j-1)/M, j/M)$ in the following way:

$$B_{i,j}^{M,N} = \begin{cases} 1 & \text{if } \exists k \text{ such that } \dfrac{k}{N} \in I_i^M \\ & \quad \text{and } f_N\left(\dfrac{k}{N}\right) \in I_j^M \\ 0 & \text{otherwise.} \end{cases} \tag{9}$$

(Observe that for $N = M$ the definitions (9) and (5) coincide.) We also define *rounding errors* on the coarse M-mesh as

$$\delta(X, M) = f(X) - f_M(X),$$
$$X \in \{0, 1/M, 2/M, \ldots, M - 1/M\}, \tag{10}$$

and their inflated versions

$$\Delta(X, M) = M\delta(X, M). \tag{11}$$

Now we will present a plausible argument showing that in the limit as $M \to \infty$ the $\Delta(X_1, M)$, $\Delta(X_2, M)$ values behave like uniformly distributed random variables on I which have arbitrarily small correlation if the inflated distance

$$D = M|X_1 - X_2| \tag{12}$$

is sufficiently large. We can regard $\Delta(X, M)$ as a function of M for fixed X, and we achieve the $M \to \infty$ limit by taking a series

$$M_i = k^i M_0. \tag{13}$$

It is easy to see that $\Delta(X, M)$, which measures the rounding errors relative to the meshsize, will grow by the same factor as the number of meshpoints, of course, modulo 1. This intuition is confirmed, as we can construct a *map* $\Delta(X, M_j) \to \Delta(X, M_{j+1})$, based on (2), (10), (11) and (13):

$$\begin{aligned}
\Delta(X&, M_{j+1}) \\
&= M_{j+1}\delta(X, M_{j+1}) \\
&= M_{j+1}(f(X) - [M_{j+1}f(X)]/M_{j+1}) \\
&= M_{j+1}(f(X) - [M_j f(X)]/M_j \\
&\quad - [M_{j+1}f(X)]/M_{j+1} + [M_j f(X)]/M_j) \\
&= M_{j+1}(\delta(X, M_j) - ([M_{j+1}f(X)] \\
&\quad - k[M_j f(X)])/M_{j+1}) \\
&= M_{j+1}(\delta(X, M_j) - [M_{j+1}f(X) \\
&\quad - k[M_j f(X)]]/M_{j+1}) \\
&= M_{j+1}(\delta(X, M_j) - [M_{j+1}\delta(X, M_j)]) \\
&= k\Delta(X, M_j) - [k\Delta(X, M_j)] \\
&= k\Delta(X, M_j) \bmod 1.
\end{aligned} \tag{14}$$

Maps of type (14) ("k-adic" maps, with k being integer) are known for being ergodic and having an absolutely continuous invariant measure which is uniform [Rényi, 1957]. For typical (irrational) initial values $\Delta(M_0)$ the iterated series (14) will produce random-like numbers, uniformly distributed on I, and the correlation between two trajectories decays exponentially with the number of iterates.

Equation (14) describes the evolution of the Δ errors on the initial $M = M_0$ grid as M is multiplied by powers of k, so the errors $\Delta(i/M_0, M_j), \Delta((i+1)/M_0, M_j)$ behave like independent, uniformly distributed random variables for typical (irrational) initial errors $\Delta(i/M_0, M_0), \Delta((i+1)/M_0, M_0)$ and sufficiently high j. If we regard the inflated distance (12) associated with two adjacent points of the M_0-mesh we obtain $D = M_j|(i+1)/M_0 - i/M_0| = M_j/M_0 = k^j$, which becomes arbitrarily large after a sufficiently high number of iterates on M, and this is an important condition guaranteeing the independence of the random variables.

Although the above argument is interesting from the theoretical point of view, its practical implementation needs further explanation. Independence is certainly not true for $\Delta(i/M_j, M_j), \Delta((i+1)/M_j, M_j)$, however, in order

to achieve closed-form solutions, we will have to assume this as well. (As we will point out later, the assumption of independence shifts our estimates towards the "safe" side.) It is not true either that the initial error $\Delta(i/M_0, M_j)$ is typically an irrational number. If f is a polynomial with rational coefficients then it will assume rational values at the meshpoints. Also, even if f is not rational, the computer simulation will truncate it to rational values. Nevertheless, we expect that rational trajectories will behave similarly to irrational ones, in the sense that the coarse-grained statistics will be similar to the discretized invariant measure. This expectation is based on the arguments in Sec. 4.3, which, in turn, rely on the main body of the paper. This seems to indicate a logical trap; however, this is not the case. We use "k-adic" maps solely to show the uniformness of the rounding errors of "typical" maps; meanwhile "k-adic" maps themselves *have identically zero rounding errors*, so there is no vicious circle in the argument.

We will further assume that M is large enough so that the derivative $f'(x)$ can be regarded as a constant on any subinterval I_i^M. Formally, this can be expressed as

$$f \approx a_i x + b_i \quad \text{if } x \in I_i^M, \tag{15}$$

where $a_i = f'(x_0), b_i = f(x_0) - a_i(x_0), x_0 = (i-1)/M$. Our next goal is to investigate (based on the previous definitions and assumptions) under which conditions $A^M = B^{M,N}$. The equivalence of the incidence matrices would signal a qualitative agreement between the discrete and the continous maps, although it does not provide immediately quantitative agreement between the invariant densities or the individual trajectories.

3. The Main Result: The Conditions for $A^M = B^{M,N}$

Assume that the interval I_i^M is mapped by f onto the intervals $I_j^M, I_{j+1}^M, \ldots, I_{j+k}^M$. The linearity assumption (15) implies that the preimages $S_{i,0}, S_{i,1}, \ldots, S_{i,k} \in I_i^M$ of $I_j^M, I_{j+1}^M, \ldots, I_{j+k}^M$ will be subintervals of equal length $1/(a_i M)$, except for the first and last ($S_{i,0}$ and $S_{i,k}$), which will be typically shorter [cf. Fig. 1(a)].

We can guarantee $A^M = B^{M,N}$ if each of these preimages contains at least one N-meshpoint. The first one ($S_{i,0}$) always does, since it contains the

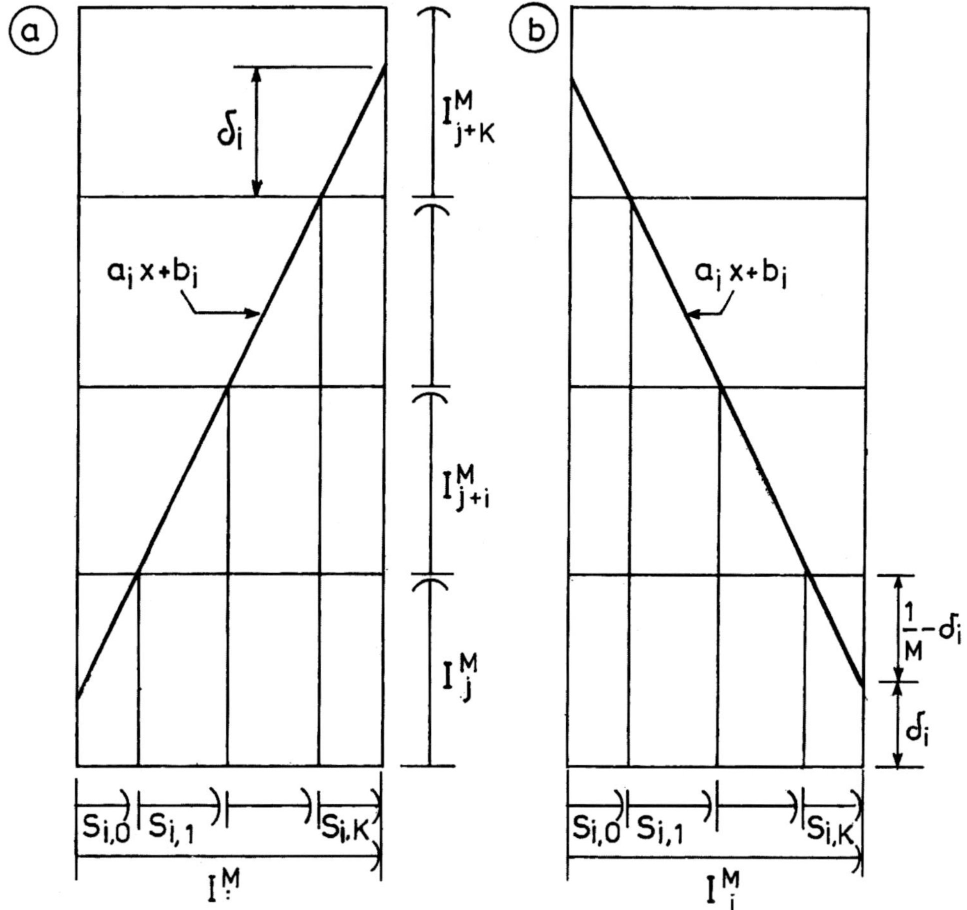

Fig. 1. Interpretation of the rounding error δ_i, the interval I_i^M, its images in the intervals $I_j^M, I_{j+1}^M, \ldots, I_{j+k}^M$ and the preimages $S_{i,0}, S_{i,1}, \ldots, S_{i,k} \in I_i^M$ of the latter in case of (a) positive and (b) negative slope.

point $x = (i-1)/M = c(i-1)/N$. If we can guarantee that

$$|S_{i,k}| > 1/N, \qquad (16)$$

then this implies $|S_{i,1}| = |S_{i,2}| = \cdots = |S_{i,k-1}| > 1/N$, so each preimage will contain at least one N-meshpoint. The length $|S_{i,k}|$ can be expressed via the rounding error δ_i (cf. Fig. 1):

$$|S_{i,k}| = \begin{cases} \delta_i/a_i & \text{if } a_i > 0 \\ (\delta_i - 1/M)/a_i & \text{if } a_i < 0. \end{cases} \qquad (17)$$

Substituting (16) into (17) and using (11) yields the following conditions for the inflated rounding error Δ_i:

$$\begin{aligned} \Delta_i &> a_i/c & \text{if } a_i > 0 \\ \Delta_i &< 1 + a_i/c & \text{if } a_i < 0. \end{aligned} \qquad (18)$$

Since we proved that the errors Δ_i behave like uniformly distributed random variables, the

probability P_i associated with the event (16) can be readily obtained via (18) as

$$P_i = P\{|S_{i,k}| > 1/N\} = \left(1 - \frac{|a_i|}{c}\right) \qquad (19)$$

Since $c \to \infty$ implies $P_i \to 1$, we are looking for the *minimal* value of c for which $A^M = B^{M,N}$ can be *expected*. We will denote this minimal value of c by the random variable ξ with distribution

$$P(\xi < c) = \prod_{i=0}^{M-1} P_i. \qquad (20)$$

Here we used the assumption that the roundoff values Δ_i are independent. Without this assumption we could not obtain a closed form solution; on the other hand, observe that the expected value $E(\xi)$ would decrease if we considered Δ_i to be correlated, so we are erring on the safe side. Since we are interested in the expected value $E(\xi)$, we will replace the individual values $|a_i| = |f'(i-1)/M|$ of the derivative by its average, which is identical to the degree D of f. So, using (19), (20), (6) we obtain

$$P(\xi < c) = \left(1 - \frac{D}{c}\right)^{\frac{N}{c}}. \qquad (21)$$

This distribution function can be differentiated to yield the density

$$\rho(c) = \left(1 - \frac{D}{c}\right)^{\frac{N}{c}}$$
$$\times \left(\frac{DN}{c^3\left(1 - \frac{D}{c}\right)} - \frac{N \log\left(1 - \frac{D}{c}\right)}{c^2}\right). \qquad (22)$$

We are interested in the asymptotic behavior for $N, M \to \infty$ (implying $N/c \gg 1$). Using the identity [Korn & Korn, 1968]

$$\lim_{n \to \infty} \left(1 + \frac{a}{n}\right)^{bn} = e^{ab} \qquad (23)$$

yields

$$\lim_{N \to \infty} \left(1 - \frac{D}{c}\right)^{\frac{N}{c}} = e^{(-DN/c^2)}. \qquad (24)$$

Taking the first term in the $D/c \ll 1$ Taylor expansion of

$$\left(\frac{DN}{c^3\left(1 - \frac{D}{c}\right)} - \frac{N \log\left(1 - \frac{D}{c}\right)}{c^2}\right) \qquad (25)$$

yields $2DN/c^3$, so the density function can be identified in this limit as

$$\lim_{N,M \to \infty} \rho(c) = \overline{\rho}(c) = \frac{2DN}{c^3} e^{\frac{-DN}{c^2}}. \qquad (26)$$

The expected value $E(\xi)$ of $\overline{\rho}(c)$ can be computed by using the identity [Korn & Korn, 1968]

$$\int_0^\infty x^{2k} e^{-ax^2} dx$$
$$= (1 \times 3 \times 5 \times \cdots \times (2k-1)\sqrt{\pi})/(2^{k+1} a^{k+1/2}) \qquad (27)$$

yielding

$$E(\xi) = \sqrt{\pi DN}. \qquad (28)$$

This formula implies that if we use a coarse grid with $c = \sqrt{\pi DN}$ (i.e. $M \approx 0.56\sqrt{N/D}$) then we can expect $A^M = B^{M,N}$.

4. Application of the Coarse-Grained Model

4.1. *Verification of formula (28)*

4.1.1. *Numerical verification*

As we pointed out earlier, since the roundoff errors are correlated, we expect to see $A^M = B^{M,N}$ already for somewhat smaller c (larger M) than predicted by (28), or, equivalently, we expect to find smaller values for the constant k in numerical experiments on individual maps.

We carried out approximately 10^3 such experiments on maps $f(x) = ax(1 - x)$ and $f(x) = (ax + b) \bmod 1$ with different values for a and b and for discretizations with $10^3 < N < 10^5$. We assumed that the relationship (28) is qualitatively true, i.e. $E(\xi) = k\sqrt{DN}$ and in each experiment we identified the value of the constant k under the condition that $A^M = B^{M,N}$ (in fact, we just checked the condition that whenever $A_{i,j}^M = 1$, this implies $B_{i,j}^{M,N} = 1$). We found values in the range $0.4 < k < 2.5$ with a mean value of roughly $k \approx 1$. This numerical result certainly confirms not only the qualitative correctness of (28), but also shows fair quantitative agreement. We can also observe that as predicted, the mean of the measured k parameters was below the theoretical expected value $k \approx 1.7724$ which indicates the correlation of the rounding errors.

4.1.2. *Theoretical verification*

We can observe that the role of N and D is symmetric in (28). This is not a coincidence; in fact it confirms the validity of the formula. We proved that the rounding errors δ_i behave like uniformly distributed random variables. This implies that their preimages under f should behave similarly; the preimages can be regarded as "rounding errors" in the x direction. We found (28) under the condition that the preimages should be larger than $1/N$, which implied that the rounding error should be larger than D/N, and we considered $M = N/c$ such independent events. We could equally concentrate on the preimages themselves, however, the factor of D would not enter into the length of the interval. On the other hand, recall that D is the degree of the map $f(x)$, so if we derive our formula based on the horizontal rounding errors then we have to consider $MD = ND/c$ independent events, so D has been smuggled back, and because of the $N - D$ symmetry of (28) we arrive at the same result.

4.2. *Templates and conservation of information*

We can associate the *oriented graphs* $G^M, G^{M,N}$ with the incidence matrices $A^M, B^{M,N}$ in a natural manner: both G^M and $G^{M,N}$ have M vertices and if $A_{i,j}^M = 1$ ($B_{i,j}^{M,N} = 1$, respectively) then the vertices i and j are connected by edge oriented towards the jth vertex. At constant M, as we increase c (and thus increase N), gradually more and more edges are added to $G^{M,N}$ until it becomes "saturated", i.e. identical to G^M. Based on the previous results we expect this to happen at $N \approx M^2$. At this "saturated" stage $G^{M,N}$ can be regarded as a template carrying all the unstable cycles of the map $f(x)$. This, of course, does not imply that one can observe all these cycles in the coarse-grained model, since only a finite number of steps s will agree, we will call these steps "topologically correct". One can give a good estimate of s based on the amount of information carried by computation. We know that the actual discrete map is working on an N-mesh, so the amount of information contained in an initial value $x_0 = i/N$ is

$$\mathcal{I}_0 = \log_2(N). \tag{29}$$

This amount of information can be utilized in different ways as we iterate x_0 forward with f_N. One part of the information, which we could call *metric* information $\mathcal{I}_m$, determines how accurately the location of any iterated value is known: obviously

$$\mathcal{I}_m = \log_2(M). \tag{30}$$

(If we choose $M = N$ then $\mathcal{I}_m = \mathcal{I}_0$ and no further information is available.) The remaining part we may call *topological* information ($\mathcal{I}_t$) indicating which vertex of the graph we are visiting. This information is equivalent to the number of "topologically correct" steps:

$$\mathcal{I}_t = s. \tag{31}$$

Since we are not creating new information in the process of the iteration we have

$$\mathcal{I}_0 = \mathcal{I}_m + \mathcal{I}_t \tag{32}$$

which yields via (6), (29) and (30)

$$\mathcal{I}_t = \log_2(c), \tag{33}$$

implying, via (31),

$$s = \log_2(c). \tag{34}$$

In the next subsection we will explore the applications of (28) and (34).

4.3. *Statistical convergence to the continuous model*

As mentioned in the Introduction, numerical experience shows that despite its fundamentally different structure, the discretized map tends to reproduce fairly accurately the invariant density of the continuous one, as long as the discretization is sufficiently fine. It may not be easy even to formulate this statement in a rigorous way, let alone to prove it. However, the idea of coarse-grained lattices coupled with the random map model may help to understand intuitively why the statistical similarities are observed.

In the previous subsection we showed that we can expect $s = \log_2(c)$ "topologically correct" steps on the coarse lattice. By choosing $c \geq \sqrt{\pi D N}$ according to (28), we can expect $s \approx \log_2 N$ steps which will agree (on the coarse-grained level) with the continuous map. If we regard the coarse lattice as a statistical sampling mesh, then these steps will be "statistically correct" on that mesh, i.e. they will be in the same sampling box as the iterated values of the continuous map. After $s = \log_2 N$ steps the discrete simulation will trail off, its global fate will be determined by the underlying fine (N-mesh) discretization. Cycle length will agree on the fine-grained (N-mesh) and coarse-grained (M-mesh) level. Although we have no direct evidence on the cycle length of the N-mesh model, the random map approach [Domokos, 1990; Lanford, 1998] suggests that for high N the cycles can be expected to be of length $L \approx \sqrt{N}$. As long as the cycle is not finished, the coarse-grained iteration can be regarded as a series of subsequent, statistically correct segments of length $s = \log_2 N$, and the random map predicts that we will have approximately $\sqrt{N}/\log_2 N$ such segments. For high N, this is more than sufficient to produce reasonably good statistical data on the $M \approx \sqrt{N}$-mesh. (In fact, since in the random map model the transients leading into the cycles have the same expected length as the cycles themselves, we can expect even more "statistically correct" segments.)

Although the above considerations are far from rigorous, they may help to develop more rigorous arguments.

4.4. *Numerical example*

Equation (28) and the measured average value $k \approx 1$ suggest the following strategies for numerical applications. If we define x as a double-precision

(52 bits) variable and carry out the iteration (3) on this arithmetic precision ($N = 2^{52}$), however, *register* the iterated values only at single precision (26 bits), regarding the latter as the coarse-grained sequence (7) ($M = 2^{26}$) then we have $M = c = \sqrt{N}$, and, assuming $D = 2$, this yields $k = 0.707$. With this arrangement we can expect the incidence matrices of the coarse lattice and the continuous map to be approximately equal, so we can explore the structure of $A^{2^{26}}$.

Based on (34) we can expect the number of correct steps s to be approximately 26, so we can hope to find cycles of length up to approximately 13. However, one could make more specific claims as well, since Eq. (34) could serve as a "digital definition" for unstable cycles. If we regard the number of correct steps s as a function of c and as we are varying c over a certain range, $s(c)$ follows the rule given in (34), then we can be more confident to have identified an unstable cycle digitally. Of course, the number of "correct" steps is

difficult to measure if the steps are not known in advance. However, assume that the cycle length L is small compared to c. As we increase c, we will observe the *identical repetition* of the same periodic pattern of integers $M\overline{X}_i \in \{0, 1, 2, \ldots, (M-1)/M\}$, (cf. (7)) on the coarse M-lattice, and each time the discrete pattern is repeated we can say that the number of correct steps has increased by L. The number of registered repeating patterns (and thus our confidence in having discovered a cycle) grows with c, our information about the *location* of the cycle decreases simultaneously. This inverse relationship is expressed formally in (6) and in its logarithmic version (32).

As we mentioned in the Introduction, the diadic map $f(x) = 2x \bmod 1$ shows extremely negative properties for $N = 2^k$ discretizations: all trajectories end after maximum k steps at $x = 0$. At first sight the discrete and the continuous maps have little in common. However, this example is an almost trivial illustration of our method: it is well-known

Table 1. $N = 2^{21}, M = 2^2$ computation of the unstable cycle $\{1/9, 2/9, 4/9, 8/9, 7/9, 5/9\}$ in the diadic map $f(x) = 2x \bmod 1$ for 21 steps. First column: serial number i of step. Second column: x_i, the iteration on the continuous map. Third column: X_i, iteration on the N-lattice. Fourth column: $|x_i - X_i|$, difference between the continuous map and the N-discretization. Fifth column: NX_i, "N-inflated" value of X_i. Sixth column: $M\overline{X}_i$, "M-inflated" value of the coarse-grained iteration. Seventh column: $M\overline{x}_i$, "M-inflated" value of the continuous iteration.

i	x_i	X_i	$|x_i - X_i|$	NX_i	$M\overline{X}_i$	$M\overline{x}_i$
0	0.1111111	0.1111106	0.0000004239	233016	0	0
1	0.2222222	0.2222213	0.0000008477	466032	0	0
2	0.4444444	0.4444427	0.0000016954	932064	1	1
3	0.8888888	0.8888855	0.0000033908	1864128	3	3
4	0.7777777	0.7777771	0.0000067817	1631104	3	3
5	0.5555555	0.5555419	0.0000135630	1165056	2	2
6	0.1111111	0.1110839	0.0000271260	232960	0	0
7	0.2222222	0.2221679	0.0000542530	465920	0	0
8	0.4444444	0.4443359	0.0001085000	931840	1	1
9	0.8888888	0.8886718	0.0002170100	1863680	3	3
10	0.7777777	0.7773437	0.0004340200	1630208	3	3
11	0.5555555	0.5546875	0.0008680500	1163264	2	2
12	0.1111111	0.1093750	0.0017361000	229376	0	0
13	0.2222222	0.2187500	0.0034722000	458752	0	0
14	0.4444444	0.4375000	0.0069444000	917504	1	1
15	0.8888888	0.8750000	0.0138888888	1835008	3	3
16	0.7777777	0.7500000	0.0277777777	1572864	3	3
17	0.5555555	0.5000000	0.0555555555	1048576	2	2
18	0.1111111	0.0000000	0.1111111111	0	0	0
19	0.2222222	0.0000000	0.2222222222	0	0	0
20	0.4444444	0.0000000	0.4444444444	0	0	1
21	0.8888888	0.0000000	0.8888888888	0	0	3

that the diadic map delivers the binary expansion of the initial value x_0 if one registers whether the iterated numbers are on the first or second continuous segment of the map. The meshpoints of an $N = 2^k$ discretization have exactly k nontrivial binary digits. In our terminology this corresponds exactly to taking $M = 2$, and (34) predicts that the initial value and the first $s = \log_2(c) = \log_2(N/M) = k-1$ binary digits of any trajectory will be computed correctly, this is the "M-inflated" series $M\overline{X}_i$, the latter defined in (7).

Below we illustrate a slightly less trivial case $M = 4$: we describe the computation of the unstable cycle $\{1/9, 2/9, 4/9, 8/9, 7/9, 5/9\}$ of length 6 in the diadic map. The "M-inflated" iterated series (8) for $M = 4$ is $\{0, 0, 1, 3, 3, 2\}$ for this cycle. We set $N = 2^{21}$ and $X_0 = [N/9]/N$ and computed the discrete sequences X_i and $\overline{X}_i$ for 21 steps. The results are summarized in Table 1. As can be observed, the "M-inflated" integer series $M\overline{x}_i$ and $M\overline{X}_i$ in the sixth and seventh columns, belonging to the continuous and the N-discretized map, respectively, agree up to 19 steps, which confirms the prediction in (34): $s = \log_2(c) = \log_2(N/M) = 19$.

5. Summary and Related Topics

In this paper we presented a "direct" approach to the main question why discretized maps resemble their continuous counterparts. We showed that in case of an N-discretization the coarse-grained model with an $M = N/c$-mesh is approaching the continuous one in the sense that the appropriate coincidence matrices become identical as we decrease c. We showed that for $c \approx \sqrt{\pi DN}$ one can expect the two matrices to have the same entries. This suggest to perform computations at double precision arithmetic but register the numbers only at single precision.

The principal benefit of the agreement between the coincidence matrices is that at that stage *all* unstable cycles of the continuous map will appear on the coarse grid temporarily. The number of correct steps will be $s \approx \log_2(N/M)$, so if we are looking for longer cycles we have to settle for less accurate information concerning their location. Although this sounds plausible, we believe that the quantitative relationship derived in this paper can be useful. Combining these results with those from the random map model [Domokos, 1990; Lanford, 1998] offered an intuitive explanation why

invariant measures can be often reliably reproduced in numerical experiments.

The relationship between discrete and continuous maps has been approached in various ways. The random map model [Lanford, 1998; Domokos, 1990] aims to predict the cycle and transient length for high N. The random perturbation methods pioneered by [Kifer, 1997; Liverani, 2001] aim to reconstruct the invariant measure associated with the continuous map based on the discrete map. (Domokos and Szász [2003] determined the minimal amount of necessary perturbation.) Although rather efficient in achieving their goals, neither of these approaches look *directly* at the discrete map (either they substitute it with a random process or they add a random process). Our approach in this paper was different since we regarded directly the discrete map, and tried to squeeze out information which is relevant when studying the continuous map. The first results consistent with this philosophy appeared in [Domokos, 1990], where the *quality* of the discrete model is defined as $Q \in [0, 1]$, and $Q = 1$ if the discrete model agrees with the continuous model in the following sense: an interval of random length and random location (both chosen uniformly) is visited with probability 1 by the discrete iterated sequence. In [Domokos, 1990] the formulae for this probability are derived, based on the incidence matrix of the discrete map. The plots showing $Q(N)$ for specific maps are interesting; they suggest that $Q \to 1$ as $N \to \infty$, but it may not be easy to prove this statement. Still another direct approach would be to study deterministic iterations on randomly generated lattices.

Acknowledgments

Several ideas in this paper originated in conversations with Tamás Tél, Mike Shub and Oscar Lanford, whom the author would like to thank. Tamás Tél also pointed out several useful formulas and helped to shape the paper. This work was supported by OTKA grant T046646 and the Bolyai Research Fellowship.

References

Bennettin, G. *et al.* [1978] "On the reliability of numerical studies of stochasticity," *Nouvo Cimento* **B44**, 183–195.

Blank, M. L. [1988] "Metric properties of epsilon-trajectories of dynamical systems with stochastic behaviour," *Ergod. Th. Dyn. Syst.* **8**, 365–378.

Cvitanovic, P. *et al.* [1999] *Classical and Quantum Chaos*, 1st edition, Niels Bohr Institute, Copenhagen, http://www.nbi.dk/ChaosBook/.

Domokos, G. [1990] "Digital modelling of chaotic motion," *Studia Sci. Math. Hung.* **25**, 323–341.

Domokos, G. & Szász, D. [2003] "Ulam's scheme revisited: Digital modeling of chaotic attractors via micro-perturbations," *Discr. Contin. Dyn. Syst.* **A4**, 859–876.

Gora, P. & Boyarsky, A. [1988] "Why computers like Lebesgue measure," *Comput. Math. Appl.* **16**, 321–329.

Keller, G. [1982] "Stochastic stability in some chaotic dynamical systems," *Monatshefte der Math.* **94**, 313–333.

Kifer, Yu. [1997] "Computations in dynamical systems via random perturbations," *Discr. Contin. Dyn. Syst.* **3**, 457–476.

Kloeden, P., Diamond, P., Klemm, A. & Pokrovski, A. [1996] "Basin of attraction of cycles of discretizations of dynamical systems with SRB invariant measures," *J. Stat. Phys.* **84**, 713–733.

Korn, G. A. & Korn, T. M. [1968] *Mathematical Handbook for Scientists and Engineers*, 2nd edition (McGraw-Hill Book Company, NY).

Lanford, O. E. [1998] "Informal remarks on the orbit structure of discrete approximations to chaotic maps," *Experim. Math.* **7**, 317–324.

Levy, Y. E. [1982] "Some remarks about computer studies of dynamical systems," *Phys. Lett.* **A88**, 1–3.

Liverani, C. [2001] "Rigorous numerical investigation of the statistical properties of piecewise expanding maps — a feasibility study," *Nonlinearity* **14**, 463–490.

Ott, E., Grebogi, C. & Yorke, J. A. [1988] "Roundoff-induced periodicity and the correlation dimension in chaotic attractors," *Phys. Rev.* **A38**, 3688–3692.

Rényi, A. [1957] "Representations of real numbers and their ergodic properties," *Acta Math. Akad. Sc. Hung.* **8**, 477–493.

Ulam, S. [1960] *Problems in Modern Mathematics* (Interscience Publishers).

MULTIPLE HELICAL PERVERSIONS OF FINITE, INTRISTICALLY CURVED RODS

G. DOMOKOS

Department of Mechanics, Materials and Structures and
Center for Applied Mathematics and Computational Physics,
Budapest University of Technology and Economics,
H-1521 Budapest, Hungary

T. J. HEALEY

Department of Theoretical and Applied Mechanics,
Cornell University, Ithaca, NY 14853-1503, USA

Received April 15, 2004; Revised June 15, 2004

We investigate mechanical spatial equilibria of slender elastic rods with intristic curvature. Our work is, to some extent, motivated by papers [Goriely & Tabor, 1998; Goriely & McMillen, 2002]. There such rods of infinite length were recently studied to quantify the behavior of botanical filaments. In particular, an adequate explanation for the existence of helical perversions (the transition between helical segments of opposite handedness) is provided in [Goriely & Tabor, 1998]. However, this theory fails to describe multiple perversions, which can be observed in Nature. In contrast we formulate a two-point boundary-value problem describing rods of *finite* length with initial curvature and clamped ends. We identify trivial solutions as straight configurations and also k-covered circles, rigorously establish the existence of local bifurcations, and then compute global solutions via the Parallel Hybrid Algorithm [Domokos & Szeberényi, 2004] to find spatially complex equilibria characterized by *multiple* perversions. Based on computational results and the White–Fuller theorem [White, 1969; Fuller, 1971; Calugareanu, 1961] we describe a heuristic global picture of the bifurcation diagram, which can serve as an explanation for the evolution of physically observable tendril shapes.

Keywords: Intristic curvature; rod theory; bifurcations; helical perversion; botanical tendrils.

1. Introduction

In this paper we examine spatially complex eqilibria of rods possessing intristic curvature. Such models have recently been employed to quantify the behavior of botanical filaments [Goriely & Tabor, 1998; Goriely & McMillen, 2002]. As observed already by Darwin [1888], the tendrils of climbing plants often assume configurations consisting of subsequent helices of opposite handedness (see Fig. 1 for Darwin's original drawing and Fig. 2 for an example in a tropical rainforest). The transition between the different helical segments is referred to as "perversion". Goriely and McMillen [2002]

give various examples of perversions occurring in long thin filamentary structures including umbilic chords. Following the ideas presented in [Goriely & Tabor, 1998], they offer a simple mechanical model predicting the occurrence of a single perversion: they study static equilibria of an infinitely long Kirchoff rod of constant intristic curvature. Regarded as an initial value problem, this approach permits the application of dynamical systems tools. In that context a perversion is represented by a heteroclinic orbit joining asymptotically two fixed points, the latter corresponding to helices with opposite handedness. While offering an adequate

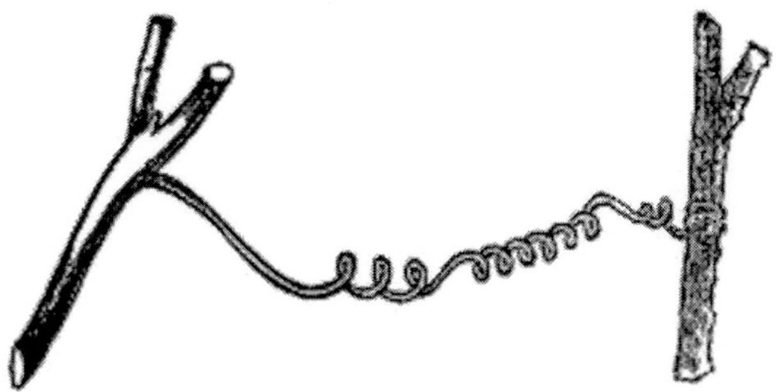

Fig. 1.   Multiple helical perversions. Drawing by Darwin [1888].

Fig. 2.   Multiple helical perversions. Daintree National Park, Queensland, Australia.

explanation for the existence of helical perversions, this theory fails to describe multiple perversions which can be observed in Nature (cf. Fig. 2) as depicted in Darwin's drawings (cf. Fig. 1). Also, due to the infinite length of the model, the genesis of helical perversions cannot be described by this model. (In [Goriely & Tabor, 1998] finite rods are also mentioned, however, just in the context of periodic solutions of the initial value problem, i.e. the boundary conditions are not specified.)

Like the approach in [Goriely & McMillen, 2002], we also look for spatial equilibria of slender, elastic rods with constant initial curvature. However, we treat boundary value problems for rods of *finite* length. Our formulation is motivated by the following hand-held experiment: Take any helical telephone cord and straighten a *finite segment* of the cord by holding it firmly with two hands and stretching it out. If the two ends are then slowly

displaced toward each other, a single perversion is born on the segment. Accordingly we consider a uniform rod with intrinsic curvature, initially occupying a straight configuration with clamped ends. The clamped ends also provide a good model for the relatively firm attachment of tendrils to their environment. We first provide a detailed local analysis of bifurcations from the straight state, as the applied end tension in the rod is relaxed. In particular, we demonstrate the birth of a single perversion that is locally stable, in accordance with the hand-held telephone cord, described above. We then combine numerical results with geometrical ideas based on the White–Fuller theorem (cf. [White, 1969]) to develop a global picture of the bifurcation diagram. In particular, we identify equilibria characterized by an *arbitrary* number of perversions with intermittent helical segments. These configurations are apparently similar to those observed on plants or on telephone cords.

In recent years, there has been considerable interest in the descriptions of spatially complex eqilibria of finite elastic rods serving as models in biology. In particular, the geometry of twisted rings has been studied as a model for DNA configurations. Global, symmetry-based analysis for the contact-free problem is carried out in [Domokos, 1995; Domokos & Healey, 2001], and the global problem with contact is solved in [Coleman *et al.*, 1995; Swigon *et al.*, 1998; Tobias *et al.*, 1994], see also [Li & Maddocks].

In the current problem of tendril perversion, the clamped–clamped boundary conditions admit an interesting family of trivial solutions, consisting of the straight rod and the series of k-covered circles. In Sec. 2 we describe the fundamental equations and analyze the straight, trivial solution. We prove that classical, planar Euler buckling modes are possible in compression. Subsequently, for rods with sufficiently high initial curvature, we identify spatial modes in tension. The first part of Sec. 3 describes our computational approach, the Parallel Simplex Algorithm and the results obtained with this method. In Sec. 3.2 we describe a heuristic, global picture of the bifurcation diagram which was confirmed both by analysis and computations. In particular, in Sec. 3.4 we apply the previous results to explain the existence and genesis of multiple tendril perversions. Our global picture is based on the application of the White–Fuller theorem [White, 1969]. In Sec. 4 we summarize our results and draws conclusions.

2. Analysis

2.1. *Formulation of the governing equations*

Let $\{\mathbf{e}_1, \mathbf{e}_2, \mathbf{e}_3\}$ denote a fixed, right-handed, orthonormal basis for $\mathbb{E}^3$. We consider a straight reference configuration parallel to $\mathbf{e}_3$. Let "s" denote the arclength coordinate (of the centerline) in the undeformed rod, and let $\mathbf{r}(s)$ denote the position vector (with respect to some fixed origin) of the material point originally at "s" in the reference configuration. We let $\mathbf{R}(s)$ denote the rotation of the cross-section spanned by $\{\mathbf{e}_1, \mathbf{e}_2\}$ at "s" in the undeformed rod. The first two unit vectors of the orthonormal field defined by

$$\mathbf{d}_i(s) = \mathbf{R}(s)\mathbf{e}_i, \quad i = 1, 2, 3, \tag{1}$$

are called *directors* in the special Cosserat theory, which we employ here. The deformed configuration of the rod is uniquely specified by the fields $\mathbf{r}(s)$ and $\mathbf{R}(s)$.

For simplicity, we consider only inextensible, unshearable rods, viz.

$$\mathbf{r}' \equiv \mathbf{d}_3. \tag{2}$$

Next, we differentiate (1) to get

$$\mathbf{d}_i' = \mathbf{R}'\mathbf{R}^T\mathbf{d}_i, \quad i = 1, 2, 3. \tag{3}$$

Since the tensor field

$$\mathbf{K} \equiv \mathbf{R}'\mathbf{R}^T \tag{4}$$

is skew-symmetric, there is a unique vector field κ such that

$$\mathbf{d}_i' = \kappa \times \mathbf{d}_i, \quad i = 1, 2, 3, \tag{5}$$

i.e. κ is the *axial vector* of $\mathbf{K}$. We then write

$$\kappa = \kappa_i \mathbf{d}_i, \tag{6}$$

where κ_1 and κ_2 are "bending curvatures" and κ_3 is the "twist".

We let $\mathbf{n}(s)$ and $\mathbf{m}(s)$ denote the internal contact force and internal contact couple, respectively, acting on the cross-section originally at "s" in the reference configuration. We write

$$\mathbf{n} = n_i \mathbf{d}_i, \quad \text{and} \quad \mathbf{m} = m_i \mathbf{d}_i. \tag{7}$$

Recall that the n_i and m_i, $i = 1, 2, 3$, are called forces and moments, respectively, cf. [Alexander & Antman, 1982]; n_1, n_2 are "shear forces", n_3 is the

"axial force", m_1, m_2 are "bending moments", and m_3 is the "torque" or "twisting moment". For a homogeneous *hyperelastic* rod, we assume the existence of a sufficiently smooth, scalar-valued *stored energy* function, $W(\kappa_1, \kappa_2, \kappa_3)$, such that

$$m_j = \frac{\partial W}{\partial \kappa_j}, \quad j = 1, 2, 3. \tag{8}$$

In accordance with the presumed curvature of the rod in a relaxed state, we assume that

$$W(\mu, 0, 0) = 0 \text{ is the global minimum of}$$
$$W(\kappa_1, \kappa_2, \kappa_3), \tag{9}$$

where $\mu \neq 0$ is the intrinsic curvature. In addition, we make the physically reasonable assumption that the Hessian matrix

$$D^2 W(\cdot) \text{ is positive definite on } \mathbb{R}^3. \tag{10}$$

We further assume that the straight rod admits two distinct transverse symmetries: A proper rotation of $180°$ about $\mathbf{e}_2$, and a reflection across the plane spanned by $\{\mathbf{e}_2, \mathbf{e}_3\}$. It is not hard to show (cf. [Healey, 2002, Sec. 7]) that these two operations induce the actions $(\kappa_1, \kappa_2, \kappa_3) \to (\kappa_1, -\kappa_2, \kappa_3)$ and $(\kappa_1, \kappa_2, \kappa_3) \to (\kappa_1, -\kappa_2, -\kappa_3)$, respectively. Accordingly, we require the stored energy function to satisfy

$$W(\kappa_1, -\kappa_2, \kappa_3) = W(\kappa_1, \kappa_2, \kappa_3), \tag{11}$$
$$W(\kappa_1, -\kappa_2, -\kappa_3) = W(\kappa_1, \kappa_2, \kappa_3). \tag{12}$$

Condition (11) implies that W is an even function of κ_2, and then (12), in turn, implies evenness of W in the argument κ_3 as well, viz.

$$W(\kappa_1, \kappa_2, \kappa_3) \equiv \Phi(\kappa_1 - \mu, \kappa_2^2, \kappa_3^2), \tag{13}$$

where Φ is some sufficiently smooth function on $\mathbb{R} \times [0, \infty) \times [0, \infty)$. From (8) we obtain

$$\begin{aligned}
m_1 &= D_1\Phi(\kappa_1 - \mu, \kappa_2^2, \kappa_3^2), \\
m_2 &= 2D_2\Phi(\kappa_1 - \mu, \kappa_2^2, \kappa_3^2)\kappa_2, \\
m_3 &= 2D_3\Phi(\kappa_1 - \mu, \kappa_2^2, \kappa_3^2)\kappa_3,
\end{aligned} \tag{14}$$

where "$D_i\Phi$" denotes the partial derviative of the function Φ with respect to its ith argument, $i = 1, 2, 3$.

The simplest example of a stored energy function fulfilling (9)–(15) is the Kirchhoff model

$$W(\kappa_1, \kappa_2, \kappa_3)$$

$$= \frac{1}{2}[A(\kappa_1 - \mu)^2 + B\kappa_2^2 + C\kappa_3^2], \qquad (15)$$

where $A, B, C > 0$ are the elastic moduli. Of course we can obtain (15) directly from (9), (10) and (13) via a truncated Taylor expansion of the latter about $(\kappa_1, \kappa_2, \kappa_3) = (\mu, 0, 0)$, where $A = D_1^2\Phi(0,0,0)$, $B = 2D_2\Phi(0,0,0), C = 2D_3\Phi(0,0,0)$.

In the absence of body forces and body couples, the well-known local forms of balance of forces and moments are

$$\mathbf{n}' = \mathbf{0}, \qquad (16)$$

$$\mathbf{m}' + \mathbf{d}_3 \times \mathbf{n} = \mathbf{0}, \qquad (17)$$

respectively. We impose "clamped" conditions at each end:

$$\mathbf{R}(0) = \mathbf{R}(1) = \mathbf{I}. \qquad (18)$$

In addition we fix the left end; we constrain the right end to move along $\mathbf{e}_3$ while prescribing the $\mathbf{e}_3$ component of the force:

$$\mathbf{r}(0) = \mathbf{0},$$

$$\mathbf{e}_\alpha \cdot \mathbf{r}(1) = 0, \quad \alpha = 1, 2, \qquad (19)$$

$$\mathbf{e}_3 \cdot \mathbf{n}(1) = \lambda. \qquad (20)$$

We now express the field equations in a convenient component form. Recalling (6) and (7), we define the triples

$$\underline{n} = (n_1, n_2, n_3), \quad \underline{m} = (m_1, m_2, m_3)$$
$$\text{and} \quad \underline{\kappa} = (\kappa_1, \kappa_2, \kappa_3), \qquad (21)$$

and we define a unique skew matrix $\underline{K}$ via

$$\underline{\kappa} \times \underline{a} = \underline{K}\underline{a} \quad \text{for all } \underline{a} \in \mathbb{R}^3. \qquad (22)$$

Using (5), we then express (17) and (17) with respect to the convected basis $\{\mathbf{d}_1, \mathbf{d}_2, \mathbf{d}_3\}$:

$$\underline{n}' + \underline{\kappa} \times \underline{n} = \underline{0}, \qquad (23)$$

$$\underline{m}' + \underline{\kappa} \times \underline{m} + (0, 0, 1) \times \underline{n} = \underline{0}. \qquad (24)$$

On the other hand, we write $\mathbf{r}$ and $\mathbf{R}$ with respect to the fixed basis:

$$\mathbf{r} = r_i\mathbf{e}_i, \bar{\mathbf{r}} \equiv (r_1, r_2, r_3), \qquad (25)$$

$$\mathbf{R} = R_{ij}\mathbf{e}_i \otimes \mathbf{e}_j, \quad \overline{\mathbf{R}} = \begin{bmatrix} R_{11} & R_{12} & R_{13} \\ R_{21} & R_{22} & R_{23} \\ R_{31} & R_{32} & R_{33} \end{bmatrix}. \qquad (26)$$

Then $(1), (2), (4), (25), (26)$ lead to

$$\bar{\mathbf{r}}' = \overline{\mathbf{R}}(0, 0, 1)^T = (R_{13}, R_{23}, R_{33}), \qquad (27)$$

$$\overline{\mathbf{R}}' = \overline{\mathbf{R}}\underline{K}. \qquad (28)$$

2.2.   *Planar configurations*

We consider the possibility of planar solutions in this section. We show that such solutions may occur only in the plane spanned by $\{\mathbf{e}_1, \mathbf{e}_2\}$. In particular, this is true in the special case (15) with $A = B$. To this end, it is convenient to first consider the description of deformation with respect to some other fixed orthonormal basis $\{\mathbf{a}_1, \mathbf{a}_2, \mathbf{a}_3\}$, where

$$\mathbf{a}_1 = \cos\psi\mathbf{e}_1 + \sin\psi\mathbf{e}_2,$$
$$\mathbf{a}_2 = -\sin\psi\mathbf{e}_1 + \cos\psi\mathbf{e}_2, \qquad (29)$$
$$\mathbf{a}_3 = \mathbf{e}_3,$$

with ψ being some fixed, but unspecified angle. Then, as in (1), we have

$$\mathbf{d}_\alpha^*(s) = \mathbf{R}(s)\mathbf{a}_\alpha, \quad \alpha = 1, 2,$$
$$\mathbf{d}_3^* = \mathbf{d}_3. \qquad (30)$$

It is not hard to show that

$$\mathbf{d}_1 = \cos\psi\mathbf{d}_1^* - \sin\psi\mathbf{d}_2^*,$$
$$\mathbf{d}_2 = \sin\psi\mathbf{d}_1^* + \cos\psi\mathbf{d}_2^*. \qquad (31)$$

Writing

$$\kappa = \kappa_1^*\mathbf{d}_1^* + \kappa_2^*\mathbf{d}_2^* + \kappa_3\mathbf{d}_3, \qquad (32)$$

we find that

$$\kappa_1 = \kappa_1^*\cos\psi - \kappa_2^*\sin\psi,$$
$$\kappa_2 = \kappa_1^*\sin\psi + \kappa_2^*\cos\psi. \qquad (33)$$

Next we seek planar solutions of the form

$$\mathbf{d}_2^* = \mathbf{a}_2,$$

$$\mathbf{d}_1^* = -\sin\theta\mathbf{e}_3 + \cos\theta\mathbf{a}_1, \tag{34}$$

$$\mathbf{d}_3^* = \mathbf{d}_3 = \cos\theta\mathbf{e}_3 + \sin\theta\mathbf{a}_1.$$

Then, as in (5), we have

$$\mathbf{d}_i^{*\prime} = \kappa \times \mathbf{d}_i^*, \quad i = 1, 2, 3, \tag{35}$$

and we readily find that

$$\kappa = \theta'\mathbf{d}_2^*, \tag{36}$$

i.e.

$$\kappa_1^* = \kappa_3 = 0, \kappa_2^* = \theta'. \tag{37}$$

Next we write $\mathbf{m} = m_i^*\mathbf{d}_i^*$ and compute m_1^*. Using (7), (14), (31), (33) and (37), we obtain

$$m_1^* = g(\theta', \psi)\cos\psi, \tag{38}$$

where

$$g(\theta', \psi)$$
$$\equiv D_1\Phi(-\theta'\sin\psi - \mu, (\theta')^2\cos^2\psi, 0)$$
$$+ 2\theta'\sin\psi D_2\Phi(-\theta'\sin\psi - \mu, (\theta')^2\cos^2\psi, 0). \tag{39}$$

A similar calculation shows that $m_3^* = m_3 = 0$. Hence, the dot product of (17) with $\mathbf{d}_3$, employing (34)–(36), reveals

$$\theta'm_1^* \equiv \theta'g(\theta', \psi)\cos\psi = 0, \tag{40}$$

i.e. either θ' or $g(\theta', \psi)$ or $\cos\psi$ vanish. If $\theta' \equiv 0$, then from (34), $\mathbf{d}_3$ is constant, and (2) and the boundary conditions (19) imply that $\mathbf{r}(s) = s\mathbf{e}_3$, i.e. the rod is in the (trivial) reference configuration. From (9), (10) and (13) and (39) we see that $g(\theta', \psi) \equiv 0$ *iff* $\psi = \pm\pi/2$ and $\theta' \equiv \mp\mu$, which is a special case of $\cos\psi = 0$. Without loss of generality, we choose $\psi = \pi/2$, in which case (30) yields $\mathbf{a}_1 = \mathbf{e}_2$ and $\mathbf{a}_2 = -\mathbf{e}_1$. From (2), (19) and (34), we then conclude:

Any planar nontrivial solution of (14), (18)–(28) *is characterized by* $\mathbf{r}(s) \in \mathrm{span}\{\mathbf{e}_2, \mathbf{e}_3\}$ *and* $\kappa(s)$ *"is an element of"* $\mathrm{span}\{\mathbf{e}_1\}$ *for all* $s \in [0, 1]$.

2.3. *Linearized problem*

In this section we obtain nontrivial solutions of the linearized problem about the straight state. Observe that for any value of the loading, $\lambda \in \mathbb{R}$, specified in (21), the straight configuration

$$\bar{\mathbf{r}} \equiv (0, 0, s),$$

$$\overline{\mathrm{R}} = I,$$

$$\underline{\kappa} \equiv \underline{0}, \tag{41}$$

$$\underline{\mathbf{m}} \equiv (\tau(\mu), 0, 0),$$

$$\underline{\mathbf{n}} \equiv (0, 0, \lambda),$$

satisfies the field equations, (14), (18)–(28), i.e. (41) characterizes the *trivial line of solutions*. Here "I" denotes the 3×3 identity matrix, and

$$\tau(\mu) \equiv D_1\Phi(-\mu, 0, 0) \tag{42}$$

is the "residual" couple maintained by the rod in the straight state (supported by the clamped ends).

In order to investigate bifurcation from the trivial line, we first obtain a consistent linearization of (14), (23), (24), (27) and (28) as follows. Set

$$\underline{\mathbf{n}} \equiv (0, 0, \lambda) + \epsilon\underline{\mathbf{N}}, \tag{43}$$

$$\bar{\mathbf{r}} \equiv (0, 0, s) + \epsilon\bar{\mathbf{u}}, \tag{44}$$

$$\overline{\mathrm{R}} = \exp(\epsilon\underline{\Theta}), \tag{45}$$

where $\underline{\mathbf{N}}, \bar{\mathbf{u}}$ are vector fields, $\underline{\Theta}$ is a skew-matrix field, and ϵ is a small parameter. From (28) we then find $\mathrm{K} = \overline{\mathrm{R}}^T\overline{\mathrm{R}}' = \epsilon\underline{\Theta}'$, from which we deduce

$$\underline{\kappa} = \epsilon\underline{\theta}' + o(\epsilon), \tag{46}$$

where $\underline{\theta}$ is the axial vector of $\underline{\Theta}$. We then substitute (43)–(46) into (14), (24), (25) and (28), compute the derivative of each with respect to "ϵ", and evaluate the resulting expressions at $\epsilon = 0$, to obtain:

$$N_1' + \lambda\theta_2' = 0,$$

$$N_2' - \lambda\theta_1' = 0,$$

$$N_3' = 0,$$

$$u_1' - \theta_2 = 0,$$

$$u_2' + \theta_1 = 0, \tag{47}$$

$$u_3' = 0,$$

$$A(\mu)\theta_1'' - N_2 = 0,$$

$$B(\mu)\theta_2'' + \tau(\mu)\theta_3' + N_1 = 0,$$

$$C(\mu)\theta_3'' - \tau(\mu)\theta_2' = 0,$$

where

$$A(\mu) = D_1^2 \Phi(-\mu, 0, 0),$$

$$B(\mu) = 2D_2 \Phi(-\mu, 0, 0), \quad \text{and} \qquad (48)$$

$$C(\mu) = 2D_3 \Phi(-\mu, 0, 0)$$

are the "instantaneous moduli" at the straight state $\kappa \equiv \underline{0}$. Finally, for compatibility with the boundary conditions (18)–(21), the "incremental" fields appearing in (43)–(45) must satisfy

$$u_\alpha(0) = u_\alpha(1) = 0, \quad \alpha = 1, 2,$$

$$u_3(0) = 0, \quad N_3(1) = 0, \qquad (49)$$

$$\underline{\theta}(0) = \underline{\theta}(1) = \underline{0}. \qquad (50)$$

A necessary condition for bifurcation is that the linearized system (47)–(50) admit nontrivial solutions. To solve the linearized problem, we first observe from $(47)_{3,6}$ and $(49)_2$ that

$$N_3 = u_3 \equiv 0. \qquad (51)$$

Next, integration of $(47)_{4,5}$, using $(49)_1$, shows that

$$\int_0^1 \theta_a = 0, \quad \alpha = 1, 2, \qquad (52)$$

and then the integration of $(47)_9$, using (50), yields

$$\theta_3(x) = \frac{\tau(\mu)}{C(\mu)} \int_0^x \theta_2(\xi)\, d\xi. \qquad (53)$$

We then integrate $(47)_{1,2}$ and substitute into $(48)_{7,8}$, employing (51) and (53), to obtain

$$\theta_1'' - \frac{\lambda}{A(\mu)} \theta_1 = \theta_1'(1) - \theta_1'(0), \qquad (54)$$

$$\theta_2'' + \frac{1}{B(\mu)}\left(\frac{(\tau(\mu))^2}{C(\mu)} - \lambda\right)\theta_2 = \theta_2'(1) - \theta_2'(0). \quad (55)$$

Both (54) and (55) are of the form

$$\theta'' + \sigma^2 \theta = \theta'(1) - \theta'(0), \qquad (56)$$

subject to

$$\theta(0) = \theta(1) = 0, \qquad (57)$$

which is equivalent to the classical eigenvalue problem associated with the planar buckling of a compressed "clamped–clamped" rod, cf. [Timoshenko & Gere, 1961]. (If we differentiate (56) and use either of $(47)_{4,5}$, we obtain the precise fourth-order formulation found in [Timoshenko & Gere, 1961].) The general solution of (56) is

$$\theta = C_1\left(\cos\sigma s - \frac{\sin\sigma}{\sigma}\right) + C_2\left(\sin\sigma s - \frac{\cos\sigma - 1}{\sigma}\right),$$
$$(58)$$

where C_1, C_2 are constants. Enforcing the boundary conditions (57), we find nontrivial solutions iff

$$\sigma\sin\sigma + 2(\cos\sigma - 1) = \sin\frac{\sigma}{2}\left(\frac{\sigma}{2}\cos\frac{\sigma}{2} - \sin\frac{\sigma}{2}\right)$$
$$= 0, \qquad (59)$$

(which agrees with Eq. (d) on p. 54 of [Timoshenko & Gere, 1961]). Accordingly, we find two families of solutions:

$$\sigma = 2n\pi; \quad \theta = \sin 2n\pi s, \qquad n = 1, 2, \ldots, \quad (60)$$

$$\frac{\sigma(m)}{2} = \tan\frac{\sigma(m)}{2};$$

$$\theta = \cos\sigma(m)s + \alpha(m)\sin\sigma(m)s - 1, \quad (61)$$

$$m = 1, 2, \ldots,$$

where $0 < \sigma(1) < \sigma(2) < \cdots$ denotes the positive solutions of the transcendental equation in (61) and

$$\alpha(m) \equiv \frac{\sigma(m) - \sin\sigma(m)}{1 - \cos\sigma(m)}. \qquad (62)$$

As discussed in [Timoshenko & Gere, 1961], the family (61) corresponds to configurations that are symmetric (reflection symmetric) about the midspan ($s = 1/2$), while (62) yields antisymmetric configurations with respect to the midspan.

From (51) and (53), we can now read off nontrivial solutions of the linearization (47), (49) and (50). There are two families of distinct solutions. The first is characterized by compressive critical loads only ($\lambda < 0$), with the linearized solutions corresponding to the planar configurations discussed in Sec. 2.2. We denote these planar solutions by P_n^I, which are reflection symmetric and P_n^{II}, which are anti-symmetric or flip-symmetric.

Planar compressive:

$$N_1 = N_3 = u_1 = u_3 = \theta_2 = \theta_3 \equiv 0; \qquad (63)$$

$$P_n^I : \begin{cases} \lambda^{2n-1} = -4n^2\pi^2 A(\mu) \\[2mm] \theta_1^{2n-1} = \sin 2n\pi s \\[2mm] u_2^{2n-1} = \dfrac{1}{2n\pi}(\cos 2n\pi s - 1) \\[2mm] N_2^{2n-1} = \lambda^{2n-1}\theta_1^{2n-1}, \quad n = 1, 2, \ldots; \end{cases} \qquad (64)$$

$$P_n^{II} : \begin{cases} \lambda^{2n} = -A(\mu)(\sigma(n))^2 \\[2mm] \theta_1^{2n} = \cos\sigma(n)s + \alpha(n)\sin\sigma(n)s - 1 \\[2mm] u_2^{2n} = s - \dfrac{1}{\sigma(n)}\sin\sigma(n)s \\[2mm] \qquad + \dfrac{\alpha(n)}{\sigma(n)}(\cos\sigma(n)s - 1) \\[2mm] N_2^{2n} = \lambda^{2n}\theta_1^{2n}(s) + A(\mu)\sigma(n)(\cos\sigma(n) - 1), \\[2mm] \qquad n = 1, 2, \ldots \end{cases}$$

$$(65)$$

The second family potentially admits tensile critical loads ($\lambda > 0$) as well as compressive critical loads, and unlike the previous family, the solutions are characterized by $\theta_3 \neq 0$. We denote these spatial solutions by S_n^I, which are reflection symmetric, and by S_n^{II}, which are flip symmetric.

Non planar tensile-compressive:

$$N_2 = N_3 = u_2 = u_3 = \theta_1 \equiv 0; \qquad (66)$$

$$S_n^I : \begin{cases} \lambda^{2n-1} = \dfrac{(\tau(\mu))^2}{C(\mu)} - 4n^2\pi^2 B(\mu) \\[2mm] \theta_2^{2n-1} = \sin 2n\pi s \\[2mm] u_1^{2n-1} = \dfrac{1}{2n\pi}(1 - \cos 2n\pi s) \\[2mm] \theta_3^{2n-1} = \dfrac{\tau(\mu)}{C(\mu)}u_1^{2n-1} \\[2mm] N_1^{2n-1} = -\lambda^{2n-1}\theta_2^{2n-1}, n = 1, 2, \ldots; \end{cases} \qquad (67)$$

$$S_n^{II} : \begin{cases} \lambda^{2n} = \dfrac{(\tau(\mu))^2}{C(\mu)} - B(\mu)(\sigma(n))^2 \\[2mm] \theta_2^{2n} = \cos\sigma(n)s + \alpha(n)\sin\sigma(n)s - 1 \\[2mm] u_1^{2n} = -s + \dfrac{1}{\sigma(n)}\sin\sigma(n)s \\[2mm] \qquad + \dfrac{\alpha(n)}{\sigma(n)}(1 - \cos\sigma(n)s) \\[2mm] \theta_3^{2n} = \dfrac{\tau(\mu)}{C(\mu)}u_1^{2n} \\[2mm] N_1^{2n} = -\lambda^{2n}\theta_2^{2n} + B(\mu)(1 - \cos\sigma(n)), \\[2mm] \qquad n = 1, 2, \ldots \end{cases} \qquad (68)$$

If the intrinsic curvature "μ" is sufficiently large, then from (8)–(10) and (42) we see that tensile "buckling loads" ($\lambda > 0$) are possible. In particular, if we specialize to the Kirchhoff model (15), viz. $A(\mu) = A, \tau(\mu) = -A\mu, B(\mu) = B, C(\mu) = C$ into (67) and (68), the characteristic equations $(67)_1$ and $(68)_1$ (non planar solutions) reduce to

$$\text{Symmetric} : \lambda^{2n-1} = \frac{A^2\mu^2}{C} - 4n^2\pi^2 B, \qquad (69)$$

and

$$\text{Antisymmetric} : \lambda^{2n} = \frac{A^2\mu^2}{C} - (\sigma(n))^2 B, \qquad (70)$$

respectively.

2.4. *Local bifurcation*

In this section, we verify the standard transversality condition insuring thatthe linearized solutions (63)–(65) and (66)–(68) correspond to actual solutions of the nonlinear problem. In order to make this precise, we need a little extra notation: Refering to (47), define the field

$$x \equiv (N_1, N_2, N_3, u_1, u_2, u_3, \theta_1, \theta_2, \theta_3), \qquad (71)$$

and for all fields u, v on $[0, 1]$, define the inner product

$$\langle u, v \rangle = \int_0^1 \sum_{i=1}^{9} u_i(s)v_i(s)\, ds. \qquad (72)$$

Next we express (47) subject to (49), (50) via differential-operator notation:

$$L(\lambda)x = 0, \tag{73}$$

If (λ^n, x^n) denotes a nontrivial solution of the linearized problem, viz.

$$L(\lambda^n)x^n = 0. \tag{74}$$

We further assume, for a given $\lambda = \lambda^n$, that (73) has only one linearly independent solution, $x = x^n$. Observe that this is the case, provided that λ^n given by (63) or (65) does not coincide with some other $\lambda^m (m \neq n)$ given by (67) or (68). Then a sufficient condition for local bifurcation in the nonlinear problem is (cf. [Crandall & Rabinowitz, 1971])

$$\langle y^n, L'(\lambda^n)x^n \rangle \neq 0, \tag{75}$$

where y^n denotes the adjoint null vector satisfying

$$L^*(\lambda^n)y^n = 0. \tag{76}$$

Here $L^*(\lambda^n)$ is the adjoint operator defined by

$$\langle L^*(\lambda^n)y, x \rangle \equiv \langle y, L(\lambda^n)x \rangle, \tag{77}$$

for all sufficiently smooth fields x, y satisfying (49), (50).

From (47) and (71) we see that

$$L'(\lambda)x = (\theta_2', -\theta_1', 0, \ldots, 0),$$

and thus, at the nontrivial solution (λ^n, x^n) of the linearized problem, we find either

$$L'(\lambda^n)x^n = \left(0, -\frac{d}{ds}\theta_1^n, 0, \ldots, 0\right), \tag{78}$$

for the "planar compressive" solutions (cf. (63)–(65)) or

$$L'(\lambda^n)x^n = \left(\frac{d}{ds}\theta_2^n, 0, \ldots, 0\right), \tag{79}$$

for the "nonplanar tensile-compressive" solutions (cf. (66)–(68)).

To compute the adjoint operator, we let

$$y \equiv (P_1, P_2, P_3, v_1, v_2, v_3, \phi_1, \phi_2, \phi_3),$$

and substitute (47) into the right side of (77). Integration by parts, using the boundary conditions (49), (50), yields the adjoint equations $L^*(\lambda)y = 0$:

$$\begin{aligned}
-P_1' + \phi_2 &= 0, \\
-P_2' - \phi_1 &= 0, \\
-P_3' &= 0, \\
-v_1' &= 0, \\
-v_2' &= 0, \\
-v_3' &= 0, \\
A(\mu)\phi_1'' + \lambda P_2' + v_2 &= 0, \\
B(\mu)\phi_2'' + \tau(\mu)\phi_3' - \lambda P_1' - v_1 &= 0, \\
C(\mu)\phi_3'' - \tau(\mu)\phi_2' &= 0,
\end{aligned} \tag{80}$$

subject to

$$P_\alpha(0) = P_\alpha(1) = 0, \quad \alpha = 1, 2, $$
$$P_3(0) = v_3(1) = 0, \tag{81}$$
$$\phi_i(0) = \phi_i(1) = 0, \quad i = 1, 2, 3. \tag{82}$$

Observe that $(80)_9$ is identical to $(47)_9$, while $(80)_{3,6}$ and $(81)_2$ yield $v_3 = P_3 \equiv 0$. Next $(80)_{1,2}$ and $(81)_1$ imply that (52) holds for $\phi_\alpha, \alpha = 1, 2$, as well. Finally, if we substitute $(80)_{1,2,4,5,9}$ into $(80)_{7,8}$, using (52) for ϕ_α, we again obtain (54) and (55) with ϕ_α in place of θ_α. We conclude that the components θ_α^n and $\phi_\alpha^n, \alpha = 1, 2$, are identical, and from $(80)_{1,2}$ and (82), we obtain the first two arguments of the adjoint null vectors as follows:

$$y^n \equiv \left(0, -\int_0^s \theta_1^n(\xi)d\xi, \ldots \right), \tag{83}$$

for the "planar compressive" solutions, and

$$y^n \equiv \left(\int_0^s \theta_2^n(\xi)d\xi, 0, \ldots \right), \tag{84}$$

for the "nonplanar tensile-compressive" solutions. Finally, we substitute either (78) and (83) into the left side of (75) or (79) and (84) into the left side of (75). For either case ($\alpha = 1$ or 2), integration by parts using (82) yields:

$$\begin{aligned}
\langle y^n, L'(\lambda^n)x^n \rangle &= \int_0^1 \frac{d}{ds}\theta_\alpha^n(s)\left(\int_0^s \theta_\alpha^n(\xi)d\xi\right)ds \\
&= -\int_0^1 (\theta_\alpha^n(s))^2\, ds \neq 0,
\end{aligned} \tag{85}$$

which verifies the transversality condition (75). Accordingly, we conclude (cf. [Crandall & Rabinowitz, 1971]) *Each of the linearized solutions given in (63)–(65) and (66)–(68), denoted $\lambda^n, \underline{N}^n, \overline{u}^n, \Theta^n$ (where $\underline{\theta}^n$ is the axial vector of $\underline{\Theta}^n$), correspond to local bifurcating solutions of the nonlinear problem in the sense that (43)–(45) are asymptotically valid, viz.*

$$\underline{n} \equiv (0,0,\lambda) + \epsilon \underline{N}^n + o(\epsilon),$$

$$\overline{r} \equiv (0,0,s) + \epsilon \overline{u}^n + o(\epsilon),$$

$$\overline{R} = I + \epsilon \underline{\Theta}^n + o(\epsilon), \tag{86}$$

$$\lambda = \lambda^n + o(\epsilon),$$

for all sufficiently small ϵ, yields a curve of nontrivial solutions of the nonlinear problem.

As suggested in $(86)_4$, each of these local bifurcations is, in fact, a so-called pitchfork. In each case, this is a consequence of Z_2 symmetry breaking. Indeed, our boundary value problem (14), (16)–(20) is equivariant under "mirror" reflections about the midplane of the straight, underformed rod perpendicular to $\mathbf{e}_3$ and also under 180° rotations or "flips" about the midpoint axes (to the undeformed rod) parallel to $\mathbf{e}_1$ and $\mathbf{e}_2$. The bifurcations associated with (64) and (67) each break a flip symmetry, while those coming from (65) and (68) break the reflection symmetry, cf. Fig. 5 in Sec. 3 Standard arguments show that each of these is necessarily a pitchfork bifurcation, [Golubitsky & Schaeffer, 1985].

In the remainder of this section we consider the Kirchhoff model (15), and we provide a more detailed analysis of the bifurcation (86) associated with (67) and (69) for $n = 1$. In particular, we assume that the intrinsic curvature μ is sufficiently large so that the "buckling load" given by (69) is positive (and hence, tensile), viz. $\lambda^1 = A^2\mu^2/C - 4\pi^2 B > 0$. This is precisely the situation in the "hand-held experiment" for a telephone cord, as discussed in the introduction. First we demonstrate that the straight (trivial) solution (42) is *stable* (the potential energy is a local minimum) for all (tensile) loading $\lambda > \lambda^1$ and *unstable* (the potential energy is not a local minimum) for all $\lambda < \lambda^1$. We start with the total potential energy functional for the rod:

$$V(\mathbf{r}, \mathbf{R}) = \int_0^1 [W(\underline{\kappa}) + \mathbf{n} \cdot (\mathbf{r}' - \mathbf{R}\mathbf{e}_3)] \, ds$$

$$- \lambda \mathbf{e}_3 \cdot \mathbf{r}(1), \tag{87}$$

where the internal contact force $\mathbf{n}$ is the Lagrange multiplier field enforcing the constraint that the rod

be inextensible and unshearable. We then compute the second variation at the trivial solution via:

$$\delta^2 V(s\mathbf{e}_3, \mathbf{I}) \equiv \frac{d^2}{d\epsilon^2} [V(s\mathbf{e}_3 + \epsilon\eta, \exp(\epsilon\Theta))]_{\epsilon=0} \tag{88}$$

A lengthy calculation leads to

$$\delta^2 V(s\mathbf{e}_3, \mathbf{I}) = \int_0^1 \{A(\theta_1')^2 + B(\theta_2')^2$$

$$+ C(\theta_3')^2 + \lambda[(\theta_1)^2 + (\theta_2)^2]$$

$$- 2A\mu\theta_3'\theta_2\} \, ds, \tag{89}$$

for all smooth test functions $\underline{\theta}(s) = (\theta_1(s), \theta_2(s), \theta_3(s))$ satisfying (50) and (52), where $\underline{\theta}(s)$ is the axial vector field corresponding to the skew-matrix field $\underline{\Theta}(s)$. By the minimum property of the smallest eigenvalue, we have the following "sharp" Poincare inequalities:

$$\int_0^1 (\theta_\alpha')^2 \, ds \geq 4\pi^2 \int_0^1 (\theta_\alpha)^2 \, ds, \quad \alpha = 1, 2, \tag{90}$$

cf. (56), (61) for $n = 1$. Also, by the arithmetic–geometric means inequality, we have

$$2 \int_0^1 |\theta_3'\theta_2| \, ds \leq \int_0^1 \left[\epsilon(\theta_2)^2 + \frac{1}{\epsilon}(\theta_3')^2 \right] \, ds, \tag{91}$$

for all numbers $\varepsilon > 0$. For any $\lambda > \lambda^1$, we choose

$$\frac{A|\mu|}{C} < \varepsilon < \frac{A|\mu|}{C} + \alpha, \tag{92}$$

where $\alpha \equiv (\lambda - \lambda^1)/A|\mu|$. It then follows from (89)–(90) that

$$\delta^2 V(s\mathbf{e}_3, \mathbf{I}) > K \int_0^1 [(\theta_1)^2 + (\theta_2)^2 + (\theta_3)^2] \, ds, \tag{93}$$

for all test functions satisfying (50), (52), where $K > 0$ is a constant.

On the other hand, suppose that $\lambda < \lambda^1$. In (89) we choose $\theta_1 \equiv 0$ and integrate by parts to get:

$$\delta^2 V(s\mathbf{e}_3, \mathbf{I}) = \int_0^1 \{[-B\theta_2'' + \lambda\theta_2 + A\mu\theta_3']\theta_2$$

$$- [C\theta_3'' + A\mu\theta_2']\theta_3\} \, ds, \tag{94}$$

for all test functions θ_2, θ_3 satisfying (50), (52). We now choose θ_2, θ_3 to coincide with the nontrivial solutions θ_2^1, θ_3^1 given in $(67)_{2,4}$, the substitution of which into (94) yields $\delta^2 V(s\mathbf{e}_3, \mathbf{I}) = (\lambda - \lambda^1)/2$.

We now determine the next nonzero term "γ" in the Taylor expansion (86) for $n = 1$, viz.

$$\lambda = \lambda^1 + \gamma\epsilon^2 + o(\epsilon^2). \tag{95}$$

To make this precise, we need a bit more notation. First we substitute (8), (13) and (15) into (24) (calling it $(24)'$) and denote the system (23), (27) and $(24)'$ via

$$F(\lambda, x) = 0. \tag{96}$$

Since we have a Z_2-symmetry-breaking pitchfork, it can be shown [Kielhöfer, 2004] that the coefficient "γ" is given by the formula

$$\gamma = -\frac{\langle y^1, D_x^3 F(\lambda^1, 0)[x^1, x^1, x^1]\rangle}{3\langle y^1, L'(\lambda^1)x^1\rangle}, \tag{97}$$

where the calculation for the numerator is facilitated by

$$D_x^3 F(\lambda^1, 0)[x^1, x^1, x^1] \equiv \frac{d^3}{d\epsilon^3}\left[F(\lambda^1, x(\epsilon))\right]_{\epsilon=0}, \tag{98}$$

with $x(\epsilon)$ is represented by (43)–(45). In particular, we use an expansion of (45) to obtain the higher-order extension of (46),

$$\underline{\kappa} = \epsilon\underline{\theta}^{1\prime} + \frac{\epsilon^2}{2}\underline{\theta}^1 \times \underline{\theta}^{1\prime} + \frac{\epsilon^3}{3!}\underline{\theta}^1$$
$$\times (\underline{\theta}^1 \times \underline{\theta}^{1\prime}) + o(\epsilon^3), \tag{99}$$

which is needed in the calculation (98). The denominator in (97) is given by (85). In the special case $A = B = C$ (cf. (15)), which we employ in our numerical work to follow, a straightforward but laborious calculation, employing (99), leads to

$$\gamma = -\frac{\mu^2(1 + \mu^2)}{3}, \tag{100}$$

i.e. the pitchfork (95) is "subcritical". By the usual exchange-of-stability-argument (cf. [Kielhöfer, 2004]), it then follows that the local bifurcating solution (95), (100) is stable. These results are summarized in Fig. 4: observe the pitchfork off branch A.

3. Global Computations and their Interpretation

3.1. *The parallel simplex algorithm and the global representation space*

In the remainder of this work, we seek a "global picture" of the solution diagram. To that end, we apply the Parallel Simplex Algorithm (PSA) which, to the best of our knowledge, is the only available code capable of determining all equilibria, connected or not, in a given domain of the solution space. We describe the method very briefly below.

The PSA, introduced in [Domokos, 1994; Domokos & Gáspár, 1995; Gáspár *et al.*, 1997] is based on some simple ideas from the theory of ordinary differential equations (ODEs), combined with the Piecewise Linear (PL) Algorithm [Allgower & Georg, 1990]. In contrast to path-continuation techniques, which deliver equilibria in sequence along solution branches, the PSA resolves simultaneously *all* equilibria (in a given domain) lying on *all* branches. The PSA can be directly applied to two-point BVPs associated with ODEs of the form

$$\dot{x}(t) = f(x(t), \lambda), \quad x \in \mathbb{R}^{2n}, \ \lambda \in \mathbb{R}^1, \ t \in [0, 2\pi]. \tag{101}$$

Let us assume that the initial ($t = 0$) conditions apply to the *first* n components ($x_i(0) = a_i, i = 1, 2, \ldots, n$) and far-end ($t = 2\pi$) conditions apply to the n components with indices $\nu_i(x_{\nu_i}(2\pi) = b_i, i = 1, 2, \ldots, n)$, where the a_i, b_i are given scalars. Let us denote the unspecified initial components by $v_{i-n} = x_i(0), i = n + 1, n + 2, \ldots, 2n$ ("variables"). The $(n + 1)$-dimensional space spanned by the variables and the parameter λ will be called the Global Representation Space (GRS) for the bifurcation problem. By using any convergent forward integrator for the Initial Value Problem (IVP), we can express the far-end values $x_{\nu_i}(2\pi), (i = 1, 2, \ldots, n)$ as *functions* of the variables v_i and the parameter $\lambda : x_{\nu_i}(2\pi) = g_i(v_1, v_2 \ldots, v_n, \lambda)$ and solve the algebraic equation system

$$g_i(v_j, \lambda) - b_i = 0,$$
$$(i, j = 1, 2, \ldots n, \ v_j \in [v_j^0, v_j^1], \ \lambda \in [\lambda^0, \lambda^1].) \tag{102}$$

by the PL algorithm [Allgower & Georg, 1990] in the prescribed $(n + 1)$-dimensional domain of the GRS (defined by the constants with superscript in (102)). Geometrically, (102) describes the intersection of n hypersurfaces in the $(n + 1)$-dimensional space, yielding typically (locally) one-dimensional solution sets, thus branches. This fact can be also expressed as

$$V = F + 1, \tag{103}$$

where V and F denote the numbers of variables and functions, respectively. These branches will appear

as polygons, due to the piecewise-linear approximation. (We remark that variables can have a far more
general interpretation in the PSA. However, the version described above is sufficient to introduce the
most important concepts.)

We now return to our problem. Due to the
clamped boundary conditions (18) we have 6 scalar
"free" initial conditions to (16), (17), viz. the values
of the components of **n** and **m** at $s = 0$, i.e. we have
$V = 6$ variables. In view of (18) and (20), observe
that $n_3(0) = \lambda$. The far-end condition (19) defines
two scalar equations, and from $(18)_2$ we deduce the
three independent scalar conditions

$$R_{12}(1) = 0$$
$$R_{13}(1) = 0 \tag{104}$$
$$R_{23}(1) = 0.$$

In addition, $(18)_2$ yields trace $\mathbf{R}(1) = R_{11} + R_{22} + R_{33} = 3$ which we also impose.

Equations (23), (24) can be integrated forward
using standard techniques; we relied on [Gáspár,
1977, 1978, 1979].

In the following subsections we will attempt to
give a partial picture of the global bifurcation diagram. Our description is partially based on computational results, partially on integer labels assigned
to the branches according to local bifurcation patterns. We will also utilize ideas connected to the
White–Fuller theorem [White, 1969] and its extension [Alexander & Antman, 1982; Heijden *et al.*,
2004].

3.2. *Classification of branches and branch labels*

The computed equilibria can be identified by the
six-dimensional vector consisting of the nonconstant initial conditions $[n_1(0), n_2(0), n_3(0), m_1(0),$
$m_2(0), m_3(0)]$, where $n_3(0) = \lambda$. For purposes
of graphical representation we will use the three-
dimensional subspace $[n_3(0) = \lambda, m_1(0), m_2(0)]$.
The computations revealed a highly complex
bifurcation diagram, consisting of a large variety
of equilibria. One portion is illustrated in Fig. 3.
All calculations were carried out for the Kirchhoff
model (15), with $A = B = C$. The bifurcation diagram in Fig. 4 has been computed with $\mu = 10$; all
other computations were carried out with $\mu = 40$.

The main thrust of this section is to gain
some (partial) understanding of the diagram. Since
our focus is the description of helical perversions,

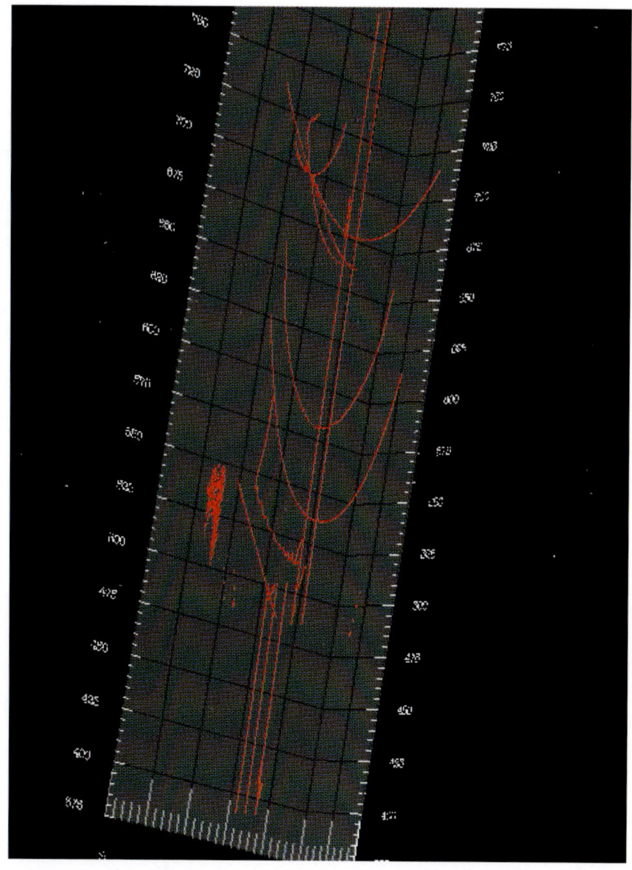

Fig. 3.   Part of the global bifurcation diagram, illustrated in
the $[n_3(0) = \lambda, m_1(0), m_2(0)]$ space.

we will concentrate on the following classes of
equilibria:

1. The primary, straight configurations, forming
   the primary trivial branch $A : [0, 0, \lambda, -\mu, 0, 0]$.
2. Planar, "classical" Euler modes P_i^I, P_i^{II}, bifurcating off the primary A branch for $\lambda < 0$,
   (compression), forming branches. The family P_i^I
   is reflection-symmetric, the family P_i^{II} is flip-
   symmetric, cf. Eqs. (64), (65).
3. Spatial modes S_i^I, S_i^{II}, bifurcating off branch
   A both for positive and negative values of the
   tension λ, forming branches. The family S_i^I
   is reflection-symmetric, the family S_i^{II} is flip-
   symmetric, cf. Eqs. (67), (68).
4. Asymptotically straight, twisted configurations b_k, located at the GRS points
   $[\infty, 0, 0, -\mu, 0, 2k\pi]$. As we will show, some
   branches approach b_k points as $\lambda \to \infty$. (In
   the case of zero initial curvature, twist is decou-
   pled from bending and one obtains branches
   of straight, twisted equilibria. In our case, the
   direction of preferred curvature changes as the

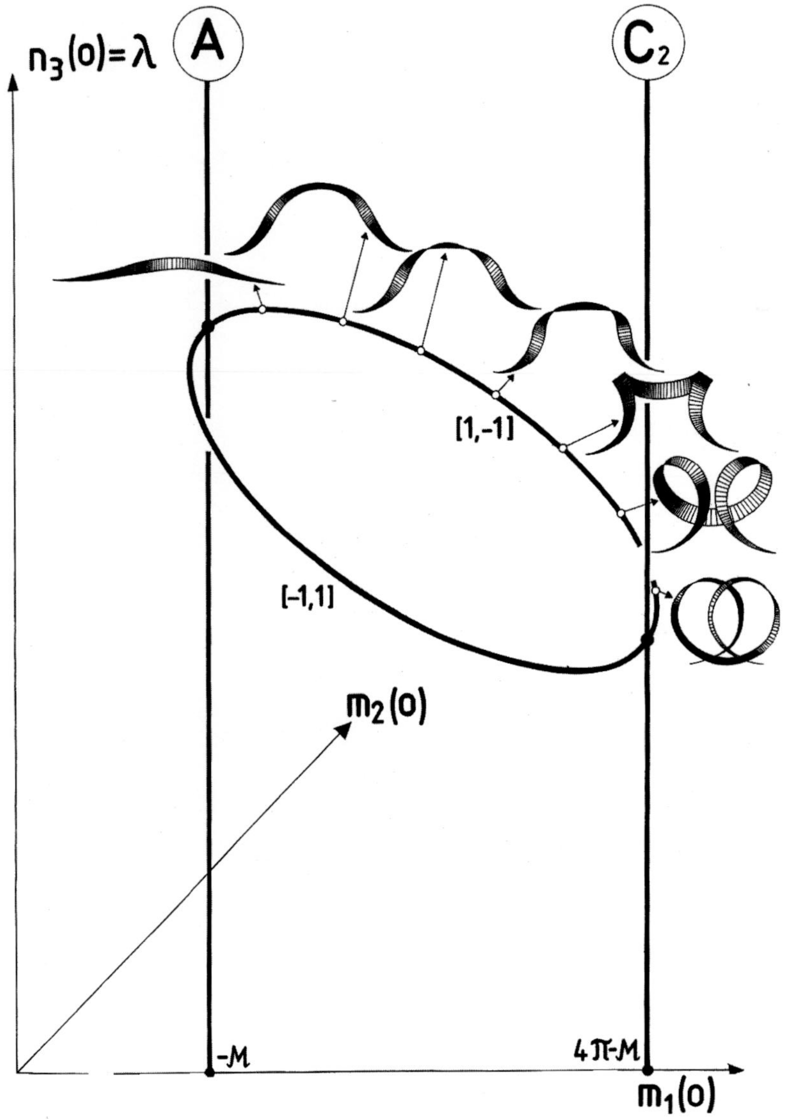

Fig. 4.   Branch S_1^I connecting the trivial A-branch to the C_2 branch. Observe perversion on the physical configurations.

cross-section is twisted, so straight, twisted equilibria can be realized only asymptotically.)

5. Planar, untwisted, self-intersecting equilibria forming the branches C_k $(k = 1, 2, \ldots)$ emerging from k-covered circles c_k at $[0, 0, 0, 2k\pi - \mu, 0, 0]$ in the GRS. Equilibria on C_k branches correspond to planar, "noninflectional" elastica lines in physical space, cf. [Love, 1927]. In the GRS description, these have the form $[0, n_2^*(k, \lambda), \lambda, m_1^*(k, \lambda), 0, 0]$, where $n_2^*(k, \lambda)$, $m_1^*(k, \lambda)$ can be expressed in closed form using Jacobian elliptic integrals. In the graphical representation these curves appear as $[\lambda, m_1^*(k, \lambda), 0]$.

6. Spatial modes bifurcating off C_k, forming branches.

7. Branches created and secondary, tertiary, etc. bifurcations.

8. Disconnected branches.

These observations are in full agreement with the findings of Sec. 2. In particular, there we show not only the existence of the branches $A, P_i^I, P_i^{II}, S_i^I, S_i^{II}$, but also the corresponding critical load parameters and eigenfunctions are given explicitly, cf. (64), (65), (67), (68), and a detailed local analysis of the branch S_1^I is provided. We can observe some of the listed equilibria in Fig. 3. All solutions shown are tensile ($\lambda > 0$). Observe that S_1^I, which connects the trivial branch A to C_2, contains "perversion" equilibiria. We illustrate this more fully in Fig. 4, displaying physical shapes as insets. Moreover, the absence of any turning points and bifurcation points along S_1^I, combined with the local stability results from Sec. 2, imply that the entire branch (excluding the bifurcation points on A and C_2) contains stable solutions. The almost parallel, almost straight lines in Fig. 3 correspond to the branches C_3, C_4, C_5, C_6 and C_7. The latter two have been computed for $\mu = 40$ on much longer segments. Observe the spatial modes bifurcating off C_4, C_6 and C_7 and branches created in secondary bifurcations. We illustrate some characteristic physical shapes in Fig. 5.

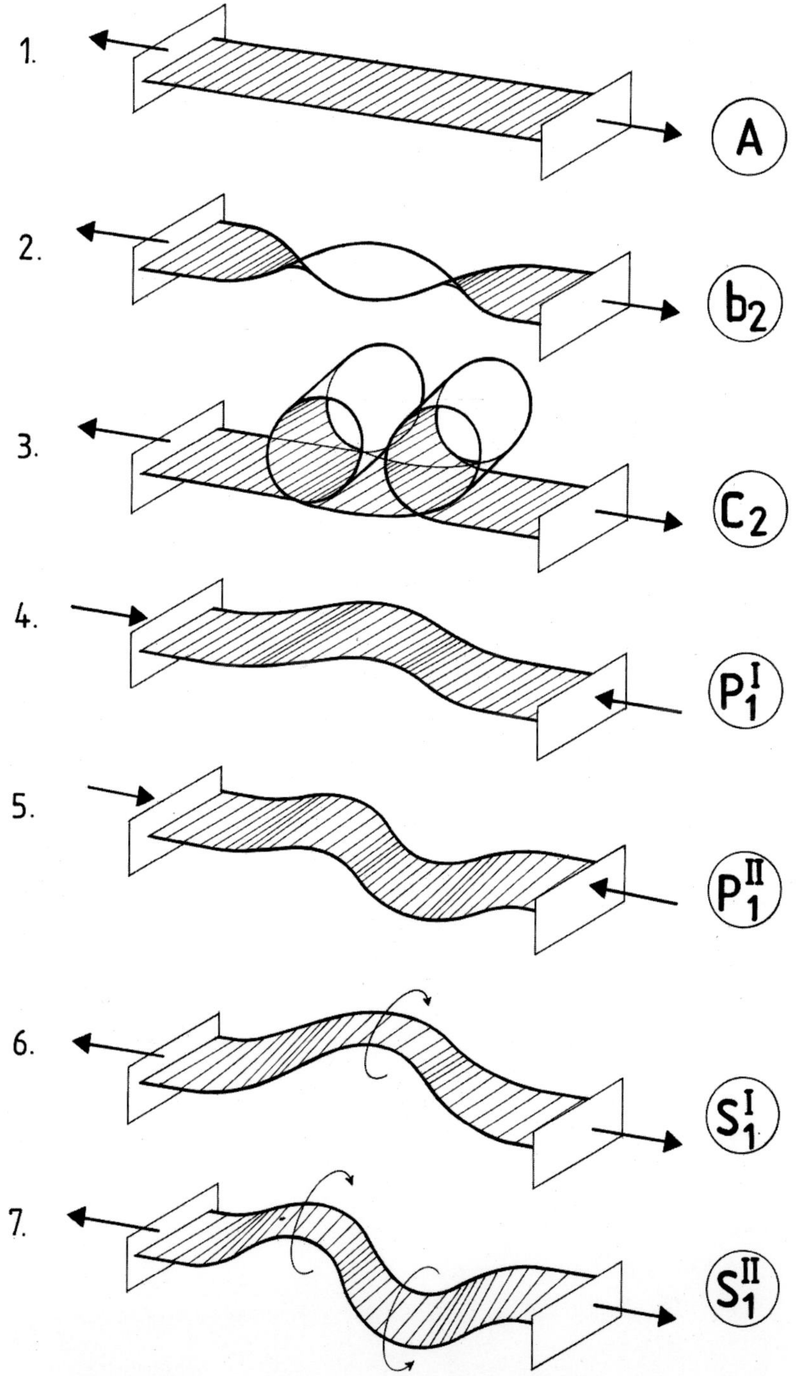

Fig. 5.   Physical shapes illustrated as ribbons. (1) The trivial eqilibria on branch A. (2) Asymptotic equilibrium b_2 at $\lambda = \infty$. (3) Equilibrium on the branch C_2. (4) Planar, compressive Euler mode P_1^I. (5) Planar, compressive Euler mode P_1^{II}. (6) Spatial (possibly tensile) mode S_1^I. (7) Spatial (possibly tensile) mode S_1^{II}.

At first sight it may be surprising that the branches C_k and the asymptotic points b_k have been included in the list; solutions on those branches are characterized by far-end clamping undergoing $2k\pi$ rotations about the various coordinate axes. Nonetheless, the boundary conditions (18) are satisfied in both cases. We will see that these solutions can be conneceted to each other and to the trivial branch as well.

Now we proceed to describe other perversions occurring among the computed equilibria. Consider the branches C_k consisting of planar equilibria in the $[2,3]$ plane. As we explain below, these planar curves have $k \leq n \leq k^2$ self-intersection points for $\lambda > 0$. For $\lambda = 0$, continuous intervals overlap, forming a k-covered circle. For small λ the clamped ends are slightly pulled apart and the k overlapping circles form k, slightly shifted loops. Each loop has two intersection points with each other loop (we will refer to these points as "inter-loop" points). In addition, each loop intersects itself once (we will call these points "bottom" points). So the total number n of self-intersections is

$$n = 2\frac{k(k-1)}{2} + k = k^2 \tag{105}$$

in this case. As λ increases and the clamped ends move further apart, the intersection points between different loops become gradually disassociated, and the self-intersection at the bottom of each loop

prevails. So we have exactly k such ("bottom") points for sufficiently high λ.

In general, one would expect that bifurcations destroy the self-intersections. Indeed, this seems to be case. However, not all self-intersections are destroyed simultaneously. In some cases several subsequent bifurcations are needed to obtain a physically relevant shape (without self-intersection). The key to identifying the bifurcations is the separation pattern.

The pattern associated with the "inter-loop" points can be rather complex for two reasons: on one hand, the number of such points is not characteristic for the branch — since it is changing with λ. On the other hand, some patterns correspond to *knotted* curves. Although such equilibria certainly exist, we do not discuss them because they are not directly related to helical perversions. One example is illustrated in Fig. 6.

For the listed reasons, we will not discuss the pattern associated with "inter-loop" points. Rather, we will assume that loops are moving independently of each other (for sufficiently high λ there are no "inter-loop" points so this assumption is certainly true). We describe only the bifurcation behavior associated with the k self-intersections of the loops ("bottom" points) surviving on the C_k branch for arbitrarily high λ. These self-intersections occur at $2k$ points $P_1, P_2, \ldots, P_{2k}$, as we follow the arclength s from 0 to 1. Pairs of points P_{2j-1}, P_{2j} $(j = 1, 2, \ldots, k)$ are coincident, forming

Fig. 6.   Knotted equilibrium shape.

the self-intersection. According to our computations, *spatial separation* of these self-intersection points, moving out of the $[x_2, x_3]$ plane, correspond to bifurcations of branches of solutions containing spatial equilibria (for $\lambda > 0$). Coincident points P_{2j-1}, P_{2j} will move either in the same or in opposite directions. In the former case they remain coincident, in the latter they separate.

We will characterize such a branch by an *integer vector label* w_i, with $i = 1, 2, \ldots, k$, $w_i \in \{-1, 0, 1\}$ defined by the sign of the initial relative displacement in the x_1 direction of the point-pair P_{2j-1}, P_{2j}. (The labels can be interpreted for the general, $k \le n \le k^2$ case as well!) The restricted labels have exactly k entries, thus we have $3^k - 1$ different labels for each value of k. The label $\{0, 0, \ldots, 0\}$ corresponds to the original branch. Bifurcating branches appear in pairs, with labels $w_i^1 = -w_i^2$, thus the labels admit $(3^k - 1)/2$ different possibilities for pairs.

In case of $k = 1$ we have only the pair $\{-1\}$ and $\{+1\}$. Based on numerical observation we believe that to each possible label-pair the corresponding branch-pair exists physically. (Others exist as well, we dealt only with k self-intersections out of the total k^2!) Labels containing $w_i = 0$ entries correspond to self-intersecting, nonphysical shapes. We will be particularly interested in physical equilibria without self-penetration, the corresponding labels do not contain zeroes: there are 2^k such labels and 2^{k-1} pairs can be identified.

3.3. *The White–Fuller theorem and global invariants*

One interesting property of the branch labels is that for shapes without self-penetration the *writhing number W* of the centerline can be obtained at the bifurcation point as

$$W = \sum_{i=1}^{n} w_i. \tag{106}$$

(We described above only the restricted $n = k$ case, however, (106) is valid in general as well.) The writhing number has been defined originally for closed curves [Calugareanu, 1961; Fuller, 1971], however, the extension to clamped–clamped end conditions is possible [Alexander & Antman, 1982; Heijden *et al.*, 2004] via a *closure*, i.e. a virtual rod segment connecting the two clamped ends. Later, we will list conditions under which this theory can

be applied; these conditions will refer both to the rod as well as to the closure. The writhing number is the number of self-intersections of planar projections of an oriented space curve, averaged over all possible projections. Each intersection is given a sign depending on whether the point with smaller or larger arclength value is closer to the plane onto which one projects. If the curve is almost planar, the writhing number is almost an integer. However, for self-intersecting curves the writhing number is not interpreted. The reason for this is that the same configuration can be approached in different limits, resulting in different (integer) writhing numbers. Our C_k branches are such curves with k self-intersection points for which the writhing number cannot be defined. As soon as these points separate, the writhing number W can be interpreted, and evidently it agrees with the sum of the w_i labels. The pairs of branches with labels $w_i^1 = -w_i^2$ illustrate why the writhing number cannot be defined for self-intersecting curves: arbitrarily close to the bifurcation point the two equilibria almost coincide, but the sign of their writhing number is different.

The White–Fuller theorem states that if we interpret the rod as a ribbon, then the linking number L of the two edges of the ribbon can be written as

$$L = T + W \tag{107}$$

where T denotes the total twist in the rod, proportional to the integral of the twist moment $m_3(s)$. Although neither W nor T are typically invariant along branches, their sum L is, as long as the following conditions are met [Alexander & Antman, 1982; Heijden *et al.*, 2004]:

1. self-penetration of the rod does not occur,
2. self-penetration of the closure does not occur,
3. the ends remain aligned.

Condition 1 has to be monitored along the branch. Conditions 2 and 3 can be guaranteed by the clamped–clamped end conditions and by admitting only positive (tensile) values for λ.

If we take any of the equilibria bifurcating off the C_k branches and increase the axial distance between the clamped ends, it is evident that the physical shape will become straight, at least asymptotically as the tension grows to infinity. In some cases there are branches carrying these equilibria. We will call the values of L, W and T close to the bifurcation point "initial values" and close to the straight shape "final values". Since L is a branch-invariant, we have

$$L_{\text{final}} = L_{\text{initial}}, \qquad (108)$$

and from the definition of the writhing number follows that

$$W_{\text{final}} = 0. \qquad (109)$$

Since the C_k branches consist of planar, untwisted equilibria,

$$T_{\text{initial}} = 0. \qquad (110)$$

From (106)–(110) it follows that

$$T_{\text{final}} = W_{\text{initial}} = \sum_{i=1}^{n} w_i. \qquad (111)$$

(We see that by using the *local* bifurcation patterns we can define a *global* invariant quantity for the branch.)

If we consider that along the primary, trivial A branch we have $L = T = W = 0$, and at the special, straight equilibria b_k we have $W = 0$, $L = T = k$, then (111) predicts two kinds of different global scenarios: (I.) If $W_{\text{initial}} = \sum_{i=1}^{n} w_i = 0$ then $T_{\text{final}} = 0$, so such a branch may be connected to the trivial branch A at finite λ. (II.) If $W_{\text{initial}} = \sum_{i=1}^{n} w_i > 0$ then $T_{\text{final}} > 0$, so such a branch may approach an equilibrium b_k as $\lambda \to \infty$.

The computations show that the two possibilities do in fact happen: we computed type I. branches *connecting* C_k branches with the trivial A branch, these connecting branches appear as loops in the global representation space, ending at two bifurcation points. We have to stress that the condition $W_{\text{initial}} = \sum_{i=1}^{n} w_i = 0$ is a *necessary* one: it simply indicates the possibility of a connecting branch. A direct connection appears to be only possible for branches with labels

$$w_j = (-1)^j, \quad j = 1, 2, \ldots, 2i, \quad k = 2i. \qquad (112)$$

These branches are *identical* with the reflection-symmetric spatial modes S_i^I. In case of other branches with $W_{\text{initial}} = \sum_{i=1}^{n} w_i = 0$ we conjecture that the connection can be established via secondary bifurcations.

We also computed type II branches, where with increasing λ the solution becomes asymptotically straight.

3.4. *The existence and genesis of simple and multiple perversions*

The branch labels w_i tell more than the scalar branch-invariant $L = W$. The exact sequence of the w_i entries defines the (approximate) shape: as long as w_i does not change sign, we have a helical segment, at the sign-change a perversion will occur. (If we include the labels defined by the "inter-loop" self-intersections, they also define the *knot type* of the solution, however, we do not investigate knotted solutions in this paper.) The simplest example are the branches bifurcating off the C_2 branch (originating in the double-covered circle at $\lambda = 0$). If we regard only physically relevant (thus non-self-intersecting) shapes, we arrive at the pattern-pairs $[1, -1], [-1, 1]$ and $[1, 1], [-1, -1]$. The former two have a sign-change consequently they contain a perversion, while the latter two labels correspond to a left-handed and a right-handed helical shape respectively, each with two total turns. Observe that for the first shapes with perversion we not only have $L = W = 0$, but the labels w_i do satisfy (112). So we expect a type I branch *connecting to the trivial branch*. In case of the helical shapes we have $L = W = \pm 2$ so we expect a type II branch converging to the b_2 equilibrium at $\lambda \to \infty$.

We computed the $[1, -1], [-1, 1]$ branch-pair and found that it actually does connect to the trivial A branch via a bifurcation point: In essence these branches form a loop in the GRS with one point connected to the C_2 branch, and one to the A branch. Figure 4 illustrates the topology of the bifurcation diagram with some physical shapes shown as insets. Observe that the perversion is most apparent at the points which are equally far both from the C_k and the A branch.

The fact that the branch-pair $[1, -1], [-1, 1]$ connects to the trivial branch is remarkable. At the end of Sec. 2.3 we demonstrated explicitly the existence of spatial buckling modes S_i^I (cf. Eq. (69)); the computations reveal that these branches can contain helical perversions.

The multiple perversions observed in nature (cf. Figs. 1 and 2) fit easily into this qualitative picture. Similarly to the C_2 branch, any branch C_k may be connected to the trivial A branch, as long as the bifurcating solution has $L = W = \sum_{i=1}^{k} w_i = 0$, and this is possible for all *even* values of k. Such

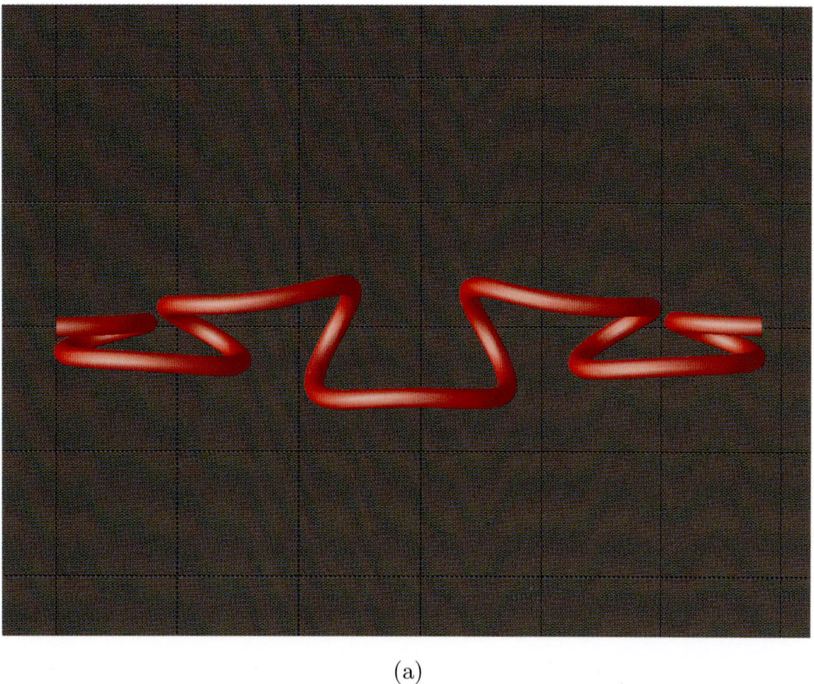

(a)

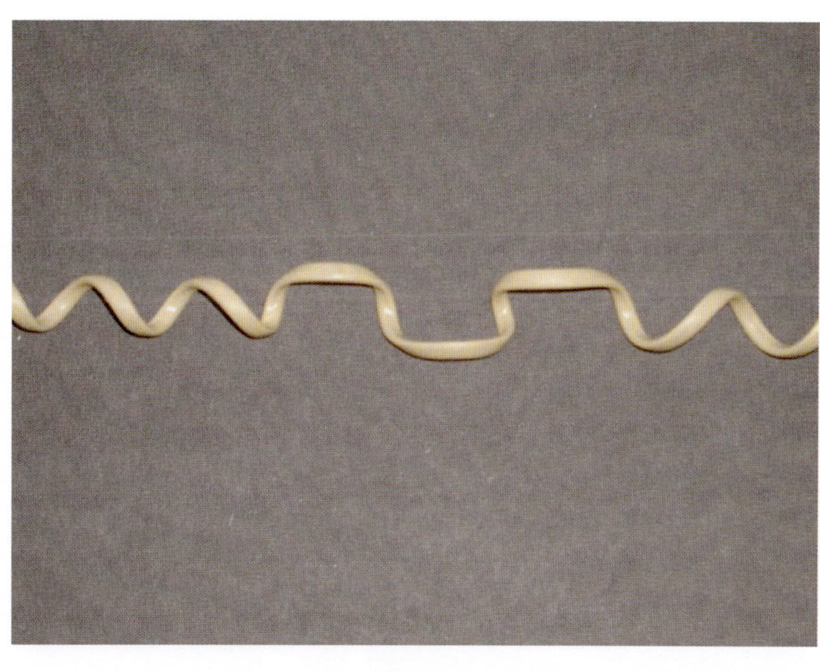

(b)

Fig. 7. Multiple perversions. (a) Configuration computed on the $[1, -1, 1, -1, 1, -1]$ branch with five perversions. (b) Telephone cord with three subsequent perversions.

an example is illustrated in Fig. 7(a), showing an equilibrium with five subsequent perversions on the $[1, -1, 1, -1, 1, -1]$ branch (also satisfying the (112) condition), which connects the C_6 to the A branch. Observe the similarity to Fig. 7(b), showing three subsequent perversions on a telephone cord. Of course, one can see multiple perversions on type II branches as well. Figure 8 illustrates such a shape with two perversions on the $[1, 1, -1, -1, 1, 1]$ branch with $L = W = \sum_{i=1}^{k} w_i = 2$, connecting the C_6 branch to the equilibrium point b_2 at $\lambda = \infty$. The C branches with odd subscript cannot be

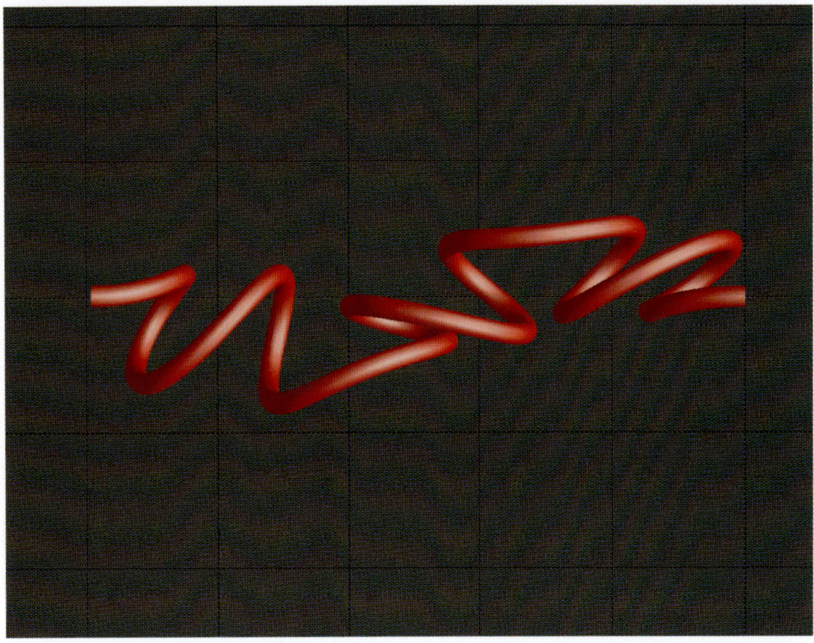

Fig. 8.   Configuration on the $[1, 1, -1, -1, 1, 1]$ branch with two perversions.

connected to the untwisted A branch, however, they may be connected to the asymptotically straight b equilibria with *odd* subscripts, i.e. with odd number of total twist. Such a shape is illustrated in Fig. 9 on the $[1, -1, -1, 1, -1, 1, -1]$ branch with $L = W = \sum_{i=1}^{k} w_i = -1$, connecting the C_7 branch to the equilibrium point b_{-1} at $\lambda = \infty$. Observe that we have five sign-changes in the last mentioned label and the physical shape exhibits exactly five perversions.

As seen, some branches bifurcating off the C_k branches connect to the trivial solution A, some

Fig. 9.   Configuration on the $[1, -1, -1, 1, -1, 1, -1]$ branch with five perversions.

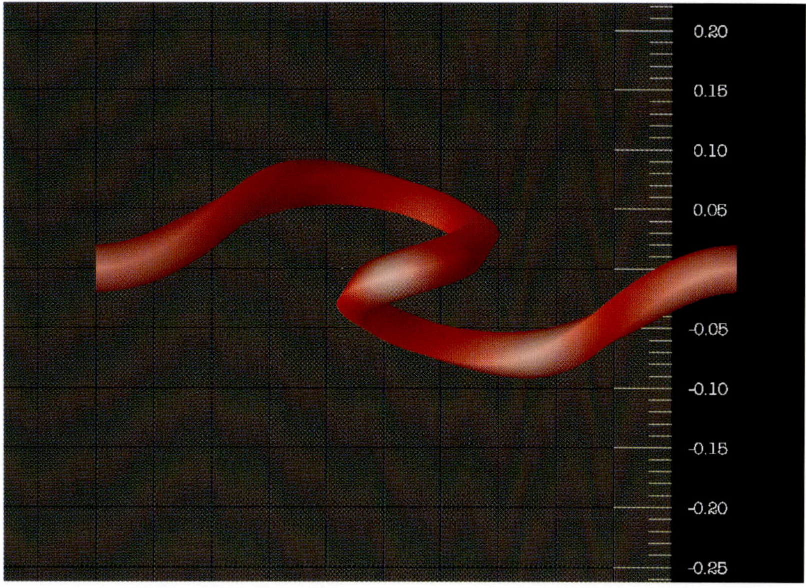

Fig. 10. The first tensile buckling mode of the trivial solution, converging to a localized loop with $L = T + W = 1 - 1 = 0$.

others do not. The inverse is also true: some buckling modes of the trivial solution (discussed in Sec. 2) connect to C_k branches, some others do not. An example for the latter is the first member of the S_i^{II} family, illustrated in Fig. 10: we can observe a branch converging to a localized loop with $L = T + W = 1 - 1 = 0$ as $\lambda \to \infty$.

4. Summary and Related Issues

In this paper we give a partial, global picture of the equilibria of intristically curved, finite, elastic rods, serving as mechanical models for things like telephone cords or botanical tendrils. In nature one can observe that the tendrils of climbing plants show spatially complex shapes consisting of several helical segments, interrupted by helical perversions, connecting two helical segments with opposite handedness.

In contrast to the works [Goriely & Tablor, 1998; Goriely & McMillen, 2002], we analyze the equilibria of *finite* rods with clamped–clamped boundary conditions. The latter are meant to model the fact that tendrils have a solid "grip" on their environment. Our main focus was the description of equilibria connected to the trivial, straight shape and the identification of equilibria with multiple perversions. (We assumed that some of the complex shapes observable in nature evolve from simple, almost straight equilibria.)

By applying analytical, computational techniques we found that the straight configuration undergoes bifurcations in tension, resulting in spatial buckling modes. Some of these branches connect to other branches originating from k-covered circles. The fact that they connect helped to identify how the geometric quantities in the White–Fuller theorem (Link, Twist, Writhe) evolve along the branch. We found that these connecting branches carry equilibria with *an arbitrary number of perversions*. We also identified branches carrying equilibria with *arbitrary number of perversions, connected by helical segments of arbitrary length*. The computed shapes correspond well to the ones observable in nature and experiments.

While the multi-covered circles and the C_k branches originating from them certainly help to understand the geometry of perversions, moreover, in some cases it might be convenient to compute perversions starting C_k branches, one has to be aware that the physical evolution is rather different since equilibria on the C_k branches are self-intersecting and thus un-physical. On the other hand, we believe that approaching a perversion from the trivial, straight solution on the bifurcation diagram is qualitatively similar to the physical evolution of equilibria, so our model can shed some light on the existence of these highly interesting, spatially complex shapes.

Acknowledgments

This work was supported by OTKA grant T046646 and the Bolyai Research Fellowship (G. Domokos), NSF grant DMS-0072514 (T. J. Healey).

References

Alexander, J. C. & Antman, S. S. [1982] "The ambigous twist of love," *Quart. Appl. Math.* **40**, 83–92.

Allgower, E. L. & Georg, K. [1990] *Numerical Continuation Methods: An Introduction* (Springer-Verlag, Berlin).

Calugareanu, G. [1961] "Sur les classes d'isotopie de noeuds tridimensionells et leurs invariants," *Czechoslovak Math. J.* **11**, 588–625.

Coleman, B. D., Tobias, I. & Swigon, D. [1995] "Theory of the influence of end conditions on self-contact in DNA loops," *J. Chem. Phys.* **103**, 9101–9109.

Crandall, M. & Rabinowitz, P. H. [1971] "Bifurcation from simple eigenvalues," *J. Funct. Anal.* **8**, 321–340.

Darwin, Ch. [1888] *The Movements and Habits of Climbing Plants* (Appleton, NY), available online at http://promo.net/pg/.

Domokos, G. [1994] "Global description of elastic bars," *Zeitschr. Angew. Math. Mech.* **74**, T289–T291.

Domokos, G. [1995] "A group-theoretic approach to the geometry of elastic rings," *J. Nonlin. Sci.* **5**, 453–478.

Domokos, G. & Gáspár, Zs. [1995] "A global, direct algorithm for path-following and active static control of elastic bar structures," *Int. J. Struct. Mach.* **23**, 549–571.

Domokos, G. & Healey, T. J. [2001] "Hidden symmetry of global solutions in twisted elastic rings," *J. Nonlin. Sci.* **11**, 47–67.

Domokos, G. & Szeberényi, I. [2004] "A hybrid parallel approach to nonlinear boundary value problems," *Comput. Assist. Mech. Eng. Sci.* **11**, 15–34.

Fuller, F. B. [1971] "The writhing number of a space curve," *Proc. Nat. Acad. Sci. USA* **68**, 815–819.

Gáspár, Zs. [1977] "The form of an ideally elastic bar with a space curve axis," *Acta Techn. Hung. Acad. Sci.* **84**, 293–306.

Gáspár, Zs. [1978] "Large deflection analzsis of bar structures," *Acta Techn. Hung. Acad. Sci.* **87**, 49–58.

Gáspár, Zs. [1979] "An exact analysis of elastic bar-structures," *Zeitschr. Angew. Math. Mech.* **59**, T179–T180.

Gáspár, Zs., Domokos, G. & Szeberényi, I. [1997] "A parallel algorithm for the global computation of elastic bar structures," *Comput. Assist. Mech. Eng. Sci.* **4**, 55–68.

Goriely, A. & McMillen, T. [2002] "Tendril perversion in intristically curved rods," *J. Nonlin. Sci.* **12**, 241–281.

Golubitsky, M. & Schaeffer, D. G. [1985] *Singularities and Groups in Bifurcation Theory*, Vol. I (Springer Verlag, NY).

Goriely, A. & Tabor, M. [1998] "The mechanics and dynamics of tendril perversion in climbing plants," *Phys. Lett.* **A250**, 311–318.

Healey, T. J. [2002] "Material symmetry and chirality in nonlinearly elastic rods," *Math. Mech. Solids* **7**, 405–420.

Heijden, G., Peletier, G. H. M. & Planque, R. [2004] "A consistent treatment of link and writhe for open rods, and their relation to end rotation," *Arch. Rat. Mech. Anal.*, submitted.

Kielhöfer, H. [2004] *Bifurcation Theory* (Springer Verlag, NY).

Li, Y & Maddocks, J. "On the computation of equilibria of elastic rods, part I: Integrals, symmetry and a Hamiltonian formulation," *J. Comput. Phys.*, submitted.

Love, A. E. H. [1927] *A Treatise on the Mathematical Theory of Elasticity* (Cambridge University Press, Cambridge, UK); Reprinted (Dover Publications, Inc. NY).

Swigon, D., Coleman, B. D. & Tobias, I. [1998] "The elastic rod model for DNA and its application to tertiary structure of DNA minicircles in mononucleosomes," *Biophys. J.* **74**, 2515–2530.

Thimoshenko, S. P. & Gere, J. M. [1961] *Theory of Elastic Stability* (McGraw-Hill, NY).

Tobias, I., Coleman, B. D. & Olson, W. [1994] "The dependence of DNA tertiary structure on end conditions: Theory and implications for topological transitions," *J. Chem. Phys.* **101**, 10990–10996.

White, J. H. [1969] "Self-linking and the gauss integral in higher dimensions," *Amer. J. Math.* **91**, 693–728.

BIFURCATIONS OF STABLE SETS IN NONINVERTIBLE PLANAR MAPS

J. P. ENGLAND, B. KRAUSKOPF and H. M. OSINGA

Bristol Centre for Applied Nonlinear Mathematics,
Department of Engineering Mathematics, University of Bristol,
Queen's Building, Bristol BS8 1TR, UK

Received May 4, 2004; Revised June 9, 2004

Many applications give rise to systems that can be described by maps that do not have a unique inverse. We consider here the case of a planar noninvertible map. Such a map folds the phase plane, so that there are regions with different numbers of preimages. The locus, where the number of preimages changes, is made up of so-called critical curves, that are defined as the images of the locus where the Jacobian is singular. A typical critical curve corresponds to a fold under the map, so that the number of preimages changes by two.

We consider the question of how the stable set of a hyperbolic saddle of a planar noninvertible map changes when a parameter is varied. The stable set is the generalization of the stable manifold for the case of an invertible map. Owing to the changing number of preimages, the stable set of a noninvertible map may consist of finitely or even infinitely many disjoint branches. It is now possible to compute stable sets with the Search Circle algorithm that we developed recently.

We take a bifurcation theory point of view and consider the two basic codimension-one interactions of the stable set with a critical curve, which we call the outer-fold and the inner-fold bifurcations. By taking into account how the stable set is organized globally, these two bifurcations allow one to classify the different possible changes to the structure of a basin of attraction that are reported in the literature. The fundamental difference between the stable set and the unstable manifold is discussed. The results are motivated and illustrated with a single example of a two-parameter family of planar noninvertible maps.

Keywords: Noninvertible map; stable set; critical curve; bifurcation; basin of attraction.

1. Introduction

One often encounters maps arising in applications that are noninvertible, by which is meant that the given map is smooth, but does not have a uniquely defined inverse. A well-referenced example of such a noninvertible system is that of a discrete-time adaptive control system [Adomaitis *et al.*, 1991; Frouzakis *et al.*, 1992; Frouzakis *et al.*, 1996]. In this example one finds multistability and the non-invertibility plays an important role in the structure of the basins of attraction of the coexisting attractors, which may consist of disconnected regions.

Generally, knowledge of the structure of the basins of attraction is key to understanding the long term evolution of the system. Other applications that give rise to noninvertible maps include models from economics [Agliari, 2000; Agliari *et al.*, 2003], radio-physics [Maistrenko *et al.*, 1996] and neural networks [Rico-Martinez *et al.*, 2000].

In this paper we focus on noninvertible maps of the plane. That is, we consider a dynamical systems that is given by a smooth planar map

$$f : \mathbb{R}^2 \mapsto \mathbb{R}^2$$

that does not have a unique inverse.

Geometrically such a noninvertible map folds the phase plane. Adopting the notation in [Nien & Wicklin, 1998] the *curve of merging preimages* (also denoted as LC_{-1}) is defined as

$$J_0 = \left\{ \mathbf{x} \in \mathbb{R}^2 \,|\, Df(\mathbf{x}) \text{ is singular} \right\},$$

and the first iterate of this curve, $J_1 = f(J_0)$, is called a *critical curve* (also denoted as LC). The dynamics of a planar map is such that the phase plane folds along the critical curves. Generically, the number of preimages of two points on either side of a fold line differs by two [Arnol'd, 1992], and points on the critical curve have two coincident preimages. We focus on this generic case of a simple fold, such that points on one side of J_1 have two more preimages than points on the other side. Common notation denotes Z_k as a region having k rank-one preimages. The simplest case of a single fold is then denoted by $(Z_0$–$Z_2)$, where points on one side of the fold have no preimages and points on the other side of the fold have two preimages. The folding of the phase plane may be more complicated, for example, for the case denoted as $(Z_1$–Z_3–$Z_1)$ there are regions with one and three preimages.

Much work has been done to investigate the dynamics of noninvertible maps and how it is related with the folding of the phase plane [Abraham *et al.*, 1997; Gumowski & Mira, 1977; 1980a, 1980b; Mira *et al.*, 1996a; Mira *et al.*, 1996b]. In particular, there has been considerable interest in bifurcations that lead to qualitative changes of basins of attraction [Agliari *et al.*, 2003; Cathala, 1998; Kitajima *et al.*, 2000; López-Ruiz & Fournier-Prunaret, 2003; Mira *et al.*, 1994]. Such basins are typically determined by computing the orbits for a large number of initial conditions. An alternative is to compute the stable set of a suitable saddle point, which forms the boundary of a given basin of attraction.

To define the stable set formally, assume that f has a saddle fixed point $\mathbf{x}_0 = f(\mathbf{x}_0)$ and that f is differentiable in a neighborhood of $\mathbf{x}_0$. The global stable set $W^s(\mathbf{x}_0)$ of $\mathbf{x}_0$ is defined as the set of points that converge to $\mathbf{x}_0$ under forward iteration of f,

$$W^s(\mathbf{x}_0) = \left\{ \mathbf{x} \in \mathbb{R}^2 \,|\, f^n(\mathbf{x}) \to \mathbf{x}_0 \text{ as } n \to \infty \right\}.$$

For an invertible map $W^s(\mathbf{x}_0)$ is an embedded manifold and one speaks of $W^s(\mathbf{x}_0)$ as the stable manifold. However, when multiple inverses exist $W^s(\mathbf{x}_0)$ may consist of disjoint pieces. In particular, this set is not an embedded manifold and this is why one speaks of $W^s(\mathbf{x}_0)$ as the global stable set [Mira *et al.*, 1996b]. Throughout this paper, the *primary manifold* is the unique connected subset of $W^s(\mathbf{x}_0)$ that contains the fixed point $\mathbf{x}_0$.

The computation of stable sets and inverse orbits is difficult due to the fact that the Jacobian may become singular and that the critical curves separate the phase plane into regions that have different numbers of preimages. Furthermore, the stable set may also consist of pieces that are disconnected from the saddle point. The recently developed Search Circle (SC) algorithm [England *et al.*, 2004] overcomes the problem of computing the primary manifold past intersections with the curve J_0 where the Jacobian is singular. It can also be used to compute disjoint pieces of the stable set. All that is needed to start a computation are the system equations themselves along with the saddle point and the stable eigenvector. All primary manifolds and stable sets in this paper have been computed with the implementation of the SC algorithm [Osinga & England, 2003] in the DsTool environment [Back *et al.*, 1992].

The SC algorithm makes it possible to find and consider codimension-one bifurcations where the stable set interacts with a critical curve. There are exactly two such codimension-one bifurcations — the outer-fold and inner-fold bifurcations that are discussed in detail in Sec. 3. (Similar interactions between unstable manifolds and attracting invariant circles have been investigated in [Frouzakis *et al.*, 1997; Frouzakis *et al.*, 2003; Maistrenko *et al.*, 1996].) Depending on the global organization of the stable set, these two basic codimension-one bifurcations may give rise to the different changes of basins of attraction that have been studied independently in the literature. This is discussed in Sec. 4. Finally, in Sec. 5 we illustrate the fundamental difference between the stable set and the unstable manifold. We discuss the codimension-one bifurcation that leads to structurally stable self-intersections of the unstable manifold, a phenomenon that cannot occur for the stable set.

2. Example

Throughout this paper we use a single example, namely the two-parameter family of planar noninvertible maps

$$Q\begin{pmatrix} x \\ y \end{pmatrix} = \begin{pmatrix} y \\ ax + bx^2 + y^2 \end{pmatrix}. \tag{1}$$

We call this map the modified Gumowski–Mira map, because Gumowski and Mira [1980a, 1980b] investigated the special case $a = 4/5$ and $b = 1$. For $b \neq -1$ this map has two fixed points. The origin is always a fixed point and it is attracting for $|a| < 1$. The other fixed point p is located at

$$p = \left(\frac{(1-a)}{(1+b)}, \frac{(1-a)}{(1+b)} \right).$$

For the special case considered in [Gumowski & Mira, 1980a, 1980b] the point p is a saddle with a negative stable eigenvalue. As is standard in such a situation, the stable set can be computed in this case by using the second iterate Q^2, but the folding by Q^2 is more complex because its Jacobian is singular both when $DQ(x)$ or $D(Q(Q(x)))$ are singular. By choosing different values for a and b this difficulty can be avoided. Specifically, we choose $a = -0.8$ throughout this paper and vary the parameter b; see also [England *et al.*, 2004]. Then the fixed point p is a saddle for $b < 3$, $b \neq -1$. The two sides of the primary manifold of the stable set of the saddle join to form a smooth closed loop. The primary manifold bounds the basin of attraction of the sink.

The map (1) is designed such that the Jacobian matrix becomes singular along a vertical line, namely

$$J_0 := \left\{ x = -\frac{a}{2b} \right\}. \tag{2}$$

The rank-one critical curve is the parabola

$$J_1 := \left\{ y = x^2 - \frac{a^2}{4b} \right\}.$$

The phase plane folds along J_0 under Q, and the image J_1 of the fold divides the plane into two distinct regions, one with two preimages and one with no preimages, denoted Z_2 and Z_0, respectively. Since a fixed point always has at least one preimage, namely itself, it must lie in Z_2. Hence, there is typically a distinct second preimage of the saddle p, which we denote by q.

The map (1) has the reflectional symmetry of a perfect fold around J_0. This means that the points

$$\left(-\frac{a}{2b} \pm x, \, y \right)$$

map to the same point under Q. In particular, the point q is the mirror image of p and has the same y-coordinate, namely

$$q = \left(-\frac{(a+b)}{b(1+b)}, \frac{(1-a)}{(1+b)} \right).$$

2.1. *Sequence of bifurcations of the stable set*

In [England *et al.*, 2004] the two particular cases $b = 0.2$ and $b = 0.1$ were used to illustrate the SC algorithm. Here we present a more detailed study of the bifurcation sequence as b is varied. Specifically, we can explain all bifurcations encountered by studying the interaction of the stable set with the critical curve J_1 in some small neighborhood.

Figure 1 shows the stable set for nine decreasing values of b. The scale of the vertical axes is the same in all panels, while the scale of the horizontal axes is adjusted, because the primary manifold is growing wider with decreasing $b > 0$. Specifically, the left point on the x-axis of each panel is fixed at $x = -3$ and the right point is chosen such that J_0 is always displayed at the center of each panel. The origin is a sink, and it is denoted by a blue triangle. In all panels the saddle p lies in the Z_2-region, above J_1, and its preimage q lies in the Z_0-region, below J_1, as indicated by green crosses. The stable sets $W^s(p)$, including the primary manifold, are shown in blue and the critical curves J_0 and J_1 are shown in gray. The red crosses are either points where the stable set is tangent to J_1, or preimages of such tangency points.

Figure 1(a) shows $W^s(p)$ for $b = 0.25$, where it only consists of the primary manifold that is connected to the saddle point. Both sides of the manifold join smoothly at q to form a closed loop. All points on $W^s(p)$ map to the segment in the Z_2-region above J_1. The situation is topologically the same for $0.189860 < b < 3$. At $b \approx 0.189860$ the primary manifold $W^s(p)$ becomes tangent to J_1; see panel (b). We observe that the point of tangency has a double preimage, which lies on J_0 (indicated by the red crosses). When decreasing b further, as is shown in panel (c) for $b = 0.14$, a new part of the stable set $W^s(p)$, namely a closed loop, is formed that is disconnected from the primary manifold. This so-called bubble has grown from the single point at the tangency and maps to the additional segment of the primary manifold that has moved above the J_1-curve. Panel (d) shows the case $b = 0.08995$, at which the disjoint bubble of the stable set is approximately tangent to J_1, giving rise to the birth of another disconnected closed

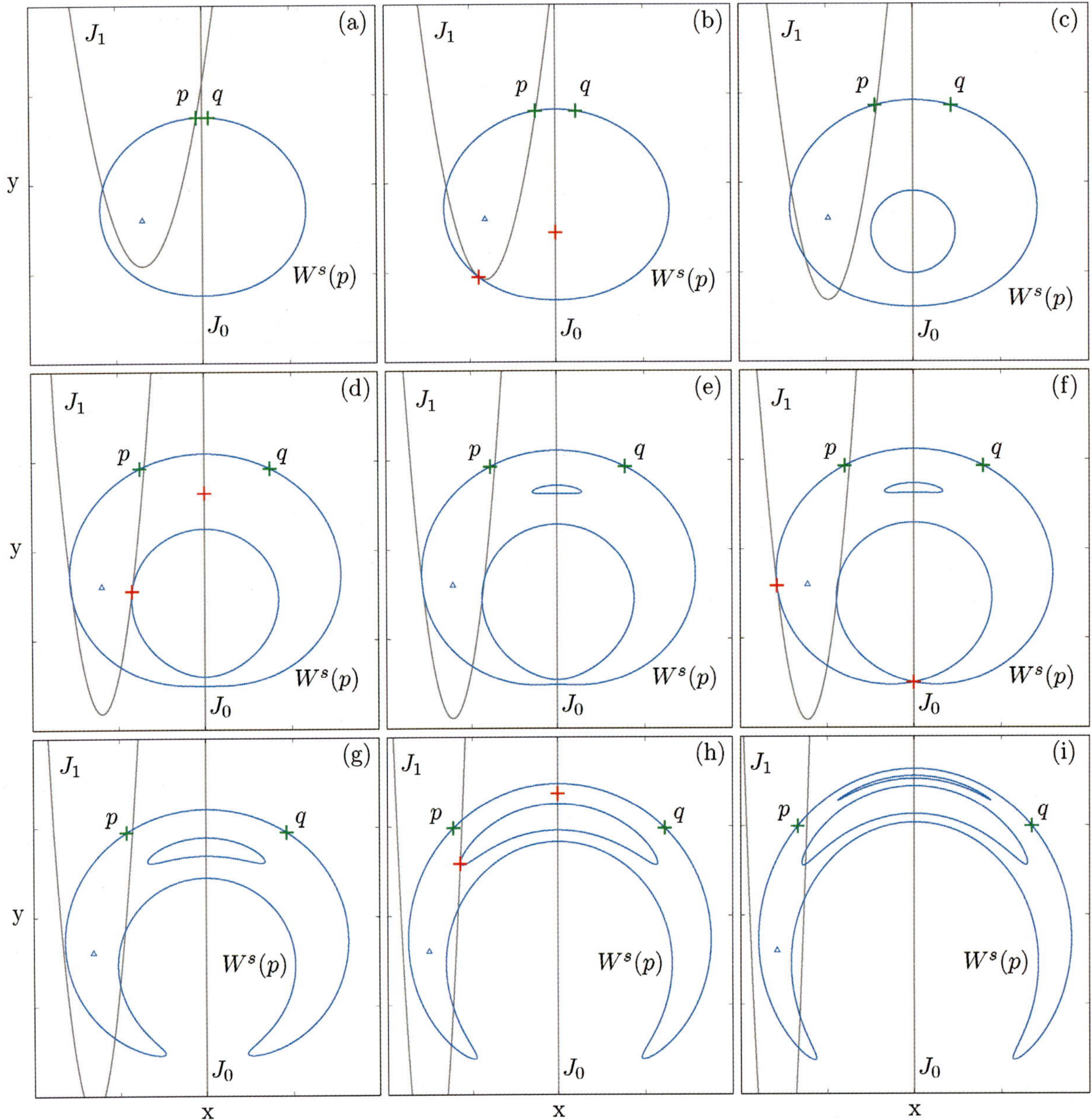

Fig. 1.   Bifurcations of the stable set $W^s(p)$ (blue curves) of the saddle p of (1) for $a = -0.8$ as b is varied. The curves J_0 and J_1 are shown in gray. The saddle point p and its preimage q are indicated by green crosses. Tangency points between $W^s(p)$ and J_1 and their preimages are indicated by red crosses. From (a) to (i) the parameter b takes the values 0.25, 0.18960, 0.14, 0.08995, 0.086, 0.0845735, 0.07, 0.04375 and 0.035.

curve, which will grow from the red cross; this is shown in panel (e) for $b = 0.086$. Approximately at $b = 0.0845735$, shown in Fig. 1(f), a tangency occurs at the left side of the picture, between the primary manifold and the curve J_1. We observe that the two separate segments of manifold that lie in the Z_2-region have joined. At the same time the original disjoint bubble of $W^s(p)$ connects with and then forms a part of the primary manifold. This connection happens at the preimage of the tangency with J_1. Decreasing b further, the primary manifold now has the shape of a horseshoe; the situation for

$b = 0.07$ is shown in panel (g). The second bubble is still disconnected from the primary manifold. As this bubble grows with decreasing b, there is a tangency between this bubble and J_1 at approximately $b = 0.04375$; see panel (h). This bifurcation gives birth to a third disjoint bubble, which is shown in panel (i) for $b = 0.035$.

In the bifurcation sequence above, there are effectively only two different bifurcations that lead to the observed changes of the stable set $W^s(p)$. Both are generic codimension-one bifurcations where there is a tangency between $W^s(p)$ and J_1.

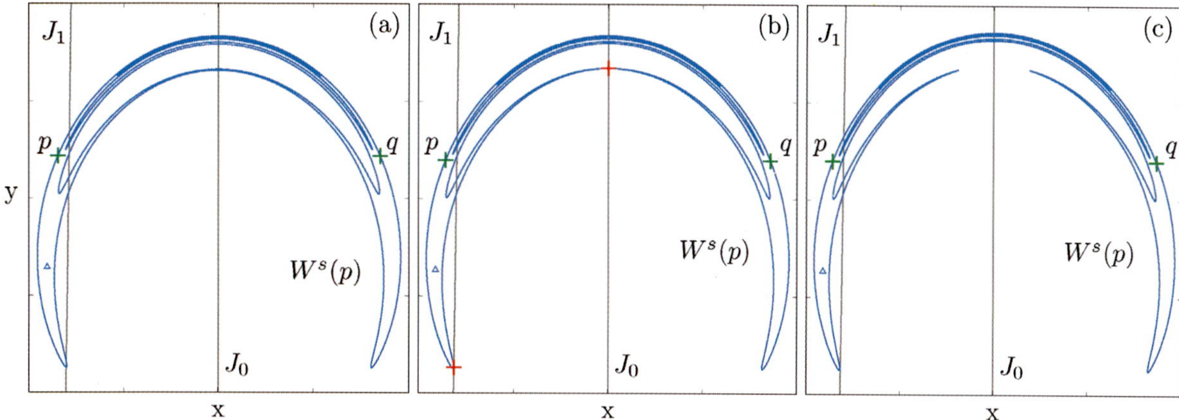

Fig. 2.   Further bifurcations of the stable set $W^s(p)$ of the saddle p of (1) for $a = -0.8$; compare Fig. 1. From (a) to (c) the parameter b takes the values 0.014, 0.01306669, and 0.0125, while the curve J_0 is at $x = 28.571$, $x = 30.612$ and $x = 32$, respectively. As $b \to \infty$ the disjoint bubbles disappear in inner-fold bifurcations; one such inner-fold bifurcation is shown in panel (b).

The first bifurcation, which we call an *outer-fold bifurcation*, results in the creation (or disappearance) of a new isolated closed curve that belongs to $W^s(p)$. The second bifurcation, which we call an *inner-fold bifurcation*, changes the local connectedness of branches of $W^s(p)$. Each of these bifurcations is discussed in detail from the point of view of bifurcation theory in the next section.

We finish this section by showing what happens if b is decreased further towards zero. Since the formula (2) for J_0 has b in the denominator, the curve J_0 tends to infinity in the limit as b approaches zero. For $b = 0$ the map (1) is actually a diffeomorphism, that is, each point has a unique preimage and the map Q is invertible. In particular, the saddle p still exists, but its second preimage q does not. Furthermore, $W^s(p)$ must be a simply connected smooth stable manifold. Hence, as b gets closer to zero all bubbles must disappear. Figure 2 gives an indication of how this happens with the three phase portraits for (a) $b = 0.014$, (b) $b = 0.01306669$ and (c) $b = 0.0125$. The bubbles are joined one by one to the primary manifold in inner-fold bifurcations; compare Fig. 2(b). As b tends to zero the left branch of $W^s(p)$ retracts further and further into the Z_2-region and, since J_0 disappears to infinity, the right branch goes off to infinity for $b = 0$.

3.   Generic Codimension-One Bifurcations of Stable Sets

The outer-fold and inner-fold bifurcations in Sec. 2 underlie various types of "basin bifurcations"

[Cathala, 1998; Kitajima *et al.*, 2000; Mira *et al.*, 1994] and "contact bifurcations" [Agliari, 2000; Agliari *et al.*, 2003; López-Ruiz & Fournier-Prunaret, 2003] that can be found in the literature. As was mentioned in the introduction, these bifurcations were not viewed in the first instance as bifurcations of a stable set, but as bifurcations of a basin of attraction. Because the underlying outer-fold or inner-fold bifurcations can change a given basin of attraction in different ways, they are associated with many different names, depending on their global flavor and which basin of attraction is under consideration. Indeed the notation in the literature is rather complicated. It is generally related to the connectedness of the basin of attraction and some papers even contain a glossary of the names that are given to the different bifurcations [López-Ruiz & Fournier-Prunaret, 2003; Mira *et al.*, 1994].

The main point of this paper is to take the point of view of bifurcation theory and singularity theory in order to provide a systematic way of classifying qualitative changes to the stable set and, hence, to basins of attraction. To this end, we first consider the *generic codimension-one bifurcations* where the stable set interacts with the critical curve J_1. The assumption that the codimension-one bifurcation be generic means, in particular, that the stable set crosses the critical curve J_1 at a generic point, that is, at an image of a regular fold point of J_0. Furthermore, genericity demands that the stable set and J_1 have a quadratic tangency. (We do not consider here the case that the stable set and J_1 already have a generic crossing, in which case the generic bifurcation would be a cubic tangency.) In other words, there are exactly two *generic codimension-one bifurcations* where the stable set crosses J_1 upon change of a parameter, namely the outer-fold and inner-fold bifurcations we already encountered in the previous section. We stress that this is true irrespective of the particular folding structure of the given map.

These two bifurcations, when seen only in a local neighborhood of the tangency, are the basic building blocks of any change in the structure of a basin boundary when a single parameter is varied. The second step is to consider the *global arrangement of the stable set* at the moment that it undergoes either an outer-fold or an inner-fold bifurcation. Since there are different possibilities of connecting branches of the stable set globally, there are different ways in which an outer-fold or inner-fold bifurcation manifests itself, for example, when one

is interested in the change of a basin of attraction. In this way, all changes to a given basin of attraction, including changes to its local connectedness, can be understood and classified in a systematic way as a combination of an outer-fold or inner-fold bifurcation with a particular global flavor.

3.1.   *Outer-fold bifurcation*

The outer-fold bifurcation occurs when a segment of the stable set becomes tangent to the J_1 curve on the outer side of the fold, so that a segment of the stable set crosses into the region with two extra preimages. The additional preimages of this segment form a disjoint bubble that is part of the stable set.

Figure 3 demonstrates how the outer-fold bifurcation creates a disjoint bubble by using data for

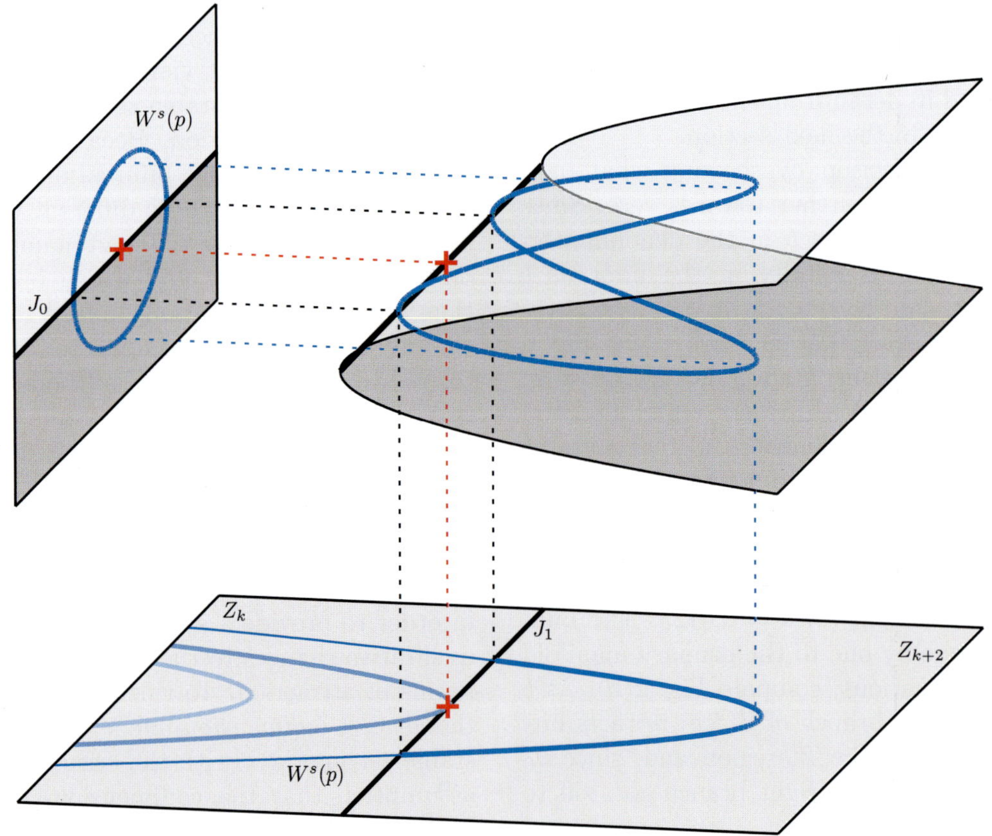

Fig. 3.   A schematic representation of the outer-fold bifurcation. The map f folds the vertical plane along J_0 and maps it to the right of J_1 onto the horizontal plane. There are two regions Z_k and Z_{k+2} to the left and right of J_1 with k and $k+2$ rank-one preimages, respectively. The stable set $W^s(p)$ does not intersect J_1 before the bifurcation, but is tangent to J_1 at the outer-fold bifurcation, and has two intersection points with J_1 after the bifurcation. After the bifurcation a part of $W^s(p)$ extends into the region Z_{k+2} where there are (locally) two extra preimages. This part of $W^s(p)$ lifts to the folded phase plane, resulting in an isolated closed curve near the preimage of the tangency point on J_0. The shown manifolds are (scaled) data of the map (1) for $a = -0.8$ and $b = 0.25$, $b = 0.189860$ and $b = 0.14$, respectively.

the map (1). The illustration is in the spirit of singularity theory and shows the folded phase plane in such a way that the action of the map f can be interpreted as a simple projection onto regions near J_0 and J_1, respectively. Indeed, a local neighborhood of J_0, shown as a vertical plane, maps to a local neighborhood of J_1, shown as a horizontal plane. It is indicated in the figure that the stable set crosses from a region Z_k with k preimages into a region Z_{k+2} with $k+2$ preimages. However, because we only consider the situation locally near the tangency point, this is (locally) equivalent to the case of a Z_0–Z_2 map, such as (1).

For $b = 0.25$, a segment of the stable set is shown in light blue. Since it does not intersect J_1 it has k preimages (outside the local neighborhood that we are interested in). For $b = 0.189860$ the stable set is shown in a darker blue and it is tangent to J_1. This means that the intersection point on $W^s(p)$ and J_1 (red cross) has one extra (double) preimage, which is illustrated by projecting the point up to the folded plane and then across, giving the double preimage (also denoted by a red cross) on the curve J_0. The stable set for $b = 0.14$ is shown in even darker blue and it crosses J_1 into the folded region, so that a whole segment of the manifold lies in the Z_{k+2}-region. If one projects this piece of manifold up to the folded plane and then across to the unfolded phase plane, it is clear how the disjoint bubble is formed around the preimage of the tangency point on J_0. Note that the piece of the stable set in the Z_{k+2}-region has k other preimages, which do not take part in the bifurcation; they correspond to the k preimages in the Z_k-region.

We remark that an outer-fold bifurcation occurs three times in the bifurcation sequence shown in Fig. 1, namely in panels (b), (d) and (h). Each case is different with respect to which part of the stable set has an outer-fold bifurcation with J_1. For example, in panel (d) there is a tangency of the first bubble with the curve J_1, leading to the creation of the second bubble. As is shown in Fig. 4, this bifurcation can be interpreted as a tangency of the primary manifold with the curve $J_2 = f(J_1)$. Indeed, all images of tangencies are again tangencies of the respective images.

3.2. *Inner-fold bifurcation*

The inner-fold bifurcation occurs when a segment of the stable set becomes tangent to the curve J_1 on the inner side of the fold, so that a segment

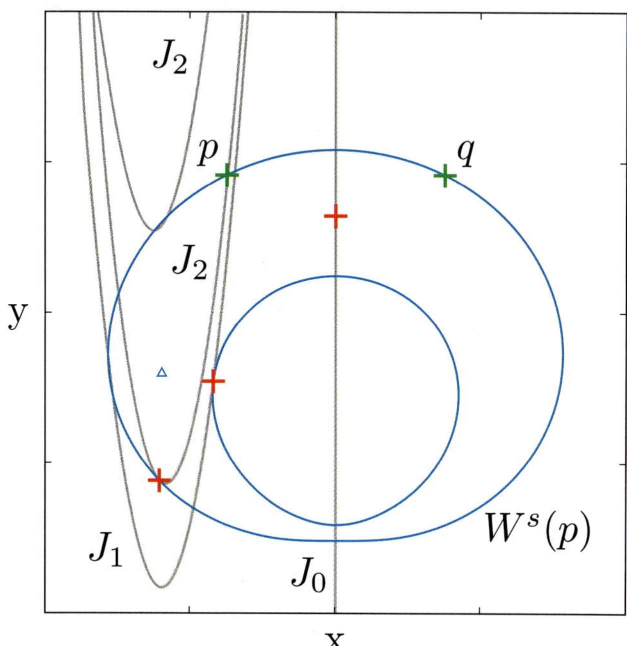

Fig. 4. The stable set $W^s(p)$ (blue curves) of (1) for $a = -0.8$ and $b = 0.08995$ shown together with the curves J_0, J_1 and $J_2 = f(J_1)$ (gray curves). The saddle point p and its preimage q are indicated by green crosses. A tangency between the disjoint bubble and J_1 must map to a tangency between the primary manifold and J_2. The preimage on J_0 of the tangency with J_1 is then the onset of a new disjoint bubble. Tangencies and their preimages are indicated by red crosses.

of the stable set crosses into the region with two less preimages. This leads to a different connectivity between the four branches of the stable set that are involved in this bifurcation. Figure 5 demonstrates this with data for the map (1). The three panels (a)–(c) show the situation before, at and after the inner-fold bifurcation in the same way as in Fig. 3. Again, a local neighborhood of J_0, shown as the vertical plane, maps to a local neighborhood of J_1, shown as the horizontal plane.

Before the tangency, in Fig. 5(a), the stable set locally extends across J_1 into the Z_k-region. The two segments in the Z_{k+2}-region each have two (local) preimages which are connected across J_0 as indicated by the vertical plane, the projection to a local neighborhood of J_0. At the inner-fold bifurcation the stable set is tangent to J_1 and the two segments in the Z_{k+2}-region connect at the tangency point with J_1, indicated again by the red cross. Therefore, their preimages are also connected at a single point on J_0. After the bifurcation, the stable set remains entirely inside the Z_{k+2}-region. Its two disjoint preimages do not connect across J_0

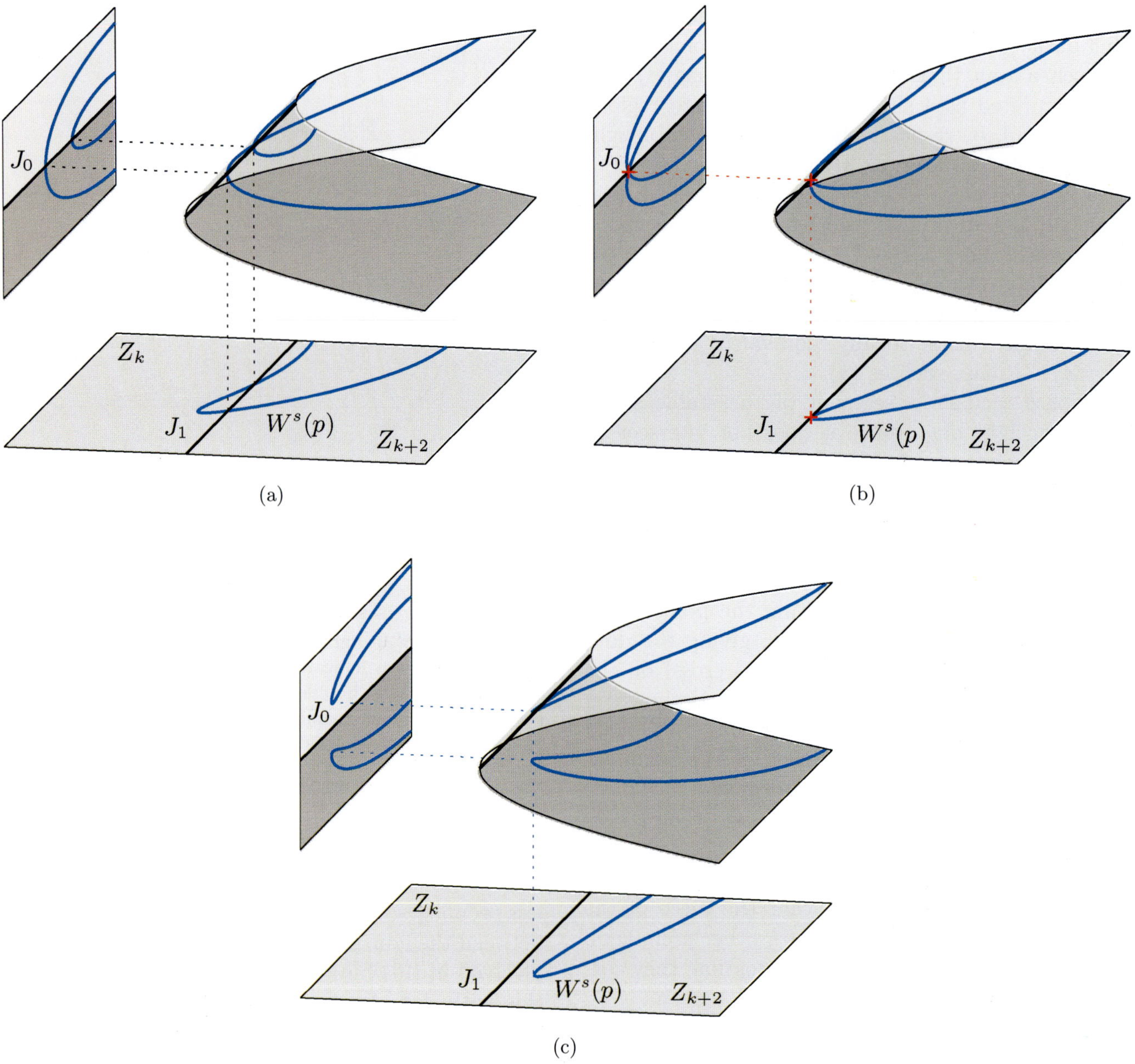

Fig. 5.   A schematic representation of the inner-fold bifurcation, illustrated in the same way as in Fig. 3. The three panels show the local phase portraits before (a), at (b), and after (c) the inner-fold bifurcation using (scaled) data from (1) for $a = -0.8$ and $b = 0.086$, $b = 0.0845735$ and $b = 0.07$, respectively.

any longer, which means that the connectivity of the branches near J_0 has changed. As before, k other preimages of the stable set do not take part in the bifurcation.

We remark that an inner-fold bifurcation occurs in Fig. 1(f). The change of the local connectivity can clearly be seen by comparing panels (e) and (g). Since the branches of the stable set involve the primary manifold and the first bubble, this inner-fold bifurcation manifests itself as a qualitative change of the primary manifold.

3.3. *Different global flavors of the inner-fold bifurcation*

The overall or global manifestation of an inner-fold bifurcation depends on which part of the stable set crosses J_1. Furthermore, it is important to know how the preimages of the two segments of the stable set, which meet at the preimage of the tangency point, are connected *outside the local neighborhood* that we consider.

Figure 6 shows two topologically different global phase portraits at the moment of an inner-fold

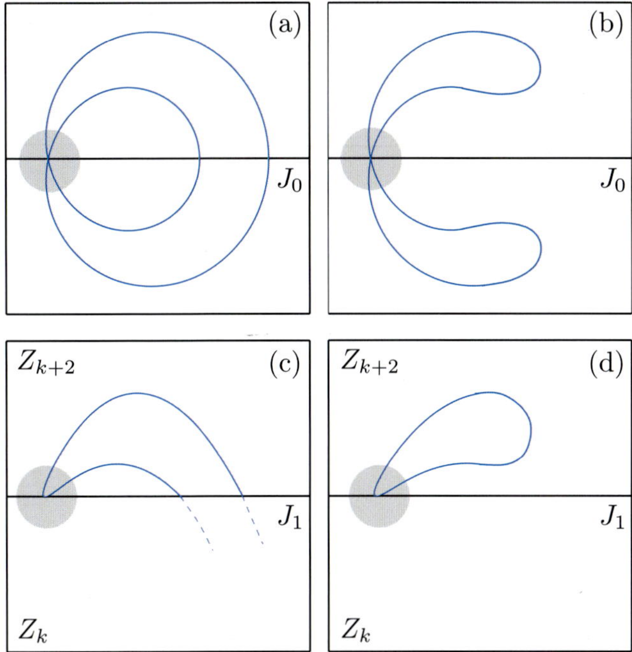

Fig. 6.   Two different global flavors of the inner-fold bifurcation. The stable sets in panels (a) and (b) are mapped to the Z_{k+2}-region as shown in panels (c) and (d), respectively. Panels (a) and (c) demonstrate the case where, at the bifurcation, the stable set intersects J_1 outside a local neighborhood of the tangency point, while in (b) and (d) the stable set does not intersect J_1 outside a local neighborhood.

bifurcation. Both agree in the small gray neighborhood, but the global structure outside this gray neighborhood is different. In panel (a) the two branches are connected across J_0 outside a neighborhood of the point where they meet. We have already seen this possibility for the inner-fold bifurcation in Figs. 1(e)–1(g). This means that the stable set must cross J_1 at two different points, as shown in the image under f in panel (c). In Fig. 6(b), on the other hand, the two branches are connected in such a way that they do not cross J_0 outside a neighborhood of the point where they meet. This means that the image, shown in panel (d), remains entirely inside the Z_{k+2}-region. In this case, the stable set in panel (b) takes the form of a pinched bubble. In the unfolding of this inner-fold bifurcation a single bubble that crosses J_0 at two nearby points pinches and then splits into two separate bubbles.

4.   Bifurcations of Basin Boundaries

The bifurcations of the stable set that we discussed so far can be interpreted directly as bifurcations of basins of attraction. To make this point, we show in Fig. 7 four instances of panels from Fig. 1 where we colored the basin of the origin in green and the basin of infinity in blue. The coloring is motivated by the literature on basins of attraction that speaks of "sea", "land", "lakes" and "islands". Indeed, as noted in [Kitajima *et al.*, 2000; Mira *et al.*, 1994], "islands" and "lakes" are equivalent, simply by exchanging the coloring of the respective basins.

Figure 7(a) shows a situation where the basin of attraction is a simply connected domain. However, as is clear from the other panels in Fig. 7, a basin of attraction of a noninvertible map need not be simply connected. While the literature speaks of the changes to the "island number" or "lake number" [Mira *et al.*, 1994], a topological classification of a basin of attraction would require one to consider its fundamental group and how it changes in bifurcations.

In this paper we argue that the notation in the literature is rather phenomenological, leading to many seemingly different cases, while the underlying bifurcation is always either an outer-fold or an inner-fold bifurcation of a particular global flavor. For example, an outer-fold bifurcation leads to the situation in Fig. 7(b) where the green basin is multiply connected (has a nontrivial fundamental group). In the literature one often finds the description that the green basin, which is an "island" or "continent" in the "sea", has a "hole" or "lake". The bifurcation itself has been called a "connected basin ↔ multiply connected basin bifurcation" when seen from the point of view of the green basin of the origin; it has also been called a "connected basin ↔ disconnected basin bifurcation", when seen from the point of view of the blue basin of infinity [Mira *et al.*, 1994]. (These two names are equivalent in the case of only two basins, because the green basin being simply connected means, by definition, that the blue basin is connected, and vice versa.) The second outer-fold bifurcation changes the connectivity of the green basin again. As shown in Fig. 7(c), there are now two "lakes" (and the fundamental group has two generators). As far as we are aware, any further increase in the connectivity of a basin is referred to in the literature as a change to either the "island number" or the "lake number" [Mira *et al.*, 1994].

Finally, in an inner-fold bifurcation with the global flavor as was shown in Figs. 6(a) and 6(c), the connectivity of the green basin changes again. As is shown in Fig. 7(d), the "lake" has now joined the

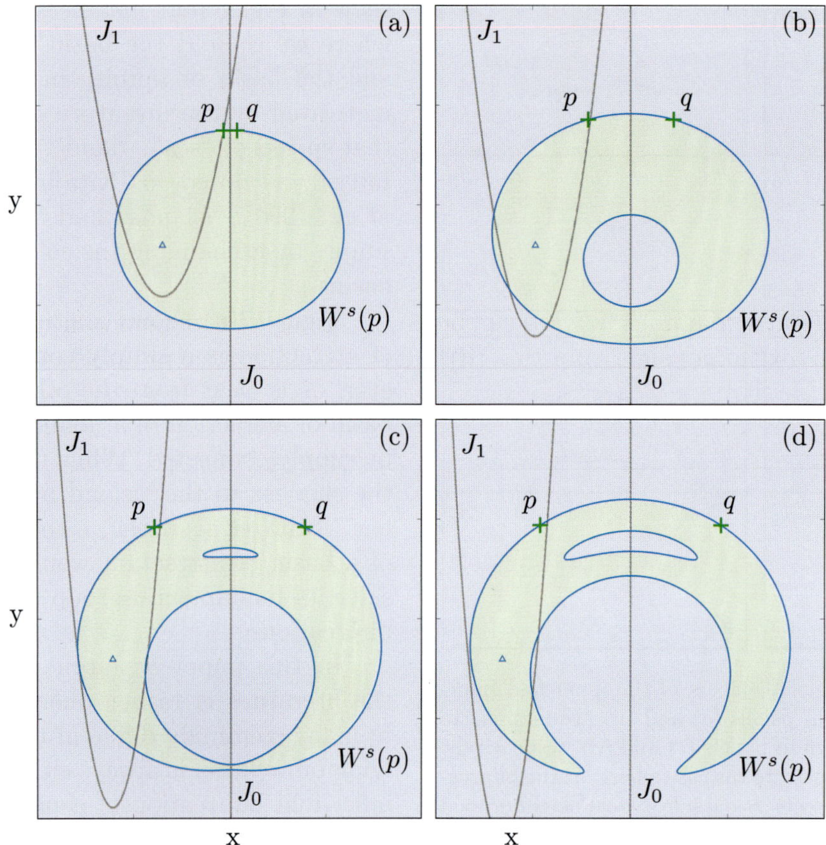

Fig. 7.   The stable sets of (1) for (a) $b = 0.25$, (b) $b = 0.14$, (c) $b = 0.086$ and (d) $b = 0.07$. The green shaded area indicates the basin of attraction for the origin (blue triangle). The blue shaded area indicates the basin of attraction of infinity.

"sea", which is also referred to as a "lake $\leftrightarrow$ roadstead" bifurcation [Mira *et al.*, 1994]. The fundamental group becomes smaller, and the topology of the basins in Fig. 7(d) is same as in Fig. 7(b).

In summary, changes to the connectivity of a basin of attraction can be classified in the spirit of bifurcation theory by only two ingredients: whether an outer-fold or an inner-fold bifurcation is involved and the global organization of the stable set at the moment of bifurcation. The latter involves information about which part of the stable set, the primary branch or a disjoint piece, undergoes the bifurcation and how branches are connected outside a neighborhood of the bifurcation point.

5.   Stable Set versus Unstable Manifold

If one considers an invertible map, that is, a diffeomorphism, then the stable manifold is the unstable manifold of the inverse f^{-1}, and vice versa. However, for a noninvertible map f as considered here this duality between stable and unstable

manifolds does not exist. Indeed, there is a fundamental difference between the stable set $W^s(\mathbf{x}_0)$ of a saddle point $\mathbf{x}_0$ and the unstable manifold $W^u(\mathbf{x}_0)$. This means that one also finds fundamentally different bifurcations when these sets interact with the curves J_0 and J_1.

By definition, the *global unstable manifold* $W^u(\mathbf{x}_0)$ consists of points that converge to $\mathbf{x}_0$ under backward iteration, that is, under application of a sequence of inverse branches of f. In terms of forward iterates this can be expressed as

$$W^u(\mathbf{x}_0) = \left\{ x \in \mathbb{R}^2 | \exists \, \{q_k\}_{k=0}^{\infty}, q_0 = x \text{ and} \right.$$

$$\left. f(q_{k+1}) = q_k, \text{ such that } \lim_{k \to \infty} q_k = \mathbf{x}_0 \right\}.$$

Because we assume that the Jacobian is nonsingular at $\mathbf{x}_0$, there exists the local unstable manifold $W^u_{\text{loc}}(\mathbf{x}_0)$, which is associated with the unique inverse branch that fixes $\mathbf{x}_0$. The global unstable manifold $W^u(\mathbf{x}_0)$ can then be expressed as

$$W^u(\mathbf{x}_0) = \bigcup_{n=1}^{\infty} f^n(W^u_{\text{loc}}(\mathbf{x}_0)). \tag{3}$$

Note that even for noninvertible f the images of $W^u_{\text{loc}}(\mathbf{x}_0)$ are unique. Indeed there may be other preimages of $W^u(\mathbf{x}_0)$, but these are not part of $W^u(\mathbf{x}_0)$ because points in these preimages do not converge to $\mathbf{x}_0$ under backward iteration. Overall, $W^u(\mathbf{x}_0)$ is generically an immersed manifold (see e.g. [Spivak, 1979]), so that it is justified to speak of $W^u(\mathbf{x}_0)$ as the *unstable manifold*. Note that $W^u(\mathbf{x}_0)$ may have generic transverse self-intersections as discussed below. The unstable manifold $W^u(\mathbf{x}_0)$ can be computed numerically with any algorithm that was developed for invertible maps.

As we discuss now, $W^u(\mathbf{x}_0)$ may have cusp points, in which case it is a piecewise immersed manifold. However, this situation is not generic and corresponds to a bifurcation of codimension at least one.

As seen, an interaction between the stable set $W^s(\mathbf{x}_0)$ and the critical curve J_1 leads to a bifurcation occurring in the neighborhood of its preimages on J_0. For the unstable manifold $W^u(\mathbf{x}_0)$, the converse is true: one needs to consider the interaction between $W^u(\mathbf{x}_0)$ and the curve J_0 where the Jacobian is singular.

A transverse intersection of $W^u(\mathbf{x}_0)$ with J_0 generically corresponds in the image to a tangency between $W^u(\mathbf{x}_0)$ and J_1. The genericity condition is that the tangent to $W^u(\mathbf{x}_0)$ at the crossing point with J_0 (which is, in fact, the eigenvector of the zero eigenvalue) does not coincide with the normal to J_0 at this point. A codimension-one bifurcation occurs when this genericity condition is violated. This is shown in Fig. 8, where panels (a) and (c) show structurally stable tangencies of $W^u(\mathbf{x}_0)$ with J_1. At the bifurcation point, as in Fig. 8(b), the tangent to $W^u(\mathbf{x}_0)$ and the normal to J_0 at the crossing point coincide. This means that (generically) $W^u(\mathbf{x}_0)$ has a cusp point where $W^u(\mathbf{x}_0)$ and J_1 meet. This bifurcation leads to the creation of a structurally stable, transverse self-intersection and a little loop of $W^u(\mathbf{x}_0)$ in a neighborhood of the image near J_1, as shown in Fig. 8(c). Note that there is a clear sense of direction along $W^u(\mathbf{x}_0)$ even when it has loops. Since $W^u(\mathbf{x}_0)$ is invariant under f, this bifurcation creates infinitely many self-intersections and loops, which are the images under f^l (for any integer $l \geq 1$) of the intersection point of $W^u(\mathbf{x}_0)$ with J_0.

Self-intersections and the associated loops of an unstable manifold have been reported in the literature; see [Lorenz, 1989], where the loops are called "antennae", and [Frouzakis *et al.*, 1997; Frouzakis *et al.*, 2003; Maistrenko *et al.*, 2003;

Mira *et al.*, 1996b]. These authors are mainly interested in bifurcations leading to the destruction of an invariant curve (also called "IC" or "torus"), which is the closure of unstable manifolds of suitable periodic points. The development of cusps and then loops of the unstable manifold is interpreted as a global bifurcation of the invariant curve. For example, Frouzakis *et al.* [2003, p. 107] reported the "destruction of the IC through a global bifurcation, appearance of loops on an unstable manifold, and the reappearance of an attractor, this time chaotic with loops". The ensuing global attractor "inherits" loops from the global unstable manifold, and this type of attractor has been called a "weakly chaotic ring" [Frouzakis *et al.*, 1997; Mira *et al.*, 1996b]. The papers [Frouzakis *et al.*, 1997; Frouzakis *et al.*, 2003] and [Maistrenko *et al.*, 2003] contain explanations of how loops are formed. In [Frouzakis *et al.*, 1997, p. 1178] the point on J_1 around which the loop forms is called a "self intersection of projection". Both [Frouzakis *et al.*, 2003] and [Maistrenko *et al.*, 2003] state the condition that the eigenvector corresponding to the zero eigenvalue coincides with the normal to J_0 at the moment when the unstable manifold develops cusps. However, none of the authors gives this codimension-one bifurcation a name or describes it in the spirit of bifurcation theory.

It does not appear to have been reported explicitly in the literature that this codimension-one bifurcation is given by a cubic tangency of $W^u(\mathbf{x}_0)$ with respect to the normal vector of J_0, and that it simply unfolds as a cusp singularity. This can be seen in the vertical neighborhood, on the left in Fig. 8, around the transverse crossing point of $W^u(\mathbf{x}_0)$ and J_0. The unfolding can be written as the normal form

$$R(s) = \mu s - s^3 \qquad (4)$$

in the (r, s)-plane, where $J_0 = \{s = 0\}$, the normal vector is $(0, 1)$, $W^u(\mathbf{x}_0) = \text{graph}(R)$ and μ is the unfolding parameter. Figure 8 was obtained by using the normal form (4) in the vertical neighborhood near J_0 and then using projections via the folded plane to the horizontal neighborhood of J_1. To avoid confusion with the codimension-two cusp bifurcation of equilibria, we call this codimension-one bifurcation of the unstable manifold the *loop bifurcation*.

We finish this section with a contrasting statement.

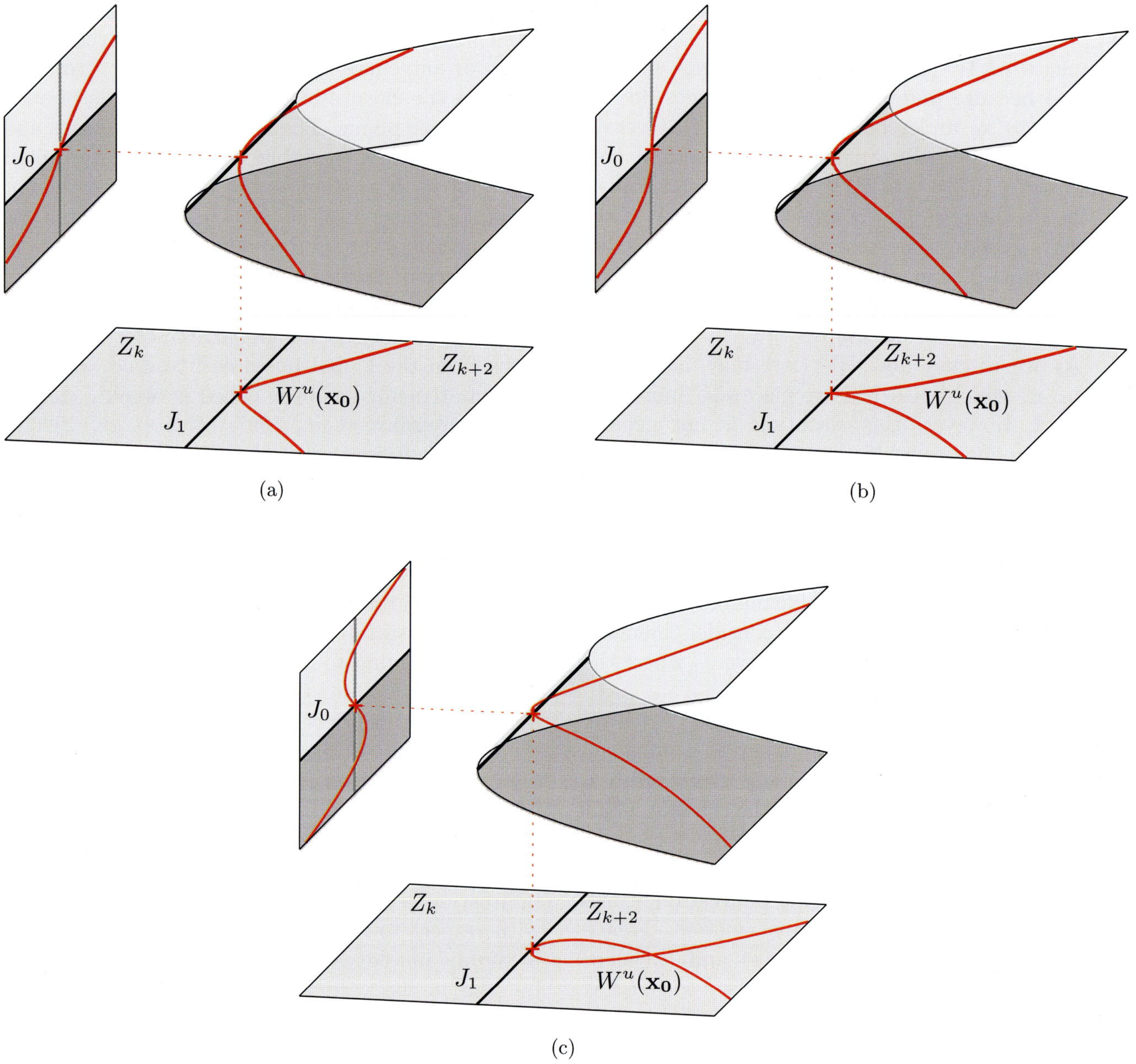

Fig. 8.   The loop bifurcation, illustrated as in Fig. 3. The three panels show the unstable manifold (red curve) in a local neighborhood before (a), at (b), and after (c) the bifurcation; the gray line in the vertical phase plane is the direction normal to J_0 at the crossing point with $W^u(\mathbf{x_0})$. The panels were obtained from the normal form (4) for $\mu = 1$, $\mu = 0$ and $\mu = -1$, respectively.

Proposition 1.   *The stable set $W^s(\mathbf{x_0})$ of a hyperbolic saddle point $\mathbf{x_0}$ of a noninvertible map $f : \mathbb{R}^n \to \mathbb{R}^n$, $n > 1$, cannot have structurally stable transverse self-intersections.*

Proof.   Since $\mathbf{x_0}$ is hyperbolic it does not lie on J_0, so that there is a small neighborhood U of $\mathbf{x_0}$ in which f is a local diffeomorphism (taking again the appropriate branch of the inverse). Hence, $W^s_{\mathrm{loc}}(\mathbf{x_0}) \subset U$ does not have self-intersections.

Consider now a transverse intersection of two branches of $W^s(\mathbf{x_0})$ in some neighborhood V. Since the intersection point $t \in V$ is on the stable set, it must map under some iterate of f, say, under f^L, into $W^s_{\mathrm{loc}}(\mathbf{x_0}) \subset U$.

Suppose now that the intersection point t does not lie on the set of preimages $\bigcup_{0 \le l \le L} f^{-l}(J_0)$ of J_0. Then all iterates f^l for $1 \le l \le L$ are diffeomorphisms on the neighborhood V. In particular, $f^L(V)$ contains a transverse intersection of two branches of $W^s(\mathbf{x}_0)$. Since $f^L(t) \in f^L(V) \cap U \neq \emptyset$ we have found a transverse intersection on $W^s_{\mathrm{loc}}(\mathbf{x}_0)$, which is a contradiction.

In other words, we must have $f^l(t) \in J_0$ for some $0 \leq l \leq L$. This shows that the self-intersection t of $W^s(\mathbf{x}_0)$ is not structurally stable, because a small perturbation of f destroys this property. $\blacksquare$

The condition that $f^l(t) \in J_0$ means that, generically, we find a codimension-one inner-fold bifurcation at $f^l(t)$ as was described in Sec. 3.2. Hence, the last self-intersection $f^l(t)$ of the stable set unfolds as shown in Fig. 5.

Note that Proposition 1 was formulated for a saddle point $\mathbf{x}_0$ for convenience and can be generalized to stable sets of other hyperbolic invariant set of saddle type. In particular, the statement holds for a saddle periodic orbit simply by means of considering an appropriate iterate of f.

6. Conclusions

Bifurcations of stable sets in noninvertible planar maps occur when the stable set interacts with a critical curve where the number of preimages changes. It is now possible to compute stable sets and find such bifurcations directly. Note that a method for computing the regions of different numbers of preimages of noninvertible maps has been developed in [Nien & Wicklin, 1998] and implemented in the program PISCES [Wicklin, 1995].

We considered a generic codimension-one bifurcation of the stable set, of which there are exactly two cases, the outer-fold and the inner-fold bifurcation. Both were illustrated with data from an example family of noninvertible maps. Furthermore, we contrasted the properties of the stable set with that of the unstable manifold. We showed how the unstable manifold may develop structurally stable self-intersections, while this is not possible for the stable set.

It must be stressed that the results presented here are valid for any noninvertible map, irrespective of its particular folding of the phase plane. This is so because the generic case is always a quadratic tangency with a critical curve along which the number of preimages changes by two. Furthermore, the outer-fold and the inner-fold bifurcations are the generic codimension-one interactions of any invariant curve (not necessarily a segment of a stable set) with the critical curve.

In other words, the unfoldings presented here are equally valid, for example, to describe bifurcations of invariant circles.

Since a basin boundary of a noninvertible planar map is typically bounded by stable sets, the outer-fold and the inner-fold bifurcations lead to changes in the structure of a given basin. In fact, we argued that all changes to basins of attraction can be classified in the spirit of bifurcation theory by the type of bifurcation in combination with its global flavor, by which is meant the global structure of the part of the stable set that undergoes the bifurcation.

Obvious future work is the consideration of generic bifurcations of stable sets (or invariant curves) of higher codimension. The codimension may be increased by interacting with the critical curve at nongeneric points, for example a cusp. Alternatively, one may look at a nonquadratic (quartic) tangency with a generic segment of the critical curve.

Finally, we mention that the general approach presented here may be used to consider noninvertible maps on higher-dimensional spaces. The next step would be to consider noninvertible maps of $\mathbb{R}^3$. This is already quite a challenge. One now deals with a two-dimensional critical manifold J_1 (instead of a critical curve) along which the number of preimages changes. Clearly, the singularity theory of smooth noninvertible maps of $\mathbb{R}^3$ is more complicated than that of smooth noninvertible maps of $\mathbb{R}^2$. Furthermore, the stable set may be up to two-dimensional.

Acknowledgments

We are grateful to Bruce Peckham for sharing his insight into the literature on noninvertible maps and for helpful comments on a draft of this paper. We thank Yuri Maistrenko for stimulating discussions. The research of J. P. England was supported by grant GR/R94572/01 from the Engineering and Physical Sciences Research Council (EPSRC).

References

Abraham, R. H., Gardini, L. & Mira, C. [1997] *Chaos in Discrete Dynamical Systems: A Visual Introduction in 2 Dimensions* (Springer-Verlag, NY).

Adomaitis, R. A. & Kevrekidis, I. G. [1991] "Noninvertibility and structure of basins of attraction in a model adaptive control system," *J. Nonlin. Sci.* **1**, 95–105.

Agliari, A. [2000] "Global bifurcations in the basins of attraction in noninvertible maps and economic applications," *Proc. Third World Congr. Nonlinear Analysts, Part 8* (Catania, 2000); [2001] *Nonlin. Anal.* **47**, 5241–5252.

Agliari, A., Gardini, L. & Mira, C. [2003] "On the fractal structure of basin boundaries in two-dimensional noninvertible maps," *Int. J. Bifurcation and Chaos* **13**, 1767–1785.

Arnol'd, V. I. [1992] *Catastrophy Theory*, 3rd, revised and expanded edition (Springer-Verlag, Berlin).

Back, A., Guckenheimer, J., Myers, M. R., Wicklin, F. J. & Worfolk, P. A. [1992] "DsTool: Computer assisted exploration of dynamical systems," *Not. Amer. Math. Soc.* **39**, 303–309.

Cathala, J. C. [1998] "Basin properties in two-dimensional noninvertible maps," *Int. J. Bifurcation and Chaos* **8**, 2147–2189.

England, J. P., Krauskopf, B. & Osinga, H. M. [2004] "Computing one-dimensional stable manifolds and stable sets of planar maps without the inverse," *SIAM J. Appl. Dyn. Syst.* **3**, 161–190.

Frouzakis, C. E., Adomaitis, R. A., Kevrekidis, I. G., Golden, M. P. & Ydstie, B. E. [1992] "The structure of basin boundaries in a simple adaptive control system," *Proc. NATO 1992, Advanced Summer Institute* (Ed. Bountis T.), pp. 195–210.

Frouzakis, C. E., Adomaitis, R. A. & Kevrekidis, I. G. [1996] "An experimental and computational study of subcriticality, hysteresis and global dynamics for a model adaptive control system," *Comput. Chem. Engin.* **120**, 1029–1034.

Frouzakis, C. E., Gardini, L., Kevrekidis, I. G., Millerioux, G., & Mira, C. [1997] "On some properties of invariant sets of two-dimensional noninvertible maps," *Int. J. Bifurcation and Chaos* **7**, 1167–1194.

Frouzakis, C. E., Kevrekidis, I. G. & Peckham, B. [2003] "A route to computational chaos revisited: Noninvertibility and the breakup of an invariant circle," *Physica* **D177**, 101–121.

Gumowski, I. & Mira, C. [1977] "Solutions chaotiques bornée d'une récurrence ou transformation ponctuelle du second ordre à inverse non-unique," *Comptes Rendus Acad. Sc. Paris* **A285**, 477–480.

Gumowski, I. & Mira, C. [1980a] *Dynamique Chaotique* (Ed. Cepadues, Toulouse).

Gumowski, I. & Mira, C. [1980b] *Recurrences and Discrete Dynamic Systems* (Springer-Verlag, NY).

Kitajima, H., Kawakami, H. & Mira, C. [2000] "A method to calculate basin bifurcation sets for a two-dimensional noninvertible map," *Int. J. Bifurcation and Chaos* **10**, 2001–2014.

López-Ruiz, R. & Fournier-Prunaret, D. [2003] "Complex patterns on the plane: Different types of basin fractalization in a two-dimensional mapping," *Int. J. Bifurcation and Chaos* **13**, 287–310.

Lorenz, E. N. [1989] "Computational chaos — a prelude to computational instability," *Physica* **D35**, 299–317.

Maistrenko, V., Maistrenko, Y. & Sushko, I. [1996] "Noninvertible two-dimensional maps arising in radiophysics," *Int. J. Bifurcation and Chaos* **4**, 383–400.

Maistrenko, V., Maistrenko, Y. & Mosekilde, E. [2003] "Torus breakdown in noninvertible maps," *Phys. Rev.* **E67**, 046215.

Mira, C., Fournier-Prunaret, D., Gardini, L., Kawakami, H. & Cathala, J. C. [1994] "Basin bifurcations of two-dimensional noninvertible maps: Fractalization of basins," *Int. J. Bifurcation and Chaos* **4**, 343–381.

Mira, C., Jean-Pierre, C., Millérioux, G. & Gardini, L. [1996a] "Plane foliation of two-dimensional noninvertible maps," *Int. J. Bifurcation and Chaos* **6**, 1439–1462.

Mira, C., Gardini, L., Barugola, A. & Cathala, J. C. [1996b] *Chaotic Dynamics in Two-Dimensional Noninvertible Maps*, Series of Nonlinear Science Series A (World Scientific, Singapore).

Nien, C. H. & Wicklin, F. J. [1998] "An algorithm for the computation of preimages in noninvertible mappings," *Int. J. Bifurcation and Chaos* **8**, 415–422.

Osinga, H. M. & England, J. P. [2003] "Global manifold 1D code, Version 2, software for use with DsTool," http://www.dynamicalsystems.org/sw/sw/detail?item=27.

Palis, J. & de Melo, W. [1982] *Geometric Theory of Dynamical Systems* (Springer-Verlag, NY/Berlin).

Rico-Martinez, R., Adomaitis, R. A. & Kevrekidis, I. G. [2000] "Noninvertibility in neural networks," *Comput. Chem. Engin.* **24**, 2417–2433.

Spivak, M. [1979] *Differential Geometry, Volume I*, 2nd edition (Publish or Perish, Houston, Texas).

Wicklin, F. J. [1995] "Pisces: a platform for implicit surfaces and curves and the exploration of singularities," Technical Report GCG #89, The Geometry Center, University of Minnesota, Minneapolis, MN. Available online at http://www.geom.uiuc.edu/~fjw/pisces/.

MULTIPARAMETRIC BIFURCATIONS IN AN ENZYME-CATALYZED REACTION MODEL

E. FREIRE, L. PIZARRO, A. J. RODRÍGUEZ-LUIS
and F. FERNÁNDEZ-SÁNCHEZ
*Department of Applied Mathematics II, E.T.S. Ingenieros, Univ. Sevilla,
Camino de los Descubrimientos s/n, 41092-Sevilla, Spain*

Received April 26, 2004; Revised June 17, 2004

An exhaustive analysis of local and global bifurcations in an enzyme-catalyzed reaction model is carried out. The model, given by a planar five-parameter system of autonomous ordinary differential equations, presents a great richness of bifurcations. This enzyme-catalyzed model has been considered previously by several authors, but they only detected a minimal part of the dynamical and bifurcation behavior exhibited by the system.

First, we study local bifurcations of equilibria up to codimension-three (saddle-node, cusps, nondegenerate and degenerate Hopf bifurcations, and nondegenerate and degenerate Bogdanov–Takens bifurcations) by using analytical and numerical techniques. The numerical continuation of curves of global bifurcations allows to improve the results provided by the study of local bifurcations of equilibria and to detect new homoclinic connections of codimension-three. Our analysis shows that such a system exhibits up to sixteen different kinds of homoclinic orbits and thirty different configurations of equilibria and periodic orbits. The coexistence of up to five periodic orbits is also pointed out. Several bifurcation sets are sketched in order to show the dynamical behavior the system exhibits. The different codimension-one and -two bifurcations are organized around five codimension-three degeneracies.

Keywords: Local bifurcations; homoclinic connections; enzyme model.

1. Introduction

In this paper we study local bifurcations of equilibria that occur in a planar five-parameter system of autonomous ordinary differential equations, arising from a *reaction–diffusion* model governed by two partial differential equations, proposed in the study of enzyme catalyzed reactions. We refer the interested reader to [Thomas, 1975; Murray, 1981a, 1981b; Kernévez *et al.*, 1979; Kernévez *et al.*, 1983; Murray, 2002, 2003]. Namely, the system we consider, that will be called *enzyme system* in the sequel, is:

$$
\begin{cases}
\dot{s} = s_0 - s - \rho\, \dfrac{sa}{1 + s + \kappa s^2}, \\[2ex]
\dot{a} = \alpha(a_0 - a) - \rho\, \dfrac{sa}{1 + s + \kappa s^2},
\end{cases}
\tag{1}
$$

where $s, a, a_0, s_0, \rho, \alpha, \kappa > 0$.

Our objective is to describe exhaustively the bifurcation behavior that the enzyme system exhibits. One of the principal methods we use is normal form theory, which provides information about the nonlinear terms essential for describing the bifurcation behavior. The analytical study of the local bifurcation behavior of a degenerate equilibrium provides knowledge of the bifurcation set in a neighborhood of the parameter space point where such degenerate equilibrium occurs (*organizing centers* in both state and parameter spaces). For a general discussion on bifurcations and normal form theory, see e.g. [Guckenheimer & Holmes, 1997; Kuznetsov, 1998; Wiggins, 2003].

The local information obtained from the unfolding of a singularity is only valid in a certain neighborhood. In practice, the size of this neighborhood is not usually very small and, therefore, the local results persist *far away* from the organizing centers. The numerical techniques of continuation are a good tool for *extending* the local results. In this way, the local analytical information gives the starting points in both state and parameter spaces.

The continuation codes we have used will not only allow to extend the local analytical results, but also to detect new bifurcation phenomena. These numerical techniques are described in [Freire *et al.*, 1999b; Freire *et al.*, 2000].

Now let us briefly describe the bifurcations we have obtained (see Table 1). Bifurcations up to codimension-three of equilibria, periodic orbits and homoclinic connections appear.

System (1) undergoes several cases of Hopf bifurcation. Let us consider the Hopf bifurcation normal form [Guckenheimer & Holmes, 1997; Kuznetsov, 1998]:

$$\dot{r} = \lambda(\eta)r + a_1 r^3 + a_2 r^5 + a_3 r^7 + \cdots, \qquad (2)$$

where $\lambda(0) = 0$ and η is the bifurcation parameter. Assuming the transversality condition $(d\lambda/d\eta)(0) \neq 0$ holds: a nondegenerate Hopf bifurcation occurs when $a_1 \neq 0$; a degenerate codimension-two Hopf bifurcation, labeled H_1, arises if $a_1 = 0$ and $a_2 \neq 0$; a degenerate codimension-three Hopf bifurcation, labeled H_2, appears if $a_1 = a_2 = 0$, $a_3 \neq 0$. When the transversality condition fails, i.e. $(d\lambda/d\eta)(0) = 0$ and $a_1 = 0$, a nontransversal degenerate codimension-three Hopf bifurcation, labeled H_T, occurs.

Let us now consider a second-order normal form of the Bogdanov–Takens bifurcation (see, for instance [Guckenheimer & Holmes, 1997; Kuznetsov, 1998]):

$$\begin{cases} \dot{x} = y, \\ \dot{y} = ax^2 + bxy. \end{cases}$$

(Note that, in this normal form, a is not the state variable of the enzyme system.) A nondegenerate Bogdanov–Takens bifurcation occurs if $a, b \neq 0$. When $a \neq 0$ and $b = 0$ a degenerate codimension-three Bogdanov–Takens bifurcation, called *cusp* of order three and labeled E, appears; if $a = 0$ and $b \neq 0$ another degenerate codimension-three Bogdanov–Takens bifurcation, labeled D, arises. These degeneracies have been studied in

Table 1. Bifurcation phenomena exhibited by the enzyme system. The symbol ($\bullet$) means that this bifurcation has been previously studied by other authors. Abbreviations are explained in the text.

Codimension	Equilibria	Periodic Orbits	Homoclinic Orbits
1	saddle-node ($\bullet$)	saddle-node ($\bullet$)	nonzero trace
	nondegenerate Hopf ($\bullet$)		central saddle-node
2	cusp	cusp	zero trace
	degenerate Hopf (H_1) ($\bullet$)		noncentral saddle-node (5 types)
	nondegenerate Bogdanov–Takens		double
3	degenerate Bogdanov–Takens (E)		CL
	degenerate Bogdanov–Takens (D)		H_{EID}
	degenerate Hopf (H_2)		
2^* *topological $\mathbb{Z}_2$-codimension	nontransversal degenerate Hopf (H_T) ($\bullet$)		

[Dumortier *et al.*, 1987] and [Dumortier *et al.*, 1991], respectively.

The different homoclinic bifurcations of Table 1 are described in Secs. 3 and 4.

The enzyme system (1) has been previously studied by several authors, but their works only describe a little part of its dynamical behavior. In this way, [Doedel *et al.*, 1991; Doedel & Kernévez, 1986; Doedel *et al.*, 1998] consider the enzyme system as an example for applying the software package AUTO, and they only provide numerical evidence of the existence of saddle-node and Hopf bifurcations of equilibria and saddle-node bifurcations of periodic orbits. Kernévez *et al.* [1983, 1985] performed a numerical study of the system (1), obtaining the same bifurcation phenomena. Hassard and Jiang devoted two papers to study the enzyme system. In [Hassard & Jiang, 1992], they detected, numerically, a nontransversal Hopf bifurcation point (they also obtained, using AUTO, curves of saddle-node bifurcation of periodic orbits which appear in an unfolding of this singularity). Note that a suitable choice of the bifurcation parameters can avoid this transversality condition failure. In [Hassard & Jiang, 1993], they carried out a detailed study of a degenerate nontransversal Hopf bifurcation point given by the vanishing of the third-order coefficient of the normal form; they located this singularity in both parameter and state spaces and they obtained some information about its unfolding.

Thus, in this work we present a more comprehensive analysis of the system (1) and we study every local and global bifurcation, up to codimension-three. The reason why our analysis improves the previous ones is that the enzyme system can be written as a low-order polynomial system by means of a suitable change of variables.

The paper is organized as follows. Section 2 is firstly devoted to the transformation of the enzyme system into a low-order polynomial system, that facilitates the application of normal form theory in order to obtain analytical results about high-codimension bifurcations of equilibria. Once the transformation is done, we study, in several subsections, bifurcations of equilibria of codimension-one and -two: saddle-node and cusp bifurcations nondegenerate and degenerate Hopf bifurcations and nondegenerate Bogdanov–Takens bifurcations. We also consider a case of nontransversal degenerate Hopf bifurcation. Two kinds of degenerate Bogdanov–Takens bifurcations and a limit case are studied in Sec. 3. In Sec. 4 we explain the numerical

methods we have used to detect and to continue the global bifurcations up to codimension-three. Note that sixteen different kinds of homoclinic orbits appear. In Sec. 5 we apply those numerical methods analyzing basically codimension-two and -three homoclinic orbits. In Sec. 6, several bifurcation sets show the way the bifurcations are organized by five local and global bifurcations of codimension-three. We finish this work with some conclusions and remarks.

2. Codimension-One and -Two Bifurcations of Equilibria

The first part of this section is devoted to transform the enzyme system (1) into a low-degree polynomial system. We note that system (1) could be defined for negative state variables, although this has not biochemical sense. In fact, it is easy to verify that if $\kappa > 1/4$ then system (1) is defined for all $(s, a) \in \mathbb{R}^2$ whereas if $\kappa \leq 1/4$ it is defined in a half-plane containing the positive quadrant $\mathbb{R}^+ \times \mathbb{R}^+$.

An equilibrium $(\overline{s}, \overline{a})$ of system (1) has to verify

$$s_0 - \overline{s} = \alpha(a_0 - \overline{a}) = \rho \frac{\overline{s}\,\overline{a}}{1 + \overline{s} + \kappa \overline{s}^2} > 0$$

and, therefore, $(\overline{s}, \overline{a}) \in (0, s_0) \times (0, a_0)$; moreover, $(\overline{s}, \overline{a})$ lies on the straight line $s_0 - s = \alpha(a_0 - a)$.

It is easy to prove that the rectangle $[0, s_0] \times [0, a_0]$ is a positively invariant set and, even more, it is an attracting set for the positive quadrant.

The following result will play a key role along this work, since it states that the enzyme system can be rewritten as a polynomial vector field whose components are of degree two and six.

Lemma 2.1. *The system (1) is, for $s > 0$, C^∞ orbital equivalent to the system*

$$\begin{cases} \dot{u} = uv, \\ \dot{v} = v^2 + F_1(u)v + F_2(u), \end{cases} \qquad (3)$$

where

$$F_1(u) = u(s_0 - u)p'(u) - (\alpha u + s_0)p(u) - \rho u^2,$$

$$F_2(u) = up(u)h(u),$$

$$p(u) = 1 + u + \kappa u^2, \qquad (4)$$

$$h(u) = h(u, a_0, s_0, \rho)$$

$$= -\rho \alpha a_0 u + (s_0 - u)[\alpha p(u) + \rho u].$$

and the symbol ′ stands for d/du.

Proof. If we make the time reparameterization $t \to (1 + s + \kappa s^2)t$ (note that $1 + s + \kappa s^2 > 0$ for $s > 0$), followed by the change of variables

$$u = s,$$
$$v = (s_0 - s)(1 + s + \kappa s^2) - \rho sa,$$

and the new time reparameterization $t \to ut$ $(u > 0)$, system (1) may be written as in (3). ∎

Now the system is written in a suitable form for its bifurcation analysis. Then, we start the study of codimension-one and -two bifurcations of equilibria of system (3) which are given by $(u_\star, 0)$, where $u_\star$ is a root of the third-degree polynomial h given in (4).

The linearization matrix of (3) at the equilibrium is

$$A = \begin{pmatrix} 0 & u_\star \\ F_2'(u_\star) & F_1(u_\star) \end{pmatrix}.$$

Since $\det A = -u_\star^2 p(u_\star) h'(u_\star)$ and the leading term coefficient in h is $-\alpha\kappa < 0$, when $h(u)$ has three roots, these equilibria correspond to a hyperbolic saddle located between two foci or nodes; in the case that $h(u)$ has only one root, it corresponds to a hyperbolic node or focus. Double and triple roots correspond to nonhyperbolic equilibria: saddle-node and cusp bifurcations of equilibria, respectively. Moreover, other bifurcations may appear: a Hopf bifurcation occurs when a focus becomes nonhyperbolic, and a Bogdanov–Takens bifurcation appears when a Hopf and a saddle-node bifurcation collide.

In Sec. 2.1 the saddle-node bifurcation of equilibria and the cusp of saddle-node bifurcation are studied. Hopf bifurcation and its degeneracies are treated in Sec. 2.2. The existence of nontransversal degenerate Hopf bifurcation is considered in Sec. 2.3. Finally, nondegenerate Bogdanov–Takens bifurcations are analyzed in Sec. 2.4.

2.1.   *Saddle-node and cusp bifurcations*

The following proposition states the existence of saddle-node and cusps bifurcations.

Proposition 2.1. *For each α and κ positive, in the (a_0, s_0, ρ)-space:*

(a) *a saddle-node bifurcation of equilibria occurs on the parameterized surface given by the rational expressions*

$$\begin{cases} s_0 = \dfrac{u_\star^2(\rho + \alpha p'(u_\star))}{\alpha(u_\star p'(u_\star) - p(u_\star))}, \\[2ex] a_0 = \dfrac{(\alpha p(u_\star) + \rho u_\star)^2}{\alpha^2(u_\star p'(u_\star) - p(u_\star))\rho}, \end{cases}$$

for $u_\star, \rho > 0$ and $h''(u_\star) \neq 0$;

(b) *the parameterized curve, given by the rational expressions*

$$\begin{cases} a_0 = \dfrac{2(u_\star p'(u_\star) - p(u_\star))[2(u_\star p'(u_\star) - p(u_\star))^2 + u_\star^2 p(u_\star)(p''(u_\star) - p'(u_\star))]}{\alpha[u_\star^2 p''(u_\star) - 2u_\star p'(u_\star) + 2p(u_\star)]p'(u_\star)[2u_\star p'(u_\star) - 2p(u_\star) - u_\star p(u_\star)]}, \\[2.5ex] s_0 = u_\star + \dfrac{2u_\star(u_\star p'(u_\star) - p(u_\star))}{u_\star^2 p''(u_\star) - 2(u_\star p'(u_\star) - p(u_\star))}, \\[2.5ex] \rho = \dfrac{2\alpha p'(u_\star)(u_\star p'(u_\star) - p(u_\star)) - \alpha u_\star p(u_\star)p''(u_\star)}{u_\star^2 p''(u_\star) - 2(u_\star p'(u_\star) - p(u_\star))}, \end{cases}$$

for $u_\star > 0$ and $h'''(u_\star) \neq 0$, is the locus where a cusp bifurcation of equilibria occurs.

Proof. Since $F_2(u_\star) = u_\star p(u_\star)h(u_\star)$ and $u_\star$, $p(u_\star) > 0$, the equilibrium $(u_\star, 0)$ undergoes a saddle-node bifurcation, for each α and κ positive, if $h(u_\star) = h'(u_\star) = 0$ and $h''(u_\star) \neq 0$. Solving the above two equations, we obtain the required expression for the surface of saddle-node bifurcation. If $h(u_\star) = h'(u_\star) = h''(u_\star) = 0$ and $h'''(u_\star) \neq 0$ the equilibrium $(u_\star, 0)$ undergoes a cusp bifurcation.

Solving these last equations, we obtain the expression for the curve of cusp bifurcation stated in the proposition. ∎

This result states where the system has one, two or three equilibria in the parameter space. The saddle-node bifurcation locus separates the regions where either one or three equilibria exist.

In Fig. 1 we observe, in the (a_0, ρ)-plane, for $s_0 = 37$, $\alpha = 0.2$ and $\kappa = 0.1$, that the two

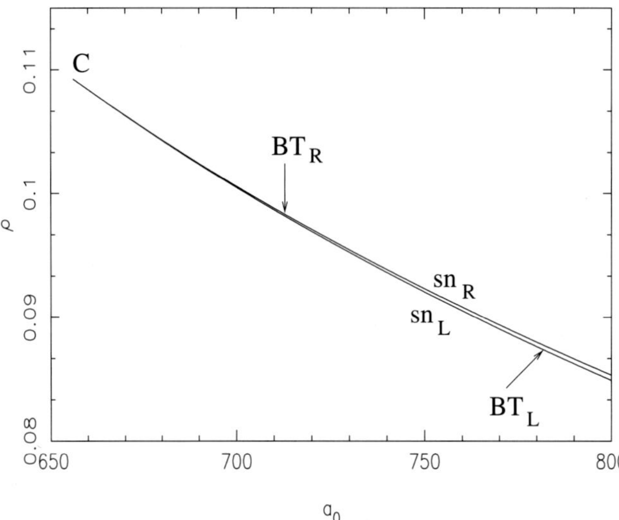

Fig. 1.   Curves of saddle-node bifurcations of equilibria, sn_R and sn_L, in the (a_0, ρ)-plane, for $s_0 = 37$, $\alpha = 0.2$ and $\kappa = 0.1$. Two Bogdanov–Takens bifurcation points, BT_R and BT_L, and a cusp bifurcation of equilibria, C, appear on those curves. The L (R) index indicates that the corresponding bifurcation is exhibited by the left (right) equilibrium.

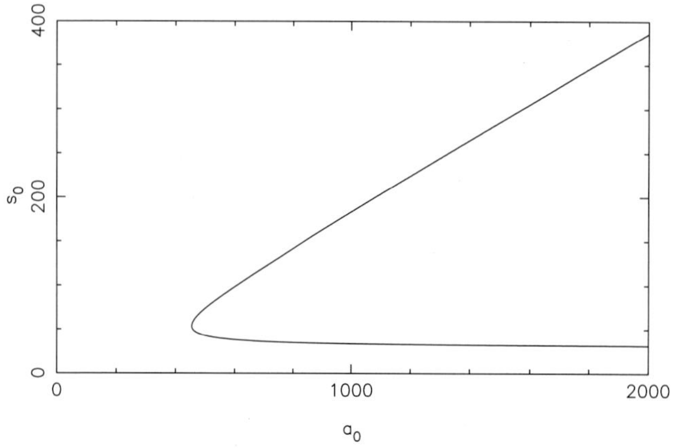

Fig. 2.   Projection, onto the (a_0, s_0)-plane, of the curve of cusp bifurcations of equilibria in the (a_0, s_0, ρ) parameter space, for $\alpha = 0.2$ and $\kappa = 0.1$.

saddle-node curves sn_L and sn_R collide in a cusp C. Two Bogdanov–Takens points (see Sec. 2.4), BT_L and BT_R, are also present. In the narrow region between sn_L and sn_R three equilibria coexist, and outside this zone the system only has one equilibrium point. The subscript L (left) in sn_L indicates that in the saddle-node bifurcation collapse the middle and the left equilibria of the system and in BT_L means that the bifurcation is exhibited by the nonhyperbolic left equilibrium. Similar comments are valid for the subscript R (right).

A projection on the (a_0, s_0)-parameter plane of the curve of cusp bifurcations of equilibria in the (a_0, s_0, ρ)-parameter space, labeled C, appears in Fig. 2, for $\alpha = 0.2$ and $\kappa = 0.1$.

2.2. *Hopf bifurcations*

The analytical and numerical study of the normal form of the Hopf bifurcation guarantees the existence of these bifurcations and some degeneracies up to codimension-three. For the next discussion, see, e.g. [Guckenheimer & Holmes, 1997; Kuznetsov, 1998].

Other cases of higher codimension Hopf bifurcation are possible when the higher-order coefficients in the Hopf bifurcation normal form vanish.

Proposition 2.2. *For each α and κ positive, in the (a_0, s_0, ρ)-space:*

(a) *a codimension-one Hopf bifurcation occurs on the parameterized surface given by the rational expressions*

$$\begin{cases} a_0 = \dfrac{(s_0 - u_\star)[u_\star p'(u_\star)(u_\star - s_0) + p(u_\star)s_0]}{\alpha[u_\star p'(u_\star)(u_\star - s_0) + p(u_\star)(\alpha u_\star + s_0)]}, \\[2ex] \rho = \dfrac{u_\star(s_0 - u_\star)p'(u_\star) - (\alpha u_\star + s_0)p(u_\star)}{u_\star^2}, \end{cases}$$

(5)

for $u_\star, s_0 > 0$ with $\sigma'(u_\star) \neq 0$ $\sigma(u_\star)$ is defined in (7), $h'(u_\star) < 0$ and

$$\Psi(u_\star) = \frac{F_1''(u_\star)}{F_1'(u_\star)} - 2\frac{p'(u_\star)}{p(u_\star)} + \frac{h''(u_\star)}{h'(u_\star)} \neq 0;$$

(b) *a degenerate Hopf bifurcation of codimension-two occurs on the curve given by the rational expressions*

$$\begin{cases} \sigma(u_\star) = 0, \\ \Psi(u_\star) = 0, \end{cases}$$

for $u_\star > 0$ with $\sigma'(u_\star) \neq 0$, $h'(u_\star) < 0$ and $a_2(u_\star) \neq 0$ (a_2 is defined in (12)).

Proof. The value of a_0 can be obtained from $h(u) = 0$ as

$$a_0 = a_0(u) = \frac{(s_0 - u)(\alpha p(u) + \rho u)}{\rho \alpha u}.$$

(6)

The linearization matrix of (3) at the equilibrium is

$$A = \begin{pmatrix} 0 & u_\star \\ \gamma(u_\star) & \sigma(u_\star) \end{pmatrix}, \qquad (7)$$

where

$$\gamma(u_\star) = u_\star p(u_\star) h'(u_\star, a_0(u_\star), s_0, \rho),$$

$$\sigma(u_\star) = F_1(u_\star, s_0, \rho).$$

Thus, the equilibrium $(u_\star, 0)$ of system (3) undergoes a Hopf bifurcation if $\sigma(u_\star) = 0$ and $\gamma(u_\star) < 0$. The transversality condition is given by

$$\mathrm{Re}\left(\frac{\partial \lambda}{\partial u_\star}\right) = \frac{1}{2}\sigma'(u_\star) \neq 0, \qquad (8)$$

where $\lambda = \eta \pm \omega i$ are the eigenvalues of (7). Note that we have taken $u_\star$ as the bifurcation parameter.

From

$$\sigma(u_\star) = u_\star(s_0 - u_\star)p'(u_\star)$$
$$-(\alpha u_\star + s_0)p(u_\star) - \rho u_\star^2 = 0, \qquad (9)$$

it is easy to obtain, for each value of ρ, α and κ, the curve of Hopf bifurcation in the $(u_\star, s_0)$-plane

$$s_0 = u_\star + \frac{u_\star}{\kappa u_\star^2 - 1}[\rho u_\star + (\alpha + 1)p(u_\star)]. \qquad (10)$$

This curve has two asymptotes: $u_\star = 1/\sqrt{\kappa}$ and $s_0 = (\alpha+2)u_\star + (\rho+\alpha+1)/\kappa$. For $u_\star \in (0, 1/\sqrt{\kappa})$, the value of s_0 is negative. Therefore, the range of $u_\star$ is the interval $(1/\sqrt{\kappa}, +\infty)$.

Combining (9) and (6) we obtain, for each α and κ positive, the parameterized surface (5). To assure the codimension-one character, we need to compute the normal form.

The translation

$$u \to u + u_\star,$$
$$v \to v,$$

followed by the rescaling transformation

$$u \to \frac{1}{\sqrt{u_\star}}\, u,$$

$$v \to \frac{1}{\sqrt{-u_\star p(u_\star)h'(u_\star)}}\, v,$$

allow to write the system (3) as

$$\begin{pmatrix} \dot{u} \\ \dot{v} \end{pmatrix} = \begin{pmatrix} 0 & -\omega_0 \\ \omega_0 & 0 \end{pmatrix}\begin{pmatrix} u \\ v \end{pmatrix} + \begin{pmatrix} f(u,v) \\ g(u,v) \end{pmatrix}, \qquad (11)$$

where

$$\omega_0 = \sqrt{-u_\star^2 p(u_\star)h'(u_\star)},$$

$$f(u,v) = \sqrt{-u_\star p(u_\star)h'(u_\star)}\,v^2 + F_1(\sqrt{u_\star}u + u_\star)v$$

$$+ \frac{F_2(\sqrt{u_\star}u + u_\star) - u_\star^{\frac{3}{2}} p(u_\star)h'(u_\star)u}{\sqrt{-u_\star p(u_\star)h'(u_\star)}},$$

$$g(u,v) = \sqrt{-u_\star p(u_\star)h'(u_\star)}\,uv.$$

In the study of the Hopf bifurcation of system (11) and its possible degeneracies, the hand calculation (as opposed to numerical evaluation) of very long expressions is required, when the corresponding bifurcation formulae are being used (see [Hassard & Jiang, 1992, 1993]). Freire *et al.* [1989] developed a recursive algorithm well suited to symbolic computation implementation, that turns out to be an efficient procedure to obtain the coefficients of the Hopf bifurcation normal form. This algorithm is based upon the use of Lie transforms; the calculations are arranged in a recursive scheme using complex variables and so the computational effort is optimized.

The application of the aforementioned algorithm, by means of a MAPLE program, to compute the coefficients a_1, a_2 and a_3 of order 3, 5 and 7, respectively, of the Hopf bifurcation normal form, provides:

$$a_1 = \frac{1}{16}\frac{A_{1,1}}{A_{1,2}}, \qquad a_2 = -\frac{1}{1152}\frac{A_{2,1}}{A_{2,2}},$$

$$a_3 = -\frac{1}{18432}\frac{A_{3,1}}{A_{3,2}}, \qquad (12)$$

with

$$A_{1,1} = u_\star F_1'' F_2' - u_\star F_2'' F_1' + 2F_1' F_2',$$

$$A_{1,2} = F_2',$$

$$A_{2,1} = 10u_\star^2 F_2' F_1''' F_1'' + 30u_\star F_2'(F_1'')^2 - 12F_1'' F_2' F_1' - 10u_\star^2 F_2''' F_1'' F_1' + 3u_\star^2 (F_1')^2 F_2^{IV}$$

$$- 16u_\star F_2' F_1''' F_1' + 4u_\star(F_1')^2 F_2'',$$

$$A_{2,2} = F_2' F_1',$$

$$A_{3,1} = 35u_\star^4 (F_1')^2 F_2^{IV}(F_2'')^3 - 35u_\star^4 F_2'' F_2^V (F_1')^2 (F_2'')^2 + 35u_\star^4 F_2''(F_2')^2 F_2^{IV} F_1''' F_1'$$
$$+ 5u_\star^4 F_2''(F_1')^2 (F_2')^2 F_2^{VI} - 140u_\star^3 F_1''' F_2' F_1'(F_2'')^3 - 315u_\star^3 F_2' F_2^{IV}(F_1')^2(F_2'')^2$$
$$+ 204u_\star^3 F_2''(F_1')^2 F_2^V (F_2')^2 - 140u_\star^3 F_2''(F_2')^3 (F_1''')^2 - 126u_\star^3 (F_2')^3 F_2^{IV} F_1''' F_1'$$
$$- 12u_\star^3 (F_1')^2 (F_2')^3 F_2^{VI} + 1260u_\star^2 F_1'''(F_2')^2 F_1'(F_2'')^2 + 942u_\star^2 (F_2')^2 F_2^{IV}(F_1')^2 F_2''$$
$$- 288u_\star^2 (F_1')^2 F_2^V (F_2')^3 + 504u_\star^2 (F_2')^4 (F_1''')^2 - 3768u_\star F_1'''(F_2')^3 F_1' F_2''$$
$$- 936u_\star (F_2')^3 F_2^{IV}(F_1')^2 + 3744 F_1'''(F_2')^4 F_1',$$
$$A_{3,2} = (F_2')^3 F_1'(5u_\star F_2'' - 12F_2'),$$

where the derivatives of the polynomial functions F_1 and F_2 are evaluated in $u_\star$.

Note that we have obtained the coefficients up to order 7 of the Hopf bifurcation normal form only in terms of the functions F_1 and F_2.

From $F_2(u) = up(u)h(u)$ an even more simplified expression for the coefficient a_1 can be provided, namely,

$$a_1 = a_1(u_\star)$$
$$= \frac{u_\star}{16}\left[F_1''(u_\star) - \left(2\,\frac{p'(u_\star)}{p(u_\star)} + \frac{h''(u_\star)}{h'(u_\star)}\right) F_1'(u_\star)\right]$$

(cf. the long expressions used for the evaluation of a_1 given in Appendix B of [Hassard & Jiang, 1992]). The nondegeneracy condition $a_1(u_\star) \neq 0$ is equivalent to $\Psi(u_\star) \neq 0$ and this proves (a).

The Hopf conditions $h(u_\star) = 0$ and $\sigma(u_\star) = 0$ along with the additional one, $a_1(u_\star) = 0$, allow to obtain, in the (a_0, s_0, ρ)-space (for each α and κ positive), the curve of degenerate codimension-two Hopf bifurcation stated in (b), assuming that it verifies the nondegeneracy condition $a_2(u_\star) \neq 0$. ∎

In Fig. 3 we show the curve of nondegenerate Hopf bifurcation (solid line) in the $(u_\star, s_0)$-plane, given by (10), for $\rho = 1$, $\alpha = 0.2$ and $\kappa = 0.1$. Its two asymptotes are also drawn (dashed lines).

By using a general purpose continuation code (PITCON 6.0, see [Rheinboldt, 1986]), we have obtained numerically for $\alpha = 0.2$ and $\kappa = 0.1$ the curve of degenerate codimension-two Hopf bifurcation. It is formed by two components, labeled H_1 and H_1', respectively, which are shown in Fig. 4. Other bifurcations shown in this figure are analyzed below.

Remarks. There exists a Hopf bifurcation point, in the (a_0, s_0, ρ)-space, labeled DH, where both

coefficients a_1 and a_2 given in (12) vanish, for each α and κ positive. This point lies on H_1' ($a_2 < 0$ for all the points of H_1) and corresponds to a degenerate codimension-three Hopf bifurcation. A curve of points DH, denoted by H_2, projected on the $(u_\star, \alpha)$-plane appears in Fig. 5 (obtained numerically for $\kappa = 0.1$ by using the aforementioned code PITCON 6.0). Local analysis provides the existence of a curve of cusp bifurcation of periodic orbits, Cu, emerging from the point DH. We have evaluated the coefficient a_3 at the parameter values at which the codimension-three Hopf bifurcation DH occurs (for $(\alpha, \kappa) \in (0, 1) \times (0, 10000)$), obtaining that a_3 is always negative. It means, on the one hand, the nonexistence of degenerate codimension-four Hopf bifurcation points and, on the other, that the curve Cu emerges from point DH by the side of the curve H_1' where the coefficient a_2 is positive (see [Takens, 1973] and Fig. 6). In Fig. 7 two qualitative

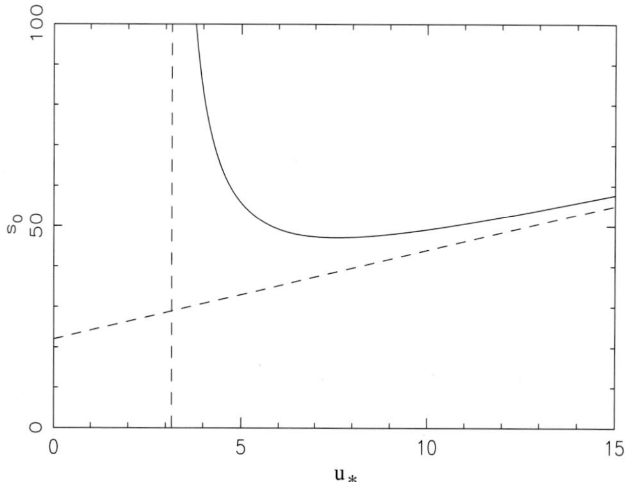

Fig. 3. Curve of nondegenerate Hopf bifurcation (solid line) and its asymptotes (dashed line) in the $(u_\star, s_0)$-plane for $\rho = 1$, $\alpha = 0.2$ and $\kappa = 0.1$.

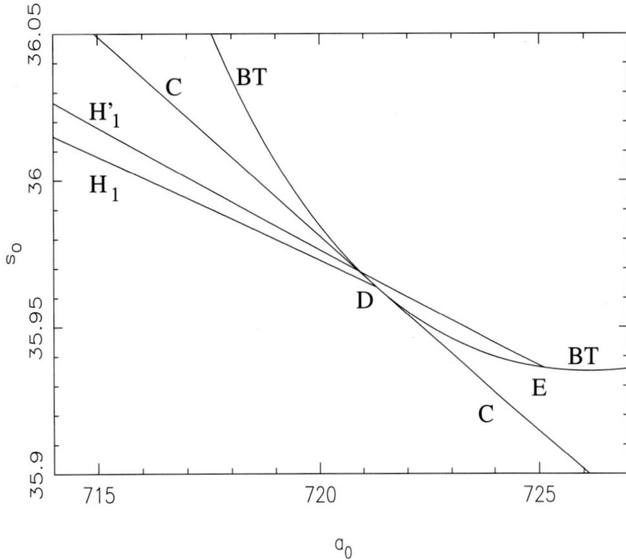

Fig. 4.   Projection, onto the (a_0, s_0)-parameter plane, for $\alpha = 0.2$ and $\kappa = 0.1$, of the curves of Bogdanov–Takens (BT), degenerate Hopf (H_1 and H_1') and cusp of equilibria (C). The degenerate Bogdanov–Takens points, E and D, are also shown.

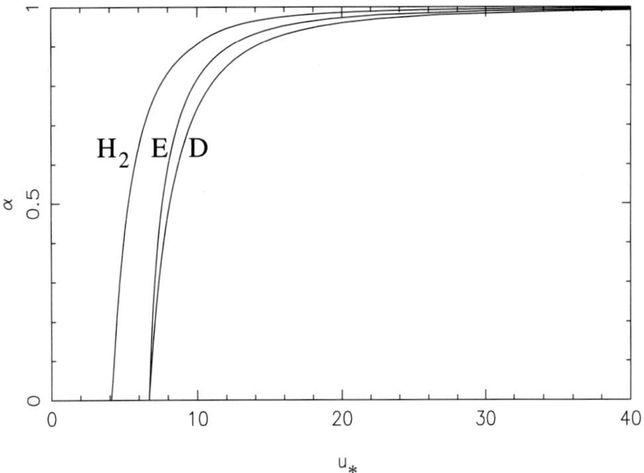

Fig. 5.   Curves of codimension-three bifurcations of equilibria in the $(u_\star, \alpha)$-plane (for $\kappa = 0.1$). E = cusps of order three; D = weak foci; H_2 = degenerate Hopf points.

pictures are represented: they show the relative position (with respect to the nondegenerate Hopf bifurcation curves), in the (a_0, ρ)-parameter plane (for α and κ constant), of the saddle-node of periodic orbits curves emerging from the codimension-two Hopf bifurcation points H_1 and H_1' for two values of s_0: one of them greater than the critical value of s_0 for which the point DH occurs (in this case, the saddle-node curves of periodic orbits

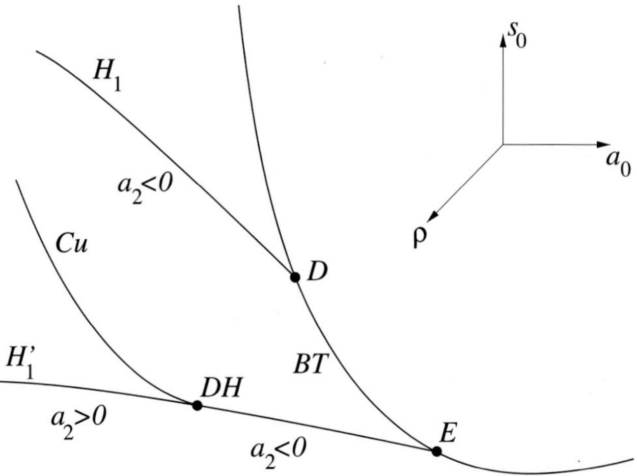

Fig. 6.   Qualitative picture, in the (a_0, s_0, ρ)-parameter space, of the curves of Bogdanov–Takens (BT), degenerate Hopf (H_1 and H_1') and cusp of periodic orbits (Cu).

coalesce in the mentioned cusp of periodic orbits) and the other value of s_0 is less than this critical value (in this case, the cusp of periodic orbits does not already exist).

2.3.   *Nontransversal degenerate Hopf bifurcations*

When a complex-conjugate pair of eigenvalues of the linearization matrix crosses the imaginary axis in a degenerate way (i.e. the crossing is nontransversal), a nontransversal Hopf bifurcation arises. If we consider the Hopf bifurcation normal form given in (2), the nontransversality condition of the Hopf bifurcation merely means that $(d\lambda/d\eta)(0) = 0$.

In this subsection we study a case of nontransversal degenerate Hopf bifurcation arising in the system (3) (this kind of degeneracy corresponds to a topological $\mathbb{Z}_2$-codimension 2 in the context of [Golubitsky & Schaeffer, 1985]). Our only aim is to show how the results of [Hassard & Jiang, 1993] may be obtained from our analysis.

We note that a first case of nontransversal degenerate codimension-two Hopf bifurcation arises when the transversality condition given in (8) fails. Hassard and Jiang [1992] studied this degeneracy. We will not consider this kind of degeneracy because of the simplicity it exhibits, in contrast to the great richness of bifurcation behavior of system (3). Thus, it will not appear in Table 1.

The case of nontransversal degenerate topological $\mathbb{Z}_2$-codimension-two Hopf bifurcations we are

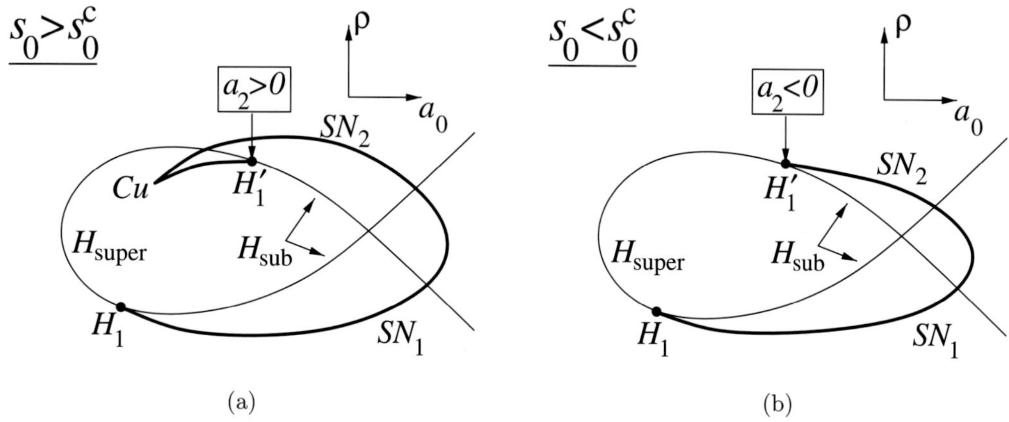

Fig. 7. Qualitative picture (for α and κ constant) of the curves of saddle-node bifurcations of periodic orbits, SN_1 and SN_2, emerging from the codimension-two Hopf bifurcation points, H_1 and H_1', for (a) $s_0 > s_0^c$ and (b) $s_0 < s_0^c$, where s_0^c is the critical value of the parameter s_0 at which the codimension-two Hopf bifurcation point DH occurs. The cusp of periodic orbits point Cu appears for a_2 (coefficient of the normal form of the Hopf bifurcation evaluated at the parameter values of H_1') positive.

interested arises when both of the following situations hold: the coefficient a_1 given in (12) vanishes and the transversality condition given in (8) fails. Such degeneracy, called H_T, can be studied in the context of the singularity theory developed by Golubitsky and Schaeffer [1985]. In [Hassard & Jiang, 1993], this degenerate point is numerically located in the (a_0, s_0, ρ)-parameter space as well as in the (s, a)-phase plane for the values $\alpha = 0.2$ and $\kappa = 0.1$.

The analytical information we have about the enzyme system allows to characterize the manifold of points H_T in the $(a_0, s_0, \rho, \alpha, \kappa)$-parameter space.

If we fix arbitrary values of the parameters α and κ, we have to find the point H_T along the curve, in the (a_0, s_0, ρ)-space, of degenerate codimension-two Hopf bifurcation given by the conditions $\sigma(u_\star) = 0$ and $a_1(u_\star) = 0$ (see paragraph (b) in Proposition 2.2). Thus, we have to obtain a solution of the system

$$\sigma(u_\star) = \sigma'(u_\star) = a_1(u_\star) = 0,$$

with $h'(u_\star) < 0$.

In order to obtain this solution, we find ρ from $h(u_\star) = 0$, $\rho = \rho(u_\star, a_0, s_0)$; this enables to get a new expression for the function σ, namely $\sigma(u_\star, a_0, s_0) = F_1(u_\star, s_0, \rho(u_\star, a_0, s_0))$.

Thus, the nontransversality condition $(\partial/\partial u_\star)\sigma(u_\star, a_0, s_0) = 0$ becomes

$$\frac{\partial}{\partial u_\star}\sigma(u_\star, a_0, s_0) = \frac{\partial F_1}{\partial u_\star} + \frac{\partial F_1}{\partial \rho}\frac{\partial \rho}{\partial u_\star} = 0. \quad (13)$$

But

$$\frac{\partial h}{\partial u_\star} = \frac{\partial h}{\partial \rho}\frac{\partial \rho}{\partial u_\star},$$

and, therefore, (13) is equivalent to

$$\frac{\partial F_1}{\partial u_\star}\frac{\partial h}{\partial \rho} - \frac{\partial F_1}{\partial \rho}\frac{\partial h}{\partial u_\star} = 0. \quad (14)$$

This is the new expression for the nontransversality condition.

Equations (14), $\sigma(u_\star) = 0$ and $a_1(u_\star, a_0, s_0, \rho(u_\star)) = 0$, determine a simple equation system whose solutions provide the manifold of points H_T. For example, for $\alpha = 0.2$ and $\kappa = 0.1$, the point H_T is given by

$$u_\star \approx 12.91003, \quad a_0 \approx 330.20156,$$

$$s_0 \approx 67.70777, \quad \rho \approx 2.30884,$$

and these parameter values are, precisely, those obtained in [Hassard & Jiang, 1993].

A projection of the curve in the $(u_\star, a_0, s_0, \alpha)$-space of points H_T obtained for $\kappa = 0.1$ appears in Fig. 8.

2.4. Bogdanov–Takens bifurcations

Finally, we consider in the following result the existence of nondegenerate Bogdanov–Takens bifurcations. A Bogdanov–Takens bifurcation arises when the linearization matrix has, for certain critical parameter values, a double-zero eigenvalue (see, e.g. [Guckenheimer & Holmes, 1997; Kuznetsov, 1998]).

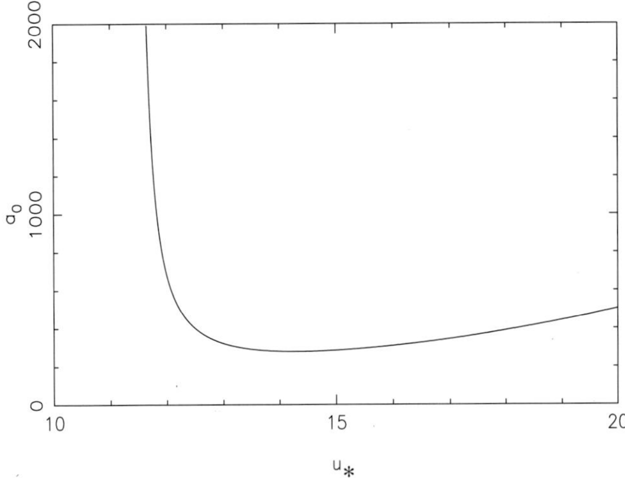

Fig. 8.   Projection, onto the $(u_\star, a_0)$-parameter plane, of the curve of degenerate Hopf bifurcation given by the failure of the transversality condition and the vanishing of the cubic term coefficient of the normal form ($\kappa = 0.1$).

Recall that system (3) has an equilibrium at $(u_\star, 0)$. It undergoes a double-zero degeneracy if the trace and the determinant of the linearization matrix (7) are equal to zero, that is,

$$\sigma(u_\star) = h'(u_\star) = 0. \tag{15}$$

Initially, we perform the rescaling $t \to t/u_\star$ which transforms (3) into

$$\begin{cases} \dot{u} = \dfrac{1}{u_\star} uv, \\[2mm] \dot{v} = \dfrac{1}{u_\star}(v^2 + F_1(u)v + F_2(u)). \end{cases} \tag{16}$$

For the critical values determined by (15), the linearization matrix of this system at $(u_\star, 0)$ is a Jordan block

$$\begin{pmatrix} 0 & 1 \\ 0 & 0 \end{pmatrix}.$$

Proposition 2.3. *For* $\alpha \in (0,1)$ *and* $\kappa > 0$, *a codimension-two nondegenerate Bogdanov–Takens bifurcation occurs on the parameterized curve given by the rational expressions*

$$\begin{cases} a_0 = \dfrac{u_\star p(u_\star)}{\alpha^2(u_\star p'(u_\star) - p(u_\star))(1 - \alpha)}, \\[3mm] s_0 = u_\star + \dfrac{u_\star p(u_\star)}{(u_\star p'(u_\star) - p(u_\star))(1 - \alpha)}, \\[3mm] \rho = \dfrac{\alpha^2 p(u_\star)}{u_\star(1 - \alpha)}, \end{cases} \tag{17}$$

for $u_\star > 0$ *with* $F_1'(u_\star) \neq 0$ *and* $h''(u_\star) \neq 0$.

Proof. The parameterized curve (17) can be easily obtained from $\sigma(u_\star) = 0$ and $h'(u_\star) = 0$, for $\alpha \in (0,1)$ and $\kappa > 0$. To guarantee that the Bogdanov–Takens bifurcation is nondegenerate we have to check that the coefficients of the second-order normal form

$$\begin{cases} \dot{u} = v, \\ \dot{v} = au^2 + buv, \end{cases} \tag{18}$$

are nonzero (see, e.g. [Guckenheimer & Holmes, 1997]). The computation of these coefficients can be done using the algorithm developed in [Freire et al., 1991] and then, we obtain the following expressions

$$a = \frac{1}{2}p(u_\star)h''(u_\star), \quad b = \frac{1}{u_\star}F_1'(u_\star),$$

that trivially lead to the nondegeneracy conditions stated in the proposition. ∎

The curve in the (a_0, s_0, ρ)-space of Bogdanov–Takens bifurcation stated in the last proposition is the organizing center for three surfaces (see [Guckenheimer & Holmes, 1997]): one of codimension-one Hopf bifurcation, another of saddle-node bifurcation of equilibria, and a third one corresponding to a nondegenerate homoclinic bifurcation.

A projection of this Bogdanov–Takens curve is drawn in Fig. 4 for $\alpha = 0.2$ and $\kappa = 0.1$. On such a curve two degenerate points (E and D) appear. Their study is performed in the next section.

3.   Degenerate Bogdanov–Takens Bifurcations

Degenerate codimension-three Bogdanov–Takens bifurcations arise when one of the quadratic coefficients in the Bogdanov–Takens normal form vanishes. Two degenerate cases may appear:

- the first one arises when $a \neq 0$ and $b = 0$ in (18) and it corresponds to a cusp of order three;
- the other case arises when $a = 0$ and $b \neq 0$ in (18) and it corresponds to a weak focus.

The existence of these degenerate Bogdanov–Takens bifurcations in the enzyme system is proved in the following subsections.

3.1. Cusp of order three case

Theorem 3.1. *For each $\kappa > 0$ there is, in the (a_0, s_0, ρ, α)-parameter space, a curve of cusps of order three given by the rational expression (17) and*

$$\alpha = \frac{1}{2}(\sqrt{9 - 4Q(u_\star)} - 1), \quad where \tag{19}$$

$$Q(u_\star) = \frac{2p(u_\star)}{(u_\star p'(u_\star) - p(u_\star))^2},$$

for $u_\star > 0$, with $h''(u_\star) \neq 0$ and $b'_4 \neq 0$ (b'_4 is defined in (21)).

Proof. If $F'_1(u_\star) = 0$ and $h''(u_\star) \neq 0$ then $a \neq 0$ and $b = 0$ in (18). A point verifying this kind of degeneracy will be labeled E in the sequel. We obtain its parameter values solving the system

$$F_1(u_\star) = F'_1(u_\star) = h(u_\star) = h'(u_\star) = 0. \tag{20}$$

From the first and second equations in (20) we get

$$s_0 = u_\star + \frac{(u_\star p'(u_\star) - p(u_\star))u_\star(\alpha + 2)}{2}.$$

Comparing this value of s_0 with the corresponding one given in (17), we obtain the following polynomial in $u_\star$

$$[2 - \alpha(1 + \alpha)]\kappa^2 u_\star^4 - 2[3 - \alpha(1 + \alpha)]\kappa u_\star^2$$
$$- 2u_\star - \alpha(1 + \alpha) = 0,$$

that leads to the required expression.

To assure codimension-three of the points E, we need to compute a fourth-order normal form (see [Dumortier *et al.*, 1987]). This can be done using the ideas of [Algaba *et al.*, 2003]. Then we obtain the following fourth-order normal form under smooth orbital equivalence:

$$\begin{cases} \dot{u} = v, \\ \dot{v} = a_2 u^2 + b'_4 u^3 v, \end{cases}$$

where

$$a_2 = \frac{1}{2}p(u_\star)h''(u_\star), \quad a_3 = \frac{1}{6}\frac{F'''_2(u_\star) + 3F''_2(u_\star)}{u_\star^2},$$

$$b'_4 = 96\kappa^5 u_\star^9 + 102\kappa^4 u_\star^8 + (24 - 268\kappa)\kappa^3 u_\star^7$$
$$- 316\kappa^3 u_\star^6 + (196\kappa - 88)\kappa^2 u_\star^5 + (264\kappa - 4)\kappa u_\star^4$$
$$+ (76 - 20\kappa)\kappa u_\star^3 + (4 - 52\kappa)u_\star^2 - (4\kappa + 12)u_\star$$

$$+ 2 + \sqrt{9\kappa^2 u_\star^4 - 26\kappa u_\star^2 - 8u_\star + 1}\left[-32\kappa^4 u_\star^7 \right.$$
$$- 34\kappa^3 u_\star^6 + (44\kappa - 8)\kappa^2 u_\star^5 + 46\kappa^2 u_\star^4$$
$$+ (8 - 24\kappa)\kappa u_\star^3 - 30\kappa u_\star^2 - 4(1 + \kappa)u_\star + 2\Big]. \tag{21}$$

The expression given in (19) corresponds to a curve of cusp of order three assuming that $b'_4 \neq 0$. ∎

The curve of cusp of order three points for $\kappa = 0.1$, projected onto the $(u_\star, \alpha)$-plane, appears in Fig. 5.

Remark. The vanishing of b'_4 provides the eventual cusps of order greater than three.

Dumortier *et al.* [1987] stated the unfolding of a cusp of order three. The intersection of the unfolding of the point E with a half-sphere with center in the parameter space point corresponding to the cusp presents the following bifurcation phenomena:

codimension-one:

1. subcritical and supercritical Hopf (H_{sub} and H_{super}, respectively);
2. saddle-node of equilibria (sn);
3. saddle-node of periodic orbits (SN);
4. left homoclinic orbit (H_L).

codimension-two:

1. degenerate Hopf (H_1);
2. Bogdanov–Takens (BT);
3. left homoclinic orbit with zero trace (H_L^D).

(Description of the different homoclinic orbits listed above appears at the end of Sec. 3.2.)

3.2. Weak focus case

Theorem 3.2. *For $\kappa > 0$, in the (a_0, s_0, ρ, α)-parameter space, there exists a curve of degenerate Bogdanov–Takens bifurcations corresponding to the vanishing of the term u^2 coefficient in the normal form. This curve is parameterized by the rational expression (17) and*

$$\alpha = 1 - \frac{p(u_\star)}{(u_\star p'(u_\star) - p(u_\star))^2} \tag{22}$$

for $u_\star > 0$, with $F'_1(u_\star) \neq 0$ and $b'_3 \neq 0$ (b'_3 is defined in (24)).

Moreover, this Bogdanov–Takens degeneracy corresponds to a weak focus, for each value of $\kappa > 0$ and $u_\star \in (0, +\infty)$. For each $\kappa > 0$, there exist 0, 2, 4 or 6 points on the curve given in (22) where the foci change their stability. These points correspond to a codimension-four degeneracy given by the vanishing of both coefficients of the terms u^2 and $u^2 v$ of the Bogdanov–Takens normal form.

Proof. If $h''(u_\star) = 0$ and $F_1'(u_\star) \neq 0$, then $a = 0$, $b \neq 0$ in the second-order normal form (18). We can obtain the parameter values of a point verifying this kind of degeneracy, labeled D in the sequel, by solving the equation system

$$F_1(u_\star) = h(u_\star) = h'(u_\star) = h''(u_\star) = 0.$$

Substituting the values of a_0, s_0 and ρ given in (17) into the fourth equation, we obtain the following polynomial in $u_\star$

$$u_\star^2 p(u_\star) p''(u_\star) - 2(u_\star p'(u_\star) - p(u_\star))$$
$$\times [(1 - \alpha) u_\star p'(u_\star) + \alpha p(u_\star)] = 0,$$

that leads to the expression stated in the theorem.

To guarantee codimension-three of the points D we need to compute a fourth-order normal form (see [Dumortier *et al.*, 1991]). As in the previous theorem, this can be done following [Algaba *et al.*, 2003]. In this manner we get the normal form

$$\begin{cases} \dot{u} = v, \\ \dot{v} = b_2 u v + a_3 u^3 + b_3' u^2 v, \end{cases} \quad (23)$$

where

$$b_2 = \frac{F_1'(u_\star)}{u_\star} = -\frac{\kappa^2 u_\star^3 - 3\kappa u_\star - 1}{u_\star p'(u_\star) - p(u_\star)},$$

$$a_3 = \frac{1}{6} p(u_\star) h'''(u_\star)$$
$$= -\frac{u_\star p(u_\star) \kappa (\kappa^2 u_\star^3 - 3\kappa u_\star - 1)}{(u_\star p'(u_\star) - p(u_\star))^2},$$

$$b_3' = -\frac{1}{5} \frac{r(u_\star)}{u_\star p(u_\star)(u_\star p'(u_\star) - p(u_\star))^2}, \quad (24)$$

where

$$r(u_\star) = 15\kappa^4 u_\star^7 + 18\kappa^3 u_\star^6 + 18\kappa u_\star + 1$$
$$- u_\star [4\kappa^3 u_\star^4 + 37\kappa^2 u_\star^3$$
$$+ (49\kappa^2 + 3\kappa) u_\star^2 + 2\kappa u_\star + 2]. \quad (25)$$

A point D can be classified into three topologically different types, depending on the values of the coefficients of the system (23): saddle (if $a_3 > 0$), focus (if $a_3 < 0$ and $b_2^2 + 8a_3 < 0$) and elliptic (if $a_3 < 0$ and $b_2^2 + 8a_3 > 0$). We are interested to know what type of point D does the system (16) have.

With this aim, let us define

$$d(\kappa, u_\star) = b_2^2 + 8a_3$$
$$= \frac{\sqrt{\kappa^2 u_\star^3 - 3\kappa u_\star - 1} - 2\sqrt{2 u_\star \kappa p(u_\star)}}{\sqrt{u_\star \kappa p(u_\star)}}.$$

If α tends to zero in the expression given in (22), it is easy to see that the abscissa $u_\star$ of the equilibrium undergoing the degeneracy D tends to a value $u_\star^0$ which is a root of the polynomial $P(u_\star) = \kappa^2 u_\star^3 - 3\kappa u_\star - 1$. Since $P'(u_\star) = 3\kappa(\kappa u_\star^2 - 1) > 0$ for all $u_\star > 1/\sqrt{\kappa}$, and $P(1/\sqrt{\kappa}) < 0$, it follows that $u_\star^0$ is the unique root of $P(u_\star)$ in $(1/\sqrt{\kappa}, +\infty)$. Thus, the curve of points D is solely defined for $u_\star \in (u_\star^0, +\infty)$ and, moreover, $P(u_\star) > 0$ for all $u_\star > u_\star^0$. Therefore, $a_3 < 0$ for all $u_\star \in (u_\star^0, +\infty)$ and for all $\kappa > 0$.

The proof of the following statements is direct:

(a) $\lim_{u_\star \to +\infty} d(\kappa, u_\star) = 1 - 2\sqrt{2}$, independently of κ;

(b) $d(\kappa, u_\star^0) = -2\sqrt{2}$, independently of κ;

(c) The function $d(\kappa, u_\star)$ is increasing in $(u_\star^0, +\infty)$, for each value of $\kappa > 0$, since

$$\frac{\partial}{\partial u_\star} d(\kappa, u_\star)$$

$$= \frac{1}{2} \frac{\kappa^2 u_\star^4 + 8\kappa^2 u_\star^3 + 6\kappa u_\star^2 + 2 u_\star + 1}{u_\star^{\frac{3}{2}} \sqrt{\kappa} p(u_\star)^{\frac{3}{2}} \sqrt{\kappa^2 u_\star^3 - 3\kappa u_\star - 1}} > 0$$

in $(u_\star^0, +\infty)$.

A sketch of the function $d(\kappa, u_\star)$ appears in Fig. 9.

Therefore, $d(\kappa, u_\star) < 0$ for each value of $\kappa > 0$ and for each value of $u_\star \in (u_\star^0, +\infty)$. Thus, the point D is always of focus type.

The stability of the focus is given by the sign of b_3', that is, the sign of $r(u_\star)$ given in (25).

Since $r(0) = 1$, $r(u_\star)$ has, at least, one negative root and, therefore, it has, at most, six positive roots. These positive roots correspond to a change of stability of the focus (note that the degeneracy condition is $b_3' = 0$) and they provide the degenerate codimension-four points stated in the theorem. ∎

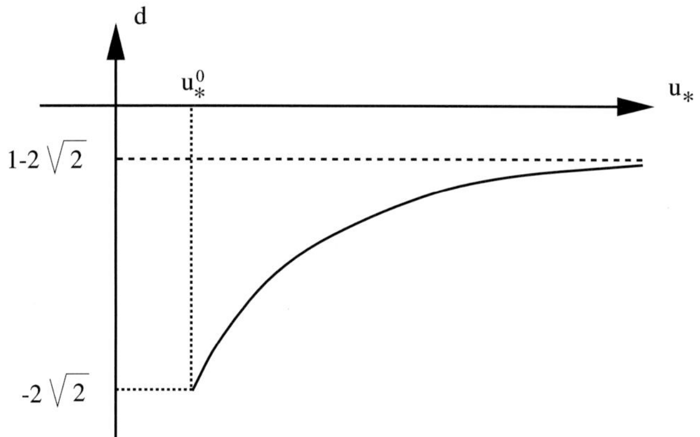

Fig. 9.   Qualitative sketch of $d(\kappa, u_\star)$ versus $u_\star$.

The curve of points D for $\kappa = 0.1$, projected onto the $(u_\star, \alpha)$-plane, appears in Fig. 5.

Dumortier *et al.* [1991] stated the unfolding of a point D in the focus case. The intersection of the unfolding of the point D in the focus case with a sphere with center in the parameter space point corresponding to the singular point presents the following bifurcation phenomena:

codimension-one:

1. subcritical and supercritical Hopf (H_{sub} and H_{super}, respectively);
2. saddle-node of equilibria (sn);
3. saddle-node of periodic orbits (SN);
4. left homoclinic orbit (H_L);
5. right homoclinic orbit (H_R);
6. lower concave homoclinic orbit (H_{LC});
7. right central saddle-node homoclinic orbit ($CSNH_R$);
8. left central saddle-node homoclinic orbit ($CSNH_L$);

codimension-two:

1. degenerate Hopf (H_1);
2. cusp of equilibria (C);
3. Bogdanov–Takens (BT);
4. lower concave homoclinic orbit with zero trace $\left(H_{LC}^D\right)$;
5. right saddle-node homoclinic orbit (SNH_R);
6. left saddle-node homoclinic orbit (SNH_L);
7. right lower concave saddle-node homoclinic orbit $\left(SNH_R^{LC}\right)$;
8. left lower concave saddle-node homoclinic orbit $\left(SNH_L^{LC}\right)$.

The equilibrium with smallest (largest) abscissa is called the left (right) equilibrium; the third equilibrium (that is always a saddle) located between the other two is called the intermediate equilibrium. So, left (right) homoclinic orbit means a homoclinic connection surrounding the left (right) equilibrium. A homoclinic orbit connecting from below (up) the intermediate equilibrium to itself and surrounding the left and right equilibria is called a lower (upper) concave homoclinic orbit.

The central saddle-node homoclinic orbit occurs when the isolated center manifold of the saddle-node point returns to it through the interior of the nodal sector; the noncentral left (right) saddle-node homoclinic orbit occurs when the isolated center manifold returns to the saddle-node point through one of the hyperbolic separatrices, surrounding the left (right) equilibrium. If the isolated center manifold returns to the saddle-node point from below (up) through one of the hyperbolic separatrices enclosing the nodal sector and the hyperbolic equilibrium is placed at the left of the nonhyperbolic one, the homoclinic connection is called a left lower (upper) concave saddle-node homoclinic orbit; if the hyperbolic equilibrium is placed at the right of the nonhyperbolic one, the homoclinic connection is called a right lower (upper) concave saddle-node homoclinic orbit. Several of the above homoclinic connections are sketched in Fig. 10.

3.3. *Additional comments*

The codimension-three degenerate Bogdanov–Takens bifurcations considered in this section have also been studied by [Medved, 1985; Guckenheimer, 1986a]. Moreover, the bifurcation diagrams for both types of the degenerate Bogdanov–Takens bifurcation have been described in [Bazykin *et al.*, 1989; Berezovskaya & Khibnik, 1985]. The latter publication contains the complete analysis leading to the bifurcation set in a neighborhood of the point E.

Points E and D appear in Fig. 4, for $\alpha = 0.2$ and $\kappa = 0.1$. Obviously, these two points are located on the Bogdanov–Takens curve BT. The curve of degenerate Hopf bifurcations (H_1) emerges from D whereas the other degenerate Hopf bifurcations (H_1') occur in a locus arising from E. Moreover, the curve of cusp bifurcations of equilibria (C) goes across D.

We end our study of degenerate Bogdanov–Takens bifurcations with two remarks. The first

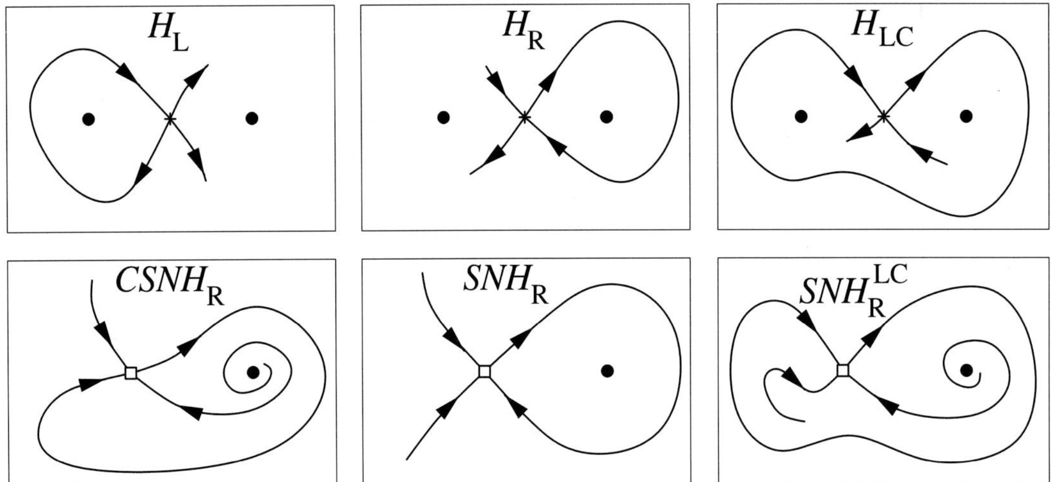

Fig. 10.   Qualitative picture of: a left homoclinic orbit, H_L; a right homoclinic orbit, H_R; a lower concave homoclinic orbit, H_{LC}; a right central saddle-node homoclinic orbit, $CSNH_R$; a right saddle-node homoclinic orbit, SNH_R; a right lower concave saddle-node homoclinic orbit, SNH_R^{LC}. Focus, saddle and nonhyperbolic saddle-node equilibria are represented, respectively, by a filled point, a cross and an empty square.

one is that both quadratic coefficients of the Bogdanov–Takens normal form cannot vanish simultaneously. From (19) and (22) such a situation would lead to

$$\frac{1}{2}(\sqrt{9 - 4Q(u_\star)} - 1) = 1 - \frac{p(u_\star)}{(u_\star p'(u_\star) - p(u_\star))^2},$$

that can occur if, and only if, $\alpha = 0$, that has no biochemical meaning.

A case where both quadratic coefficients of the Bogdanov–Takens normal form vanish appears in [Dangelmayr & Guckenheimer, 1987].

The second remark refers to the limit case $\alpha = 1$. It is easy to verify that the curves of points E and D obtained in Theorems 3.1 and 3.2, respectively, tend to $\alpha = 1$ as the parameter $u_\star$ tends to infinity. An evaluation of the curve of degenerate Hopf points H_2 provides an analogous result (see Fig. 5). This fact is justified by the following reasonings.

If we consider the system (1) for $\alpha = 1$,

$$\begin{cases} \dot{s} = s_0 - s - \rho \dfrac{sa}{1 + s + \kappa s^2}, \\[2mm] \dot{a} = a_0 - a - \rho \dfrac{sa}{1 + s + \kappa s^2}, \end{cases} \tag{26}$$

and we make the change of variables $w = s - a$, this new variable verifies the differential equation

$$\dot{w} = w_0 - w, \tag{27}$$

where $w_0 = s_0 - a_0$.

The equilibrium of Eq. (27), $w = w_0$, corresponds to the straight line $s - a = s_0 - a_0$, that is invariant for the flow of (26). Thus, the equilibria of system (26) lie on the line $s - a = s_0 - a_0$ and, therefore, the existence of limit cycles, homoclinic connections and equilibria of focus type is impossible. We remark that, in spite of the richness of the dynamic and bifurcation behavior of the system (1) for all values of parameter α arbitrarily close to the critical value $\alpha = 1$, there are no more limit sets than equilibria for the value $\alpha = 1$, and therefore the only bifurcations that remain are saddle-nodes and cusps of equilibria.

In short, the limit sets that exist for $\alpha < 1$ suffer a *stretching* when α tends to 1 and they disappear for $\alpha = 1$ (see Fig. 5).

4. Homoclinic Orbits and Their Numerical Continuation

When dealing with parameterized systems of autonomous ordinary differential equations, the presence of a homoclinic orbit (that is, a trajectory which is bi-asymptotic to the same stationary point in both forwards and backwards time) may reveal the existence of other bifurcations (see e.g. [Wiggins, 2003]).

In autonomous planar systems, under certain *nondegeneracy* conditions, a homoclinic bifurcation simply creates or destroys a single periodic orbit. Roughly speaking, it is a bifurcation of a periodic orbit to "infinite period". Nevertheless, the presence of degenerate homoclinic connections will lead to the existence of several bifurcations where more than one periodic orbit is involved (for example, saddle-node and cusp of saddle-node bifurcations). Then, it will be of importance to combine analytical tools with numerical methods that detect and continue degenerate homoclinic connections, since they act as important organizing centers in the dynamical behavior of systems.

The techniques to study homoclinic orbits in planar vector fields were well developed by the 1920's in the works of Dulac. The fundamental idea is that the recurrent behavior near a connecting orbit should be studied in a fashion similar to that used in studying periodic orbits via a Poincaré return map. But there are some additional complications in the study of homoclinic orbits compared to that of periodic orbits which significantly complicate the analysis (see e.g. [Guckenheimer & Worfolk, 1993]).

There are many types of codimension-two bifurcations of connecting orbits. Failure of one of the conditions that characterize a generic homoclinic orbit will lead to a degenerate bifurcation. This can occur, in planar systems, for eigenvalues degeneracies and for multiple connecting orbits [Guckenheimer & Worfolk, 1993]. In the system we consider, as we are going to see, two kinds of eigenvalue degeneracies may occur. The first one appears for homoclinic orbits to nonhyperbolic equilibria (a zero eigenvalue). The second one is present when a nonresonant condition is violated (zero trace). On the other hand, a double homoclinic connection appears for certain values of the parameters. Furthermore, the presence of two kinds of codimension-three homoclinic orbits will be pointed out.

Among the numerical continuation methods proposed in the literature there exist basically two groups: *boundary-value* and *shooting* methods. The *boundary-value* methods truncate the homoclinic problem to a finite time interval and impose certain boundary conditions at the end points of that interval (see e.g. [Beyn, 1990; Friedman & Doedel, 1993]). The second technique uses shooting, that is, the numerical integration of orbits in the stable and unstable manifolds of the equilibrium and the computation of a distance between them (see e.g. [Rodríguez-Luis *et al.*, 1990]).

The above methods detect the homoclinic orbit and provide the curve, in a parameter plane, where the global codimension-one bifurcation

occurs. As global bifurcations may exhibit degeneracies, numerical techniques are also needed for these higher codimension situations. In this direction, [Champneys & Kuznetsov, 1994, 1996], have developed a continuation code for several cases of codimension-two homoclinic bifurcations. This code, called HomCont, has been included in the AUTO97 continuation and bifurcation software [Doedel *et al.*, 1998].

In the following we briefly describe the numerical shooting methods used along this work for the continuation of homoclinic orbits. Moreover, we will give information about some theoretical results on the homoclinic orbits that appear in this system. In particular, in Sec. 4.3, we will emphasize on cuspidal loops, a codimension-three homoclinic connection.

First, in Sec. 4.1, we deal with nondegenerate homoclinic connections and, in Sec. 4.2, we consider the cases of degenerate homoclinic orbits. The basic idea of the numerical method we use is to establish a correspondence, under the adequate hypothesis, between the homoclinic connections and the zeros of a certain function. Then, the continuation of homoclinic connections will be equivalent to the continuation of the zeros of such a function.

4.1. *Continuation of nondegenerate homoclinic connections*

We consider the one-parameter autonomous planar system

$$\dot{x} = X(x, \mu), \quad x = (x_1, x_2) \in \mathbb{R}^2, \quad \mu \in I \subseteq \mathbb{R},$$

where $X \in C^\infty \left(\mathbb{R}^2 \times I; \mathbb{R}^2\right)$ is a family of vector fields and I is some neighborhood of $\mu_0 \in \mathbb{R}$, for which value a homoclinic orbit occurs.

Suppose the origin, $x = 0$, is a hyperbolic equilibrium, $X(0, \mu) = 0$ for all $\mu \in I$, of saddle-type. Without loss of generality we may suppose that the linearization matrix has the form

$$D_x X(0, \mu) = \begin{pmatrix} -\lambda_1(\mu) & 0 \\ 0 & \lambda_2(\mu) \end{pmatrix},$$

where $\lambda_1(\mu)$ and $\lambda_2(\mu)$ are positive scalars, for $\mu \in I$.

Under the adequate hypothesis [Freire *et al.*, 1999b], the existence of a nondegenerate homoclinic connection corresponds to a regular zero of a certain scalar function $G_1(\mu)$ that measures, on an adequate transversal section, the distance between the stable and the unstable manifolds of the equilibrium. If now the system is bi-parametric, $\mu \in \mathbb{R}^2$, under the adequate hypothesis, a curve of nondegenerate homoclinic orbits may be continued solving $G_1(\mu) = 0$, that is, the continuation in the parameter plane of the homoclinic connections locus is a problem equivalent to tracing zeros of a function of one component and two independent variables. If there are more parameters, $\mu \in \mathbb{R}^m$, $m > 2$, to continue curves of degenerate homoclinic bifurcations it is enough to add the appropriate test functions defining the degeneracy in question.

4.2. *Continuation of degenerate homoclinic connections*

For $\mu \in \mathbb{R}^2$, to detect a codimension-two point along the homoclinic curve we monitor a test function Ψ_1 (for instance, the test function that detects the vanishing of the trace is simply $\Psi_1(\mu) = -\lambda_1(\mu) + \lambda_2(\mu)$). If Ψ_1 changes sign we first accurately locate its zero. We can then continue numerically the curve of codimension-two homoclinic orbits in three parameters by restarting from the detected zero of Ψ_1, freeing an additional parameter ($\mu \in \mathbb{R}^3$) and appending the extra algebraic constraint $\Psi_1(\mu) = 0$ to $G_1(\mu) = 0$. It is clear that this strategy may be applied to compute curves of codimension-three points as four parameters are allowed to vary and to continue curves of codimension-four points when $\mu \in \mathbb{R}^5$. Obviously, some extra transversality assumption has to be satisfied to guarantee that we have a regular zero of the corresponding test functions (see details of this detection and continuation strategy in [Champneys & Kuznetsov, 1994]).

A first eigenvalue degeneracy (codimension-two) appears when the homoclinic orbit connects an hyperbolic equilibrium point with zero trace, the so-called neutral resonant saddle case. This situation was studied completely by [Nozdrachova, 1982] for two-dimensional systems. A curve of fold (saddle-node) bifurcations of periodic orbits emerges from this codimension-two bifurcation point.

In the bi-parametric case $\mu = (\mu_1, \mu_2)$, the curve of homoclinic connections is defined by $(\mu_1(s), \mu_2(s))$, where s adequately parameterizes such a curve. To guarantee a regular zero of the test function $\Psi_1(\mu(s))$, at $s = s_0$ say, we have to add the extra transversality assumption $d\Psi_1/ds|_{s=s_0} \neq 0$.

In the zero-trace case, the stability of the homoclinic orbit is determined by the integral of the divergence of the vector field along the homoclinic orbit. In fact, the homoclinic orbit Γ is asymptotically stable (resp. unstable) if, and only if, $\int_\gamma \operatorname{div} X < 0$ (resp. $\int_\gamma \operatorname{div} X > 0$), where $\gamma(t)$, $t \in (-\infty, \infty)$, parameterizes Γ. Then, an additional degeneracy (codimension-three) appears when $\int_\gamma \operatorname{div} X = 0$. As it is easier to compute the exponential of the integral of the divergence EID [Freire *et al.*, 1999b],

$$\text{EID} \overset{\text{def}}{=} e^{\int_\gamma \operatorname{div} X}, \tag{28}$$

this codimension-three homoclinic singularity has simultaneously zero trace and EID $= 1$. Its numerical continuation will be done using the test functions $\Psi_1 = -\lambda_1 + \lambda_2$ and $\Psi_2 = \text{EID} - 1$, with the corresponding transversality assumptions to guarantee the regularity of such zeros. From this codimension-three point a curve of cusps of saddle-node of periodic orbits will emerge.

The stability of this codimension-three homoclinic orbit is governed by a new resonant local coefficient RES which may be computed taking advantage of the duality between the Hopf bifurcation (and its degeneracies) and the homoclinic bifurcation (and its degeneracies in the case of zero trace). (See details in [Freire *et al.*, 1999b; Joyal, 1988].) The first result used to look for an expression of RES is the following:

Proposition 4.1. *Let u_0 be a hyperbolic saddle point of the planar system*

$$\dot{u} = X(u), \quad u = (x, y) \in \mathbb{R}^2, \tag{29}$$

with $\operatorname{div} X(u_0) = 0$. Under these conditions, system (29) is C^∞ orbitally equivalent to

$$\begin{cases} \dot{x} = x, \\ \dot{y} = -y + \displaystyle\sum_{k=1}^{n} \alpha_{k+1} x^k y^{k+1} + O\big(|x, y|^{2n+3}\big). \end{cases}$$

System (29) can be written as

$$\begin{pmatrix} \dot{x} \\ \dot{y} \end{pmatrix} = \begin{pmatrix} 0 & 1 \\ 1 & 0 \end{pmatrix} \begin{pmatrix} x \\ y \end{pmatrix} + \begin{pmatrix} f(x, y) \\ g(x, y) \end{pmatrix},$$

where $f(x, y), g(x, y) = O(|x, y|^2)$, and we have assumed that $u_0 = 0$ is a hyperbolic equilibrium of (29).

It is then possible to obtain that RES is given by

$$\mathrm{RES} = (f_{yy}g_{yy} + f_{yy}f_{xy} + g_{yy}g_{xy} - f_{xy}f_{xx}$$
$$- g_{xy}g_{xx} - f_{xx}g_{xx} - g_{yyy}$$
$$- f_{xyy} + g_{xxy} + f_{xxx})/16. \tag{30}$$

When this coefficient RES vanishes, a codimension-four homoclinic singularity appears. In this situation (we have numerically checked that this does not occur in the enzyme system), a curve of swallowtail singularities of periodic orbits will emerge in a four-parameter space from such a codimension-four point.

A second eigenvalue degeneracy (codimension two) appears when the homoclinic curve (in a bi-parametric space) reaches a curve of saddle-node bifurcations of equilibria. This situation, known as the saddle-node separatrix-loop bifurcation, was analyzed by [Schecter, 1987]. He showed that the homoclinic curve meets the fold curve with a quadratic tangency.

When the curve of nondegenerate homoclinic connections is approaching the fold curve, it is better to take, in the numerical method, the abscissa of the equilibrium as continuation parameter [Freire *et al.*, 2000]. This is the way to detect such a codimension-two point.

To continue the curve of these degenerate homoclinic connections in a three-parameter space we have to adapt our strategy to the presence of a nonhyperbolic equilibrium. We take a linear approximation for the hyperbolic manifold (stable or unstable) and a quadratic approximation for the center manifold. In this way we have a continuation problem with a function of three variables (the three parameters) and two components (the first one is the distance, on a transversal section, between the orbits integrated from the approximations of the center and the hyperbolic manifolds; the second one is the condition of saddle-node bifurcation of equilibria).

On the other hand, a double homoclinic connection (degeneracy for multiple connecting orbits) is easily detected looking at the crossing of two homoclinic curves, one corresponding to left homoclinic orbits, H_L, and the other one to right homoclinic orbits, H_R. Its continuation in a three-parameter space is performed looking at the zeros of a two-component function (each component corresponds to the condition of existence of one homoclinic connection). Other curves of homoclinic orbits (namely,

of lower concave, H_{LC}, and of upper concave, H_{UC}) emerge from such a double homoclinic point, HH. An example of this situation appears, for instance, in [Freire *et al.*, 1996].

4.3. *Cuspidal loops*

In this subsection we will summarize some results about cuspidal loops and the numerical method for the continuation of these planar codimension-three homoclinic orbits [Freire *et al.*, 2000]. A cuspidal loop occurs when the separatrices of an equilibrium of cusp type intersect and a cusp point is a nonhyperbolic equilibrium with a double-zero eigenvalue (Bogdanov–Takens bifurcation).

Let X be a planar vector field, $X \in C^\infty(\mathbb{R}^2)$, and

$$\begin{cases} \dot{x} = X_1(x, y), \\ \dot{y} = X_2(x, y), \end{cases} \tag{31}$$

a dynamical system, with an equilibrium at the origin of cusp type, that is, the equilibrium has stable and unstable local separatrices forming a cusp. It is well known (see e.g. [Guckenheimer & Holmes, 1997]) that system (31) is C^∞ orbitally equivalent to a system in the form

$$\begin{cases} \dot{x} = y + O(|x, y|^{k+1}), \\ \dot{y} = \sum_{j=2}^{k} a_j x^j + b_j x^{j-1} y + O(|x, y|^{k+1}). \end{cases} \tag{32}$$

When $a_2 \neq 0$, the topological type of (32) is determined by the truncated second-order system

$$\begin{cases} \dot{x} = y, \\ \dot{y} = ax^2 + bxy, \end{cases} \tag{33}$$

(where $a = a_2$ and $b = b_2$). We assume that the separatrices intersect forming a cuspidal loop.

We also assume that the cuspidal loop is parameterized by the function $\gamma(t)$, for $t \in (-\infty, +\infty)$. The stability of the cuspidal loop is governed by the integral of the divergence of the vector field along the homoclinic loop, $\int_\gamma \operatorname{div} X$, in the case that this quantity does not vanish, as is stated in [Dumortier *et al.*, 1997] (the cuspidal loop is an attracting singular cycle if $\int_\gamma \operatorname{div} X < 0$, and it is a repelling one if $\int_\gamma \operatorname{div} X > 0$).

To carry out the analysis of the local stability of a cuspidal loop and to establish the different unfoldings of such a singularity, two local transversal sections to the loop are taken and three maps are considered [Freire *et al.*, 1999a]:

- The Dulac map D, that provides local information of the behavior in the vicinity of the cusp equilibrium point.
- The regular transition map along the homoclinic orbit, R.
- The Poincaré map, P, given by the composition of R and D.

Two cases appear, depending on the signs of a and b:

1. $a > 0, b > 0$ (topologically equivalent to the case $a < 0, b < 0$). The slope of D, in this case, is greater than 1, that indicates that the local behavior, in the vicinity of the cusp point, is asymptotically unstable. The global behavior, along the regular arc of the homoclinic orbit, and, therefore, the stability of the cuspidal loop, is given by the sign of ω (the slope of R is $1 + \omega$):

   (i) If $\omega < 0$ the cuspidal loop is an attracting cycle;
   (ii) if $\omega > 0$ the cuspidal loop is a repelling cycle.

   The limit situation, $\omega = 0$, corresponds, therefore, to a codimension-four degeneracy, since the stability of the homoclinic orbit has changed.

   The case (ii) corresponds to the simplest type of cuspidal loop. In this case the homoclinic orbit that rises from the Bogdanov–Takens point and the cuspidal loop have the same stability (this kind of cuspidal loop occurs in the enzyme system). The case (i) is more complex, since both stabilities are now opposite, producing, in the corresponding unfolding, a much richer dynamical behavior (an example of this kind of cuspidal loop appears in the *continuous flow stirred tank reactor*, CSTR [Guckenheimer, 1986b]). In Fig. 11 the unfolding of the first type of (unstable) cuspidal loop is shown [Dumortier *et al.*, 1997]. This figure has been obtained intersecting the unfolding of the cuspidal loop with a sphere with center in the parameter space point corresponding to the cuspidal loop. Numbers in this figure make reference to the different phase portraits displayed at the bottom of it. In these phase portraits, the solid (dotted) line represents a stable (unstable) periodic orbit; the

symbols: •, ∘ and + mean stable, unstable and saddle equilibrium, respectively.

The bifurcation phenomena that appear in the vicinity of a cuspidal loop can be classified by their codimension:

codimension-one:

(a) (subcritical) Hopf (H_{sub});
(b) saddle-node of equilibria (sn_L);
(c) saddle-node of periodic orbits (SN);
(d) left homoclinic orbit (H_L);
(e) right homoclinic orbit (H_R);
(f) lower concave homoclinic orbit (H_{LC});
(g) upper concave homoclinic orbit (H_{UC});
(h) right central saddle-node homoclinic orbit ($CSNH_R$);

codimension-two:

(a) cusp of periodic orbits (Cu);
(b) Bogdanov–Takens (BT_L);
(c) right homoclinic orbit with zero trace $\left(H_R^D\right)$;
(d) lower concave homoclinic orbit with zero trace $\left(H_{LC}^D\right)$;
(e) double homoclinic orbit (HH);
(f) right saddle-node homoclinic orbit (SNH_R);
(g) right lower concave saddle-node homoclinic orbit $\left(SNH_R^{LC}\right)$;
(h) right upper concave saddle-node homoclinic orbit $\left(SNH_R^{UC}\right)$.

Recall the description of the different homoclinic orbits given at the end of Sec. 3.2. Moreover, all the above homoclinic connections are sketched in Fig. 23.

The subscript L (left) in sn_L indicates that in the saddle-node bifurcation collapse the middle and the left equilibria of the system and BT_L means that the bifurcation is exhibited by the nonhyperbolic left equilibrium.

2. $a > 0, b < 0$ (topologically equivalent to the case $a < 0, b > 0$). Similarly, $p < 1$ is obtained (p is the slope of D), which means asymptotically stable local behavior, and the following results can be summarized:

   (1) $\omega < 0$: asymptotically stable global behavior; the cuspidal loop is asymptotically stable. Simple case.
   (2) $\omega > 0$: asymptotically unstable global behavior; the cuspidal loop is asymptotically stable. Complex case.

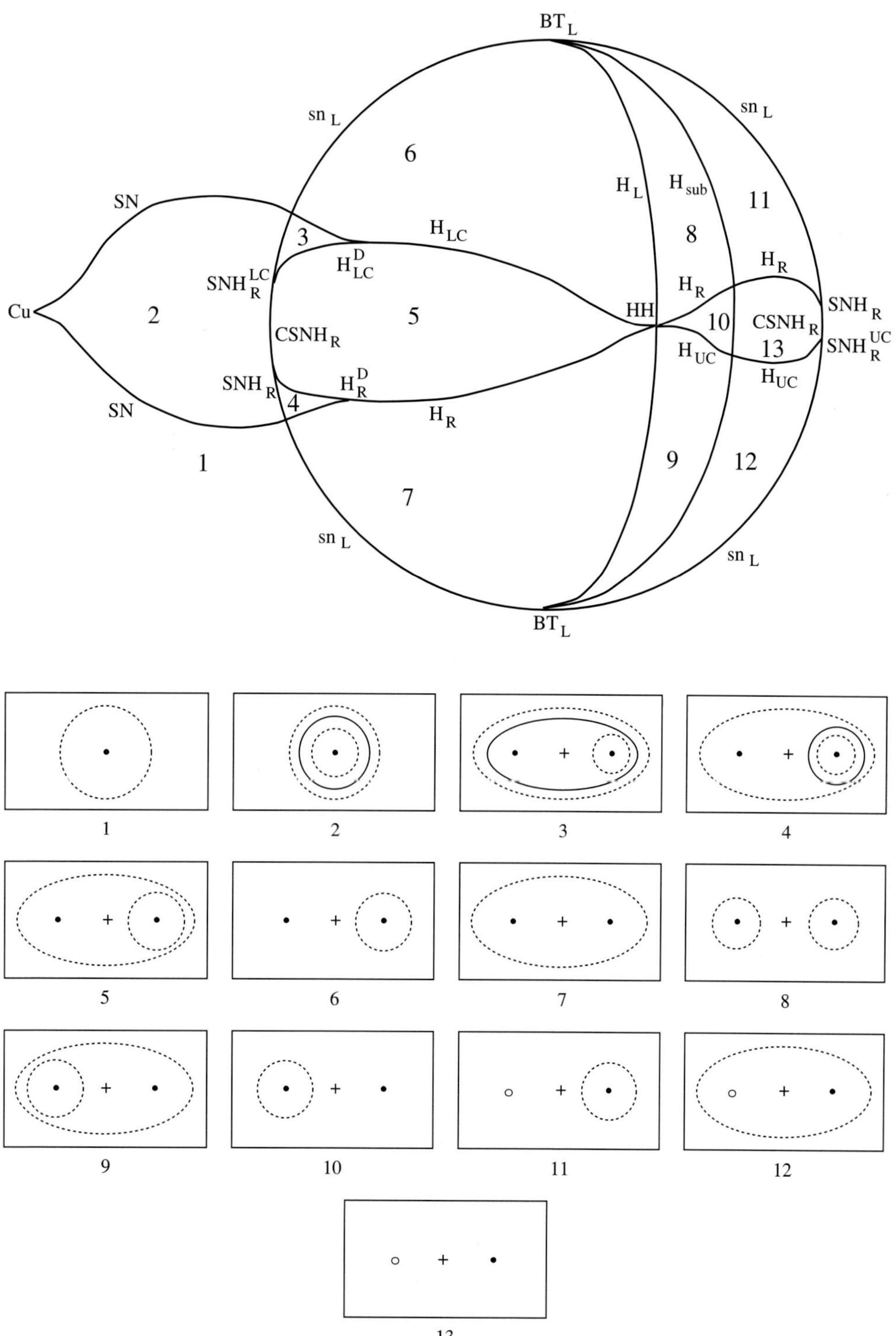

Fig. 11.   Unfolding of the simplest type of (unstable) cuspidal loop. The different phase portraits appearing in the unfolding of this type of cuspidal loop are also sketched. A solid (dotted) line represents a stable (unstable) periodic orbit; the symbols •, ◦ and + mean stable, unstable and saddle equilibrium, respectively.

With respect to the numerical continuation of cuspidal loops, the separatrices of the nonhyperbolic equilibrium can be approximated by means of the semi-cubic $y^2 = ax^3$ (if the cusp point is at the origin). This is deduced from (33), as in the vicinity of the origin (cusp point) the separatrices have horizontal tangent and then the term bxy is negligible, in first approximation, with respect to the ax^2 term [Freire *et al.*, 2000].

5. Homoclinic Bifurcations in the Enzyme System

We start this section adapting the numerical methods for homoclinic continuation, summarized above, to the system under study. The information we get on homoclinic connections allows to show, in Sec. 6, eight representative bifurcation sets with the dynamical behavior exhibited by this system. These bifurcation sets include the previous information we obtained by using analytical methods concerning local bifurcations as well as the global bifurcations arising in enzyme system not yet considered. These are the upper concave homoclinic orbit (H_{UC}), the right upper concave saddle-node homoclinic orbit (SNH_R^{UC}), the double homoclinic orbit (HH), the cuspidal loop (CL) and the lower concave homoclinic orbit with simultaneously zero trace and coefficient $EID = 1$ (H_{EID}). In this way, we obtain two objectives: to show the unfoldings of the different codimension-three bifurcation phenomena as well as the transition among these important organizing centers of the dynamics exhibited by the enzyme system.

We are now interested in how an orbit in the a–s phase plane appears in the u–v plane and vice versa (see Fig. 12). First, note that the curve $\dot{s} = 0$ is mapped into the u-axis and the straight line labeled as r (namely, $s_0 - s = \alpha(a_0 - a)$) is transformed into the $v = h(u)$ curve. Secondly, the region where $\dot{s} < 0$ maps into the region $v < 0$. Thus, the equilibria of system (1), that appeared on the straight line r, occur now on the u-axis, in system (4).

In general, an orbit in the region $\dot{s} < 0$ will move from right to left (as t increases) whereas the corresponding orbit in the u–v plane will move from left to right in the $v < 0$ zone. Analogously, the orbits in the $\dot{s} > 0$ region, that move from left to right, correspond to orbits moving from right to left in the $v > 0$ region.

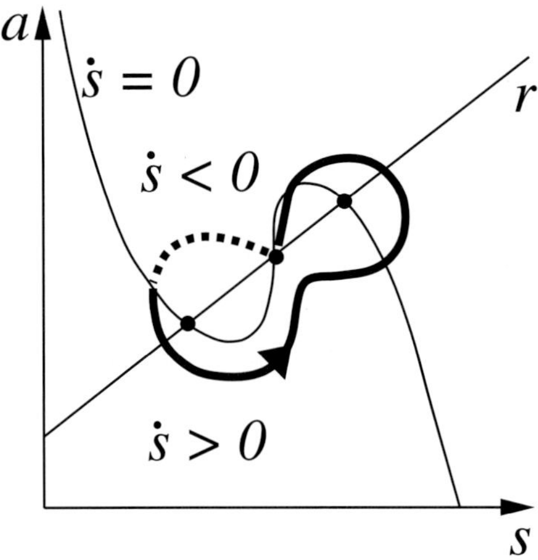

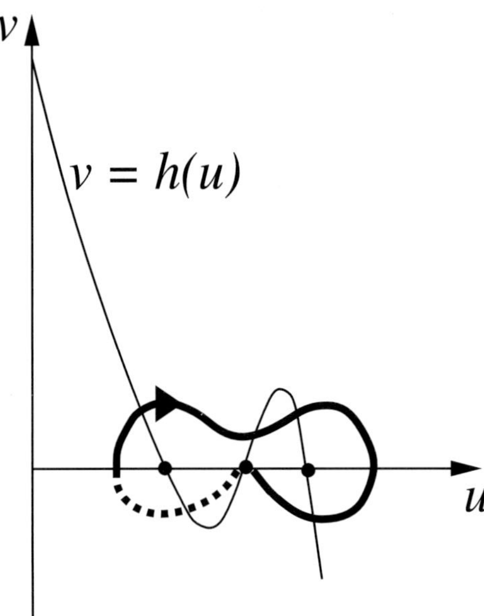

Fig. 12. Sketch to understand how the orbits in the s–a plane map into the u–v plane. We have represented, in the plane s–a, the nullcline corresponding to $\dot{s} = 0$ and the straight line r where the equilibria appear. In the plane u–v we have drawn the equilibria, as the intersection points between the u-axis and the curve $v = h(u)$. A homoclinic orbit also appears in both planes: it is lower concave in the s–a plane and upper concave in the u–v plane.

In particular, the lower concave homoclinic orbit drawn in the s–a plane is transformed into the upper concave homoclinic orbit in the u–v plane. Thus, all kinds of homoclinic connections (except the concave ones) keep their shape in both planes but the orbits are described in opposite senses.

Note that the names and labels of the homoclinic orbits along this work correspond to their shape in the u–v plane.

Now we describe the strategy followed in the continuation of the homoclinic connection loci. In all cases of homoclinic orbits to a hyperbolic saddle point, we have taken linear approximations to the stable and unstable manifolds of the equilibrium. The addition of the test function $\Psi_1 = -\lambda_1 + \lambda_2$ allows to detect degenerate homoclinic connections with zero trace (codimension-two). If we consider the test functions Ψ_1 along with $\Psi_2 = \mathrm{EID} - 1$ we may continue the codimension-three homoclinic orbits given by the vanishing of both trace and integral of the divergence (see algorithm EID developed in [Freire *et al.*, 1999b]).

In principle, there are four kinds of homoclinic orbits that may become degenerate due to the vanishing of the trace: H_L, H_R, H_{UC} and H_{LC}. However, we have checked that this degeneration only appears in two cases (labeled as H_L^D and H_{LC}^D). Moreover, the following degeneration ($\mathrm{EID} = 1$) only occurs for the H_{LC}^D homoclinic connections. We

might call H_{LC}^{EID} to this codimension-three homoclinic orbits but, for simplicity, we denote them as H_{EID}.

For the reference values found in the literature $\alpha = 0.2$ and $\kappa = 0.1$ (we will use along all this section), we have located a homoclinic connection H_{EID} (it is a degeneration point in the curve of lower concave homoclinic orbit with zero trace H_{LC}^D) for the parameter values

$$u_\star \approx 7.4884, \quad a_0 \approx 717.2305,$$
$$s_0 \approx 36.122, \quad \rho \approx 0.0939.$$

We remark that the other homoclinic connections with zero trace do not present this additional degeneracy ($\mathrm{EID} = 1$).

We have also computed the resonant coefficient RES given in (30), that determines the stability of the homoclinic with zero trace and $\mathrm{EID} = 1$, H_{EID}:

$$\mathrm{RES} = \frac{1}{16} \frac{u_\star \left[F_1'(u_\star) F_2''(u_\star) - F_1''(u_\star) F_2'(u_\star) \right] - 2 F_1'(u_\star) F_2'(u_\star)}{\sqrt{u_\star} \sqrt{F_2'(u_\star) F_2'(u_\star)}}.$$

We have verified that, for $\kappa = 0.1$, it does not vanish (in fact, it remains always positive). Therefore, no codimension-four homoclinic orbit with zero-trace, $\mathrm{EID} = 1$ and $\mathrm{RES} = 0$ arises.

When a curve of left homoclinic orbits intersects with a curve of right homoclinic orbits, a double homoclinic connection occurs. In this situation, the equilibrium point is hyperbolic, but other codimension-one homoclinic curves appear. In Fig. 13, we show the details of the tangency between the curves of right homoclinic orbits (H_R) and lower concave homoclinic orbits (H_{LC}) and between the curves of left homoclinic orbits (H_L) and upper concave homoclinic orbits (H_{UC}). Note that if we superimpose both figures, the curve H_{UC} would be imperceptible with respect to H_{LC}, as the parameter region shown in (a) is approximately ten times the region drawn in (b).

At this moment we think it is interesting to have a realistic idea of the phase portrait of the homoclinic connections exhibited by the enzyme system. In Fig. 14 we represent the phase portraits of two homoclinic orbits, obtained with Dstool [Guckenheimer & Kim, 1992], for the following values of the parameters: $s_0 = 37$, $\alpha = 0.2$ and $\kappa = 0.1$. The first one, a right homoclinic connection H_R,

occurs for $a_0 \approx 683.39886$ and $\rho \approx 0.1036702$. Its phase portrait in the original s–a plane appears in Fig. 14(a) whereas its phase portrait in the u–v plane is drawn in Fig. 14(b). The second one, a lower concave homoclinic connection H_{LC} in the u–v plane (and an upper concave homoclinic connection in the s–a), exists for $a_0 \approx 683.46591$ and $\rho \approx 0.1036624$. We show its phase portrait in the original s–a plane in Fig. 14(c) and in the u–v plane in Fig. 14(d).

Two conclusions follow from this picture. First, it is evident that the u–v plane is more convenient for the representation of the orbits (they are easier to see in such a plane). Secondly, small variations of the parameters imply important changes in the dynamic behavior of the system (compare the parameter values of the two homoclinic orbits). Then this shows clearly the importance of analytical and very precise numerical results to understand the full dynamical behavior this system exhibits. Therefore, the narrow interval of the parameters where the phenomena occur (that has a parallelism in the narrow region of the phase plane where the important orbits are) makes completely useless the utilization of a brute-force simulation strategy of the system.

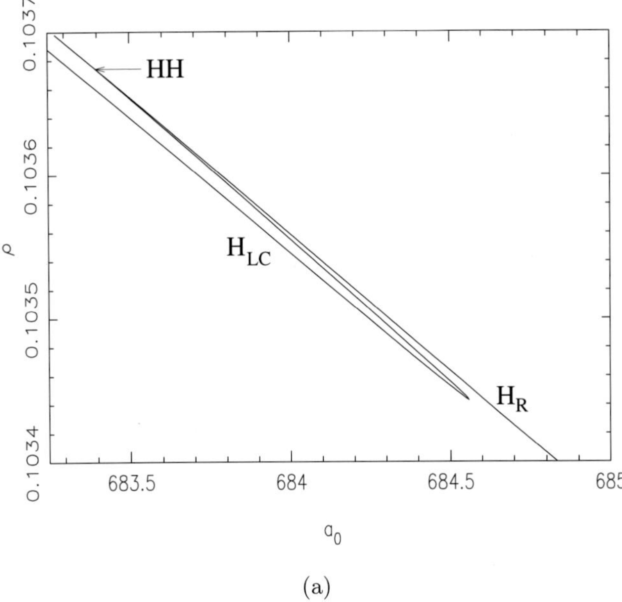

(a)

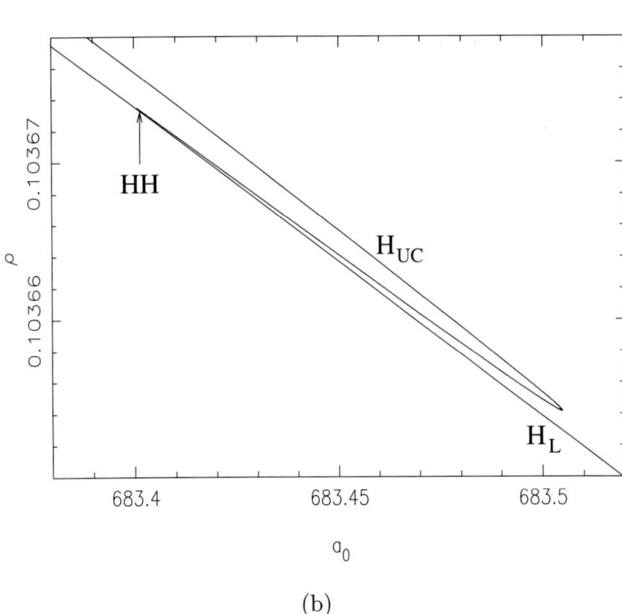

(b)

Fig. 13.   Detail of the tangency between the curves of: (a) right homoclinic orbits (H_R) and lower concave homoclinic orbits (H_{LC}); (b) left homoclinic orbits (H_L) and upper concave homoclinic orbits (H_{UC}). In both cases the point of tangency corresponds to a double homoclinic orbit (HH) ($s_0 = 37$, $\alpha = 0.2$ and $\kappa = 0.1$).

Another important case of codimension-two homoclinic connections appears when the homoclinic curve meets a curve of saddle-node bifurcation of equilibria. In these cases of saddle-node homoclinic orbits (there are five in the enzyme system, corresponding to SNH_R, SNH_L, SNH_R^{LC}, SNH_L^{LC}, SNH_L^{UC}) we have taken a quadratic approximation

for the center manifold and a linear approximation for the hyperbolic manifold. For that, we have translated the nonhyperbolic equilibrium of (3) to the origin, followed by a Taylor expansion in a neighborhood of the origin, obtaining the following second-order truncated system:

$$
\begin{cases}
\dot{u} = vu_\star + uv, \\
\dot{v} = F_1(u_\star)v + v^2 + F_1'(u_\star)uv + \dfrac{1}{2}F_2''(u_\star)u^2.
\end{cases}
\tag{34}
$$

The linear change of variables given by

$$
\begin{pmatrix} u \\ v \end{pmatrix} = \begin{pmatrix} 1 & u_\star \\ 0 & F_1(u_\star) \end{pmatrix} \begin{pmatrix} x \\ y \end{pmatrix}
$$

uncouples the linear part of (34), obtaining in the new variables x and y the following system

$$
\begin{pmatrix} \dot{x} \\ \dot{y} \end{pmatrix} = \begin{pmatrix} 0 & 0 \\ 0 & F_1(u_\star) \end{pmatrix} \begin{pmatrix} x \\ y \end{pmatrix} + \begin{pmatrix} f(x,y) \\ g(x,y) \end{pmatrix},
\tag{35}
$$

where

$$
\begin{aligned}
f(x,y) = -u_\star &\left[F_1(u_\star)y^2 + F_1'(u_\star)y(u_\star y + x) \right. \\
&\left. + \frac{F_2''(u_\star)}{2F_1(u_\star)}(u_\star y + x)^2 \right] \\
&+ F_1(u_\star)y(u_\star y + x), \\
g(x,y) = F_1(u_\star)&y^2 + F_1'(u_\star)y(u_\star y + x) \\
&+ \frac{F_2''(u_\star)}{2F_1(u_\star)}(u_\star y + x)^2.
\end{aligned}
$$

The origin, that is a semi-hyperbolic equilibrium of (35), has a center manifold with tangent space on the OX axis. This center manifold is given, up to second order, by the equation $y = ax^2$, for a certain value of a. Differentiating and identifying coefficients, we obtain, in the original variables, the following expression for the center manifold

$$
v = -\frac{F_2''(u_\star)}{2F_1(u_\star)}\left[u^2 - \frac{2u_\star}{F_1(u_\star)}uv + \frac{u_\star^2}{F_1(u_\star)^2}v^2 \right].
$$

With respect to the location and the continuation of cuspidal loops, we have approximated the separatrices of the nonhyperbolic equilibrium by means of the semi-cubic $(u - u_\star)^3 = av^2$, where $u_\star$ is the abscissa of the equilibrium and a is a parameter to be determined. Let us consider the system (3) written in the Bogdanov–Takens normal form, in a

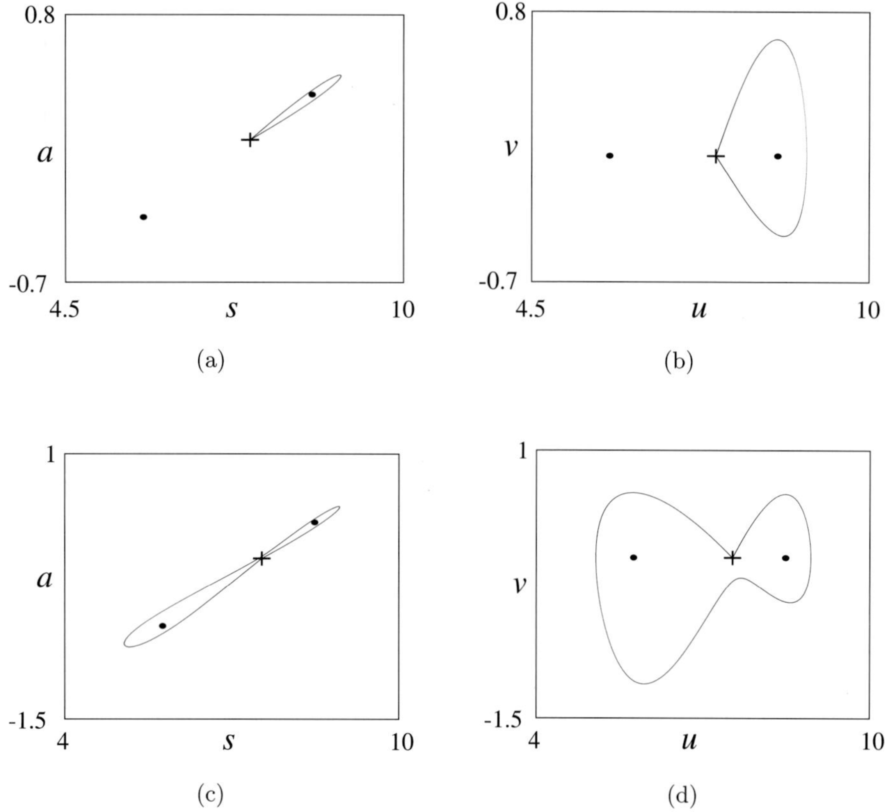

Fig. 14.   Phase portraits of two homoclinic orbits for $s_0 = 37$, $\alpha = 0.2$ and $\kappa = 0.1$. The first one is a right homoclinic connection $\mathrm{H_R}$ that occurs for $a_0 \approx 683.39886$ and $\rho \approx 0.1036702$: (a) plane s–a; (b) plane u–v. The second one is a lower concave homoclinic connection $\mathrm{H_{LC}}$ that occurs for $a_0 \approx 683.46591$ and $\rho \approx 0.1036624$: (c) plane s–a; (d) plane u–v.

neighborhood $(u_\star, 0)$, up to second order but avoiding the uv term:

$$\begin{cases} \dot{u} = v, \\ \dot{v} = \dfrac{1}{2}\dfrac{F_2''(u_\star)}{u_\star}(u - u_\star)^2. \end{cases} \qquad (36)$$

Differentiating with respect to t in the semi-cubic, and from (36), we get the following value

$$a = \frac{3u_\star}{F_2''(u_\star)}.$$

To start the continuation, we have to previously locate a cuspidal loop. It has been an easy task proceeding in the following way. Firstly, we have detected a parameter space point corresponding to a simultaneous Hopf bifurcation in an equilibrium and a Bogdanov–Takens in the other equilibrium. Next, the periodic orbit arising from the Hopf bifurcation evolves, as the parameters run over the Bogdanov–Takens curve, towards the cuspidal loop.

Let the parameter-space point $(a_0, s_0, \rho, \alpha, \kappa)$ be such that, simultaneously, the equilibrium $(u_1^\star, 0)$ undergoes a Hopf bifurcation and the equilibrium $(u_2^\star, 0)$ undergoes a Bogdanov–Takens bifurcation. Since both $u_1^\star$ and $u_2^\star$ are zeros of the equation $F_1(u) = 0$, it is easily deduced that

$$\alpha = \frac{(u_1^\star p'(u_1^\star) - p(u_1^\star))^2 - p(u_1^\star)}{(u_1^\star p'(u_1^\star) - p(u_1^\star))^2 + p(u_1^\star)}, \qquad (37)$$

that is an expression, for κ constant, of the value of α in terms of the abscissa of the equilibrium undergoing the Bogdanov–Takens bifurcation.

Since $u_2^\star$ is a double root and $u_1^\star$ is a single root of the cubic polynomial $h(u)$ given in (4), it is deduced, using the Cardano relations, that

$$u_2^\star = \frac{\kappa(u_1^\star)^2 + \alpha}{(1 + \alpha)\kappa u_1^\star} = \frac{1}{2}\frac{\kappa^2(u_1^\star)^3 + \kappa u_1^\star + 1}{\kappa(\kappa(u_1^\star)^2 - 1)}. \qquad (38)$$

From (37) and (38), and for $\alpha = 0.2$ and $\kappa = 0.1$, we obtain the parameter values where a Hopf bifurcation and a Bogdanov–Takens bifurcation simultaneously occur:

$$u_1^\star \approx 6.4560, \quad u_2^\star \approx 7.4798, \quad a_0 \approx 716.0002,$$
$$s_0 \approx 36.1198, \quad \rho \approx 0.094084.$$

If we continue the periodic orbit arising from this Hopf bifurcation point (as the parameters run over the Bogdanov–Takens curve), we obtain the parameter values for which a cuspidal loop exists:

$$(u_\star, a_0, s_0, \rho)$$
$$\approx (7.5602, 715.2247, 36.1692, 0.094415).$$

In Fig. 15 we show, using simulation with Dstool, the evolution of the phase portraits of system (3), for $\alpha = 0.2$ and $\kappa = 0.1$, as the parameters run over the Bogdanov–Takens curve. We focus on the transition in the vicinity of a cuspidal loop point. In a first moment, only a stable large-amplitude periodic orbit exists (i.e. a periodic orbit surrounding all the equilibria) [see Fig. 15(a)]. An unstable small-amplitude periodic orbit appears (i.e. a periodic orbit surrounding only one equilibrium point) in a Hopf bifurcation exhibited by the left equilibrium [see Fig. 15(b)]. When this orbit grows, it collapses with the cusp point giving rise to an unstable cuspidal loop [see Fig. 15(c)]. This codimension-three global connection does not destroy the periodic orbit but it allows the transition between a small- and a large-amplitude periodic orbit [see Fig. 15(d)]. Finally, this unstable periodic orbit disappears in a saddle-node bifurcation together with the stable large-amplitude periodic orbit that has been present in all this sequence [see Fig. 15(e)].

Now we compute, from the expressions given in Sec. 3, the parameter values where the three codimension-three bifurcations of equilibria occur, for $\alpha = 0.2$ and $\kappa = 0.1$. On the one hand, the parameter values for the degenerate Bogdanov–Takens points D and E are, respectively

$$(u_\star, a_0, s_0, \rho)$$
$$\approx (7.111385, 721.304124, 35.963550, 0.09258790)$$

and

$$(u_\star, a_0, s_0, \rho)$$
$$\approx (6.931252, 725.121350, 35.936106, 0.091869965).$$

On the other hand, the codimension-three degenerate Hopf bifurcation point H_2, occurs when

$$(u_\star, a_0, s_0, \rho)$$
$$\approx (4.4721298, 458.8523045, 51.6934269, 0.3533743).$$

This point lies on the codimension-two degenerate Hopf bifurcation curve H_1' arising from the codimension-three degenerate Bogdanov–Takens point E.

Moreover, from the analytical expressions obtained we conclude that no codimension-four bifurcations of equilibria (Hopf and Bogdanov–Takens) may occur. In the case $\kappa = 0.1$, we have checked that the seventh-order coefficient a_3 of the normal form of the Hopf bifurcation (2) remains always negative for all the degenerate Hopf bifurcation points H_2; thus, no codimension-four Hopf bifurcation arises.

Such a codimension-four point would lead to the existence of a curve of swallowtail singularities of periodic orbits (codimension-three). Although we have not numerically found a swallowtail singularity (*a cusp of cusps*) of periodic orbits, its existence would not be strange (there are several codimension-three bifurcations of equilibria and homoclinic orbits).

The sign of the seventh-order Hopf bifurcation coefficient a_3 determines the relative position of the curve of cusp bifurcations of periodic orbits arising from the point H_2 with respect to the degenerate Hopf H_1'. Analogously, the sign of the resonant coefficient RES determines the relative position of the curve of cusp bifurcations of periodic orbits, that emerges from the point H_{EID}, with respect to the curve of lower concave homoclinic orbit with zero trace $H_{\mathrm{LC}}^{\mathrm{D}}$. In Fig. 16, a qualitative picture of all of these curves are sketched. As predicted by the theory (see [Takens, 1973]), the cusp curve Cu emerges from H_2 by the side of the degenerate Hopf curve H_1' where the coefficient a_2 is positive. In a similar manner [Dumortier *et al.*, 1994], the cusp curve Cu emerges from H_{EID} by the side of the degenerate homoclinic curve $H_{\mathrm{LC}}^{\mathrm{D}}$ where the coefficient EID is less than one. This second cusp curve Cu ends at the cuspidal loop point CL. In this figure the degenerate Bogdanov–Takens points E and D are also shown because they are the starting points, respectively, of the curves H_1' and $H_{\mathrm{LC}}^{\mathrm{D}}$.

To put in evidence the usefulness of both the analytical results obtained in Secs. 2 and 3 and the numerical methods for homoclinic connections, we

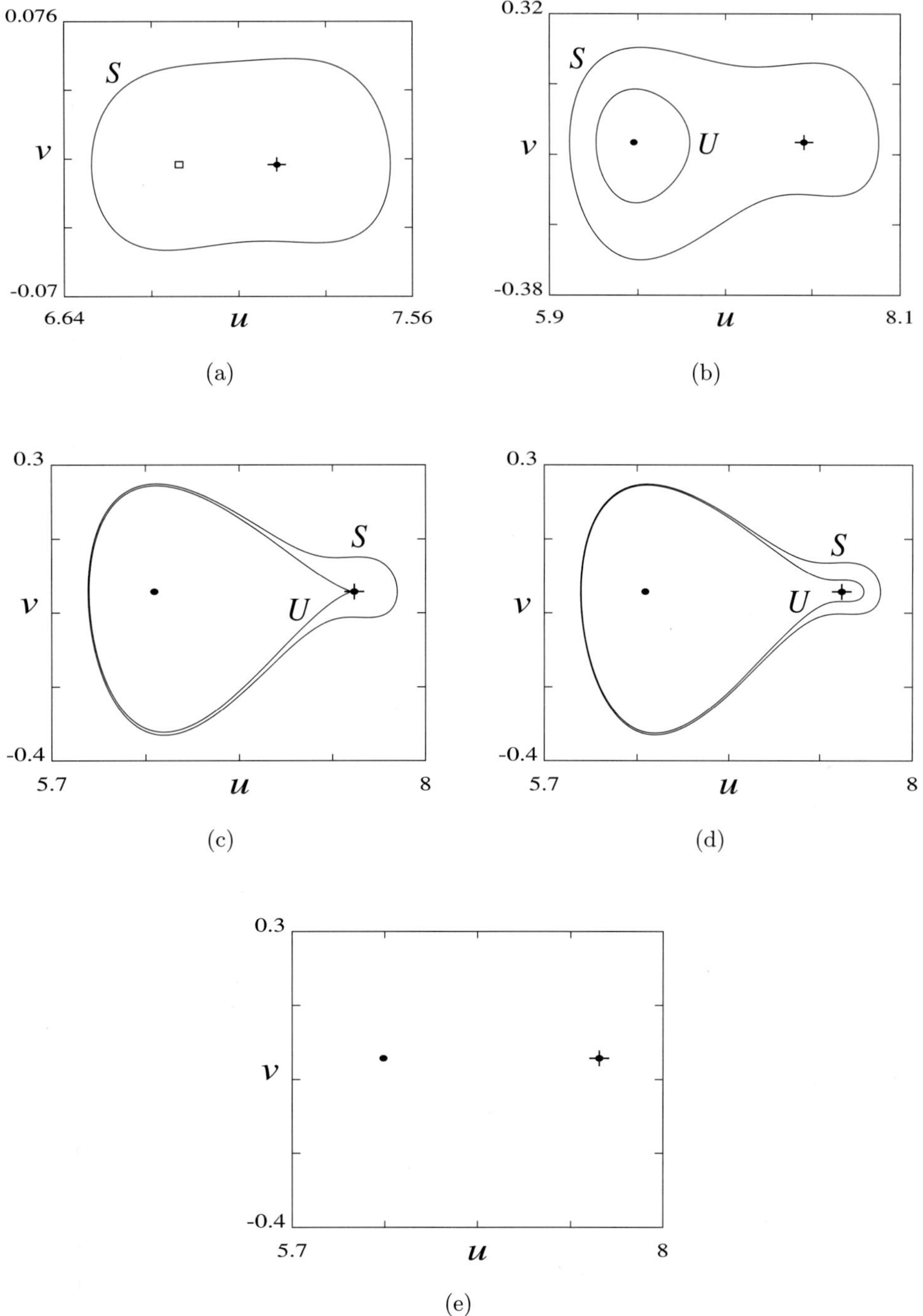

Fig. 15. Evolution of the phase portraits along the Bogdanov–Takens bifurcation in the (u, v)-plane for $\alpha = 0.2$ and $\kappa = 0.1$: (a) $u_\star = 7.2$; (b) $u_\star = 7.5$; (c) $u_\star = 7.56022$; (d) $u_\star = 7.561$ and (e) $u_\star = 7.6$. The meaning of the symbols is the following: S = Stable periodic orbit; U = Unstable periodic orbit; $\bullet$ = Stable equilibrium; $\square$ = Unstable equilibrium; $+$ = Saddle equilibrium. Note that the right equilibrium is nonhyperbolic (marked by superimposing a cross and a filled circle).

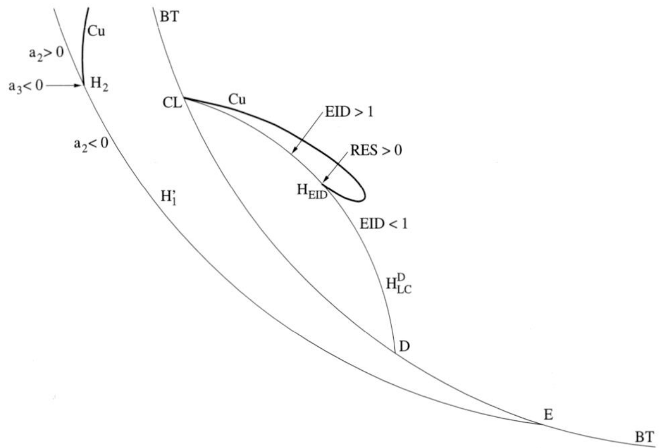

Fig. 16. Relative position of the curves of cusp bifurcations of periodic orbits (Cu) with respect to the curves of degenerate Hopf H_1' and of lower concave homoclinic orbit with zero trace H_{LC}^D. This scheme corresponds to a projection from the (a_0, s_0, ρ)-parameter space onto the (a_0, s_0)-plane, for $\alpha = 0.2$ and $\kappa = 0.1$.

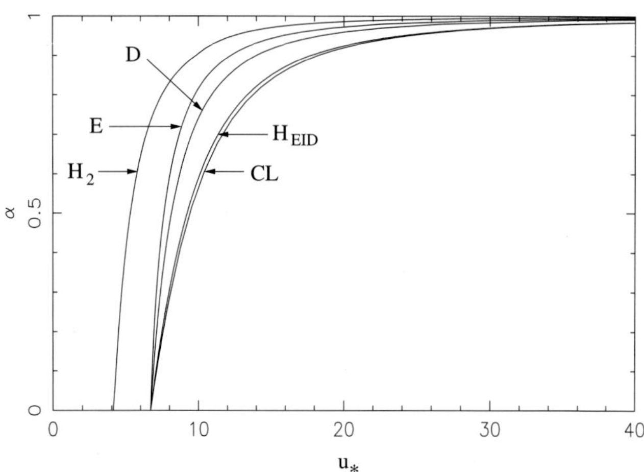

Fig. 17. Curves of codimension-three bifurcations, in the $(u_\star, \alpha)$-plane, for $\kappa = 0.1$: E = cusps of order three (Bogdanov–Takens bifurcation with degeneracy in the uv term of its normal form); D = foci (Bogdanov–Takens bifurcation with degeneracy in the u^2 term of its normal form); H_2 = Hopf bifurcation with degeneracy in both order three and five of its normal form); CL = cuspidal loops and H_{EID} = homoclinic orbits with zero trace and coefficient EID equal to one.

will now consider the (a_0, s_0, ρ, α)-parameter space, for $\kappa = 0.1$. In this situation, the codimension-three bifurcations the enzyme system exhibits occur on curves we are able to compute from the analytical results or with the numerical continuation procedures. In Fig. 17, all the curves corresponding to codimension-three bifurcations are represented in the $(u_\star, \alpha)$-plane [recall that $u_\star$ determines uniquely a point (a_0, s_0, ρ)]: the cusps of order three (E), the foci (D), the degenerate Hopf (H_2), the cuspidal loops (CL) and the lower concave homoclinic orbits with simultaneously zero trace and zero integral of the divergence (H_{EID}). Observe that all the curves approach asymptotically to $\alpha = 1$ as for this value all the dynamics disappear as was pointed out in Sec. 3.3. On the other extreme, four of the curves collapse for $\alpha = 0$. This would be a codimension-four point if this parameter value had biological meaning, but this is not the case.

Now we are interested in the stability of the cuspidal loop, to determine if some additional degeneracy, that is, a change in its stability (that will imply a codimension-four bifurcation) may be present. In the case $\alpha = 0.2$ and $\kappa = 0.1$, we have verified that both the cuspidal loop and the small-amplitude periodic orbits arising from the homoclinic orbit associated with the Bogdanov–Takens bifurcation are unstable. This means that this cuspidal loop lies in the simplest case of cuspidal loops considered in Sec. 4.3.

To detect possible changes in the stability of the cuspidal loops, we have performed a numerical study for $\alpha \in (0, 0.9994705)$ and $\kappa = 0.1$. This change would give rise to the richest case of cuspidal loops. The result of this study has been negative. We have not detected any change in the stability of such cuspidal loops. However, we have verified that for $\alpha > \alpha_0 \approx 0.98$, a greater richness in the dynamics of the enzyme system occurs. It is due to the appearance of a degenerate Hopf bifurcation over the curve of points where both a Hopf bifurcation in one equilibrium and a Bogdanov–Takens in the other one simultaneously occur (this curve is of coexistence of Hopf and Bogdanov–Takens bifurcations, HT).

From this degenerate point a curve of saddle-node of small periodic orbits, sn, arises, that coexists with the curve of saddle-node of large-amplitude periodic orbits, SN, aforementioned. For $\alpha \in (0, \alpha_0)$ the Hopf bifurcation is subcritical and for $\alpha \in (\alpha_0, 1)$ the Hopf bifurcation is supercritical. In Fig. 18 we show qualitatively the relative positions, in the $(u_\star, \alpha)$-plane, of the curve of cuspidal loops (CL) and the curve of coexistence of Hopf and Bogdanov–Takens bifurcations (HT). We have verified that both curves intersect at the value $\alpha_1 \approx 0.99$, exchanging their relative positions. There exists, moreover, a value $\alpha_2 \approx 0.994$ for which both curves HT and SN intersect.

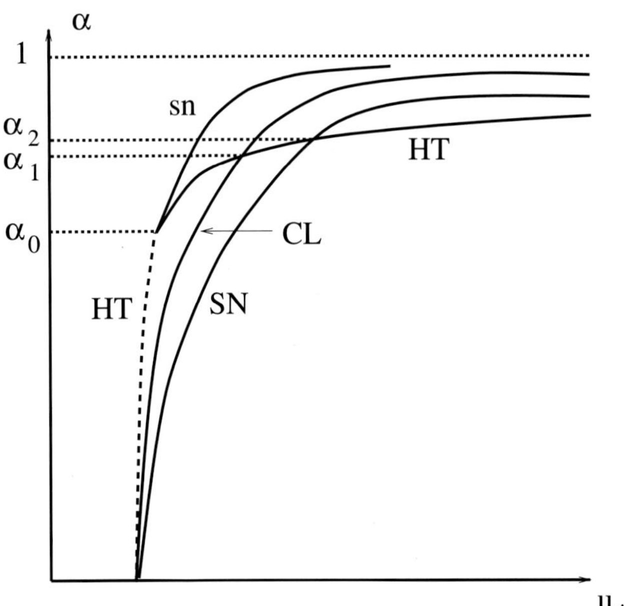

Fig. 18. Qualitative representation, for $\kappa = 0.1$, of the curves of: cuspidal loops (CL); coexistence of Hopf and Bogdanov–Takens bifurcations (HT); saddle-node of small-amplitude periodic orbits (sn); saddle-node of large-amplitude periodic orbits (SN). The solid line (resp. dashed) in the HT curve means that the Hopf bifurcation is super-critical (resp. subcritical).

The above bifurcation set will give rise to four different bifurcation diagrams depending on the α value, for $\kappa = 0.1$ (see Fig. 19). In the first case, when $\alpha \in (0, \alpha_0)$, the sequence of the bifurcation points is HT-CL-SN. For $\alpha \in (\alpha_0, \alpha_1)$, a point of saddle-node bifurcation of small-amplitude periodic orbits (sn) appears as consequence of the degenerate Hopf bifurcation. For $\alpha \in (\alpha_1, \alpha_2)$, the bifurcations HT and CL have changed their relative positions. And finally, for $\alpha \in (\alpha_2, 1)$, HT and SN interchange their position. Note that the cuspidal loop CL always occurs on the unstable branch: it connects an unstable small-amplitude periodic orbit with an unstable large-amplitude periodic orbit.

In Fig. 20 we show, using again simulation with Dstool, the evolution of the phase portraits of system (3), for $\alpha = 0.9971544168$ and $\kappa = 0.1$, as the parameters run over the Bogdanov–Takens bifurcation curve. In a first moment only a stable large-amplitude periodic orbit is present [see Fig. 20(a)]. Later, two small-amplitude periodic orbits emerge from a saddle-node bifurcation [see Fig. 20(b)]. The unstable small-amplitude periodic orbit grows and collapses with the cusp point giving rise to a cuspidal loop [see Fig. 20(c)]. After the

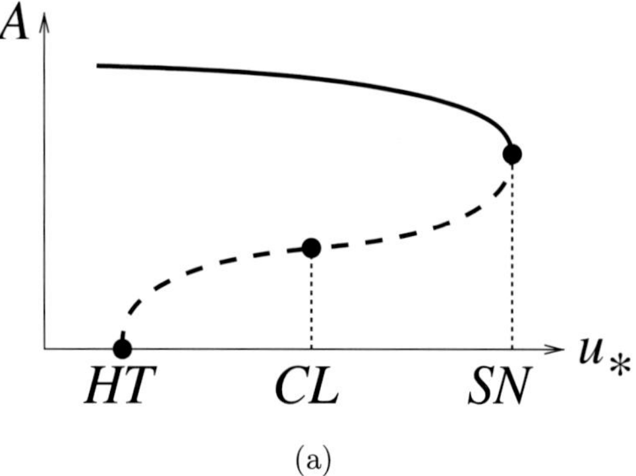

(a)

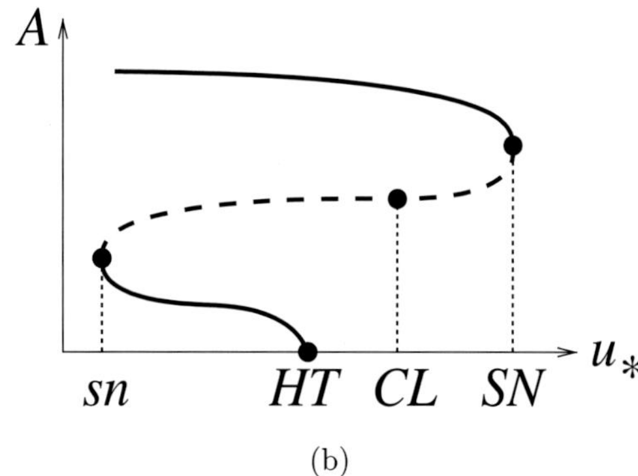

(b)

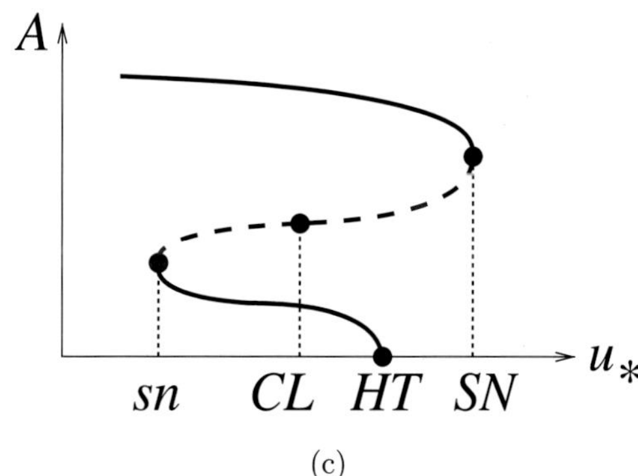

(c)

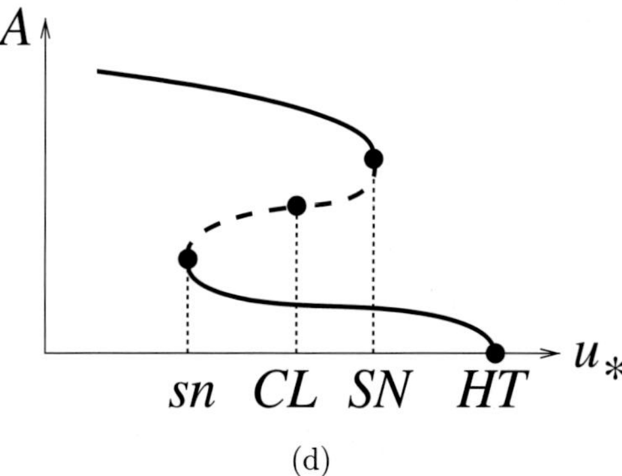

(d)

Fig. 19. Qualitative bifurcation diagrams of a family of periodic orbits near the cuspidal loop, for $\kappa = 0.1$: (a) $\alpha \in (0, \alpha_0)$; (b) $\alpha \in (\alpha_0, \alpha_1)$; (c) $\alpha \in (\alpha_1, \alpha_2)$; (d) $\alpha \in (\alpha_2, 1)$. The ordinate, A, represents the amplitude of the periodic orbit. A solid (resp. dashed) line means stable (resp. unstable) periodic orbit.

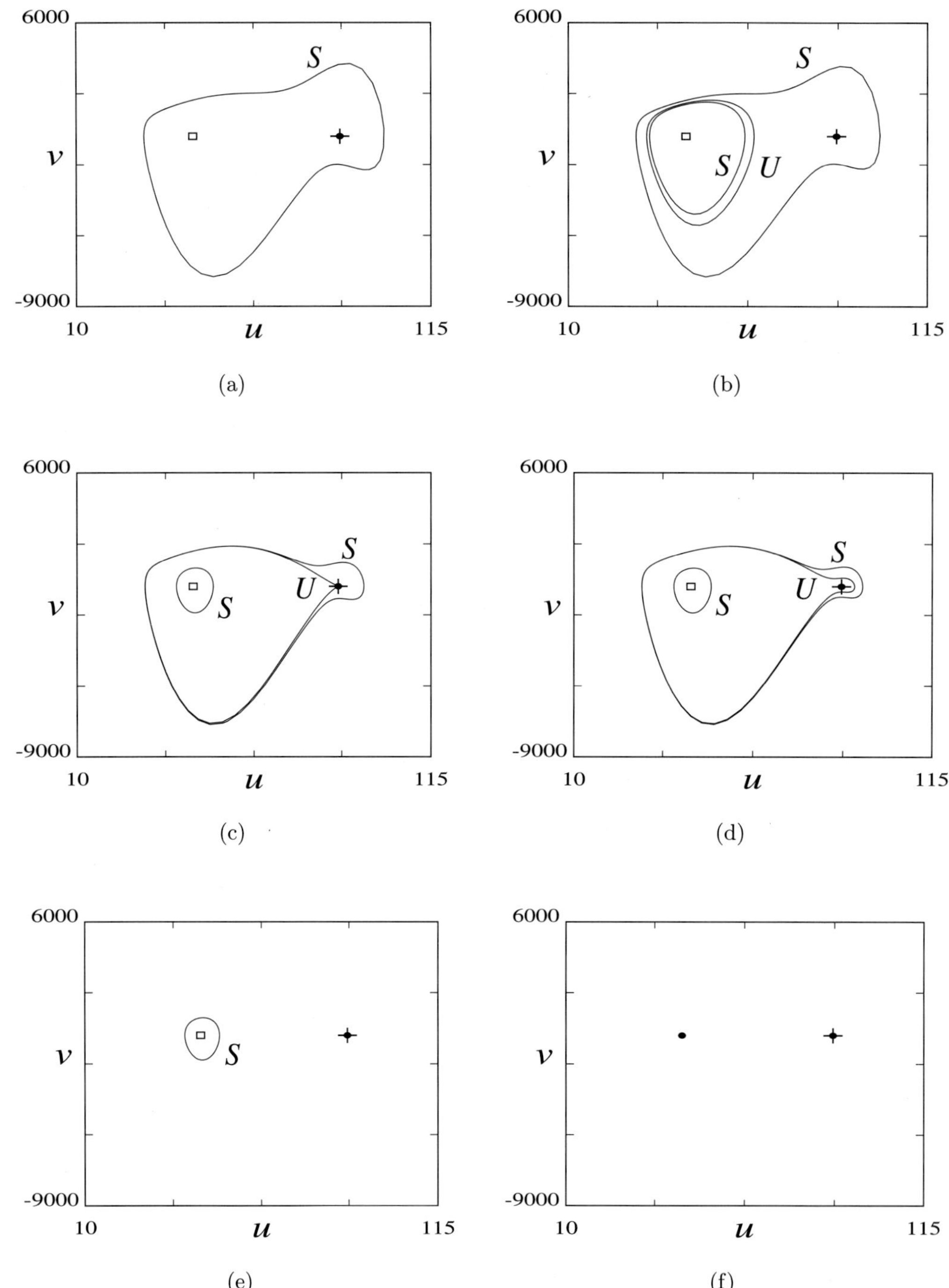

Fig. 20.   Evolution of the phase portraits in the (u, v)-plane for $\alpha = 0.9971544168$ and $\kappa = 0.1$: (a) $u_\star = 88.3$; (b) $u_\star = 88.32$; (c) $u_\star = 88.4715$; (d) $u_\star = 88.473$; (e) $u_\star = 88.48$; (f) $u_\star = 88.8$. S = Stable limit cycle; U = Unstable limit cycle; $\bullet$ = Stable equilibrium; $\square$ = Unstable equilibrium; $+$ = Equilibrium of saddle type. Note that the right equilibrium is nonhyperbolic (marked by superimposing a cross and a filled circle).

breaking of the cuspidal loop, an unstable large-amplitude periodic orbit appears [see Fig. 20(d)]. The two large-amplitude periodic orbits collapse and disappear in a saddle-node bifurcation [see Fig. 20(e)]. Finally, the stable small-amplitude periodic orbit disappears in a supercritical Hopf bifurcation and only the two equilibria are present [see Fig. 20(f)].

To complete with the special attention we have devoted to the cuspidal loop, we now show two such global connections in the (s, a)-phase plane (see Fig. 21). The first one corresponds to the values of the parameters $\alpha = 0.2$ and $\kappa = 0.1$ whereas the second one occurs for $\alpha \approx 0.97552$ and $\kappa = 0.1$. We pay attention to several things. First, the orbits are better observed in the (u, v)-plane than in the (s, a)-plane [compare these cuspidal loops with the other two represented in Figs. 15(c) and 20(c)]. Second, the cuspidal loop of Fig. 21(b) allows to see the behavior the system exhibits when α tends to 1: we observe how it approaches the line $s - a = s_0 - a_0$. Recall that for $\alpha = 1$, the straight line $s - a = s_0 - a_0$ is invariant for the enzyme system and even all the equilibria lie on such a line (see Sec. 3.3). Thus, the existence of limit cycles, homoclinic connections and foci is not possible: there are no more limit sets than equilibria for the value $\alpha = 1$, and therefore

the only bifurcations that remain are saddle-nodes and cusps of equilibria.

To have an idea of the rich homoclinic behavior exhibited by the enzyme system, we show in Fig. 22, for $\alpha = 0.2$ and $\kappa = 0.1$, the projection on the (a_0, s_0)-parameter plane of all codimension-two homoclinic orbits curves (that are in the (a_0, s_0, ρ)-parameter space). We remark that these curves of codimension-two homoclinic orbits start and/or end at the codimension-three points D, E, CL and $\mathrm{H_{EID}}$. The two degenerate Bogdanov–Takens points (D and E) as well as the cuspidal loop (CL) appear on the Bogdanov–Takens curve BT whereas the $\mathrm{H_{EID}}$ point is on the curve of lower concave homoclinic orbits with zero trace $\mathrm{H_{LC}^D}$.

Note that six codimension-two homoclinic connections are related to the cuspidal loop point CL. One of them, the curve of left saddle-node homoclinic orbits $(\mathrm{SNH_L})$ exists on both sides of CL. The other five curves, corresponding to double homoclinic orbits (HH), left upper concave saddle-node homoclinic orbits $(\mathrm{SNH_L^{UC}})$, lower concave homoclinic orbits with zero trace $(\mathrm{H_{LC}^D})$, left lower concave saddle-node homoclinic orbits $(\mathrm{SNH_L^{LC}})$ and left homoclinic orbits with zero trace $(\mathrm{H_L^D})$, emerge from CL. Three of these six curves end at the degenerate Bogdanov–Takens point D: $\mathrm{H_{LC}^D}$, $\mathrm{SNH_L^{LC}}$ and $\mathrm{SNH_L}$, whereas the curve $\mathrm{H_L^D}$ ends at the other degenerate Bogdanov–Takens point E. Moreover, two homoclinic curves, corresponding to right saddle-node homoclinic orbit $(\mathrm{SNH_R})$ and right lower concave saddle-node homoclinic orbit $(\mathrm{SNH_R^{LC}})$, emerge from D.

Some comments are now in order. First, in the unfolding of the cuspidal loop shown in Fig. 11, several of the homoclinic connections correspond to the right equilibrium whereas in Fig. 22 the same kinds of homoclinic orbits connect the left equilibrium. The reason is that, in the unfolding shown, the cuspidal loop occurs at the left equilibrium whereas in the enzyme system the cusp point connected by a loop is the right one.

Secondly, remark that some of these curves are almost undistinguishable and then the need of a qualitative picture (see the bottom sketch of Fig. 22). Finally, the tangential behavior a homoclinic curve presents when it approaches a curve of saddle-node of equilibria (in the so-called saddle-node separatrix-loop bifurcation, see Sec. 4.2) is also present in the curves $\mathrm{SNH_R}$, $\mathrm{SNH_R^{LC}}$, $\mathrm{SNH_L^{LC}}$ and $\mathrm{SNH_L}$ when they approach the degenerate Bogdanov–Takens point D.

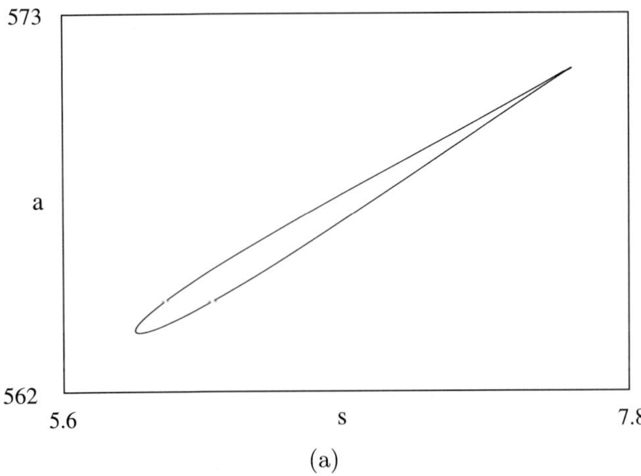

(a)

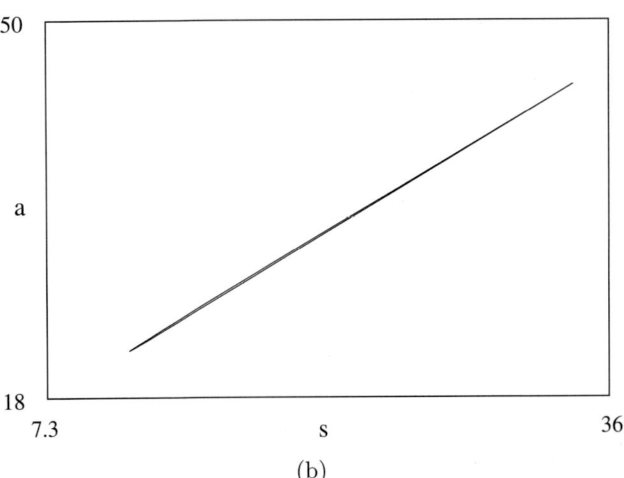

(b)

Fig. 21. Cuspidal loops in the (s, a)-phase plane, for $\kappa = 0.1$ and: (a) $\alpha = 0.2$, $a_0 \approx 715.2247$, $s_0 \approx 36.1692$ and $\rho \approx 0.09441$; (b) $\alpha \approx 0.97552$, $a_0 \approx 1881.2647$, $s_0 \approx 1823.2647$ and $\rho \approx 168.8043$.

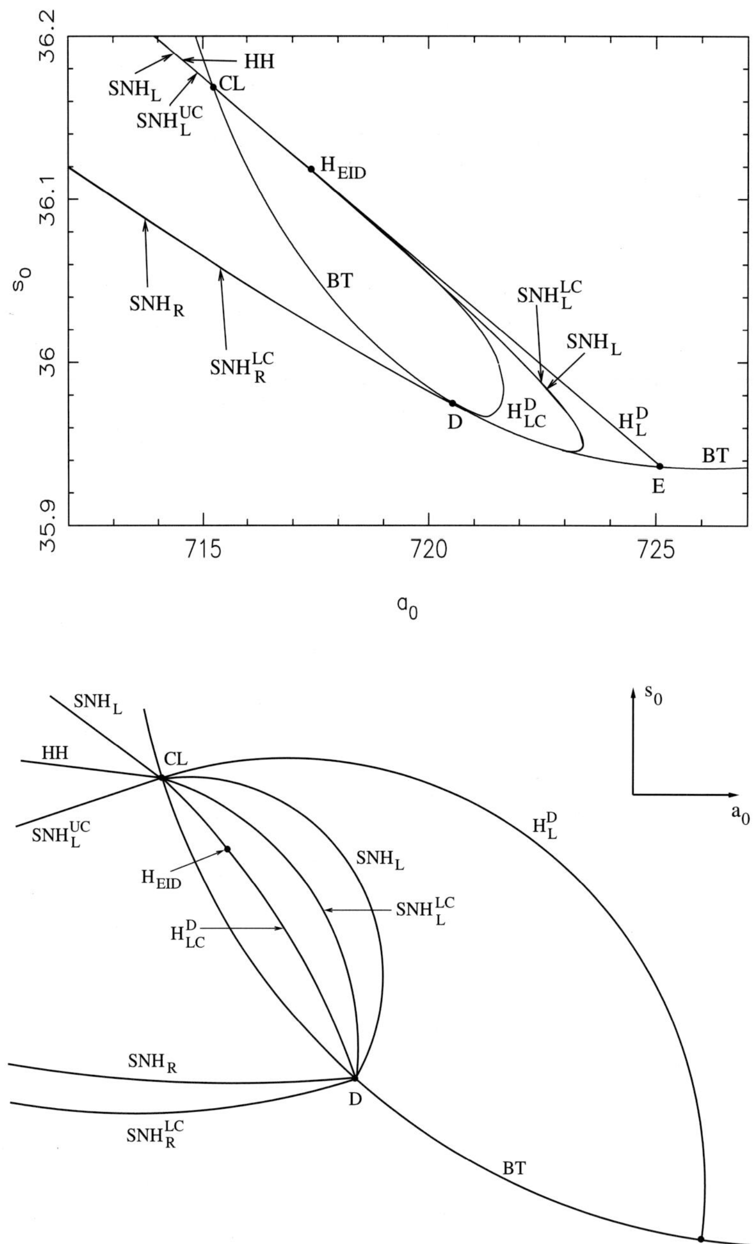

Fig. 22.   Projection, onto the (a_0, s_0)-plane, of the curves of codimension-two homoclinic orbits arising from the codimension-three points E, D, CL and H_{EID}. These numerical (top) and qualitative (bottom) pictures are obtained for $\alpha = 0.2$ and $\kappa = 0.1$.

Due to the great variety of homoclinic connections exhibited by the enzyme system, we consider useful to schematize as much information as possible in Fig. 23. We then devote a window for each of the sixteen kinds of homoclinic connections. In each window we show the following information about a homoclinic orbit: in the upper left corner we write the label used for the homoclinic connection along the text; its codimension is indicated in the upper right corner; a qualitative picture of its phase portrait, with information about the equilibria involved (center); finally, at the bottom of the window we have put the bifurcations of higher codimension in whose unfolding appears such a homoclinic connection.

The convention we have used in such a table is now indicated. Hyperbolic equilibria are represented by a filled point, except in the case of zero trace where a star is used. Saddle-node equilibria appear as an empty point and cusp points

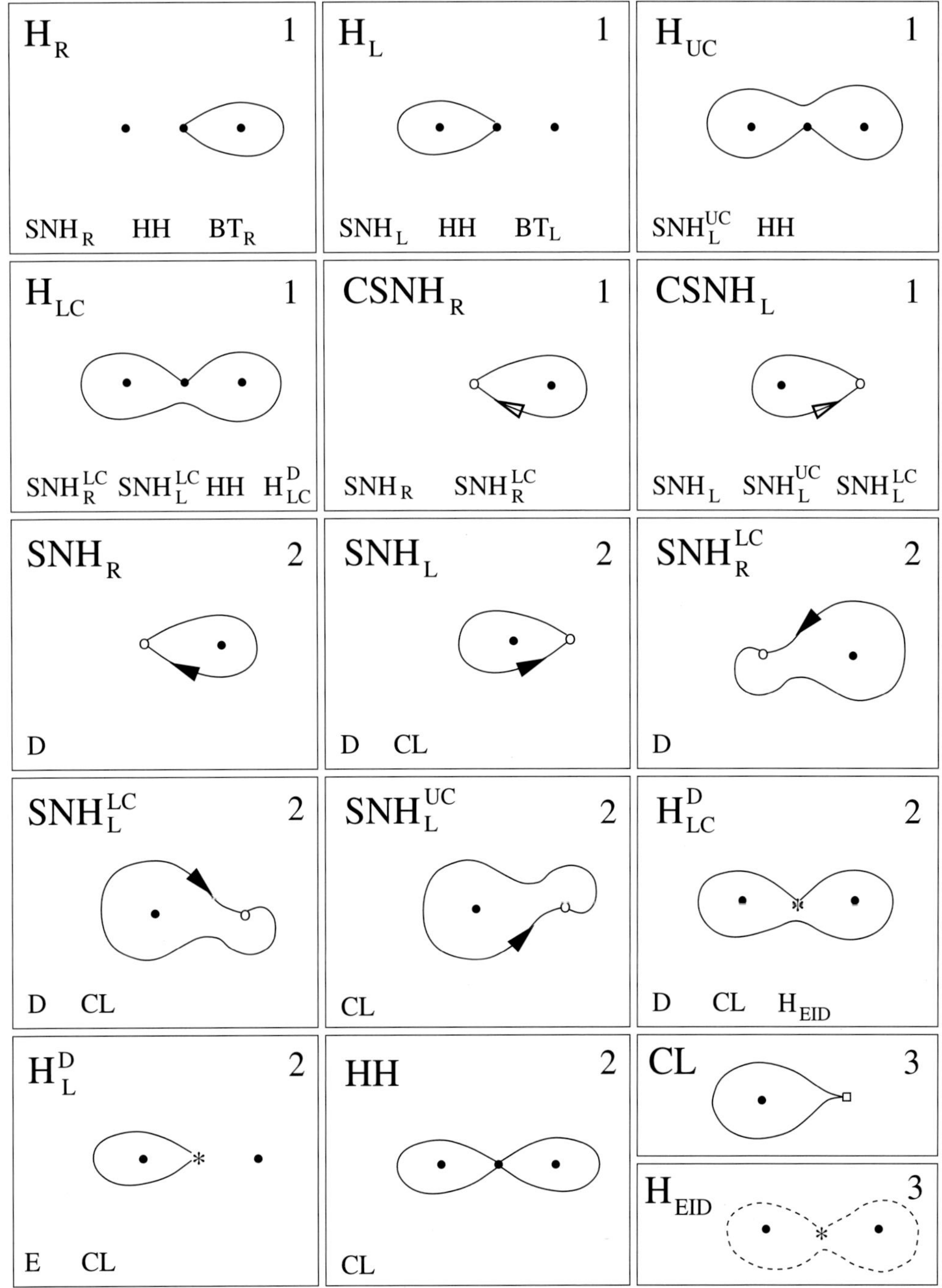

Fig. 23. Table of the sixteen kinds of homoclinic connections exhibited by the enzyme system. In each window we show the following information about a homoclinic orbit: the label used for the homoclinic connection along the text (upper left corner); its codimension (upper right corner); a scheme of its phase portrait (center); the bifurcations of higher codimension related to the homoclinic (bottom). The following convention has been used: filled point (nonresonant hyperbolic equilibrium); empty point (nonhyperbolic saddle-node equilibrium); star (hyperbolic equilibrium with zero trace); empty square (cusp point); filled head of the arrow (noncentral homoclinic orbit); empty head of the arrow (central homoclinic orbit); dashed orbit (EID = 1).

as an empty square. The presence of a central homoclinic orbit (codimension-one) is denoted by an empty head of the arrow whereas a noncentral homoclinic orbit (codimension-two) is indicated by a filled head of the arrow. Finally, the vanishing of the integral of the divergence along the homoclinic orbit (EID = 1) is marked with a dashed orbit.

6. Bifurcation Sets

Our aim in this section is to show all the information obtained from the previous analytical and/or numerical study of the bifurcations of codimension-one, -two and -three. In this way, we get a complete picture of the nice and complex dynamics exhibited by the enzyme system.

The need of qualitative diagrams is made evident looking at the following pictures. First, see again Fig. 1, that shows the curves of saddle-node bifurcations of equilibria, $\mathrm{sn_L}$ and $\mathrm{sn_R}$, in the a_0–ρ parameter plane, for $s_0 = 37$, $\alpha = 0.2$ and $\kappa = 0.1$. In the narrow area between both curves three equilibria exist whereas outside this region the system only has one equilibrium point. We observe how these two curves collapse in a cusp C and also the presence of two Bogdanov–Takens bifurcations, $\mathrm{BT_R}$ and $\mathrm{BT_L}$. In fact, all the homoclinic bifurcation phenomena occur inside the zone of three equilibria, delimited by the curves $\mathrm{sn_L}$ and $\mathrm{sn_R}$, and between the cusp of equilibria point C and the Bogdanov–Takens point $\mathrm{BT_L}$. Thus, the different curves numerically obtained would not be distinguishable among them.

In Fig. 24(a) we show, for the same values of the parameters, the two Hopf curves H that emerge from the Bogdanov–Takens points $\mathrm{BT_R}$ and $\mathrm{BT_L}$. The presence of two degenerate Hopf points, $\mathrm{H_1}$ and $\mathrm{H_1'}$, indicates the existence of saddle-node bifurcations of periodic orbits $\mathrm{SN_1}$ and $\mathrm{SN_2}$ that we have numerically continued. We have drawn the Hopf curves as dashed lines to distinguish them from the curves $\mathrm{SN_1}$ and $\mathrm{SN_2}$.

Even in the zoom of the region where the saddle-node curves exist, shown in Fig. 24(b), it would be easy to believe that both curves collapse in a cusp. This *optical illusion* is due to the proximity of both curves (see Fig. 27 to see the correct bifurcation behavior). On the other hand, it is evident that if we superimpose the curves of Figs. 1 and 24(a) a not very useful bifurcation picture would appear.

Our next objective is to perform a complete study of the bifurcation sets related to the five codimension-three points D, E, $\mathrm{H_{EID}}$, $\mathrm{H_2}$ and CL, as well as of the transitions among them. For that, we have intersected the (a_0, s_0, ρ)-parameter space (for $\alpha = 0.2$ and $\kappa = 0.1$) with different planes $s_0 = $ constant, in the eight situations represented in Fig. 25.

The bifurcation sets intersected by the aforementioned planes appear in Figs. 26–33. In all these

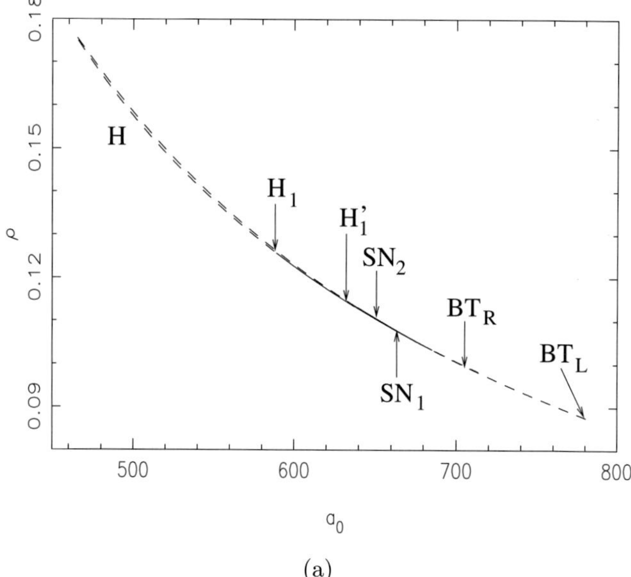

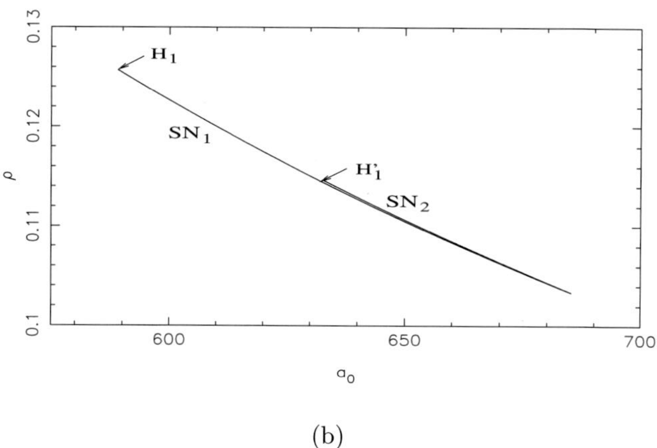

Fig. 24. (a) Curves of Hopf bifurcations of equilibria (dashed lines) and curves of saddle-node bifurcations of periodic orbits (solid lines), in the (a_0, ρ)-plane, for $s_0 = 37$, $\alpha = 0.2$ and $\kappa = 0.1$. The degenerate Hopf points ($\mathrm{H_1}$ and $\mathrm{H_1'}$) and the Bogdanov–Takens points ($\mathrm{BT_R}$ and $\mathrm{BT_L}$) are marked. (b) Zoom of the region where the curves of saddle-node bifurcations of periodic orbits exist.

cases we have performed the numerical location and continuation of all the bifurcation phenomena appearing in the aforementioned figures. We recall that we have the analytical expression for several of the bifurcation curves and points which appear in Figs. 26–33, namely saddle-node bifurcation of equilibria ($\mathrm{sn_L}$ and $\mathrm{sn_R}$), cusp bifurcation of equilibria (C), supercritical and subcritical Hopf bifurcation ($\mathrm{H_{super}}$ and $\mathrm{H_{sub}}$, respectively), codimension-two Hopf bifurcation ($\mathrm{H_1}$ and $\mathrm{H_1'}$) and cusp of order three and foci bifurcations (E and D, respectively). The saddle-node bifurcation curves of periodic

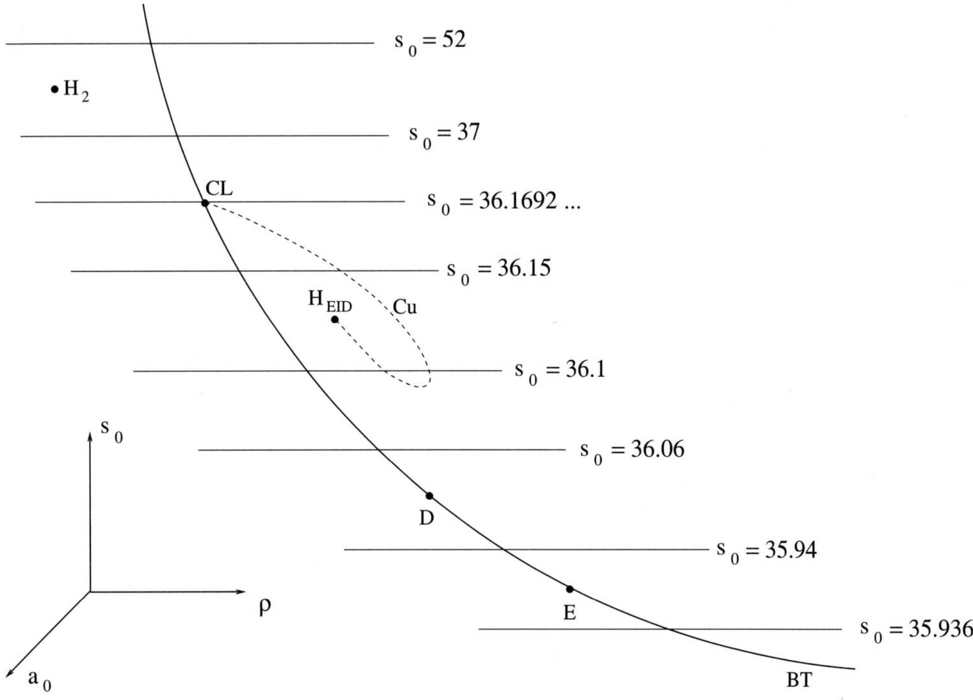

Fig. 25.   Relative position, with respect to the codimension-three points, of the different planes $s_0 = $ constant for which the bifurcation sets of Figs. 26–33, for $\alpha = 0.2$ and $\kappa = 0.1$, are obtained.

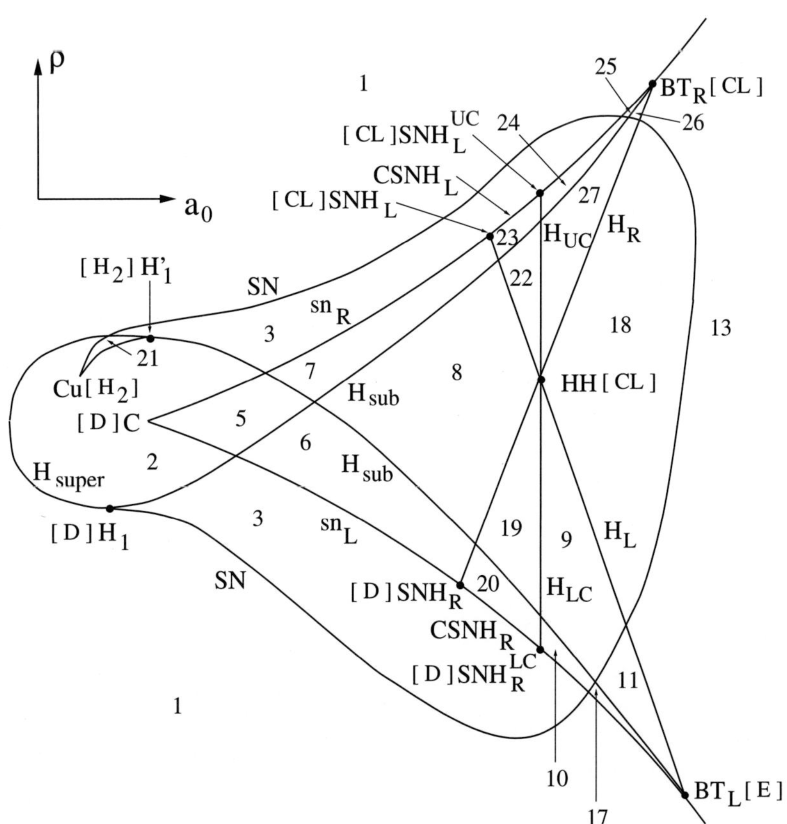

Fig. 26.   Intersection of the (a_0, s_0, ρ)-space with the plane $s_0 = 52$.

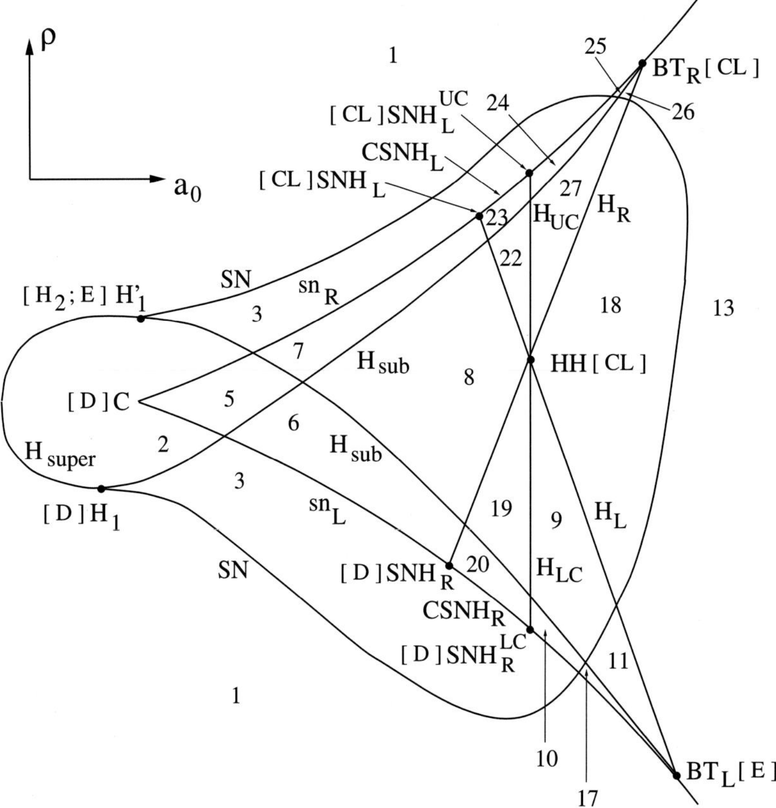

Fig. 27.   Intersection of the (a_0, s_0, ρ)-space with the plane $s_0 = 37$.

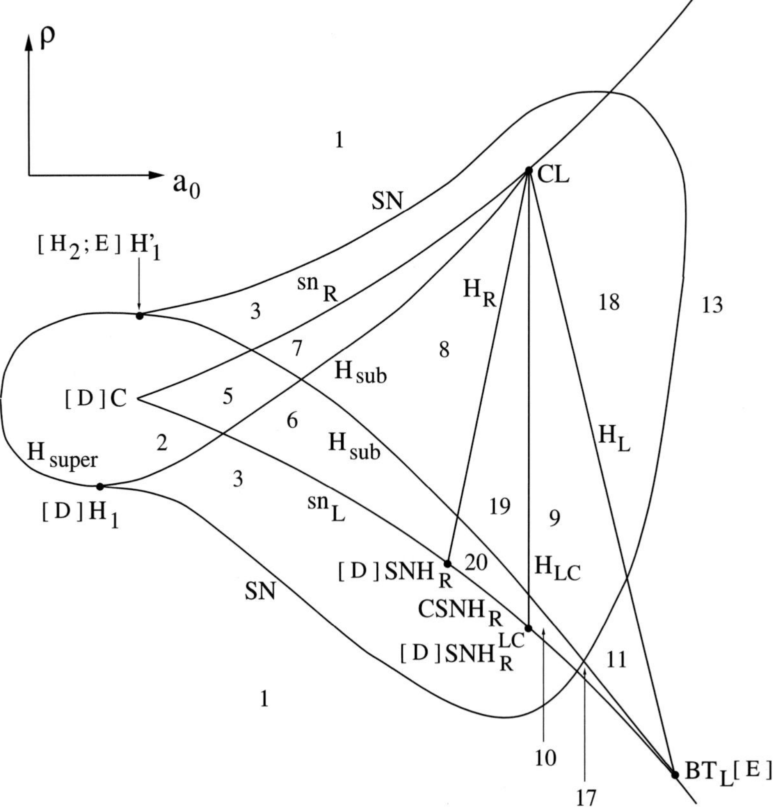

Fig. 28.   Intersection of the (a_0, s_0, ρ)-space with the plane $s_0 = s_0^{CL} \approx 36.1692$.

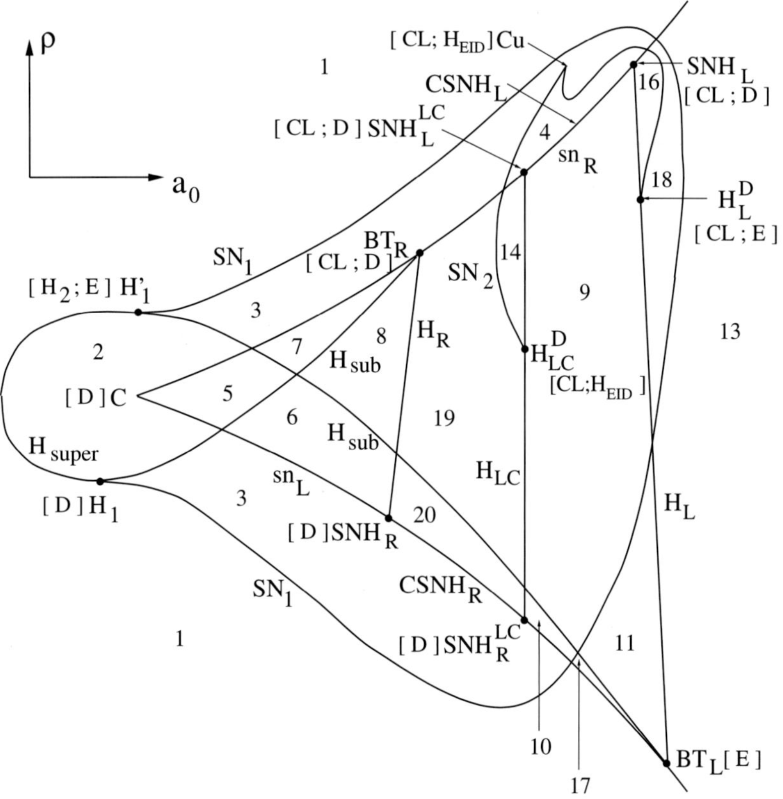

Fig. 29.   Intersection of the (a_0, s_0, ρ)-space with the plane $s_0 = 36.15$.

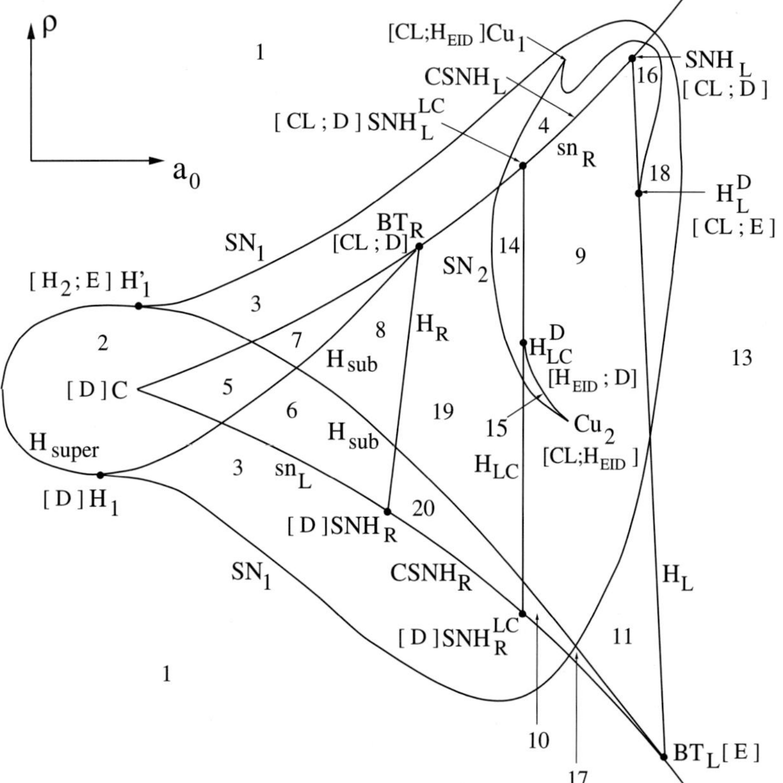

Fig. 30.   Intersection of the (a_0, s_0, ρ)-space with the plane $s_0 = 36.1$.

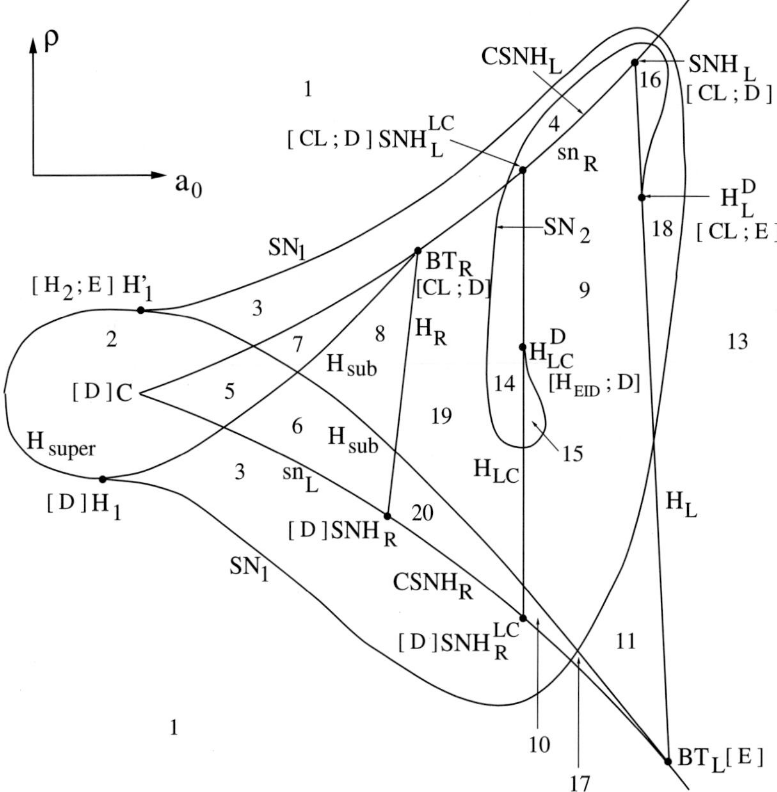

Fig. 31.   Intersection of the (a_0, s_0, ρ)-space with the plane $s_0 = 36.06$.

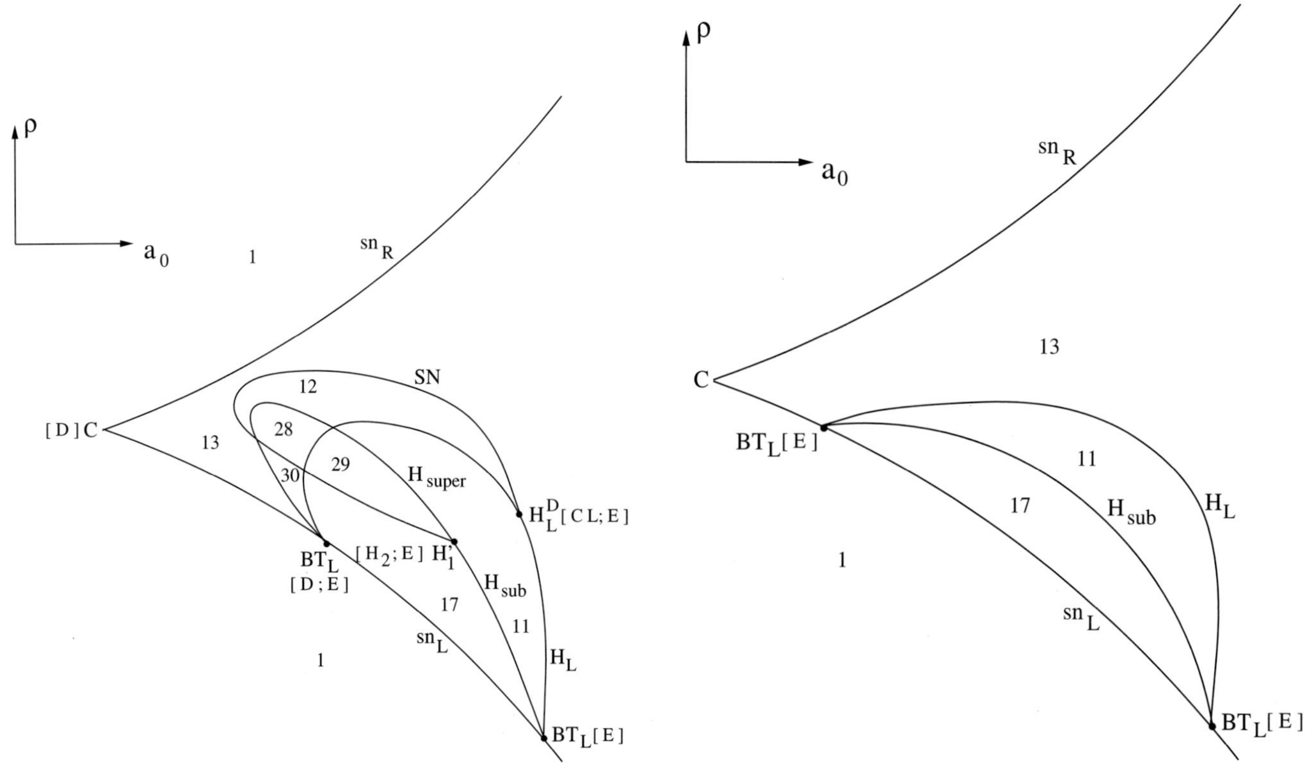

Fig. 32.   Intersection of the (a_0, s_0, ρ)-space with the plane $s_0 = 35.94$.

Fig. 33.   Intersection of the (a_0, s_0, ρ)-space with the plane $s_0 = 35.936$.

orbits (SN_1 and SN_2) have been continued using AUTO97 (see [Doedel *et al.*, 1998]). The numerical continuation of all kinds of homoclinic orbits has been performed with the methods developed in [Freire *et al.*, 1999b] and [Freire *et al.*, 2000] that have been summarized in Sec. 4 and, when it has been possible, the corresponding loci have also been computed with HomCont [Doedel *et al.*, 1998].

In Figs. 26–33, the symbol $[X;Y]$ beside a codimension-two point P means that point P tends to the codimension-three point X as the parameter s_0 increases and P tends to the codimension-three point Y if s_0 decreases; if the symbol $[Z]$ is beside P, it means that P tends to the codimension-three point Z as s_0 either increases or decreases.

In Fig. 34 we have represented the thirty phase portraits corresponding to the different regions bounded by curves of codimension-one arising in Figs. 26–33.

The first bifurcation set, sketched in Fig. 26, corresponds to $s_0 = 52$. There are five codimension-two bifurcations of equilibria (Bogdanov–Takens, BT_R and BT_L; cusp C; degenerate Hopf points, H_1 and H_1'), five codimension-two global connections (double homoclinic orbit, HH; left upper concave saddle-node homoclinic orbit, SNH_L^{UC}; left saddle-node homoclinic orbit, SNH_L; right lower concave saddle-node homoclinic orbit, SNH_R^{LC}; right saddle-node homoclinic orbit, SNH_R) and one codimension-two bifurcation of periodic orbits (cusp, Cu).

From BT_R, that is placed on the curve of saddle-node of equilibria sn_R, a subcritical Hopf curve (H_{sub}) emerges as well as a curve of right homoclinic orbits, H_R. The Hopf curve has a first degeneracy point (H_1) where it becomes supercritical. Later, in a new degeneracy point (H_1'), it becomes again subcritical and finally it ends at the point BT_L, that is on the curve sn_L. From H_1, a curve of saddle-node of periodic orbits appears (SN). This curve is connected with the saddle-node curve emerged from H_1', but in the vicinity of this degenerate Hopf point a cusp (Cu) appears, due to the proximity in the parameter space with the double-degenerate Hopf point H_2. Precisely, the points Cu and H_1' collapse at H_2 (in other words, the curves of cusps Cu and of degenerate Hopf H_1', in the (a_0, s_0, ρ)-space, emerge from H_2). The homoclinic curve H_R finishes, on the sn_L curve, at SNH_R. Analogously, the left homoclinic curve H_L, started at BT_L, finishes at SNH_R, on the sn_R curve. From the HH point, the cross-point between

H_R and H_L, a curve of upper concave homoclinic orbits (H_{UC}) and a curve of lower concave homoclinic orbits (H_{LC}) emerge. The first one ends at SNH_L^{UC} and the second one at SNH_R^{LC}. Note that on a portion of sn_R, delimited by SNH_L and SNH_L^{UC}, a curve of central saddle homoclinic connections ($CSNH_L$) appears. Analogously, on a portion of sn_L, delimited by SNH_R and SNH_R^{LC}, a curve of central saddle homoclinic orbits exists ($CSNH_R$). Note that, for simplicity, we have not drawn in all these qualitative figures the homoclinic curves touching tangentially the saddle-node curves sn_L and sn_R, as it really occurs.

In Fig. 27, the bifurcation set for $s_0 = 37$ is drawn. In the meantime, the system has exhibited a double-degenerate Hopf bifurcation H_2 (it occurs for $s_0 \approx 51.6934269$). For this reason, the cusp of periodic orbits Cu has disappeared in the curve of saddle-node bifurcations of periodic orbits SN, that connect the degenerate Hopf points H_1 and H_1'. All the other bifurcations are present for $s_0 = 37$.

Decreasing s_0, the codimension-two points SNH_L, SNH_L^{UC}, HH and BT_R approach the cuspidal loop point CL and collapse at such a point. This codimension-three global bifurcation occurs for $s_0 \approx 36.1692$. The bifurcation set for this parameter value appears in Fig. 28.

Evidently, between $s_0 = 37$ and $s_0 \approx 36.1692$, the relative position of the curve SN and the point BT_R has changed. We have not drawn this situation in a different bifurcation set for the sake of brevity and because the only consequence this change will have is the disappearance of regions 25 and 26.

On the other side of the CL point, for instance, for $s_0 = 36.15$ (see Fig. 29) important changes appear. The three curves H_R, H_L and H_{LC} do not intersect (and then HH do not exist). Moreover, a degenerate zero-trace left homoclinic orbit (H_L^D) appears on the curve H_L, a degenerate zero-trace lower concave homoclinic orbit (H_{LC}^D) appears on the curve H_{LC} and a cusp of periodic orbits Cu appears on the saddle-node curve of periodic orbits SN_2 that connects the points H_L^D and H_{LC}^D. Now, the central saddle-node homoclinic connection curve ($CSNH_L$), on sn_R, is bounded by SNH_L^{LC} and SNH_L.

As parameter s_0 decreases and reaches the critical value $s_0 \approx 36.122$, the coefficient EID, given in (28), equals one and the codimension-two degenerate homoclinic orbit H_{LC}^D becomes a codimension-three degenerate homoclinic orbit H_{EID}. From this point a new cusp of periodic orbits Cu_2 arises as s_0 decreases, as appears in Fig. 30.

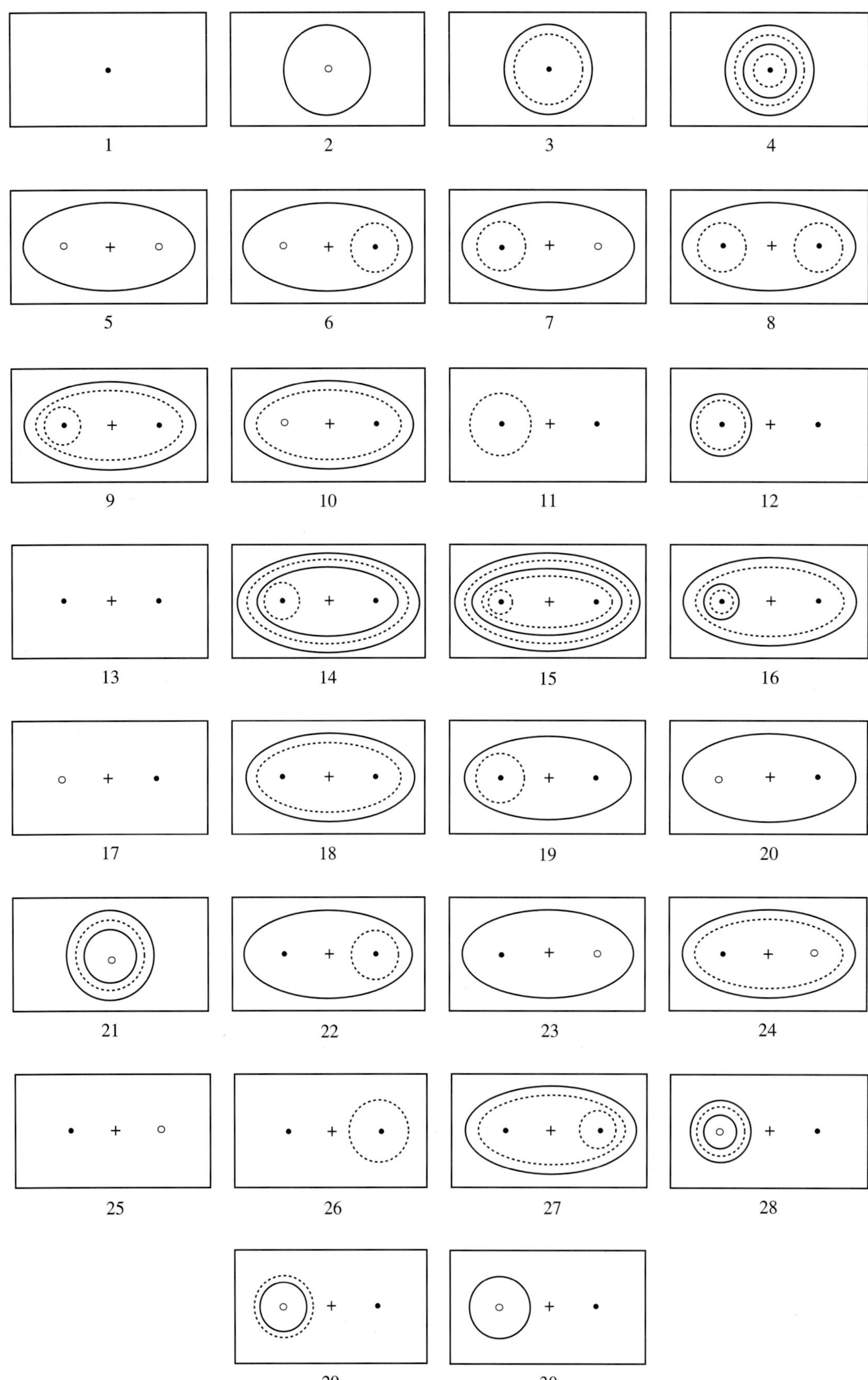

Fig. 34.   Phase portraits for Figs. 26–33. A solid (dotted) line represents a stable (unstable) periodic orbit; the symbols ●, ○ and + mean stable, unstable and saddle equilibrium, respectively.

The cusp of periodic orbits curve connecting points CL and H_{EID} has a minimal point with respect to the s_0 axis (see Fig. 25). Thus, as s_0 decreases, points Cu_1 and Cu_2 appearing in Fig. 30 coalesce and disappear, obtaining a situation such as shown in Fig. 31.

Between the two situations shown in Figs. 31 and 32 a codimension-three degenerate Bogdanov–Takens point D occurs (at that moment the BT_R point is just on the cusp point C). For the critical value of s_0 corresponding to point D, namely, $s_0 \approx 35.9635498$, the codimension-two points BT_R, SNH_L^{LC}, SNH_L, H_{LC}^D, SNH_R^{LC}, SNH_R and H_1 coalesce in the cusp of equilibria point C. This produces the vanishing of the following curves: the arc of subcritical Hopf bifurcation H_{sub} connecting the points BT_R and H_1, the right homoclinic orbit curve H_R, the lower concave homoclinic orbit curve H_{LC}, and both central saddle-node homoclinic orbit curves $CSNH_L$ and $CSNH_R$. Notice that in Fig. 32 both Bogdanov–Takens points BT_R and BT_L lie now on the saddle-node of equilibria curve sn_L.

As parameter s_0 decreases and reaches the critical value $s_0 \approx 35.9361063$, the Bogdanov–Takens point BT_R shown in Fig. 32 degenerates into a cusp of order three point E. At this critical value the codimension-two points H_1' and H_L^D coalesce at E and disappear as s_0 decreases. Thus, the curve SN connecting the points H_1' and H_L^D also disappears. The situation corresponding to a value of s_0 smaller than the aforementioned critical value is shown in Fig. 33. At this level, only the saddle-node of equilibria curves sn_L and sn_R are present, the cusp of equilibria point C, the Bogdanov–Takens points BT_R and BT_L and the two curves connecting them, namely, the subcritical Hopf bifurcation curve H_{sub} and the left homoclinic orbit curve H_L.

The Bogdanov–Takens bifurcation curve in the (a_0, s_0, ρ)-space has a minimum point for the value $s_0 \approx 35.9350685$. Therefore, for values of s_0 below this critical value, the Bogdanov–Takens points have disappeared, as well as both Hopf bifurcation H_{sub} and left homoclinic orbit H_L curves. At this moment, the only bifurcation phenomena that still persist are the saddle-node of equilibria bifurcations sn_L and sn_R and the cusp bifurcation of equilibria C. Then, in this case the only configurations of equilibria present are 1 and 13 of Fig. 34.

In fact, we have chosen the (a_0, ρ)-plane to represent the bifurcation sets (for $s_0 = $ constant) because of the shape of the Bogdanov–Takens curve we perfectly know analytically. For this reason,

we have seen the bifurcation sets in Figs. 26–33 with two Bogdanov–Takens points, and then the interaction between all the curves related to this bifurcation has been made evident.

Our bifurcation analysis shows the presence of thirty different regimes (phase portraits) of dynamical behavior of the enzyme system depending on the parameters. Multistability and oscillatory regimes are present in different combinations. We can classify the phase portraits of the system by the number of attractors (stable equilibrium and stable limit cycle), in a similar way as is done in [Bazykin, 1998]. In this manner, there are eight different groups:

  (i) a single equilibrium, in regions 1, 17 and 25;

  (ii) a single limit cycle, in regions 2 and 5;

  (iii) one equilibrium and one periodic orbit, in regions 3, 6, 7, 10, 20, 23, 24, 29 and 30;

  (iv) two equilibria, in regions 11, 13 and 26;

  (v) two equilibria and one limit cycle, in regions 8, 9, 12, 16, 18, 19, 22 and 27;

  (vi) two equilibria and two periodic orbits, in regions 14 and 15;

  (vii) one equilibrium and two limit cycles, in regions 4 and 28.

(viii) two limit cycles, in region 21.

Note that when there are three equilibria, the middle one is a saddle and then its manifolds will play an important role in the delimitation of the basin of attraction of the corresponding attractors. On the other hand, unstable periodic orbits act as boundaries of the basin of attraction.

As we can see, the system has one globally attracting equilibrium in regions 1, 17 and 25. The distinction in the system behavior between region 1 and the other two is the transitional processes of getting back to the equilibrium after the system has been perturbed.

In cases where more than one attractor is present, the initial condition will determine in which of them the system will end up. Observe this situation, for instance in region 15. The three unstable periodic orbits and the manifolds of the saddle equilibrium mark the boundaries of the basins of attraction for the four coexisting attractors.

Note that, for instance, in region 3 we may find *hard generation* of oscillations (also called *abrupt excitation* of oscillations). In this case the phase portrait includes a stable equilibrium with a basin of attraction that is bounded by an unstable limit cycle. For small perturbations, damped oscillations restore the equilibrium, but the system goes into oscillations for rather strong perturbations.

We now proceed by describing some events that may occur in the enzyme system with respect to attractors if we gradually vary parameters. When parameters change, stable equilibria may show different types of behavior:

(1) Jump from one equilibrium to another. This hysteresis phenomenon between equilibria occurs, for instance, when the system is in the right equilibrium of region 13 and changing the parameters it crosses the curve sn_R (where such an equilibrium disappears) and enters in region 1. Then the system jumps to the other stable equilibrium.

(2) Gradual excitation of oscillations. When the parameters cross from region 1 into region 2 (supercritical Hopf bifurcation, H_{super}) oscillations are gradually excited around the unique equilibrium.

(3) Abrupt excitation of oscillations. When the parameters cross, for example, from region 3 into 2 or from region 6 into 5 or from region 8 (if the system is initially on the right equilibrium) into 7 (subcritical Hopf bifurcation, H_{sub}) we find this hysteresis phenomenon between equilibrium and limit cycle. In the aforementioned transitions the initial values of the variables at the equilibrium are within the range where the oscillations (abruptly excited) exist. This situation also occurs in the transition from region 14 into region 4 (saddle-node of equilibria, sn_R) provided the system was initially at the right equilibrium.

(4) Jump from an equilibrium to a distant limit cycle. When the parameters cross from region 16 into region 4 (saddle-node of equilibria, sn_R) the system, if it is initially in the right equilibrium, moves to an oscillatory regime after the equilibrium disappears. The difference from the previous case is that the initial values of the variables at the equilibrium lie generally beyond the range that they have in the new stable oscillations.

Let us comment some events that may occur if the system is in an oscillatory regime and the parameters are changed.

- Gradual decay of oscillations. This phenomenon, reverse to the phenomenon of gradual excitation

of oscillations, occurs for parameters crossing from region 2 into region 1.

- Abrupt termination of oscillations. This phenomenon, reverse to the phenomenon of abrupt excitation of oscillations, occurs, for example, for parameters crossing from region 10 into region 17 and from region 3 into region 1.

- Breaking up of oscillations in a homoclinic loop. This occurs when the parameters cross from region 16 into region 9, provided that for parameters from region 16 the system was in the oscillatory regime. As the parameters approach the bifurcation curve H_L, the amplitude increases and the oscillations change to relaxation type. On the other hand, note that other crossing of curves H_L, H_R, H_{LC} and H_{UC} gives rise to the appearance/disappearance of an unstable limit cycle.

In this system, the appearance of oscillations from a saddle-node loop (and then its reverse phenomenon, termination of oscillations in a saddle-node loop bifurcation) is not directly observable because the periodic orbits that emerge/disappear when crossing the curves $CSNH_R$ and $CSNH_L$ are unstable (see transitions from regions 20 and 23 into region 3). The presence of the unstable limit cycle may be detected looking at the basin of attraction.

Note that the twelve different phase plane portraits that appear for cubic autocatalysis with decay (see Fig. 8.14 of [Gray & Scott, 1990]) are included between the thirty phase portraits that the enzyme system exhibits for $\alpha = 0.2$ and $\kappa = 0.1$.

7. Conclusions

We have shown how the local bifurcation theory may provide important analytical information about the organizing centers of the dynamical behavior of the five-parameter enzyme system considered along this work. Sometimes, as in this case, it is very useful to rewrite the system in a more convenient way for the application of the Bifurcation Theory tools.

The complete study of codimension-one, -two and -three bifurcations of equilibria (and the proof that there are no local bifurcations of higher codimension) indicates the presence of a very rich dynamical scenario for a planar system: for instance, the emergence of up to three periodic orbits from degenerate Hopf bifurcations and the presence of several degenerate codimension-two homoclinic connections (that are also related to periodic orbits).

However, numerical methods are needed to complete the analysis. On the one hand, several analytical expressions are rather cumbersome and their numerical evaluation (and/or the numerical continuation of the locus where a bifurcation occurs) will be needed to understand all the information they have inside.

On the other hand, the results on global connections (homoclinic orbits in this system) are only first-order approximations, that need to be *extended* with the help of numerical methods. The information is very useful to guarantee the existence of such homoclinic connections as well as to help in their detection and continuation, as occurs in the case of the Bogdanov–Takens bifurcation.

While the use of a brute-force simulation strategy of the system would provide very few results (the narrow interval of the parameters where the phenomena occur makes very difficult to find them without an analytical previous information), an exhaustive description of the dynamical behavior this system exhibits has been carried out along our study. However, although the presented analysis is rather detailed, we cannot exclude the existence of closed curves related to limit cycle bifurcations away from the studied codimension-one, -two and -three points.

In the case of the Bogdanov–Takens bifurcation, the theoretical study of the corresponding unfolding provides information about a lot of codimension-one and -two bifurcations, but when numerical methods extend these local results, new bifurcations theoretically unexpected may appear (this is the case of the cuspidal loop, CL, and of the lower concave homoclinic orbit, H_{EID}). These numerical results may open new research frontiers in the theoretical field, and an interesting feedback process may provide advances in both theoretical and numerical areas.

The presence of hysteresis behavior between equilibria and/or periodic orbits is one of the features that may be deduced from the results achieved. Other characteristics of excitable media are present (trigger mechanism, threshold phenomena, slow–fast motions,...; see, for instance, [Murray, 2002, 2003]) and would be easy to find in the five-parameter space. We have not emphasized on these topics for the sake of brevity.

The mathematical results we have obtained about the enzyme system provide a deep insight of the model and will be useful for biochemist-mathematicians to check its validity and limitations

comparing them with the reality it tries to model. For example, they have to evaluate if some evolutionary factors can force the system to operate in the narrow domains where the complicated phase portraits occur (idea suggested by [Bazykin, 1998]).

Acknowledgments

This work has been partially supported by the *Ministerio de Ciencia y Tecnología, fondos FEDER* in the frame of the project BFM2001-2608 and by the *Consejería de Educación de la Junta de Andalucía* (TIC-0130). The authors wish to thank the comments of A. R. Champneys and E. Gamero on a draft of this paper.

References

Algaba, A., Freire, E. & Gamero, E. [2003] "Computing simplest normal forms for the Takens-Bogdanov singularity," *Qual. Th. Dyn. Syst.* **3**, 377–435.

Bazykin, A. D., Kuznetsov, Yu. A. & Khibnik, A. I. [1989] *Bifurcation Portraits: Bifurcation Diagrams of Dynamical Systems on the Plane*, Series in Mathematics and Cybernetics, Vol. 89 (Znanie, Moscow) (in Russian).

Bazykin, A. D. [1998] *Nonlinear Dynamics of Interacting Populations*, World Scientific Series on Nonlinear Science, Series A, Vol. 11 (World Scientific, Singapore).

Berezovskaya, F. S. & Khibnik, A. I. [1985] "Bifurcations of a dynamical second-order system with two zero eigenvalues and additional degeneracy," in *Methods of Qualitative Theory of Differential Equations* (Gorkii State University, Gorkii) (in Russian), pp. 128–138.

Beyn, W.-J. [1990] "The numerical computation of connecting orbits in dynamical systems," *IMA J. Numer. Anal.* **9**, 379–405.

Champneys, A. R. & Kuznetsov, Yu. A. [1994] "Numerical detection and continuation of codimension-two homoclinic bifurcations," *Int. J. Bifurcation and Chaos* **4**, 785–822.

Champneys, A. R., Kuznetsov, Yu. A. & Sandstede, B. [1996] "A numerical toolbox for homoclinic bifurcation analysis," *Int. J. Bifurcation and Chaos* **6**, 867–888.

Dangelmayr, G. & Guckenheimer, J. [1987] "On a four parameter family of planar vector fields," *Arch. Rat. Mech. Anal.* **97**, 321–352.

Doedel, E. J. & Kernévez, J. P. [1986] "AUTO: Software for continuation and bifurcation problems in ordinary differential equations," Applied Mathematics Report, California Institute of Technology.

Doedel, E. J., Keller, H. B. & Kernévez, J. P. [1991] "Analysis and control of bifurcation problems, Part I: Bifurcation in finite dimensions," *Int. J. Bifurcation and Chaos* **1**, 493–520.

Doedel, E. J., Champneys, A. R., Fairgrieve, T. F., Kuznetsov, Yu. A., Sandstede, B. & Wang, X. [1998] "AUTO97: Continuation and bifurcation software for ordinary differential equations (with HomCont), User's Guide," Concordia University, Montreal, Canada.

Dumortier, F., Roussarie, R. & Sotomayor, J. [1987] "Generic 3-parameter families of vector fields on the plane, unfolding a singularity with nilpotent linear part," *Ergod. Th. Dyn. Syst.* **7**, 375–413.

Dumortier, F., Roussarie, R. & Sotomayor, J. [1991] *Generic 3-Parameter Families of Planar Vector Fields, Unfoldings of Saddle, Focus and Elliptic Singularities with Nilpotent Linear Parts*, Lecture Notes in Mathematics, Vol. 1480 (Springer, Berlin).

Dumortier, F., Roussarie, R. & Sotomayor, J. [1994] "Elementary graphics of ciclicity 1 and 2," *Nonlinearity* **7**, 1001–1043.

Dumortier, F., Roussarie, R. & Sotomayor, J. [1997] "Bifurcations of cuspidal loops," *Nonlinearity* **10**, 1369–1408.

Fernández-Sánchez, F., Freire, E., Pizarro, L. & Rodríguez-Luis, A. J. [1996] "Analytical and numerical study of a van der Pol-Duffing oscillator," in *NDES '96: Fourth Int. Workshop on Nonlinear Dynamics of Electronic Systems* (Centro Nacional de Microelectrónica, Sevilla), pp. 321–326.

Freire, E., Gamero, E. & Ponce, E. [1989] "An algorithm for symbolic computation of Hopf bifurcation," in *Computers and Mathematics*, eds. Kaltofen, E. & Watt, S. M. (Springer, NY), pp. 109–118.

Freire, E., Pizarro, L. & Rodríguez-Luis, A. J. [1999a] "Examples of non-degenerate and degenerate cuspidal loops in planar systems," *Dyn. Stab. Syst.* **14**, 129–161.

Freire, E., Pizarro, L. & Rodríguez-Luis, A. J. [1999b] "Numerical continuation of degenerate homoclinic orbits in planar systems," *IMA J. Numer. Anal.* **19**, 51–75.

Freire, E., Pizarro, L. & Rodríguez-Luis, A. J. [2000] "Numerical continuation of homoclinic orbits to nonhyperbolic equilibria in planar systems," *Nonlin. Dyn.* **23**, 353–375.

Friedman, M. J. & Doedel, E. J. [1993] "Computational methods for global analysis of homoclinic and heteroclinic orbits: A case study," *J. Dyn. Diff. Eqs.* **5**, 37–57.

Gamero, E., Freire, E. & Ponce, E. [1991] "On the normal forms for planar systems with nilpotent linear parts," in *Bifurcation and Chaos: Analysis, Algorithms, Applications*, eds. Seydel, R., Schneider, F. W., Küpper, T. & Troger, H., International Series of Numerical Mathematics, Vol. 97 (Birkhäuser, Basel), pp. 123–127.

Golubitsky, M. & Schaeffer, D. G. [1985] *Singularities and Groups in Bifurcation Theory, Vol. I*, Applied Mathematics Science Series, Vol. 51 (Springer, Berlin).

Gray, P. & Scott, S. K. [1990] *Chemical Oscillations and Instabilities. Non-Linear Chemical Kinetics*, International Series of Monographs on Chemistry, Vol. 21 (Clarendon Press, Oxford).

Guckenheimer, J. [1986a] "Multiple bifurcation problems for chemical reactors," *Physica* **D20**, 1–20.

Guckenheimer, J. [1986b] *Global Bifurcations in Simple Models of a Chemical Reactor*, Lectures in Applied Mathematics, Vol. 24, pp. 163–174.

Guckenheimer, J. & Kim, S. [1992] "Dstool: A dynamical system toolkit with an interactive graphical interface, User's Guide," Center for Applied Mathematics, Cornell University, Ithaca, NY.

Guckenheimer, J. & Worfolk, P. [1993] "Dynamical systems: some computational problems," in *Bifurcations and Periodic Orbits of Vector Fields*, ed. Schlomiuk, D., NATO ASI Series, Series C, Vol. 408 (Kluwer, Dordrecht), pp. 241–277.

Guckenheimer, J. & Holmes, P. J. [1997] *Nonlinear Oscillations, Dynamical Systems, and Bifurcations of Vector Fields*, Applied Mathematical Science Series, Vol. 42 (Springer, Berlin).

Hassard, B. & Jiang, K. [1992] "Unfolding a point of degenerate Hopf bifurcation in an enzyme-catalyzed reaction model," *SIAM J. Math. Anal.* **23**, 1291–1304.

Hassard, B. & Jiang, K. [1993] "Degenerate Hopf bifurcation and isolas of periodic solutions in an enzyme-catalyzed reaction model," *J. Math. Anal. Appl.* **177**, 170–189.

Joyal, P. [1988] "Generalized Hopf bifurcation and its dual generalized homoclinic bifurcation," *SIAM J. Appl. Math.* **48**, 481–496.

Kernévez, J. P., Joly, G., Duban, M. C., Bunow, B. & Thomas, D. [1979] "Hysteresis, oscillations, and pattern formation in realistic immobilized enzyme systems," *J. Math. Biol.* **7**, 41–56.

Kernévez, J. P., Doedel, E., Duban, M. C., Hervagault, J. F., Joly, G. & Thomas, D. [1983] "Spatio-temporal organization in immobilized enzyme systems," in *Rhythms in Biology and Other Fields: Deterministic and Stochastic Approaches*, eds. Demongeot, J. & Le Breton, A., Lecture Notes in Biomathematics, Vol. 49 (Springer, Berlin), pp. 50–70.

Kernévez, J. P., Doedel, E. & Thomas, D. [1985] "Mathematical modeling of immobilized enzyme systems," *Biomed. Biochim. Acta 44* **6**, 993–1003.

Kuznetsov, Yu. A. [1998] *Elements of Applied Bifurcation Theory*, Applied Mathematical Science Series, Vol. 112 (Springer, Berlin).

Medved, M. [1985] "The unfoldings of a germ of vector fields in the plane with a singularity of codimension 3," *Czech. Math. J.* **35**, 1–42.

Murray, J. D. [1981a] "On pattern formation mechanism for lepidopteran wing pattern and mammalian coat markings," *Phil. Trans. Roy. Soc.* **B295**, 473–496.

Murray, J. D. [1981b] "A pre-pattern formation mechanism for animal coat markings," *J. Theor. Biol.* **88**, 161–199.

Murray, J. D. [2002] *Mathematical Biology. I: An Introduction*, Interdisciplinary Applied Mathematics, Vol. 17 (Springer, Berlin).

Murray, J. D. [2003] *Mathematical Biology. II: Spatial Models and Biomedical Applications*, Interdisciplinary Applied Mathematics, Vol. 18 (Springer, Berlin).

Nozdrachova, V. [1982] "Bifurcation of a noncourse separatrix loop," *Diff. Eqs.* **18**, 1098–1104.

Rheinboldt, W. C. [1986] *Numerical Analysis of Parametrized Nonlinear Equations*, The University of Arkansas Lecture Notes in the Mathematical Science, Vol. 7 (John Wiley, NY).

Rodríguez-Luis, A. J., Freire, E. & Ponce, E. [1990] "A method for homoclinic and heteroclinic continuation in two and three dimensions," in *Continuation and Bifurcations: Numerical Techniques and Applications*, eds. Roose, D., de Dier, B. & Spence, A. NATO ASI Series, Series C, Vol. 313 (Kluwer, Dordrecht), pp. 197–210.

Schecter, S. [1987] "The saddle-node separatrix-loop bifurcation," *SIAM J. Math. Anal.* **18**, 1142–1156.

Takens, F. [1973] "Unfoldings of certain singularities of vectorfields: Generalized Hopf bifurcations," *J. Diff. Eqs.* **14**, 476–493.

Thomas, D. [1975] "Artificial enzyme membranes, transport, memory, and oscillatory phenomena," in *Analysis and Control of Immobilized Enzyme Systems*, eds. Thomas, D. & Kernévez, J. P. (Springer, Berlin), pp. 115–150.

Wiggins, S. [2003] *Introduction to Applied Nonlinear Dynamical Systems and Chaos*, Texts in Applied Mathematics, Vol. 2 (Springer, Berlin).

STRAIGHTFORWARD COMPUTATION OF SPATIAL EQUILIBRIA OF GEOMETRICALLY EXACT COSSERAT RODS

T. J. HEALEY

Theoretical & Applied Mechanics and Center for Applied Mathematics,
Cornell University, Ithaca, NY 14850, USA

P. G. MEHTA*

Center for Applied Mathematics,
Cornell University, Ithaca, NY 14850, USA

Received July 9, 2004; Revised July 28, 2004

In this paper, we present a well posed "force" based formulation for nonlinearly elastic Cosserat rods with general boundary conditions enabling straightforward, efficient computation of spatial equilibria. We illustrate the ease and utility of our approach in four example problems, each exhibiting large spatial buckling, employing the path-following software AUTO.

Keywords: Elastic Cosserat rods; geometrically exact; computation of spatial equilibria.

1. Introduction

In this paper we consider the problem of computation of spatial equilibria of nonlinear elastic Cosserat rods, cf. [Antman, 1995]. At first glance this appears innocuous — merely the solution of a nonlinear two-point boundary value problem is required. However, in the special Cosserat theory, which we consider here, and which contains the classical Kirchhoff theory [Love, 1934] as a special case, the kinematical description of the rod requires the determination of the rotation field of the cross-sections. Herein lies the main difficulty — use of a standard (two-point boundary-value) solver with Newton type iteration will typically lead to "drift" in the rotation field. Namely, the rotations belong to a set of SO(3)-valued mappings, which is not a linear space. In the absence of explicit constraint equations, this is generally at odds with an additive iteration scheme. In this paper, we present a consistent formulation of the rod equations in the setting of a *linear* space, free of algebraic constraints, enabling the computation of equilibria via standard numerical techniques for two-point boundary value problems.

Simo and Vu-Quoc [1986] proposed a solution algorithm for statical Cosserat rod problems featuring a multiplicative updating procedure — in essence, a Newton solver on the differentiable manifold. There, the rotations are parametrized via unit quaternions (Euler parameters), and the incremental rotations (coming from the solution of a linearized problem) are efficiently exponentiated via the so-called Rodrigues formula. Their approach is geometrically natural and correct, if a bit formidable, and their reported numerical results are certainly very good. However, from a practical point of view, their methodology is not compatible for use with standard numerical packages, and consequently, it has not been widely adopted.

More recently the numerical implementation of a Hamiltonian formulation for the statical rod

---

*Current address: United Technologies Research Center, 411 Silver Lane, East Hartford, CT 06018, USA.

equations (treating the arc-length of the rod as a "time-like" variable) has been proposed [Li & Maddocks, 1996; Dichmann *et al.*, 1996]. The rotation field is explicitly parametrized via unit quaternions or Euler parameters. Of course, this introduces a different nonlinear space — a set of S^3-valued mappings (instead of SO(3)-valued mappings), where S^3 denotes the unit sphere in $\mathbb{R}^4$. A four-vector field, conjugate to the quaternion field, arises in lieu of the couple field in [Li & Maddocks, 1996; Dichmann *et al.*, 1996]. In particular, this leads to an extra differential equation in the system replacing the balance-of-moments equation. The role of this extra equation and, in particular, the consistent assignment of boundary conditions within a general class of problems is not addressed.

In this work, we propose a "force" formulation, based directly upon the convective form of the balance laws (force and moment balance). We too explicitly employ unit quaternions for the rotation field. This formulation reveals an inconsistency in the assignment of boundary conditions: For example, the prescription of a boundary rotation requires the assignment of the four components of the quaternion, whereas a prescribed couple at the boundary has only three components. Of course in the former case the four components are not independent — the quaternion field must have unit length, and a consistent formulation would generally require the inclusion of that algebraic constraint. Instead, we exploit the fact that the "unit-constraint equation" is actually a conservation law, and we eliminate its explicit appearance from the field equations via an approach similar to that employed in the proof of the Liapunov Center Theorem — sometimes referred to as "vertical Hopf bifurcation", e.g. cf. [Ambrosetti & Prodi, 1993]. We note that this same type of approach has been used successfully in the numerical computation of periodic solutions of the three-body problem in [Doedel, 2000; Doedel *et al.*, 2003], and more generally, for periodic solutions of conservative and Hamiltonian systems in [Munoz-Almaraz *et al.*, 2003]. For the convenience of the reader, we now summarize the well-known theorem underlying the method:

Consider a system of differential equations of the form

$$\dot{\mathbf{x}} = \mathbf{f}(\mathbf{x}), \qquad (1)$$

where $\mathbf{f} \colon \mathbb{R}^n \to \mathbb{R}^n$ is sufficiently smooth. Further we presume the existence of a real-valued conservation law. Specifically, for some differentiable function $E \colon \mathbb{R}^n \to \mathbb{R}$, we have

$$E(\mathbf{x}) \equiv C \qquad (2)$$

on all solutions of (1), where C is a constant, viz.

$$\frac{d}{dt}E(\mathbf{x}) = \langle \nabla E(\mathbf{x}), \dot{\mathbf{x}} \rangle = \langle \nabla E(\mathbf{x}), \mathbf{f}(\mathbf{x}) \rangle \equiv 0, \qquad (3)$$

where $\langle \cdot, \cdot \rangle$ denotes the standard Euclidean inner product on $\mathbb{R}^n$. The following is the cornerstone of our approach:

Proposition 1.1. *Consider the augmented system*

$$\dot{\mathbf{x}} = \mathbf{f}(\mathbf{x}) + \mu \nabla E(\mathbf{x}), \quad a < t < b, \qquad (4)$$

where $\mu \in \mathbb{R}$ is an unspecified parameter. Then any solution of (4) satisfying the end conditions

$$E(\mathbf{x}(a)) = E(\mathbf{x}(b)), \qquad (5)$$

is also a solution of (1) and (2).

The proof of Proposition 1.1 is simple: Substituting (4) into the first equation in (3), while using the last equality in (3), we see that

$$\frac{dE}{dt} = \mu \left\| \nabla E \right\|^2. \qquad (6)$$

In view of (5) we conclude that $\mu = 0$, i.e. (2) holds and (4) coincides with (1).

From a computational point of view, one advantage of treating (4), (5) is clear: The accuracy of *any* reasonable method employed in solving (4), (5) (supplemented by appropriate initial or boundary conditions) is carried over automatically to (2). The free parameter μ is simply a "dummy" unknown in (4) — its computed value is an extremely small number in practice. On the other hand, even a highly accurate solver for (1) will be inconsistent, in general, with a conservation law (2). Thus, working directly with (1) requires either that (2) be carried along as an algebraic constraint or that a discretization scheme be found for (1) that automatically fulfills the conservation law (2), the latter approach of which may be neither convenient nor practical. Moreover, as discussed below, the presence of the free parameter μ in the rod equations enables a consistent prescription of boundary conditions for a general class of boundary value problems.

The outline of the paper is as follows: In Sec. 2, we present the equilibrium equations solely in terms of force and moment fields via a complementary-energy formulation. These, in turn, are coupled to kinematical equations for the displacements and rotations. Here we see plainly the difficulty in working directly with rotation matrices — several constraint equations arise. We eliminate all but one constraint equation in the usual way via the Euler parameters. We then eliminate the unit constraint equation via Proposition 1.1. Here we see another advantage of treating (4), (5): In rod problems, the multiplier μ provides an extra unknown, which enables the consistent prescription of boundary conditions. For example, it turns out that for a prescribed rotation at an end-point, one of the quantities (5) is necessarily equal to unity; when a couple is prescribed at a boundary point, we *impose* one of (5) to be unity as a boundary condition at that location. In this way we always have the same number of boundary conditions appropriate for the number of unknowns in the field equations. This is true for a general class of "mixed" boundary conditions as well. Thus we always have well-posed two-point boundary value problems (in the absence of other symmetries). We also point out another advantage of our force-based formulation (over displacement-based formulations, e.g. [Simo & Vu-Quoc, 1986]), viz. the various classical constraints, such as inextensibility and/or unshearability, are readily incorporated with only minor modifications and without the need of Lagrange multipliers. In Sec. 3 we present four numerical examples of large, spatial buckling of elastic rods using the software package AUTO [Doedel, 2000], demonstrating the ease and utility of our formulation. In the first example we consider large lateral buckling of an end-loaded cantilevered rod in the shape of a thin ruler. Next we obtain large helical buckled states of a compressed hemitropic rod in the absence of external twist, cf. [Papadopoulos, 1999; Healey, 2002]. Third we consider a boundary value problem governing the spatial equilibria of a finite rod with intrinsic curvature (cf. [Domokos & Healey, 2005] for a systematic study). In particular, we obtain large helical solutions and so-called helical "perversions" [McMillen & Goriely, 2002] bifurcating from the straight rod in tension. Finally we consider again a long thin "ruler" with one end clamped while the other end is twisted via a hinged connection, i.e. the orientation of the cross-section is only partially prescribed. The second and third examples illustrate the utility of our approach in the presence of "mixed" boundary conditions.

2. Formulation

Let $\{\mathbf{e}_1, \mathbf{e}_2, \mathbf{e}_3\}$ denote a fixed, right-handed, orthonormal basis for $\mathbb{E}^3$. We consider a straight rod of unit length occupying a reference configuration parallel to $\mathbf{e}_3$. Let $s \in [0,1]$ denote the arclength coordinate (of the centerline) in the undeformed rod, and let $\mathbf{r}(s)$ denote the position vector (with respect to some fixed origin) of the material point originally at "s" in the reference configuration. We let $\mathbf{R}(s)$ denote the rotation of the cross-section spanned by $\{\mathbf{e}_1, \mathbf{e}_2\}$ at "s" in the undeformed rod. The first two unit vectors of the orthonormal field defined by

$$\mathbf{d}_i(s) = \mathbf{R}(s)\mathbf{e}_i, \quad i = 1, 2, 3, \tag{7}$$

are called *directors* in the special Cosserat theory, which we employ here. The deformed configuration of the rod is uniquely specified by the fields $\mathbf{r}(s)$ and $\mathbf{R}(s)$.

Differentiation of (1) yields

$$\mathbf{d}_i' = \mathbf{R}'\mathbf{R}^T\mathbf{d}_i, \quad i = 1, 2, 3. \tag{8}$$

Since the tensor field

$$\mathbf{K} \equiv \mathbf{R}'\mathbf{R}^T \tag{9}$$

is skew-symmetric, there is a unique vector field κ such that

$$\mathbf{d}_i' = \kappa \times \mathbf{d}_i, \quad i = 1, 2, 3, \tag{10}$$

i.e. κ is the *axial vector* of $\mathbf{K}$. We write

$$\mathbf{r}' = \nu_i\mathbf{d}_i, \quad \text{and} \quad \kappa = \kappa_i\mathbf{d}_i. \tag{11}$$

The numbers ν_i, κ_i are the "strains" in this theory, cf. [Antman, 1995]; ν_1, ν_2 are "shears", ν_3 is the "stretch", κ_1, κ_2 are "curvatures", and κ_3 is the "twist".

We let $\mathbf{n}(s)$ and $\mathbf{m}(s)$ denote the internal contact force and internal contact couple, respectively, acting on the cross-section originally at "s" in the reference configuration. We write

$$\mathbf{n} = n_i\mathbf{d}_i, \quad \text{and} \quad \mathbf{m} = m_i\mathbf{d}_i. \tag{12}$$

Recall that the n_i and m_i, $i = 1, 2, 3$, are called forces and moments, respectively, cf. [Antman, 1995;

n_1, n_2 are "shear forces", n_3 is the "axial force", m_1, m_2 are "bending moments", and m_3 is the "torque" or "twisting moment". For a *hyperelastic* rod, we assume the existence of a twice-differentiable, scalar-valued *stored energy* function, $W(\nu_1, \nu_2, \nu_3, \kappa_1, \kappa_2, \kappa_3, s)$, such that

$$n_j = \frac{\partial W}{\partial \nu_j} \quad \text{and} \quad m_j = \frac{\partial W}{\partial \kappa_j}, \quad j = 1, 2, 3. \quad (13)$$

If we define the triples $\underline{n} = (n_1, n_2, n_3)$, $\underline{m} = (m_1, m_2, m_3)$, $\underline{v} = (\nu_1, \nu_2, \nu_3)$, and $\underline{k} = (\kappa_1, \kappa_2, \kappa_3)$, and define $W(\underline{v}, \underline{k}, s) = W(\nu_1, \nu_2, \nu_3, \kappa_1, \kappa_2, \kappa_3, s)$, then (13) takes the compact form

$$\underline{n} = \frac{\partial W}{\partial \underline{v}}, \quad \text{and} \quad \underline{m} = \frac{\partial W}{\partial \underline{k}}. \quad (14)$$

We make the physically reasonable assumption that the Hessian $D^2 W(\cdot)$ is positive-definite matrix for each of its arguments on $\mathbb{R}^2 \times (0, \infty) \times \mathbb{R}^3$. Consequently, there is a *complementary energy function* (the Legendre transform of W), denoted by $\Upsilon(\underline{n}, \underline{m}, s)$, such that

$$\underline{v} = \frac{\partial \Upsilon}{\partial \underline{n}}, \quad \text{and} \quad \underline{k} = \frac{\partial \Upsilon}{\partial \underline{m}}. \quad (15)$$

Next we assume that the rod is subjected to a distributed, external body force per unit undeformed length, $\mathbf{b}(s)$, and a distributed, external body couple per unit undeformed length, $\mathbf{g}(s)$. Then the well-known local forms of balance of forces and moments are given by (cf. [Antman, 1995])

$$\mathbf{n}' + \mathbf{b} = \mathbf{0}, \quad (16)$$

and

$$\mathbf{m}' + \mathbf{r}' \times \mathbf{n} + \mathbf{g} = \mathbf{0}, \quad (17)$$

respectively.

Finally, we must specify boundary conditions at the two ends of the rod, $s = 0$ and $s = 1$. For boundary conditions of *place*, we specify the configuration $(\mathbf{r}, \mathbf{R})$ at an endpoint, while boundary conditions of *force* entail the specification of $(\mathbf{n}, \mathbf{m})$ at a boundary point. Of course, various "mixed" combinations can also be imposed, e.g. $(\mathbf{r}, \mathbf{m})$ could be specified at an endpoint.

3. Numerical Implementation

We first rewrite Eqs. (16) and (17) with respect to the convected basis $\{\mathbf{d}_1, \mathbf{d}_2, \mathbf{d}_3\}$, employing (10)–(12):

$$\underline{n}' + \underline{k} \times \underline{n} + \underline{b} = \underline{0}, \quad (18)$$

$$\underline{m}' + \underline{k} \times \underline{m} + \underline{v} \times \underline{n} + \underline{g} = \underline{0}, \quad (19)$$

where $\mathbf{b} = b_i \mathbf{d}_i$ and $\underline{b} \equiv (b_1, b_2, b_3)$, etc., as in (14). On the other hand, we write $\mathbf{r}$ and $\mathbf{R}$ with respect to the fixed basis:

$$\mathbf{r} = r_i \mathbf{e}_i, \quad \overline{r} \equiv (r_1, r_2, r_3), \quad (20)$$

$$\mathbf{R} = R_{ij} \mathbf{e}_i \otimes \mathbf{e}_j, \quad \overline{R} = \begin{bmatrix} R_{11} & R_{12} & R_{13} \\ R_{21} & R_{22} & R_{23} \\ R_{31} & R_{32} & R_{33} \end{bmatrix}. \quad (21)$$

Then (7), (9) and (11) lead to

$$\overline{r}' = \overline{R}\underline{v}, \quad (22)$$

$$\overline{R}' = \overline{R}\underline{K}, \quad (23)$$

where $\underline{K}$ is uniquely defined by $axial(\underline{K}) \equiv \underline{k}$.

Next we employ (15) in (18), (19), (22) and (23), to obtain the following system of first-order ODEs:

$$\underline{n}' = \underline{n} \times \frac{\partial \Upsilon}{\partial \underline{m}}, \quad (24)$$

$$\underline{m}' = \underline{m} \times \frac{\partial \Upsilon}{\partial \underline{m}} + \underline{n} \times \frac{\partial \Upsilon}{\partial \underline{n}}, \quad (25)$$

$$\overline{r}' = \overline{R}\frac{\partial \Upsilon}{\partial \underline{n}}, \quad (26)$$

$$\overline{R}' = \overline{R}\hat{\underline{K}}(\underline{n}, \underline{m}), \quad (27)$$

where $\hat{\underline{K}}(\underline{n}, \underline{m})$ denotes the skew-matrix-valued function uniquely defined by

$$axial(\hat{\underline{K}}(\underline{n}, \underline{m})) \equiv \frac{\partial \Upsilon}{\partial \underline{m}}. \quad (28)$$

The main difficulty with the numerical implementation of formulation (24)–(27) is that $\overline{R}(s) \in \mathrm{SO}(3)$, the latter of which is *not* a linear space, i.e.

$$\det(\overline{R}(s)) \equiv 1 \quad \text{and} \quad \overline{R}^T \overline{R} \equiv \overline{I} \quad (29)$$

must be imposed as constraints.

In an effort to reduce the number of constraints, we look to a well-known, singularity-free parametrization of SO(3). For any rotation $\overline{R}$, Euler's theorem asserts the existence of an axis of rotation, which corresponds to an eigenvector $\bar{a}$ satisfying

$$\bar{R}\bar{a} = \bar{a}, \quad |\bar{a}| = 1. \tag{30}$$

Let $\theta \in \mathbb{R}(\mathrm{mod}\, 2\pi)$ denote the counterclockwise rotation angle (according to the right-hand rule about $\bar{a}$) also given by Euler's theorem. We then introduce the quantities

$$q_0 \equiv \cos\left(\frac{\theta}{2}\right), \quad (q_1, q_2, q_3) \equiv \sin\left(\frac{\theta}{2}\right)\bar{a}, \tag{31}$$

and the four-vector

$$\mathsf{q} \equiv (q_0, q_1, q_2, q_3). \tag{32}$$

We then observe

$$\langle \mathsf{q}, \mathsf{q} \rangle \equiv q_0^2 + q_1^2 + q_2^2 + q_3^2 = 1, \tag{33}$$

viz. q is a unit *quaternion*; the scalars q_0, q_1, q_2, q_3 are typically called the *Euler parameters*. Here $\langle \cdot, \cdot \rangle$ denotes the standard Euclidean inner-product on $\mathbb{R}^4$. It can then be shown [Darboux, 1972] that

$$\begin{aligned}
\bar{R} \\
= \hat{R}(\mathsf{q}) \\
\equiv 2 \begin{bmatrix} q_0^2 + q_1^2 - \dfrac{1}{2} & q_1 q_2 - q_0 q_3 & q_1 q_3 + q_0 q_2 \\[2mm] q_1 q_2 + q_0 q_3 & q_0^2 + q_2^2 - \dfrac{1}{2} & q_2 q_3 - q_0 q_1 \\[2mm] q_1 q_3 - q_0 q_2 & q_2 q_3 + q_0 q_1 & q_0^2 + q_3^2 - \dfrac{1}{2} \end{bmatrix},
\end{aligned} \tag{34}$$

and

$$\mathsf{q}' = A(\mathsf{q})\underline{\mathsf{k}}, \tag{35}$$

where

$$A(\mathsf{q}) \equiv \frac{1}{2} \begin{bmatrix} -q_1 & -q_2 & -q_3 \\ q_0 & -q_3 & q_2 \\ q_3 & q_0 & -q_1 \\ -q_2 & q_1 & q_0 \end{bmatrix}. \tag{36}$$

We can now use (15), (34) and (35) to replace (26) and (27) in our earlier formulation:

$$\bar{r}' = \hat{R}(\mathsf{q}) \frac{\partial \Upsilon}{\partial \underline{n}}(\underline{n}, \underline{m}), \tag{37}$$

$$\mathsf{q}' = A(\mathsf{q}) \frac{\partial \Upsilon}{\partial \underline{m}}(\underline{n}, \underline{m}). \tag{38}$$

Our new system comprises (24), (25), (37) and (38), subject to the constraint (33).

In spite of the drastic reduction in the number of constraints (cf. (29) versus (33)), system (24), (25), (37) and (38) possesses a slight inconsistency that is revealed through the prescription of boundary conditions. For example, for a placement boundary condition, $(\bar{r}, \mathsf{q})$ is prescribed at an endpoint, which entails seven quantities. On the other hand, a force boundary condition entails only six specified quantities, e.g. $(\underline{n}, \underline{m})$. Of course, the four components of q in the former must also satisfy (33), which, in some sense, reconciles the actual count. At any rate, we would like to eliminate (33) altogether, which would enable the use of standard two-point boundary-value problem solvers.

Observe from (36) that

$$A^T(\mathsf{q})\mathsf{q} \equiv \mathsf{o} \quad \text{for all } \mathsf{q}. \tag{39}$$

In view of (35) and (39), we find that

$$\frac{d}{ds}\langle \mathsf{q}, \mathsf{q} \rangle = 2\langle \mathsf{q}, \mathsf{q}' \rangle = 2\langle A^T(\mathsf{q})\mathsf{q}, \underline{\mathsf{k}} \rangle \equiv 0,$$

i.e. $E \equiv \langle \mathsf{q}, \mathsf{q} \rangle = 1$ is a conservation law for (35). The following is a special case of Proposition 1.1:

Proposition 3.1. *Consider the augmented system (24), (25), (37), with (38) replaced by*

$$\mathsf{q}' = A(\mathsf{q})\frac{\partial \Upsilon}{\partial \underline{m}}(\underline{n}, \underline{m}) + \mu\mathsf{q}, \tag{40}$$

where $\mu \in \mathbb{R}$ is a free parameter, in the absence of (33). If (33) is satisfied at the endpoints, $s = 0$ and $s = 1$, then any solution of the augmented system, (24), (25), (37) and (40), is also a solution of the algebraic-differential system (24), (25), (33), (37) and (38).

Proposition 3.1 has important practical ramifications for numerical implementation: The augmented system (24), (25), (37), and (40) can be solved without explicitly enforcing (33) pointwise, provided that the latter is satisfied at the endpoints. We accommodate this as follows: For boundary conditions of placement, for which the orientation of the cross-section is prescribed, (33) is naturally satisfied at the boundary points. For force boundary conditions, (33) must be prescribed as a boundary condition as well. In either case, we always end up with seven boundary conditions at each boundary point, which agrees with the fourteen unknowns inherent in the augmented system, viz. $\underline{n}, \underline{m}, \bar{r}, \mathsf{q}$ and μ. This is also true for (well-posed) mixed boundary value problems, as demonstrated in the next section.

Another major advantage of our "force-based" approach, is that we can easily accommodate the common constrained rod theories automatically — without Lagrange multipliers. We summarize the three most common cases. Recall that a rod is said to be *inextensible* if $\nu_3 \equiv 1$ is imposed as a constraint. A rod is said to be *unshearable* if $\nu_1 = \nu_2 \equiv 0$ are imposed as constraints. Of course, all three of these may be imposed, in which case the rod is said to be *inextensible and unshearable*. In each of these cases, we simply replace the vector-valued function $\underline{v} = (\partial\Upsilon/\partial\underline{n})(\underline{n},\underline{m})$ in (25) and (37) with the following expression:

$$\text{Inextensible: } \underline{v} = \left(\frac{\partial\Upsilon}{\partial n_1}, \frac{\partial\Upsilon}{\partial n_2}, 1\right). \tag{41}$$

$$\text{Unshearable: } \underline{v} = \left(0, 0, \frac{\partial\Upsilon}{\partial n_3}\right). \tag{42}$$

$$\text{Inextensible, unshearable: } \underline{v} = (0,0,1). \tag{43}$$

4. Examples

In this section, we apply the above framework to obtain computational bifurcation and continuation results for four different example problems. We use the software package AUTO [Doedel, 2000] to carry out the computations. AUTO has the capability to continue and locate bifurcations of the solutions of general two-point boundary value problems (BVP). A well posed BVP (with a correct count of boundary conditions) is discretized in AUTO using the method of orthogonal collocation. A solution curve for the resulting square system of algebraic equations is continued in a single parameter using the method of arc length continuation employing Newton iteration [Keller, 1977]. The discretized *state*, say $\mathbf{x}$ and the parameter λ are parametrized as $(\mathbf{x}(s), \lambda(s))$ where s denotes the arc length continuation parameter. For this purpose, an initial solution $\mathbf{x}(0)$ for some given parameter value $\lambda(0)$ is needed for the continuation to begin.

In the examples presented below, there is always an "extra" boundary condition relative to the number of (first order) ODEs and the parameter μ is then the "extra" unknown which makes the discretized algebraic system well-posed (square). The parametrization for the arc length continuation is then given as $(x(s), \lambda(s), \mu(s))$ where s as before denotes the arc length continuation parameter. In consonance with Proposition 3.1, all computed

values of μ in each of the four examples presented below is observed to be *numerically* zero, i.e. on the order of 10^{-14}.

4.1. *Large lateral buckling of a "ruler"*

Consider an unshearable rod with one end clamped and with the other end subjected to a "dead" transverse force λ, as shown in Fig. 1. The assumed constitutive laws are summarized in Table 1. Observe that one bending stiffness is ten times the other. Accordingly, we call such a rod a "ruler," as suggested by the depiction in Fig. 1. The boundary conditions at the clamped end $(s = 0)$ are

$$\mathbf{r}(0) = \mathbf{0}, \tag{44}$$

$$q_0(0) = 1, \quad (q_1, q_2, q_3)(0) = \mathbf{0}, \tag{45}$$

and at the end $(s = 1)$ where normal force λ is applied, we impose

$$\mathbf{m}(1) = \mathbf{0}, \tag{46}$$

$$\mathbf{n}(1) = \lambda\mathbf{e}_1, \tag{47}$$

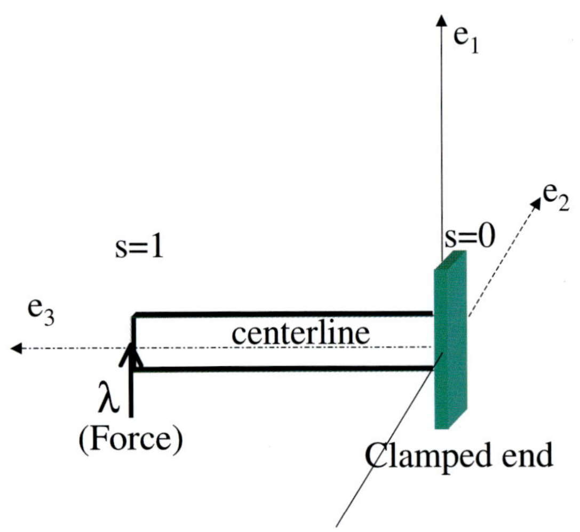

Fig. 1.   Schematic of the ruler showing reference configuration (where $\lambda = 0$) and boundary conditions at the two ends.

Table 1.   Constitutive laws for ruler.

Unshearable	$\nu_1 = 0, \ \nu_2 = 0$
Axial force	$n_3 = 20\log(\nu_3)$
Bending moments	$m_1 = \kappa_1, \ m_2 = 10\kappa_2$
Twisting moment	$m_3 = \kappa_3$

for a total of thirteen boundary conditions. In the convective coordinates used for computations, the boundary condition (47) is expressed as

$$n_1(1) = 2\lambda(q_1 q_2 + q_0 q_3), \tag{48}$$

$$n_2(1) = 2\lambda(q_0^2 + q_2^2 - 0.5), \tag{49}$$

$$n_3(1) = 2\lambda(q_2 q_3 - q_0 q_1). \tag{50}$$

We are interested in computing the solutions of this problem as the force parameter λ is increased from zero (for which the ruler is in its reference configuration). The boundary conditions (44)–(46), (48)–(50) together with the system of differential Eqs. (24), (25), (37), and (38) describe a well-posed continuation problem in the single parameter λ. However, on account of numerical errors (as λ is increased) implicit in any boundary value solver, the solution as it is continued may "move off" the admissible manifold of solutions defined by constraint (32). For this purpose, and as discussed in the previous section, we consider the augmented system of Eqs. (24), (25), (37), and (39) and introduce an additional boundary condition at $s = 1$

$$(q_0^2 + q_1^2 + q_2^2 + q_3^2)(1) = 1. \tag{51}$$

The solution of the system of Eqs. (24), (25), (37), and (39) together with the boundary condition Eqs. (44)–(46), (48)–(51) is continued in two parameters (λ, μ). As the applied force λ is increased from zero, the tip of the rod (at $s = 1$) moves in the $\mathbf{e}_2 - \mathbf{e}_3$ plane in the direction of the applied force. At a critical value, the planar solution buckles thereby resulting in a bifurcated nonplanar solution. Figure 2 plots the centerline of planar solutions for increasing values of λ together with the centerline of a single bifurcated nonplanar solution.

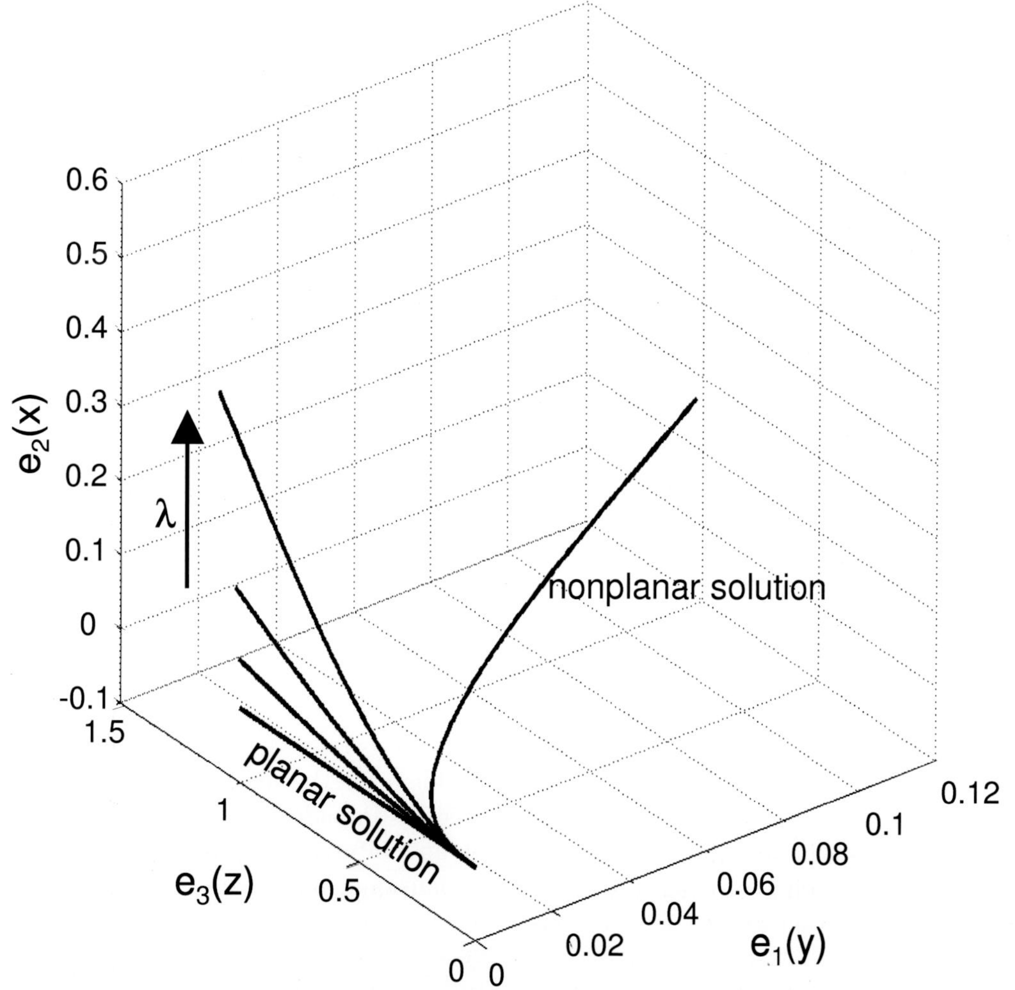

Fig. 2. Centerline of the ruler as force at the tip — λ is increased.

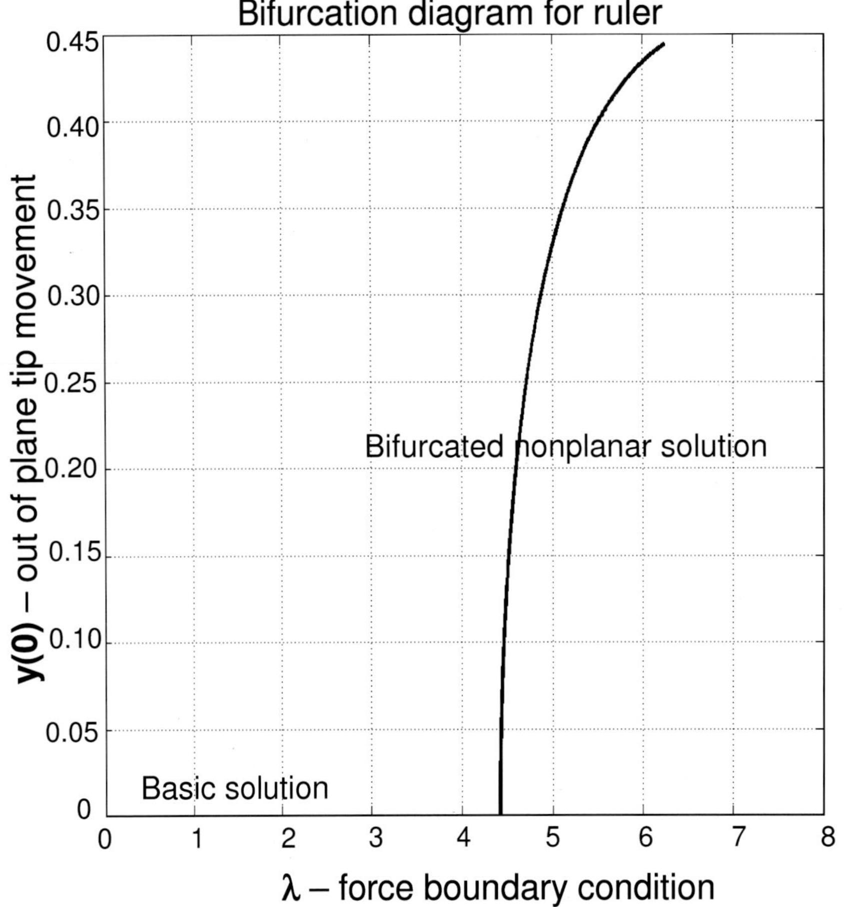

Fig. 3.   Bifurcation diagram for ruler.

Figure 3 plots the bifurcation diagram as a function of λ and Fig. 4 plots a typical bifurcated nonplanar solution showing the buckling experienced by the ruler as the normal force exceeds the critical value. We note that the load at bifurcation agrees well with that predicted buckling load in [Timoshenko & Gere, 1961, Eqs. (6)–(23)], the latter of which is only approximate, given that an infinitesimal pre-buckled configuration is presumed.

4.2. *Nonplanar solutions of a compressed "cable" or "DNA strand"*

Consider an unshearable hemitropic rod (see Table 2 for the constitutive laws assumed). Hemitropy is a natural model of long filaments having a helical micro-structure in the relaxed state, cf. [Healey, 2002]. The two ends of the rod are "clamped" against rotation and transverse displacements, while the axial displacements of the

two endpoints are prescribed (and equal):

$$r_\alpha(-1) = 0, \quad \alpha = 1, 2, \quad r_3(-1) = -1 + \lambda, \quad (52)$$

$$\mathsf{q}(-1) = (1, 0, 0, 0), \quad (53)$$

$$r_\alpha(1) = 0, \quad \alpha = 1, 2, \quad r_3(1) = 1 - \lambda, \quad (54)$$

$$\mathsf{q}(1) = (1, 0, 0, 0) \quad (55)$$

imposed at the two ends $s = -1$ and $s = 1$ as shown in Fig. 5.

We are interested in computing the solutions of this problem as the displacement parameter λ is increased from zero (for which the rod is assumed to be in its reference configuration). The bifurcation problem for the unshearable case has been considered in [Papadopoulos, 1999] and the linearization at any bifurcation point shown to possess a two-dimensional null space. This two-dimensional null space arises due to the presence of the symmetry group $O(2) \subset \mathrm{SO}(3)$: $\mathrm{SO}(2)$ due to rotations of the rod about $\mathbf{e}_3$ (the centerline), and Z^2 corresponding to $180°$ rotations of the rod about any perpendicular

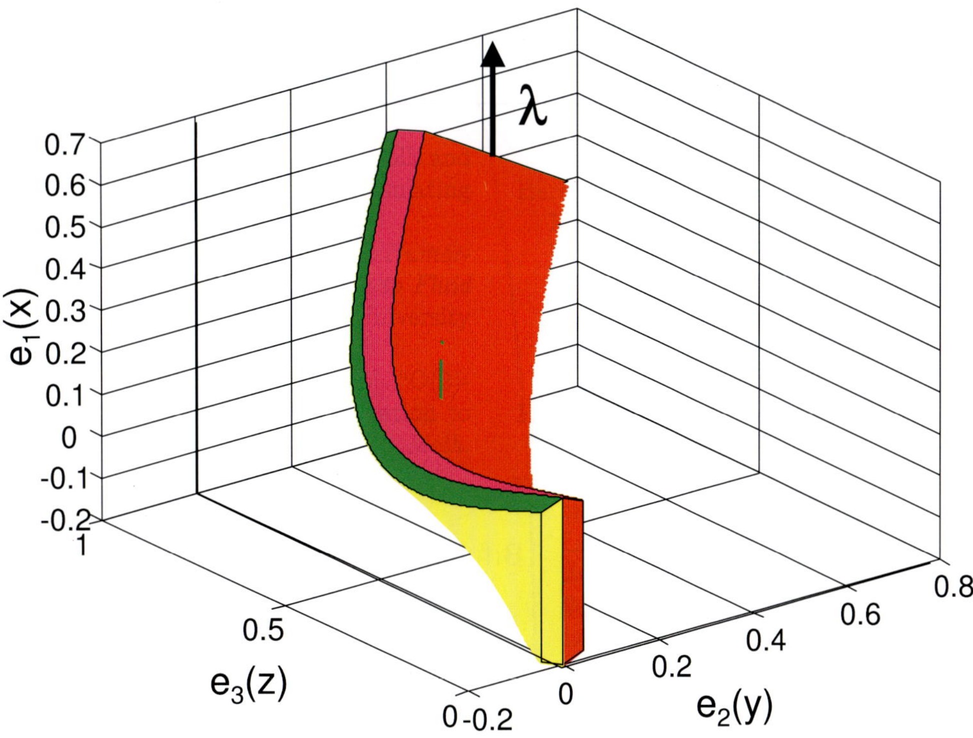

Fig. 4.   A typical 3D nonplanar solution along the bifurcated branch for the ruler.

Table 2.   Constitutive laws for hemitropic rod (see [Healey, 2002] for the definition of hemitropic rod).

Unshearable	$\nu_1 = 0, \quad \nu_2 = 0$
Axial Force	$n_3 = 10(\nu_3 - 1) - 10\kappa_3$
Bending moments	$m_1 = \kappa_1, \quad m_2 = \kappa_2$
Twisting moment	$m_3 = \kappa_3 - 10(\nu_3 - 1)$

bisector of the centerline. We obtain solutions by working in a suitable fixed-point space corresponding to solutions which are symmetric with respect to $180°$ rotations of the rod about $\mathbf{e}_2$ at $s = 0$ — the so-called Z^2 isotropy subgroup, cf. [Papadopoulos, 1999]. In this fixed-point space, the boundary conditions at $s = 1$ remain as before [Eqs. (52) and (53)] while the boundary conditions are now imposed at the midpoint $s = 0$ and are given as

$$n_2(0) = 0, \tag{56}$$

$$m_2(0) = 0, \tag{57}$$

$$r_1(0) = 0, \tag{58}$$

$$r_3(0) = 0, \tag{59}$$

$$q_1(0) = 0, \tag{60}$$

$$q_3(0) = 0. \tag{61}$$

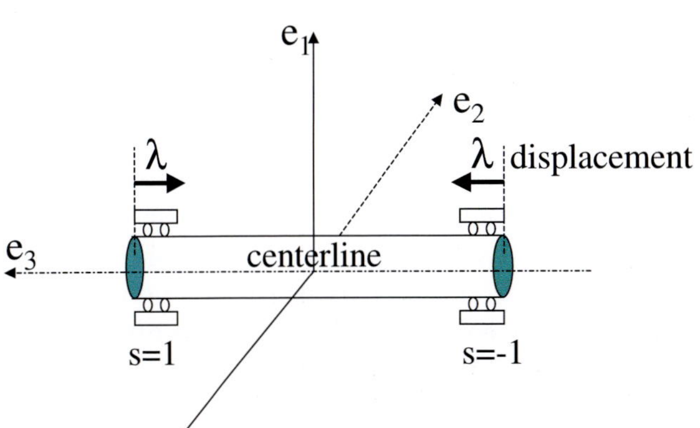

Fig. 5.   Schematic of the rod showing reference configuration (where $\lambda = 0$) and boundary conditions at the two ends.

Once the solution is obtained in the fixed-point space for $s \in [0, 1]$, the solution for $s \in [-1, 0]$ is obtained by $180°$ rotation. Moreover, an entire orbit of solutions may be obtained by rotating the

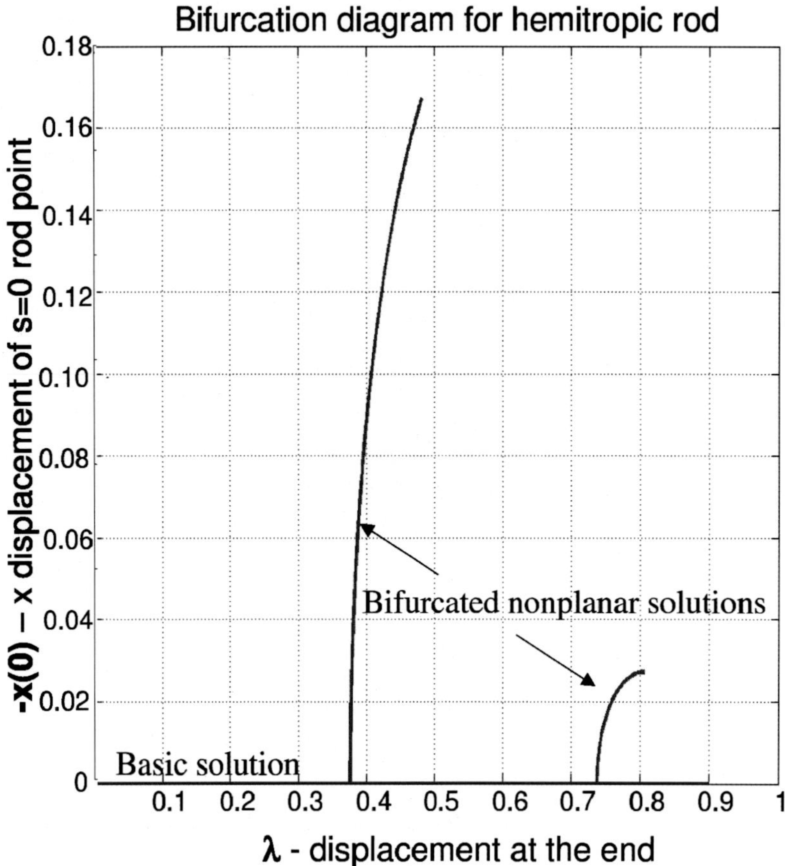

Fig. 6.   Bifurcation diagram for the hemitropic rod.

above solution (applying the SO(2) quotient group). Physically though, these solutions are all the same as they correspond to rigid rotation of the rod about its centerline.

As in the case of the ruler, the thirteen boundary conditions (52), (53), (56)–(61) in the fixed-point space are augmented with an extra boundary condition

$$(q_0^2 + q_1^2 + q_2^2 + q_3^2)(0) = 1 \qquad (62)$$

in order to satisfy the constraint. In view of the two boundary conditions (60) and (61), instead of Eq. (62),

$$(q_0^2 + q_2^2)(0) = 1 \qquad (63)$$

is actually used in carrying out the computations. The solution of the augmented system of Eqs. (24), (25), (37), and (39) together with the augmented set of fourteen boundary condition Eqs. (52), (53), (56)–(61) and (63) is continued in two parameters (λ, μ). Figure 6 plots the bifurcation diagram of the obtained solutions showing two bifurcation points for the problem.

These bifurcation points agree with the analysis in [Papadopoulos, 1999; Papadopoulos & Healey, 2004] and Figs. 7 and 8 plot two typical nonplanar solutions along the resulting bifurcated branches.

4.3.   *Perversions of a "telephone cord"*

Next we compute helical and so-called perversion or helical-reversal solutions exhibited by a rod of finite length with intrinsic curvature, e.g. a telephone cord. We refer to [McMillen & Goriely, 2002] for an analytical study of the perversion solutions for infinite rods and to [Domokos & Healey, 2005] for a systematic study of the class of finite-length rod problems considered here. For our computational study, we assume an unshearable, inextensible rod with initial curvature κ_0 about the $\mathbf{e_1}$ direction; the constitutive laws are summarized in Table 3. The initial curvature κ_0 is related to the length L of the rod via

$$N2\pi\frac{1}{\kappa_0} = L, \qquad (64)$$

A typical solution along first bifurcated branch

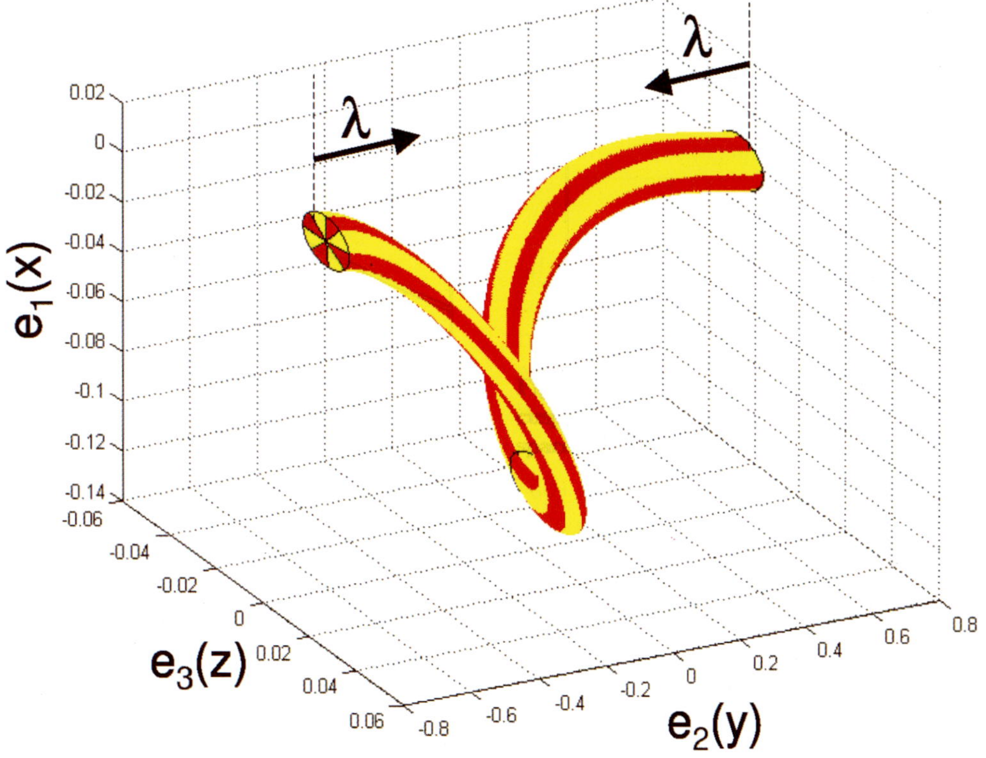

Fig. 7.   A typical 3D nonplanar solution along the first bifurcated branch for the hemitropic rod.

A typical solution along second bifurcated branch

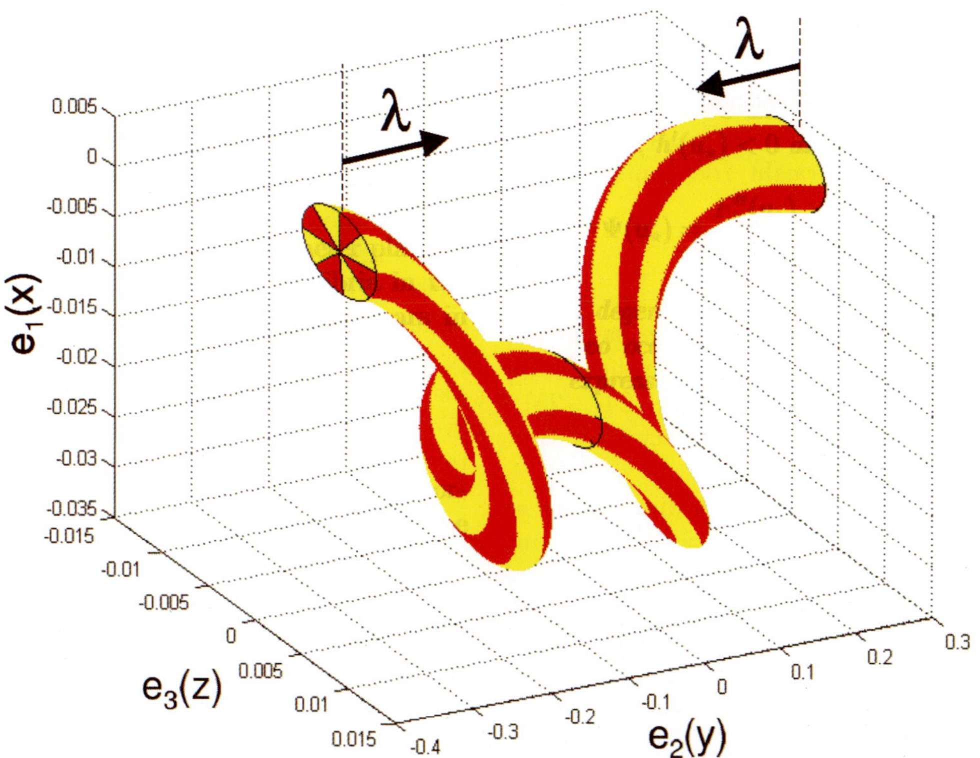

Fig. 8.   A typical 3D nonplanar solution along the second bifurcated branch for the hemitropic rod.

Table 3. Constitutive laws for telephone cord: κ_0 is the initial curvature.

Unshearable	$\nu_1 = 0,\ \nu_2 = 0$
Inextensible	$\nu_3 = 1$
Bending moments	$m_1 = \kappa_1,\ \ m_2 = \kappa_2 - \kappa_0$
Twisting moment	$m_3 = \kappa_3$

where N corresponds to the number of turns of the telephone cord being modeled and $1/\kappa_0$ is the radius of curvature. At $s = 0$ the rod is clamped; the end at $s = 1$ is clamped against rotation and transverse displacements. We also impose axial tension at the end $s = 1$ given by

$$\mathbf{n}(1) \cdot \mathbf{e}_3 = \lambda, \qquad (65)$$

where λ denotes the magnitude of the imposed tensile force. Since the rotation at $s = 1$ is completely constrained, Eq. (65) becomes

$$n_3(1) = \lambda. \qquad (66)$$

The rest of the boundary conditions correspond to the placements

$$\mathbf{r}(0) = \mathbf{0}, \qquad (67)$$

$$\mathsf{q}(0) = (1, 0, 0, 0), \qquad (68)$$

$$r_1(1) = 0, \qquad (69)$$

$$r_2(1) = 0, \qquad (70)$$

$$\mathsf{q}(1) = (1, 0, 0, 0), \qquad (71)$$

for a total of fourteen boundary conditions. In particular, the rotations at the two ends are fixed, which ensure that the straight rod configuration with bending moment

$$m_1(s) = -\kappa_0 \qquad (72)$$

is a basic solution for all values of the loading λ. We compute the solution as the tensile load λ on the cord is increased from its initial zero value. Note that the problem with differential Eqs. (24), (25), (37), and (39) and the fourteen boundary condition Eqs. (66)–(71) is well-posed.

Figure 10 plots the bifurcation diagram as a function of λ for the choice of $N = 1.5$ turns in Eq. (64). As the parameter λ is increased from its zero value, the basic solution undergoes a bifurcation into a nonplanar solution shown in Fig. 11. As the parameter λ is further increased, the basic solution undergoes a second bifurcation resulting in a so-called perversion as shown in Fig. 12. The computed shapes depicted in Figs. 11 and 12 were first obtained using the Parallel Hybrid

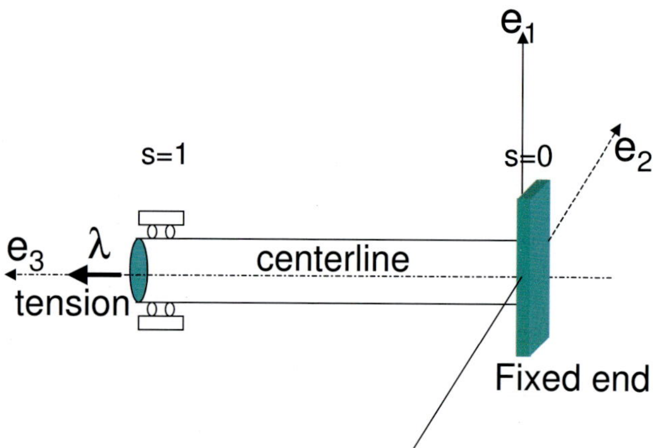

Fig. 9. Schematic of the telephone cord showing reference configuration (where $\lambda = 0$) and boundary conditions at the two ends.

Algorithm [Domokos & Szeberenyi, 2004] in [Domokos & Healey, 2005]. The latter paper also contains a detailed local analysis, with which our computational results agree well. The computational results in [Domokos & Healey, 2005] go well beyond those presented in this section. In particular, solutions characterized by more than one helical reversal along the length of the rod are obtained.

4.4. *Multiple solutions for a twisted "ruler"*

Consider finally a buckling problem for a classical unshearable, inextensible rod with constitutive laws summarized in Table 4. The rod is clamped at one end $(s = 0)$ and is subjected to a tensile load at the other end $(s = 1)$. In addition, the rod is attached to a movable hinge at $s = 1$. The rod is free to rotate about the axis of the hinge, initially aligned with $\mathbf{e}_1$, while the orientation or twist of the hinge about $\mathbf{e}_3$, through a counter-clockwise angle "α", is prescribed, as illustrated in Fig. 13. The transverse displacements of the rod at $s = 1$ are constrained. We are interested in computing the solutions of the problem in the presence of prescribed tensile end load λ, as in the telephone cord problem, and in the presence of twist α. The twist boundary condition is specified by the rotation imposed at the cross-section at $s = 1$:

$$\mathbf{R}(1)\mathbf{e}_1 = \cos(\alpha)\mathbf{e}_1 + \sin(\alpha)\mathbf{e}_2, \qquad (73)$$

and in terms of coordinates used for computations

$$2q_0^2(1) + 2q_1^2(1) - 1 = \cos(\alpha), \qquad (74)$$

$$q_1(1)q_3(1) - q_0(1)q_2(1) = 0. \qquad (75)$$

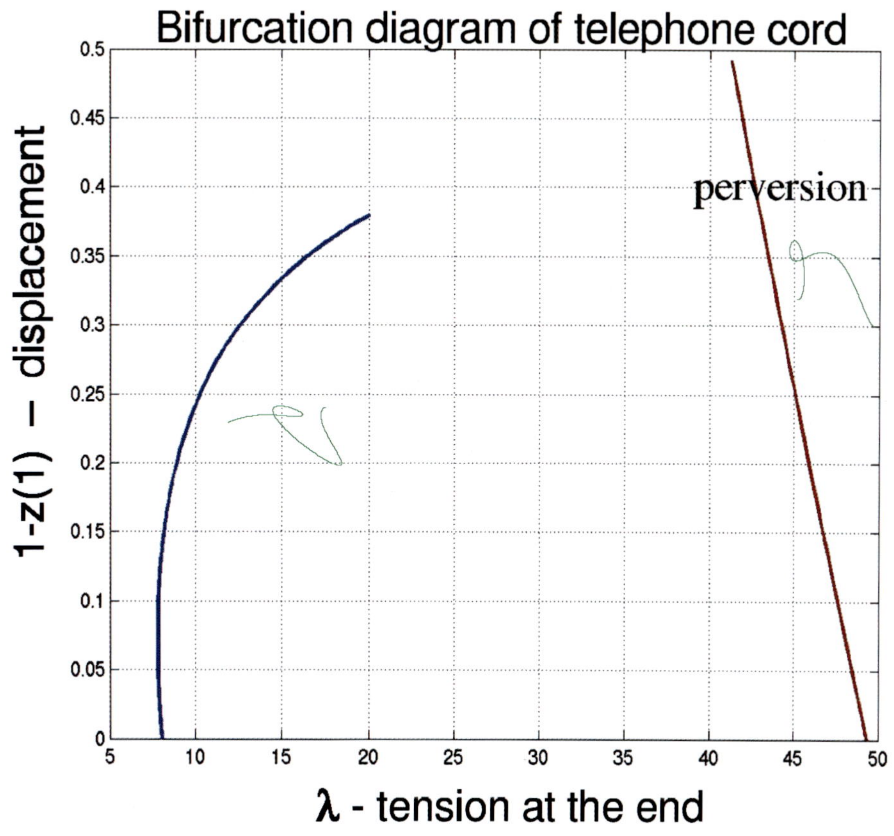

Fig. 10.   Bifurcation diagram for telephone cord.

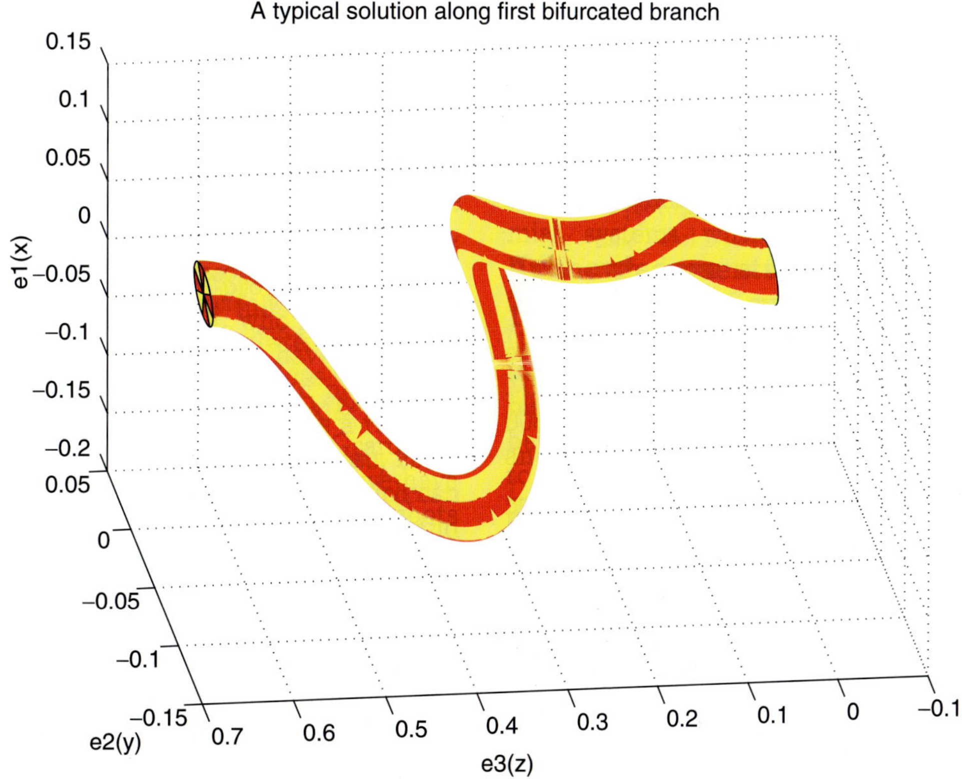

Fig. 11.   A typical nonplanar solution along the first bifurcated branch for the telephone cord.

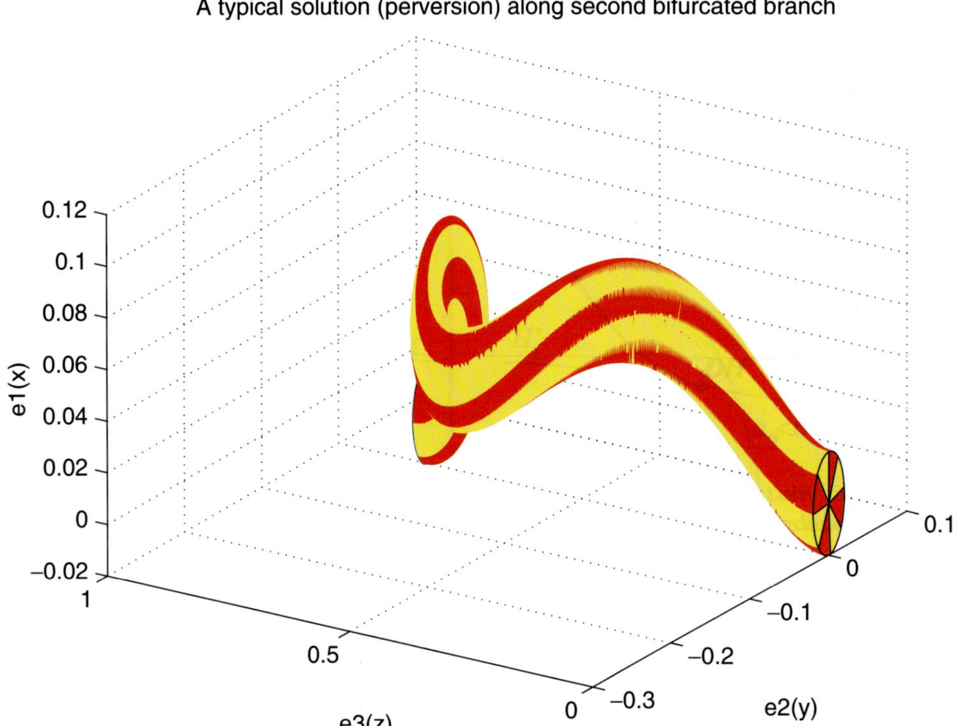

Fig. 12.   A typical perversion along the second bifurcated branch for the telephone cord.

Table 4.   Constitutive laws for the twisted ruler problem.

Unshearable	$\nu_1 = 0, \ \nu_2 = 0$
Inextensible	$\nu_3 = 1$
Bending moments	$m_1 = \kappa_1, \ m_2 = 10\kappa_2$
Twisting moment	$m_3 = \kappa_3$

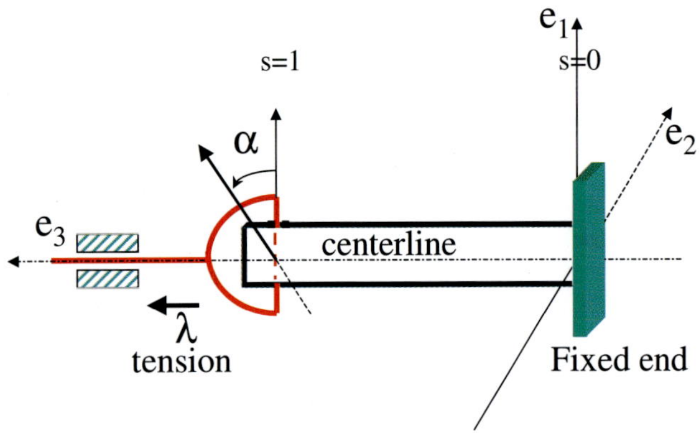

Fig. 13.   Schematic of the ruler showing reference configuration and boundary conditions at the two ends: λ denotes the tensile load and α denotes the angle of twist.

The rest of the boundary conditions arise as

$$\mathbf{r}(0) = \mathbf{0}, \tag{76}$$

$$\mathbf{q}(0) = (1, 0, 0, 0), \tag{77}$$

$$r_1(1) = 0, \tag{78}$$

$$r_2(1) = 0, \tag{79}$$

$$m_1(1) = 0, \tag{80}$$

and a tensile boundary condition as in Eq. (65)

$$\left[2\left(q_3 q_1 - q_0 q_2\right) n_1 + 2\left(q_3 q_2 + q_0 q_1\right) n_2 \right. $$
$$\left. + 2\left(q_0^2 + q_3^2 - \frac{1}{2}\right) n_3 \right]_{s=1} = \lambda, \tag{81}$$

for a total of thirteen boundary conditions. These thirteen boundary conditions are then augmented by the extra boundary condition

$$(q_0^2 + q_1^2 + q_2^2 + q_3^2)(1) = 1, \tag{82}$$

in accordance with Proposition 1.1.

We next describe some of the continuation and bifurcation results obtained with this example. With no twist ($\alpha = 0$), the primary response is governed by a standard planar buckling problem. Figure 14 depicts this branch together with a branch corresponding to the case in which a small amount of twist ($\alpha = 30°$) is applied. A more interesting situation arises where the buckling load is kept at a constant value and twist applied. Figure 15

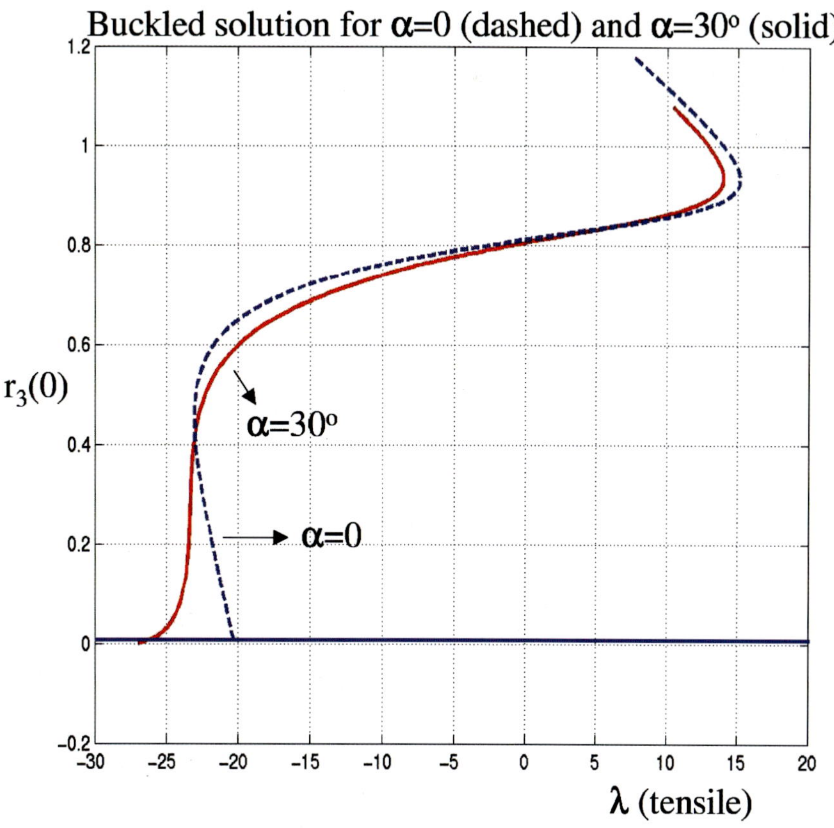

Fig. 14.   Bifurcation diagram (in parameter λ) for twisted ruler.

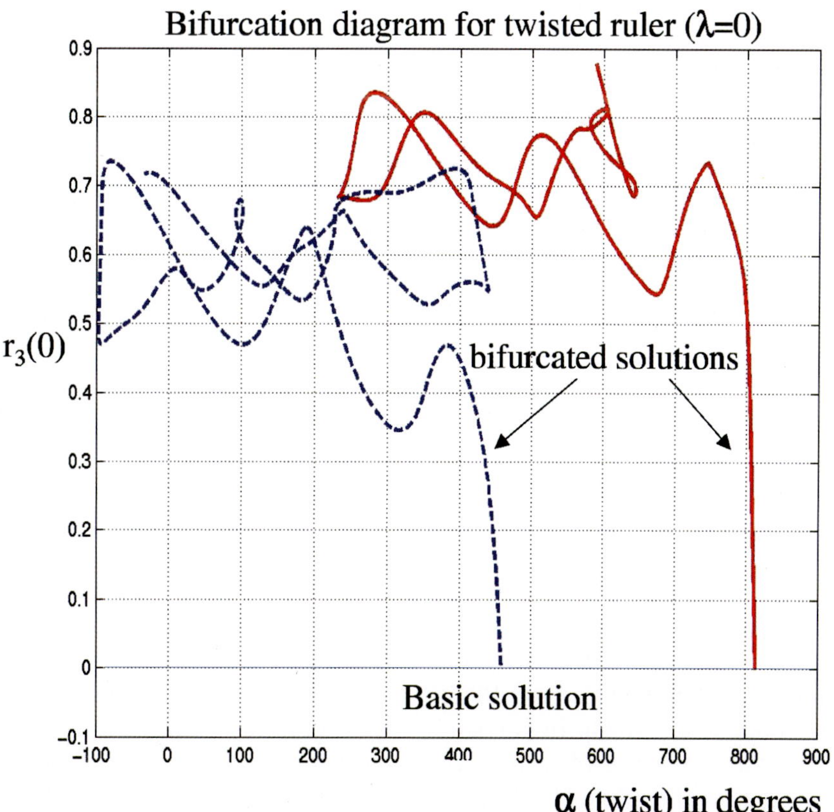

Fig. 15.   Bifurcation diagram (in parameter α) for twisted ruler.

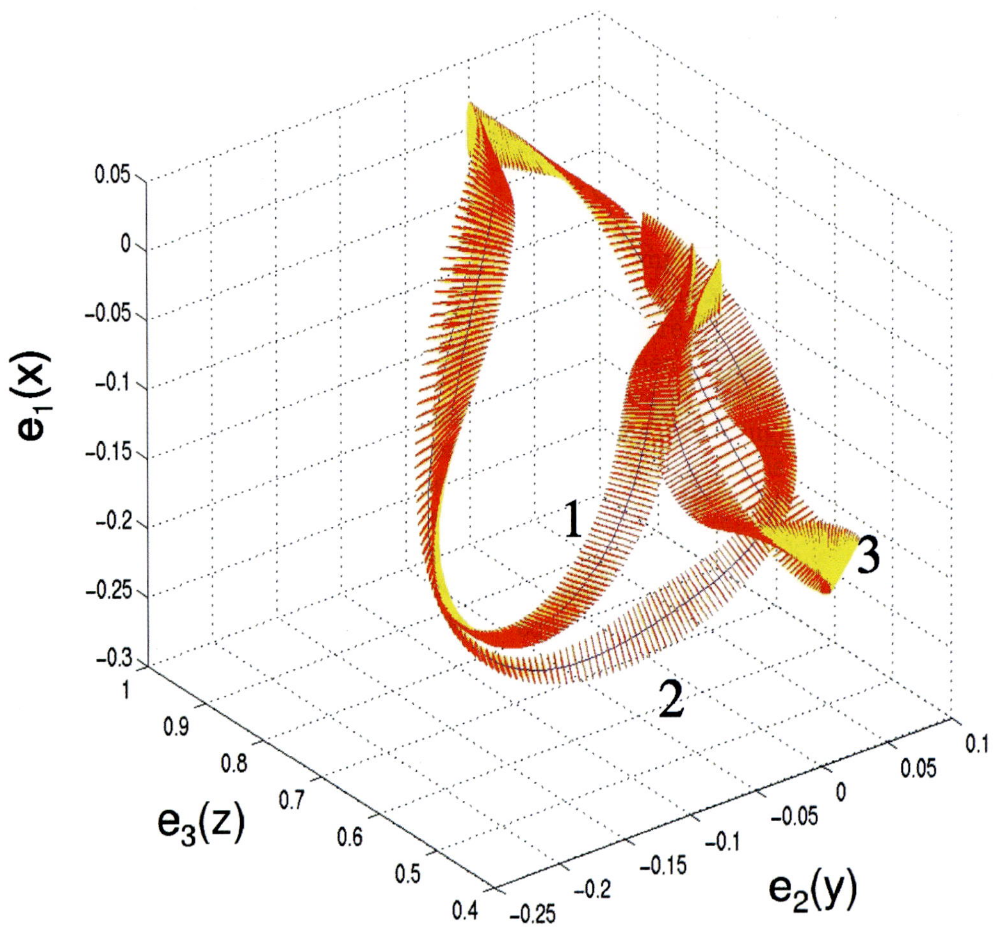

Fig. 16.   Multiple solutions for parameter values $\lambda = \alpha = 0$ (no loading, no twist).

plots the bifurcation diagram for this case where $\lambda = 0$ and parameter α is varied. Continuation here shows the intricate nature of multiple buckled nonplanar solutions for any given choice of angle. For example, Fig. 16 plots three distinct solutions for $\alpha = 0$ along the first bifurcating branch.

5.   Concluding Remarks

The use of quaternions in rigid body dynamics and its applications, including their use in simulations, is well known, e.g. [Darboux, 1972; Mitchell & Rodgers, 1965; Spurrier, 1978; Kane *et al.*, 1983; Taylor & Paul, 1990; Cooke *et al.*, 1994]. In particular, in [Taylor & Paul, 1990] the direct integration of (38) (where $\underline{k}$ (cf. (15)) is now the triple of the components of the angular velocity in body coordinates) is advocated provided that "periodic normalization

to unit magnitude is accomplished" — this is simply an *ad hoc* version of Proposition 3.1. Of course, Proposition 3.1 can be used in rigid-body dynamical simulations, but this requires the use of implicit methods, given that the unit condition is specified at the end of a time step or sequence of time steps. We plan to pursue such ideas elsewhere. We also point out that in [Cooke *et al.*, 1994] equations of the form (40) are advocated (presumably for explicit methods of numerical integration), but there $\mu \neq 0$ is imposed as an "integration drift correction gain".

John Maddocks has kindly pointed out to us that the discretization scheme employed in the package AUTO [Doedel, 2000] preserves quadratic conservation laws, e.g. (33). In the first example above, this means that we could directly tackle (38) without imposing (33) as a pointwise constraint and without the need for (51). Indeed one then has the thirteen boundary conditions (44)–(46) and thirteen (scalar) unknowns, $\underline{n}, \underline{m}, \bar{r}$, and $\mathfrak{q}$ — so with one loading parameter, we expect solution branches. Indeed, we carried this procedure out, which yields results virtually identical to those obtained above. At the very least, this is a nice check on our results. On the other hand, if we try this procedure for any problem like the third example (telephone cord), for which the orientation at both ends is prescribed, we end up with fourteen boundary conditions for the thirteen unknowns — with one loading parameter we will not be able to compute solution branches. As pointed out earlier, all such inconsistencies are avoided by our approach based upon Proposition 3.1. We *always* prescribe fourteen boundary conditions for the fourteen scalar unknowns, $\underline{n}, \underline{m}, \bar{r}, \mathfrak{q}$ and μ, and with one loading parameter (and in the absence of other symmetries) we get solution branches. Moreover, our approach automatically insures the same accuracy for the conservation law as that employed in the solution of the differential equations — *independent* of the particular integration scheme chosen. In particular, problems for elastic frameworks of Cosserat rods (with more than two rods coming together at a joint) are beyond the capabilities of any two-point boundary value problem solver like AUTO. Here a more general finite-element formulation is required, for which our method is very attractive. We will pursue such work elsewhere.

Acknowledgment

This work was supported in part by the National Science Foundation through grant DMS-0072514.

References

Ambrosetti, A. & Prodi, G. [1993] *A Primer of Nonlinear Analysis* (Cambridge University Press, Cambridge).

Antman, S. [1995] *Nonlinear Problems of Elasticity* (Springer-Verlag, NY).

Cooke, J. M., Zyda, M. J., Pratt, D. R. & McGhee, R. B. [1994] "Npsnet: Flight simulation dynamic modeling using quaternions," *Presence* **1**, 404–420.

Darboux, G. [1972] *Lecons Sur La Théorie Générale Des Surfaces, Premiere Partie* (Chelsea Publishing Company, NY).

Dichmann, D., Li, Y. & Maddocks, J. [1996] "Hamiltonian formulation and symmetries in rod mechanics," in *Mathematical Approaches to Biomolecular Structure and Dynamics*, ed. Mesirov, J. (Springer-Verlag, NY), pp. 71–113.

Doedel, E. J. [2000] "AUTO2000: Continuation and bifurcation software for ordinary differential equations."

Doedel, E. J., Paffenroth, R. C., Keller, H. B., Dichmann, D. J., Galan-Vioque, J. & Vanderbauwhede, A. [2003] "Computation of periodic solutions of conservative systems with application to the 3-body problem," *Int. J. Bifurcation and Chaos* **12**, 1353–1381.

Domokos, G. & Szeberenyi, I. [2004] "A hybrid parallel approach to nonlinear boundary value problems," *Comp. Ass. Mech. Eng. Sci.* **11**, 15–34.

Domokos, G. & Healey, T. J. [2005] "Multiple helical perversions of finite, intristically curved rods," *Int. J. Bifurcation and Chaos* **15**, 871–890.

Healey, T. J. [2002] "Material symmetry and chirality in nonlinearly elastic rods," *Math. Mech. Solids* **7**, 405–420.

Kane, T. R., Likins, P. W. & Levinson, D. A. [1983] *Spacecraft Dynamics* (McGraw-Hill Book Company).

Keller, H. B. [1977] "Numerical solution of bifurcation and nonlinear eigenvalue problems," in *Applications of Bifurcation Theory*, ed. Rabinowitz, P. H. (Academic Press), pp. 359–384.

Li, Y. & Maddocks, J. [1996] "On the computation of equilibria of elastic rods. Part I: Integrals symmetry and a Hamiltonian formulation," manuscript.

Love, A. [1934] *A Treatise on the Mathematical Theory of Elasticity*, 4th edition (Cambridge University Press, Cambridge).

McMillen, T. & Goriely, A. [2002] "Tendril perversion in intrisically curved rods," *J. Nonlin. Sci.* **12**, 169–205.

Mitchell, E. E. & Rodgers, A. E. [1965] "Quaternion parameters in the simulation of a spinning rigid body," *Simulation* **18**.

Munoz-Almaraz, F., Freire, E., Galan, J., Doedel, E. & Vanderbauwhede, A. [2003] "Continuation of periodic orbits in conservative and Hamiltonian systems," *Physica* **D181**, 1–38.

Papadopoulos, C. M. [1999] "Nonplanar buckled states of hemitropic rods," Ph.D. thesis, Cornell University.

Papadopoulos, C. & Healey, T. J. [2004] "Large buckling of a compressed hemitropic rod," manuscript.

Simo, J. & Vu-Quoc, L. [1986] "A three-dimensional finite-strain rod model. Part II: Computational aspects," *Comput. Meth. Appl. Mech. Engin.* **58**, 79–116.

Spurrier, R. A. [1978] "Comment on singularity-free extraction of a quaternion from a direction-cosine matrix," *J. Spacecraft* **15**, p. 255.

Taylor, R. H. & Paul, R. P. [1990] "On homogeneous transforms, quaternions, and computational efficiency," *IEEE Trans. Robot. Autom.* **6**, 382–387.

Timoshenko, S. & Gere, J. [1961] *Theory of Elastic Stability* (McGraw-Hill, NY).

MULTIPARAMETER PARALLEL SEARCH BRANCH SWITCHING

MICHAEL E. HENDERSON

IBM Research Division, T. J. Watson Research Center,
Yorktown Heights, NY 10598, USA
mhender@watson.ibm.com

Received March 10, 2004; Revised June 8, 2004

A continuation method (sometimes called path following) is a way to compute solution curves of a nonlinear system of equations with a parameter. We derive a simple algorithm for branch switching at bifurcation points for multiple parameter continuation, where surfaces bifurcate along singular curves on a surface. It is a generalization of the parallel search technique used in the continuation code AUTO, and avoids the need for second derivatives and a full analysis of the bifurcation point.

The one parameter case is special. While the generalization is not difficult, it is nontrivial, and the geometric interpretation may be of some interest. An additional tangent calculation at a point near the singular point is used to estimate the tangent to the singular set.

Keywords: Numerical continuation; continuation methods; multiparameter branch switching; implicitly defined manifolds.

1. Background and Basic Result

A continuation method (sometimes called path following) is a way to compute solution curves of a nonlinear system of equations with a parameter. For an introduction to these methods see, for example [Allgower & Georg, 2003; Garcia & Zangwill, 1981] and more recently [Govaerts, 2000; Beyn *et al.*, 2002], and papers [Doedel, 1997] and [Seydel, 1997].

In [Henderson, 2002] the author described a generalization of these methods to problems with more than one parameter, where the solution manifolds are surfaces instead of curves. One practical issue that was not addressed there is how to generalize the second-derivative-free parallel search branch switching algorithm that is used in codes like AUTO [Keller, 2001; Keller & Doedel, 2003]. The one parameter case is special. While the generalization is not difficult, it is nontrivial, and the geometric interpretation may be of some interest.

Suppose Γ_0 is a regular connected component of the solution manifold of

$$\mathbf{F}(\mathbf{u}) = 0, \quad \mathbf{u} \in \mathbb{R}^n \quad \mathbf{F} : \mathbb{R}^n \to \mathbb{R}^{n-k}$$

containing the initial point $\mathbf{u}_0$ and restricted to some computational domain $\Omega \subset \mathbb{R}^n$. That is, a point $\mathbf{v}$ is in Γ_0 if there is a continuous curve $\mathbf{u}(s)$, $s \in [0, 1]$, of regular solutions of $\mathbf{F} = 0$ connecting $\mathbf{v}$ to $\mathbf{u}_0$ through Ω (see Fig. 1)

$$\mathbf{F}(\mathbf{u}(s)) = 0, \quad \mathbf{u}(s) \subset \Omega,$$

$$\mathrm{rank}(\mathbf{F}_{\mathbf{u}}(\mathbf{u}(s))) = n - k.$$

$$\mathbf{u}(0) = \mathbf{u}_0, \quad \mathbf{u}(1) = \mathbf{v}.$$

If $\mathbf{F}$ is smooth Γ_0 is a k-dimensional manifold with a boundary, and the boundary is made up of $(k-1)$-dimensional manifolds (again with boundaries) which either lie on $\delta\Omega$, or are such that the Jacobian $\mathbf{F}_{\mathbf{u}}$ is of rank $n - k - 1$.

Consider a point $\mathbf{u}^*$ on the singular boundary of Γ_0 (see Fig. 2). This point can be found by monitoring an indicator function $\chi(\mathbf{u})$, which changes

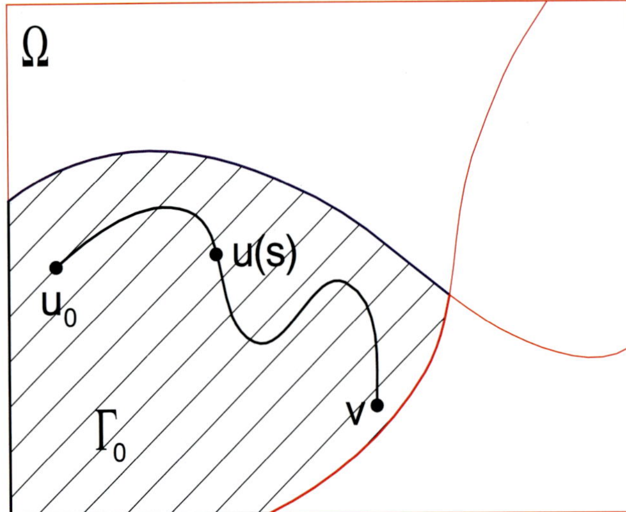

Fig. 1.   A regular connected component Γ_0 of $\mathbf{F} = 0$ in Ω. For every point $\mathbf{v}$ in Γ_0 there is a path $\mathbf{u}(s)$ of regular solutions connecting it to $\mathbf{u}_0$.

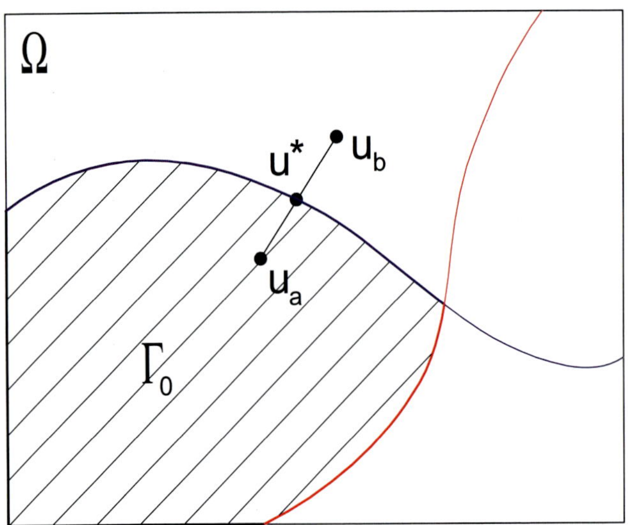

Fig. 2.   The singular boundary of a regular connected component. Points on this boundary, like $\mathbf{u}^*$, are found by monitoring a test function which changes between points on opposite sides of the boundary ($\mathbf{u}_a$ and $\mathbf{u}_b$). Bisection or a root finding algorithm may be used to locate $\mathbf{u}^*$.

sign or jumps when evaluated for points on opposite sides of a singular curve. (See [Beyn *et al.*, 2002] for a description of indicator functions for various bifurcations.) Bisection or a root finding algorithm may then be used to locate $\mathbf{u}^*$ in the interval $[\mathbf{u}_a, \mathbf{u}_b]$ where the indicator function changes. The aim of branch switching is to find points near $\mathbf{u}^*$ that are interior to the other regular connected components containing $\mathbf{u}^*$ (Fig. 3).

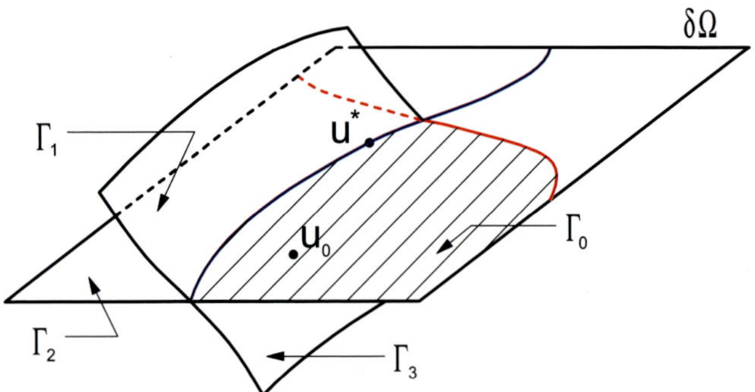

Fig. 3.   The regular connected components sharing a point $\mathbf{u}^*$ on their singular boundaries.

1.1.   *The geometry of the solution manifold near a singular point*

The tangent space of Γ_0 at the singular boundary point can be found by interpolation between the tangent spaces at $\mathbf{u}_a$ and $\mathbf{u}_b$ (which are regular points and have unique tangent spaces). We can therefore find an orthonormal basis $\{\phi_0, \ldots, \phi_{k-1}\}$ for the k-dimensional tangent space of Γ_0 at $\mathbf{u}^*$

$$\mathbf{F_u}(\mathbf{u}^*)\phi_i = 0$$

$$\phi_i^T \phi_j = \delta_{ij}$$

If $\mathbf{u}^*$ is interior to the singular boundary, the rank of $\mathbf{F_u}(\mathbf{u}^*)$ will be $n - k - 1$, and so there is a right null vector $\phi_k \in \mathbb{R}^n$, and left null vector $\psi \in \mathbb{R}^{n-k}$ of the augmented Jacobian

$$\mathbf{F_u}(\mathbf{u}^*)\phi_k = 0 \quad \psi^T \mathbf{F_u}(\mathbf{u}^*) = 0$$

$$\phi_i^T \phi_k = 0 \ (i \neq k)$$

$$\phi_k^T \phi_k = 1 \quad \psi^T \psi = 1$$

To study the geometry of the bifurcation we use a Lyapunov–Schmidt decomposition. Let

$$\mathbf{u} = \mathbf{u}^* + \sum_{i=0}^{k} \phi_i s^i + \eta, \quad \phi_i^T \eta = 0,$$

and consider first the projection of $\mathbf{F}$ onto the range of the Jacobian

$$(I - \psi\psi^T)\mathbf{F}\left(\mathbf{u}^* + \sum_{i=0}^{k} \phi_i s^i + \eta\right) = 0$$

$$\phi_i^T \eta = 0$$

as a system for η. The Jacobian at the solution $s^i = 0$, $\eta = 0$ is nonsingular, so using the Implicit Function Theorem (IFT) there is a unique function $\eta(s^0, \ldots, s^k)$ which satisfies the projected equations in a neighborhood of $s^i = 0$. At $s^i = 0$ we have

$$\eta = 0, \quad \eta_{s^i} = 0$$

$$(I - \psi\psi^T)(\mathbf{F_u}\eta_{s^i,s^j} + \mathbf{F_{uu}}\phi_i\phi_j) = 0$$

$$\phi_l^T \eta_{s^i,s^j} = 0$$

(Similar equations can be written for the higher derivatives of η by repeated differentiation.)

To satisfy $\mathbf{F} = 0$, one further scalar equation must be satisfied (the Bifurcation Equation Eq. (1))

$$\psi^T \mathbf{F}\left(\mathbf{u}^* + \sum_{i=0}^{k} \phi_i s^i + \eta(s^0, \ldots, s^k)\right) = 0. \quad (1)$$

The linearization of this is zero at $\mathbf{s} = 0$, and we can remove this (so that the IFT can be used), by introducing a small parameter ϵ –

$$\frac{1}{\epsilon^2}\psi^T \mathbf{F}\left(\mathbf{u}^* + \epsilon\sum_{i=0}^{k} \phi_i s^i + \eta(\epsilon s^0, \ldots, \epsilon s^k)\right) = 0,$$

$$s^i s^i = 1.$$

A Taylor series (in ϵ) of this begins:

$$\frac{1}{\epsilon^2}\psi^T \mathbf{F} = \sum_{i,j}\psi^T \mathbf{F_{uu}}\phi_i\phi_j s^i s^j$$

$$+ \epsilon\sum_{i,j,l}\left(\psi^T \mathbf{F_{uuu}}\phi_i\phi_j\phi_l s^i s^j s^l\right.$$

$$\left. + \psi^T \mathbf{F_{uu}}\phi_l\eta_{s^i,s^j}s^l\right) + \cdots$$

Suppose that the Algebraic Bifurcation Equation (ABE) Eq. (2)

$$\sum_{i,j}\psi^T \mathbf{F_{uu}}\phi_i\phi_j s^i s^j = 0 \quad (2)$$

is satisfied and the first-order term is nonzero

$$\sum_{i,j,l}\psi^T \mathbf{F_{uuu}}\phi_i\phi_j\phi_l s^i s^j s^l + \sum_{i,j,l}\psi^T \mathbf{F_{uu}}\phi_l\eta_{s^i,s^j}s^l \neq 0.$$

Using the IFT, a set of functions $s^i(\epsilon)$ with $s^i(0) = s^i$ exists in a neighborhood of $\epsilon = 0$. Each solution of the ABE therefore corresponds to a curve (parameterized by ϵ) on the solution surface through $\mathbf{u}^*$. Varying the s^i subject to the ABE traces out the surface.

We know one set of solutions — any vector s with $s^k = 0$. (This is because we chose the first k null vectors to be a basis for the tangent space of Γ_0.) The ABE is therefore of the form

$$s^k\left(\sum_{i=0}^{k}\psi^T \mathbf{F_{uu}}\phi_i\phi_k s^i\right) = 0, \quad \text{or}$$

$$s^k\left(\sum_{i=0}^{k} N_i s^i\right) = 0.$$

Therefore $\mathbf{N} \in \mathbb{R}^{k+1}$ is orthogonal to the bifurcating branch, and $N_i = \psi^T \mathbf{F_{uu}}\phi_i\phi_k$. The tangent space of the singular boundary is

$$s^k = 0, \quad \sum_{i=0}^{k} N_i s^i = 0$$

Let $\{\sigma_0, \ldots, \sigma_{k-2}\}$ be an orthonormal basis for this $(k - 1)$-dimensional tangent space. The tangent space to the bifurcating sheet includes the additional vector $\sigma_{k-1} = (N_k N_0, \ldots, N_k N_{k-1}, -\sum_0^{k-1} N_i N_i)$. This is orthogonal to both $\mathbf{N}$ and the other σ_i. (It is not normalized.)

1.2. *Special case: $k = 1$*

When $k = 1$ the singular set is a point (see Fig. 4). We have

$$(N_0, N_1) = (\psi^T \mathbf{F_{uu}}\phi_0\phi_1, \psi^T \mathbf{F_{uu}}\phi_1\phi_1),$$

and the tangent (not normalized) of the bifurcating branch in $\mathbf{s}$-space is

$$\sigma_{k-1} = N_0(N_1, -N_0)$$

1.3. *Special case: $k = 2$*

The vector (s^0, s^1, s^2) gives a point in the $k + 1 = 3$ dimensional null space of $\mathbf{F_u}(\mathbf{u}^*)$ (in the

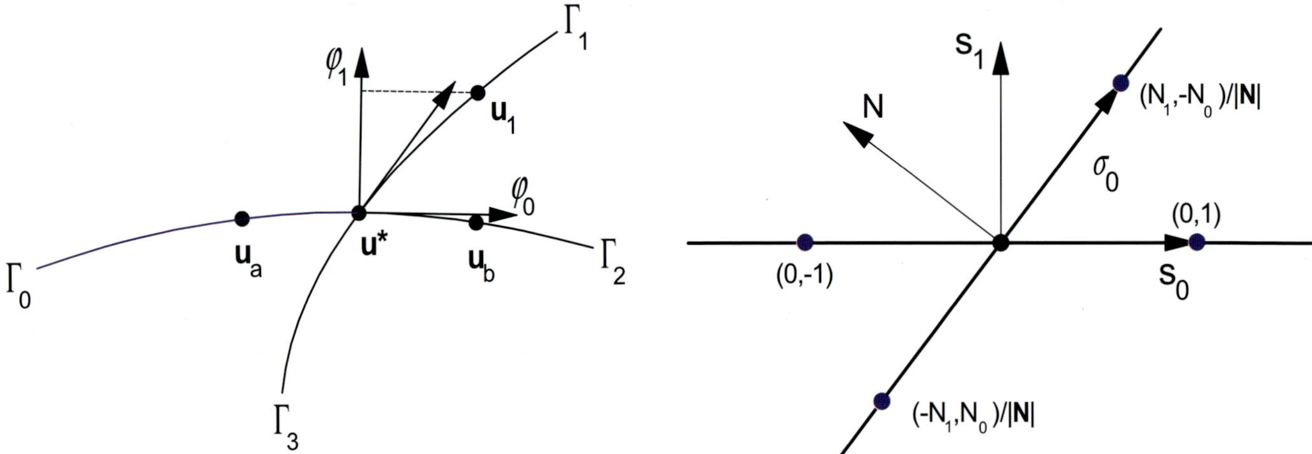

Fig. 4.   (Left) The parallel search branch switching algorithm used in AUTO. A bifurcation point $\mathbf{u}^*$ is located between $\mathbf{u}_a$ and $\mathbf{u}_b$, and the null vector of the augmented Jacobian ϕ_1 is used as a tangent for the next step. (Right) A sketch of the corresponding s-space, showing the four roots of the ABE.

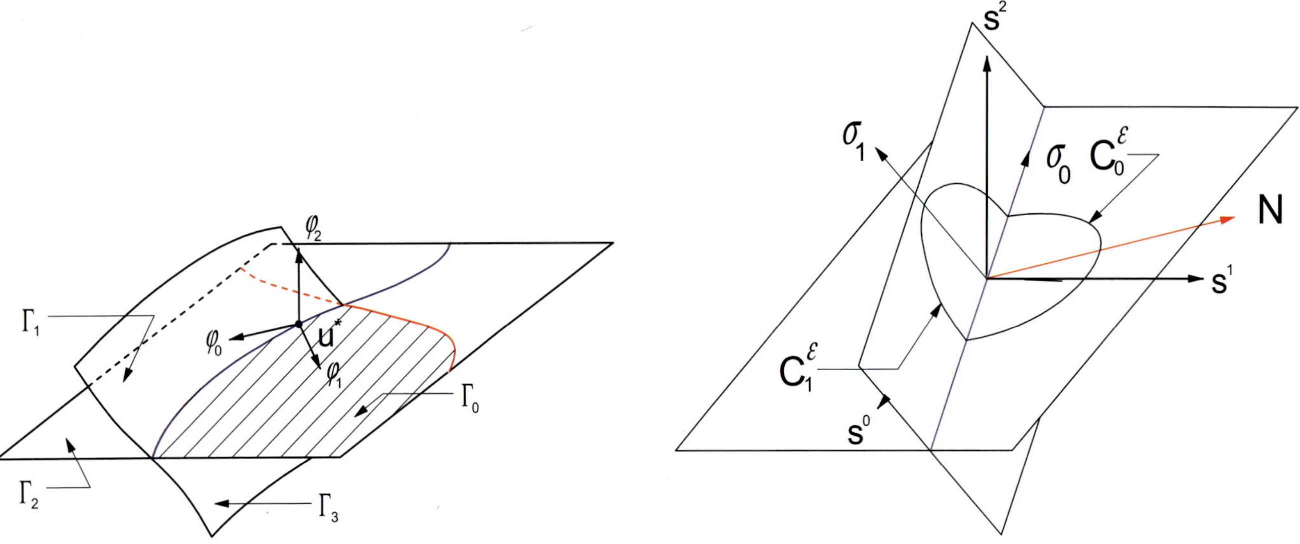

Fig. 5.   (Left) The basis for the right null space of $\mathbf{F_u(u^*)}$. The first two $(k-1)$ basis vectors lie in the tangent plane of Γ_0 and Γ_2. The third is orthogonal to the first two. (Right) Solutions to the ABE's in s-space. Circle C_0 is $s^2 = 0$, $s^0 s^0 + s^1 s^1 = 1$. Circle C_1 lies in the plane $N.s = 0$, where $N = (\psi^T \mathbf{F_{uu}}\phi_0\phi_2, \psi^T \mathbf{F_{uu}}\phi_1\phi_2, \psi^T \mathbf{F_{uu}}\phi_2\phi_2)$. The singular set is $N.s = 0$, $s^k = 0$.

basis ϕ_0, ϕ_1, ϕ_2) (Fig. 5). We have

$$(N_0, N_1, N_2)$$
$$= (\psi^T \mathbf{F_{uu}}\phi_0\phi_2, \psi^T \mathbf{F_{uu}}\phi_1\phi_2, \psi^T \mathbf{F_{uu}}\phi_2\phi_2).$$

The branch corresponding to Γ_0 (and Γ_2) is $s^2 = 0$

$$\mathbf{u}^* + \epsilon \left(\phi_0 s^0 + \phi_1 s^1\right) + \eta(\epsilon s^0, \epsilon s^1, 0),$$
$$s^0 s^0 + s^1 s^1 = 1$$

The other branches (Γ_1 and Γ_3) are

$$\mathbf{u}^* + \epsilon \left(\phi_0 s^0 + \phi_1 s^1 + \phi_2 s^2\right) + \eta(\epsilon s^0, \epsilon s^1, \epsilon s^2),$$
$$N_0 s^0 + N_1 s^1 + N_2 s^2 = 0$$
$$s^0 s^0 + s^1 s^1 + s^2 s^2 = 1$$

See Fig. 5. The tangent to the singular set is, in s-space and $\mathbb{R}^n$

$$\sigma_0 = (-N_1, N_0, 0) \quad \leftrightarrow \quad -N_1\phi_1 + N_0\phi_0$$

and the tangent vector (not normalized, and in s-space) to the bifurcating branch orthogonal to the singular set is

$$\sigma_1 = (N_0 N_2, N_1 N_2, -N_0 N_0 - N_1 N_1)$$

1.4. *Parallel search branch switching*

These quantities can be computed, and the tangent to the bifurcating components found directly. However, in many instances the second derivatives are not available, and we need a branch switching algorithm which does not assume they are.

The goal is to find a point on the bifurcating component which can be used as an initial point to compute the component. To project a point $\mathbf{v}$ onto a regular component, a system of the form

$$\mathbf{F}(\mathbf{u}) = 0$$
$$\tilde{\Phi}^T(\mathbf{u} - \mathbf{v}) = 0$$

is used. As long as the projection of the k vectors $\tilde{\Phi}$ (the columns) onto the null space of the Jacobian at $\mathbf{u}$ spans the null space this is a nonsingular system. As the name parallel search implies, we choose $\tilde{\Phi}$ orthogonal to the tangent to Γ_0. The condition that the augmented Jacobian be nonsingular is that the bifurcation be transverse to Γ_0.

For $k = 1$, ϕ_1 has a nonzero projection onto the tangent to the bifurcating curve $N_1\phi_0 - N_0\phi_1$ if

$N_0 \neq 0$ (the nontransverse case). So a point on the bifurcating branch may be found by solving

$$\mathbf{F}(\mathbf{u}) = 0$$
$$\phi_1^T(\mathbf{u} - (\mathbf{u}^* + \Delta s\phi_1)) = 0$$

This is the technique used in AUTO (described in [Beyn *et al.*, 2002] and [Keller, 2001]).

For $k > 1$ we need to find a k-dimensional subspace whose projection onto the tangent space of the bifurcating sheet spans that tangent space. For $k = 1$ we projected orthogonal to ϕ_k. This defines a $(k-1)$-dimensional curve on the bifurcating sheet, and so if $k \neq 1$ we need additional constraints to define a unique point on the bifurcating sheet. The tangent to the singular set lies in both tangent spaces, and we need to project orthogonal to that as well.

To estimate the tangent to the singular set we use a perturbation

$$\tilde{\mathbf{F}}(\mathbf{u}) = \mathbf{F}(\mathbf{u}) - \mathbf{F}(\mathbf{u}^* + \epsilon\Delta\phi_k).$$

For this (as before $N_i = \psi^T \mathbf{F}_{\mathbf{uu}}\phi_i\phi_k$)

$$\frac{1}{\epsilon^2}\psi^T \tilde{\mathbf{F}} = \sum_{i=0}^{k} N_i s^i s^k - N_k \Delta^2 + O(\epsilon).$$

In s-space solutions of this perturbed equation are a pair of hyperbolic sheets which asymptote to the solutions of the unperturbed equation (Fig. 6).

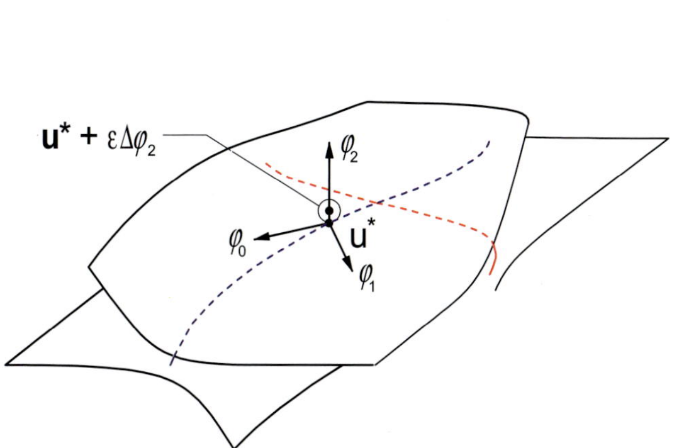

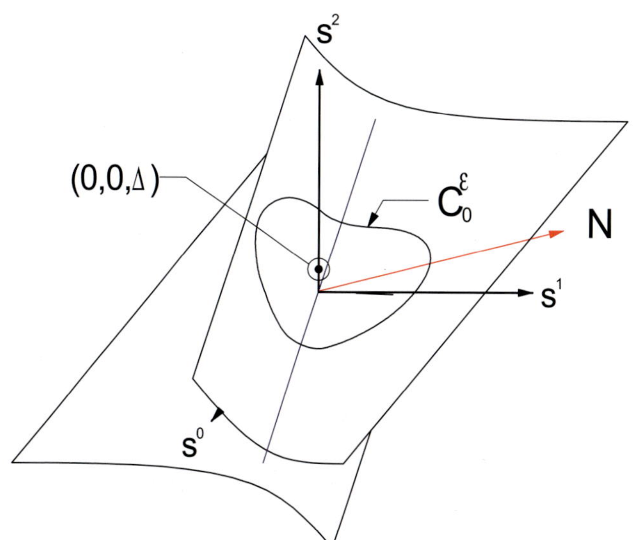

Fig. 6. (Left) The perturbed surface $\mathbf{F}(\mathbf{u}) = \mathbf{F}(\mathbf{u}^* + \epsilon\Delta\phi_k)$. (Right) The same surface in the s-space defined by the unperturbed problem.

Suppose σ is any vector in the tangent space of the singular set. That is

$$\sum_{i=0}^{k} N_i \sigma^i = 0 \quad \sigma^k = 0$$

By construction, we know one point on the perturbed surface, $\mathbf{u}^* + \epsilon\phi_k$, which corresponds to the point $(0, \ldots, 0, \Delta)$ in the s-space since it is a solution of

$$\sum_{i=0}^{k} N_i s^i s^k = \Delta^2 N_k.$$

This equation is invariant to a shift in the σ direction, so σ lies in the tangent space of the perturbed system at the known point $(0, \ldots, 0, \Delta)$. This gives us a way to compute the tangent space of the singular set: it is the common $(k-1)$-dimensional subspace of the tangent to Γ_0 and the tangent of the perturbed system (the null space of $\mathbf{F}_\mathbf{u}(\mathbf{u}^* + \epsilon\phi_k)$).

2.   Statement of the Algorithm

Detection — Locate a pair of points $\mathbf{u}_a$ and $\mathbf{u}_b$ on $\mathbf{F} = 0$ such that $\chi(\mathbf{u}_a) \neq \chi(\mathbf{u}_b)$.

Location — Using bisection, or a root finding method locate the point $\mathbf{u}^*$ on $\mathbf{F} = 0$ in the interval at which $\chi(\mathbf{u})$ changes. Use the tangent spaces at $\mathbf{u}_a$ and $\mathbf{u}_b$ to interpolate $(\phi_0, \ldots, \phi_{k-1})$, an orthonormal approximation to a basis for the tangent space of Γ_0 at $\mathbf{u}^*$.

Branch Switching —

(1) Find the right null vector ϕ_k, $\phi_k^T \phi_k = 1$

$$\mathbf{F}_\mathbf{u}(\mathbf{u}^*)\phi_k = 0$$

$$\phi_i^T \phi_k = 0, \quad i = 0, \ldots, k-1$$

(2) Find an orthonormal basis $\tilde{\Phi}$ of the k-dimensional null space of

$$\mathbf{F}_\mathbf{u}(\mathbf{u}^* + \epsilon\Delta\phi_k)\tilde{\phi}_i = 0$$

$$\tilde{\phi}_i^T \tilde{\phi}_j = \delta_{ij}$$

(3) Find an orthonormal basis $\{\sigma_0, \ldots, \sigma_{k-2}\}$ for the common subspace of Φ and $\tilde{\Phi}$. Since ϕ_k is orthogonal to Φ this is the subspace $\phi_k^T \tilde{\Phi} = 0$, so the Gram–Schmidt algorithm can be used on the set of $k+1$ vectors $\{\phi_k, \tilde{\phi}_0, \ldots, \tilde{\phi}_{k-1}\}$ to find the subspace (the first will be ϕ_k, and the last will be zero. The ones in between are the σ_i).

(4) Find points on the bifurcating sheets by solving

$$\mathbf{F}(\mathbf{u}) = 0$$

$$\sigma_i^T(\mathbf{u} - (\mathbf{u}^* + \Delta s\phi_k)) = 0, \quad i = 0, \ldots, k-2$$

$$\phi_k^T(\mathbf{u} - (\mathbf{u}^* + \Delta s\phi_k)) = 0$$

With $\Delta s > 0$ we get a point on Γ_1, and $\Delta s < 0$ gives a point on Γ_3. For the point on Γ_2 we can use $\mathbf{u}_b$, which was found in the detection step.

Notes:

- Δ controls the shape of the hyperbola in s-space, and ϵ is small relative to the norm of $\mathbf{u}^*$. Therefore $\epsilon\Delta$ should be something like $10^{-3}|\mathbf{u}^*|$.
- There is a technique, described in [Allgower & Georg, 2003] which perturbs the problem in order to switch branches. This approach does that in some sense by using the tangent of a perturbation.
- For Hopf and other bifurcations the null vector ϕ_k at the singular point is of a different class than the other null vectors. For example, $\phi_0, \ldots, \phi_{k-1}$ may be in $\mathbb{R}^n$, while ϕ_k is in $\mathbb{R}^n \times S^1$. All this means is that the other null vectors must be promoted to the larger space, since the tangent space $\tilde{\Phi}$ at the perturbed point is in the larger space.

3.   Examples

3.1.   *Cusp*

Our first example is a complexified cusp [Henderson & Keller, 1990].

$$(x + iy) \cdot ((x + iy)^2 + \lambda) = \mu.$$

This is $n = 4$, $k = 2$, and we can easily find an initial solution $x_0 = y_0 = 0$ at $\mu_0 = 0$, $\lambda_0 = 1$. Figure 7 shows $x + y$ as a function of (λ, μ). The single initial point, with the branch switching algorithm described in the preceding section was sufficient to compute the four regular connected components. Note that the blue components ($y = 0$) are the cusp catastrophe.

3.2. $(2, 4)$ *cell interaction model*

Our second example is a model of the $(2, 4)$ cell mode interaction in Taylor–Couette flow [Meyer-Spasche, 1991, pp. 106–110]. It is based on an analysis of a two eigenvalue bifurcation by

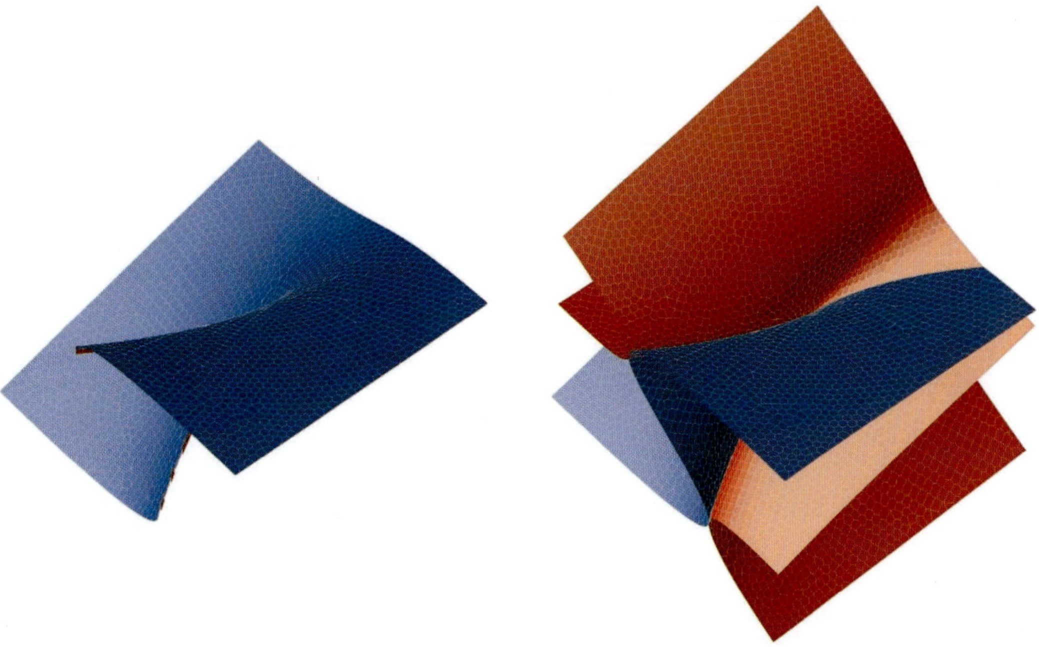

Fig. 7.   A computation of solutions of $u(u^2 - \lambda) = \mu$. The projection used for rendering is $(\mu, \lambda, x + y)$. (Left) Γ_0, the regular component connected to the initial point. (Right) All components ($y \neq 0$ is red, $y = 0$ is blue).

Andreichikov [1979], with coefficients computed by Bolstad [1992]. The computation of the coefficients is as described in [Ramaswamy & Keller, 1995]. At radius ratio $\eta = 0.615$ and a 12×48 grid the bifurcation point was found to be at Reynolds number $R = 78.53836$, aspect ratio $\lambda = 2.881799$. The model is

$$x(x^2 + a_1 y^2 - f_1 + b_1 y) = 0$$
$$y(a_2 x^2 + y^2 - f_2) + b_2 x^2 = 0$$

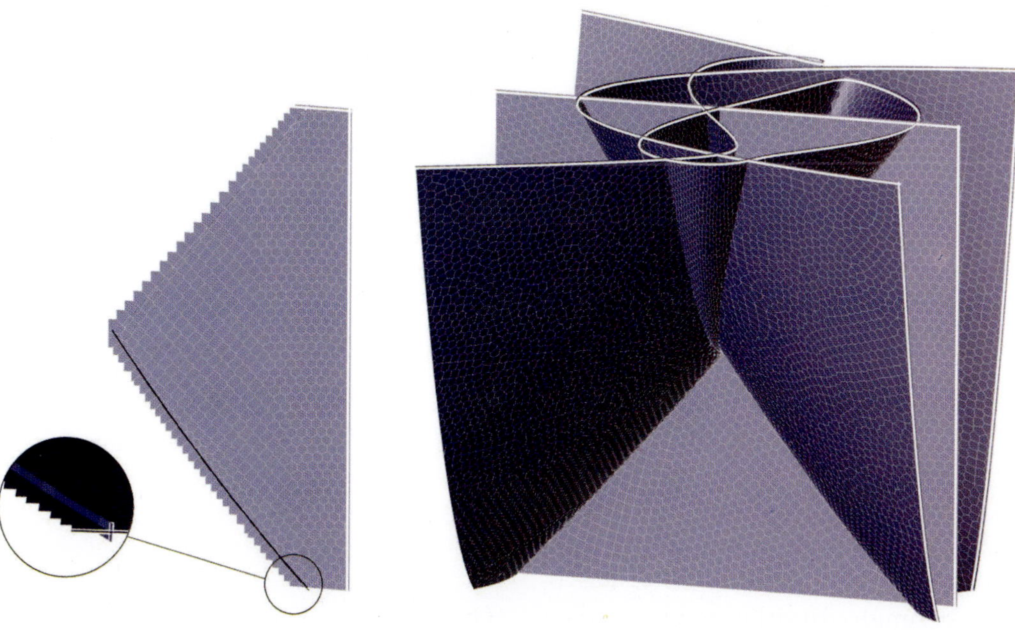

Fig. 8.   A computation of solutions of a Model for the (2,4) mode interaction in Taylor–Couette flow. The projection used for rendering is $(\Delta R, \Delta \lambda, x + y)$. (Left) Γ_0, the regular component connected to the initial point. The inset shows tiles created using $\{\sigma_0, \ldots, \sigma_{k-2}, \phi_k\}$ (right) all components.

where

$$a_1 = 3.67$$

$$b_1 = -0.0975 - 0.00392\Delta R + 0.0543\Delta\lambda$$

$$f_1 = 0.00117\Delta R - 0.0137\Delta\lambda - 0.00000427\Delta R^2$$
$$- 0.000407\Delta R\Delta\lambda + 0.00106\Delta\lambda^2$$

$$a_2 = 1.19$$

$$b_2 = 0.0331 + 0.000476\Delta R - 0.00546\Delta\lambda$$

$$f_2 = 0.000681\Delta R + 0.00955\Delta\lambda - 0.000002605^2$$
$$- 0.000216\Delta R\Delta\lambda - 0.004925\Delta\lambda^2.$$

The solution manifold consists of three pieces with different symmetries (Fig. 8):

$x = 0, \quad y = 0$ — The trivial solutions.

$x = 0, \quad y \neq 0$ — The 4-cell solutions.

$$y^2 - f_2(\Delta R, \Delta\lambda) = 0.$$

$x \neq 0, \quad y \neq 0$ — The mixed 2-cell/4-cell solutions.

References

Allgower, E. L. & Georg, K. [2003] *Introduction to Numerical Continuation Methods*, Classics in Applied Mathematics, Vol. 45 (SIAM, Philadelphia).

Andreichikov, I. P. [1979] "Branching of secondary modes in the flow between rotating cylinders," *Fluid Dyn.* **12**, 38–43.

Beyn, W.-J., Champneys, A., Doedel, E., Govarets, W., Kuznetsov, U. A., Yu, A. & Sandstede, B. [2002] *Numerical Continuation, and Computation of Normal Forms*, Handbook of Dynamical Systems, Vol. 2 (Elsevier Science).

Bolstad, J. [1992] Private communication.

Doedel, E. J. [1997] "Nonlinear numerics," *Int. J. Bifurcation and Chaos* **7**, 2127–2143.

Garcia, C. B. & Zangwill, W. I. [1981] *Pathways to Solutions, Fixed Points and Equilibria* (Prentice-Hall).

Govaerts, W. J. F. [2000] *Numerical Methods for Bifurcations of Dynamical Equilibria* (SIAM, Philadelphia).

Henderson, M. E. & Keller, H. B. [1990] "Complex bifurcation from real paths," *SIAM J. Appl. Math.* **50**, 460–482.

Henderson, M. E. [2002] "Multiple parameter continuation: Computing implicitly defined k-manifolds," *Int. J. Bifurcation and Chaos* **12**, 451–476.

Keller, H. B. [2001] "Continuation and bifurcations in scientific computation," *Math. TODAY* **76**, 493–520.

Keller, H. B. & Doedel, E. J. [2003] "Path following in scientific computing and its implementation in AUTO," in *Sourcebook of Parallel Computing*, eds. Dongarra, J., Foster, I., Fox, G., Gropp, W., Kennedy, K. Torczon, L. & White A. (Morgan Kaufman, San Francisco), Chap. 23, pp. 670–700.

Meyer-Spasche, R. [1991] *Pattern Formation in Viscous Flows* (Springer-Verlag, NY).

Ramaswamy, M. & Keller, H. B. [1995] "A local study of a double critical point in Taylor-Couette flow," *Acta Mech.* **109**, 27–39.

Seydel, R. [1997] "Nonlinear computation," *Int. J. Bifurcation and Chaos* **7**, 2105–2126.

EQUATION-FREE, EFFECTIVE COMPUTATION FOR DISCRETE SYSTEMS: A TIME STEPPER BASED APPROACH

J. MÖLLER and O. RUNBORG

Department of Numerical Analysis and Computer Science, KTH,
10044 Stockholm, Sweden

P. G. KEVREKIDIS

Department of Mathematics and Statistics, University of Massachusetts,
Amherst, MA 01003, USA

K. LUST

Departement Computerwetenschappen, Katholieke Universiteit Leuven,
Celestijnenlaan 200A, B-3001 Heverlee, Belgium

I. G. KEVREKIDIS

Department of Chemical Engineering,
Program for Applied and Computational Mathematics,
Department of Mathematics,
Princeton University, Princeton, NJ 08544, USA

Received May 5, 2004; Revised August 3, 2004

We propose a computer-assisted approach to studying the effective continuum behavior of spatially discrete evolution equations. The advantage of the approach is that the "coarse model" (the continuum, effective equation) need not be explicitly constructed. The method only uses a time-integration code for the discrete problem and judicious choices of initial data and integration times; our bifurcation computations are based on the so-called Recursive Projection Method (RPM) with arc-length continuation [Shroff & Keller, 1993]. The technique is used to monitor features of the genuinely discrete problem such as the pinning of coherent structures and its results are compared to quasi-continuum approaches such as the ones based on Padé approximations.

Keywords: Equation-free methods; homogenization; discrete problems; bifurcation; pinning condition.

1. Introduction

In contemporary science and engineering modeling many situations arise in which the physical system consists of a lattice of discrete interacting units. The role of discreteness in modifying the behavior of solutions of continuum nonlinear PDEs has recently been increasingly appreciated. The relevant physical contexts can be quite diverse, ranging from the calcium burst waves in living cells [Dawson

et al., 1999] to the propagation of action potentials through the tissue of the cardiac cells [Keener, 1991] and from chains of chemical reactions [Laplante & Erneux, 1992] to applications in superconductivity and Josephson junctions [Ustinov *et al.*, 1993], nonlinear optics and waveguide arrays [Christodoulides & Joseph, 1988], complex electronic materials [Swanson *et al.*, 1999], the dynamics of neuron chains or lattices [Rinzel *et al.*, 1998; McLaughlin *et al.*, 2000] or the local denaturation of the DNA double strand [Peyrard & Bishop, 1989].

Whether the phenomenon in question is the propagation of an excitation wave along a neuron lattice, the electric field envelope in an optical waveguide array, or the behavior of a tissue consisting of an array of individual cells, we would often like to model the system through a "coarse level" effective continuum evolution equation that retains the essential features of the actual (discrete) problem. Typically computational modeling of such systems involves two steps: the derivation of effective continuum equations, followed by their analysis through traditional numerical tools. In this paper we attempt to circumvent the derivation of explicit (closed) continuum effective equations, and analyze the effective behavior directly. This is accomplished through short, appropriately initialized simulations of the detailed discrete process, a procedure that we call the "coarse time stepper". These simulations provide estimates of the quantities (residuals, action of Jacobians, time derivatives, Fréchet derivatives) that would be directly evaluated from the effective equation, had such an equation been available. The estimated quantities are processed by a higher level numerical procedure (in this case, the Recursive Projection Method, RPM, of [Shroff & Keller, 1993]) which computes the effective, macroscopic behavior (in this case, traveling waves and their coarse bifurcations). A more general discussion of the combination of coarse time stepping with continuum numerical techniques beyond RPM can be found in [Gear *et al.*, 2002; Kevrekidis *et al.*, 2003]. We have recently demonstrated such an approach to the computation of the effective behavior (in some sense, homogenization) of spatially heterogeneous problems [Runborg *et al.*, 2002]. This paper constitutes an extension of this idea to spatially discrete problems.

The paper is organized as follows: We begin with a brief review of the coarse time stepper for spatially discrete problems. We then discuss our illustrative problem (a front in a discrete reaction–diffusion system) and its properties. A description of our implementation of the coarse time stepper for the bifurcation analysis of this particular problem is then presented, followed by numerical results. We conclude with a discussion of an alternative approach that involves the derivation of an explicit effective evolution equation (based on Padé approximations), and of the scope and applicability of our method.

2. A Coarse Time Stepper for Discrete Systems

Consider a discrete system where each unknown is associated with a point on a lattice in space. In the discussion here, we consider a one-dimensional regular lattice for simplicity. Higher dimensional and/or possibly irregular, lattices can be treated in a similar way. We denote the unknowns $\{u_\ell\}$, with $\ell \in \mathbb{Z}$, and the corresponding points $\{x_\ell\}$, such that $x_\ell = \ell \Delta x$, where Δx is the lattice spacing. We assume that the system is governed by the ordinary differential equations

$$\frac{du_\ell}{dt} = F(t, u_{\ell-n}, \ldots, u_{\ell+n}), \quad \ell \in \mathbb{Z}, \qquad (1)$$

where $n > 0$ is an integer representing the range of interaction between lattice points. We want to describe this discrete system dynamics through a continuous function $v(t, x)$ that models the "coarse" behavior of the unknowns on the lattice:

$$u_\ell(t) \approx v(t, x_\ell), \quad \forall\, t, \ell,$$

in some appropriate sense. We denote v as the *coarse continuous solution* of (1) and we assume that n is not large and that there exists an effective, spatially continuous evolution equation for $v(x, t)$ of the form

$$v_t = P\left(t, v, \partial_x v, \ldots, \partial_x^M v\right), \qquad (2)$$

for some P and integer M. Such an effective equation for v should "average over" the detailed discrete structure of the medium; if there are no macroscopic variations of the discrete medium, this equation should therefore be translationally invariant; for the moment, we will confine ourselves to this case. In terms of (1), we can express this as: if F does not depend on ℓ, and if v and $\tilde{v}$ are two solutions to the effective equation (2) satisfying $v(0, x) = \tilde{v}(0, x + s)$ for all x, then $v(t, x) = \tilde{v}(t, x + s)$ for all time $t > 0$, all x, and all shifts s.

It is interesting to consider what the result of integrating such an effective equation with a particular, continuum initial condition $v_0(x)$, would physically mean. There clearly exists an uncertainty in how such a continuum initial condition would be imparted to (sampled by) the lattice. One way would be to set $u_\ell(0) = v_0(x_\ell)$, for all ℓ, but we could equally well set $u_\ell(0) = v_0(x_\ell + s)$ for any $s \in [0, \Delta x)$. There exists, therefore, a one-parameter uncertainty parametrized by a continuous shift s.

Simulations resulting from different lattice samplings of the same continuum initial condition could be quite different. This is best illustrated by thinking of a single-peaked function as the continuum initial condition: the peak may lie precisely at a lattice point, or could fall in-between lattice points. It is reasonable to consider as an useful effective continuum equation one which takes into account all possible shifts of the initial condition within a cell; in analogy with our earlier work [Runborg *et al.*, 2002], we would like to analyze an effective equation that would describe the expected result — taken over all possible shifts — of sampling the initial condition by the lattice.

We will use the coarse time stepper approach to simulate an effective equation like (2). In this setting, we approximate $v(t, x)$ by the coarse time stepper solution $\tilde{u}(t, x)$ at discrete times nT, where T is the *time horizon* of the coarse time stepper. Using the terminology of this framework, we take the following steps, starting from a continuous initial condition $v_0(x) = \tilde{u}(0, x)$.

- *Lifting.* This initial data $v_0(x)$ is "lifted" to an ensemble of N_c different initial states of (1) by sampling,

$$u_\ell^j(0) = v_0(x_\ell + j\Delta s), \quad \Delta s = \Delta x / N_c,$$

$$j = 0, \ldots, N_c - 1. \tag{3}$$

Setting $\mathbf{u}_j = \{u_\ell^j\}$, we write this symbolically as

$$\mathbf{u}_j(0) = \mu_j v_0,$$

where $\{\mu_j\}$ are called the lifting operators. In this case they simply sample a continuous function.
- *Evolve.* Each ensemble of initial data is evolved till time T according to the "true dynamics" (1),

$$\mathbf{u}_j(T) = \mathcal{T}_T \mathbf{u}_j(0), \quad j = 0, \ldots, N_c - 1. \tag{4}$$

where $\mathcal{T}_\tau$ is the solution operator of (1) evolving $\mathbf{u}(t)$ to $\mathbf{u}(t + \tau)$. This step thus generates an ensemble of solutions $\mathbf{u}_j(T)$ at time T.
- *Restrict.* Via the restriction operator $\mathcal{M}$, the ensemble of solutions is brought back to a continuous function.

$$\tilde{u}(T, x) = \mathcal{M}\{\mathbf{u}_j(T)\}, \quad j = 0, \ldots, N_c - 1. \tag{5}$$

To ensure consistency we require that $\mathcal{M}\{\mu_j\} = I$. The restriction operator $\mathcal{M}$ is typically defined

as follows. The solutions $\mathbf{u}_j(T)$ are thought of as sample values of a function $\overline{u}$ such that $\overline{u}(x_\ell + j\Delta s) = u_\ell^j$. The function $\overline{u}$ is recovered by interpolating the sample values and the restriction $\tilde{u}(x, T) = \mathcal{M}\{\mathbf{u}_j(T)\}$ is finally given as a coarse scale filtering of $\overline{u}(x)$.

These steps are illustrated in Fig. 1. For $n > 0$ we define $\tilde{u}(nT, x)$ recursively by applying the same construction. Hence,

$$\tilde{u}(nT, x) = \mathcal{M}\{\mathcal{T}_T \mu_j\}\tilde{u}((n - 1)T, x). \tag{6}$$

The hope is that the coarse time stepper solution $\tilde{u}(nT, x)$, at these discrete points in time, can be obtained from a closed evolution equation like (2) whose solution, $v(t, x)$ (defined for all t), agrees, at least approximately, with the coarse solution obtained from the procedure above, at the discrete points in time, $v(nT, x) \approx \tilde{u}(nT, x)$. We will refer to the procedure as the *coarse time stepper.*

In order to approximate v numerically, we must use a finite representation of $\tilde{u}(nT, x)$. We let $\mathbf{v}^n = \{v_k^n\}_{k=0}^{M-1}$, be this representation at time $t = nT$. The elements $\{v_k^n\}$ could be nodal values, cell averages or, more generally, coefficients for finite elements or other basis functions. Let Π be the operator realizing the function from the finite representation, $(\Pi \mathbf{v}^n)(x) = \tilde{u}(nT, x)$. We also require that the restriction operator projects on the subspace spanned by the finite representation, and we can redefine it to also convert the projected function to this representation. Symbolically, we then write the coarse time stepping

$$\mathbf{v}^{n+1} = \mathcal{M}\{\mathcal{T}_T \mu_j\}\Pi \mathbf{v}^n =: G(\mathbf{v}^n). \tag{7}$$

Note that we may not be able to write down the explicit expression for G or Eq. (2) for $v(t, x)$, but our definition of $\tilde{u}(t, x)$ allows us to realize its time-T map numerically in a straightforward fashion.

Applied directly to the simulation, the coarse time stepper does nothing to reduce the cost of detailed computation with the discrete dynamics. It is only in conjunction with other techniques (like projective integration [Gear & Kevrekidis, 2003], or matrix-free fixed point techniques) that the coarse time stepper may provide computational or analytical benefits. Here we will make use of the coarse time stepper in conjunction with the Recursive Projection Method (RPM), to perform stability and

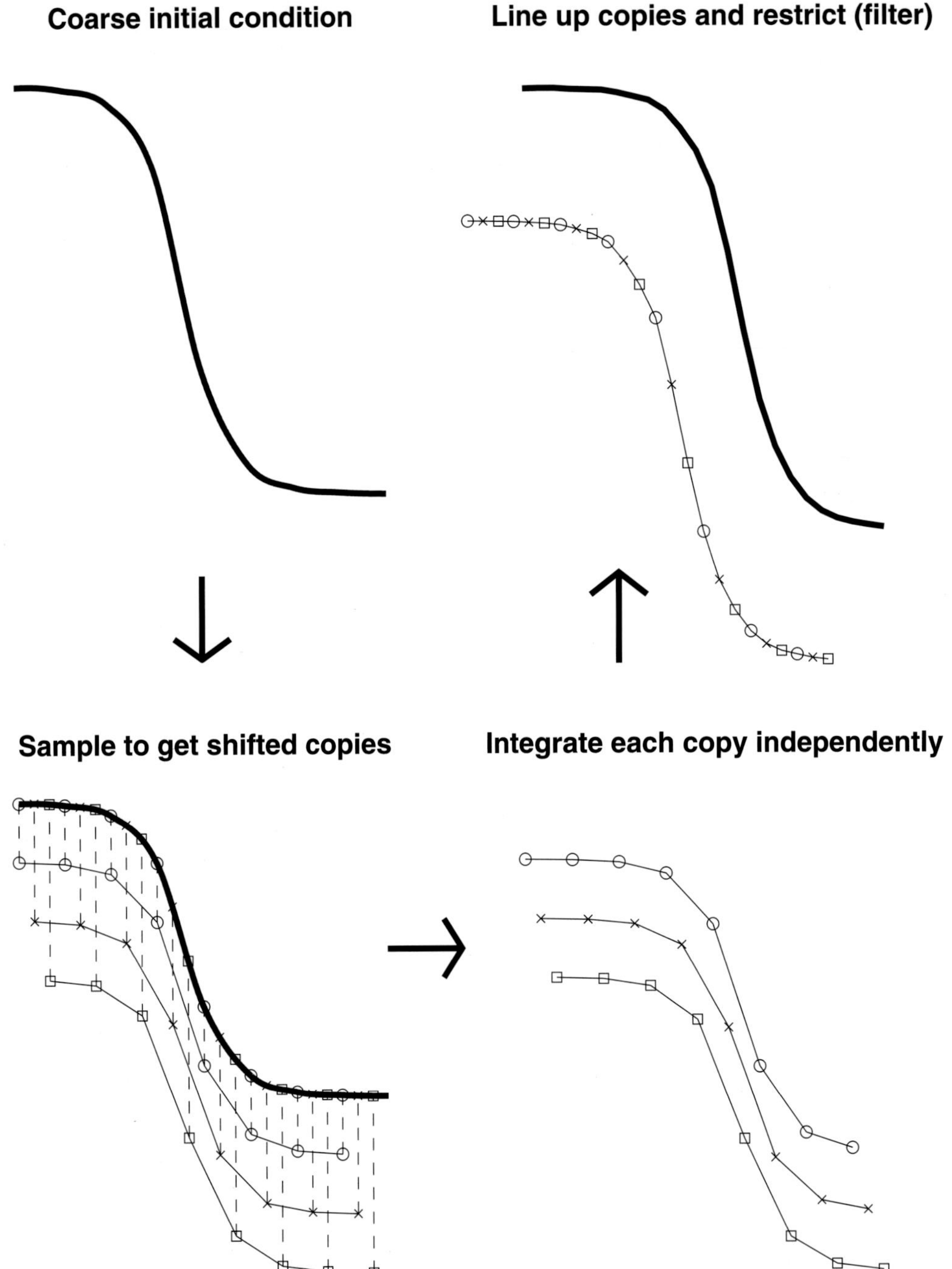

Fig. 1.   The coarse time stepper: Starting from a coarse initial condition $v_0(x)$, lift it by sampling to an ensemble of initial data, $\{\mathbf{u}_j(0)\}$, $j = 0, \ldots, N_c - 1$, for the system and evolve each set for time T. Line up solutions at time T and interpolate to get $\overline{u}(x)$. Finally, filter $\overline{u}(x)$ to get $\tilde{u}(T, x)$, the result of the coarse time stepper at $t = T$.

bifurcation analysis of certain types of solutions of the (unavailable) coarse evolution equation. For a schematic illustration of the coarse time stepper with RPM, see Fig. 2.

RPM helps locate fixed points, allows us to trace fixed point branches and locate their local bifurcations; when the bifurcations in (7) that we are interested in do not involve fixed points, G has to be reformulated. How this is done depends on the application; for the type of solutions considered here (traveling fronts), the appropriate modification is discussed in Sec. 3.2.

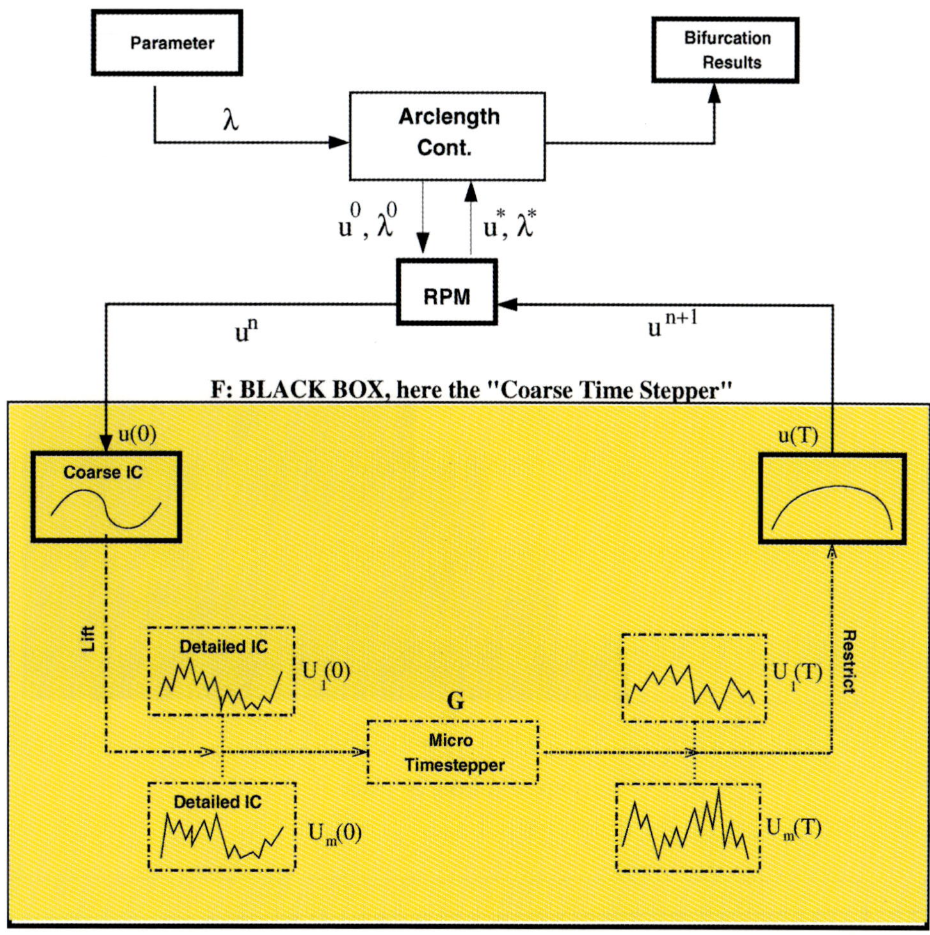

Fig. 2. An overview of the coarse time stepper with RPM.

3. A Discrete Traveling Front Example

The effects of discreteness on the propagation of traveling wave solutions have been documented and analyzed in many different settings over the last two decades. From the pinning of traveling waves in discrete arrays of coupled torsional pendula and Hamiltonian models [Ishimori & Munakata, 1982; Peyrard & Kruskal, 1984], to the trapping of coherent structures in dissipative lattices of coupled cells [Keener, 2000; Keener & Sneyd, 1998; Fath, 1998] (see also references therein), the role of spatial discreteness has triggered a large interest in a diverse host of settings. Recent studies have addressed rather extensively the possibility for stable, traveling wave fronts to exist in discrete reaction–diffusion systems; see, e.g. [Zinner, 1991, 1992; Zinner *et al.*, 1993], as well as the more recent work of [Bates *et al.*, 2003; Beyn & Thümmler, 2003].

Herein we focus on an alternative viewpoint (with respect to the above works), namely the one of effective equations. Such models, if capable of

describing the nature of the solutions of discrete problems, should successfully capture the effects of discreteness on the traveling wave shape and speed. More importantly, they should be capable of accurately predicting qualitative transitions (bifurcations) that are *inherently due to the discreteness*. The most prominent of those is probably the pinning of traveling waves and fronts often observed when the lattice spacing becomes sufficiently large. To illustrate the performance of our proposed coarse equation in capturing such a front pinning, we have chosen what is arguably a prototypical spatially discrete problem capable of exhibiting it: a one-dimensional lattice with scalar bistable on-site kinetics and nearest neighbor diffusive coupling between lattice sites. Our test problem is, therefore, a discrete reaction–diffusion system described by

$$\frac{du_\ell}{dt} = \frac{1}{(\Delta x)^2}(u_{\ell-1} - 2u_\ell + u_{\ell+1}) + f(u_\ell), \quad \ell \in \mathbb{Z}$$

$$(8)$$

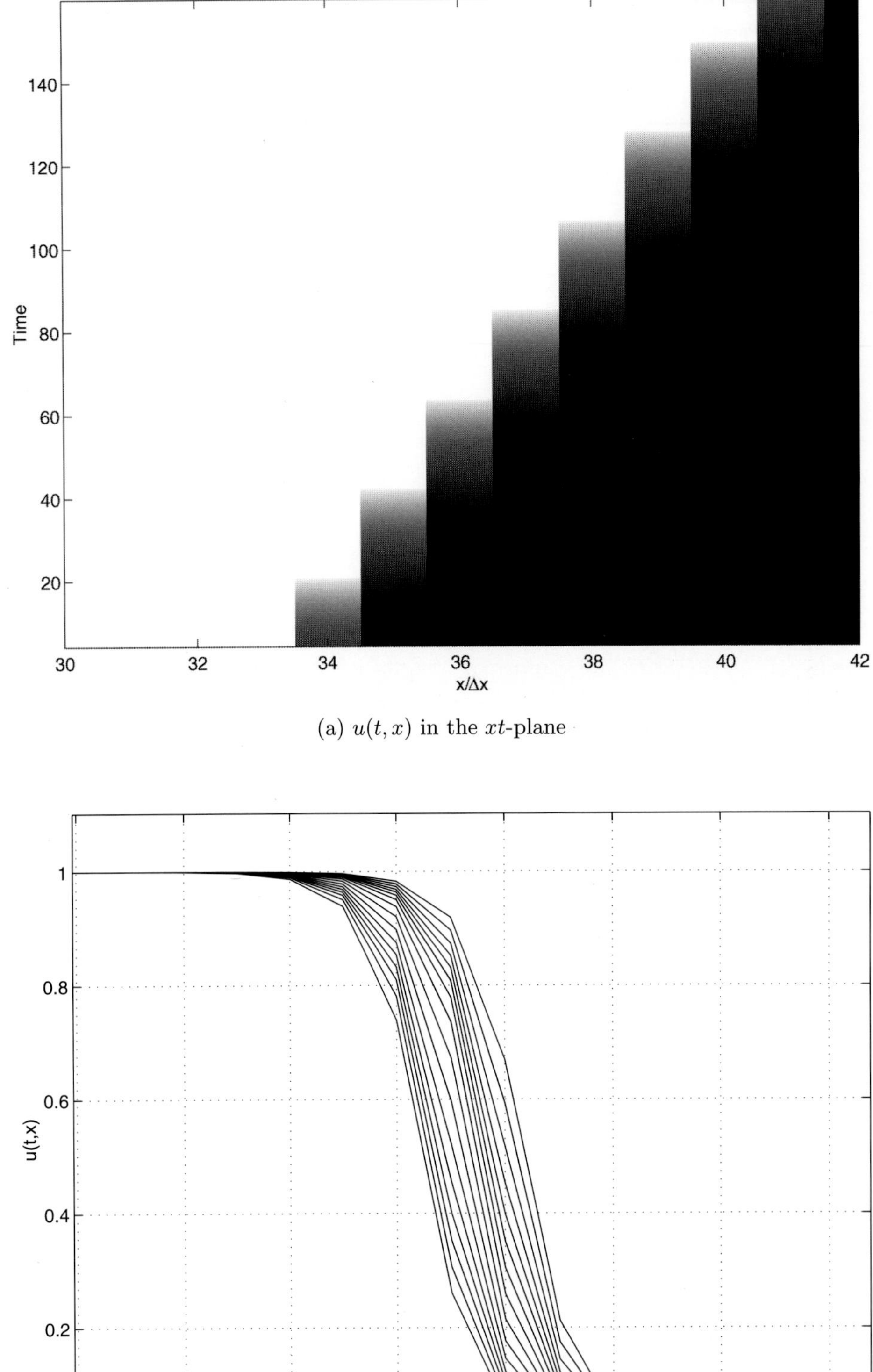

(a) $u(t, x)$ in the xt-plane

(b) $u(t, x)$, $t = 0, 2.5, 5, \ldots, 40$

Fig. 3.   The plot illustrates how the front advances when $\Delta x = 1.75$. In (a) the front in the xt-plane is shown; the grayscale is proportional to the solution $u(t, x)$. In (b) the solution as a function of x at different time levels is shown. The time interval is $t \in [0, 40]$. Looking at the spacing between the solution instances, we can see how the front speed varies in a lurching manner.

with

$$f(u) = 2u(u-1)(\eta - u), \quad \eta = 0.45. \tag{9}$$

This can serve as a model of e.g. individual cells in the cardiac tissue which are resistively coupled through gap junctions (see e.g. [Keener, 2000] and references therein). In this case the solution u_ℓ, would correspond to the electrical potential of the cells. For small Δx the system possesses solutions that can be characterized as *discrete traveling fronts*: see Fig. 3. These solutions have a near constant shape and travel in a "lurching" manner. When Δx becomes sufficiently large, front propagation fails (front pinning). In our example, this happens at $\Delta x = \Delta x^* \approx 2.3$, see Fig. 4. The front speed for an infinite lattice approaches the asymptotic "PDE speed" value 0.1 as the lattice size tends to zero.

We will examine how faithful the coarse time stepper is to the properties of the solutions of the full discrete model (8). Our numerical simulations are restricted to a finite domain, using $N = 64$ grid points. At the boundaries, we prescribe Neumann-type conditions

$$u_N - u_{N-1} = 0,$$
$$u_0 - u_{-1} = 0.$$

This should model the full problem accurately as long as the (relatively narrow) front is positioned sufficiently far from the boundary.

3.1. *Construction of the coarse time stepper*

In this section we detail the procedures associated with the coarse time stepper applied to the test problem (8, 9) on the finite interval $I = [0, L]$, where $L = N\Delta x$ and the cell locations are $x_j = j\Delta x$, with $j = 0, \ldots, N-1$.

Our choice of finite representation of the coarse solution are M nodal values $\mathbf{v}^n = \{v_k^n\}$, $k = 0, \ldots, M-1$, evaluated at $t = nT$ and $y_k = k\Delta y$, with $M\Delta y = N\Delta x$.

For many solution shapes Fourier interpolation would be a natural interpolation operator realizing the coarse solution $\tilde{u}(nT, x)$ from $\mathbf{v}^n$. We denote direct Fourier interpolation by Π^f. We could then define the corresponding lifting operators μ_j^f via the *shifting* operator $\mathcal{S}_s^f : \mathbb{R}^M \to \mathbb{R}^N$,

$$\mu_j^f \mathbf{u} := \mathcal{S}_{j\Delta s}^f \mathbf{u}, \quad (\mathcal{S}_s^f \mathbf{u})_\ell := (\Pi^f \mathbf{u})(x_\ell + s), \quad s \geq 0,$$

where Π^f uses $\{y_k\}$ as interpolation nodes. In our case, however, the solution is not periodic on I and

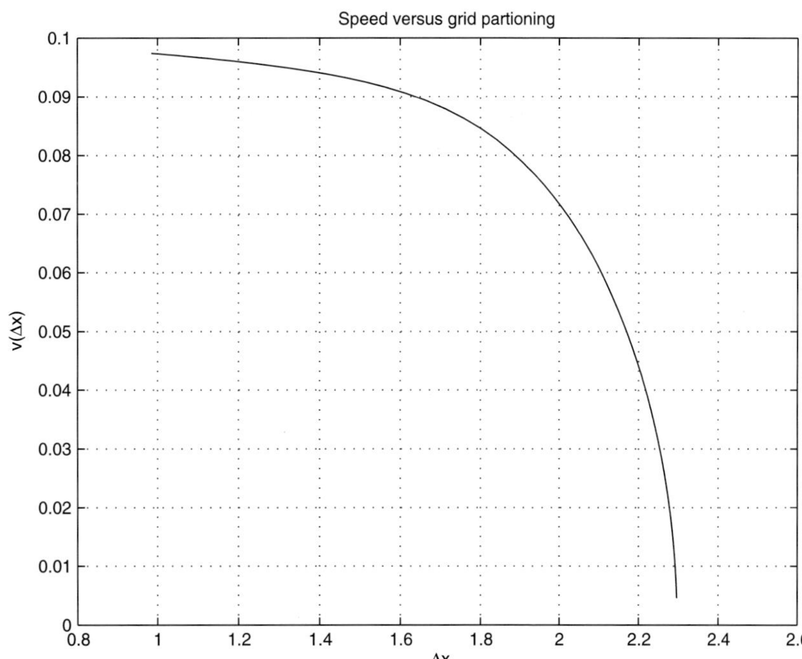

Fig. 4.   The speed of the front as a function of Δx. As the lattice spacing is increased, the speed v approaches zero; the front stops at $\Delta x^* \approx 2.3$.

we get large errors if we use $\mathcal{S}_s^f$ directly. Instead we apply Fourier interpolation to the *differences* of the $\mathbf{v}^n$ sequence. We thus use the modified shifting operator $\mathcal{S}_s : \mathbb{R}^M \to \mathbb{R}^N$ given by

$$\mathcal{S}_s\mathbf{u} := C\mathcal{S}_s^f D\mathbf{u},$$

$$(C\mathbf{u})_\ell := 1 + \sum_{j=0}^{\ell} u_j, \tag{10}$$

$$(D\mathbf{u})_\ell := \begin{cases} u_0 - 1, & \ell = 0, \\ u_\ell - u_{\ell-1}, & \ell > 0. \end{cases}$$

We then define the lifting operator $\mu : \mathbb{R}^M \to \mathbb{R}^{N \times N_c}$ (acting directly on $\mathbf{v}^n$) as

$$\mu\mathbf{v}^n = \{\mu_j\mathbf{v}^n\}, \quad \mu_j\mathbf{v}^n := \mathcal{S}_{j\Delta s}\mathbf{v}^n, \quad \Delta s = \frac{\Delta y}{N_c},$$

where $j = 0, \ldots, N_c - 1$.

The restriction operator $\mathcal{M} : \mathbb{R}^{N \times N_c} \to \mathbb{R}^M$ is also defined using the shifting operators, but now with negative shifts,

$$\mathcal{S}_{-s}^f : \mathbb{R}^N \to \mathbb{R}^M, \quad (\mathcal{S}_{-s}^f\mathbf{u})_k := (\Pi^f \mathbf{u})(y_k - s),$$

$$s \geq 0,$$

where Π^f uses $\{x_\ell\}$ as interpolation nodes. We then set $\mathcal{S}_{-s} = C\mathcal{S}_{-s}^f D$ and let

$$\mathcal{M}\{\mathbf{u}_j\} := \frac{1}{N_c} \sum_{j=0}^{N_c-1} \mathcal{S}_{-j\Delta s}\mathbf{u}_j.$$

Note that these choices of μ and $\mathcal{M}$ are consistent when $N \geq M$. Then, by the sampling theorem $\mathcal{S}_{-s}^f \mathcal{S}_s^f = I$ on $\mathbb{R}^M$. Moreover, it is easy to see that $CD = DC = I$. Therefore, we also have

$$\mathcal{S}_{-s}\mathcal{S}_s = C\mathcal{S}_{-s}^f DC\mathcal{S}_s^f D = C\mathcal{S}_{-s}^f \mathcal{S}_s^f D = CD = I,$$

on $\mathbb{R}^M$ and consequently,

$$\mathcal{M}\mu\mathbf{v}^n = \frac{1}{N_c} \sum_{j=0}^{N_c-1} \mathcal{S}_{-j\Delta s}\mu_j\mathbf{v}^n$$

$$= \frac{1}{N_c} \sum_{j=0}^{N_c-1} \mathcal{S}_{-j\Delta s}\mathcal{S}_{j\Delta s}\mathbf{v}^n$$

$$= \frac{1}{N_c} \sum_{j=0}^{N_c-1} \mathbf{v}^n = \mathbf{v}^n.$$

We should also remark here that, in the special case when $N = M$, we have

$$\left(\frac{1}{N_c} \sum_{j=0}^{N_c-1} \mathcal{S}_{-j\Delta s}^f \mathbf{u}_j \right)_\ell = (P_N \Pi^f \overline{\mathbf{u}})(x_\ell),$$

$$\overline{\mathbf{u}} = \{\overline{u}_r\}, \quad \overline{u}_{\ell+jN} = u_\ell^j,$$

where P_N is a projection on the N lowest Fourier modes. Hence, if we used direct Fourier interpolation and $M = N$, then our definition of $\mathcal{M}$ is equivalent to lowpass filtering of $\overline{\mathbf{u}}$, the lined up copies described in Fig. 1, top right. When we replace $\mathcal{S}_s^f$ by $\mathcal{S}_s$ we do not retain exactly this property, and a definition of $\mathcal{M}$ based on simple lowpass filtering is no longer consistent. However, our procedure still corresponds to a type of lowpass filtering, although a more complicated one.

For the time integration of (8) we use the Crank–Nicolson method, treating the nonlinear term explicitly. Thus, with $\mathbf{w}^0 = \{w_\ell^0\} \in \mathbb{R}^N$,

$$\mathcal{T}_T\mathbf{w}^0 := \mathbf{w}^{N_T} = \left\{w_\ell^{N_T}\right\}, \quad N_T\Delta t = T,$$

where $\{w_\ell^n\}$ are given iteratively by

$$w_\ell^{n+1} - \frac{\Delta t}{2(\Delta x)^2}\left(w_{\ell-1}^{n+1} - 2w_\ell^{n+1} + w_{\ell+1}^{n+1}\right)$$

$$= w_\ell^n + \frac{\Delta t}{2(\Delta x)^2}\left(w_{\ell-1}^n - 2w_\ell^n + w_{\ell+1}^n\right) + \Delta t f(w_\ell^n),$$

for $\ell = 0, \ldots, N-1$, together with the free boundary conditions

$$w_{-1}^n - w_0^n = 0,$$

$$w_N^n - w_{N-1}^n = 0.$$

In our computations we use the time step $\Delta t = 0.01$.

3.2.   *Steady state formulation*

The coarse solution $\tilde{u}(nT, x)$ as we have defined it is a (practically) constant shape moving front. In order to convert this moving state into a stationary state, we can factor out the movement through a procedure based on *template fitting* ([Rowley & Marsden, 2000; Runborg *et al.*, 2002], see also [Chen & Goldenfeld, 1995]) which pins the traveling front at a fixed x-coordinate. This is performed by a "pinning-shift" operator, which we denote as $\mathcal{P}$. Our coarse time stepping is then modified from (7) to

$$\mathbf{v}^{n+1} = \mathcal{P}\mathcal{M}\{\mathcal{T}_T\mu_j\}\Pi\mathbf{v}^n =: G(\mathbf{v}^n). \tag{11}$$

This formulation has a steady state at the constant shape moving front.

Let us start from the basic, Fourier based, pinning-shift operator $\mathcal{P}^f : \mathbb{R}^M \to \mathbb{R}^M$. After introducing a template function $S(x)$, we define

$$\mathcal{P}^f \mathbf{w} := \mathcal{S}_c^f \mathbf{w},$$
$$c = \arg\max_{c' \in \mathbb{R}} \int_0^L (\Pi^f \mathbf{w})(x + c') S(x)\, dx. \tag{12}$$

Hence, $\mathcal{P}^f \mathbf{w}$ is the shifted version of $\mathbf{w}$ that best fits the template $S(x)$, in the sense that it maximizes the L_2-inner product between its Fourier interpolant and S. Upon convergence, the effective front speed v can be deduced from the converged value of c and the time reporting horizon T simply by taking $v = c/T$. With the template $S(x) = 1 - \cos(2\pi x/L)$ we can compute the inner product in (12) explictly,

$$\frac{1}{L} \int_0^L (\Pi^f \mathbf{w})(x + c') S(x)\, dx = \hat{w}_0 - \Re(\hat{w}_1 e^{ic'}),$$
$$\tag{13}$$

where $\hat{w}_k$ are the Fourier coefficients of $\mathbf{w}$. Hence, since $\hat{w}_0$ is real c in (12) should be chosen such that $\hat{w}_1 e^{ic}$ is real and negative. This is easily implemented numerically together with the Fourier shift $\mathcal{S}_c^f$.

For the same reasons as in the implementation of the coarse time stepper, we would like to avoid direct Fourier interpolation of the solution, since it is not periodic. Therefore, we modify $\mathcal{P}^f$ to operate on differences instead. In the same spirit as in Sec. 3.1, we let

$$\mathcal{P} := C\mathcal{P}^f D,$$

with C and D defined in (10). We still use the effective propagation speed given by $\mathcal{P}^f$.

An important property of the Fourier based pinning shift operator is that it satisfies $(\mathcal{P}^f)^2 = \mathcal{P}^f$, which follows from the sampling theorem [Runborg *et al.*, 2002]. For other types of interpolation, such as piecewise polynomial interpolation, the pinning shift operator will not have this property and a steady moving coarse shape may not translate into a fixed point for (11). Our modification still has this property though, since

$$\mathcal{P}^2 = C\mathcal{P}^f DC\mathcal{P}^f D = C(\mathcal{P}^f)^2 D = C\mathcal{P}^f D = \mathcal{P},$$

where we used the fact that $DC = I$.

3.3. *The RPM with pseudo-arclength continuation*

RPM is an iterative procedure which can accelerate the location of fixed points of processes; under certain conditions it can help locate steady states of dynamic processes (in particular, discretized parabolic PDEs). It can be an acceleration technique for the solution of nonlinear equations, and a stabilizer of unstable numerical procedures (as first presented [Shroff & Keller, 1993]). Consider the fixed point problem

$$F(u; \lambda) = u, \tag{14}$$

and let J be the Jacobian of F.

- Like the Newton method, RPM can converge rapidly to the fixed point solution u^* provided the initial guess is good enough; the convergence occurs even if $J(u^*)$ has a few eigenvalues larger than one. The computational cost and convergence rate depend on the eigenvalues of J. Optimally there should be a clear gap in the spectrum between small and large (near the unit circle) eigenvalues and a limited number of large (in norm) eigenvalues for RPM to perform well.
- J never needs to be evaluated directly, only F. We can therefore apply RPM to any "black box" code that defines a function F; it is a "matrix-free" method.
- As a by-product, RPM also computes approximations of the largest eigenvalues of J. This gives approximate stability information about the fixed point.

When RPM is used for the computer-assisted bifurcation analysis of steady states of (usually dissipative evolution) PDEs, the function F represents a *time stepper*: a subroutine that takes initial data and reports the solution of the PDE after some fixed time (the reporting horizon T). A fixed point then satisfies (14). The conventional way of finding the steady state using a time stepper would be to call it many times in succession — in effect, to integrate the PDE for a long time, corresponding to solving (14) by simple fixed point (Picard) iteration.

RPM can improve this approach in two important respects. First, the convergence can be significantly accelerated. The nature of many transport PDEs usually encountered in engineering modeling (the action of viscosity, heat conduction, diffusion, and the resulting spectra) dictates that there exists a separation of time-scales, which translates into an eigenvalue gap in the spectrum of J at the steady state. Second, RPM converges even if the steady state is slightly unstable, i.e. when J has a few eigenvalues outside the unit circle. It may thus be possible to compute (mildly) unsteady branches of the bifurcation diagram using forward integration (but in a nonconventional way, dictated by the RPM protocol). RPM still retains the simplicity of the fixed point iteration, in the sense that no more information is needed than just the time-integration code. This code, which may be a legacy code, and can incorporate the best physics and modeling available for the process, is used by RPM as a black box.

RPM can be seen as a modified version of fixed point iteration. It adaptively identifies the subspace corresponding to large (in norm) eigenvalues of J, hence the directions of slow or unstable time-evolution in phase space. In these directions the fixed point iteration is replaced by (approximate) Newton iteration. More precisely, suppose $F : \mathbb{R}^N \times \mathbb{R} \to \mathbb{R}^N$ in (14). Let $\mathbb{P}$ be the maximal invariant subspace of J corresponding to the m largest eigenvalues and let $\mathbb{Q}$ be its orthogonal complement in $\mathbb{R}^N$. The solution u is decomposed as $u = p + q = Pu + Qu$, where P and Q, are the projection operators in $\mathbb{R}^N$ on $\mathbb{P}$ and $\mathbb{Q}$. These are constructed from an orthogonal basis V_p

$$P = V_p V_p^T,$$

$$Q = I - V_p V_p^T.$$

In a pseudo-arclength continuation context the solution $u = u(s)$ and $\lambda = \lambda(s)$, where s parameterizes the bifurcation curve. In addition to (14) we then use an algebraic equation to be able to handle turning points,

$$S(u, \lambda, \Delta s) = \frac{\|u(s) - u(s - \Delta s)\|^2}{\Delta s}$$

$$+ \frac{|\lambda(s) - \lambda(s - \Delta s)|^2}{\Delta s} - \Delta s$$

$$= 0, \tag{15}$$

where $u(s - \Delta s)$ and $\lambda(s - \Delta s)$ refers to the converged solution at the previous point on the continuation curve.

The solution is advanced using a predictor–corrector method. Via extrapolation from previous points $u_i = u(s_i)$, $\lambda_i = \lambda(s_i)$ and $\Delta s_i = s_{i+1} - s_i$, the predictor-solution is obtained. Comparing a

first-order extrapolation,

$$\lambda^* = \lambda_i + \frac{\lambda_i - \lambda_{i-1}}{\Delta s_{i-1}} \Delta s_i,$$

$$u^* = u_i + \frac{u_i - u_{i-1}}{\Delta s_{i-1}} \Delta s_i,$$

with a second-order extrapolation,

$$\lambda^{**} = \lambda^* + \frac{1}{2} \frac{\lambda_i(1 - \gamma) - 2\lambda_{i-1} + (1 + \gamma)\lambda_{i-2}}{\Delta s_{i-1}\Delta s_{i-2}} \Delta s_i^2,$$

$$u^{**} = u^* + \frac{1}{2} \frac{u_i(1 - \gamma) - 2u_{i-1} + (1 + \gamma)u_{i-2}}{\Delta s_{i-1}\Delta s_{i-2}} \Delta s_i^2$$

$$\gamma = \frac{\Delta s_{i-1} - \Delta s_{i-2}}{\Delta s_{i-1} + \Delta s_{i-2}},$$

and requiring that

$$\max(\|u^{**} - u^*\|, |\lambda^{**} - \lambda^*|) < \epsilon \tag{16}$$

the stepsize is determined. Here ϵ is a user specified tolerance. As the corrector method, we use RPM with pseudo-arclength continuation, see [Shroff & Keller, 1993; Lust, 1997]. Starting from $u^0 = u^{**}$ and $\lambda^0 = \lambda^{**}$, the iterative scheme is given by

$$q^{n+1} = QF(u^n, \lambda^n),$$

$$\begin{bmatrix} (V_p^T J V_p - I) & V_p^T F_\lambda \\ S_u^T V_p & S_\lambda \end{bmatrix} \begin{bmatrix} \Delta p \\ \Delta \lambda \end{bmatrix}$$

$$= -\begin{bmatrix} V_p^T F\left(p^n + q^{n+1}, \lambda^n\right) - p^n \\ S(p^n + q^{n+1}, \lambda^n) \end{bmatrix},$$

$$u^{n+1} = p^n + V_p \Delta p^n + q^{n+1},$$

$$\lambda^{n+1} = \lambda^n + \Delta \lambda^n,$$

where the left-hand side consists of partial derivatives of S in (15) and of F in (14) with respect to u and λ. The iterates $u^n = p^n + q^n$ will converge to the solution of (14) under the assumptions discussed above. If the number of large norm eigenvalues, m, is limited, the dimension of $\mathbb{P}$ and the projected Jacobian in the Newton iteration, $V_p^T J V_p - I$, remains small. Only this small matrix needs to be inverted. For a more complete description of RPM we refer to [Shroff & Keller, 1993].

4. Numerical Results

In this section we present some numerical results using the coarse time stepper and the procedure described above to simulate an effective equation for the discrete problem in (8). We will start by discussing the "exact" bifurcation diagram of the discrete system, which we attempt to approximate. We will then show results obtained through the coarse time stepper, and discuss the effect of time stepper "construction parameters" like the reporting time horizon, T (the time to which (1) is integrated within the coarse time stepper), and the number of different initial shifted copies, N_c.

Figure 5 shows the bifurcation diagram of the discrete problem as a function of the parameter Δx, the lattice spacing, in the regime close to the onset of pinning. For lattice spacings smaller than $\Delta x^* \approx 2.3$ the system has, as we discussed, an attracting, front-like solution that travels; its motion is *modulated* as it "passes over" the lattice points. For an infinite lattice, this modulated traveling solution possesses a discrete translational invariance: $u_{\ell+1}(t + \tau) = u_\ell(t)$. The shape of the modulating front is shifted by one (resp. $2, 3, \ldots, n$) lattice spacing after time τ (resp. $2\tau, 3\tau, \ldots, n\tau$); this helps us define its effective speed $v(\Delta x) \equiv \Delta x/\tau$ (see Fig. 4). As Δx approaches zero, for an infinite lattice, the discrete front approaches the continuum front of the PDE, and its speed (the period of the modulation divided by Δx approaches the PDE front speed, 0.1 (see Fig. 4)).

If we identify shapes shifted by one lattice constant, the attractor appears as a limit cycle with period τ. As the lattice spacing approaches the critical value Δx^* the speed of propagation approaches zero (the period of the "limit cycle" approaches infinity); asymptotically, $v(\Delta x) \approx |\Delta x - \Delta x^*|^{0.5}$. As discussed in [Kevrekidis *et al.*, 2001; Carpio & Bonilla, 2003a, 2003b] what occurs is a Saddle-Node Infinite Period (SNIPER) bifurcation: a saddle-node bifurcation where both new fixed points appear "on" the limit cycle. For larger values of Δx the "saddle" and the "node" move away from each other, and what used to be the limit cycle is now comprised from the saddle, the node, and both sides of the one-dimensional unstable manifold of the saddle, which asymptotically approaches the node.

The saddle and the node are, of course, stationary fronts. A pair of them exists for every "unit cell": all "node fronts" are shifts of each other by one lattice spacing, and all "saddle fronts" are also shifts of each other by one lattice spacing. Since the medium has a discrete translational invariance, this makes sense — if an initial condition gives rise to a front eventually pinned at some location in the discrete medium, the shift of this initial condition by one lattice spacing will eventually get trapped one lattice spacing further. This saddle-node bifurcation can be seen in Fig. 5(a); linearizing around the saddle front will give a positive eigenvalue λ_s, while the corresponding eigenvalue λ_n for the node front would be negative. Since we look at the problem in discrete time, what is plotted is the *multiplier* $\mu_{n,s} = \exp(\lambda_{n,s}T)$, where T is the reporting horizon. The saddle front has a multiplier larger than 1, while the corresponding multiplier for the stable node is less than 1; both multipliers asymptote to 1 at the SNIPER (Δx^*).

Figure 5(b) shows the bifurcation diagram in terms of the front traveling speed. Since both the saddle and the node fronts are pinned (have zero speed) they both fall on the zero axis; we plotted their eigenvalues in Fig. 5(a) to distinguish between them. The true traveling speed (broken line) is compared with the effective traveling speed predicted by a coarse time stepper using $N_c = 5$ copies within each unit cell, and a reporting horizon of $T = 32$. The coarse time stepper speed is a byproduct of fixed point computation and continuation with it; short bursts of detailed simulation are used in the RPM framework to construct a contraction mapping that converges to a fixed point of the time stepper. The final shift upon convergence (from the pinning-shift computation), divided by the time stepper reporting horizon gives us an estimate of the "effective speed". Inspection of Fig. 5(b) indicates that the coarse time stepper never predicts a speed that is exactly zero; yet it gives a good approximation of the effective speed, all the way from small Δx to the near neighborhood of the pinning transition, when the effective speed becomes small.

We will return to discussing this issue of "small residual motion" for the coarse time stepper shortly. To give an indication of when the procedure stops being quantitative, we have included the $\Delta x/T$ curve in Fig. 5(b): disagreement starts well in the regime where the effective movement is *less* than one unit cell per observation period. In the next section we will compare the "goodness of approximation" of our coarse time stepper to the effective speed predicted by the Padé approach to extracting effective continuum equations. It is interesting that the coarse time stepper sometimes predicts a small hysteresis loop at low speeds, relatively close to "true pinning"; notice in Fig. 5(a) the unstable (larger than one) multipliers for the brief saddle

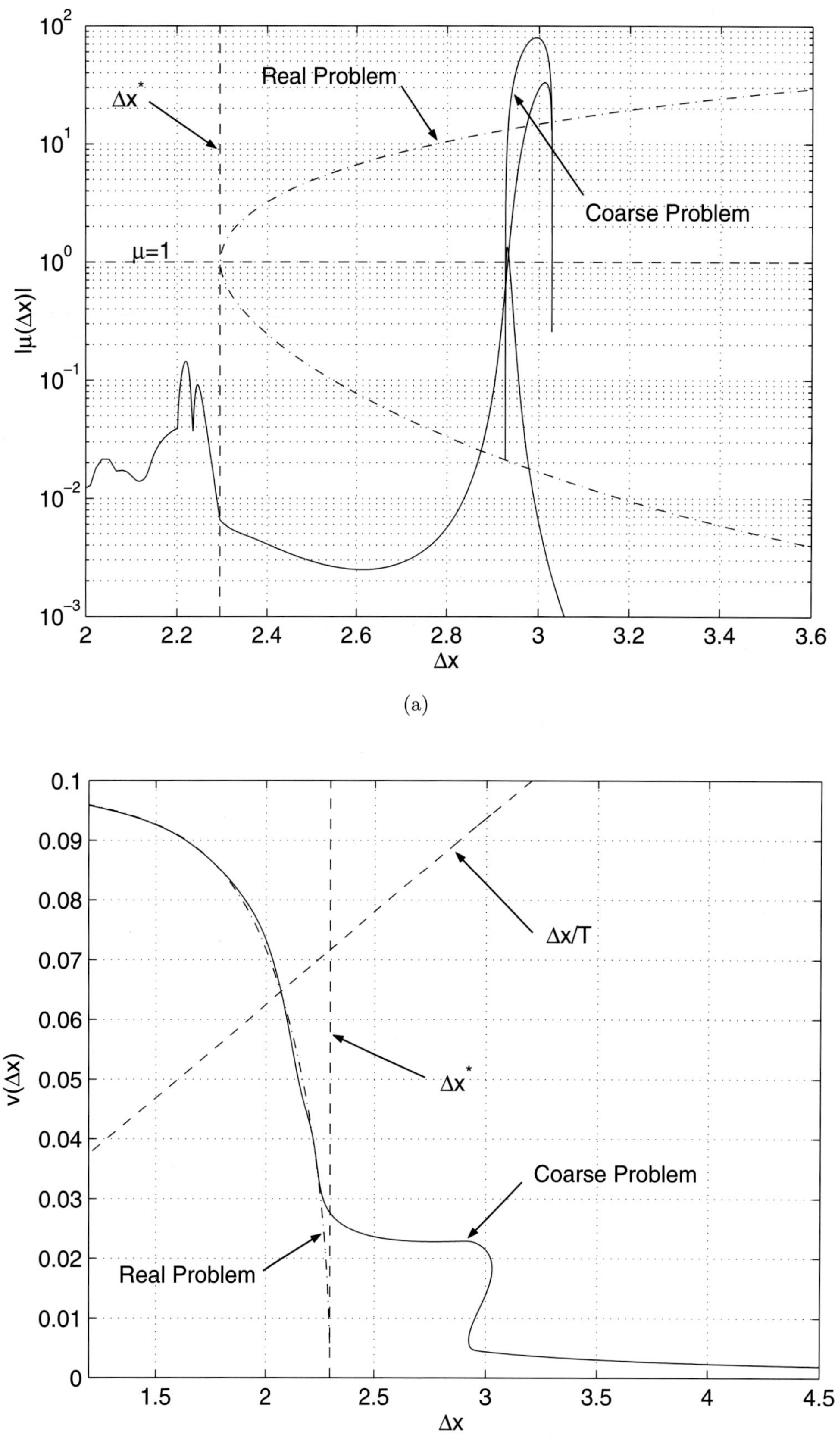

Fig. 5.   Detailed bifurcation diagram and coarse time stepper bifurcation diagram with parameters $N_c = 5, T = 32$. (a) Multipliers versus Δx, (b) effective front speed versus Δx.

part of this loop. We will discuss a tentative rationalization of this below.

Figure 6 illustrates the effects of "time stepper construction" parameters on the effective behavior predicted by the time stepper: the reporting time-horizon, for two different sets of shifted copies ($N_c = 3$ and $N_c = 10$) as well as the effect of the number of copies for a fixed time horizon ($T = 16$). Augmenting the time stepper reporting horizon is shown in Figs. 6(a) and 6(b); clearly, in both cases, extending the time stepper reporting horizon extends the region over which its effective speed agrees with the true problem closer to Δx^*. Larger numbers of copies ($N_c = 5, 10, 20$) also perform slightly better than smaller numbers ($N_c = 3$). In all cases the qualitative behavior is the same: (a) successful approximation of the effective speed until reasonably close to true pinning; (b) all differences occur when the average front motion is significantly less than one unit cell per reporting horizon; (c) there is always a slight residual motion, which — possibly

after a small hysteresis loop close to true pinning — eventually becomes negligible.

We now turn to the discussion of the slight residual motion of the coarse time stepper at large Δx beyond Δx^*. For an infinite domain, the saddle and node pinned fronts appearing there are invariant to translations by one lattice spacing; for a large enough computational domain we still see two pinned front solutions per cell. When we "sprinkle" initial conditions along the cell, depending on their location with respect to the saddle front, the trajectories may either be attracted to the stable node "to the right" or to the one "to the left" of the saddle. It is instructive to represent these solutions as in Fig. 7(a), in a way that identifies the "right" node front with the "left" one; here translation along the lattice corresponds roughly to rotation along the circle. The node is denoted by a black circle, and the saddle by a white one. The small squares represent the initial positions of our initial condition "copies". The fate of our distribution of initial conditions is

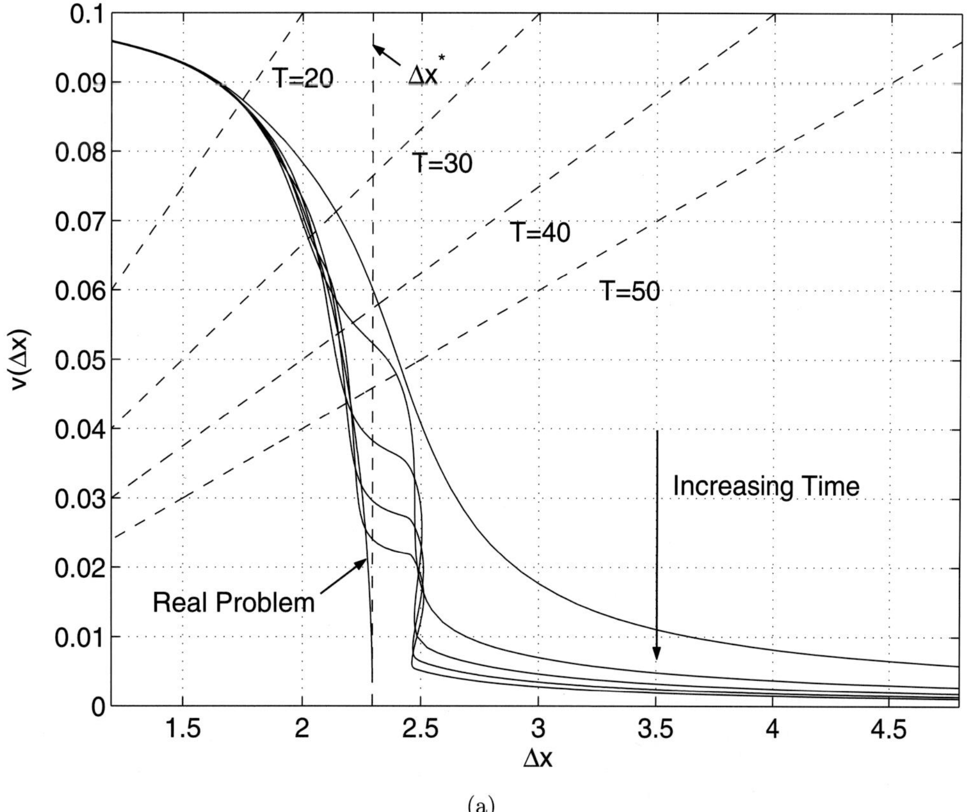

Fig. 6.   Effective front speed versus Δx and the effect of varying the time horizon T and the number of copies N_c. (a) Varying time horizon, $T = 10, \ldots, 50$, with fixed $N_c = 3$. Dashed lines show $\Delta x/T$, i.e. speed required to traverse one cell, (b) varying time horizon, $T = 5, \ldots, 16$, with fixed $N_c = 10$, (c) varying number of copies, $N_c = 3, 5, 10, 20$, with fixed $T = 16$.

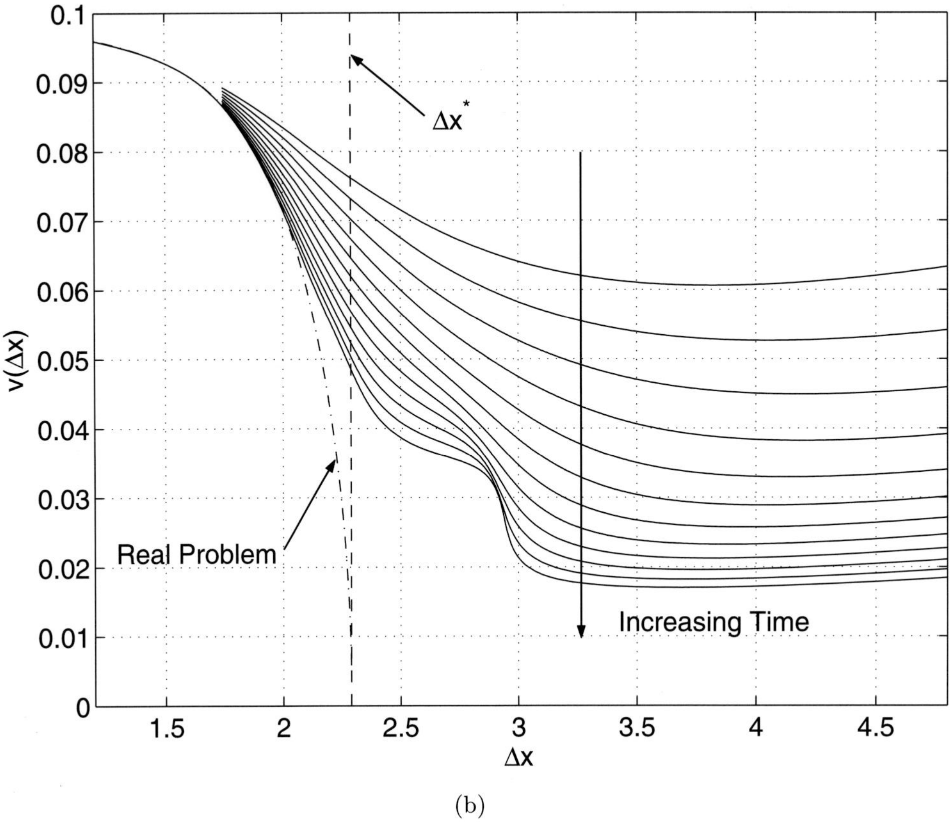

(b)

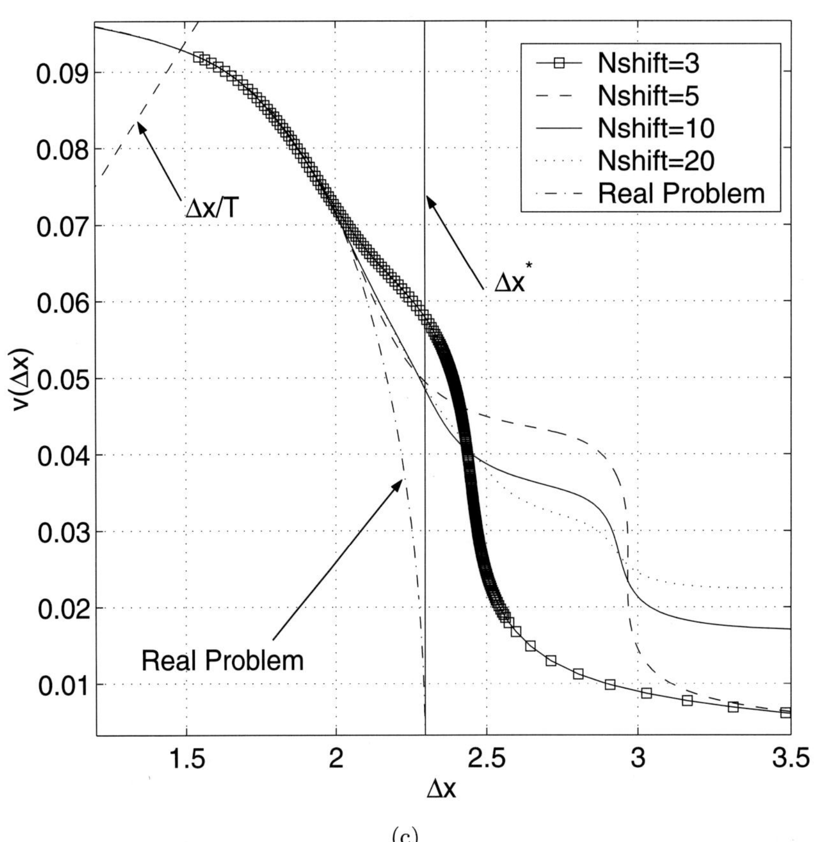

(c)

Fig. 6.   (*Continued*)

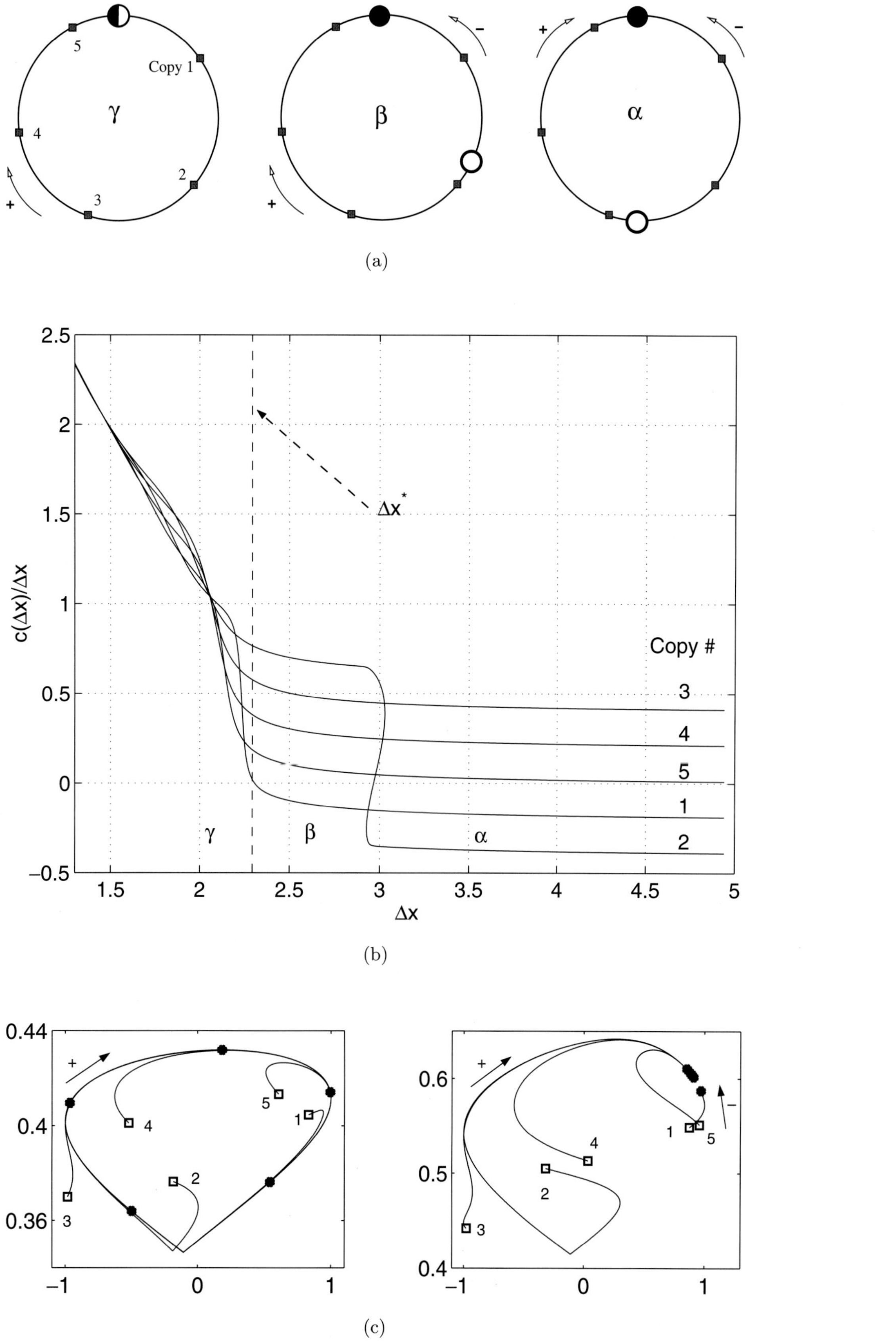

(a)

(b)

(c)

Fig. 7.   Movement of the individual copies, for $N_c = 5$, $T = 32$. (a) Schematic movement of copies in phase space, (b) distance traversed by copies, (c) real movement of copies in phase space for $\Delta x = 1.6$ (left) and $\Delta x = 2.3$ (right). (See text for specification of axes.)

governed by their initial "angle" on the circle — as our time horizon grows all initial conditions will asymptote to a stable front, either the left one (moving counterclockwise on the circle) or the right one (clockwise movement). We now see clearly the physical reason behind the net residual motion for any finite time horizon for the coarse time stepper. An initial condition that is put down "at random" in a unit cell deep in the pinned regime, even if it never exits this unit cell, will gradually traverse the part of the circle separating it from the closest node front.

When the critical parameter value is approached from the pinned side, the saddle and the node fronts approach each other on the circle, on their way to coalescing at the SNIPER bifurcation point [Kevrekidis *et al.*, 2001; Carpio & Bonilla, 2003a, 2003b]. Figure 7(b) shows how this process becomes manifest in the coarse time stepper computations, using the problem in Fig. 5 as our example. Deep in the pinning regime (high Δx, marked α) the relative "phase" of the saddle and the node pinned fronts on the circle remains roughly constant. The distance each member of our ensemble of initial conditions has traversed during one time horizon can be deduced from Fig. 7(b): the copy with the largest negative movement is the one closest to the saddle but on its left (copy number two). One can similarly rationalize the labeling of the remaining curves in Fig. 7(b). When Δx is reduced approaching the onset of pinning, at some point the saddle front starts moving appreciably towards the node front. As part of this movement, it "sweeps" the circle counterclockwise; at $\Delta x \approx 2.8$ it has its first encounter with one of our initial conditions — the closest one on the left. When the saddle "moves past" it into the regime marked β, this copy, which was responsible for the largest negative displacement now approaches asymptotically the node front on the right, performing the largest *positive* displacement (and so on for the remaining copies). Eventually, in the propagating regime, marked γ, and for long enough reporting horizons, the initial "phase" difference (a fraction of a cell) becomes negligible compared to the net displacement of each point (several cells).

The real movement in phase space is shown in Fig. 7(c) for two different Δx. In these subfigures, the x-axis represents $\sin(2\pi x_c)$ where x_c corresponds to the location of the front, more specifically $x_c = \sum_\ell (D\mathbf{u})_\ell \ell$. The y-axis represents $\max_\ell |(D\mathbf{u})_\ell|$. The initial positions of the copies are indicated by small squares and their locations at $t = T$, the time horizon, are marked by filled circles. The labels refer to the same copies as in Fig. 7(b).

As the reporting time horizon of the time stepper goes to infinity, it is clear that one can compute the average residual movement from the asymptotic position of the saddle front, i.e. from the relative extent of the circle "to the right" and "to the left" of the saddle front. The most reasonable point to "declare" as an estimate of the true pinning from coarse time-stepper computations would come from a polynomial extrapolation of the "successful" regime (close to the tip of the "apparent parabola" in Fig. 5); alternatively, a value of Δx where the speed is small enough (well below one unit cell per time horizon) and its variation with number of copies and time horizon is below a user-prescribed tolerance, would also serve this purpose. While there is no well-defined pinning bifurcation for the coarse time stepper (since pinning is an inherently nontranslationally invariant bifurcation), the procedure can provide a good approximation of the effective shape and speed of the traveling fronts, as well as "common sense" ways of numerically estimating the true pinning.

5. An Alternative Continuum Approach: Padé Approximations

In this section, we propose an alternative scheme for capturing effects of discreteness, by means of a (now explicit) continuum equation. This PDE is obtained by means of Padé approximations [Cabannes, 1976; Elphick *et al.*, 1990] which can be used to approximate discreteness in a quasi-continuum way, through the use of pseudo-differential operators. In particular, starting from the Taylor expansion for analytic functions, see e.g. [Christiansen *et al.*, 2001],

$$u(x + m) = \exp(m\partial_x)u(x),$$

one can then express spatial discreteness as

$$u_{\ell+1} + u_{\ell-1} - 2u_\ell$$
$$\equiv (\exp(\Delta x \partial_x) + \exp(-\Delta x \partial_x) - 2)u(x)$$
$$\equiv 4\sinh^2\left(\frac{\Delta x \partial_x}{2}\right)u(x,t).$$

Expanding $\exp(\pm\Delta x\partial_x)$ [Elphick *et al.*, 1990], one then obtains

$$\exp(\pm\Delta x\partial_x) - 1 = \frac{1}{2}\Delta x^2\left(1 + \frac{\Delta x^2}{12}\partial_x^2 + \cdots\right)\partial_x^2$$

$$\pm\Delta x\left(1 + \frac{1}{6}\Delta x^2\partial_x^2 + \cdots\right)\partial_x$$

Finally, regrouping the terms in the manner of Padé [Cabannes, 1976; Elphick *et al.*, 1990] yields

$$\exp(\pm\Delta x\partial_x) - 1 \approx \frac{1}{2}\frac{\Delta x^2\partial_x^2}{1 - \frac{\Delta x^2}{12}\partial_x^2} \pm \frac{\Delta x\partial_x}{\left(1 - \frac{\Delta x^2}{12}\partial_x^2\right)^2} \tag{17}$$

We now use the pseudo-differential operator approximation in (17) to convert the discrete model in (8) into the PDE approximation of the form:

$$u_t = \frac{\partial_x^2}{1 - \frac{\Delta x^2}{12}\partial_x^2}u + f(u). \tag{18}$$

Such approaches were introduced and used extensively by Rosenau and collaborators [Rosenau, 1986, 1987, 1989, 1992; Doering *et al.*, 1987] to regularize nonlinear wave equations, particularly of the Klein–Gordon type.

Equation (18) clearly emulates the discrete setting in some key aspects of the relevant spectral operator properties (i.e. the discrete Laplacian in comparison with the pseudo-differential operator of (18)). For example, considering plane wave solutions of the form $\exp(\lambda t - ikx)$, we obtain in the discrete case the linearized dispersion relation (around a uniform state $u = u_{\text{hom}}$)

$$\lambda = \frac{2}{\Delta x^2}\left(\cos(k\Delta x) - 1\right) + f'(u_{\text{hom}}).$$

In the case of (18), the corresponding equation becomes

$$\lambda = -\frac{k^2}{1 + \frac{\Delta x^2}{12}k^2} + f'(u_{\text{hom}}).$$

Apart from sharing the continuum limit, the two dispersion relations share another qualitative feature which is particularly important [Rosenau, 1986, 1987, 1989, 1992; Doering *et al.*, 1987]; namely, the presence of a lower bound in the continuous spectrum. Notice, however, that the two lower

bounds are different ($f'(u_{\text{hom}}) - 4/\Delta x^2$ in the discrete case versus $f'(u_{\text{hom}}) - 12/\Delta x^2$ in the Padé approximation).

It would then be of interest to alleviate this spectral discrepancy, as well as to match the discrete operator (if possible) to a higher order in the Taylor expansion

$$\frac{u_{n+1} + u_{n-1} - 2u_n}{\Delta x^2} = \sum_{j=1}^{\infty}\frac{2\Delta x^{2j-2}}{(2j)!}u^{(2j)}$$

$$= u_{xx} + \frac{\Delta x^2}{12}u_{xxxx}$$

$$+ \frac{\Delta x^4}{360}u_{6x} + O(\Delta x^6).$$

This can be achieved by a natural generalization in the form of a continued fraction such as e.g.

$$\frac{\partial_x^2}{1 - \frac{A\partial_x^2}{1 - \frac{B\partial_x^2}{1 - C\partial_x^2}}}. \tag{19}$$

In order to use (19) in practice (i.e. for computational purposes), we convert the three fractions into one of the form

$$\frac{\partial_x^2(1 + \alpha\Delta x^2\partial_x^2)}{1 + (\alpha + \beta)\Delta x^2\partial_x^2 + \gamma\Delta x^4\partial_x^4}, \tag{20}$$

where a simple (algebraic) reduction of A, B, C to α, β, γ has been used. We then use Taylor expansion of the denominator to convert the expression of (20) into one resembling (19). By matching up to $O(h^6)$ the exact Taylor expansion, we obtain three algebraic equations for α, β and γ. In this way, we obtain a set of solutions for α, β and γ. We use here the set $\alpha = -0.007, 912$, $\beta = -1/12$, $\gamma = 0.002, 056$. An additional benefit (to the matching of the Taylor expansion up to correction terms of $O(h^8)$) that should be highlighted here is the value $\alpha/(\gamma\Delta x^2) = 3.848/\Delta x^2$ of the lower bound expression for λ, which is much closer to the theoretical lower bound of $4/\Delta x^2$ than the prediction $12/\Delta x^2$ of the leading order approximation presented previously. The resulting evolution equation will then read:

$$u_t = \frac{\partial_x^2(1 + \alpha\Delta x^2\partial_x^2)}{1 + (\alpha + \beta)\Delta x^2\partial_x^2 + \gamma\Delta x^4\partial_x^4} + f(u) \tag{21}$$

Both (18) and (21) can be numerically implemented in a straightforward manner, by means of the spectral techniques described in [Kevrekidis *et al.*, 2002]. We have performed numerical simulations of the front propagation, using 1024 modes in the spectral decomposition of (18) and (21). We will refer to these equations as the (Padé) models A and B, respectively. A fourth order Runge–Kutta algorithm has been used for the time integration. For each value of Δx, we identify the position x_c of the front as the point where the ordinate of the front acquires the value $u = 1/2$. The linear interpolation scheme suggested in [Boesch *et al.*, 1989] has been implemented and has proved to be an efficient front tracking algorithm in all the examined cases.

Our results of this quasi-continuum approach to the discrete problem can be summarized in Figs. 8 and 9. Figure 8 shows the speed of the fronts in Padé models A and B, respectively. We can observe that the critical value of Δx beyond which trapping of the front occurs is significantly displaced from the actual one of $\Delta x^* \approx 2.3$, for $\eta = 0.45$. In particular, for model A, $\Delta x^* \approx 6.4$, while for model B, the corresponding critical value is $\Delta x^* \approx 3.8$. We can deduce that the latter model is closer to the actual physical reality, even though the relevant prediction is still considerably higher than its actual value for the discrete model.

In part at least, these results (and the discrepancy from the actual discrete case) can be justified by observing Fig. 9. The bottom panel of the figure suggests that the *only* way in which the front can stop in these quasi-continuum Padé approximations is by becoming practically a vertical shock-like structure. In this case, the "mass" of the front which is given by $\int_{-\infty}^{\infty} u_x^2\, dx$ (see e.g. [Boesch *et al.*, 1989] and references therein) becomes practically infinite. This means that the inertia of the front becomes too big for the front to move and hence "pinning" occurs. However, notice that this process of pinning is significantly different than the details of the discrete structure of the problem (such as e.g. the saddle-node bifurcation and the transition to pinned solutions). The translationally invariant quasi-continuum Padé approximations of models A and B do not "see" such features. Instead, they incorporate the well-known feature of front steepening for stronger discreteness [Peyrard & Kruskal, 1984] and the criticality of the latter feature eventually leads to pinning.

An additional pointer to the fact that such (pseudo-differential operator) models are "eligible" to pinning is that they are devoid of some of the important symmetries that are inherently related

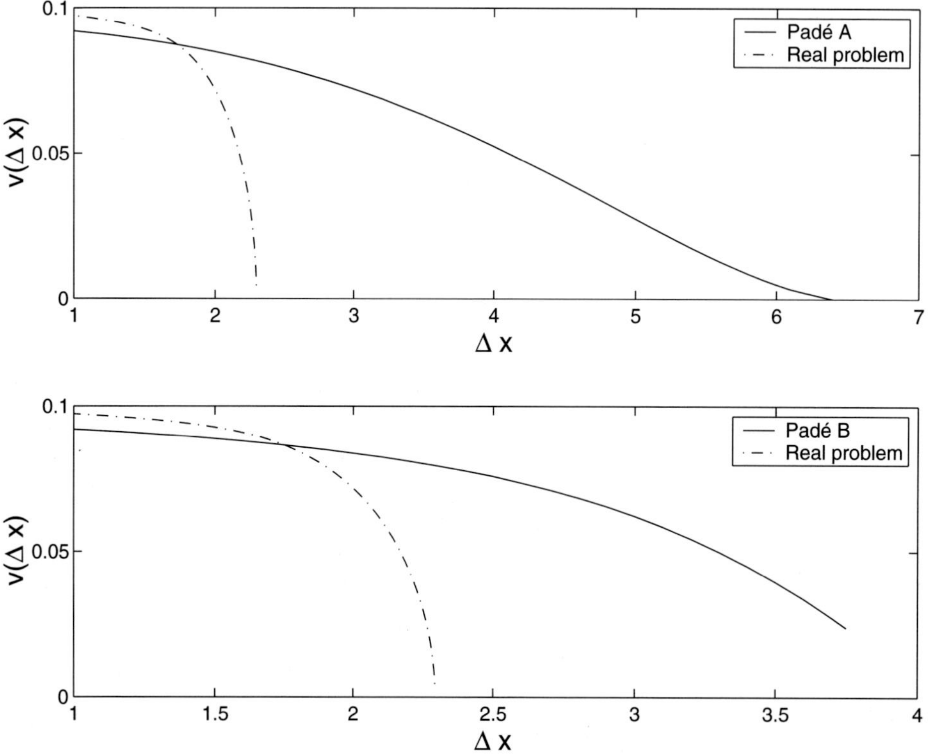

Fig. 8.   Effective front speed as a function of Δx, for the Padé model A (top panel) and model B (bottom panel).

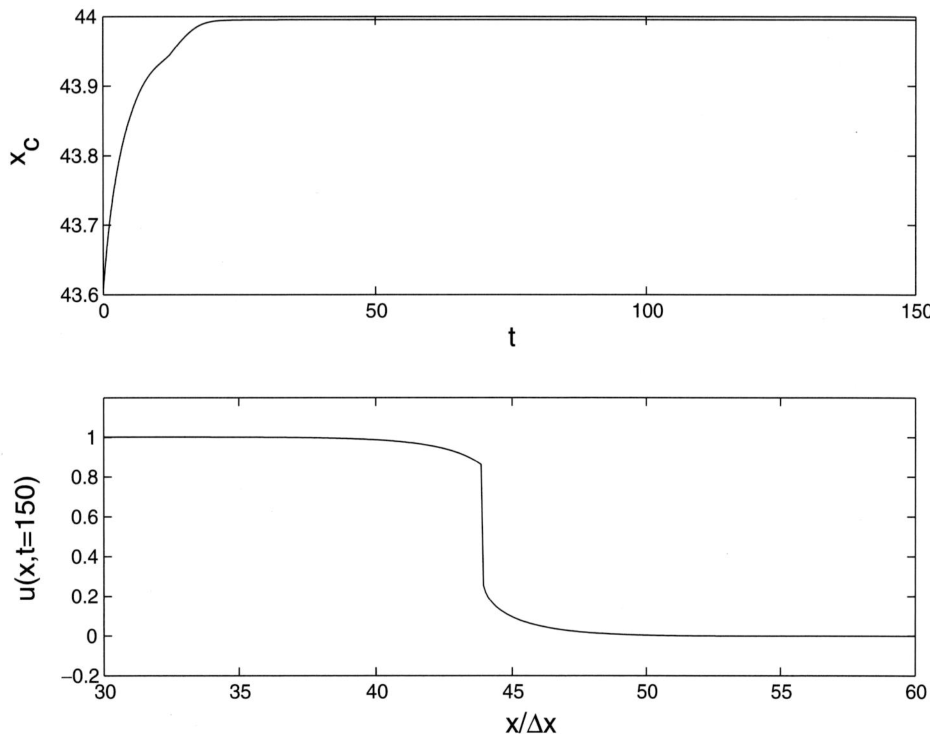

Fig. 9.   The figure shows the time evolution of the front for model A and for $\Delta x = 6.4$. The top panel shows the time evolution of the front center which eventually leads to trapping. The bottom panel shows the final front configuration of the numerical simulation at $t = 150$.

to traveling such as the Galilean invariance in the case of continuum bistable equation or the Lorentz invariance of its Hamiltonian (nonlinear Klein–Gordon) analog.

6. Summary and Discussion

We presented a computer-assisted approach for the *solution* of effective, translationally invariant equations for spatially discrete problems without deriving these equations in closed form. Assuming that such an equation exists, its time-one map is approximated through the coarse time stepper, constructed through an ensemble of appropriately initialized simulations of the detailed discrete problem. Combining the coarse time stepper with matrix-free based numerical analysis techniques, e.g. contraction mappings such as RPM, can then help analyze the unavailable effective equation. We are currently exploring the use of our coarse time stepper with coarse projective integration [Gear & Kevrekidis, 2003; Gear *et al.*, 2002; Kevrekidis *et al.*, 2003; Rico-Martinez *et al.*, 2004]. Matrix-free eigenanalysis techniques should also be explored, especially since they can help test the "fast slaving"

hypothesis underlying the existence of a closed effective equation (see, for example, the discussion in [Makeev *et al.*, 2002; Hummer & Kevrekidis, 2003]).

We also presented initial computational results exploring the effect of certain "construction parameters" of the approach: the number of shifted copies in the ensemble of initial conditions, as well as the time-horizon used. We included a comparison between our approach and a particular way of obtaining explicit approximate translationally invariant evolution equations for such a problem (the Padé approximation). More work is necessary along these lines, exploring the relation of our approach with traditional homogenization methods at small lattice spacings. A discrete problem whose detailed solution can be obtained explicitly (perhaps a piecewise-linear kinetics problem) or at least approximated very well analytically over short times, would be the ideal context in which to study these issues.

Several extensions of the approach can be envisioned, and might be interesting to explore. A time stepper based approach can be applied without modification to hybrid discrete-continuum media, e.g. continuum transport with a lattice of sources or sinks, such as cells secreting ligands into and binding them back from a liquid solution, [Pribyl *et al.*, 2003]. It is clear that it can be tried in more than one dimensions, and for regular lattices of different geometry. For irregular lattices the averaging "over all shifts" we performed here for periodic media can be substituted with a Monte Carlo sampling over the distribution of possible lattices that takes into account what we know about the statistical geometry of the lattices. In this paper we assumed that an equation existed and closed for the *expected shape* of the solution. Conceivably one can attempt to develop time steppers not only for the expectation (the first moment of a distribution of possible results), but, say, for the expectation *and* the standard deviation of possible results; the lifting operator would then have to be appropriately modified. Finally, our time stepper here was built on short simulations of the *entire* detailed discete system in space. Hybrid simulations, where a known, explicit effective equation is accurate over *part* of the physical domain can be done; an "overall hybrid coarse" time stepper (explicit equation over part of the domain, and the coarse time stepper in this paper over the rest of the domain) will then be used. In a multiscale context, we have proposed "gaptooth" and "patch dynamics" simulations [Gear *et al.*, 2003; Kevrekidis *et al.*, 2003], where the present coarse time stepper integrations are performed not over the entire domain, but over a mesh of small computational "boxes". Both hybrid and "gaptooth" simulations, if possible, require careful boundary conditions for the "handshaking" between the continuum equation and the discrete simulations, or the discrete simulations in distant boxes, effectively implementing smoothness of the solution of the unavailable effective equation (e.g. [Kevrekidis *et al.*, 2003; Li *et al.*, 1998a, 1983b; Shenoy *et al.*, 1999; E & Huang, 2001]).

We close with a discussion of the "onset of pinning", the transition around which our test example of the coarse time stepper was focused. Continuum effective equations such as the ones discussed here through the numerical time-stepping procedure do *not*, strictly speaking, possess a bifurcation at the critical point of the genuinely discrete problem. In this effective process, the bifurcation

is smeared out and rendered a "continuum transition" (see, for example, materials science models of the onset of movement of a front [Cahn, 1962; Maroudas & Brown, 1991]). On the other hand, one might argue that this is an acceptable, and possibly optimal way for a continuum equation to represent the discrete bifurcation to pinning. We can see that other procedures, such as the discreteness-emulating Padé type ones, lose a lot of the quantitative structure of the relevant transition. On the other hand, if a continuum *differential* (as opposed to pseudo-differential) equation was constructed to "model" this transition, the latter would possess other artificial features such as a topologically mandated, unstable branch of traveling wave solutions. See e.g. [Kness *et al.*, 1992] and references therein. It is conceivable that the short hysteresis loop sometimes predicted by the coarse time stepper close to pinning conditions is a "vestige" of this unstable branch that translationally invariant equations would necessarily predict. In conclusion, it can be appreciated that genuinely discrete problems and continuum ones have inherent differences[1] that cannot be fully captured by emulating (or "summarizing") the one context through the other. Nevertheless, the approach proposed here, combined with a "common sense" interpretation of its results with respect to the genuinely discrete problem, performs in a satisfactory way for the modeler, even for the "most different" features between discrete and continuum models.

Acknowledgments

Part of the research for this paper was carried out while Olof Runborg held a post-doctoral appointment with the Program for Applied and Computational Mathematics at Princeton University, supported by NSF KDI grant DMS-9872890. Panayotis G. Kevrekidis gratefully acknowledges support from a UMass FRG, NSF-DMS-0204585 and from the Eppley Foundation for Research. Kurt Lust is a postdoctoral fellow of the Fund for Scientific Research–Flanders. This paper presents research results of the Belgian Programme on Interuniversity Poles of Attraction, initiated by the Belgian State, Prime Minister's Office for Science, Technology and Culture. The scientific

---

[1]A similar example can be found in the comparison of discrete and *periodic* continuum problems, where the former ones possess a single permissible band of excitations, while the latter possess an infinity of such bands and hence allow for interband transitions [Alfimov *et al.*, 2002].

responsibility rests with its authors. Ioannis G. Kevrekidis gratefully acknowledges the support of AFOSR (Dynamics and Control) and an NSF-ITR grant.

References

Alfimov, G. L., Kevrekidis, P. G., Konotop, V. V. & Salerno, M. [2002] "Wannier functions analysis of the nonlinear Schrödinger equation with a periodic potential," *Phys. Rev.* **E66**, 046608-6.

Bates, P. W., Chen, X. F. & Chmaj, A. J. J. [2003] "Traveling waves of bistable dynamics on a lattice," *SIAM J. Math. Anal.* **35**, 520–546.

Beyn, W.-J. & Thümmler, V. [2004] "Freezing solutions of equivariant evolution equations," *SIAM J. Appl. Dyn. Syst.* **3**, 85–116.

Boesch, R., Willis, C. R. & El-Batanouny, M. [1989] "Spontaneous emission of radiation from a discrete sine-Gordon kink," *Phys. Rev.* **B40**, 2284–2296.

Cabannes, H. (ed.) [1976] *Padé Approximants Method and its Applications to Mechanics* (Springer-Verlag, Berlin).

Cahn, J. W. [1962] "The impurity drag effect in grain boundary motion," *Acta Metall.* **10**, 789–798.

Carpio, A. & Bonilla, L. [2003a] "Oscillatory wave fronts in chains of coupled nonlinear oscillators," *Phys. Rev.* **E67**, 056621-11.

Carpio, A. & Bonilla, L. L. [2003b] "Depinning transitions in discrete reaction-diffusion equations," *SIAM J. Appl. Math.* **63**, 1056–1082.

Chen, L.-Y. & Goldenfeld, N. [1995] "Numerical renormalization-group calculations for similarity solutions and traveling waves," *Phys. Rev.* **E51**, 5577–5581.

Christiansen, P. L., Gaididei, Y. G., Mertens, F. G. & Mingaleev, S. F. [2001] "Multi-component structure of nonlinear excitations in systems with length-scale competition," *Eur. Phys. J.* **B19**, 545–553.

Christodoulides, D. N. & Joseph, R. I. [1988] "Discrete self-focusing in nonlinear arrays of couped waveguides," *Opt. Lett.* **13**, 794–796.

Dawson, S. P., Keizer, J. & Pearson, J. E. [1999] "Fire-diffuse-fire model of dynamics of intracellular calcium waves," *Proc. Natl. Acad. Sci. USA* **96**, 6060–6063.

Doering, C. R., Hagan, P. S. & Rosenau, P. [1987] "Random-walk in a quasi-continuum," *Phys. Rev.* **A36**, 985–988.

E, W. & Huang, Z. [2001] "Matching conditions in atomistic-continuum modeling of materials," *Phys. Rev. Lett.* **87**, 135501-4.

Elphick, C., Meron, E. & Spiegel, E. A. [1990] "Patterns of propagating Pulses," *SIAM J. Appl. Math.* **50**, 490–503.

Fath, G. [1998] "Propagation failure of traveling waves in a discrete bistable Medium," *Physica* **D116**, 176–190.

Gear, C. W., Kevrekidis, I. G. & Theodoropoulos, C. [2002] "'Coarse' integration/bifurcation analysis via microscopic simulators: Micro-Galerkin methods," *Comp. Chem. Eng.* **26**, 941–963.

Gear, C. W. & Kevrekidis, I. G. [2003] "Projective methods for stiff differential equations: Problems with gaps in their eigenvalue spectrum," *SIAM J. Sci. Comput.* **24**, 1091–1106 (electronic).

Gear, C. W., Li, J. & Kevrekidis, I. G. [2003] "The gap-tooth method in particle simulations," *Phys. Lett.* **A316**, 190–195.

Hummer, G. & Kevrekidis, I. G. [2003] "Coarse molecular dynamics of a peptide fragment: Free energy, kinetics and long time dynamics computations," *J. Chem. Phys.* **118**, 10762–10773.

Ishimori, Y. & Munakata, T. [1982] "Kink dynamics in the discrete sine-Gordon system: A perturbational approach," *J. Phys. Soc. Jpn.* **51**, 3367–3374.

Keener, J. & Sneyd, J. [1998] *Mathematical Physiology* (Springer-Verlag, NY).

Keener, J. P. [1991] "The effects of discrete gap junction coupling on propagation in myocardium," *J. Theor. Biol.* **148**, 49–82.

Keener, J. P. [2000] "Homogenization and propagation in the bistable equation," *Physica* **D136**, 1–17.

Kevrekidis, I. G., Gear, C. W., Hyman, J. M., Kevrekidis, P. G., Runborg, O. & Theodoropoulos, C. [2003] "Equation-free, coarse-grained multiscale computation: Enabling microscopic simulators to perform system-level analysis," *Commun. Math. Sci.* **1**, 715–762; Original version can be found as physics/0209043 at arXiv.org.

Kevrekidis, P. G., Kevrekidis, I. G. & Bishop, A. R. [2001] "Propagation failure, universal scaling and Goldstone modes," *Phys. Lett.* **A279**, 361–369.

Kevrekidis, P. G., Kevrekidis, I. G., Bishop, A. R. & Titi, E. S. [2002] "Continuum approach to discreteness," *Phys. Rev.* **E65**, 046613-13.

Kness, M., Tuckermann, L. S. & Barkley, D. [1992] "Symmetry-breaking bifurcations in one-dimensional excitable media," *Phys. Rev.* **A46**, 5054–5062.

Laplante, J. P. & Erneux, T. [1992] "Propagation failure in arrays of coupled bistable chemical reactors," *J. Phys. Chem.* **96**, 4931–4934.

Li, J., Liao, D. & Yip, S. [1998a] "Coupling continuum to molecular-dynamics simulation: Reflecting particle method and the field estimator," *Phys. Rev.* **E57**, 7259–7267.

Li, J., Liao, D. & Yip, S. [1998b] "Imposing field boundary conditions in MD simulations of fluids: Optimal particle controller and buffer zone feedback," *Mat. Res. Soc. Symp. Proc.* **538**, 473–478.

Lust, K. [1997] "Numerical bifurcation analysis of periodic solutions of partial differential equations," Ph.D. thesis, Katholieke Universiteit Leuven.

Makeev, A. G., Maroudas, D. & Kevrekidis, I. G. [2002] "'Coarse' stability and bifurcation analysis using stochastic simulators: Kinetic Monte Carlo examples," *J. Chem. Phys.* **116**, 10083–10091.

Maroudas, D. & Brown, R. A. [1991] "Model for dislocation locking by oxygen gettering in silicon crystals," *Appl. Phys. Lett.* **58**, 1842–1844.

McLaughlin, D., Shapley, R., Shelley, M. & Wielaard, D. J. [2000] "A neuronal network model of macaque primary visual cortex (v1): Orientation tuning and dynamics in the input layer $4C\alpha$," *Proc. Natl. Acad. Sci. USA* **97**, 8087–8092.

Peyrard, M. & Kruskal, M. D. [1984] "Kink dynamics in the highly discrete sine-Gordon system," *Physica* **D14**, 88.

Peyrard, M. & Bishop, A. R. [1989] "Statistical mechanics of a nonlinear model for DNA denaturation," *Phys. Rev. Lett.* **62**, 2755–2758.

Pribyl, M., Muratov, C. B. & Shvartsman, S. [2003] "Discrete models of autocrine cell communication in epithelial layers," *Biophys. J.* **84**, 3624–3635.

Rico-Martinez, R., Gear, C. W. & Kevrekidis, I. G. [2004] "Coarse projective kMC integration: Forward/reverse initial and boundary value problems," *J. Comp. Phys.* **196**, 474–489.

Rinzel, J., Terman, D., Wang, X.-J. & Ermentrout, B. [1998] "Propagating activity patterns in large-scale inhibitory neuronal networks," *Science* **279**, 1351–1355.

Rosenau, P. [1986] "Dynamics of nonlinear mass-spring chains near the continuum-limit," *Phys. Lett.* **A118**, 222–227.

Rosenau, P. [1987] "Dynamics of dense lattices," *Phys. Rev.* **B36**, 5868–5876.

Rosenau, P. [1989] "Extending hydrodynamics via the regularization of Chapman-Enskog expansion," *Phys. Rev.* **A40**, 7193–7196.

Rosenau, P. [1992] "Tempered diffusion: A transport process with propagating fronts and inertial delay," *Phys. Rev.* **A46**, R7371–R7374.

Rowley, C. W. & Marsden, J. E. [2000] "Reconstruction equations and the Karhunen-Loeve expansion for systems with symmetry," *Physica* **D142**, 1–19.

Runborg, O., Theodoropoulos, C. & Kevrekidis, I. G. [2002] "Effective bifurcation analysis: a time-stepper based approach," *Nonlinearity* **15**, 491–511.

Shenoy, V. B., Miller, R., Tadmor, E. B., Rodney, D., Phillips, R. & Ortiz, M. [1999] "An adaptive finite element approach to atomic-scale mechanics — The quasicontinuum method," *J. Mech. Phys. Solids* **47**, 611–642.

Shroff, G. M. & Keller, H. B. [1993] "Stabilization of unstable procedures: The recursive projection method," *SIAM J. Numer. Anal.* **30**, 1099–1120.

Swanson, B. L., A. Brozik, J., Love, S. P., Strouse, G. F., Shreve, A. P., Bishop, A. P., Wang, W.-Z. & Salkola, M. I. [1999] "Observation of intrinsically localized modes in a discrete low-dimensional material," *Phys. Rev. Lett.* **82**, 3288–3291.

Ustinov, A. V., Doderer, T., Vernik, I. V., Pedersen, N. F., Huebener, R. P. & Oboznov, V. A. [1993] "Experiments with solitons in annular Josephson junctions," *Physica* **D68**, 41–44.

Zinner, B. [1991] "Stability of traveling wave-fronts for the discrete Nagumo Equation," *SIAM J. Math. Anal.* **22**, 1016–1020.

Zinner, B. [1992] "Existence of traveling wave-front solutions for the discrete Nagumo equation," *J. Diff. Eqs.* **96**, 1–27.

Zinner, B., Harris, G. & Hudson, W. [1993] "Traveling wave-fronts for the discrete Fisher's equation," *J. Diff. Eqs.* **105**, 46–62.

MODEL REDUCTION FOR FLUIDS, USING BALANCED PROPER ORTHOGONAL DECOMPOSITION

C. W. ROWLEY

Department of Mechanical and Aerospace Engineering,
Princeton University, Princeton, NJ 08544, USA

Received May 15, 2004; Revised June 7, 2004

Many of the tools of dynamical systems and control theory have gone largely unused for fluids, because the governing equations are so dynamically complex, both high-dimensional and nonlinear. Model reduction involves finding low-dimensional models that approximate the full high-dimensional dynamics. This paper compares three different methods of model reduction: proper orthogonal decomposition (POD), balanced truncation, and a method called balanced POD. Balanced truncation produces better reduced-order models than POD, but is not computationally tractable for very large systems. Balanced POD is a tractable method for computing approximate balanced truncations, that has computational cost similar to that of POD. The method presented here is a variation of existing methods using empirical Gramians, and the main contributions of the present paper are a version of the method of snapshots that allows one to compute balancing transformations directly, without separate reduction of the Gramians; and an output projection method, which allows tractable computation even when the number of outputs is large. The output projection method requires minimal additional computation, and has *a priori* error bounds that can guide the choice of rank of the projection. Connections between POD and balanced truncation are also illuminated: in particular, balanced truncation may be viewed as POD of a particular dataset, using the observability Gramian as an inner product. The three methods are illustrated on a numerical example, the linearized flow in a plane channel.

Keywords: Model reduction; proper orthogonal decomposition; balanced truncation; snapshots.

1. Introduction

The past several decades have produced major advances in techniques for analyzing dynamical systems, both analytically and numerically. However, despite continuing improvements in computing power, many systems of interest remain out of reach of these tools, because of their high dimension. For instance, the mechanisms by which a fluid flow transitions from laminar to turbulent are still not fully understood: at this point, it is not even clear whether the mechanisms are fundamentally nonlinear [Holmes *et al.*, 1996] or linear [Farrell & Ioannou, 1993; Bamieh & Daleh, 2001]. The full nonlinear partial differential equations that describe a fluid flow are too complex to be analyzed directly, so in order to answer questions such as these, lower-dimensional models that approximate the full system are desirable.

The problem of obtaining a lower-dimensional approximation to a high-dimensional dynamical system is known as *model reduction*. This paper reviews two well-known approaches to model reduction, and presents a method which compares favorably with both of these. The method of *proper orthogonal decomposition* (POD) and Galerkin projection is popular in the fluids community, and in this method, one obtains a lower-dimensional approximation by projecting the full nonlinear system onto a set of basis functions determined from empirical data. However, the POD/Galerkin method can yield unpredictable results, and is sensitive to details such as the empirical data used [Rathinam & Petzold, 2003], and the choice of inner product [Colonius & Freund, 2002]. POD/Galerkin models near stable equilibrium points can even be unstable [Smith, 2003].

A related method known as *balanced truncation* was developed in the control theory community for stable, linear, input–output systems, and does not suffer the same limitations as the POD method. Most notably, balanced truncation has error bounds that are close to the lowest error possible from any reduced-order model. In addition, this method has recently been extended to nonlinear systems using two distinct approaches [Lall *et al.*, 2002; Scherpen, 1993]. Balanced truncation has been used on some fluid problems [Cortelezzi & Speyer, 1998], but becomes computationally intractable for systems of very large dimension (e.g. 10 000 states or more), and so is not practical for many fluids systems.

This paper presents a method we refer to as *balanced* proper orthogonal decomposition, which combines ideas from POD and balanced truncation. The goal is to compute balanced truncations, or approximations to these, with computational cost similar to POD. Several previous methods have combined ideas from POD and balanced truncation, including the original work of Moore [1981]. The method presented here relies heavily on the work of Lall *et al.* [1999, 2002], who used empirical Gramians to generalize balanced truncation to nonlinear systems. Our goal is to use empirical Gramians to compute balancing transformations for very large systems. Previous works have addressed this problem as well, notably the work of Willcox and Peraire [2002], which used POD to compute low-rank approximations to the Gramians, from which the balancing transformation was computed using an efficient solver to find the eigenvectors of their product. However, this method has several drawbacks. In particular, it becomes intractable when the number of outputs is large, as a separate adjoint simulation is required for each output. Furthermore, in reducing the rank of the controllability and observability of Gramians before the balancing is performed, one risks prematurely truncating states that are poorly observable yet very strongly controllable, which can lead to less accurate models, as we shall see in the numerical example shown in Figs. 7 and 8.

The present method overcomes this latter drawback using a different method of snapshots, described in Sec. 3.1, in which one computes the balancing transformation directly from the snapshots, without individual reduction of the Gramians, and without a separate eigenvector solve. Furthermore, we describe an output projection method in Sec. 3.2, which allows the empirical observability Gramian to be computed even when the number of outputs is large, using many fewer adjoint simulations. This output projection is optimal in an L_2 sense, involves very little extra computation, and comes with an *a priori* error bound which can guide the rank of the output projection used [Eq. (27)]. Like balanced truncation, the present method is limited to stable, linear systems. However, because our method uses many of the same ideas as Lall *et al.* [1999] (in particular, empirical Gramians constructed from impulse responses), it is likely that similar computational techniques may be applied to nonlinear systems as well.

The paper is outlined as follows: in Sec. 2, we review the methods of POD/Galerkin projection and balanced truncation; we present our method in Sec. 3; and in Sec. 4, we compare the three methods on a example, the linearized flow in a plane channel.

2. Background on Model Reduction

The model reduction methods discussed in this paper fall in the category of *projection methods*, in that they involve projecting the equations of motion onto a subspace of the original phase space. The methods of POD/Galerkin and balanced truncation are briefly reviewed here, both for comparison with balanced POD, and also because our method uses ideas from both POD and balanced truncation. There are many other methods available for reducing both linear and nonlinear systems, and several of these are reviewed in [Antoulas *et al.*, 2001].

2.1. *Proper orthogonal decomposition*

Proper orthogonal decomposition, also known as principal component analysis, or the Karhunen–Loéve expansion, has been used for some time in developing low-dimensional models of fluids [Lumley, 1970; Sirovich, 1987; Holmes *et al.*, 1996]. The idea is, given a set of data that lies in a vector space $\mathcal{V}$, to find a subspace $\mathcal{V}_r$ of fixed dimension r

such that the error in the projection onto the subspace is minimized. Here, for simplicity, we will consider the case where $V = \mathbb{R}^n$. For a fluid, V will be infinite-dimensional, consisting of functions on some spatial domain (for instance, velocity and pressure everywhere), but we will assume that the equations have already been discretized in space, for instance by a finite-difference or spectral method, so that V has finite dimension n (e.g. for a finite-difference simulation, n is the number of grid-point times the number of flow variables). For the infinite-dimensional case, see [Holmes *et al.*, 1996; Rowley *et al.*, 2004].

Suppose we have a set of data given by $x(t) \in \mathbb{R}^n$, with $0 \leq t \leq T$. We seek a projection $P_r : \mathbb{R}^n \to \mathbb{R}^n$ of fixed rank r, that minimizes the total error

$$\int_0^T \|x(t) - P_r x(t)\|^2 \, dt. \tag{1}$$

To solve this problem, introduce the $n \times n$ matrix

$$R = \int_0^T x(t)x(t)^* \, dt, \tag{2}$$

where * denotes the transpose, and find the eigenvalues and eigenvectors of R, given by

$$R\varphi_k = \lambda_k \varphi_k, \quad \lambda_1 \geq \cdots \geq \lambda_n \geq 0. \tag{3}$$

Since R is symmetric, positive-semidefinite, all the eigenvalues λ_k are real and non-negative, and the eigenvectors φ_k may be chosen to be orthonormal. The main result of POD is that the optimal subspace of dimension r is spanned by $\{\varphi_1, \ldots, \varphi_r\}$, and the optimal projection P_r is then given by

$$P_r = \sum_{k=1}^r \varphi_k \varphi_k^*. $$

The vectors φ_k are called *POD modes*.

2.1.1. *Galerkin projection*

One can then form reduced order models using *Galerkin projection* onto this subspace. Suppose the dynamics of a system are described by

$$\dot{x}(t) = f(x(t)). \tag{4}$$

Galerkin projection specifies dynamics of a variable $x_r(t) \in \text{span}\{\varphi_1, \ldots, \varphi_r\}$ by $\dot{x}_r(t) = P_r f(x_r(t))$, that is, simply projecting the original vector field f onto the r-dimensional subspace. Writing

$$x_r(t) = \sum_{j=1}^r a_j(t)\varphi_j, \tag{5}$$

substituting into the equations, and multiplying by φ_k^*, one obtains

$$\dot{a}_k(t) = \varphi_k^* f(x_r), \quad k = 1, \ldots, r, \tag{6}$$

a set of r ODEs that describe the evolution of $x_r(t)$.

2.1.2. *Method of snapshots*

To compute the POD modes, one must solve an $n \times n$ eigenvalue problem (3). For a discretization of a fluid problem, the dimension n often exceeds 10^6, so direct solution of this eigenvalue problem is not often feasible. If the data is given as "snapshots" $x(t_j)$ at discrete times $t_1, \ldots, t_m$, then one can transform the $n \times n$ eigenvalue problem (3) into an $m \times m$ eigenvalue problem [Sirovich, 1987]. In this case, the integral in (3) becomes a sum

$$R = \sum_{j=1}^m x(t_j)x(t_j)^* \delta_j \tag{7}$$

where δ_j are quadrature coefficients. Assembling the data into an $n \times m$ matrix

$$X = \begin{bmatrix} x(t_1)\sqrt{\delta_1} & \cdots & x(t_m)\sqrt{\delta_m} \end{bmatrix} \tag{8}$$

the sum (7) may be written $R = XX^*$. In the method of snapshots, one then solves the $m \times m$ eigenvalue problem

$$X^* X u_k = \lambda_k u_k, \quad u_k \in \mathbb{R}^m, \tag{9}$$

where the eigenvalues λ_k are the same as in (3). The eigenvectors u_k may be chosen to be orthonormal, and the POD modes are then given by $\varphi_k = X u_k / \sqrt{\lambda_k}$. In matrix form, with $\Phi = [\varphi_1 \ \cdots \ \varphi_m]$, and $U = [u_1 \ \cdots \ u_m]$, this becomes

$$\Phi = XU\Lambda^{-1/2}. \tag{10}$$

The $m \times m$ eigenvalue problem (9) is more efficient than the $n \times n$ eigenvalue problem (3) when the number of snapshots m is smaller than the number of states n.

2.1.3. *Remarks and limitations*

A physical explanation of POD modes is that they maximize the average energy in the projection of the data onto the subspace spanned by the modes. This is equivalent to minimizing the error (1), since

$$\arg \min_{\{\varphi_k\}} \langle \|x - P_r x\|^2 \rangle = \arg \max_{\{\varphi_k\}} \langle \|P_r x\|^2 \rangle$$

where $\langle \cdot \rangle$ is the average over the data ensemble (this follows from the Pythagorean theorem, since P_r is an orthogonal projection). In particular, the energy in the projection is given by

$$\int_0^T \|P_r x(t)\|^2 \, dt = \sum_{k=1}^r \lambda_k. \tag{11}$$

Though POD modes are very effective (indeed optimal) at approximating a given dataset, they are not necessarily the best modes for describing the *dynamics* that generate a particular dataset, since low-energy features may be critically important to the dynamics. For instance, in a fluid flow where acoustic resonances occur, acoustic waves play a crucial role, even though they have much smaller energy than hydrodynamic pressure fluctuations. In practice, one sometimes neglects some of the higher-energy POD modes in forming reduced-order models [Smith, 2003], in favor of lower-energy modes that are more dynamically important. In fact, adding more POD modes can even make dynamical models worse [Rowley *et al.*, 2004]. These are undesirable characteristics of a model reduction procedure, and part of the motivation behind balanced POD is to improve on these limitations.

2.2. *Balanced truncation*

Balanced truncation is a method of model reduction for stable, linear input–output systems, introduced by [Moore, 1981]. Consider a stable linear input–output system

$$\begin{aligned} \dot{x} &= Ax + Bu \\ y &= Cx \end{aligned} \tag{12}$$

where $u(t) \in \mathbb{R}^p$ is a vector of inputs, $y(t) \in \mathbb{R}^q$ is a vector of outputs, and $x(t) \in \mathbb{R}^n$ is the state vector.

One begins by defining *controllability* and *observability Gramians*, which are symmetric, positive-semidefinite matrices defined by

$$W_c = \int_0^\infty e^{At} BB^* e^{A^*t} \, dt$$

$$W_o = \int_0^\infty e^{A^*t} C^* C e^{At} \, dt, \tag{13}$$

usually computed by solving the Lyapunov equations

$$\begin{aligned} AW_c + W_c A^* + BB^* &= 0 \\ A^* W_o + W_o A + C^* C &= 0. \end{aligned} \tag{14}$$

The controllability Gramian W_c measures to what degree each state is excited by an input. For two states x_1 and x_2 with $\|x_1\| = \|x_2\|$, if $x_1^* W_c x_1 > x_2^* W_c x_2$, then state x_1 is "more controllable" than x_2 (i.e. it takes a smaller input to drive the system from rest to x_1 than to x_2). The Gramian W_c is positive-definite if and only if all states are reachable with some input $u(t)$.

Conversely, the observability Gramian W_o measures to what degree each state excites future outputs. For an initial state x_0, and with zero input, one has $\|y\|_2^2 = x_0^* W_o x_0$, where $\|\cdot\|_2$ denotes the $L_2[0, \infty)$ norm. States which excite larger output signals are called "more observable," and in this sense are more dynamically important than states that are less observable.

The Gramians depend on the coordinates, and under a change of coordinates $x = Tz$, they transform as

$$W_c \mapsto T^{-1} W_c (T^{-1})^*, \quad W_o \mapsto T^* W_o T.$$

Balancing refers to changing to coordinates in which the controllability and observability properties are balanced — more precisely, the transformed Gramians are equal and diagonal:

$$T^{-1} W_c (T^{-1})^* = T^* W_o T = \Sigma = \mathrm{diag}(\sigma_1, \ldots, \sigma_n). \tag{15}$$

The diagonal elements $\sigma_1 \geq \cdots \geq \sigma_n \geq 0$ are called the *Hankel singular values* of the system, and are independent of the coordinate system. A basic result is that a balancing transformation T exists as long as the system is both controllable and observable (i.e. $W_c, W_o > 0$). The transformation is found by computing appropriately scaled eigenvectors of the product $W_c W_o$ (in particular, $W_c W_o T = T\Sigma^2$). In the balanced coordinates, the states that are least influenced by the input also have the least influence on the output. Balanced truncation involves first changing to these coordinates, and then truncating the least controllable/observable states, which have little effect on the input–output behavior.

2.2.1. *Error bounds*

A useful property of balanced truncation is that one has *a priori* error bounds that are close to the lower bound achievable by any reduced-order model. To understand these error bounds, consider the transfer function

$$\hat{G}(s) = C(sI - A)^{-1} B,$$

which relates the Laplace transform of the input to the Laplace transform of the output ($\hat{y}(s) = \hat{G}(s)\hat{u}(s)$). The L_2-induced operator norm of $\hat{G}$ is defined by

$$\max_u \frac{\|Gu\|_2}{\|u\|_2} = \|G\|_\infty \equiv \max_\omega \sigma_1(\hat{G}(i\omega)), \qquad (16)$$

where $\sigma_1(M)$ denotes the maximum singular value of the matrix M. The following error bounds are standard results [Dullerud & Paganini, 1999]: first, any reduced order model G_r with r states must satisfy

$$\|G - G_r\|_\infty > \sigma_{r+1}, \qquad (17)$$

where σ_{r+1} is the first neglected Hankel singular value of G. This is a fundamental limitation for any reduced order model. Balanced truncation also guarantees an upper bound of the error:

$$\|G - G_r\|_\infty < 2 \sum_{j=r+1}^{n} \sigma_j, \qquad (18)$$

which is usually close to the lower bound (17), if the Hankel singular values drop off quickly. Balanced truncation is not optimal, in the sense that there may be other reduced-order models with smaller error norms, but *a priori* guarantees and strong heuristic justification make it a popular and effective technique.

2.2.2. *Empirical Gramians*

Instead of computing the Gramians by solving Lyapunov equations (14), one may compute them from data from numerical simulations. This was the original approach used by [Moore, 1981], and was

used in [Lall *et al.*, 1999, 2002] to extend balanced truncation to nonlinear systems.

2.2.3. *Controllability Gramian*

To compute the controllability Gramian for a system with p inputs, writing $B = [b_1, \ldots, b_p]$, one forms the state responses to unit impulses

$$x_1(t) = e^{At}b_1$$
$$= \text{response to impulsive input } u_1(t) = \delta(t)$$
$$\vdots$$
$$x_p(t) = e^{At}b_p$$
$$= \text{response to impulsive input } u_p(t) = \delta(t)$$

Then the controllability Gramian is given by

$$W_c = \int_0^\infty (x_1(t)x_1(t)^* + \cdots + x_p(t)x_p(t)^*)dt. \qquad (19)$$

Note the similarity between the expression above and the operator in (2) that arises in POD of the dataset $\{x_1(t), \ldots, x_p(t)\}$. In fact, the POD modes for this dataset of impulse responses are just the largest eigenvectors of W_c, or, in other words, the most controllable modes of the realization. Note that since the Gramian matrices depend on the coordinate system, so do the POD modes of this dataset.

If data from simulations is used to find the impulse responses, then it is usually given at discrete times $t_1, \ldots, t_m$, and the integral above becomes a quadrature sum, as in (7), and we may stack the snapshots as columns of a matrix

$$X = [x_1(t_1)\sqrt{\delta_1} \quad \cdots \quad x_1(t_m)\sqrt{\delta_m} \quad \cdots \quad x_p(t_1)\sqrt{\delta_1} \cdots x_p(t_m)\sqrt{\delta_m}], \qquad (20)$$

where again δ_j are quadrature coefficients. The quadrature approximation to (19) is then

$$W_c = XX^*. \qquad (21)$$

2.2.4. *Observability Gramian*

The procedure for computing the empirical observability Gramian proceeds analogously: we compute impulse responses of the adjoint system

$$\dot{z} = A^*z + C^*v.$$

If q is the number of outputs and $C^* = (c_1, \ldots, c_q)$, then let

$$z_1(t) = e^{A^*t}c_1$$
$$= \text{response to impulsive input } v_1(t) = \delta(t)$$
$$\vdots$$
$$z_q(t) = e^{A^*t}c_q$$
$$= \text{response to impulsive input } v_q(t) = \delta(t),$$

from which the observability Gramian is given by

$$W_o = \int_0^\infty (z_1(t)z_1(t)^* + \cdots + z_q(t)z_q(t)^*)dt.$$

One then forms the data matrix Y, as in (20), and writes the Gramian as

$$W_o = YY^*.$$

Note that this method requires q integrations of the adjoint system, where q is the number of outputs. Thus, this method is not feasible when the number of outputs is large, for instance, if the output is the full state. The empirical Gramian may also be computed from n simulations of the primal system $\dot{x} = Ax$, where n is the number of states (as is done in [Lall *et al.*, 2002]), but clearly this is also not feasible when the number of states is large. This difficulty is the motivation behind the output projection method to be discussed in Sec. 3.2.

3. Balanced POD

The main idea of balanced POD is to obtain an approximation to balanced truncation that is computationally tractable for large systems. The present method involves two components: computing the balancing transformation directly from snapshots of empirical Gramians, without needing to compute the Gramians themselves; and an output projection method to enable tractable computation even when the number of outputs is large. The method has deep connections with POD: it may be viewed as POD with respect to a particular inner product, or as a biorthogonal decomposition, as discussed in Sec. 3.4.

3.1. *Balanced truncation using the method of snapshots*

Suppose the controllability and observability Gramians may be factored as

$$W_c = XX^*, \quad W_o = YY^*, \tag{22}$$

where W_c and W_o are $n \times n$ square matrices, but X and Y may be rectangular, with differing dimensions. For instance, X and Y may be data matrices used to form empirical Gramians, as described in the previous section. In the method of snapshots used here, the balancing modes are computed by forming the singular value decomposition (SVD) of

the matrix Y^*X:

$$
\begin{aligned}
Y^*X &= U\Sigma V^* \\
&= \begin{bmatrix} U_1 & U_2 \end{bmatrix} \begin{bmatrix} \Sigma_1 & 0 \\ 0 & 0 \end{bmatrix} \begin{bmatrix} V_1^* \\ V_2^* \end{bmatrix} \\
&= U_1\Sigma_1 V_1^*
\end{aligned}
\tag{23}
$$

where $\Sigma_1 \in \mathbb{R}^{r \times r}$ is invertible, r is the rank of Y^*X, and $U_1^*U_1 = V_1^*V_1 = I_r$. Define the matrices $T_1 \in \mathbb{R}^{n \times r}$ and $S_1 \in \mathbb{R}^{r \times n}$ by

$$T_1 = XV_1\Sigma_1^{-1/2}, \quad S_1 = \Sigma_1^{-1/2}U_1^*Y^*. \tag{24}$$

A proposition proved in the appendix establishes that if $r = n$ (that is, the Gramians are full rank), then the matrix Σ_1 contains the Hankel singular values, T_1 determines the balancing transformation, and S_1 is its inverse. Furthermore, if $r < n$, then the columns of T_1 form the first r columns of the balancing transformation, and the rows of S_1 form the first r rows of the inverse transformation.

Remark. The major advantage of the above method for computing the balancing transformation is that the Gramians themselves never need to be computed. Only one SVD is needed, of a matrix with dimension $N_p \times N_d$, where N_p is the number of primal snapshots (columns of X), and N_d is the number of dual snapshots (columns of Y). If the number of snapshots is much smaller than the number of states n, as is typical for a problem in fluids, then this represents considerable savings. In particular, the size of the SVD is independent of n, and once the snapshots are computed, the entire method scales linearly with n. Thus, the overall computation time is similar to POD (compare (23)–(24) with (9)–(10)), except that here one also needs to compute adjoint snapshots, which do not arise in POD.

The method above is also similar to a well-known method for computing balancing transformations from the Cholesky factorization of the Gramians [Laub *et al.*, 1987]. The present method differs in that the factorization (22) need not be the Cholesky factorization, and neither of the Gramians needs to be full-rank. (In particular, the system does not need to be controllable or observable.) The present method does share the same desirable numerical characteristics as the method in [Laub *et al.*, 1987], in particular that the Gramians never need to be "squared up," and thus the method is less sensitive to numerical round-off than methods that involve computing the full Gramians W_c and W_o, rather than a factorization.[1]

3.2. *Output projection*

Recall from Sec. 2.2 that in order to compute data for the observability Gramian, one requires q simulations of the adjoint system, where q is the number of outputs. This procedure is clearly not feasible if the number of outputs is large. The idea of this section is to alleviate this problem by projecting the output onto an appropriate subspace, in such a way that the input–output behavior is almost unchanged. Instead of the system (12), consider the related system

$$\dot{x} = Ax + Bu$$
$$y = P_r C x \tag{25}$$

where P_r is an orthogonal projection with rank r. Such a projection allows us to compute the empirical observability Gramian using only r simulations of the adjoint system, rather than q simulations. To see this, write the projection P_r as the product $P_r = \Phi_r \Phi_r^*$, where Φ_r is a $q \times r$ matrix, with $\Phi_r^* \Phi_r = I_r$ (this can always be done for any orthogonal projection). The observability Gramian (13) then becomes

$$W_o = \int_0^\infty e^{A^* t} C^* \Phi_r \Phi_r^* C e^{At}\, dt$$

and so may be computed from r simulations of the adjoint system

$$\dot{z}(t) = A^* z + C^* \Phi_r v$$

where $v \in \mathbb{R}^r$. When the number of outputs q is large, the reduction in computational cost is substantial.

We would like to choose P_r such that the input–output behavior of (25) is as close as possible to the input–output behavior of (12). We can measure this input–output behavior by considering the *impulse response matrix* $G(t)$, whose element $G_{ij}(t)$ is the output component $y_i(t)$ corresponding to an impulsive input $u_j(t) = \delta(t)$. The impulse response completely determines the input–output behavior of a linear system. If $G(t)$ is the impulse response of (12), then the impulse response of (25) is

$P_r G(t)$, and we seek a projection P_r that minimizes the error

$$\int_0^\infty \|G(t) - P_r G(t)\|^2\, dt \tag{26}$$

with respect to some norm on matrices. If we use a norm induced by an inner product, for instance the Frobenius norm $\|A\|_F^2 = \mathrm{Tr}(A^* A)$, which is induced by the inner product $\langle A, B \rangle = \mathrm{Tr}(A^* B)$, then *the projection P_r that minimizes the error (26) is the projection onto the first r POD modes of the dataset $G(t)$.* For instance, if $\Phi_r = [\varphi_1 \ \cdots \ \varphi_r]$ is a matrix containing the first r POD modes of $G(t)$, then $P_r = \Phi_r \Phi_r^*$ is the projection that minimizes (26).

A convenient numerical feature of this method for computing P_r is that the necessary snapshots for computing the POD modes of $G(t)$ have already been computed, for the empirical controllability Gramian. To compute the snapshots for W_c, as in Sec. 2.2, we compute impulse responses $x_1(t), \ldots, x_p(t)$, for each of the p inputs. The dataset required for computing P_r is simply $C x_1(t), \ldots, C x_p(t)$, so we need only to multiply each of our snapshots by the output matrix C.

3.2.1. *Error bounds*

One can also quantify the error for the projected system. In particular, if $\lambda_1, \ldots, \lambda_m$ denote the POD eigenvalues of the dataset $\{C x_1(t), \ldots, C x_p(t)\}$, then

$$\|G - P_r G\|_2^2 = \sum_{j=r+1}^{m} \lambda_j, \tag{27}$$

where m is the number of outputs, and the 2-norm is given by

$$\|G\|_2^2 = \int_0^\infty \mathrm{Tr}(G(t)^* G(t))\, dt. \tag{28}$$

The proof follows immediately from a variant of (11). This result gives us guidance in choosing the number of modes to keep in the projection, based on the desired accuracy of the reduced-order model, and the POD eigenvalues computed from the impulse response data.

---

[1]As one reviewer remarked, POD modes may also be computed by a SVD of the snapshot matrix X from (8). This approach also has better roundoff properties than computing the eigenvalue decomposition of $X^* X$ as in (9), although it requires more computation.

3.3. *Summary*

To summarize, the steps in the balanced POD method are as follows:

1. Integrate solutions $x_1(t), \ldots, x_p(t)$ of the system $\dot{x} = Ax$, with initial conditions $x_k(0) = b_k$, where b_k denotes the kth column of the B matrix in (12).
2. Compute POD modes φ_k of the dataset $\{Cx_1(t), \ldots, Cx_p(t)\}$, and choose a projection rank r such that the error (27) is acceptable.
3. Integrate solutions $z_1(t), \ldots, z_r(t)$ of the adjoint system $\dot{z} = A^* z$, with initial conditions $z_k(0) = C^* \varphi_k$.
4. Form the data matrices X and Y for the primal and dual solutions, as in (20).
5. Compute the SVD of $Y^* X$, and the balanced POD modes are given by (24).

If the number of outputs is small, then one may skip step 2 and in step 3 use initial conditions $z_k(0) = c_k^*$, where c_k is the kth row of C.

Reduced-order models may then be formed by transforming to balanced coordinates and projecting. Note that there is no need to transform all of the states: if we write

$$x(t) = Tz(t) = \begin{bmatrix} T_1 & T_2 \end{bmatrix} \begin{bmatrix} z_1(t) \\ z_2(t) \end{bmatrix}$$

$$= T_1 z_1(t) + T_2 z_2(t),$$

where $z_1(t)$ are states to be retained and $z_2(t)$ are states to be truncated, then the transformed equations are

$$\dot{z}_1 = S_1 A T_1 z_1 + S_1 A T_2 z_2 + S_1 B u$$

$$\dot{z}_2 = S_2 A T_1 z_1 + S_2 A T_2 z_2 + S_2 B u$$

$$y = C T_1 z_1 + C T_2 z_2,$$

where $S = T^{-1}$. Setting $z_2 = 0$ gives the truncated model

$$\dot{z}_1 = S_1 A T_1 z_1 + S_1 B u$$

$$y = C T_1 z_1$$

Thus, to compute a reduced-order model of order r, all we need is the first r columns of T and the first r rows of S, given by (24). Note, however, that this is not the same as orthogonal projection onto the subspace spanned by the first r columns of T, since the columns of T are not orthogonal.

3.4. *Relation to POD*

There are deep connections between the POD/Galerkin method and balanced truncation, which are elucidated by the balanced POD procedure. For instance, balanced truncation may be viewed as a biorthogonal decomposition, instead of the orthogonal decomposition given by POD. Alternatively, balanced truncation may be viewed as a special case of POD, using a particular dataset (impulse responses), and using the observability Gramian as an inner product. The former point of view is useful for numerics, and the latter is useful for analysis, as it yields a guarantee that if balanced POD is used, then Galerkin projections of stable nonlinear systems are guaranteed to be stable as well.

3.4.1. *Biorthogonal decomposition*

In the POD/Galerkin procedure, one finds a sequence of orthogonal basis functions $\{\varphi_j\}$, for projection of the dynamics. Balanced truncation can be viewed in the same way, but using a sequence of biorthogonal functions $\{\varphi_j\}, \{\psi_j\}$. Let the matrices T_1 and S_1 from (24) be written

$$T_1 = \begin{bmatrix} \varphi_1 & \cdots & \varphi_r \end{bmatrix}, \quad S_1 = \begin{bmatrix} \psi_1^* \\ \vdots \\ \psi_r^* \end{bmatrix},$$

with $\varphi_j, \psi_j \in \mathbb{R}^n$. Then since $S_1 T_1 = I_r$, we have $\psi_i^* \varphi_j = \delta_{ij}$, so the sequences are biorthogonal. Now, approximate $x(t)$ as in (5), as

$$x_r(t) = \sum_{j=1}^{r} a_j(t) \varphi_j, \quad a_j(t) = \psi_j^* x(t).$$

Substituting into the equation $\dot{x} = f(x)$, multiplying by ψ_k^* and using biorthogonality now gives

$$\dot{a}_k = \psi_k^* f(x),$$

which is identical to (6), but using the *adjoint modes* ψ_k for the projection. Of course, one needs a linear system to define Gramians or adjoint equations, but the idea is that even for a nonlinear system, one may compute balancing modes $\{\varphi_j\}, \{\psi_j\}$ using a linearization, or a method similar to that in [Lall *et al.*, 2002], and then project the nonlinear system $\dot{x} = f(x)$ without having to transform the entire state before truncating.

3.4.2. *Observability Gramian as an inner product*

One of the difficulties with the POD/Galerkin method is that the inner product used for computing POD modes and projecting the dynamics is arbitrary. Sometimes, an appropriate inner product is obvious, as for incompressible flow [Holmes *et al.*, 1996], but other times, as for compressible flow, a suitable inner product is not obvious [Rowley *et al.*, 2004], and different choices can give dramatically different results [Colonius & Freund, 2002]. Perhaps the deepest connection between POD/Galerkin and balanced truncation is that for a stable linear system, balanced truncation may be viewed as a *special case* of POD, using impulse responses for a dataset (i.e. the matrix X in (20)), and using the observability Gramian as an inner product.

To see this, first define an inner product on $\mathbb{R}^n$ by

$$\langle a, b \rangle_{W_o} = a^* W_o b \qquad (29)$$

where W_o is the observability Gramian (which is positive definite as long as the system is observable). As mentioned in Sec. 2.2, W_o measures states of large "dynamical importance," so this inner product weights dynamically important states more heavily. The POD modes of the dataset X with respect to this inner product are eigenvectors of $R = XX^* W_o$ (see [Rowley *et al.*, 2004] for an explanation of POD with respect to an arbitrary inner product). These eigenvectors will be orthogonal with respect to the inner product (29), though not with respect to the standard inner product.

POD modes are normalized balancing modes. Since the dataset X was produced such that $XX^* = W_c$, the POD modes are just the eigenvectors of $R = W_c W_o$: in other words, they are the balancing modes, normalized differently. Furthermore, the eigenvalues of R are the squares of the Hankel singular values. If we compute the POD modes using the method of snapshots as in (9), we form the SVD $X^* W_o X = V_1 \Sigma_1^2 V_1^*$, and the POD modes are columns of

$$\Phi = [\tilde\varphi_1 \quad \cdots \quad \tilde\varphi_r] = X V_1 \Sigma^{-1}.$$

Note that these modes are the same as columns of T_1 in (24), with a different scaling. If we define "adjoint modes" $\tilde\psi_j = W_o \tilde\varphi_j$, then

$$\langle \tilde\varphi_i, \tilde\varphi_j \rangle_{W_o} = \tilde\varphi_i^* W_o \tilde\varphi_j = \tilde\psi_i^* \tilde\varphi_j = \delta_{ij}$$

so these adjoint modes may be viewed as a biorthogonal decomposition with respect to the standard inner product $\langle \psi, \varphi \rangle = \psi^* \varphi$, as in the previous section. These adjoint modes are also rescaled versions of the rows of S_1 in (24), since one easily checks that, with $W_o = YY^*$, and $X^* Y = U_1 \Sigma_1 V_1^*$,

$$\tilde S_1 := \begin{bmatrix} \tilde\psi_1^* \\ \vdots \\ \tilde\psi_r^* \end{bmatrix} = \Phi^* W_o = \Sigma_1^{-1} V_1^* X^* YY^* = U_1^* Y^*,$$

a rescaling of S_1 in (24).

3.4.3. *Guaranteed stability*

A useful consequence of using the observability Gramian as an inner product for Galerkin projection is that in this case, the reduced-order model preserves the stability of an equilibrium point at the origin, even if the full model is nonlinear. It is well-known that balanced truncations of stable linear systems are stable, but POD/Galerkin models of nonlinear systems may be unstable even if the nonlinear system is linearly stable at the origin [Smith, 2003].

The stability result follows from a result in [Rowley *et al.*, 2004]: if the norm induced by an inner product is a Lyapunov function for a nonlinear system with a stable equilibrium point at the origin, then orthogonal projection of the dynamics onto *any* subspace will also be stable at the origin. One sees from (14) that $V(x) = \langle x, x \rangle_{W_o}$ is a Lyapunov function of the linearized system $\dot x = Ax$, with $\dot V(x) = -C^* C \leq 0$. If the nonlinear system $\dot x = f(x)$ has a linearly stable equilibrium point at the origin, with $Df(0) = A$, then $V(x)$ is also a Lyapunov function for the nonlinear system, and so Galerkin projections using $\langle \cdot, \cdot \rangle_{W_o}$ will also be stable.

4. Example: Linearized Channel Flow

In order to compare the effectiveness of the three model reduction methods considered in this paper, we consider the problem of fluid flow in a plane channel. In particular, we use linearized equations with a coarse enough discretization that conventional balanced truncation is still computationally tractable. Since balanced POD is meant to approximate balanced truncation, we may evaluate how close the approximation is, and compare the resulting models to those formed with the standard POD/Galerkin method. Focusing on linearized equations allows us to use operator norms to objectively compare the errors in the reduced order models.

4.1.   *Equations of motion*

Consider the problem of a fluid flowing in a plane channel, as depicted in Fig. 1. We focus on the linearized case, considering small perturbations about a steady, laminar flow. The flow is assumed periodic in the x- and z-directions, with no-slip boundary conditions at the walls $y = \pm 1$. We force the flow with a body force given by $B(y,z)f(t)$, acting in the wall-normal direction (here $B(y,z)$ specifies the spatial distribution of the force, and $f(t)$ is regarded as an input). We restrict ourselves to streamwise-constant perturbations (no variations in the x-direction), and for this case the equations are given by

$$\frac{\partial v}{\partial t} = \frac{1}{R}\nabla^2 v + Bf$$

$$\frac{\partial \eta}{\partial t} = \frac{1}{R}\nabla^2 \eta - U'\frac{\partial v}{\partial z}$$

where v is the wall-normal velocity and $\eta = u_z - w_x$ is the perturbation in wall-normal vorticity. Numerical investigations indicate that the laminar velocity profile $\mathbf{u} = (U(y),0,0)$, with $U(y) = 1 - y^2$, is linearly stable for Reynolds numbers $R < 5772$ [Drazin & Reid, 1981], so the infinite-time Gramians will be well defined.

For the numerical examples considered here, we consider $R = 100$, on the domain $z \in [0, 2\pi]$, and discretize the problem using 16 Chebyshev modes in the y-direction, and 16 Fourier modes in the z-direction. The forcing $B(y,z)$ is zero everywhere except in a small region at the center of the domain ($y = 0$, $z = \pi$). We take the output to be the entire state, that is, the values of (v, η) everywhere in space. The total number of states is $2 \cdot 16 \cdot 15 = 480$,

which is small enough that we may compute the full Gramians exactly, for comparison with our approximate methods.

4.2.   *Results*

4.2.1.   *Hankel singular values*

We begin by comparing the Hankel singular values σ_j, shown in Fig. 2. Here, the exact values for balanced truncation are compared to the approximate values for balanced POD, for both five-mode and ten-mode output projections P_r. Also shown are the POD eigenvalues λ_j, computed from (9), and observe that the eigenvalues fall off quite rapidly. The first five POD modes capture 95.6% of the energy, while the first ten modes capture 99.8% of the energy. Thus, one expects that five-mode and ten-mode output projections should closely match the full input–output system.

In Fig. 2, the exact Hankel singular values are computed using the algorithm in [Laub *et al.*, 1987], while the approximate versions are computed from (23). Both the primal and dual solutions were computed using 1000 snapshots equally spaced within time $0 \le t \le 200$, by which time transients have decayed to a maximum value of 0.0002, from a maximum value of 1 at the initial time.

For the five-mode output projection, the first five singular values match closely, while for the ten-mode output projection, the first ten singular values match. Though there is no guarantee that for an output projection of rank r, the first r singular values will be approximated well, empirically this seems to be the case, at least for the channel flow problem.

4.2.2.   *Modes*

The first three modes are plotted in Figs. 3–5, which compare modes from exact balanced truncation, balanced POD with a five-mode output projection, and conventional POD. As explained in Sec. 3.4, for exact balanced truncation and balanced POD, the kth mode is the kth column of the transformation T, from (15) and (24), respectively. The POD modes are the eigenvectors from (3), also columns of the matrix Φ from (10).

The modes from balanced POD are nearly identical to those from exact balanced truncation, even for the five-mode output projection. For the ten-mode output projection, the modes also look

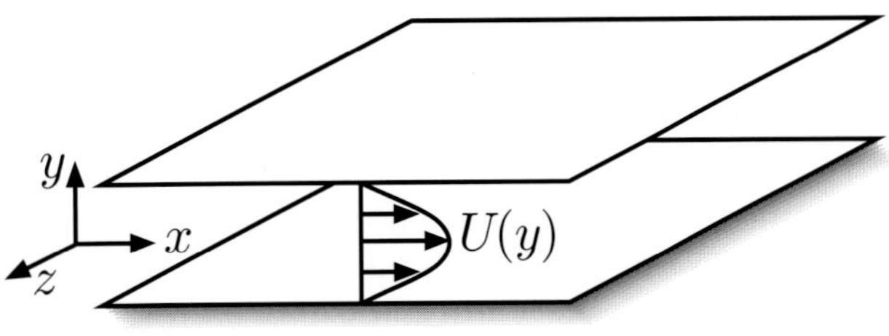

Fig. 1.   Schematic of channel flow example.

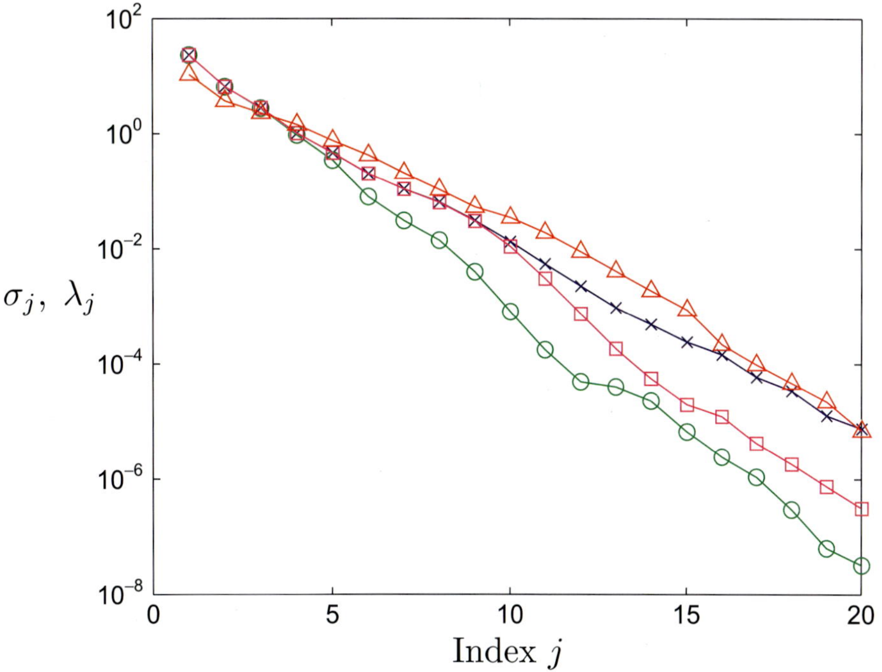

Fig. 2.   Hankel singular values σ_j for linearized channel flow: balanced truncation ($\times$), balanced POD with five-mode output projection ($\circ$), ten-mode output projection ($\square$); and POD eigenvalues λ_j ($\triangle$).

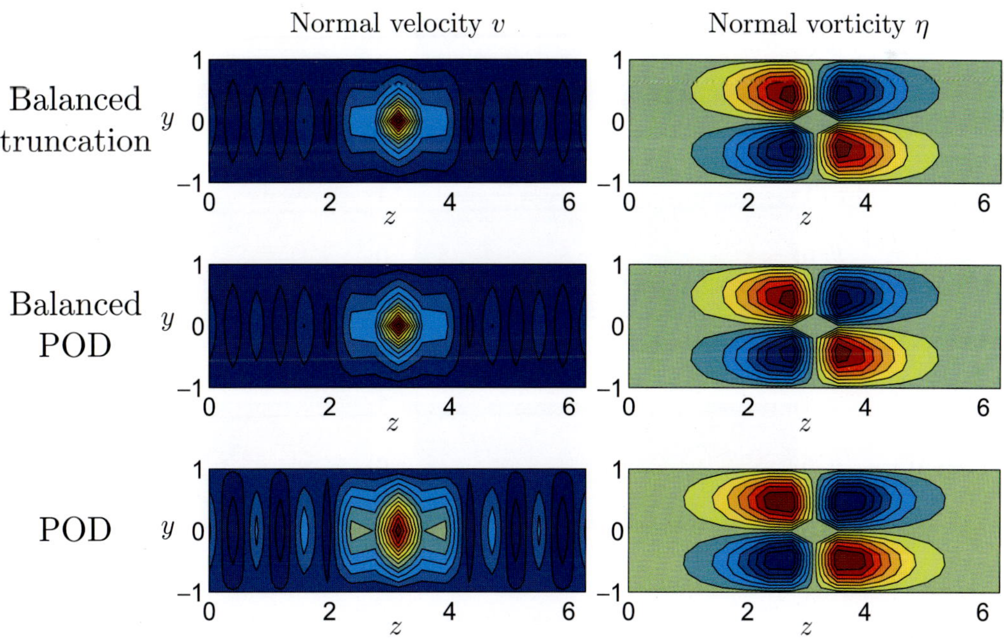

Fig. 3.   Mode 1 for channel flow.

visually identical, so these are not shown. The conventional POD modes look similar in general structure, especially mode 1, but there are distinct differences in modes 2 and 3. Of course, we would not expect the POD modes to be the same as the balancing modes, unless the observability Gramian Y is the identity, so it is interesting that the POD modes look so similar.

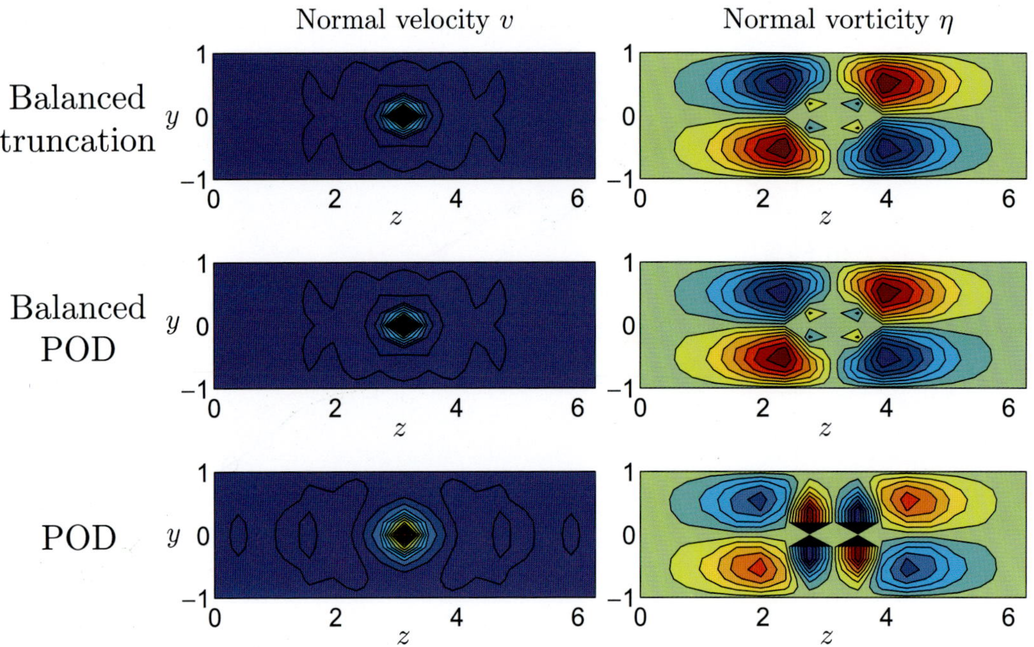

Fig. 4.   Mode 2 for channel flow.

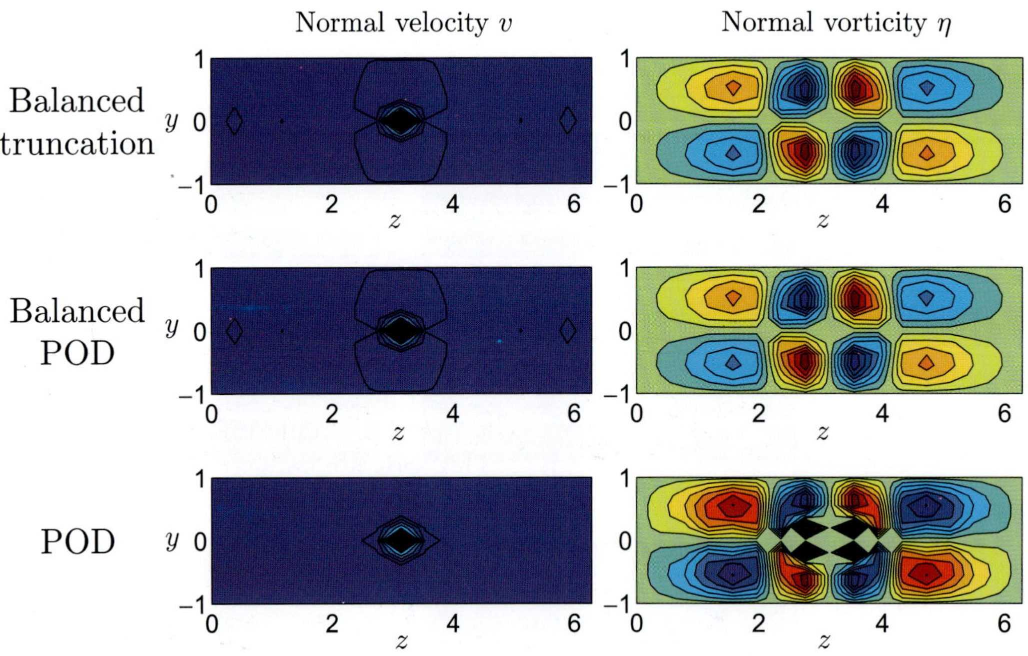

Fig. 5.   Mode 3 for channel flow.

4.2.3.   *Adjoint modes*

The corresponding adjoint modes for balanced POD are shown in Fig. 6. These look visually identical to the adjoint modes from balanced truncation (i.e. the first three rows of S_1 in (24)), so these are not shown. Recall that the POD modes are orthogonal, not biorthogonal, so the "adjoint modes" for POD are the same as the primal modes shown in Figs. 3–5. The adjoint modes in Fig. 6 look quite different from the primal modes or the POD modes,

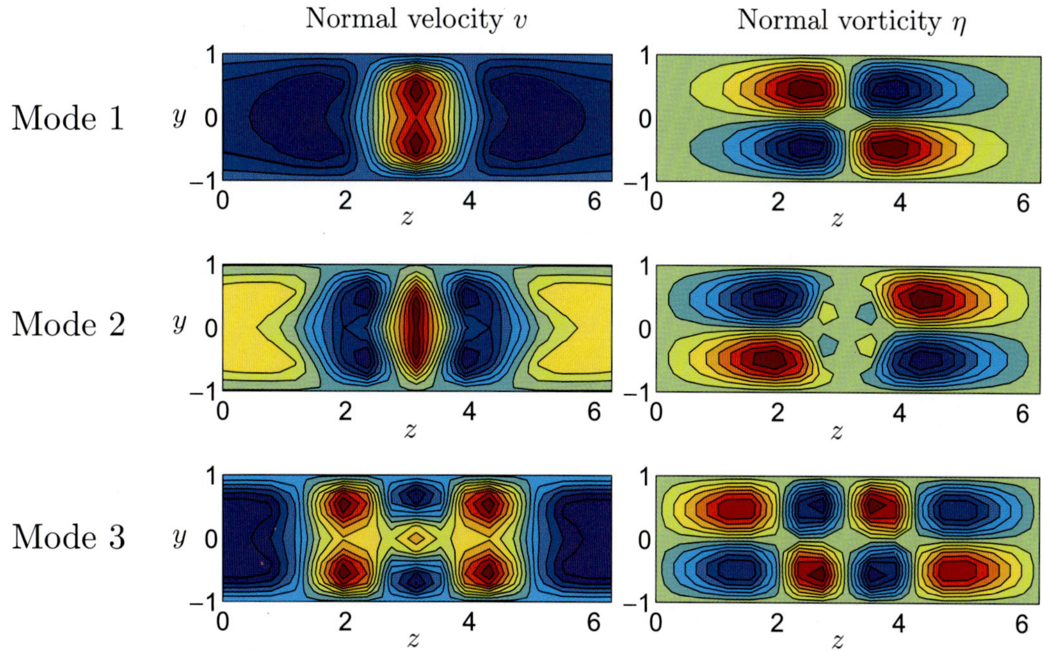

Fig. 6.   Adjoint modes 1–3 for balanced POD. The adjoint modes for balanced truncation are nearly identical, and the adjoint modes for POD are the same as the primal modes.

so it is reasonable to say that, for this problem, the main difference between balanced POD and conventional POD is the choice of inner product used for the projection.

4.2.4. *Error norms*

The main reason for using a linear system to compare these model reduction procedures is to have an objective measure of how effective the various reduced-order models are at approximating the full-order system. For linear systems, we have norms which enable such an objective comparison. Perhaps the most intuitive norm is the H_2 norm, defined by (28). Since we have a single input, the impulse response matrix $G(t)$ is a column vector $g(t)$, and so

$$\|G\|_2^2 = \int_0^\infty \|g(t)\|^2 \, dt,$$

just the regular $L_2[0,\infty)$ norm of the impulse response vector. We can think of the error norm $\|G - G_r\|_2$ as being the RMS error between a simulation of the reduced-order model G_r and a simulation of the full model G, where the simulation begins with $v(x,z,0) = \eta(x,z,0) = 0$, and the forcing is $f(t) = \delta(t)$. This error is shown in Fig. 7,

as the order r varies from 1 to 10. Notice that the error norms for balanced POD with both five-mode and ten-mode output projections are virtually the same as for balanced truncation, while POD is significantly worse for models of dimension six or smaller. For models of dimension greater than six, the error norms become smaller and all methods perform about the same.

Also shown is the error from an approximate balanced truncation in which the exact Gramians are computed, and then separately approximated by low-rank projections (to rank 30) using SVD. This separate reduction of Gramians is performed in the method of snapshots used in [Willcox & Peraire, 2002], although here their method of snapshots was not literally used, since it would require 480 adjoint simulations (the exact Gramians were computed by solving (14) instead). The balancing transformations are then found from the low-rank Gramians, and the L_2 errors of the resulting models are plotted in Fig. 7. One sees that the errors are significantly increased. It is interesting that if only the controllability Gramian is reduced to rank 30, while the exact observability Gramian is retained, then the results are similar to full balanced truncation or balanced POD (though these results are not shown in the figure). Thus, in truncating the observability Gramian, one is removing states that are almost unobservable, but apparently strongly controllable, and this causes increased errors in the resulting models. This illustrates one of the advantages of our method of snapshots (Sec. 3.1), which does not require separate reduction of the Gramians.

The differences between balanced truncation and POD become even more apparent when one

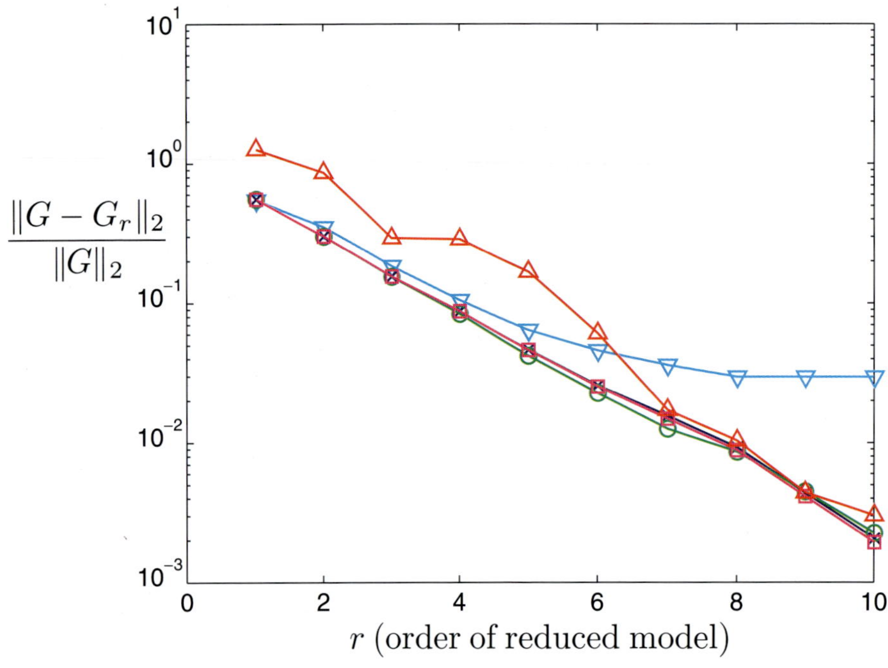

Fig. 7.   Error $\|G - G_r\|_2/\|G\|_2$, for balanced truncation ($\times$), balanced POD with five-mode and ten-mode output projection ($\circ$ and $\square$), POD ($\triangle$), and approximate balanced truncation with separate reduction of Gramians to rank 30 ($\triangledown$).

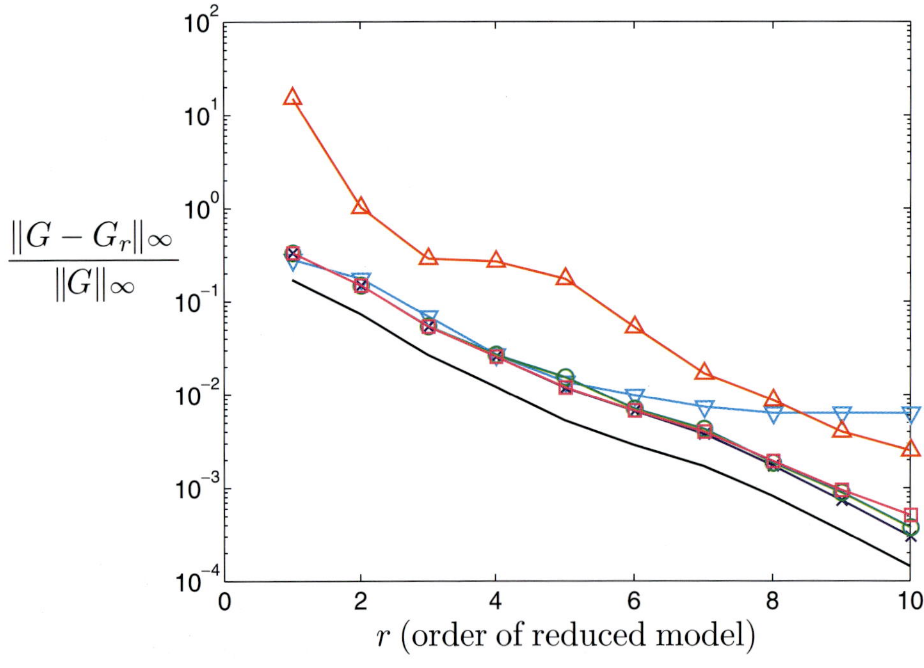

Fig. 8.   Error $\|G - G_r\|_\infty/\|G\|_\infty$, for balanced truncation ($\times$), balanced POD with five-mode and ten-mode output projection ($\circ$ and $\square$), POD ($\triangle$) approximate balanced truncation with separate reduction of Gramians to rank 30 ($\triangledown$), and lower bound for any model reduction scheme (—).

considers the H_∞ norm $\|G - G_r\|_\infty$, defined by (16). This norm is perhaps the most useful, because it is an induced norm, and measures the maximum error over all possible inputs, not just an impulsive input. Figure 8 shows the error $\|G - G_r\|_\infty$ for the various reduced-order models G_r. Again, the norms for balanced POD are almost identical to the norms for exact balanced truncation, for both five-mode and ten-mode output projections. Here, the norms for POD are about an order of magnitude higher, for all models considered. The error from an approximate balanced truncation using a rank-30 reduction of the exact Gramians is also shown, and again results in larger errors for the more accurate models. Also shown in this figure is the lower bound (17) achievable by any reduced-order model of dimension r, and the balanced POD norms are indeed very close to this lower bound.

5. Conclusions

The balanced POD method described here is not the first to use empirical Gramians to compute approximate balanced truncations using simulation data. These empirical Gramians were used by Moore [1981] in his original development of balancing, and by others in extending balancing to nonlinear systems [Lall *et al.*, 1999, 2002], and computing balancing transformations for large systems [Willcox & Peraire, 2002].

This work addresses computing balancing transformations (or approximations of them) for very large systems with, e.g. millions of states, as arise in discretizations of problems in fluids. Standard methods for computing balanced truncations involve singular value decompositions of the empirical Gramians, which are full $n \times n$ matrices (where n is the number of states), which is not feasible when n is large. Previous computational methods for large systems [Willcox & Peraire, 2002] involve separate reduction of the Gramians, which can lead to less accurate models, as we have seen (Figs. 7 and 8). The method of snapshots described in Sec. 3.1 allows computing balanced truncations from SVDs of much smaller matrices, with dimension $N_p \times N_d$, where N_p and N_d are numbers of snapshots in a dataset of primal and dual solutions, respectively, without separate reduction of the Gramians.

Furthermore, previous methods as in [Lall *et al.*, 1999] and [Willcox & Periare, 2002] are not tractable for systems with large numbers of outputs,

as one must integrate an adjoint solution for each output. Section 3.2 describes an output projection method that approximates full balanced truncation with guaranteed error bounds, and dramatically reduces the number of adjoint solutions necessary. In the example shown, integration of five adjoint solutions produced models that were virtually indistinguishable in the H_∞ norm from full balanced truncations, which would have required 480 adjoint simulations using previous methods.

The formulation of balanced POD also clarifies some connections between balanced truncation and POD, most importantly that for a linear system, balanced truncation is a special case of POD. In particular, one uses a dataset consisting of responses to unit impulses (one for each input), and uses the observability Gramian for the inner product. This inner product weight states of large "dynamical importance," as opposed to POD, which retains only the most energetic modes. This suggests that even for a nonlinear system, the observability Gramian from a linearization might be a good choice of inner product for POD, if reduced-order models are desired. The balanced POD procedure not only removes subjectivity in the choice of inner product for POD, but also guarantees that a Galerkin projection of a nonlinear system with a stable equilibrium point at the origin will also have a stable equilibrium point at the origin.

Although many of the developments in this paper are restricted to stable, linear systems, Sec. 3.4 suggests how many of these ideas might be extended to large-scale nonlinear systems as well, following the approaches in [Lall *et al.*, 2002].

Acknowledgments

This work was partially supported by the NSF, grant CMS-0347239, under program manager M. Tomizuka; and by AFOSR, grant F49620-03-1-0081, under program managers B. King, S. Heise and J. Schmisseur.

References

Antoulas, A. C., Sorensen, D. C. & Gugercin, S. [2001] "A survey of model reduction methods for large-scale systems," *Contemp. Math.* **280**, 193–219.

Bamieh, B. & Daleh, M. [2001] "Energy amplification in channel flows with stochastic excitation," *Phys. Fluids* **13**, 3258–3269.

Colonius, T. & Freund, J. B. [2002] "POD analysis of sound generation by a turbulent jet," AIAA Paper 2002-0072.

Cortelezzi, L. & Speyer, J. L. [1998] "Robust reduced-order controller of laminar boundary layer transitions," *Phys. Rev.* **E58**, 1906–1910.

Drazin, P. G. & Reid, W. H. [1981] *Hydrodynamic Stability* (Cambridge University Press).

Dullerud, G. E. & Paganini, F. [1999] *A Course in Robust Control Theory: A Convex Approach*, Texts in Applied Mathematics, Vol. 36 (Springer-Verlag).

Farrell, B. F. & Ioannou, P. J. [1993] "Stochastic forcing of the linearized Navier–Stokes equations," *Phys. Fluids* **A5**, 2600–2609.

Holmes, P., Lumley, J. L. & Berkooz, G. [1996] *Turbulence, Coherent Structures, Dynamical Systems and Symmetry* (Cambridge University Press).

Lall, S., Marsden, J. E. & Glavaški, S. [1999] "Empirical model reduction of controlled nonlinear systems," in *Proc. IFAC World Congress*, Vol. F, pp. 473–478.

Lall, S., Marsden, J. E. & Glavaški, S. [2002] "A subspace approach to balanced truncation for model reduction of nonlinear control systems," *Int. J. Robust Nonlin. Contr.* **12**, 519–535.

Laub, A. J., Heath, M. T., Page, C. C. & Ward, R. C. [1987] "Computation of balancing transformations and other applications of simultaneous diagonalization algorithms," *IEEE Trans. Automat. Contr.* **32**, 115–122.

Lumley, J. L. [1970] *Stochastic Tools in Turbulence* (Academic Press).

Moore, B. C. [1981] "Principal component analysis in linear systems: Controllability, observability, and model reduction," *IEEE Trans. Automat. Contr.* **26**, 17–32.

Rathinam, M. & Petzold, L. R. [2003] "A new look at proper orthogonal decomposition," *SIAM J. Numer. Anal.* **41**, 1893–1925.

Rowley, C. W., Colonius, T. & Murray, R. M. [2004] "Model reduction for compressible flow using POD and Galerkin projection," *Physica* **D189**, 115–129.

Scherpen, J. M. A. [1993] "Balancing for nonlinear systems," *Syst. Contr. Lett.* **21**, 143–153.

Sirovich, L. [1987] "Turbulence and the dynamics of coherent structures, parts I–III," *Q. Appl. Math.* **XLV**, 561–590.

Smith, T. R. [2003] "Low-dimensional models of plane Couette flow using the proper orthogonal decomposition," PhD thesis, Princeton University.

Willcox, K. & Peraire, J. [2002] "Balanced model reduction via the proper orthogonal decomposition," *AIAA J.* **40**, 2323–2330.

Appendix A
Theorems on Computing Balancing Transformations

Here, we consider empirical Gramians defined by (22), with balancing transformations T_1 and S_1 defined by (23)–(24). The following theorem establishes that if one takes enough snapshots that the empirical Gramians W_c and W_o have full rank n (clearly, at least n snapshots are required, and the system must be both controllable and observable), then Σ_1 contains the Hankel singular values (square roots of the eigenvalues of the product $W_c W_o$), and T_1 is the balancing transformation that simultaneously diagonalizes W_c and W_o.

Proposition 1. *Let W_c and W_o be empirical Gramians defined by (22), and suppose $Y^* X$ has rank $r = n$. Then the matrix T_1 is square and invertible, with inverse S_1, and*

$$S_1 W_c S_1^* = T_1^* W_o T_1 = \Sigma_1.$$

Proof. To show $S_1 = T_1^{-1}$, we have

$$S_1 T_1 = \Sigma_1^{-1/2} U_1^* Y^* X V_1 \Sigma_1^{-1/2}$$
$$= \Sigma_1^{-1/2} \Sigma_1 \Sigma_1^{-1/2} = I_n.$$

Also,

$$S_1 W_c S_1^* = \Sigma_1^{-1/2} U_1^* Y^* X X^* Y U_1 \Sigma^{-1/2}$$
$$= \Sigma_1^{-1/2} (\Sigma_1 V_1^*)(V_1 \Sigma_1) \Sigma_1^{-1/2} = \Sigma_1,$$

and a similar calculation shows $T_1^* W_o T_1 = \Sigma_1$. ∎

Of course, our main interest is in large systems for which the number of snapshots, and hence the rank of W_c, W_o is much smaller than n. The following theorem establishes that in this case, Σ_1 also contains all nonzero Hankel singular values, and T_1 contains the first r columns of the balancing transformation.

Proposition 2. *Suppose $Y^* X$ has rank $r < n$. Then there exist matrices $S_2, T_2 \in \mathbb{R}^{n \times (n-r)}$ such that for*

$$T = [T_1 \quad T_2], \quad S = \begin{bmatrix} S_1 \\ S_2 \end{bmatrix},$$

T is invertible with $T^{-1} = S$, and

$$SW_c W_o T = \begin{bmatrix} \Sigma_1^2 & 0 \\ 0 & 0 \end{bmatrix}, \tag{A.1}$$

and furthermore,

$$SW_cS^* = \begin{bmatrix} \Sigma_1 & 0 \\ 0 & M_1 \end{bmatrix},$$

$$T^*W_oT = \begin{bmatrix} \Sigma_1 & 0 \\ 0 & M_2 \end{bmatrix}, \qquad (A.2)$$

where M_1 and M_2 are matrices in $\mathbb{R}^{(n-r)\times(n-r)}$.

Proof. As in the proof of Theorem 1, $S_1T_1 = I_r$. Choose T_2 such that its columns form a basis for the nullspace of S_1 (an $(n-r)$-dimensional subspace of $\mathbb{R}^n$). Then $S_1T_2 = 0$, and T is invertible, since its columns are linearly independent. Define S_2 as the last $n-r$ rows of T^{-1}, and it follows that $S_2T_1 = 0$.

First, we show

$$T^*W_oT = \begin{bmatrix} T_1^*W_oT_1 & T_1^*W_oT_2 \\ T_2^*W_oT_1 & T_2^*W_oT_2 \end{bmatrix} = \begin{bmatrix} \Sigma_1 & 0 \\ 0 & M_2 \end{bmatrix}.$$

As in the proof of Theorem 1, $T_1^*W_oT_1 = \Sigma_1$. Next,

$$T_1^*W_oT_2 = \Sigma_1^{-1/2}V_1^*X^*YY^*T_2$$

$$= \Sigma_1^{-1/2}(\Sigma_1U_1^*)Y^*T_2 = \Sigma_1S_1T_2 = 0_{r\times(n-r)},$$

and thus $T_2^*W_oT_1 = (T_1^*W_oT_2)^* = 0_{(n-r)\times r}$. The results for SW_cS^* and SW_cW_oT follow similarly, using $S_2T_1 = 0$.   ∎

BIFURCATION TRACKING ALGORITHMS AND SOFTWARE FOR LARGE SCALE APPLICATIONS

A. G. SALINGER*, E. A. BURROUGHS[†], R. P. PAWLOWSKI,
E. T. PHIPPS and L. A. ROMERO
*Sandia National Laboratories, Albuquerque,
NM 87185-1111, USA*
*agsalin@sandia.gov

Received March 30, 2004; Revised June 28, 2004

We present the set of bifurcation tracking algorithms which have been developed in the LOCA software library to work with large scale application codes that use fully coupled Newton's method with iterative linear solvers. Turning point (fold), pitchfork, and Hopf bifurcation tracking algorithms based on Newton's method have been implemented, with particular attention to the scalability to large problem sizes on parallel computers and to the ease of implementation with new application codes. The ease of implementation is accomplished by using block elimination algorithms to solve the Newton iterations of the augmented bifurcation tracking systems. The applicability of such algorithms for large applications is in doubt since the main computational kernel of these routines is the iterative linear solve of the same matrix that is being driven singular by the algorithm. To test the robustness and scalability of these algorithms, the LOCA library has been interfaced with the MPSalsa massively parallel finite element reacting flows code. A bifurcation analysis of an 1.6 Million unknown model of 3D Rayleigh–Bénard convection in a $5 \times 5 \times 1$ box is successfully undertaken, showing that the algorithms can indeed scale to problems of this size while producing solutions of reasonable accuracy.

Keywords: Bifurcation analysis; continuation; flow stability.

1. Introduction

Bifurcation analysis is an important and powerful tool for performing computational design of modeled systems. Identifying bifurcations in parameter space is important since they represent a discontinuous change in a system's behavior with respect to changes in parameter. This behavior is often an undesirable phenomenon to be designed away from, such as the onset of flow instabilities in a chemical vapor deposition reactor [Pawlowski *et al.*, 2001] or the buckling of a structure [Fujii *et al.*, 2000]. It can also be desired, such as the onset of oscillations in an resonant tunnelling diode [Lasater *et al.*, 2004] or symmetry breaking in morphogenesis [Muratov

& Shvartsman, 2003]. In any event, the computational design process of numerous systems can be aided by the availability of software with efficient and robust algorithms for tracking bifurcations. In this work we present the algorithms implemented in the LOCA library that have been developed for large-scale applications, such as those arising from the discretizations of PDEs in multiple dimensions. (Our definition of "large-scale" for Newton-based applications are those that use approximate iterative methods for solving the linear system in Newton's methods, and that are likely parallel.)

Certainly, bifurcation analysis software exists. A partial list includes the AUTO code of Doedel

---

[†]Current address: Department of Mathematics, Humboldt State University, Arcata CA.

et al. [1997], CONTENT by Kuznetsov and Levitin [1995–1997], the MATCONT package for use within Matlab by Dhooge *et al.* [2003], and DDE-Biftool for delay differential equations by Engelborghs *et al.* [2002]. As a generalization, these software packages are aimed at applications consisting of sets of ODEs, including those that come from discretizations of 1D PDEs. For these problems, the developers of these codes have implemented bifurcation analysis capabilities that go well beyond the generic one-parameter bifurcations that are the focus of this paper, including the tracking of periodic orbits, heteroclinic orbits, bifurcations of delay equations, and tracking of higher co-dimension bifurcations. It is our understanding that the only general purpose bifurcation analysis software for large-scale systems is the PDEcont code of Lust (e.g. [Lust *et al.*, 1998]), which uses a Newton–Picard algorithm and is aimed at transient-based simulation codes.

The development of algorithms for larger problems, such as those coming from multi-dimensional PDEs, is not new. A thorough treatment is presented in the book by Govaerts [2000]. Also, an excellent review of the theory, algorithms, and applications to problems in fluid mechanics was published in 2000 by Cliffe *et al.* [2000a]. What distinguishes our present work from previous work (such as using the Entwife code [Cliffe *et al.*, 2000b]) is that we have worked towards developing a general purpose software library for these problems, and therefore maintained a separation between the bifurcation library and the application code. Because of this, we have refrained from major modifications to the application codes, such as to explicitly form the augmented systems for distinguishing bifurcation points or to compute analytic derivatives for additional quantities needed in the bifurcation analysis. Furthermore, we have targeted very large systems where direct solvers are no longer a scalable option. In this respect, our present application is most closely related to the methods of Tuckermann and coworkers (e.g. [Mamun & Tuckerman, 1995; Nore *et al.*, 2003; Xin & Le Quéré, 2002]) who use a matrix-free Newton–Krylov approach to solve for fixed points and to perform stability analysis of a time-stepper for 2D and 3D fluid mechanics applications.

In our development of the LOCA software, we have targeted codes that use a Newton-based solution algorithm and iterative linear solvers to reach equilibrium solutions. (While the algorithms presented here do work for problems with direct solvers, they do not take advantage of such capabilities as convenient monitoring of the sign of the determinant.) This is a different set of application codes than those targeted by the PDEcont code and therefore a complementary approach. Due to the fact that there are numerous linear solver algorithms which are tailored to different physics, data structures, and even discretizations, and that this is an active area of research and development, we have chosen not to **own** this computation in the continuation and bifurcation library. Instead, we have implemented block elimination algorithms, sometimes referred to as bordering algorithms, that use the solve of a linear system with the Jacobian matrix ($\mathbf{J}^{-1}\mathbf{v}$) as the main computational kernel. (The Hopf tracking algorithm is an exception.)

The ramifications of this approach, which is motivated for reasons to do with implementation and software, and not numerics, are many. On the positive side, this approach renders the library readily usable by any Newton-based code, which must by definition already possess this inversion capability. The library can be written with no knowledge of the matrix and its (parallel) data structures or solution algorithm. On the negative side, the bifurcation tracking algorithms are numerically unstable, using the linear solve of the Jacobian matrix as part of the iteration process to drive that same matrix singular. This will be seen clearly in the presentations of the algorithms in Sec. 2, and the effect will be documented in a numerical experiment in Sec. 3.4.

To demonstrate and evaluate the algorithms, we present in Sec. 3 results for tracking secondary bifurcations in the classical Rayleigh–Bénard problem. This problem involves natural convection flows and the discretization of five coupled PDEs in three dimensions. A brief description of the problem and PDE solution algorithms are presented in Sec. 3.1, followed by bifurcation tracking results in Secs. 3.2 and 3.3.

The bifurcation tracking algorithms presented and demonstrated in this paper are included in the LOCA software library along with complementary capabilities of parameter continuation and a linear stability analysis capability. The parameter continuation routines include the pseudo-arclength continuation algorithm [Keller, 1977] and multiparameter continuation using the *multifario* code of Henderson [2002]. We have previously reported (see

[Lehoucq & Salinger, 2001a; Burroughs *et al.*, 2001, 2004]) on our approach to large-scale eigenvalue approximation using the generalized Cayley transformation and then Arnoldi iterations using the ARPACK code [Lehoucq *et al.*, 1998; Maschhoff & Sorensen, 1996]. We have found that the eigensolver exhibits even better scalability than the steady state solution algorithm since the matrix requiring inversion in the Cayley transform is better conditioned than the Jacobian matrix.

The application presented in this paper is the largest we have analyzed with LOCA and serves to demonstrate the scalability of the algorithms on a familiar problem. Other large-scale applications that have been analyzed include natural convection flows in 2D enclosures [Salinger *et al.*, 2002a; Burroughs *et al.*, 2004], flows in chemical reactors [Pawlowski *et al.*, 2001], and density functional theory calculations of capillary condensation of confined fluids [Salinger & Frink, 2003; Frink & Salinger, 2003] and polymer self-assembly [Frischknecht *et al.*, 2002]. Current work includes the release of a completely new version of LOCA as part of a larger solver framework effort [Heroux *et al.*, 2003], and the development and implementation of alternative algorithms for mitigating or removing the solves of the nearly singular systems in the current algorithms.

2. Bifurcation Tracking Algorithms

In this section we describe the methods implemented in the LOCA library for locating three common instabilities exhibited in nonlinear systems: turning point, pitchfork and Hopf bifurcations. Each of the algorithms solves simultaneously for the steady state solution vector $\mathbf{x}$ of length n, the parameter at which the bifurcation occurs, λ, and the null vector $\mathbf{w} = \mathbf{y} + i\mathbf{z}$, which is the eigenvector associated with the eigenvalue that has zero real part. The bifurcations are tracked as a function of a second parameter to generate the loci of bifurcation points in two-parameter space.

It is assumed that the application code uses a fully coupled Newton method to solve for steady states of a set of nonlinear equations. In this paper, the equations are the n residual equations $\mathbf{R}$ of the finite element discretization of the PDEs that govern fluid flow and heat transfer. The steady state problem is written as

$$\mathbf{R}(\mathbf{x}, \lambda) = \mathbf{0}, \qquad (1)$$

which, given an initial guess for $\mathbf{x}$, is solved iteratively with Newton's method,

$$\mathbf{J}\Delta\mathbf{x} = -\mathbf{R}; \quad \mathbf{x}^{\text{new}} = \mathbf{x} + \Delta\mathbf{x}, \qquad (2)$$

where the Jacobian matrix $\mathbf{J} = \partial\mathbf{R}/\partial\mathbf{x}$. The iteration on $\mathbf{x}$ converges when $\|\Delta\mathbf{x}\|$ and/or $\|\mathbf{R}\|$ decrease below some tolerances. For scalability to large applications, the matrix equation (2) must be solved iteratively.

Steady state solution branches are tracked using continuation algorithms. Zero order continuation (natural parameter continuation using the previous solution as the initial guess), first order continuation (natural continuation with an Euler predictor requiring an extra matrix solve), and pseudo arclength continuation algorithms [Keller, 1977] have all been implemented in the LOCA library. Details of these methods can be found elsewhere [Cliffe *et al.*, 2000a; Salinger *et al.*, 2002b], and include code to automatically balance the scaling between the solution and parameter components of the arclength constraint and step size control algorithms.

The stability of the steady solutions to small perturbations can be ascertained through linear stability analysis. Linearization of the transient equations, which can be written generally as $\mathbf{R}(\dot{\mathbf{x}}, \mathbf{x}, \lambda) = 0$, around the steady state, leads to a generalized eigenvalue problem of the form

$$\mathbf{J}\mathbf{w} = \gamma\mathbf{B}\mathbf{w}, \qquad (3)$$

where $\mathbf{B} = -\partial\mathbf{R}/\partial\dot{\mathbf{x}}$ is the matrix of coefficients of time-dependent terms, γ is an eigenvalue of the system (generally complex), and $\mathbf{w}$ is the associated eigenvector, which can be written in terms of real value vectors $\mathbf{w} = \mathbf{y} + i\mathbf{z}$. If any eigenvalue has positive real part, then perturbations with any component in the direction of the associated eigenvector will grow exponentially, and the steady state solution is deemed unstable. A system loses stability, and experiences a bifurcation, when a stable steady state solution branch, as parameterized by a system parameter λ, passes through a point where $\text{Real}(\gamma) = 0$.

We have developed a robust linear stability analysis capability for large scale problems that accurately approximates leading eigenvalues of the system in Eq. (3). A detail of the method is found in [Lehoucq & Salinger, 2001b] while benchmarking and application of the method to incompressible flows are found in [Burroughs *et al.*, 2001, 2004; Salinger *et al.*, 1999].

2.1. *The turning point (fold) tracking algorithm*

The turning point (fold) tracking algorithm in LOCA uses Newton's method to converge to a turning point and simple zero order continuation to track it as a function of a second parameter. At a turning point bifurcation (or fold), there is a single eigenvalue $\gamma = 0$ with an associated real null vector $\mathbf{y}$. We use the formulation of Moore and Spence [1980] to characterize the turning point:

$$\mathbf{R} = \mathbf{0}, \tag{4}$$

$$\mathbf{Jy} = \mathbf{0}, \tag{5}$$

$$\phi \cdot \mathbf{y} - 1 = 0. \tag{6}$$

Here ϕ is a constant vector. The first vector equation (which is n scalar equations) specifies that the solution be on the steady state solution branch, the second vector equation specifies that a real-valued eigenvector $\mathbf{y}$ exists that corresponds to a zero eigenvalue, and the last scalar equation pins the length of the null vector at length 1 (and removes the trivial solution $\mathbf{y} = 0$). This set of $2n + 1$ equations uniquely specifies the values of $\mathbf{x}$, $\mathbf{y}$, and λ given a nondegenerate turning point, as long as ϕ is chosen to be any vector such that $\phi \cdot \mathbf{y} \neq 0$.

A full Newton's method applied to this system requires linear solves of the form

$$\begin{bmatrix} \mathbf{J} & \mathbf{0} & \dfrac{\partial \mathbf{R}}{\partial \lambda} \\[2mm] \dfrac{\partial \mathbf{Jy}}{\partial \mathbf{x}} & \mathbf{J} & \dfrac{\partial \mathbf{Jy}}{\partial \lambda} \\[2mm] \mathbf{0} & \phi^T & 0 \end{bmatrix} \begin{bmatrix} \Delta \mathbf{x} \\ \Delta \mathbf{y} \\ \Delta \lambda \end{bmatrix} = - \begin{bmatrix} \mathbf{R} \\ \mathbf{Jy} \\ \phi \cdot \mathbf{y} - 1 \end{bmatrix}, \tag{7}$$

It would be desirable to formulate this system and send it to an efficient linear solver, but this is not practical with many large-scale engineering simulation codes. One hurdle would be the formulation of $(\partial \mathbf{Jy}/\partial \mathbf{x})$, which is a matrix formed by the derivative of the vector $\mathbf{Jy}$ with respect to the vector $\mathbf{x}$ (the same as the notation $\mathbf{J} = \partial \mathbf{R}/\partial \mathbf{x}$). The computation of this matrix requires derivatives not normally calculated in an engineering code and does not lend itself well to efficient numerical differentiation. The second issue is the work involved in determining the sparse matrix storage for iterative linear solvers and partitioning and load balancing for applications sent to parallel computers. The last row and column are not in general sparse and so matrix-vector multiplications

would require global communications between all processors. The sparsity of the matrix $\mathbf{J}$ coming from many PDE solution methods (e.g. finite element, finite difference, finite volume) limits communications in the linear solver to only local communications between a processor and ~ 10 of its neighbors.

To reduce the effort in implementing the bifurcation algorithms with application codes, block elimination algorithms are used to solve the system of equations in (7). The solution to (7) can be equivalently formulated with four linear solves of the matrix $\mathbf{J}$ [Eqs. (8)–(11)] and some simple algebra:

$$\mathbf{Ja} = -\mathbf{R}, \tag{8}$$

$$\mathbf{Jb} = -\frac{\partial \mathbf{R}}{\partial \lambda}, \tag{9}$$

$$\mathbf{Jc} = -\frac{\partial \mathbf{Jy}}{\partial \mathbf{x}} \mathbf{a}, \tag{10}$$

$$\mathbf{Jd} = -\frac{\partial \mathbf{Jy}}{\partial \mathbf{x}} \mathbf{b} - \frac{\partial \mathbf{Jy}}{\partial \lambda}, \tag{11}$$

$$\Delta \lambda = \frac{1 - \phi \cdot \mathbf{c}}{\phi \cdot \mathbf{d}}, \tag{12}$$

$$\Delta \mathbf{x} = \mathbf{a} + \Delta \lambda \mathbf{b}, \tag{13}$$

$$\Delta \mathbf{y} = \mathbf{c} + \Delta \lambda \mathbf{d} - \mathbf{y}. \tag{14}$$

The variables $\mathbf{a}$, $\mathbf{b}$, $\mathbf{c}$ and $\mathbf{d}$ are temporary vectors of length n. Each of the four linear solves of $\mathbf{J}$ are performed by the application code, in the same way that this matrix is solved in Newton iteration (2). Work is saved in the second, third and fourth solves, by reusing the preconditioner for an preconditioned iterative solver (and the factorization for a direct solver). The algorithm requires initial guesses for $\mathbf{x}$ and λ, which usually come from a steady solution near the turning point as located by an arclength continuation run. The initial guess for the null vector is chosen to be a scaled version of the $\mathbf{b}$ vector from Eq. (9)

$$\mathbf{y}^{\text{init}} = \frac{\mathbf{b}}{\phi \cdot \mathbf{b}}. \tag{15}$$

The logic for this choice is based upon the realization that if $\mathbf{J}$ is nearly singular, then $\mathbf{J}^{-1}(\partial \mathbf{R}/\partial \lambda)$ should have a large component in the direction of the null vector. For coupled PDE applications, we have found that a good choice for the scaling vector ϕ is the vector given by the inverse of the average of the solution values for each PDE variable. This tends to make each variable's contribution to $\phi \cdot \mathbf{y}$ to be of the same relative magnitude.

The derivatives on the right-hand side of Eqs. (9)–(11) are all calculated with first-order finite differences and directional derivatives. The following formulas are used:

$$\frac{\partial \mathbf{R}}{\partial \lambda} \approx \frac{\mathbf{R}(\mathbf{x}, \lambda + \varepsilon_1) - \mathbf{R}(\mathbf{x}, \lambda)}{\varepsilon_1}, \qquad (16)$$

$$\frac{\partial \mathbf{J} \mathbf{y}}{\partial \mathbf{x}} \mathbf{a} \approx \frac{\mathbf{J}(\mathbf{x} + \varepsilon_2 \mathbf{a}, \lambda) \mathbf{y} - \mathbf{J}(\mathbf{x}, \lambda) \mathbf{y}}{\varepsilon_2}, \qquad (17)$$

$$\frac{\partial \mathbf{J} \mathbf{y}}{\partial \mathbf{x}} \mathbf{b} + \frac{\partial \mathbf{J} \mathbf{y}}{\partial \lambda} \approx \frac{1}{\varepsilon_3} \mathbf{J}(\mathbf{x} + \varepsilon_3 \mathbf{b}, \lambda) \mathbf{y}$$

$$+ \frac{1}{\varepsilon_1} \mathbf{J}(\mathbf{x}, \lambda + \varepsilon_1) \mathbf{y}$$

$$- \left(\frac{1}{\varepsilon_3} + \frac{1}{\varepsilon_1} \right) \mathbf{J}(\mathbf{x}, \lambda) \mathbf{y}. \qquad (18)$$

The robustness and accuracy of the algorithm is dependent on the choice of the perturbations ε. The following choices have been found to work well on sample applications for $\delta = 10^{-6}$:

$$\varepsilon_1 = \delta(|\lambda| + \delta), \qquad (19)$$

$$\varepsilon_2 = \delta \left(\frac{\|\mathbf{x}\|}{\|\mathbf{a}\|} + \delta \right), \qquad (20)$$

$$\varepsilon_3 = \delta \left(\frac{\|\mathbf{x}\|}{\|\mathbf{b}\|} + \delta \right). \qquad (21)$$

After convergence to a turning point, a slight modification of simple zero order continuation is often used to converge to the next turning point at the next value of second parameter. We have found more robust convergence when the solution vector $\mathbf{x}$ was perturbed off the singularity by a small random perturbation of relative magnitude 10^{-5}. The initial guesses for λ and $\mathbf{y}$ are the converged values at the previous turning point. The constant vector ϕ may be recomputed by recalculating the average of the solution values across the mesh, although we typically find this unnecessary.

It should be pointed out that the minimally augmented system

$$\mathbf{R} = \mathbf{0}, \qquad (22)$$

$$\sigma = 0. \qquad (23)$$

is recommended in the literature for use as the set of defining equations for a turning point bifurcation in place of Eqs. (4)–(6) listed above [Govaerts, 2000]. Here σ is a scalar measure of the singularity of J that is implicitly defined through additional matrix equations. An algorithm based on this approach should be more robust than the current method, yet requires solves with the transpose of the Jacobian matrix. This would preclude its use by codes that do not have the capability to solve $\mathbf{J}^{-T} v$, such as those that use matrix-free Newton–Krylov methods.

2.2. *The pitchfork tracking algorithm*

An algorithm for tracking pitchfork bifurcations has been developed that requires little modifications to the application code and model. Pitchfork bifurcations occur when a symmetric solution loses stability to a pair of asymmetric solutions. In this algorithm, we require that the user defines the symmetry by supplying a constant vector, ψ, that is antisymmetric with respect to the symmetry being broken. We specify the pitchfork by the following set of $(2n + 2)$ coupled equations:

$$\mathbf{R} + \sigma \psi = \mathbf{0}, \qquad (24)$$

$$\mathbf{J} \mathbf{y} = \mathbf{0}, \qquad (25)$$

$$\langle \mathbf{x}, \psi \rangle = 0, \qquad (26)$$

$$\phi \cdot \mathbf{y} - 1 = 0. \qquad (27)$$

The variable not previously defined in the turning point algorithm is the scalar variable σ that is a slack variable representing the asymmetry in the problem. This additional unknown is associated with the additional equation (26), which enforces that the solution vector is orthogonal to the antisymmetric vector. The notation in this equation represents an inner product. For a symmetric model, σ will go to zero at the solution. This approach for generating a regular system has been presented by Govaerts [2000] as an alternative to the approach of Werner and Spence [1984].

There are a few assumptions that were made to ease the implementation of the pitchfork tracking algorithm, yet can make it trickier to use. First, we require that any odd symmetry in the variables is about zero so that the inner product of the solution vector with the antisymmetric vector is zero. For instance, the cold and hot temperatures in a thermal flow problem should be set at -0.5 and 0.5 instead of 0 and 1. Second, our current implementation uses a dot product of the vectors to calculate the inner product $\langle \mathbf{x}, \psi \rangle$; however, this strictly should be an integral over the computational domain. For instance, if the discretization (i.e. finite element mesh) is not symmetric with respect to the symmetry, then the dot product of the solution vector and antisymmetric coefficient vectors would not be zero. We allow the users of the LOCA library to supply the integrated inner product, yet in our applications we have replaced it with the vector dot product. If the mesh is not symmetric with respect to the symmetry in the PDEs that is being broken at the pitchfork bifurcation, the discretized system will exhibit an imperfect bifurcation. The algorithm presented here will converge to a point that is a reasonable approximation of the pitchfork bifurcation. However, at this point $\sigma \neq 0$ and therefore we will not have $\mathbf{R} = 0$.

To start the algorithm, we require the user to supply the vector ψ. The null vector $\mathbf{y}$ has the antisymmetry that we are requiring of ψ. We calculate ψ and the initial guess for $\mathbf{y}$ by first detecting the pitchfork bifurcation with an eigensolver. The eigenvector associated with the eigenvalue that is passing through zero at the pitchfork is used for ψ and the initial guess for $\mathbf{y}$. For problems that have multiple pitchfork bifurcations in the same region of parameter space, which is often the case when the system can go unstable to different modes, the pitchfork algorithm can be started multiple times with different ψ vectors to track each pitchfork separately. We choose $\sigma = 0$ as an initial guess and we rarely see it increase past 10^{-10} throughout the iterations. The constant vector ϕ is chosen to be the scaling vector as in the turning point algorithm.

As with the turning point algorithm, we use a fully coupled Newton method to converge to the pitchfork bifurcation and a block elimination algorithm to simplify the solution of the Newton iteration. The Newton iteration for this system is

$$
\begin{bmatrix}
\mathbf{J} & 0 & \psi & \dfrac{\partial \mathbf{R}}{\partial \lambda} \\[2ex]
\dfrac{\partial \mathbf{Jy}}{\partial \mathbf{x}} & \mathbf{J} & 0 & \dfrac{\partial \mathbf{Jy}}{\partial \lambda} \\[2ex]
\dfrac{\partial \langle \mathbf{x}, \psi \rangle}{\partial x} & 0 & 0 & 0 \\[2ex]
0 & \phi^T & 0 & 0
\end{bmatrix}
\begin{bmatrix}
\Delta \mathbf{x} \\ \Delta \mathbf{y} \\ \Delta \sigma \\ \Delta \lambda
\end{bmatrix}
= -
\begin{bmatrix}
\mathbf{R} + \sigma \psi \\ \mathbf{Jy} \\ \langle \mathbf{x}, \psi \rangle \\ \phi \cdot \mathbf{y} - 1
\end{bmatrix} .
\tag{28}
$$

It can be solved using a mathematically (but not numerically) equivalent block elimination algorithm:

$$\mathbf{Ja} = -\mathbf{R}, \tag{29}$$

$$\mathbf{Jb} = -\frac{\partial \mathbf{R}}{\partial \lambda}, \tag{30}$$

$$\mathbf{Jc} = -\psi, \tag{31}$$

$$\mathbf{Jd} = -\frac{\partial \mathbf{Jy}}{\partial \mathbf{x}}\mathbf{a}, \tag{32}$$

$$\mathbf{Je} = -\frac{\partial \mathbf{Jy}}{\partial \mathbf{x}}\mathbf{b} - \frac{\partial \mathbf{Jy}}{\partial \lambda}, \tag{33}$$

$$\mathbf{Jf} = -\frac{\partial \mathbf{Jy}}{\partial \mathbf{x}}\mathbf{c}, \tag{34}$$

$$
\Delta \sigma = -\sigma \\
+ \frac{(\langle \mathbf{x}, \psi \rangle + \langle \mathbf{a}, \psi \rangle)\phi \cdot \mathbf{e} + \langle \mathbf{b}, \psi \rangle (1 - \phi \cdot \mathbf{d})}{\langle \mathbf{b}, \psi \rangle \phi \cdot \mathbf{f} - \langle \mathbf{c}, \psi \rangle \phi \cdot \mathbf{e}},
\tag{35}
$$

$$\Delta \lambda = \frac{1 - \phi \cdot \mathbf{d} - \phi \cdot \mathbf{f}(\Delta \sigma + \sigma)}{\phi \cdot \mathbf{e}}, \tag{36}$$

$$\Delta \mathbf{x} = \mathbf{a} + \Delta \lambda \mathbf{b} + (\Delta \sigma + \sigma)\mathbf{c}, \tag{37}$$

$$\Delta \mathbf{y} = \mathbf{d} + \Delta \lambda \mathbf{e} + (\Delta \sigma + \sigma)\mathbf{f} - \mathbf{y}. \tag{38}$$

This algorithm has six temporary vectors $(\mathbf{a}, \mathbf{b}, \mathbf{c}, \mathbf{d}, \mathbf{e}, \text{and } \mathbf{f})$, each of which is the result of a linear solve with the same matrix $\mathbf{J}$. Again the preconditioner is only calculated once. The use of a block solver, where solves for $\mathbf{a}, \mathbf{b}$, and $\mathbf{c}$ are performed simultaneously (as are $\mathbf{d}, \mathbf{e}$, and $\mathbf{f}$), would be advantageous. The right-hand sides of these six linear systems are mostly the same as for the turning point algorithm, and so reuse the same routines and differencing schemes and perturbations presented above [Eqs. (16) and (19)].

2.3. *The Hopf tracking algorithm*

The algorithm for tracking Hopf bifurcations, where a complex pair of eigenvalues have zero real part, is similar to the above turning point and pitchfork tracking algorithms. It is however more complicated in that it involves complex numbers. The purely imaginary eigenvalues at the bifurcation point can be written $\gamma = \pm i\omega$ with complex eigenvectors $\mathbf{w} = \mathbf{y} + i\mathbf{z}$. The following set of equations, presented by Griewank and Reddien [1983], specify the Hopf bifurcation,

$$\mathbf{R} = \mathbf{0}, \tag{39}$$

$$\mathbf{J}\mathbf{y} + \omega \mathbf{B}\mathbf{z} = 0, \tag{40}$$

$$\mathbf{J}\mathbf{z} - \omega \mathbf{B}\mathbf{y} = 0, \tag{41}$$

$$\phi \cdot \mathbf{y} - 1 = 0, \tag{42}$$

$$\phi \cdot \mathbf{z} = 0. \tag{43}$$

This system of $3n + 2$ equations and unknowns solves for the solution vector $\mathbf{x}$, $\mathbf{y}$, $\mathbf{z}$, ω, and λ. The first vector equation specifies that we are on the solution branch, the next two equations specify that we are at a place where there is a purely imaginary eigenvalue, and the last two scalar equations set the phase and amplitude of the eigenvectors (which are otherwise free). The same Hopf bifurcation can admit a second solution to this system of equations at $(\mathbf{x}, \mathbf{y}, -\mathbf{z}, -\omega, \lambda)$.

One Newton iteration for the fully coupled solution of this system is the linear system,

$$
\begin{bmatrix}
\mathbf{J} & \mathbf{0} & \mathbf{0} & \mathbf{0} & \dfrac{\partial \mathbf{R}}{\partial \lambda} \\[2ex]
\dfrac{\partial \mathbf{J}\mathbf{y}}{\partial \mathbf{x}} + \omega \dfrac{\partial \mathbf{B}\mathbf{z}}{\partial \mathbf{x}} & \mathbf{J} & \omega \mathbf{B} & \mathbf{B}\mathbf{z} & \dfrac{\partial \mathbf{J}\mathbf{y}}{\partial \lambda} + \omega \dfrac{\partial \mathbf{B}\mathbf{z}}{\partial \lambda} \\[2ex]
\dfrac{\partial \mathbf{J}\mathbf{z}}{\partial \mathbf{x}} - \omega \dfrac{\partial \mathbf{B}\mathbf{y}}{\partial \mathbf{x}} & -\omega \mathbf{B} & \mathbf{J} & -\mathbf{B}\mathbf{y} & \dfrac{\partial \mathbf{J}\mathbf{z}}{\partial \lambda} - \omega \dfrac{\partial \mathbf{B}\mathbf{y}}{\partial \lambda} \\[2ex]
\mathbf{0} & \phi^T & \mathbf{0} & 0 & 0 \\[1ex]
\mathbf{0} & \mathbf{0} & \phi^T & 0 & 0
\end{bmatrix}
\begin{bmatrix}
\Delta \mathbf{x} \\[1ex] \Delta \mathbf{y} \\[1ex] \Delta \mathbf{z} \\[1ex] \Delta \omega \\[1ex] \Delta \lambda
\end{bmatrix}
= -
\begin{bmatrix}
\mathbf{R} \\[1ex] \mathbf{J}\mathbf{y} + \omega \mathbf{B}\mathbf{z} \\[1ex] \mathbf{J}\mathbf{z} - \omega \mathbf{B}\mathbf{y} \\[1ex] \phi \cdot \mathbf{y} - 1 \\[1ex] \phi \cdot \mathbf{z}
\end{bmatrix}. \tag{44}
$$

In this derivation we have allowed for $\partial \mathbf{B}/\partial \mathbf{x} \neq 0$ and $\partial \mathbf{B}/\partial \lambda \neq 0$. While in many situations these terms can be neglected, the matrix $\mathbf{B}$ can depend on the solution vector through dependence of the inertial coefficients (e.g. density and heat capacity) on the local state vector. The matrix $\mathbf{B}$ will depend on the parameter very strongly when λ is a geometric parameter that moves the mesh locations.

Again we solve this linear system by a block elimination algorithm that breaks it into simpler linear solves. It is not possible to solve this system by solves of just the matrix $\mathbf{J}$, but also requires solves of the complex matrix $\mathbf{J} + i\omega \mathbf{B}$. The block elimination algorithm for the Newton iteration of the Hopf tracking algorithm, written in terms of real-valued variables, is,

$$\mathbf{J}\mathbf{a} = -\mathbf{R}, \tag{45}$$

$$\mathbf{J}\mathbf{b} = -\frac{\partial \mathbf{R}}{\partial \lambda}, \tag{46}$$

$$
\begin{bmatrix} \mathbf{J} & \omega \mathbf{B} \\ -\omega \mathbf{B} & \mathbf{J} \end{bmatrix}
\begin{bmatrix} \mathbf{c} \\ \mathbf{d} \end{bmatrix}
= \begin{bmatrix} \mathbf{B}\mathbf{z} \\ -\mathbf{B}\mathbf{y} \end{bmatrix}, \tag{47}
$$

$$
\begin{bmatrix} \mathbf{J} & \omega \mathbf{B} \\ -\omega \mathbf{B} & \mathbf{J} \end{bmatrix}
\begin{bmatrix} \mathbf{e} \\ \mathbf{f} \end{bmatrix}
=
\begin{bmatrix}
-\left(\dfrac{\partial \mathbf{J}\mathbf{y}}{\partial \mathbf{x}} + \omega \dfrac{\partial \mathbf{B}\mathbf{z}}{\partial \mathbf{x}} \right)\mathbf{a} \\[2ex]
-\left(\dfrac{\partial \mathbf{J}\mathbf{z}}{\partial \mathbf{x}} - \omega \dfrac{\partial \mathbf{B}\mathbf{y}}{\partial \mathbf{x}} \right)\mathbf{a}
\end{bmatrix}, \tag{48}
$$

$$
\begin{bmatrix} \mathbf{J} & \omega \mathbf{B} \\ -\omega \mathbf{B} & \mathbf{J} \end{bmatrix}
\begin{bmatrix} \mathbf{g} \\ \mathbf{h} \end{bmatrix}
$$
$$
=
\begin{bmatrix}
-\left(\dfrac{\partial \mathbf{J}\mathbf{y}}{\partial \mathbf{x}} + \omega \dfrac{\partial \mathbf{B}\mathbf{z}}{\partial \mathbf{x}} \right)\mathbf{b} - \left(\dfrac{\partial \mathbf{J}\mathbf{y}}{\partial \lambda} + \omega \dfrac{\partial \mathbf{B}\mathbf{z}}{\partial \lambda} \right) \\[2ex]
-\left(\dfrac{\partial \mathbf{J}\mathbf{z}}{\partial \mathbf{x}} - \omega \dfrac{\partial \mathbf{B}\mathbf{y}}{\partial \mathbf{x}} \right)\mathbf{b} - \left(\dfrac{\partial \mathbf{J}\mathbf{z}}{\partial \lambda} - \omega \dfrac{\partial \mathbf{B}\mathbf{y}}{\partial \lambda} \right)
\end{bmatrix}, \tag{49}
$$

$$\Delta \lambda = \frac{(\phi \cdot \mathbf{c})(\phi \cdot \mathbf{f}) - (\phi \cdot \mathbf{e})(\phi \cdot \mathbf{d}) + (\phi \cdot \mathbf{d})}{(\phi \cdot \mathbf{d})(\phi \cdot \mathbf{g}) - (\phi \cdot \mathbf{c})(\phi \cdot \mathbf{h})}, \tag{50}$$

$$\Delta \omega = \frac{(\phi \cdot \mathbf{h})\Delta \lambda + (\phi \cdot \mathbf{f})}{\phi \cdot \mathbf{d}}, \tag{51}$$

$$\Delta \mathbf{x} = \mathbf{a} + \Delta \lambda \mathbf{b}, \tag{52}$$

$$\Delta \mathbf{y} = \mathbf{e} + \Delta \lambda \mathbf{g} - \Delta \omega \mathbf{c} - \mathbf{y}, \tag{53}$$

$$\Delta \mathbf{z} = \mathbf{f} + \Delta \lambda \mathbf{h} - \Delta \omega \mathbf{d} - \mathbf{z}. \tag{54}$$

Table 1. This table summarizes the functionality that an application code must supply in order to access each of the stability analysis algorithms. The requirement for the Hopf tracking is in addition to those listed above for the other methods.

Method	Requirements	Description
Parameter continuation	$\mathbf{R}$	Residual calculation
Turning point tracking	$\mathbf{Jv}$	Jacobian-vector multiply
Pitchfork tracking	$\mathbf{J}^{-1}\mathbf{v}$	Solve with Jacobian
	set λ	Set parameters
Eigensolve	$\mathbf{Bv}$	Mass matrix-vector multiply
	$(\mathbf{J} - \sigma\mathbf{B})^{-1}\mathbf{v}$	Solve with shifted Jacobian
Hopf tracking	$(\mathbf{J} - i\omega\mathbf{B})^{-1}\mathbf{v}$	Solve with complex matrix

This algorithm has eight temporary vectors $\mathbf{a}$ through $\mathbf{h}$, which are solved with two solves of the $\mathbf{J}$ matrix and three solves of the $2n \times 2n$ matrix $\begin{bmatrix} \mathbf{J} & \omega\mathbf{B} \\ -\omega\mathbf{B} & \mathbf{J} \end{bmatrix}$. This algorithm differs from the turning point and pitchfork tracking algorithms which only require solution of the steady state Jacobian $\mathbf{J}$, a capability already possessed by codes using Newton's method. Since the location of the non-zeros in the sparse matrix $\mathbf{B}$ is typically a subset of those for the matrix $\mathbf{J}$, a parallel iterative solver for the $2n \times 2n$ matrix can use the same local communication maps as used for solves of $\mathbf{J}$. An algorithm for solving complex matrix equations with a real-valued sparse iterative solver has been published [Day & Heroux, 2001] and implemented in the Komplex extension to the Aztec library of preconditioned iterative Krylov solvers. This algorithm also requires the formulation of the $\mathbf{B}$ matrix, for which a code performing linear stability analysis of Eq. (3) will already have a routine.

To initialize the routine, we assume that an initial Hopf bifurcation has been detected with an eigensolver, by having the real part of a complex pair of eigenvalues pass through zero with successive steps in the parameter. This gives good starting values for all the unknowns in the Hopf tracking algorithm. Also, the constant vector ϕ is chosen as before.

2.4. *Interface requirements*

By using block elimination algorithms to solve the Newton iterations of the augmented systems describing the bifurcations and using finite differencing to compute all the derivatives except for the Jacobian matrix, the requirements on the application code to access these algorithms are rather small. The requirements are summarized in Table 1.

3. Bifurcation Analysis of Rayleigh–Bénard Convection in a $5 \times 5 \times 1$ Box

As a demonstration of the algorithms presented above, we choose to study the secondary bifurcations in the Rayleigh–Bénard problem, which consists of a fluid that is heated from below and cooled from above, so that the thermal expansion of the fluid in the presence of gravity produces a destabilizing density gradient. In particular, we compute the convection rolls arising from the first symmetry breaking bifurcation and then analyze the loss of stability of these rolls in two-parameter space. This system is controlled by two dimensionless groups (which will be defined in the following section): the Rayleigh number Ra which is a measure of the destabilizing buoyancy effect compared to the stabilizing diffusive effects, and the Prandtl number Pr, which is a property of the fluid comparing the relative diffusive strengths of momentum and heat. We have chosen our parameters and boundary conditions so that we get all three bifurcations generic to a one-parameter system with symmetry.

Of particular interest in the past is the onset of oscillatory instabilities. The stability of the rolls has been considered both experimentally [Willis & Deardorff, 1970] and numerically [Busse &

Clever, 1979; Clever & Busse, 1995; Tangborn *et al.*, 1995; Nakamura, 1997; Sone *et al.*, 1997; Cox & Matthews, 2000]. In their paper, Busse and Clever [1979] numerically analyzed the stability of the convection rolls in the absence of any side walls. Their equilibrium solution is a two-dimensional solution periodic in the direction perpendicular to the axis of the rolls. They analyze the stability of this solution by Fourier transforming the disturbances and looking for the most unstable wavelength. Their results show that as the Prandtl number goes to zero, the rolls have an oscillatory instability at a Rayleigh number close to the critical Rayleigh number of the first bifurcation. Although their results are in qualitative agreement with experiments, quantitatively their results predict that bifurcation occurs closer to the original bifurcation than the experiments do. They argue that this is most likely a result of ignoring the side walls.

In a technical report [Burroughs *et al.*, 2001], we computed the steady convective rolls in a $5 \times 5 \times 1$ box, and analyzed their stability using the eigensolver. For a fluid with $\mathrm{Pr} = 0.01$, the onset of an oscillatory instability was estimated to be at $\mathrm{Ra}_{\mathrm{cr}} = 1910$ on the finest finite element discretization of 16 Million unknowns. The appeal of bifurcation tracking for this problem is to calculate the whole curve of $\mathrm{Ra}_{\mathrm{cr}}(\mathrm{Pr})$ and further determine under what conditions the convective rolls undergo an oscillatory instability.

3.1. *Model and methods*

The governing partial differential equations for the Rayleigh–Bénard problem are the incompressible Navier–Stokes equations for momentum transport, the continuity equation for mass conservation, and the heat equation. The Boussinesq approximation is used, which allows for a linear dependence of density on temperature in the body force term, yet assumes constant density in all other terms. In nondimensional form, the equations are:

$$\frac{\partial \mathbf{u}}{\partial t} + \mathbf{u} \cdot \nabla \mathbf{u} + \nabla P = \nabla^2 \mathbf{u} + \mathrm{Ra}\, \mathrm{Pr}\, T \mathbf{e}_z, \quad (55)$$

$$\nabla \cdot \mathbf{u} = 0, \quad (56)$$

$$\frac{\partial T}{\partial t} + \mathbf{u} \cdot \nabla T = \mathrm{Pr}^{-1} \nabla^2 T. \quad (57)$$

Here $\mathbf{u} = u\mathbf{e}_x + v\mathbf{e}_y + w\mathbf{e}_z$, P and T are the unknown velocity, pressure and temperature fields. The vector $\mathbf{e}_z$ is the unit vector in the direction of the gravitational acceleration. The two dimensionless groups controlling the system are the Rayleigh number

$$\mathrm{Ra} = \frac{\rho^2 C_p g \beta \Delta T L^3}{k \mu},$$

and the Prandtl number

$$\mathrm{Pr} = \frac{\mu C_p}{k}.$$

Here ρ is the density at reference temperature $T = 0$, C_p is the heat capacity of the fluid, g is the gravitational acceleration, β is the coefficient of thermal expansion, ΔT is the temperature difference over the box, L is the height of the box, k is the thermal conductivity, and μ is the viscosity. In this formulation, distances are made dimensionless with respect to L, the velocities with respect to $\mu/\rho L$, the pressure with respect to $\mu^2/\rho L^2$, the temperature with respect to ΔT, and time with respect to $\mu/\rho L^2$.

Even though we only solve for steady solutions in this paper, the time-dependent versions of the equations are written since these terms come to play in the linear stability analysis and the Hopf tracking algorithm. Note that for incompressible flow there are no time derivatives of the pressure field so the mass matrix $\mathbf{B}$ is 20% rank deficient.

In the following sections we analyze the flow stability for two closely related systems, that differ in the boundary conditions on the side walls. The first is the closed box, with no-slip boundary conditions on the side walls for the entire velocity vector,

closed box:

$$\mathbf{u}(0, y, z) = \mathbf{u}(5, y, z) = \mathbf{u}(x, 0, z) = \mathbf{u}(0, 5, z) = 0.$$

The second is the symmetric box with symmetry boundary conditions on the side walls, including no normal flow and no shear stress,

symmetric box:

$$u(0, y, z) = u(5, y, z) = v(x, 0, z) = v(x, 5, z) = 0,$$

$$\frac{\partial u}{\partial y}(x, 0, z) = \frac{\partial u}{\partial y}(x, 5, z)$$

$$= \frac{\partial w}{\partial y}(x, 0, z) = \frac{\partial w}{\partial y}(x, 5, z) = 0,$$

$$\frac{\partial v}{\partial x}(0, y, z) = \frac{\partial v}{\partial x}(5, y, z)$$

$$= \frac{\partial w}{\partial x}(0, y, z) = \frac{\partial w}{\partial x}(5, y, z) = 0.$$

The rest of the boundary conditions are the same for both systems, including adiabatic conditions on the side walls,

$$\frac{\partial T}{\partial x}(0, y, z) = \frac{\partial T}{\partial x}(5, y, z)$$

$$= \frac{\partial T}{\partial y}(x, 0, z) = \frac{\partial T}{\partial y}(x, y, z) = 0.$$

with no-slip boundary conditions and a hot temperature on the bottom surface,

$$\mathbf{u}(x, y, 0) = 0, \quad T(x, y, 0) = 0.5,$$

and no-slip boundary conditions and a cold temperature on the top surface,

$$\mathbf{u}(x, y, 1) = 0, \quad T(x, y, 1) = -0.5.$$

The temperature boundary conditions are chosen so that the solution is symmetric about zero, as required by the pitchfork algorithm.

The above system of five coupled PDEs and boundary conditions are solved for unknowns u, v, w, P, T with the MPSalsa code. MPSalsa uses a Galerkin/least-squares finite element method [Hughes *et al.*, 1989a, 1989b; Shadid, 1999] to discretize these equations over the spatial domain. This stabilization procedure allows for the use of equal order linear FE basis functions for all variables while avoiding spurious pressure oscillations for incompressible flows. For these calculations we did not need to include the streamline upwind Petrov-Galerkin (SUPG) type methodology for controlling oscillations due to convective effects. An additional important aspect of this stabilization procedure is that a fully-implicit (for time-dependent systems) and a direct-to-steady-state solution procedure using Newton–Krylov methods can be implemented [Shadid, 1999]. The Aztec package of preconditioned iterative Krylov methods is used to solve the linear systems [Hutchinson *et al.*, 1995]. In this work, we have used the ILUT domain decomposition preconditioner, where each processor owns one domain. We chose 1 level of overlap between domains and a fill factor of 1.5, which allows for the preconditioner to have 1.5 as many nonzeroes as the Jacobian itself. The GMRES linear solver was used without restarts, and orthogonality is maintained with the modified Graham–Schmidt algorithm. A typical linear solve for the solution of a steady state used a relative tolerance of 10^{-3} and built a Krylov space of size 220, while a solve for a bifurcation tracking run used a relative tolerance of 10^{-8} and built a Krylov space of size 400.

The MPSalsa code is designed for general unstructured meshes in 2D and 3D, and runs on massively parallel computers. The majority of the results in this paper were calculated for a $100 \times 100 \times 30$ mesh of eight-node trilinear hexahedral elements, which corresponds to 316231 nodes and over 1.58 Million equations and unknowns.

Fig. 1.   A visualization of the partition of the 316231 node mesh for 48 processors is shown. The colored patches are elements (in the finite element discretization) whose nodes are all owned by a given processor, while the red strips are elements whose nodes are owned by multiple processors. Inter-processor communication is needed only across these elements for performing the finite element method.

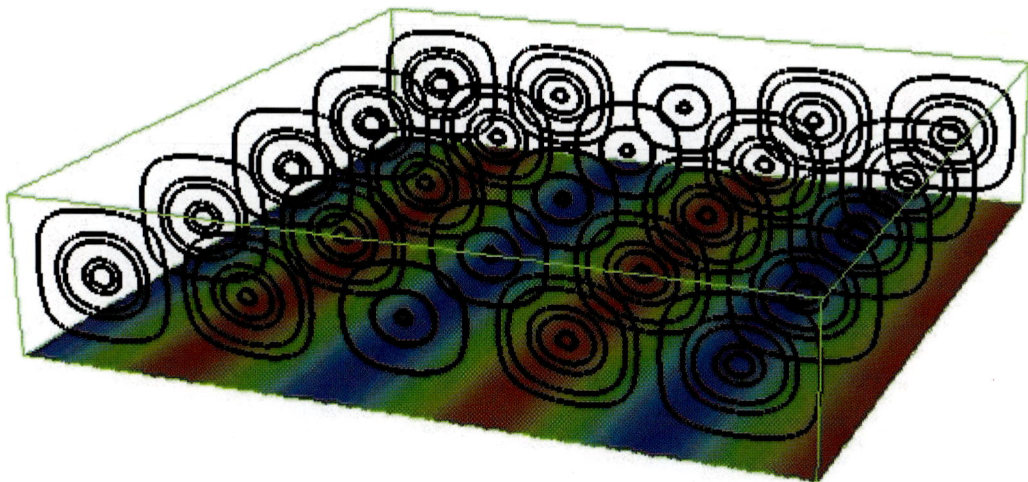

Fig. 2. Visualization of the stable convective flow state for the symmetric box at Ra = 3328.1 and Pr = 1.0. The black circles are streamlines, and the color contours are heat flux through the bottom surface.

The mesh is produced using the CUBIT software [Shepherd, 2000] and decomposed for parallel solution using the Chaco graph partitioning package [Hendrickson & Leland, 1995a, 1995b]. The partitioner assigns each node to a processor in a way to evenly distribute the work load while minimizing interprocessor communication. The decomposition of the mesh for 48 processors is visualized in Fig. 1. Finite elements with all eight corner nodes owned by the same processor are given a color unique to that processor. Elements broken over multiple processors are colored red (and form jagged lines) and are representative of the amount of information that needs to be communicated to perform a matrix fill or matrix-vector multiply.

A steady-state solution of the convective roll cells in the symmetric box is shown in Fig. 2. Note that the solution is two-dimensional, with the streamlines showing five tubes of circulating flow. The color contours show heat flux through the bottom of the box, where the high red values correspond to regions of downward flow, and low blue to regions correspond to upward flow.

3.2. *Results for closed box*

Our first model system is the closed box, with no-slip boundary conditions on all walls. A bifurcation diagram with respect to the Rayleigh number Ra for fixed Pr = 1.0 is shown in Fig. 3. The no-flow, conduction solution was calculated to bifurcate to the convective rolls solution at Ra = 1774.0. This calculation involved the computation of the eigenvector

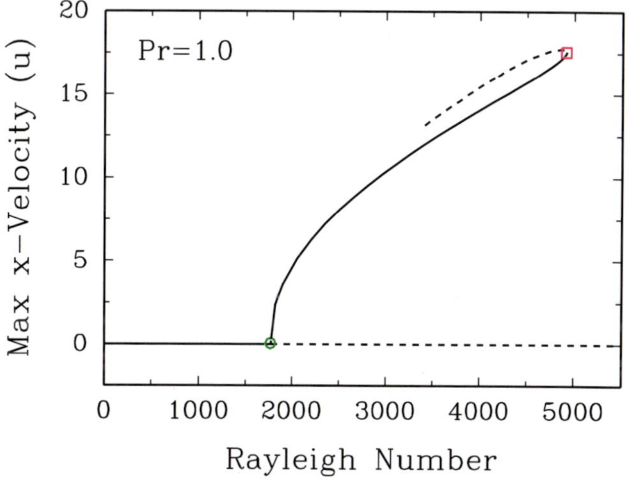

Fig. 3. Plot of steady solution branches in the closed box as a function of Rayleigh number for Pr = 1.0, showing a pitchfork bifurcation from the stationary solution to a branch of convective rolls. A turning point (fold) is seen on the second branch, which was subsequently calculated to occur at Ra = 4915.2.

near this singularity, followed by a solve with the pitchfork tracking algorithm using the eigenvector as the ψ antisymmetric vector and the initial guess for the null vector.

This convective flow branch has predominantly two-dimensional profile, with roll cells resembling those in Fig. 2, yet with 3D effects due to the no-slip walls. Since we were not able to adequately visualize the flow field, we instead show the heat flux through the bottom surface in Fig. 4(a). The

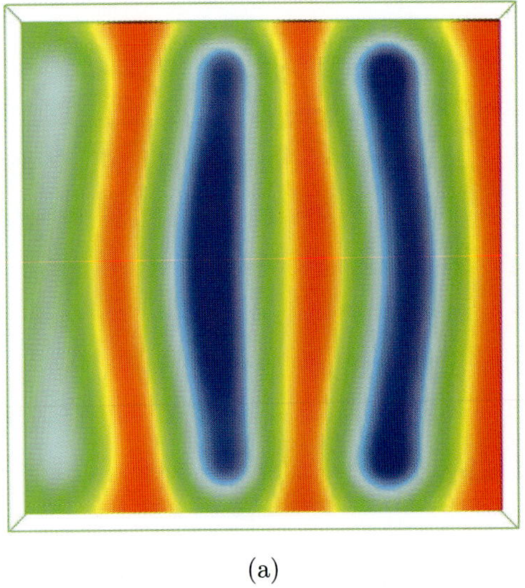

(a)

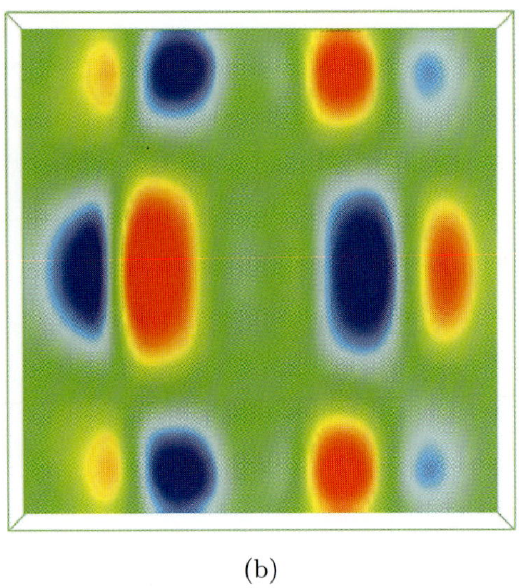

(b)

Fig. 4.   The heat flux through the bottom of the closed box is shown for (a) the solution and (b) the null vector, at the turning point at Ra = 4915.2 and Pr = 1.0. The red and blue regions correspond to downward and upward flow.

red and blue regions correspond to downward and upward flow. The effect of the no-slip wall can be clearly seen by comparison with the results for the symmetric box, which was visible in Fig. 2 and reproduced in the same format in Fig. 7(a).

The solution branch was tracked with pseudo arclength continuation where it encountered a turning point near Ra = 4900. (The unstable branch, shown as the dotted line, was followed back through other bifurcations to another turning point near Ra = 3400, where linear stability analysis revealed nine eigenvalues with positive real parts.) The turning point algorithm of Sec. 2.1 was then used to converge to the bifurcation at Pr = 1.0 and Ra = 4915.2. The null vector is visualized in Fig. 4(b) and seen to have significant variation in the y-direction, indicating an end to the flow branch consisting of the tubular roll cells. (The convergence details of this calculation are the subject of the numerical experiments in Sec. 3.4.)

The turning point was then tracked with decreasing Pr. The results of this tracking are shown in Fig. 5. The pitchfork bifurcation from the trivial branch, known to be independent of Pr, is drawn in as well. The symbols at Pr = 1 correspond to the similarly marked solutions in Fig. 3. Regions of no-flow, stable convective roll cells of predominantly 2D flow, and the region where the cells are no longer a stable solution are delineated by these curves of bifurcations. The calculation of the curve of turning points involved 12 consecutive turning

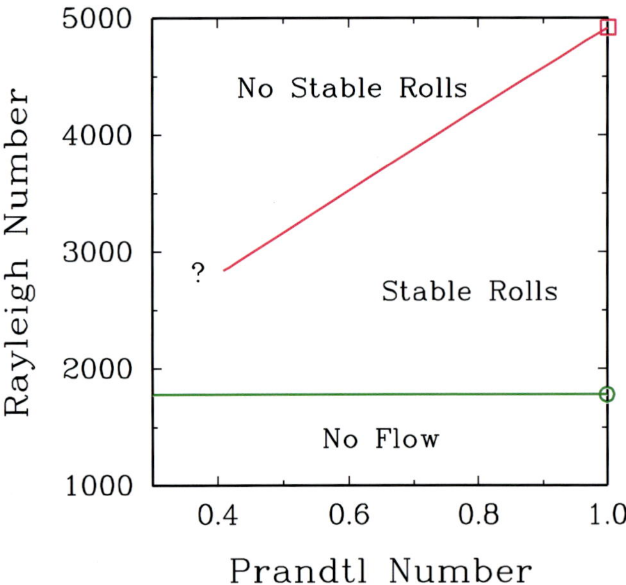

Fig. 5.   The results of tracking the turning point bifurcation seen in Fig. 3 (at Pr = 1.0 and Ra = 4915.2) as a function of Prandtl number is shown. This bifurcation represents the limit of the nearly two-dimensional roll cells in a closed box. The branch of turning point bifurcations ends in what is presumably a cusp near Pr = 0.4075, and the stability behavior in the region of the ? symbol remain uninvestigated. The pitchfork bifurcation signalling the onset of flow is drawn in as well.

point calculations, each requiring about 45 min on a cluster of 48 3.0 GHz processors. The last solution on this branch is calculated at Pr = 0.4075 and Ra = 2831.7, below which the branch ends, presumably in a cusp. Further investigation of the stability behavior at lower Pr is beyond the scope of this study.

3.3. *Results for symmetric box*

A similar set of calculations were performed on the symmetric box, where symmetry boundary conditions were placed on all side walls. This linear stability of this system has been previously probed, including the detection of a Hopf bifurcation from the convective rolls solution for $Pr = 0.01$ in the range of $Ra = 1900 - 1950$ [Burroughs *et al.*, 2001].

A parameter continuation study in Ra was performed on this system at $Pr = 1.0$, and is presented in Fig. 6. Again, a pitchfork bifurcation from the trivial no-flow, conduction solution is found. The critical Rayleigh number is found to be at $Ra = 1703.7$, about 4% lower than for the closed box where the side walls stabilize the no-flow solution. The two-dimensional convective rolls solution is continued until the linear stability analysis detects a secondary pitchfork bifurcation near $Ra = 3300$. This singularity represents the end of the stable two-dimensional solution. The solution at the pitchfork bifurcation was visualized in Fig. 2. The heat flux through the bottom of the box is again shown for the solution and null vector in Figs. 7(a) and 7(b). The null vector shows that the convective roll solution destabilizes to 3D disturbances.

(a)

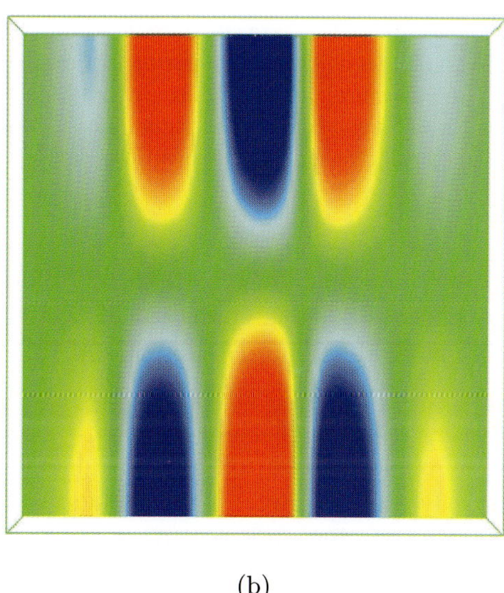

(b)

Fig. 7. The heat flux through the bottom of the symmetric box is shown for (a) the solution and (b) the null vector, at the pitchfork bifurcation at $Ra = 3338.1$ and $Pr = 1.0$. The red and blue regions correspond to downward and upward flow.

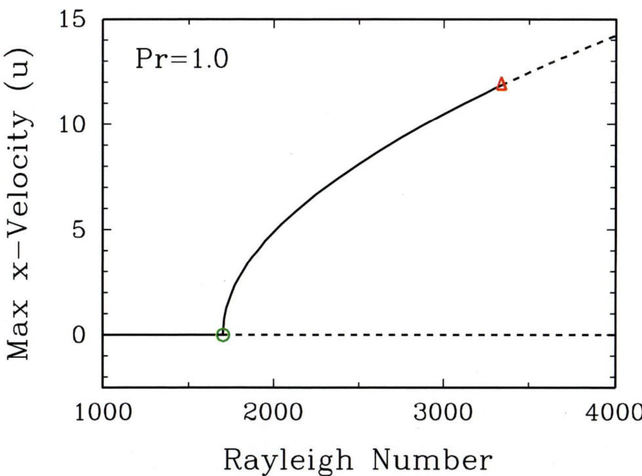

Fig. 6. Plot of steady solution branches in a symmetric box as a function of Rayleigh number for $Pr = 1.0$, showing a pitchfork bifurcation from the stationary solution to a branch of convective rolls. A second pitchfork bifurcation is seen on the asymmetric branch at $Ra_{cr} = 3338.1$.

The pitchfork tracking algorithm in Sec. 2.2 was launched using an eigenvector calculated with the linear stability analysis capability as the ψ vector. This same vector was used as the initial guess for the null vector $\mathbf{y}$. This algorithm located the pitchfork at $Pr = 1.0$ to be at $Ra = 3338.1$ for the mesh of 1.58 Million unknowns. About 30 solutions were calculated along the branch, calculated down to $Pr = 0.037$, as shown in Fig. 8. The symbols at $Pr = 1$ correspond to the similarly marked solutions in Fig. 6. Each solution required 40 min on average on a cluster of 48 3.0 GHz processors. The pitchfork bifurcation corresponding to the initial bifurcation

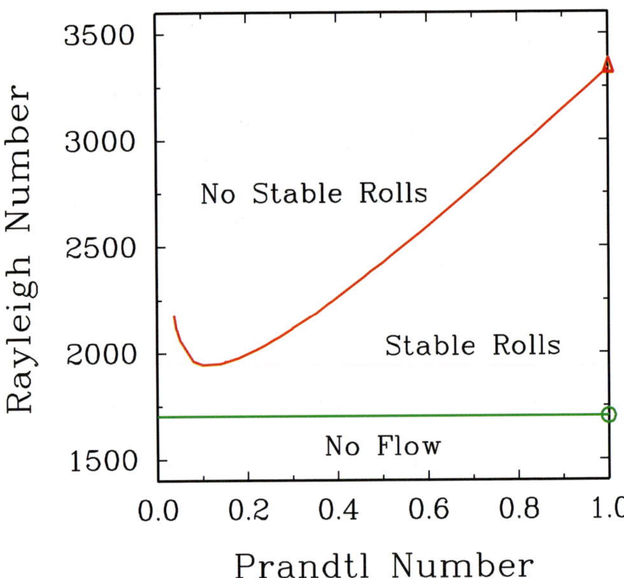

Fig. 8.   The results of tracking the pitchfork bifurcation seen in Fig. 6 in a symmetric box as a function of Prandtl number is shown. Furthermore, a Hopf bifurcation branch is computed, since this is the bifurcation signaling the loss of stability of the convective rolls solution at low Pr.

from the trivial solution, known to be independent of Prandtl number, is drawn in as well. The different flow regions are delineated by curves of bifurcation points.

Linear stability calculations at Pr = 0.04 revealed a complex conjugate pair of eigenvalues with positive real part. This indicates that a Hopf bifurcation had overtaken the pitchfork as

the first destabilizing mode, confirming the results of previous works that oscillatory instabilities destabilize the convective rolls at low Pr. Since we have not developed algorithms for directly locating higher co-dimension bifurcations, the coincidence of the Hopf and Pitchfork bifurcations was found, by repeated stability analysis calculations along the curve of pitchfork bifurcation, to occur near Pr = 0.0434 and Ra = 2106. The real and imaginary parts of the eigenvector corresponding to the Hopf bifurcation are visualized in Fig. 9. These solutions show five cells developing in the y-direction, where previously there was no variation.

Starting from this point, and using the real and imaginary eigenvectors as initial guesses, the Hopf tracking algorithm was launched. Since the solutions of the complex matrix (or rank $2n$ real valued matrix) require considerable extra memory and time, the solution and eigenvector were first interpolated to a coarser mesh. Since the solution with these boundary conditions has no variation in the y-direction (although the eigenvectors do), the mesh was coarsened only in this dimension. By reducing from 100 to 32 elements in this dimension, a mesh corresponding to 516 K unknowns was produced.

The results of the Hopf tracking runs are shown (along with the pitchfork curves from Fig. 8) in Fig. 10, with the Prandtl number axis switched to a log scale. The Rayleigh number of the Hopf bifurcation appears to be approaching a low Prandtl

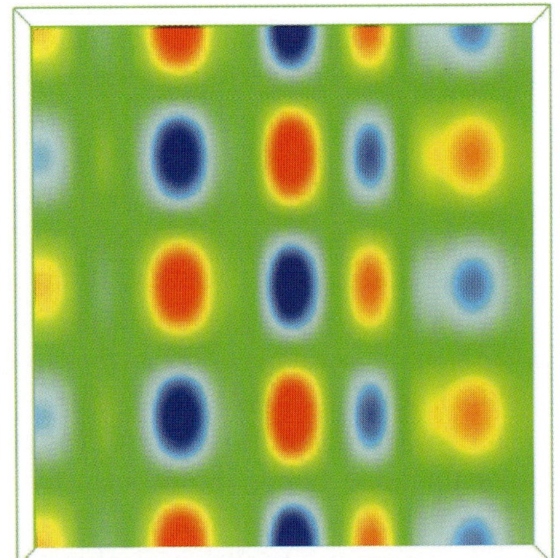 

Fig. 9.   The heat flux through the bottom of the symmetric box is shown for real and imaginary parts of the null vector for the Hopf bifurcation, in the neighborhood of the higher codimension bifurcation at Pr = 0.0434 and Ra = 2106. The red and blue regions correspond to downward and upward flow.

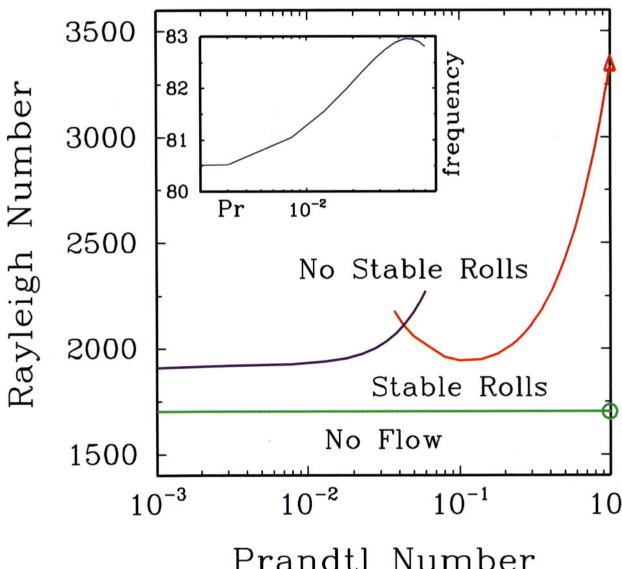

Fig. 10. The results of tracking the pitchfork bifurcation seen in Fig. 6 in a symmetric box as a function of Prandtl number is shown. Furthermore, a Hopf bifurcation branch is computed, since this is the bifurcation signaling the loss of stability of the convective rolls solution at low Pr.

number limit around Ra = 1900. The inset figure showing the frequency of the Hopf bifurcation shows that this quantity (with our choice of the nondimensionalization of time) is also becoming insensitive to the Prandtl number. The calculation of 13 points along the Hopf curve required on average 50 min per solution on a cluster of 48 3.0 GHz processors.

3.4. *Effect of linear solver tolerance*

In this section we present a simple numerical experiment designed to inform both on the accuracy and robustness of the bifurcation tracking algorithms described in this paper. One would expect that the algorithms, which use iterative linear solves of the matrix being driven singular, would continue to converge towards the singularity until the condition number of the matrix being inverted multiplied by the error in the linear solve was order one. Therefore a tighter tolerance on the iterative linear solves would give a more accurate solution.

We performed a set of four computations using the turning point tracking algorithm to converge to the turning point at Ra = 4915.2 (and Pr = 1.0), starting from a converged steady state solution at Ra = 4875. Each computation was forced to run

for 12 Newton iterations, by setting an unreachable convergence tolerance for the nonlinear system. This was repeated for four different tolerances for the reduction in the residual for the iterative linear solves: $10^{-4}, 10^{-6}, 10^{-8}, 10^{-10}$. Figure 11 shows a plot of the norm — the residual of the turning point equations (4) (which are dominated by $|\mathbf{Jy}|$) as a function of Newton iteration for each of these four tolerances. (We should note that several of the linear solves did not reach their requested tolerances.)

The results can be interpreted both in terms of robustness in accuracy. When looking at robustness, the numerical instability of the algorithms becomes apparent. After reaching a level of convergence to the singularity, the inexact solves of the nearly singular matrix can lead to bad Newton steps. This is particularly noticeable in the 10^{-4} run. This results in unacceptably large increases in the residual, and we have seen occurrences where the code has not recovered from these lapses. (This behavior can be mitigated by damping or other globalizations of Newton's method [Pawlowski *et al.*, 2004].) In practice, we set a tight tolerance on the iterative linear solves (e.g. 10^{-8}),

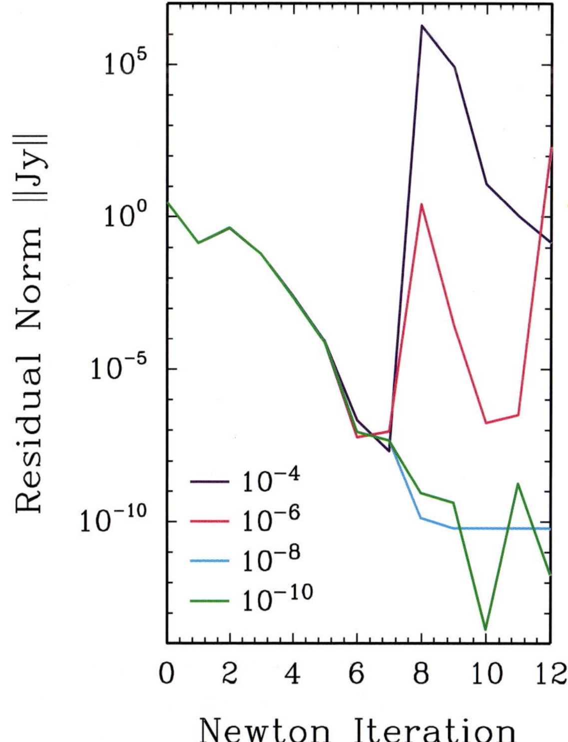

Fig. 11. Convergence history of the turning point algorithm is plotted as a function of a Newton iteration for four different linear solver tolerances.

and a moderate tolerance on the nonlinear system, and the robustness issues do not usually come into play. This does limit the scalability of the algorithms since this requirement puts a larger burden on the preconditioners and linear solver algorithms than the steady state solve.

With regard to accuracy, it can be seen that a tighter linear solver tolerance leads to a more accurate solution of the bifurcation. All runs succeeded in dropping the residual seven orders of magnitude after six Newton iterations, and the 10^{-10} curve reaches a very low residual near 10^{-14}. All predicted the same parameter value of the bifurcation parameter to five digits, to Ra = 4915.2, and only the 10^{-4} run moved away from this value with a bad step. To put this in perspective, the same turning point calculated on a mesh of $208\,\mathrm{K}$ unknowns, corresponding to half as many elements in each direction, was located at Ra = 5167.8, a full 5% difference, and implying that the solution on our current mesh still has a 1–2% discretization error. Furthermore, in many applications the modeling error, such as knowing the true value for the viscosity, even swamps the discretization error, further decreasing the importance of locating the bifurcation point to high accuracy.

One final point regarding the results in Fig. 11 comes from the observation that the four curves overlap for the first five Newton iterations. This suggests that an inexact Newton algorithm, where the linear solver convergence tolerance starts out very loose and is dynamically tightened at later Newton iterations, would be appropriate for these calculations. The savings would be significant, since the typical linear solver time to reach a 10^{-4} tolerance was **half** that needed to make the tightest 10^{-10} tolerance.

4. Summary and Conclusions

In this paper we present a set of bifurcation tracking algorithms used in the LOCA software library and aimed at large scale applications, such as those coming from discretizations of PDEs in multiple dimensions. The augmented systems defining the bifurcations are solved with a Newton method. The linear solves within the Newton methods are solved with block elimination, resulting in a numerically unstable procedure involving the linear solve of the same Jacobian matrix being driven singular. This choice, however, leads to a simple interface to existing Newton-based application codes and frees

the bifurcation library from needing any information about the storage or solution of the linear systems.

To demonstrate the scalability of the algorithms, a bifurcation analysis of a three-dimensional natural convection flow application was undertaken. The limit of stability of convective roll cells in the Rayleigh–Bénard problem was investigated as a function of the Rayleigh number and Prandtl number. Turning point, pitchfork, and Hopf bifurcations indicating the limit of stability of the convective roll solutions were successfully tracked, with no failures in the continuation process. The first two were tracked on a mesh corresponding to 1.58 Million unknowns, and the third on a mesh of 0.51 Million unknowns.

The accuracy and robustness of the algorithms were shown in a numerical experiment to be a strong function of the tolerance of the iterative linear solver used to invert the Jacobian matrix. While the accuracy of this approach was found to be more than adequate for this problem — finding the parameter value of the bifurcation to several digits — the current algorithms lack robustness when trying to solve the bifurcation problem to high accuracy. Work is underway to improve the robustness by looking at reformulations of the linear solves, to implement algorithms based on minimally augmented systems [Govaerts, 2000], and to look at more invasive approaches.

Acknowledgments

The authors would like to thank those that contributed code, advice and support for this work, including John Shadid, Rich Lehoucq, David Day, Ray Tuminaro, Ed Wilkes, David Womble and Sudip Dosanjh. Funding for this work came from the US DOE MICS and ASCI programs. Sandia is a multiprogram laboratory operated by Sandia Corporation, a Lockheed Martin Company, for the United States Department of Energy under Contract DE-AC04-94AL85000.

References

Burroughs, E. A., Romero, L. A., Lehoucq, R. B. & Salinger, A. G. [2001] "Large scale eigenvalue calculations for computing the stability of buoyancy driven flows," Technical Report SAND2001–0113, Sandia National Laboratories, Albuquerque, NM.

Burroughs, E. A., Romero, L. A., Lehoucq, R. B. & Salinger, A. G. [2004] "Linear stability of flow in a differentially heated cavity via large-scale eigenvalue calculations," *Int. J. Numer. Meth. Heat Fluid Flow* **14**, 803–822.

Busse, F. H. & Clever, R. M. [1979] "Instabilities of convection rolls of moderate Prandtl number," *J. Fluid Mech.* **91**, 319–335.

Clever, R. M. & Busse, F. H. [1995] "Convection rolls and their instabilities in the presence of a nearly insulating upper boundary," *Phys. Fluids* **7**, 92–97.

Cliffe, K., Spence, A. & Tavener, S. [2000a] *The Numerical Analysis of Bifurcation with Application to Fluid Mechanics*, Acta Numerica (Cambridge University Press), pp. 39–131.

Cliffe, K., Spence, A. & Tavener, S. [2000b] "O(2)-symmetry breaking bifurcation: With application to the flow past a sphere in a pipe," *Int. J. Numer. Meth. Fluids* **32**, 175–200.

Cox, S. M. & Matthews, P. C. [2000] "Instability of rotating convection," *J. Fluid Mech.* **403**, 153–172.

Day, D. & Heroux, M. [2001] "Solving complex-valued linear systems via equivalent real formulations," *SIAM J. Sci. Comp.* **23**, 480–498.

Dhooge, A., Govaerts, W. & Kuznetsov, Y. [2003] "MATCONT: A MATLAB package for numerical bifurcation analysis of ODEs," *ACM Trans. Math. Softw.* **29**, 141–164.

Doedel, E. J., Champneys, A. R., Fairgrieve, T. F., Kuznetsov, Y. A., Sandstede, B. & Wang, X. J. [1997] "AUTO: Continuation and bifurcation software with ordinary differential equations (with homcont), user's guide," Technical Report, Concordia University, Montreal, Canada.

Engelborghs, K., Luzyanina, T. & Roose, D. [2002] "Numerical bifurcation analysis of delay differential equations using DDE-BIFTOOL," *ACM Trans. Math. Softw.* **28**, 1–21.

Frink, L. & Salinger, A. [2003] "Rapid analysis of phase behavior with density functional theory, part II: Capillary condensation in disordered porous media," *J. Chem. Phys.* **118**, 7466–7476.

Frischknecht, A., Weinhold, J., Salinger, A., Curro, J., Frink, L. & McCoy, J. [2002] "Density functional theory of inhomogeneous polymer systems: I. Numerical methods," *J. Chem. Phys.* **117**, 10385–10397.

Fujii, F., Noguchi, H. & Ramm, E. [2000] "Static path jumping to attain postbuckling equilibria of a compressed circular cylinder," *Comput. Mech.* **26**, 259–266.

Govaerts, W. [2000] *Numerical Methods for Bifurcations of Dynamic Equilibria* (SIAM, Philadelphia, PA).

Griewank, A. & Reddien, G. [1983] "The calculation of Hopf points by a direct method," *IMA J. Numer. Anal.* **3**, 295–303.

Henderson, M. [2002] "Multiple parameter continuation: Computing implicitly defined k-manifolds," *Int. J. Bifurcation and Chaos* **12**, 451–476.

Hendrickson, B. & Leland, R. [1995a] "The Chaco user's guide: Version 2.0," Technical Report SAND94–2692, Sandia National Labs, Albuquerque, NM.

Hendrickson, B. & Leland, R. [1995b] "An improved spectral graph partitioning algorithm for mapping parallel communications," *SIAM J. Sci. Comput.* **16**, 452–469.

Heroux, M., Bartlett, R., Howle, V., Hoekstra, R., Hu, J., Kolda, T., Lehoucq, R., Long, K., Pawlowski, R., Phipps, E., Salinger, A., Thornquist, H., Tuminaro, R., Willenbring, J. & Williams, A. [2003] "An overview of Trilinos," Technical Report SAND2003–2927, Sandia National Labs, Albuquerque, NM.

Hughes, T. J. R., Franca, L. P. & Balestra, M. [1989a] "A new finite element formulation for computational fluid dynamics: V. Circumventing the Babuska–Brezzi condition: A stable Petrov-Galerkin formulation of the Stokes problem accommodating equal-order interpolation," *Comput. Meth. Appl. Mech. Engin.* **59**, 85–99.

Hughes, T. J. R., Franca, L. P. & Hulbert, G. M. [1989b] "A new finite element formulation for computational fluid dynamics: VII. the Galerkin/Least-Squares method for advective-diffusive equation," *Comput. Meth. Appl. Mech. Engin.* **73**, 173–189.

Hutchinson, S. A., Shadid, J. N. & Tuminaro, R. S. [1995] "Aztec user's guide: Version 1.0," Technical Report SAND95-1559, Sandia National Laboratories, Albuquerque, New Mexico 87185.

Keller, H. B. [1977] "Numerical solution of bifurcation and nonlinear eigenvalue problems," in *Applications of Bifurcation Theory*, ed. Rabinowitz, P. H. (Academic Press, NY), pp. 159–384.

Kuznetsov, Y. A. & Levitin, V. V. [1995–1997] "CONTENT: A multiplatform environment for analyzing dynamical systems," Dynamical Systems Laboratory, CWI, Amsterdam, The Netherlands.

Lasater, M. S., Kelley, C. T., Salinger, A. G., Woolard, D. L. & Zhao, P. [2004] "Parallel solution of the Wigner–Poisson equations for RTDs," *Proc. 2004 Int. Symp. Distributed Computing and Applications to Business, Engineering, and Science.*

Lehoucq, R. & Salinger, A. [2001a] "Large-scale eigenvalue calculations for stability analysis of steady flows on massively parallel computers," *Int. J. Numer. Meth. Fluids* **36**, 309–327.

Lehoucq, R. B. & Salinger, A. G. [2001b] "Large-scale eigenvalue calculations for stability analysis of steady flows on massively parallel computers," *Int. J. Numer. Meth. Fluids* **36**, 309–327.

Lehoucq, R. B., Sorensen, D. C. & Yang, C. [1998] *ARPACK USERS GUIDE: Solution of Large Scale Eigenvalue Problems with Implicitly Restarted Arnoldi Methods* (SIAM, Philadelphia, PA).

Lust, K., Roose, D., Spence, A. & Champneys, A. [1998] "An adaptive Newton-Picard algorithm with subspace iteration for computing periodic solutions," *SIAM J. Sci. Comput.* **19**, 1188–1209.

Mamun, C. & Tuckerman, L. [1995] "Asymmetric and Hopf bifurcation in spherical Couette flow," *Phys. Fluids* **7**, 80–91.

Maschhoff, K. J. & Sorensen, D. C. [1996] "P_ARPACK: An efficient portable large scale eigenvalue package for distributed memory parallel architectures," in *Applied Parallel Computing in Industrial Problems and Optimization*, eds. Wasniewski, J., Dongarra, J., Madsen, K. & Olesen, D., Lecture Notes in Computer Science. Vol. 1184 (Springer-Verlag, Berlin).

Moore, G. & Spence, A. [1980] "The calculation of turning points of nonlinear equations," *SIAM J. Numer. Anal.* **17**, 567–576.

Muratov, C. & Shvartsman, S. [2003] "An asymptotic study of the inductive pattern formation mechanism in drosophila egg development," *Physica* **D186**, 93–108.

Nakamura, Y. [1997] "Spatio-temporal dynamics of forced periodic flows in a confined domain," *Phys. Fluids* **9**, 3275–3287.

Nore, C., Tuckerman, L., Daube, O. & Xin, S. [2003] "The $1:2$ mode interaction in exactly counter-rotating von Karman swirling flow," *J. Fluid Mech.* **477**, 51–88.

Pawlowski, R. P., Salinger, A. G., Romero, L. A. & Shadid, J. N. [2001] "Computational design and analysis of MPOVPE reactors," *J. Phys. IV* **11**, 197–204.

Pawlowski, R. P., Simonis, J. P., Shadid, J. N. & Walker, H. F. [2004] "Globalization techniques for Newton–Krylov methods and applications to the fully-coupled solution of the Navier–Stokes equations," Technical Report SAND2004, Sandia National Labs, Albuquerque, NM.

Salinger, A. & Frink, L. [2003] "Rapid analysis of phase behavior with density functional theory, part I: Novel numerical methods," *J. Chem. Phys.* **118**, 7457–7465.

Salinger, A., Lehoucq, R., Pawlowski, R. & Shadid, J. [2002a] "Computational bifurcation and stability studies of the $8:1$ cavity problem," *Int. J. Numer. Meth. Fluids* **40**, 1059–1073.

Salinger, A. G., Bou-Rabee, N., Pawlowski, R. P., Wilkes, E. D., Burroughs, E. A., Lehoucq, R. B. & Romero, L. A. [2002b] "LOCA 1.0: Library of continuation algorithms — Theory and implementation manual," Technical Report SAND2002-0396, Sandia National Laboratories, Albuquerque, New Mexico 87185.

Salinger, A. G., Shadid, J. N., Hutchinson, S. A., Hennigan, G. L., Devine, K. D. & Moffat, H. K. [1999] "Analysis of gallium arsenide deposition in a horizontal chemical vapor deposition reactor using massively parallel computations," *J. Cryst. Growth* **203**, 516–533.

Shadid, J. N. [1999] "A fully-coupled Newton-Krylov solution method for parallel unstructured finite element fluid flow, heat and mass transport," *IJCFD* **12**, 199–211.

Shepherd, J. F. [2000] "CUBIT mesh generation toolkit," Technical Report SAND2000-2647, Sandia National Laboratories, Albuquerque, New Mexico 87185.

Sone, Y., Aoki, K. & Sugimoto, H. [1997] "The Bénard problem for a rarefied gas: Formation of steady flow patterns and stability of array of rolls," *Phys. Fluids* **9**, 3898–3914.

Tangborn, A. V., Zhang, S. Q. & Lakshminarayanan, V. [1995] "A three-dimensional instability in mixed convection with streamwise periodic heating," *Phys. Fluids* **7**, 2648–2658.

Werner, B. & Spence, A. [1984] "The computation of symmetry-breaking bifurcation points," *SIAM J. Numer. Anal.* **21**, 388–399.

Willis, G. E. & Deardorff, J. W. [1970] "The oscillatory motions of Rayleigh convection," *J. Fluid Mech.* **44**, 661–672.

Xin, S. & Le Quéré, P. [2002] "An extended Chebyshev pseudo-spectral benchmark for the $8:1$ differentially heated cavity," *Int. J. Numer. Meth. Fluids* **40**, 981–998.

AN ALGORITHM FOR FINDING INVARIANT ALGEBRAIC CURVES OF A GIVEN DEGREE FOR POLYNOMIAL PLANAR VECTOR FIELDS

GRZEGORZ ŚWIRSZCZ

IBM Watson Research Center, Yorktown Heights NY 10598, USA

Institute of Mathematics, University of Warsaw,
02–097 Warsaw, Banacha 2, Poland
swirszcz@us.ibm.com
swirszcz@mimuw.edu.pl

Received March 10, 2004; Revised June 10, 2004

Given a system of two autonomous ordinary differential equations whose right-hand sides are polynomials, it is very hard to tell if any nonsingular trajectories of the system are contained in algebraic curves. We present an effective method of deciding whether a given system has an invariant algebraic curve of a given degree. The method also allows the construction of examples of polynomial systems with invariant algebraic curves of a given degree. We present the first known example of a degree 6 algebraic saddle-loop for polynomial system of degree 2, which has been found using the described method. We also present some new examples of invariant algebraic curves of degrees 4 and 5 with an interesting geometry.

Keywords: Invariant algebraic curve; symbolic computations; linear algebra.

1. Introduction and Preliminary Definitions

Since Darboux [1878] had found connections between algebraic geometry and the existence of first integrals of polynomial systems (polynomial planar vector fields), algebraic invariant curves have been a central object in the theory of integrability of polynomial systems in $\mathbb{R}^2$. Today, after more than a century of investigations, the theory of invariant algebraic curves is still full of open questions. One of the reasons for this is the fact that examples of polynomial systems with invariant algebraic curves are extremely hard to find. The calculations required to find such examples exceed human abilities. Recent development of the theory of integrability would have been impossible without the use of automatic computations. Thanks to pioneering works of von Neumann [1946, 1958] and Turing [1950] we have now at our disposal sound foundations of methods performing symbolic computations and providing us with reliable results. The visionary ideas of von Neumann about the architecture of computers (called today "von Neumann Architecture") [von Neumann, 1945], his concepts of "code" and data processing have laid the foundations to modern computer science and its applications to pure and applied mathematics. Thanks to his ideas we know now how to obtain "reliable answers from unreliable computer components" and with the aid of computers we are able to develop proofs and theories which are strictly correct from the mathematical point of view.

Dynamical systems are one of the fields of mathematics where the combination of pure science, modeling and computational methods have led to amazing results. In the theory of iterations of maps, thanks to the use of computers, we have beautiful visualizations of fractals (see for example the famous book [Mandelbrot, 1982]). The computer simulations have also provided useful tools and intuitions for many mathematical proofs, see for example [Lanford, 1982]. In the theory of vector fields the computer assisted methods have a wide array of applications. The methods of modeling are used to obtain the approximate phase portraits for systems of differential equations, but probably even more important application is the use of symbolic arithmetics. The possibility to perform in a relatively short time extremely complex symbolic operations have led in the last years to the discovery of new examples of invariant algebraic curves for polynomial systems and to a much better understanding of the theory of their integrability. Nevertheless, even with the help of computers it is far from obvious how to look for such examples. In the present paper we propose an approach based on symbolic computations and methods of linear algebra, which turned out to be very effective for low-degree polynomial systems. It allows to reduce the problem of finding invariant algebraic curves to the problem of finding zeroes of a set of relatively simple polynomial equations. This gives a link between the theory of integrability of polynomial systems and a classical chapter in the computer assisted mathematics — the theory of Gröbner bases. Before we proceed with the introduction we present some definitions.

A *polynomial system* of a degree k in $\mathbb{R}^2$ is a system of two autonomous differential equations

$$\dot{x} = p(x,y),$$
$$\dot{y} = q(x,y), \tag{1}$$

where p, q are coprime polynomials of degree k, that is,

$$p(x,y) = \sum_{i,j=0}^{k} p_{i,j}x^i y^j, \quad q(x,y) = \sum_{i,j=0}^{k} q_{i,j}x^i y^j.$$

We say that the algebraic curve is an *invariant algebraic curve of degree n* if it is contained in the union of trajectories of (1) and it is given by zeroes of a polynomial φ of a degree n

$$\varphi(x,y) = \sum_{i,j=0}^{n} \varphi_{i,j}x^i y^j.$$

From basic properties of polynomials follows the fundamental fact that the algebraic curve $\varphi(x,y) = 0$ is an invariant algebraic curve of system (1) if and only if there exists a polynomial $\kappa = \kappa(x,y)$ satisfying

$$p\frac{\partial \varphi}{\partial x} + q\frac{\partial \varphi}{\partial y} - \kappa\varphi = 0. \tag{2}$$

The polynomial κ is called a *cofactor* of the curve $\varphi = 0$. Of course, the degree of the cofactor can be at most $k-1$, so

$$\kappa(x,y) = \sum_{i,j=0}^{k-1} k_{i,j}x^i y^j. \tag{3}$$

An invariant algebraic curve $\varphi = 0$ is called *irreducible* if the polynomial φ is irreducible. In the rest of the paper all the invariant algebraic curves are assumed to be irreducible unless stated otherwise.

A trajectory γ of system (1) is a *limit cycle* if it is nonconstant periodic and there are no other periodic trajectories in some neighborhood of γ. The orbit γ is an *algebraic limit cycle* of system (1) if it is a limit cycle and if it is contained in some irreducible algebraic invariant curve $\varphi = 0$ of system (1). An algebraic saddle-loop is defined analogously.

Polynomial system (1) that has enough invariant algebraic curves must be integrable. Let $\varphi_i(x,y)$ be polynomials defining invariant algebraic curves of system (1). We say that first integral H of polynomial system (1) is in Darboux form if it satisfies for some $K \in \mathbb{N}$, and some $\alpha_i \in \mathbb{C}$

$$H(x,y) = \prod_{i=1}^{K} \varphi_i^{\alpha_i}(x,y).$$

We say that integrating factor M of polynomial system (1) is in Darboux form if it satisfies for some $L \in \mathbb{N}$ and some $\beta_i \in \mathbb{C}$

$$M(x,y) = \prod_{i=1}^{L} \varphi_i^{\beta_i}(x,y).$$

The classical result of Darboux is

Theorem 1.1. (Darboux) *If polynomial system (1) of degree k has more than $k(k+1)/2$ irreducible invariant algebraic curves $\varphi_i(x,y) = 0$, then there exist constants α_i such that the product*

$$\prod_{i=1}^{\frac{k(k+1)}{2}} \varphi_i^{\alpha_i}(x,y)$$

is a first integral in Darboux form. When the number of invariant algebraic curves is equal to $k(k+1)/2$, the system has an integrating factor in Darboux form.

Nevertheless, the above conditions are too strong in general. This motivates the following problem.

Problem 1. What are the connections between the possible degrees and numbers of invariant algebraic curves of a polynomial system of degree k and the existence and type of its first integral?

Understanding the significance of invariant algebraic curves [Poincaré, 1891, 1897] has formulated a slightly different question: Estimate the greatest possible degree $n = n(k)$ of an invariant algebraic curve for a polynomial system of degree k. In this formulation the question has a simple answer, the system

$$\dot{x} = nx, \qquad \dot{y} = y \qquad (4)$$

has the invariant algebraic curve $x - y^n = 0$, therefore even $n(1)$ is unbounded. Nevertheless, system (4) has a rational first integral xy^{-n}, so each of its trajectories is contained in some algebraic curve. Therefore, Problem 1 is often referred to as the "Poincaré's problem". Another approach is to look for "nontrivial" examples of invariant algebraic curves of high degrees, like algebraic limit cycles or algebraic saddle-loops.

One of the main problems in the development of the theory of invariant algebraic curves is the fact that there are not many examples known. Even for systems of degree 2 the structure of invariant algebraic curves turned out to be much more complex than has been expected. For example it has been conjectured that

Conjecture (Lins–Neto). *There exists a number* $N(2)$ *such that, if a quadratic system has an invariant algebraic curve of a degree* $n > N(2)$, *then the system has a rational first integral.*

This conjecture has been proved to be false by [Christopher & Llibre, 2002], who have found a class of quadratic systems that can have an invariant algebraic curve of any degree, and not have a rational first integral. Their example has a rational integrating factor. Later [Chavarriga & Grau, 2002] have found a family of quadratic systems which can have an invariant algebraic curve of any degree and without a rational integrating factor. It has an integrating factor in Darboux form and it is still an open question if the following conjecture is true.

Conjecture (Weakened Lins–Neto). *There exists a number* $N(2)$ *such that if a quadratic system has an invariant algebraic curve of degree greater than* $N(2)$, *then the system has an integrating factor in Darboux form.*

The problem of classification of algebraic limit cycles for quadratic system is also open, for almost 30 years there have been only three known examples, one of degree 2 [Qin, 1958] and two of degree 4 [Yablonskii, 1966; Filiptsov, 1973]. It was also known [Evdokimenco, 1970, 1974, 1979] that there are no quadratic systems with algebraic limit cycles of degree 3. Then in the year 2000 two more families of quadratic systems with algebraic limit cycles of degree 4 have been found; see [Chavarriga *et al.*, 2001; Chavarriga *et al.*, 2000]. It has also been proved by [Chavarriga *et al.*, 2000] that there are no other families of quadratic systems with algebraic limit cycles of degree 4. The question if there exist quadratic systems with algebraic limit cycles of degree greater than 4 remained open until recently two new examples, one of degree 5 and one of degree 6, have been found [Christopher *et al.*, 2003]. Also [Christopher *et al.*, 2003] presents the first example of an algebraic saddle-loop of degree 5. In Sec. 4.3 we give the first example of an algebraic saddle-loop of degree 6.

Another simple and interesting class of polynomial systems for which one may ask a question about the existence of algebraic limit cycles are Liénard systems $\dot{x} = y$, $\dot{y} = -F_k(x)y - G_m(x)$ (F_k, G_m-polynomials of degrees n and m, respectively). In this case, the question has been answered by Żołądek [1998] for all values of k and m except for $k = 1$, $m = 3$, for which the question still remains open.

These, and many more similar questions, motivate the need for an efficient algorithm to efficiently find examples of families of polynomial systems with invariant algebraic curves. Until now, most attempts were based on looking for algebraic curves in some special form (usually hyperelliptic) for the sake of simplifying the calculations. However successful this simple approach was in many cases, it is far from being general and fails completely when one tries to look for invariant curves of a high degree. This is the reason that there have been practically no known examples of invariant algebraic curves of degrees higher than 4. For quadratic systems even the invariant algebraic curves of degree 4 are not well investigated. As one of the examples of the application of the presented algorithm, we give in Sec. 4 two examples of invariant algebraic curves of degree 4 with an interesting geometry, which to our knowledge have not been known before.

With the method described in the present paper we have been able to successfully investigate some families of quadratic systems with invariant algebraic curves of degrees as high as 14.

2. The Problem of Invariant Algebraic Curves from the Point of View of Linear Algebra

The method we present is based on the observation that the problem of existence and finding a solution to Eq. (2) is a purely linear problem. To be more precise, we look for a polynomial $\varphi(n)$ of degree less or equal to n. Such polynomials form a linear space $\mathcal{V}_n$ of dimension $(n+1)(n+2)/2$. Given a polynomial system (1) of degree k and a polynomial $\kappa(x,y)$ of degree $k-1$ we define an operator $\Xi : \mathcal{V}_n \to \mathcal{V}_{n+k-1}$ as

$$\Xi[\varphi] = p\frac{\partial\varphi}{\partial x} + q\frac{\partial\varphi}{\partial y} - \kappa\varphi.$$

Of course, Ξ is a linear operator. An obvious consequence of the definition is:

Proposition 2.1. *Polynomial system (1) has an invariant algebraic curve φ of degree less or equal to n with cofactor κ if and only if the operator Ξ has a nontrivial kernel.*

To investigate the kernel of Ξ we shall use the language of matrices. We introduce the following basis $\mathcal{B}$ in $\mathcal{V}_n$:

$$x^i y^j = e_{\mu(i,j)},$$

where $\mu(i,j) = (i+j)(i+j+1)/2 + i$. This comes from linearly ordering homogenous monomials in the following way: $x^i y^j > x^k y^l$ if and only if $i+j > k+l$ or $i+j = k+l$ and $i > k$.

Remark 2.2. Note that the function μ is a bijection from $\mathbb{N} \times \mathbb{N} \to \mathbb{N}$, so it has an inverse function. Therefore, it makes sense to use both $\mu = \mu(i,j)$ and $i = i(\mu)$, $j = j(\mu)$.

Every polynomial $\varphi \in \mathcal{V}_n$ has a unique representation as a vector in the basis $\mathcal{B}$ — its coordinates are simply the coefficients of the polynomial φ. Now the operator Ξ is represented in the basis $\mathcal{B}$ by a $[(n+k)(n+k+1)/2] \times [(n+1)(n+2)/2]$ matrix $A = (a_{IJ})$. The terms a_{IJ} satisfy

$$\begin{aligned}
a_{IJ} = {} & i(J)p_{i(I)-i(J)+1,j(I)-j(J)} \\
& + j(J)q_{i(I)-i(J),j(I)-j(J)+1} \\
& - k_{i(I)-i(J),j(I)-j(J)},
\end{aligned} \tag{5}$$

where $i(I), j(I), i(J), j(J)$ are the unique numbers satisfying $\mu(i(I),j(I)) = I$, and $\mu(i(J),j(J)) = J$ (see Remark 2.2). We apply the convention that we set p_{ij}, q_{ij}, k_{ij} equal to 0 if (i,j) is out of the range of definition, i.e. i or j is negative, or their sum is greater than the degree of the polynomial of their coefficients.

Matrix A has the following block-multidiagonal form

$A =$

<table>
<tr><td>B_n^{n+k-1}</td><td></td><td></td><td></td><td></td><td></td><td></td><td></td><td></td><td></td><td></td><td></td><td></td><td></td></tr>
<tr><td>B_n^{n+k-2}</td><td>B_{n-1}^{n+k-2}</td><td></td><td></td><td></td><td></td><td></td><td></td><td></td><td></td><td></td><td></td><td></td><td></td></tr>
<tr><td>$\vdots$</td><td>$\vdots$</td><td>$\ddots$</td><td></td><td></td><td></td><td></td><td></td><td></td><td></td><td></td><td></td><td></td><td></td></tr>
<tr><td>B_n^n</td><td>B_{n-1}^n</td><td>$\cdots$</td><td>B_{n-k+1}^n</td><td></td><td></td><td></td><td></td><td></td><td></td><td></td><td></td><td></td><td></td></tr>
<tr><td>B_n^{n-1}</td><td>B_{n-1}^{n-1}</td><td>$\cdots$</td><td>B_{n-k+1}^{n-1}</td><td>B_{n-k}^{n-1}</td><td></td><td></td><td></td><td></td><td></td><td></td><td></td><td></td><td></td></tr>
<tr><td></td><td>B_{n-1}^{n-2}</td><td>$\cdots$</td><td>B_{n-2}^{n-2}</td><td>B_{n-3}^{n-2}</td><td>B_{n-4}^{n-2}</td><td></td><td></td><td></td><td></td><td></td><td></td><td></td><td></td></tr>
<tr><td></td><td></td><td>$\ddots$</td><td>$\vdots$</td><td>$\vdots$</td><td>$\vdots$</td><td>$\ddots$</td><td></td><td></td><td></td><td></td><td></td><td></td><td></td></tr>
<tr><td></td><td></td><td></td><td></td><td></td><td></td><td>$\vdots$</td><td></td><td></td><td></td><td></td><td></td><td></td><td></td></tr>
<tr><td></td><td></td><td></td><td></td><td></td><td></td><td>$\ddots$</td><td>$\vdots$</td><td>$\vdots$</td><td>$\vdots$</td><td>$\ddots$</td><td></td><td></td><td></td></tr>
<tr><td></td><td></td><td></td><td></td><td></td><td></td><td></td><td>B_{k+2}^{k+1}</td><td>B_{k+1}^{k+1}</td><td>B_k^{k+1}</td><td>$\cdots$</td><td>B_2^{k+1}</td><td></td><td></td></tr>
<tr><td></td><td></td><td></td><td></td><td></td><td></td><td></td><td></td><td>B_{k+1}^k</td><td>B_k^k</td><td>$\cdots$</td><td>B_2^k</td><td>B_1^k</td><td></td></tr>
<tr><td></td><td></td><td></td><td></td><td></td><td></td><td></td><td></td><td></td><td>B_k^{k-1}</td><td>$\cdots$</td><td>B_2^{k-1}</td><td>B_1^{k-1}</td><td>B_0^{k-1}</td></tr>
<tr><td></td><td></td><td></td><td></td><td></td><td></td><td></td><td></td><td></td><td></td><td>$\ddots$</td><td>$\vdots$</td><td>$\vdots$</td><td>$\vdots$</td></tr>
<tr><td></td><td></td><td></td><td></td><td></td><td></td><td></td><td></td><td></td><td></td><td></td><td>B_2^1</td><td>B_1^1</td><td>B_0^1</td></tr>
<tr><td></td><td></td><td></td><td></td><td></td><td></td><td></td><td></td><td></td><td></td><td></td><td></td><td>B_1^0</td><td>B_0^0</td></tr>
</table>

where each of the blocks B_i^j is a $(i+1) \times (j+1)$ matrix.

Let M_0 denote the set of all the minors of maximum dimension (determinants of $(n+1)(n+2)/2 \times (n+1)(n+2)/2$ submatrices) of the matrix A. M_0 is a set of polynomials in the variables p_{ij}, q_{ij} and k_{ij}. The number of polynomials in the set M_0 is equal to

$$\binom{\dfrac{(n+k)(n+k+1)}{2}}{\dfrac{(n+1)(n+2)}{2}}$$

and each of its elements depends, in general, on $(n+1)(3n+4)/2$ variables. From fundamental facts of linear algebra follows

Theorem 2.3. *Polynomial system (1) has an invariant algebraic curve φ of degree less or equal to n with cofactor κ if and only if all the polynomials in M_0 vanish simultaneously.*

Theorem 2.3 suggests the following algorithm. If we want to find a polynomial system of a given degree k with an invariant algebraic curve of degree less or equal to n, we calculate the corresponding matrix A for the system 1, and the corresponding set M_0. Next we try to solve the equation $M_0 = 0$. (In the language of algebraic geometry this means that we look for a simple description of the algebraic set $V(M_0)$.) Methods for solving systems of polynomial equations are very well developed. The theory of Gröbner bases and multipolynomial resultants can be applied here; see for example [Cox *et al.*, 1998]. Nevertheless, one can immediately see that, if we try to use this straightforward approach, we end up with an enormous number of equations in many variables.

Fortunately, when we look for the examples of polynomial systems with invariant algebraic curves, we usually consider certain families, depending only on a few parameters. Therefore, the number of variables is usually not too big.

The key to reducing the number of equations is a standard linear-algebra approach. First we note that, if there is a row i in the matrix A containing only a single nonzero constant term $a_{i,j}$, then each of the vectors in the kernel of A must have 0 at the jth coordinate. Therefore, we can remove the column j from the matrix A, limiting our considerations to a certain subspace of the

space $\mathcal{V}_n$. Moreover, after the removal there can appear more rows with only one nonzero constant term in them, so sometimes the size of the matrix A can be reduced significantly in that way. We can also remove all the rows containing only zeroes. We obtain the *reduced matrix B*.

Once we have found the matrix B we apply Gauss–Jordan elimination. We should note here that applicability of numerical methods to Gauss elimination is the subject of a fundamental paper by [Goldstine & von Neumann, 1947]. When the polynomial system is expressed in a normal form, one may expect the matrix B to have a lot of terms which are constants, that is, they do not depend on the parameters of the system and the coefficients of the cofactor.

3. The Algorithm

We get the following algorithm. Given a family of polynomial systems

$$\dot{x} = \sum_{i,j=0}^{k} p_{i,j} x^i y^j$$

$$\dot{y} = \sum_{i,j=0}^{k} q_{i,j} x^i y^j \tag{6}$$

whose coefficients $p_{i,j}$, $q_{i,j}$ depend on some parameters $p_1, \ldots, p_s$ and an integer n we want to find those values of the parameters for which the system has an invariant algebraic curve of degree n.

The procedure

1. We use changes of variables to transform simultaneously the system (6) and the potential cofactor $\kappa(x, y) = \sum_{i,j=0}^{k-1} k_{i,j} x^i y^j$ to the simplest form. Usually we strive to make as many of the coefficients $p_{i,j}$, $q_{i,j}$, $k_{i,j}$ as possible zero or equal to constants, all other coefficients are treated as the parameters of the family. We shall call the family obtained in this way the *simplified family*.

2. We find the matrix A for the simplified family.

3. We generate a vector $\tilde{W} \in K[x,y]^{\frac{(n+1)(n+2)}{2}}$, whose ith coordinate is a monomial $e_{\mu(i)}$, i.e. $\tilde{W} = (x^n, x^{n-1}y, x^{n-2}y^2, \ldots, y^n, x^{n-1}, x^{n-2}y, \ldots, x, y, 1)$. We create an extended matrix $\tilde{A}$ obtained by adding the vector $\tilde{W}$ as the last row to the matrix A. This is done only to make the transformation of the obtained

vector-solution into a corresponding polynomial more convenient.

4. We perform the preliminary simplification of the extended matrix $\tilde{A}$: if there is any row i in the matrix $\tilde{A}$ containing only a single nonzero constant term $a_{i,j}$, we remove the jth column from the matrix $\tilde{A}$. We keep repeating this process till there are no more rows with only one nonzero constant term. Then from the obtained matrix we remove all the rows with only zeroes in them. We denote the *extended reduced matrix* matrix we have obtained by $\tilde{B}$.

5. We denote the last row of the matrix $\tilde{B}$ by W. We remove it. The matrix we obtain is the reduced matrix B for the simplified family.

6. We apply the process of Gauss–Jordan elimination to the matrix B, using only nonzero constant terms. Namely, starting from the leftmost column we pick a nonzero constant term and use row reduction to make all the other terms in that column equal to zero. Then we proceed to the next column. If there is a column with all the terms in it depending on the parameters, we skip it in the process. We denote the obtained matrix by C.

7. We apply the process described in step 4 to the matrix C. In other words, this means that we remove all columns with precisely only constant term in them, and then we remove all rows with only zeroes in them. The matrix we obtain is denoted by D.

8. We calculate the set M_1 of minors of maximum dimension of the matrix D. From the standard facts of linear algebra it follows that M_0 vanishes if and only if M_1 vanishes.

9. We try to solve the system of equations $M_1 = 0$. We find a set of solutions $\{S_1, S_2, \ldots, S_d\}$.

10. For each S_i we substitute it to the matrix B, obtaining a matrix $B_i = B|_{S_i}$. Next, we solve the linear system of equations $B_i \cdot X = 0$. Of course, each of the matrices B_i is a degenerate matrix, so for each i we have a nonempty set of l_i solutions $\{X_i^l\}_{l=1}^{l_i}$, $l_i \geq 1$. Note that in most cases B_i is a family of matrices — after the substitution of the solution S_i, B usually still depends on some parameters, and so does each of the corresponding vectors X_i^l. Therefore, we shall refer to each of X_i^l as to a family of solutions, although in some cases it can be a constant family (see Sec. 4.2).

11. For each pair (i, l), the family of polynomials $\varphi_i^l(x, y) = W \cdot X_i^l$ defines a family of invariant

algebraic curves for the subfamily of the simplified family (6) defined by the conditions S_i. Note that S_i usually contains some equations that must be satisfied by the coefficients of the cofactor, as well as the coefficients of the system.

Remark 3.1. One may notice that steps 3–5 of our algorithm seem unnecessary. Indeed, one could apply Gauss–Jordan elimination immediately to the matrix A. Nevertheless, the form of the vector W and the simplified matrix B contain some information about the structure of the invariant algebraic curve we are trying to find. This is particularly helpful when we try to determine if the family of systems we are investigating is a good candidate. Sometimes it can suggest how to change the family. Another advantage is that performing this preliminary reduction makes the elimination process run faster.

Remark 3.2. In most cases, the system of linear equations $B_i \cdot X = 0$ in step 10 of our algorithm has only one solution X_i^1. In case $l_i > 1$ the polynomial system corresponding to S_i has a rational first integral. Indeed, invariant algebraic curves φ_i^1 and φ_i^2 have the same cofactor κ, so

$$\left(\frac{\varphi_i^1}{\varphi_i^2} \right) = \frac{\kappa \varphi_i^1 \varphi_i^2 - \varphi_i^1 \kappa \varphi_i^2}{(\varphi_i^2)^2} = 0$$

Remark 3.3. To solve/simplify the system of polynomial equations $M_1 = 0$ the methods of applied algebraic geometry can be used; see [Cox *et al.*, 1998]. In many cases standard packages using Gröbner bases are efficient, in other cases the combination of those and the use of resultants turned out to be very effective.

4. Examples

4.1. *Degree 4 invariant algebraic curves for a certain family of quadratic systems*

We look for invariant algebraic curves of degree 4 within the family of quadratic systems

$$\dot{x} = x + y + xy,$$

$$\dot{y} = Kx + Ly + \alpha x^2 + \beta xy + 2y^2$$

with cofactor $4y$. This family depends on the four parameters $\{K, L, \alpha, \beta\}$. We perform steps 1–3 of our algorithm. The extended matrix $\tilde{A}$ for the system is

$$
\left[\begin{array}{ccccccccccccccc}
0 & \alpha & 0 & 0 & 0 & 0 & 0 & 0 & 0 & 0 & 0 & 0 & 0 & 0 & 0 \\
0 & \beta & 2\alpha & 0 & 0 & 0 & 0 & 0 & 0 & 0 & 0 & 0 & 0 & 0 & 0 \\
0 & 1 & 2\beta & 3\alpha & 0 & 0 & 0 & 0 & 0 & 0 & 0 & 0 & 0 & 0 & 0 \\
0 & 0 & 2 & 3\beta & 4\alpha & 0 & 0 & 0 & 0 & 0 & 0 & 0 & 0 & 0 & 0 \\
0 & 0 & 0 & 3 & 4\beta & 0 & 0 & 0 & 0 & 0 & 0 & 0 & 0 & 0 & 0 \\
0 & 0 & 0 & 0 & 4 & 0 & 0 & 0 & 0 & 0 & 0 & 0 & 0 & 0 & 0 \\
4 & K & 0 & 0 & 0 & 0 & \alpha & 0 & 0 & 0 & 0 & 0 & 0 & 0 & 0 \\
4 & 3+L & 2K & 0 & 0 & -1 & \beta & 2\alpha & 0 & 0 & 0 & 0 & 0 & 0 & 0 \\
0 & 3 & 2+2L & 3K & 0 & 0 & 0 & 2\beta & 3\alpha & 0 & 0 & 0 & 0 & 0 & 0 \\
0 & 0 & 2 & 1+3L & 4K & 0 & 0 & 1 & 3\beta & 0 & 0 & 0 & 0 & 0 & 0 \\
0 & 0 & 0 & 1 & 4L & 0 & 0 & 0 & 2 & 0 & 0 & 0 & 0 & 0 & 0 \\
0 & 0 & 0 & 0 & 0 & 3 & K & 0 & 0 & 0 & \alpha & 0 & 0 & 0 & 0 \\
0 & 0 & 0 & 0 & 0 & 3 & 2+L & 2K & 0 & -2 & \beta & 2\alpha & 0 & 0 & 0 \\
0 & 0 & 0 & 0 & 0 & 0 & 2 & 1+2L & 3K & 0 & -1 & 2\beta & 0 & 0 & 0 \\
0 & 0 & 0 & 0 & 0 & 0 & 0 & 1 & 3L & 0 & 0 & 0 & 0 & 0 & 0 \\
0 & 0 & 0 & 0 & 0 & 0 & 0 & 0 & 0 & 2 & K & 0 & 0 & \alpha & 0 \\
0 & 0 & 0 & 0 & 0 & 0 & 0 & 0 & 0 & 2 & 1+L & 2K & -3 & \beta & 0 \\
0 & 0 & 0 & 0 & 0 & 0 & 0 & 0 & 0 & 0 & 1 & 2L & 0 & -2 & 0 \\
0 & 0 & 0 & 0 & 0 & 0 & 0 & 0 & 0 & 0 & 0 & 0 & 1 & K & 0 \\
0 & 0 & 0 & 0 & 0 & 0 & 0 & 0 & 0 & 0 & 0 & 0 & 1 & L & -4 \\
0 & 0 & 0 & 0 & 0 & 0 & 0 & 0 & 0 & 0 & 0 & 0 & 0 & 0 & 0 \\
\hline
x^4 & x^3y & x^2y^2 & xy^3 & y^4 & x^3 & x^2y & xy^2 & y^3 & x^2 & xy & y^2 & x & y & 1
\end{array}\right]
$$

The reduced matrix is

$$
B = \left[\begin{array}{ccccccccc}
4 & 0 & \alpha & 0 & 0 & 0 & 0 & 0 & 0 \\
4 & -1 & \beta & 0 & 0 & 0 & 0 & 0 & 0 \\
0 & 3 & K & 0 & \alpha & 0 & 0 & 0 & 0 \\
0 & 3 & 2+L & -2 & \beta & 2\alpha & 0 & 0 & 0 \\
0 & 0 & 2 & 0 & -1 & 2\beta & 0 & 0 & 0 \\
0 & 0 & 0 & 2 & K & 0 & 0 & \alpha & 0 \\
0 & 0 & 0 & 2 & 1+L & 2K & -3 & \beta & 0 \\
0 & 0 & 0 & 0 & 1 & 2L & 0 & -2 & 0 \\
0 & 0 & 0 & 0 & 0 & 0 & 1 & K & 0 \\
0 & 0 & 0 & 0 & 0 & 0 & 1 & L & -4
\end{array}\right]
$$

and the monomial vector is

$$
W = (x^4, x^3, x^2y, x^2, xy, y^2, x, y, 1).
$$

We proceed to step 7 of our algorithm and get

$$
D = \begin{bmatrix}
2\alpha - 2\beta + 3\alpha\beta - 3\beta^2 - 2L + 3\alpha L - 6\beta L - 2KL - L^2 & 2 - 2\alpha + 5\beta + 2K + L \\[4pt]
3\alpha\beta - 3\beta^2 - \beta K + \alpha L - 3\beta L - KL & -\alpha + 3\beta + K \\[4pt]
2\alpha - 2\beta + 3\alpha\beta - 3\beta^2 + 2K - 4L + 3\alpha L - 6\beta L - 3L^2 & 4 - 3\alpha + 6\beta + 3K + 3L
\end{bmatrix}.
$$

Now we are ready to calculate M_1. It consists of three terms that, after multiplication by a constant, are equal to

$$
(\alpha - \beta - K)(-2\alpha + 3\alpha\beta - 6\beta^2 - 2\beta K - \alpha L),
$$

$$
(\alpha - \beta - K)(-4 + 2\alpha - 8\beta + 3\alpha\beta - 3\beta^2 - 4K - 8L + 3\alpha L - 12\beta L - 6KL - 3L^2),
$$

$$
(\alpha - \beta - K)(-2\alpha - 6\beta + 6\alpha\beta - 9\beta^2 - 2K - 3\beta K - 9\beta L - 3KL).
$$

The set of equations $M_1 = 0$ can be solved explicitly, and we have the following solutions

$$
S_1 = \{\alpha = \beta + K\},
$$

$$
S_2 = \{K = -1 \wedge L = -1 \wedge \alpha = -2 \wedge \beta = -1\},
$$

$$
S_3 = \left\{K = -1 \wedge L = -1 \wedge \beta = \frac{1}{3}\right\},
$$

$$
S_4 = \{\beta = -1 \wedge K = 2\alpha + 3 \wedge K \neq -1 \wedge L = -1\},
$$

$$
S_5 = \left\{(3\beta + 2\sqrt{-2 - 3L} = 2 + 3L \vee 2 + 2\sqrt{-2 - 3L} + 3L = 3\beta) \wedge\right.
$$

$$
\left. L \neq -1 \wedge 2\alpha + \frac{(1 + \beta)(1 + K)}{1 + L} = 5 + 3\beta + 3K + 3L\right\}.
$$

The kernel of $B_1 = B|_{S_1}$ is generated by the vector $X_1^1 = ((\beta + K)^2, 4K(\beta + K), -4(\beta + K), 2K(\beta + 3K) - 2(\beta + K)L, -8K, 4, 4K(K - L), -4K + 4L, (K - L)^2)^T$. Therefore an invariant algebraic curve $W \cdot X_1^1 = (L - \beta x^2 - K(1 + x)^2 + 2y)^2$, which is reducible, corresponds to S_1.

Similarly, the reducible invariant algebraic curve $(x + x^2 + y)^2 = 0$ corresponds to S_2.

The invariant curve $18x^2 + 4x^3 - 12\alpha x^3 - 3\alpha x^4 + 36xy + 12x^2 y + 18y^2 = 0$ corresponding to S_3, has for $\alpha < 2/3$ a form of a cuspidal loop (see Fig. 1.) containing all three singular points of the system.

Corresponding to S_4 is the invariant curve $(x^2 + 2y)(x(4 + 3x) + 2y) - Kx^2(2 + x)^2 = 0.$

The solution S_5 corresponds, in fact, to several families of algebraic invariant curves. Here we present only one example, belonging to a two-parameter family

$$
a = \frac{K(2 - \sqrt{-2 - 3L} + 3L) + (2 + 3L)(4 + \sqrt{-2 - 3L} + 3L)}{3(1 + L)},
$$

$$
\beta = \frac{2}{3}(1 + \sqrt{-2 - 3L}) + L
$$

with an invariant algebraic curve

$$
\varphi_{0,0} + \varphi_{1,0}x + \varphi_{2,0}x^2 + \varphi_{3,0}x^3 + \varphi_{4,0}x^4 + \varphi_{0,1}y
$$

$$
+ \varphi_{1,1}xy + \varphi_{2,1}x^2 y + \varphi_{0,2}y^2 = 0
$$

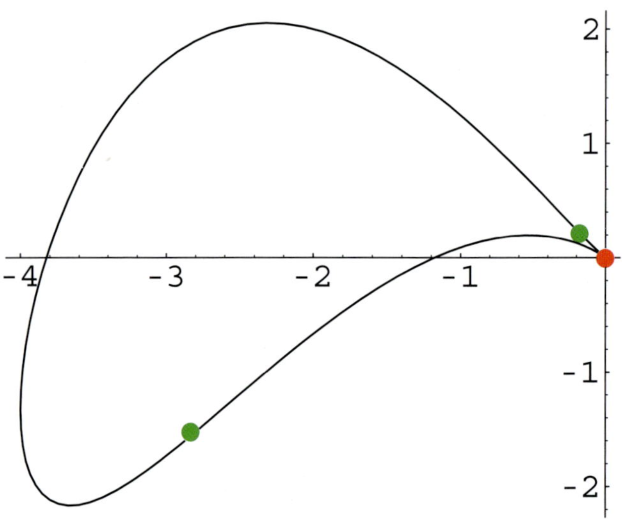

Fig. 1.   The curve corresponding to $K = -1$, $L = -1$, $\alpha = -4/3$, $\beta = 1/3$.

$$\varphi_{0,0} = 27(K - L)(1 + L)^3$$

$$\varphi_{1,0} = 108K(1 + L)^3$$

$$\varphi_{2,0} = 18(1 + L)((4 + \sqrt{-2 - 3L} + 3L)$$
$$\times (2 + 5L + 3L^2) + K(8 - \sqrt{-2 - 3L}$$
$$+ 18L + 9L^2))$$

$$\varphi_{3,0} = 4(2 + 3L)(6(4 + \sqrt{-2 - 3L})$$
$$+ 17(4 + \sqrt{-2 - 3L})L$$
$$+ (60 + 9\sqrt{-2 - 3L})L^2 + 18L^3$$
$$+ K(10 - 2\sqrt{-2 - 3L} + 21L + 9L^2))$$

$$\varphi_{4,0} = (2 + 3L)K(10 - 2\sqrt{-2 - 3L} + 21L + 9L^2)$$
$$+ (2 + 3L)^2(14 + 8\sqrt{-2 - 3L}$$
$$+ 3(7 + 2\sqrt{-2 - 3L})L + 9L^2)$$

$$\varphi_{0,1} = -108(1 + L)^3$$

$$\varphi_{1,1} = -36(1 + L)(6 + (13 + \sqrt{-2 - 3L})L + 6L^2)$$

$$\varphi_{2,1} = -12(4 + \sqrt{-2 - 3L} + 3L)(2 + 5L + 3L^2)$$

$$\varphi_{0,2} = 18(1 + \sqrt{-2 - 3L})(1 + L)$$

The shape of the curve for $k = -17/3$, $c = -8/5$ is presented in Fig. 2.

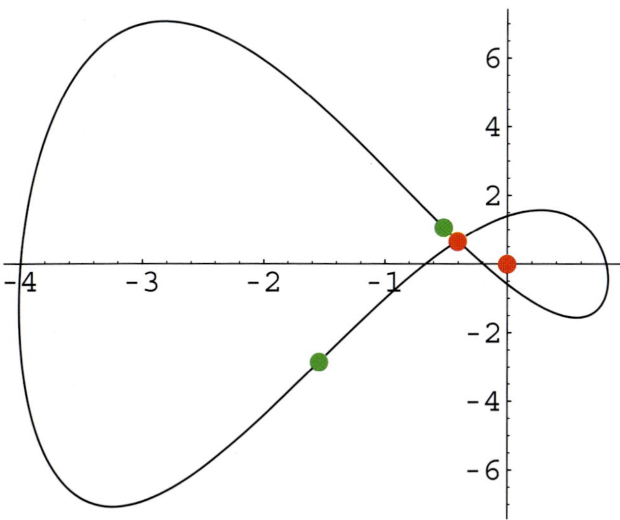

Fig. 2.   The curve corresponding to $K = -17/3$, $L = -8/5$, $\alpha = -(1/135)(1358 + 43\sqrt{70}), \beta = (2/15)(\sqrt{70} - 7)$.

4.2. Degree 5 invariant algebraic curves for quadratic systems

We present two examples of degree 5 invariant algebraic curves for quadratic systems. They have been found by applying our algorithm to the family of quadratic systems

$$\dot{x} = x + y + xy,$$

$$\dot{y} = Kx + Ly + \alpha x^2 + \beta xy + \frac{5}{4}y^2$$

with cofactor $5y$. The set of minors M_1 consists of four polynomials depending on four variables K, L, α, β that, after multiplication by constants are equal to

$$80\alpha^2(909\beta + 86L - 25) - 5(4\beta + K)(22176\beta^3$$
$$+ 245K(5 + 6L) + 36\beta^2(1725 + 1078L)$$
$$+ \beta(4490K + 3(5 + 6L)(745 + 462L)))$$
$$+ 2\alpha(133056\beta^3 - 33(5 + L)(5 + 6L)(10 + 7L)$$
$$+ 48\beta^2(924L - 1585) - 5K(2665 + 953L)$$
$$+ \beta(490K - 2(85175 + 22L(2570 + 63L)))),$$

$$400\alpha^2(9 + 606\beta + 83L) + 20\alpha(44352\beta^3$$
$$+ 48\beta^2(462L - 307) - 2K(1940 + 1039L)$$
$$+ 2\beta(-26245 + 109K - 20338L + 1386L^2)$$
$$- 3(1025 + L(2460 + 1501L)))$$
$$- (4\beta + K)(310464\beta^3 + 504\beta^2(1945 + 1562L)$$
$$+ 7(5 + 6L)(520K + 3(5 + 11L)(5 + 16L))$$
$$+ 2\beta(106175 + 25780K$$
$$+ 12L(19795 + 11088L))),$$

$$40\alpha^2(1704\beta - 83L - 625) - 55\beta(4\beta + K)(100\,K$$
$$+ 3\beta(365 + 672\beta + 168L))$$
$$+ 2\alpha(94248\beta^3 - 10K(380 + 181L)$$
$$- 6\beta^2(38245 + 3696L) - \beta(39125$$
$$+ 19610K + L(25885 + 2772L))),$$

$$- 210\beta^3(4\beta + K) + \alpha\beta(-95K + 2\beta(-505$$
$$+ 168\beta - 63L)) + 12\alpha^2(11\beta - L - 5).$$

The system of equations $M_1 = 0$ has many solutions, some of them being isolated points in $\mathbb{C}^4$. The examples we present correspond to two of these isolated solutions, namely

$$K = \frac{375(8836\sqrt{21} - 1828897)}{722131963},$$

$$L = \frac{5(170\sqrt{21} - 41951)}{219961},$$

$$\alpha = \frac{46875(748\sqrt{21} - 2331)}{2888527852},$$

$$\beta = \frac{375(9\sqrt{21} + 182)}{439922}$$

and $K = 189$, $L = -11$, $\alpha = 405/4$, $\beta = -27/2$. Therefore these examples are isolated, not belonging to any families of quadratic systems with invariant algebraic curve of degree 5. We have:

The system

$$\dot{x} = x + y + xy,$$

$$\dot{y} = \frac{375(8836\sqrt{21} - 1828897)}{722131963}x$$
$$+ \frac{5(170\sqrt{21} - 41951)}{219961}y$$
$$+ \frac{46875(748\sqrt{21} - 2331)}{2888527852}x^2$$
$$+ \frac{375(9\sqrt{21} + 182)}{439922}xy + \frac{5}{4}y^2$$

has the invariant algebraic curve

$$-3.1973 \cdot 10^{57} + 2.06748 \cdot 10^{60}x - 3.7594 \cdot 10^{62}x^2$$
$$+ 1.32337 \cdot 10^{64}x^3 - 1.46055 \cdot 10^{64}x^4$$
$$+ 2.21 \cdot 10^{62}x^5 + 2.22619 \cdot 10^{60}y$$
$$- 8.09555 \cdot 10^{62}xy + 4.27331 \cdot 10^{64}x^2y$$
$$- 6.20874 \cdot 10^{64}x^3y - 4.36964 \cdot 10^{62}y^2$$
$$+ 4.63717 \cdot 10^{64}xy^2 - 1.09432 \cdot 10^{65}x^2y^2$$
$$+ 1.69051 \cdot 10^{64}y^3 - 9.20718 \cdot 10^{64}xy^3$$
$$- 3.02394 \cdot 10^{64}y^4 = 0$$

with cofactor $5y$. We present the coefficients in numerical form because the exact formula is over two pages long. The curve is presented in Fig. 3.

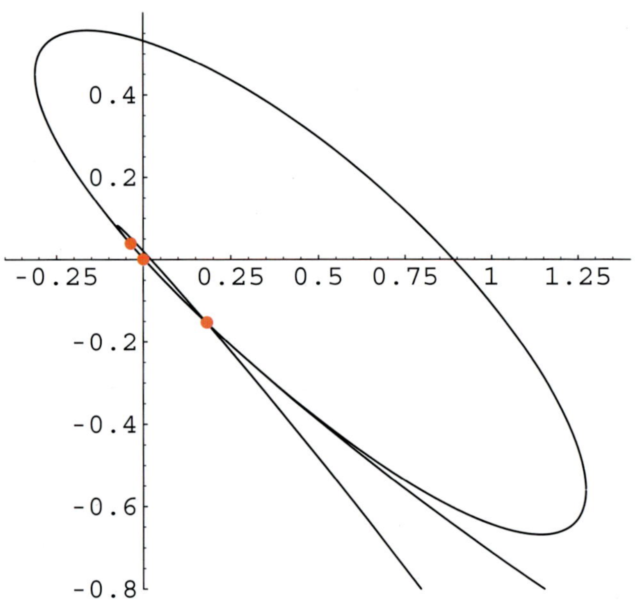

Fig. 3.   Invariant algebraic curve of degree 5.

The system

$$\dot{x} = x + y + xy,$$

$$\dot{y} = 189x - 11y + \frac{405}{4}x^2 - \frac{27}{2}xy + \frac{5}{4}y^2$$

has the invariant algebraic curve

$$25600000 + 120960000x + 224272800x^2$$
$$+ 203163552x^3 + 89367381x^4 + 15116544x^5$$
$$- 640000y - 2030400xy - 2137104x^2y$$
$$- 746496x^3y + 16800y^2 + 35136xy^2$$
$$+ 18306x^2y^2 - 208y^3 - 216xy^3 + y^4 = 0$$

with cofactor $5y$. This curve is presented in Fig. 4.

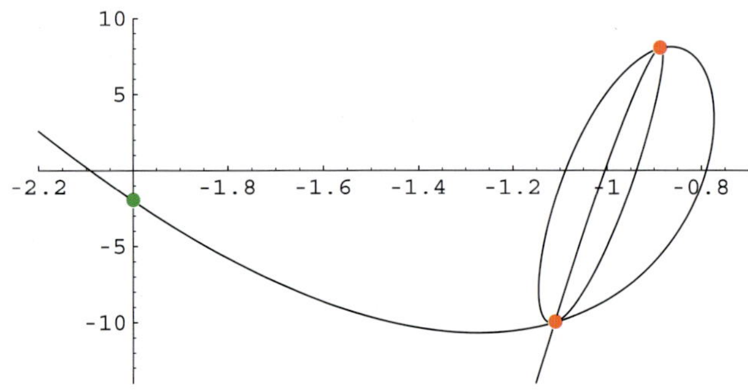

Fig. 4.   Another invariant algebraic curve of degree 5.

4.3. *Degree 6 invariant algebraic curve containing a saddle-loop for a certain family of quadratic systems*

Application of our algorithm to the family of systems

$$\dot{x} = 1 + x + xy,$$

$$\dot{y} = (K - \alpha) + Kx + Ly + \alpha x^2 + \beta xy + 2y^2$$

with cofactor $6y$ and $n = 6$ leads to the discovery of a degree 6 algebraic saddle-loop. As far as we know, this is the first known example of an algebraic saddle-loop of degree greater than 5 for quadratic systems.

Theorem 4.1. *The system*

$$\dot{x} = 1 + x + xy,$$

$$\dot{y} = \frac{-22 - 47L - 21L^2}{10} - \frac{34 + 87L + 60L^2 + 9L^3}{10}x$$

$$+ Ly - \frac{(3 + L)(2 + 3L)^2}{10}x^2$$

$$+ \frac{(3 + L)(2 + 3L)}{10}xy + 2y^2$$

has an invariant algebraic curve defined by

$$\varphi_{0,0} + \varphi_{1,0}x + \varphi_{2,0}x^2 + \varphi_{3,0}x^3 + \varphi_{4,0}x^4 + \varphi_{5,0}x^5$$
$$+ \varphi_{6,0}x^6 + \varphi_{0,1}y + \varphi_{1,1}xy + \varphi_{2,1}x^2y$$
$$+ \varphi_{3,1}x^3y + \varphi_{4,1}x^4y + \varphi_{0,2}y^2 + \varphi_{1,2}xy^2$$
$$+ \varphi_{2,2}x^2y^2 + \varphi_{0,3}y^3 = 0$$

where

$$\varphi_{0,0} = -200(192 + 1104L + 2184L^2$$
$$+ 1732L^3 + 463L^4)$$

$$\varphi_{1,0} = -2400(88 + 474L + 937L^2 + 834L^3$$
$$+ 325L^4 + 42L^5)$$

$$\varphi_{2,0} = -60(2 + 3L)^2(1296 + 3160L + 2506L^2$$
$$+ 701L^3 + 47L^4)$$

$$\varphi_{3,0} = 20(2 + 3L)^3(-884 - 1496L - 615L^2$$
$$+ 4L^3 + 19L^4)$$

$$\varphi_{4,0} = 12(2 + 3L)^4(-138 - 109L + 33L^2$$
$$+ 24L^3 + 2L^4)$$

$$\varphi_{5,0} = 6(L - 2)(3 + L)^3(2 + 3L)^5$$

$$\varphi_{6,0} = (L - 2)(3 + L)^3(2 + 3L)^6$$

$$\varphi_{0,1} = -8000(12 + 29L + 33L^2 + 6L^3)$$

$$\varphi_{1,1} = -1200(192 + 584L + 544L^2$$
$$+ 152L^3 + 3L^4)$$

$$\varphi_{2,1} = -600(2 + 3L)^2(28 + 20L - 11L^2 - 3L^3)$$

$$\varphi_{3,1} = -120(L - 2)^2(3 + L)(2 + 3L)^3$$

$$\varphi_{4,1} = 60(L - 2)(3 + L)^2(2 + 3L)^4$$

$$\varphi_{0,2} = 120000L(1 + L)$$

$$\varphi_{1,2} = 24000(6 + 17L + 14L^2 + 3L^3)$$

$$\varphi_{2,2} = -600(4 + 4L - 3L^2)^2$$

$$\varphi_{0,3} = 80000$$

with cofactor $6y$. For $1 < L < 2$ this curve contains a saddle-loop.

The shape of the curve for $L = 11/7$ is presented in Fig. 5.

Remark 4.2. Most examples presented in the paper belong to a very special class of quadratic systems. There are certain conditions that must be satisfied for a quadratic system to have an invariant algebraic curve of a high degree. Such quadratic systems have been studied by Llibre and Świrszcz [2003] and all quadratic systems admitting high-degree limit cycles have been classified. In particular, the family

$$\dot{x} = x + y + xy,$$

$$\dot{y} = Kx + Ly + \alpha x^2 + \beta xy + \gamma y^2$$

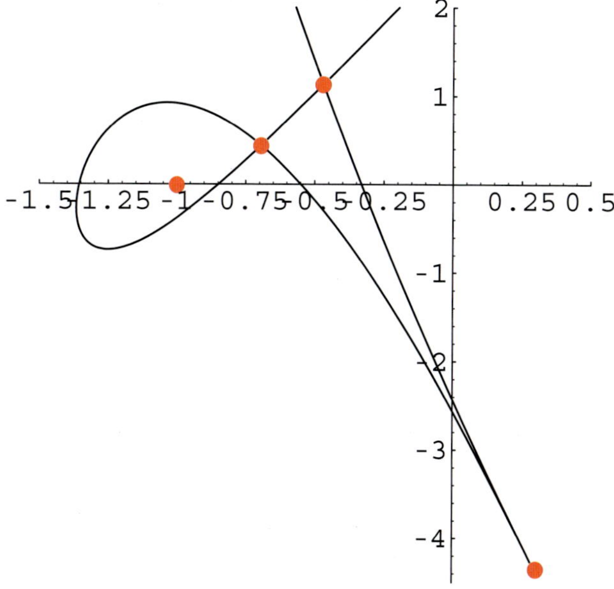

Fig. 5.   Degree 6 algebraic saddle-loop for $L = 11/7$.

with cofactor ny (denoted by S_n^n in [Llibre & Świrszcz, 2003]) is a very promising class of systems. Many other examples of quadratic systems with invariant algebraic curves have been found using the described algorithm, but they usually do not have such interesting geometry. Similar conditions to some of those presented in [Llibre & Świrszcz, 2003] have been found for polynomial systems (not necessarily quadratic) by [Chavarriga *et al.*, 2003].

Acknowledgment

This paper has been partially supported by Polish KBN Grant 2PO3A 01022.

References

Chavarriga, J., Llibre, J. & Sorolla, J. [2000] "Algebraic limit cycles of quadratic systems," preprint.

Chavarriga, J., Giacomini, H. & Llibre, J. [2001] "Uniqueness of algebraic limit cycles for quadratic systems," *J. Math. Anal. Appl.* **261**, 85–99.

Chavarriga, J. & Grau, M. [2002] "A family of non Darboux–integrable quadratic polynomial differential systems with algebraic solutions of arbitrarily high degree," *Appl. Math. Lett.* **16**, 833–837.

Chavarriga, J., Giacomini, H. & Grau, M. [2003] "Necessary conditions for the existence of invariant algebraic curves for planar polynomial systems," *Bull. Sci. Math.*, to appear.

Christopher, C. & Llibre, J. [2002] "A family of quadratic polynomial differential systems with invariant algebraic curves of arbitrarily high degree without rational first integrals," *Proc. Amer. Math. Soc.* **130**, 2025–2030.

Christopher, C., Llibre, J. & Świrszcz, G. [2003] "Invariant algebraic curves of large degree for quadratic systems," *J. Math. Anal. Appl.*, to appear.

Cox, D., Little, J. & O'Shea, D. [1998] *Using Algebraic Geometry*, Graduate Texts in Mathematics, Vol. 185 (Springer-Verlag, NY).

Darboux, G. [1878] "Mémoire sur les équations différentielles algébriques du premier ordre et du premier degré (Mélanges)," *Bull. Sci. Math. 2ème série* **2**, 60–96, 123–144, 151–200.

Evdokimenco, R. M. [1970] "Construction of algebraic paths and the qualitative investigation in the large of the properties of integral curves of a system of differential equations," *Diff. Eqs.* **6**, 1349–1358.

Evdokimenco, R. M. [1974] "Behavior of integral curves of a dynamic system," *Diff. Eqs.* **9**, 1095–1103.

Evdokimenco, R. M. [1979] "Investigation in the large of a dynamic system," *Diff. Eqs.* **15**, 215–221.

Filiptsov, V. F. [1973] "Algebraic limit cycles," *Diff. Eqs.* **9**, 983–986.

Goldstine, H. H. & von Neumann, J. [1947] "Numerical inversion of matrices of high order," *Bull. AMS*, 1021–1099.

Lanford, O. E. [1982] "A computer–assisted proof of the Feigenbaum conjectures," *Bull. Amer. Math. Soc.* **6**, 427–434.

Llibre, J. & Świrszcz, G. [2003] "Classification of quadratic systems admitting the existence of an algebraic limit cycle," preprint.

Mandelbrot, B. [1982] *The Fractal Geometry of Nature* (W.H. Freeman and Company, NY).

Poincaré, H. [1981] "Sur l'intégration des équations différentielles du premier ordre et du premier degré I and II," *Rendiconti del Circolo Matematico di Palermo* **5** (1891), 161–191; [1987] **11** (1897), 193–239.

Qin, Y.-X. [1958] "On the algebraic limit cycles of second degree of the differential equation $dy/dx = \sum_{0 \leq i+j \leq 2} a_{ij} x^i y^j / \sum_{0 \leq i+j \leq 2} b_{ij} x^i y^j$," *Acta Math. Sin.* **8**, 23–35.

Turing, A. [1950] "Computing machinery and intelligence," *Mind* **49**, 433–460.

von Neumann, J. [1945] "First Draft of a Report on the EDVAC," Contract No. W-670-ORD-492, Moore School of Electrical Engineering, Univ. of Penn., Philadelphia (30 June 1945); Reprinted (in part): Randell, B., *Origins of Digital Computers: Selected Papers* (Springer–Verlag, Berlin, Heidelberg), pp. 383–392.

von Neumann, J. [1946] *The Principles of Large-Scale Computing Machines*; Reprinted, *Ann. Hist. Comp.* **3**, 263–273.

von Neumann, J. [1958] *The Computer and the Brain*, (Yale University Press, New Haven).

Yablonskii, A. I. [1966] "Limit cycles of a certain differential equations," *Diff. Eqs.* **2**, 335–344 (in Russian).

Żołądek, H. [1998] "Algebraic invariant curves for the Lienard equation," *Trans. Amer. Math. Soc.* **4**, 1681–1701.

AUTHOR INDEX